PROCEEDINGS

SPIE—The International Society for Optical Engineering

Optical Methods for Time- and State-Resolved Chemistry

Cheuk-Yiu Ng
Chair/Editor

23–25 January 1992
Los Angeles, California

Sponsored and Published by
SPIE—The International Society for Optical Engineering

Volume 1638

PROCEEDINGS
SPIE—The International Society for Optical Engineering

Optical Methods for Time- and State-Resolved Chemistry

Cheuk-Yiu Ng
Chair/Editor

23–25 January 1992
Los Angeles, California

Sponsored and Published by
SPIE—The International Society for Optical Engineering

Volume 1638

SPIE (The Society of Photo-Optical Instrumentation Engineers) is a nonprofit society dedicated to the advancement of optical and optoelectronic applied science and technology.

The papers appearing in this book comprise the proceedings of the meeting mentioned on the cover and title page. They reflect the authors' opinions and are published as presented and without change, in the interests of timely dissemination. Their inclusion in this publication does not necessarily constitute endorsement by the editors or by SPIE.

Please use the following format to cite material from this book:
  Author(s), "Title of paper," *Optical Methods for Time- and State-Resolved Chemistry*, Cheuk-Yiu Ng, Editor, Proc. SPIE 1638, page numbers (1992).

Library of Congress Catalog Card No. 92-80040
ISBN 0-8194-0784-4

Published by
SPIE—The International Society for Optical Engineering
P.O. Box 10, Bellingham, Washington 98227-1101 USA
Telephone 206/676-3290 (Pacific Time) • Fax 206/647-1445

Copyright ©1992, The Society of Photo-Optical Instrumentation Engineers.

Copying of material in this book for internal or personal use, or for the internal or personal use of specific clients, beyond the fair use provisions granted by the U.S. Copyright Law is authorized by SPIE subject to payment of copying fees. The Transactional Reporting Service base fee for this volume is $4.00 per article (or portion thereof), which should be paid directly to the Copyright Clearance Center (CCC), 27 Congress Street, Salem, MA 01970. Other copying for republication, resale, advertising or promotion, or any form of systematic or multiple reproduction of any material in this book is prohibited except with permission in writing from the publisher. The CCC fee code is 0-8194-0784-4/92/$4.00.

Printed in the United States of America.

Contents

vii   *Conference Committee*
ix   *Introduction*

SESSION 1   CONDENSED-PHASE ULTRAFAST SPECTROSCOPY, PHOTOCHEMISTRY, AND TIME-RESOLVED DYNAMICS

2   **Photodissociation dynamics in crystalline solids: many-body dynamics** [1638-01]
V. A. Apkarian, P. Ashjian, N. Blake, A. Danylichev, M. Hill, G. J. Hoffman, D. Imre, W. G. Lawrence, E. Sekreta, R. Zadoyan, J. Zoval, Univ. of California/Irvine.

12   **Structural relaxation in liquids and glasses by transient hole burning** [1638-02]
M. Berg, J. Yu, P. Earvolino, Univ. of Texas/Austin.

21   **Ultrafast studies on electron transfer in the betaines: evidence for local heating** [1638-38]
N. E. Levinger, A. E. Johnson, G. C. Walker, E. Åkesson, P. F. Barbara, Univ. of Minnesota.

29   **Solution photochemistry of OClO: excited-state dissociation and isomerization** [1638-05]
R. C. Dunn, B. N. Flanders, J. D. Simon, Univ. of California/San Diego.

36   **Structure-reactivity effects in the picosecond dynamics of isolated clusters** [1638-06]
J. A. Syage, J. Steadman, The Aerospace Corp.

SESSION 2   SPECTROSCOPY AND REACTION DYNAMICS OF CLUSTERS

44   **Size- and state-selected cluster chemistry** [1638-07]
Y. Achiba, T. Moriwaki, H. Shiromaru, Tokyo Metropolitan Univ. (Japan).

56   **Photodissociation of mass-selected clusters: solvated metal ions** [1638-08]
J. M. Farrar, S. G. Donnelly, Univ. of Rochester.

63   **Stimulated Raman spectroscopy of complexes and clusters in supersonic molecular beams** [1638-09]
B. F. Henson, G. V. Hartland, V. A. Venturo, P. M. Maxton, P. M. Felker, Univ. of California/Los Angeles.

74   **Photoinduced intracluster electron transfer reactions of captured intermediates in gas-phase S_N2 reactions** [1638-10]
D. M. Cyr, G. A. Bishea, C. Han, L. A. Posey, M. A. Johnson, Yale Univ.

85   **Photo- and collision-induced dissociation of Ar cluster ions** [1638-11]
T. Kondow, T. Nagata, S. Nonose, Univ. of Tokyo (Japan).

92   **Vibrational spectroscopy of size-selected metal-ion-solvent clusters** [1638-12]
J. M. Lisy, Univ. of Illinois/Urbana-Champaign.

102   **Laser desorption jet cooling spectroscopy of organic clusters** [1638-43]
M. S. de Vries, H. E. Hunziker, G. Meijer, H. R. Wendt, IBM/Almaden Research Ctr.

(continued)

106 **Gas-phase study of the group VI transition metal tricarbonyl complexes by negative-ion photoelectron spectroscopy (Poster Paper)** [1638-45]
A. A. Bengali, S. M. Casey, C. Cheng, J. P. Dick, P. T. Fenn, P. W. Villalta, D. G. Leopold, Univ. of Minnesota.

118 **Resonant MPI spectrum of allyl radicals (Poster Paper)** [1638-46]
D. W. Minsek, J. A. Blush, P. Chen, Harvard Univ.

127 **High-resolution inner-valence UV photoelectron spectra of the O_2 molecule and CI calculations of $^2\Pi_u$ states between 20 and 26 eV (Poster Paper)** [1638-47]
P. Baltzer, B. Wannberg, L. Karlsson, M. Carlsson-Göthe, Uppsala Univ. (Sweden); M. Larsson, Royal Institute of Technology (Sweden).

132 **CO(v,J) product-state distributions from the reaction $O(^3P) + OCS \rightarrow CO + SO$ (Poster Paper)** [1638-39]
S. L. Nickolaisen, Univ. of Southern California; H. E. Cartland, D. Veney, U.S. Military Academy; C. Wittig, Univ. of Southern California.

SESSION 3 — PHOTODISSOCIATION DYNAMICS

144 **Translational spectroscopy of H atom photofragments** [1638-13]
M. N. Ashfold, I. R. Lambert, D. H. Mordaunt, G. P. Morley, C. M. Western, Univ. of Bristol (UK).

156 **Site-specific photochemistry** [1638-14]
B. D. Koplitz, J. L. Brum, S. Deshmukh, X. Xu, Z. Wang, Y. Yen, Tulane Univ.

164 **C-H dissociation rate constants of alkylbenzenes from hot molecule formed by 158-nm (F_2 laser) irradiation** [1638-15]
N. Nakashima, T. Shimada, Institute of Laser Engineering/Osaka Univ. (Japan); Y. Ojima, Institute for Laser Technology (Japan); Y. Izawa, Institute of Laser Engineering/Osaka Univ. (Japan); C. Yamanaka, Institute for Laser Technology (Japan).

170 **How to photograph a chemical reaction** [1638-40]
K. R. Leopold, S. W. Reeve, M. A. Dvorak, W. A. Burns, Univ. of Minnesota; R. S. Ford, Scott Community College; F. J. Lovas, R. D. Suenram, National Institute of Standards and Technology.

179 **Primary processes involved in the photodissociation of saturated hydrocarbons at 157 nm** [1638-16]
K. A. Prather, Y. T. Lee, Lawrence Berkeley Lab. and Univ. of California/Berkeley.

185 **Photodissociation dynamics of ethylene sulfide and allene at 193 nm studied by time-of-flight mass spectroscopy** [1638-17]
K. Tabayashi, K. Shobatake, Institute for Molecular Science (Japan).

197 **157-nm photodissociation dynamics of CO_2 via photofragment-translational spectroscopy** [1638-18]
A. Stolow, Y. T. Lee, Univ. of California/Berkeley and Lawrence Berkeley Lab.

SESSION 4 — PHOTOIONIZATION, PHOTOELECTRON, AND ION PHOTODISSOCIATION SPECTROSCOPY

210 **Laser spectroscopy of radicals and carbenes** [1638-30]
H. Clauberg, D. W. Minsek, P. Chen, Harvard Univ.

216   High-brightness cm^{-1}-resolution threshold photoelectron spectroscopic technique [1638-19]
K. Kimura, M. Takahashi, Institute for Molecular Science (Japan).

228   Studies of doubly charged molecular ions by means of ion photofragment spectroscopy [1638-20]
M. Larsson, G. Sundström, Royal Institute of Technology (Sweden); L. Broström, S. Mannervik, Manne Siegbahn Institute of Physics (Sweden).

234   Intracavity laser photoelectron spectroscopy of Cr_2^-, Cr_2H^-, and Cr_2D^- [1638-41]
S. M. Casey, D. G. Leopold, Univ. of Minnesota.

245   Dynamics of S production in the 193-nm photodissociation of CH_3SCH_3, CH_3SSCH_3, CH_3SH, and H_2S [1638-21]
BC. Liao, C. Hsu, C. Ng, Ames Lab./USDOE and Iowa State Univ.

254   Electronic spectroscopy of ionic clusters [1638-22]
E. J. Bieske, A. M. Soliva, A. Friedmann, J. P. Maier, Univ. Basel (Switzerland).

264   Experimental methods for probing structure and dynamics of gas-phase molecular dications [1638-23]
K. Yokoyama, D. M. Szaflarski, A. S. Mullin, W. C. Lineberger, Joint Institute for Laboratory Astrophysics, National Institute of Standards and Technology, and Univ. of Colorado/Boulder.

273   Rotationally resolved threshold photoionization of H_2S [1638-24]
R. T. Wiedmann, M. G. White, Brookhaven National Lab.

282   Ultrafast Raman echo experiments in the liquid phase (Poster Paper) [1638-48]
M. Berg, L. J. Muller, D. Vanden Bout, Univ. of Texas/Austin.

291   Photophysics of rigidized 7-aminocoumarin laser dyes (Poster Paper) [1638-49]
M. R. Sahyun, 3M Graphic Research Lab.; D. K. Sharma, Concordia Univ. (Canada).

SESSION 5   OPTICAL TECHNIQUES AND MOLECULAR SPECTROSCOPY

298   Population transfer by stimulated Raman scattering with delayed pulses: the concept, some problems, and experimental demonstration [1638-25]
K. Bergmann, S. Schiemann, A. Kuhn, Univ. Kaiserslautern (FRG).

310   Orientation and molecular orbital dependences in electronic relaxation collisions studied through van der Waals complexes [1638-26]
L. Lapierre, P. Y. Cheng, S. S. Ju, Y. M. Hahn, Univ. of Pennsylvania; H. Dai, Univ. of Pennsylvania (USA) and National Taiwan Univ. (Taiwan).

322   Advances in high-repetition-rate ultrashort gigawatt laser systems for time-resolved spectroscopy [1638-27]
L. F. DiMauro, Brookhaven National Lab.

333   Ultraviolet resonance Raman spectroscopy: studies of depolarization dispersion and strong vibronic coupling [1638-28]
B. S. Hudson, Univ. of Oregon.

(continued)

345 **Vibrational spectroscopy and picosecond dynamics of gaseous trienes and tetraenes in S_1 and S_2 electronic states** [1638-29]
H. Petek, A. J. Bell, Institute for Molecular Science (Japan); R. L. Christensen, Bowdoin College; K. Yoshihara, Institute for Molecular Science (Japan).

357 **Coherent effects in laser chemistry** [1638-44]
M. Shapiro, Weizmann Institute of Science (Israel); P. Brumer, Univ. of Toronto (Canada).

368 **Vibrational energy transfer in the chemical energy regime using stimulated emission pumping** [1638-31]
X. Yang, E. H. Kim, A. M. Wodtke, Univ. of California/Santa Barbara.

SESSION 6   STATE-SELECTED AND STATE-TO-STATE REACTION DYNAMICS

386 **Reactive scattering of "hot" H atoms with CO_2 and OCS** [1638-32]
H. E. Cartland, U.S. Military Academy; S. L. Nickolaisen, C. Wittig, Univ. of Southern California.

399 **Chemiluminescent reactions of laser-generated metal atoms with oxidants** [1638-33]
K. Chen, National Sun Yat-sen Univ. (Taiwan).

416 **State-to-state collision dynamics of molecular free radicals** [1638-34]
R. G. Macdonald, K. Liu, Argonne National Lab.

423 **State-selected ion-neutral reactive scattering** [1638-35]
J. E. Pollard, The Aerospace Corp.

431 **Investigation of the roles of vibrational excitation and collision energy in the ion-molecule reaction $NH_3^+(v_2) + ND_3$** [1638-42]
L. A. Posey, Vanderbilt Univ.; R. D. Guettler, R. N. Zare, Stanford Univ.

441 **State-to-state studies of intramolecular dynamics** [1638-36]
T. R. Rizzo, X. Luo, R. D. Settle, Univ. of Rochester.

453 **Photoelectron spectroscopy with cm^{-1} resolution: VO^+, $benzyl^+$, $toluene^+$** [1638-37]
J. C. Weisshaar, Univ. of Wisconsin/Madison.

464 Addendum
465 Author Index

Conference Committee

Conference Chair

Cheuk-Yiu Ng, Ames Laboratory/USDOE and Iowa State University

Cochairs

Paul F. Barbara, University of Minnesota
Edward R. Grant, Purdue University
John W. Hepburn, University of Waterloo (Canada)
Andrew H. Kung, Lawrence Berkeley Laboratory
Robert L. Whetten, University of California/Los Angeles

Session Chairs

Session 1—Condensed-Phase Ultrafast Spectroscopy, Photochemistry, and Time-Resolved Dynamics
Paul F. Barbara, University of Minnesota

Session 2—Spectroscopy and Reaction Dynamics of Clusters
Robert L. Whetten, University of California/Los Angeles

Session 3—Photodissociation Dynamics
Andrew H. Kung, Lawrence Berkeley Laboratory

Session 4—Photoionization, Photoelectron, and Ion Photodissociation Spectroscopy
John W. Hepburn, University of Waterloo (Canada)

Session 5—Optical Techniques and Molecular Spectroscopy
Edward R. Grant, Purdue University

Session 6—State-Selected and State-to-State Reaction Dynamics
Cheuk-Yiu Ng, Ames Laboratory/USDOE and Iowa State University

---

Conference 1638, *Optical Methods for Time- and State-Resolved Chemistry,* was part of a four-conference program on Laser Spectroscopy held at SPIE's OE/LASE '92 Lasers, Sensors, and Spectroscopy Symposium, 19–25 January 1992, in Los Angeles, California. The other conferences were:

Conference 1636, *Applied Spectroscopy in Materials Science*
Conference 1637, *Environmental and Process Monitoring Technologies*
Conference 1639, *Scanning Probe Microscopies*

Program Chair: **Joseph R. Lakowicz,** University of Maryland

Introduction

Optical techniques involving the use of lasers have had a profound influence on the fundamental study of chemical processes. This conference emphasized some recent developments and applications of laser techniques for time- and state-resolved studies of molecular dynamics in gaseous and condensed phases. Sophisticated laser techniques for the spectroscopic characterization of molecules, ions, and clusters were also included. Papers presented at this meeting represent state-of-the-art experimental and theoretical research in chemical physics.

A major activity in cluster research is the investigation of the transition from gaseous state to condensed phase. It was the intent of Sessions 1 and 2 to compare results of the time-resolved ultrafast studies in the condensed phase with those of the state- or energy-resolved experiments of gaseous clusters and cluster ions.

Session 3 was primarily concerned with recent investigations of molecular photodissociation dynamics using ultraviolet and vacuum ultraviolet (VUV) lasers. The advantages and limitations of the high-resolution time-of-flight (TOF) techniques for detection of H atom fragments and the conventional TOF mass spectrometric method were addressed.

The theme of Session 4 was the application of laser techniques, including laser ion photofragmentation, resonance-enhanced multiphoton ionization, and VUV photoionization for spectroscopic and energetic studies of molecular cations and anions, cluster cations, and doubly charged cations. Threshold photoelectron spectra with cm^{-1} resolution can now be obtained routinely using the delayed, pulsed-field ionization technique.

Session 5 focused on new coherent laser techniques for the study of state- and product-selective chemistry. Experiments presented in this session also included the application of coherent Raman and stimulated emission pumping for spectroscopic and intra- and intermolecular dynamics studies.

Session 6 discussed state-selected and state-to-state reaction dynamics. Papers presented in this session concerned the novel applications of laser techniques for reactant-state preparation and for product-state identification. The array of papers included detailed studies of neutral-neutral, ion-molecule, and unimolecular reactions. In addition to the six oral sessions, two poster sessions were held.

The success of this meeting is a tribute to the cochairs: Paul F. Barbara, University of Minnesota; Robert L. Whetten, University of California/Los Angeles; Andrew H. Kung, Lawrence Berkeley Laboratory; John W. Hepburn, University of Waterloo (Canada); and Edward R. Grant, Purdue University. I wish to thank each of them for their efforts.

Cheuk-Yiu Ng
Ames Laboratory/USDOE and Iowa State University

OPTICAL METHODS FOR
TIME- AND STATE-RESOLVED CHEMISTRY

Volume 1638

SESSION 1

Condensed-Phase Ultrafast Spectroscopy, Photochemistry, and Time-Resolved Dynamics

Chair
Paul F. Barbara
University of Minnesota

Photodissociation Dynamics in Crystalline Solids: Many-Body Dynamics

V. A. Apkarian, P. Ashjian, N. Blake, A. Danylichev, M. Hill, G. J. Hoffman, D. Imre, W. G. Lawrence, E. Sekreta, R. Zadoyan, J. Zoval

University of California, Irvine, Department of Chemistry
Irvine, CA 92717.

Understanding many-body dynamics on a molecular level is a major aim in condensed phase photodynamical research. Much can be learned about this general field through studies of molecular photodissociation in model systems, namely crystalline rare gas solids. The aim of this presentation is to illustrate this proposition by highlights drawn from a variety of related investigations. Under the title of photodissociation in solids, several related processes can be categorized: charge transfer induced dissociation,[1] radiative dissociation,[2] atomic photomobility,[3] are examples.

The experimental paradigm in its simplest form would consist of: isolating a diatomic molecule in an ordered host of known structure, optically inducing a bound-to-free electronic transition, and following the subsequent dynamics of all atoms, on all relevant time scales. Iteration of this experiment in the laboratory and on computer, through Molecular Dynamics simulations (MD) is used to converge to a first principles understanding of the many-body interactions.

The starting point of the laboratory experiments is the preparation of doped crystalline samples. There are no general recipes for this step. Free standing crystals grown near thermodynamic equilibrium conditions, temperature or pressure annealed thin films produced by either pulsed deposition or by slow spray-on, are the main techniques utilized. Indirect spectroscopic information is used to characterize structure. Does *long-range order* effect photodissociation dynamics? A dramatic example arguing in the affirmative to this natural question was recently discovered in photodissociation studies of O_2. Excitation of O_2 at 193 nm, over the Schumann-Runge band, leads to the $O_2(B)$ state, which in the gas phase predissociates to form $O(^3P)+O(^3P)$. In crystalline Ar, Kr and Xe, this channel is blocked by a strong cage effect, as evidenced by the c→a, C→X, b→X recombinant emissions which can be followed in solid Ar and Kr (quenched in Xe). In doubly doped O_2:Xe/Kr, efficient photodissociation is observed at 24K only. The process can be followed by monitoring the formation of O atoms, which can be monitored via O/Xe and O/Kr charge transfer transitions. At all temperatures, a photochemical steady-state between O and O_2 can be established due to the *photomobility* of excited O atoms. The XeO concentration as a function of temperature clearly shows a lambda type phase transition, which is further verified by light scattering experiments (see figure 1). The transition is believed to be impurity induced fcc to hcp, martensitic type, which are well known in doped rare gas crystals.[4] In this case, it is clear that O_2 dissociates at the lambda point, in the disordered state created by the slip of the stacking plane. The absence of O_2 dissociation, despite O atom photomobilities can be understood by making reference to Rg-O pair potentials: $O(^3P)$ is too "big", while both $O(^1D)$ and O^- are much "smaller", and therefore subject to small cage barriers. Note, although in most of the chosen systems for study the relevant pair potentials are well characterized, in the case of open shell fragments the relation between pair and many-body potentials is less than direct. Spectroscopies that probe fragment-host interactions are particularly useful in this respect. Spectroscopy in O /Rg solids have revealed different portions of the cage potential. Emission spectra for both centro-symmetric and eccentric trapping sites can be obtained. An example of the latter is the emission from the charge transfer states of O/Xe, in which the early time cage dynamics is probed.

In general, photodissociation is expected to be most sensitive to the *short-range order*, namely: the structure of the local cage of isolation. In fact, quite commonly, photodissociation is defined as the cage exit probability of photofragments. In this respect a categorization is possible by making reference to the dynamics of the fragment and the immediate cage as: a) impulsive/sudden, b) delayed/sudden, c) delayed/accommodated (relaxed).

To the first category belongs F_2 in crystal Ar and Kr, in which satisfactory convergence between experiment and simulation is thought to be reached.[5] Among the observables used in the comparisons are the quantum yields of dissociation versus excess energy, an inverse temperature dependence ascribed to molecule-exit window *pre-alignment*,[6] and long-range migration of atoms. The agreement is surprisingly good in view of the fact that the multiplicity of electronic surfaces is not properly taken into account.[5] A more rigorous test of our understanding of this system is provided by pressure dependent studies. The direct photodissociation channel is blocked at 40 kbar, even at excess energies of 2.8 eV for which $\Phi \sim 1$ at zero pressure. This dramatic effect would not be expected by a naive summation of pair potentials. A proper account of nonadiabatic dynamics is essential. In a related study of Cl_2/Xe, $\Phi \sim 0$ at 40 kbar (100 K), in-cage geminate recombination can be followed and is observed to retain at least partial memory of its initial *orientation*, as evidenced by polarization of the recombinant emission (see figure 2).[7]

The extent of control exercised on potentials in diamond anvil cells (in the range 0-200 kbar) is quite impressive, hence the utility of this tool in dynamical studies. At high pressure, both in the case of F_2/Kr and Cl_2/Xe, although the direct dissociation channel is blocked, dissociation via two photon induced access of ionic states proceeds efficiently.[8] Dynamics over the ionic surfaces is very poorly understood, here one needs to proceed beyond nearest neighbor interactions. Charge transfer spectroscopy and dynamics provides a glimpse of the essential ingredients for a molecular level of interpretations.

Due to the fact that pairwise additivity is a reliable construct in the case of $H(^1S) + Rg(^1S)$ interactions, some of the most reliable insights can be obtained from the scrutiny of photodissociation involving the ejection of H atoms: hydrogen halides, H_2O, and H_2S. Early simulations depicted a delayed cage exit scenario, after complete accommodation of the cage.[9] The most complete experimental studies to date are on H_2O.[10] The experiments suggest a sudden cage exit process. A statistical model for sudden cage exit is possible to construct, and is quite revealing. The cage exit barrier in fcc crystals correspond to the D_{3h} window. This corresponds to a tight site for H atoms, ~ 0.4 eV up on the repulsive H-Ar pair-potential. Given the steep gradient of the potential at these distances, a large distribution in barrier heights can be expected due to the thermal fluctuations of the lattice even at 5K (see figure 3). It is possible to estimate window cross sections analytically, and in the sudden cage limit, to derive expressions for cage exit probabilities including multiple tries(delay). The results, at least qualitatively vindicate the experimental conclusion. Moreover, they point out the weaknesses of simulations: choice of potentials, absence of recombination(multiple surface dynamics), and the need for smart sampling.

More recent results from H_2S studies most clearly indicate the need for careful treatment of the in-cage dynamics. Photodissociation in this case is followed by laser induced fluorescence of the SH fragment (see figure 4). Large cross sections are observed below the threshold region determined in the H_2O studies, moreover clear evidence for the $HS+H \rightarrow H_2+S$ channel is observed. The latter by thermoluminescence of S_2 upon warm-up of the solids.

Charge transfer transitions in rare gas solids reveal a variety of limiting behaviors, from fully delocalized extended states to strictly molecular type transitions.[11] The dynamics associated with these transitions is quite rich: charge separation,[12] self-trapping,[12] and formation of new ionic species[13] are phenomena associated with delocalization. In the case of molecular charge transfer states, the host response (solvation) dynamics is probed. An example of the latter is obtained from studies of XeF doped crystal Ar. Optical excitation of XeF leads to a full charge transfer creating an excited state dipole of ~10D. The host response on a fs time scale is probed by excitation spectra, which indicate a strong nonlinear coupling between the dipole and Ar motions. Longer time scales can be probed by transient gain measurements. Two types of atomic configurations are observed: one in which the system is locked in the solvated geometry, and one in which the system remains on the neutral surface minimum. Relaxation in the locked geometry can be described as vibrational cascading that proceeds on the time scale of 2-200 ps. Electronic relaxation in the ionic surface of the neutral geometry is complete in 56 ps, however the required atomic rearrangement for full solvation cannot be completed during the excited state lifetime of 6 ns. Given the simplicity of the system, it should be possible to clearly identify the collective coordinates involved in the observed dynamics of this system.

Acknowledgements

This research is supported by the US Air Force Phillips Laboratory under contract, S04611-90-K-0035, the National Science Foundation by a grant, ECS-8914321, and a fellowship to V. A. A. from the Alfred P. Sloan Foundation.

References

1. M. E. Fajardo, R. Whithnall, J. Feld, F. Okada, W. Lawrence, L. Wiedeman, and V. A. Apkarian, "Condensed Phase Laser Induced Harpoon Reactions," *Laser Chem.* 9, 1(1988).

2. H. Kunttu, J. Feld, R. Alimi, and V. A. Apkarian, "Charge Transfer and Radiative Dissociation Dynamics in Fluorine Doped Solid Krypton and Argon," *J. Chem. Phys.*, 94, 6671(1990).

3. J. Feld, H. Kunttu, and V. A. Apkarian, "Photodissociation of F_2 and Mobility of F Atoms in Crystalline Argon," *J. Chem. Phys.* 93, 1009(1990).

4. L. Meyer, "Phase Transitions In Van Der Waal's Lattices," *Advances In Chemical Physics,* Vol 16, pp. 343, 1970.

5. R. Alimi, R. B. Gerber, and V. A. Apkarian, "Photodissociation Dynamics of F_2 in Solid Kr: Theory versus Experiment,"*Phys. Rev. Lett.* 66, 1295(1991); "Dynamics of Molecular Reactions in Solids: Photodissociation of F_2 in Crystalline Argon", *J. Chem. Phys.* 92, 3551(1990).

6. H. Kunttu and V. A. Apkarian, "Photodissociation of F_2 in Crystalline Krypton: Effect of Molecule-Lattice Prealignment," *Chem. Phys. Lett.* 171, 423(1990).

7. G. J. Hoffman, E. Sekreta, and V. A. Apkarian, "Oriented Geminate Recombination of Cl_2 in Solid Xenon at High Pressure," *Chem. Phys. lett.* (accepted, 1992).

8. A. I. Katz and V. A. Apkarian, "Photodynamics in Cl_2-Doped Xenon under High Pressures: A Diamond Anvil Cell Study," *J. Phys. Chem.* 94, 6671(1990).

9. R. Alimi, R. B. Gerber, and V. A. Apkarian, "Dynamics of Molecular Reactions in Solids: Photodissociation of HI in Crystalline Xe," *J. Chem. Phys.*, 89, 174(1988).

10. a) R. Schriever, M. Chergui, H. Kunz, V. Stepanenko, and N. Schwentner, "Cage Effect for the Abstraction of H from H_2O in Ar Matrices," *Chem. Phys.* 91, 4128(1989).

    b) R. Schriever, M. Chergui, O. Unal, N. Schwentner, V. Stepanenko, "Threshold and Cage Fffect for Dissociation of H_2O and D_2O in Ar and Kr Matrices," *J. Chem. Phys.* 93, 3245(1990).

    c) R. Schriever, M. Chergui, and N. Schwentner, "Absolute Photodissociation Quantum Yield of H_2O in Ar Matrices," *J. Chem. Phys.* 93, 2499(1990).

11. a) M. E. Fajardo and V. A. Apkarian, "Charge Transfer Photodynamics in Halogen Doped Xenon Matrices II: Photoinduced Harpooning and the Delocalized Charge Transfer States of Solid Xenon Halides (F, Cl, Br, I)," *J. Chem. Phys.* 89, 4102(1988).

b) N. Schwentner, M. E. Fajardo, and V. A. Apkarian, "Rydberg Series of Charge Transfer Excitations in Halogen Doped Rare Gas Crystals," *Chem. Phys. Lett.* _154_, 237(1989).
12. M. E. Fajardo and V. A. Apkarian, "Energy Storage and Thermoluminescence in Halogen Doped Solid Xenon II: Photodynamics of Charge Separation, Self-Trapping and Ion-hole Recombination," *J. Chem. Phys.* _89_, 4124(1988).
13. H. Kunttu, J. Seetula, M. Rasanen, and V. A. Apkarian, "Photogeneration of Ions via Delocalized Charge Transfer States I: Xe_2H^+ and Xe_2D^+ in Solid Xe," *J. Chem. Phys.* (in print 1992).

Figure Captions

Figure 1: The photochemical steady-state concentration of XeO as a function of temperature shows a lambda-type phase transition (A) in crystalline Kr originally doped with O_2 and Xe. The phase transition can be observed by light scattering measurements, transmission of a He:Ne laser versus T (B). In this rapid scan hysteresis is evident in the warm-up and cool down cycles. The O_2 and O atom concentrations are clearly anticorrelated during heat cycles under constant irradiation at 193 nm (C).

Figure 2: The $Cl_2(A' \rightarrow X)$ recombinant emission in Xe is observed to be polarized (polarization ratio of 0.95 ± 0.02) at 40kbar, and 100K. The experiment is conducted in a diamond anvil cell (shown above), using a pair of polarizers as selector and analyzer.

Figure 3: Barrier height distributions for sudden cage exit in H/Ar, at 5K and 30K.

Figure 4: Excitation (top) and emission (middle) spectra of SH trapped in solid Ar. Photodissociation is probed by monitoring the growth of the SH signal as a function of irradiation time (lower panel).

Figure 1

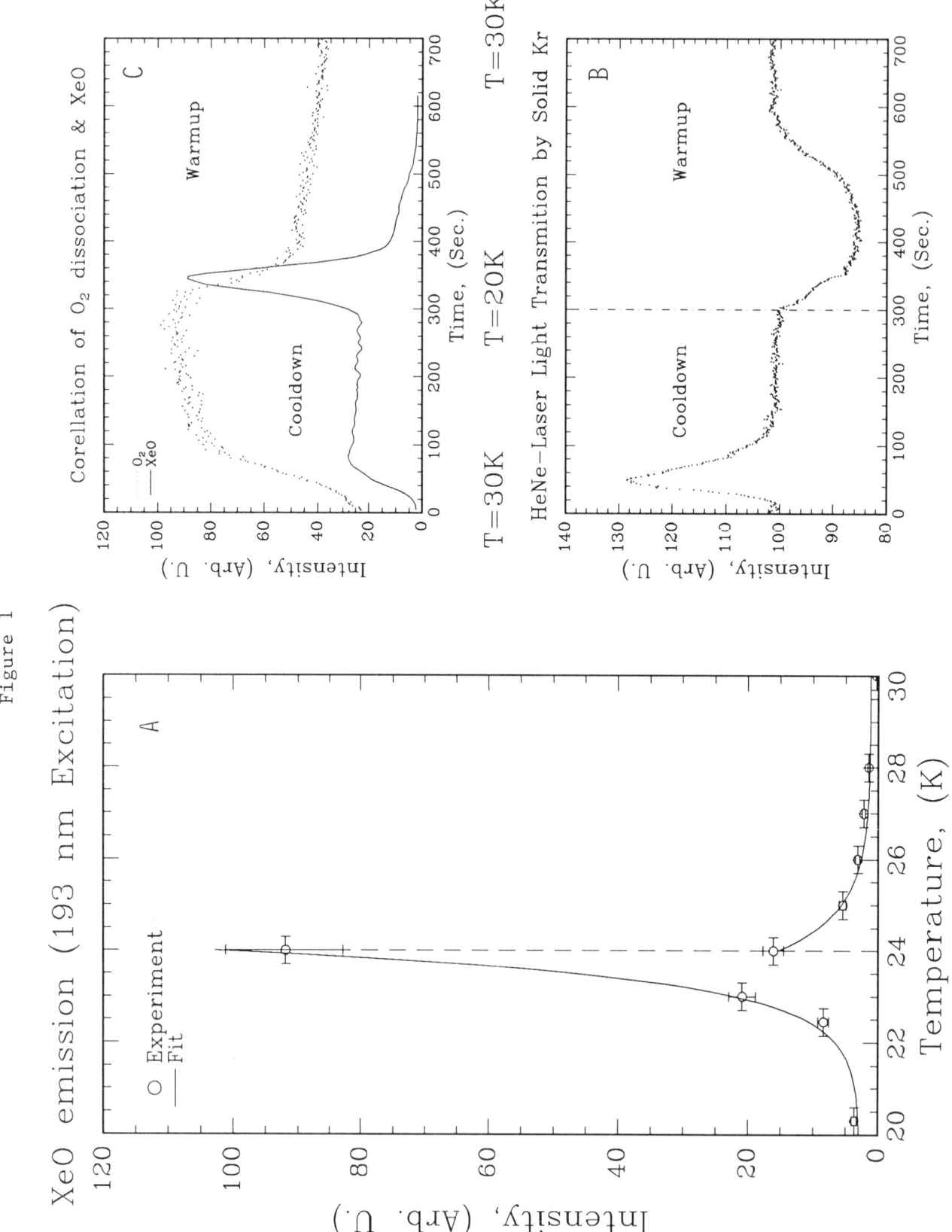

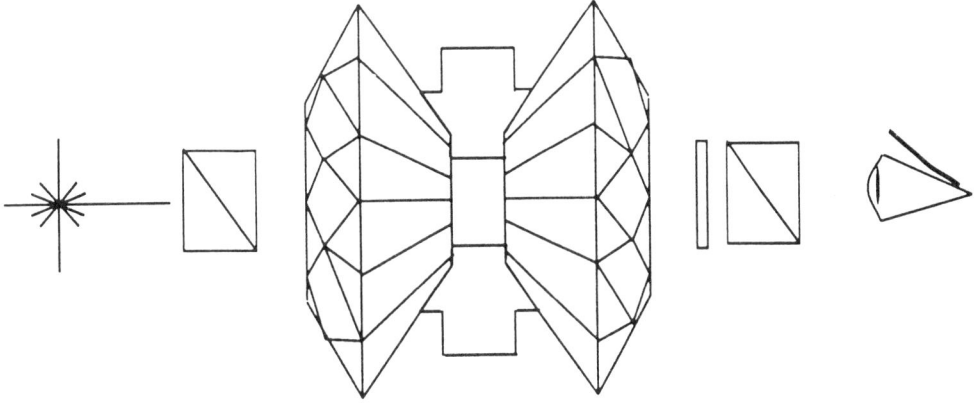

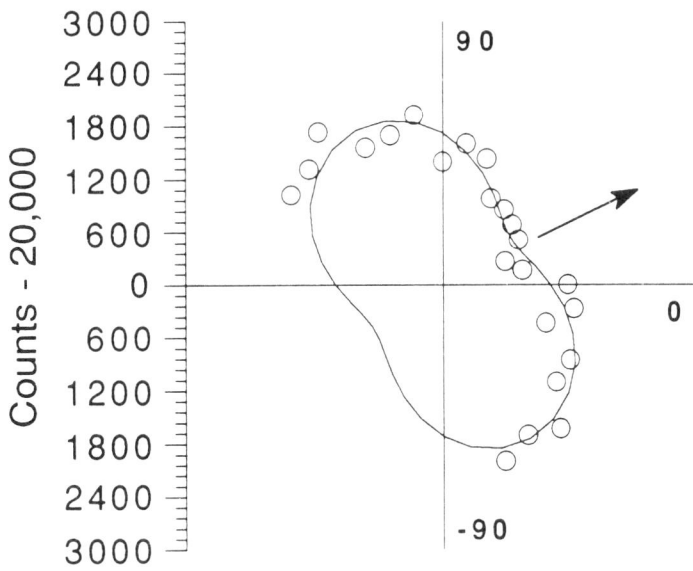

Figure 2

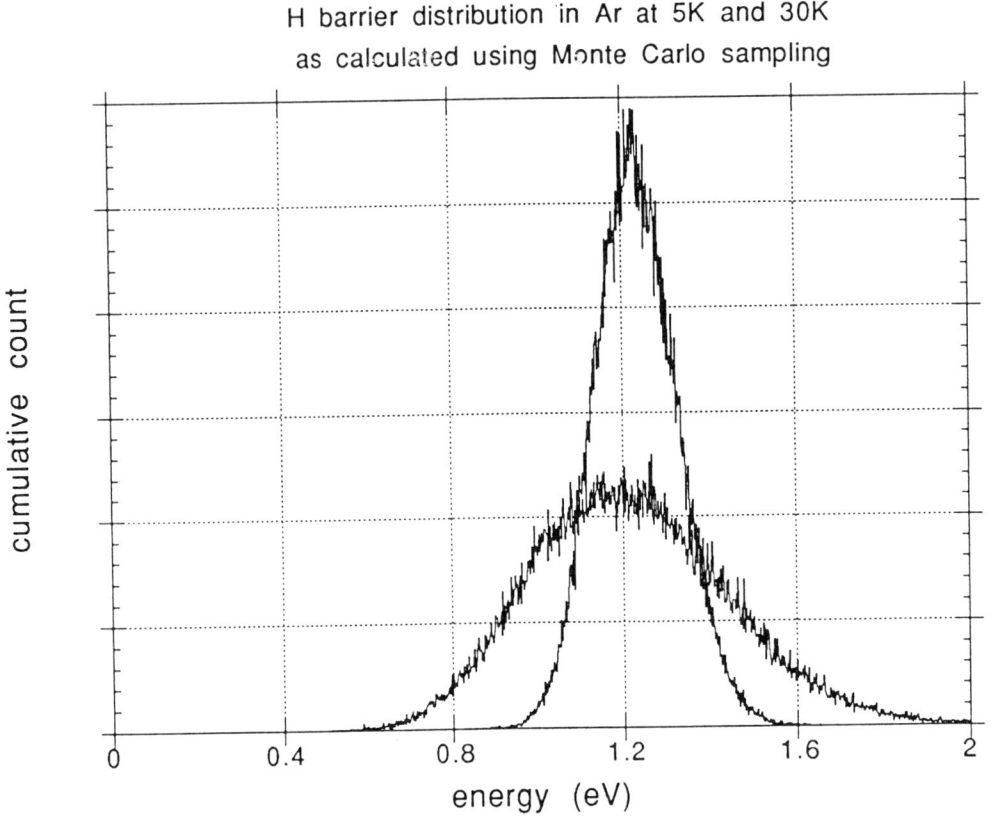

Figure 3

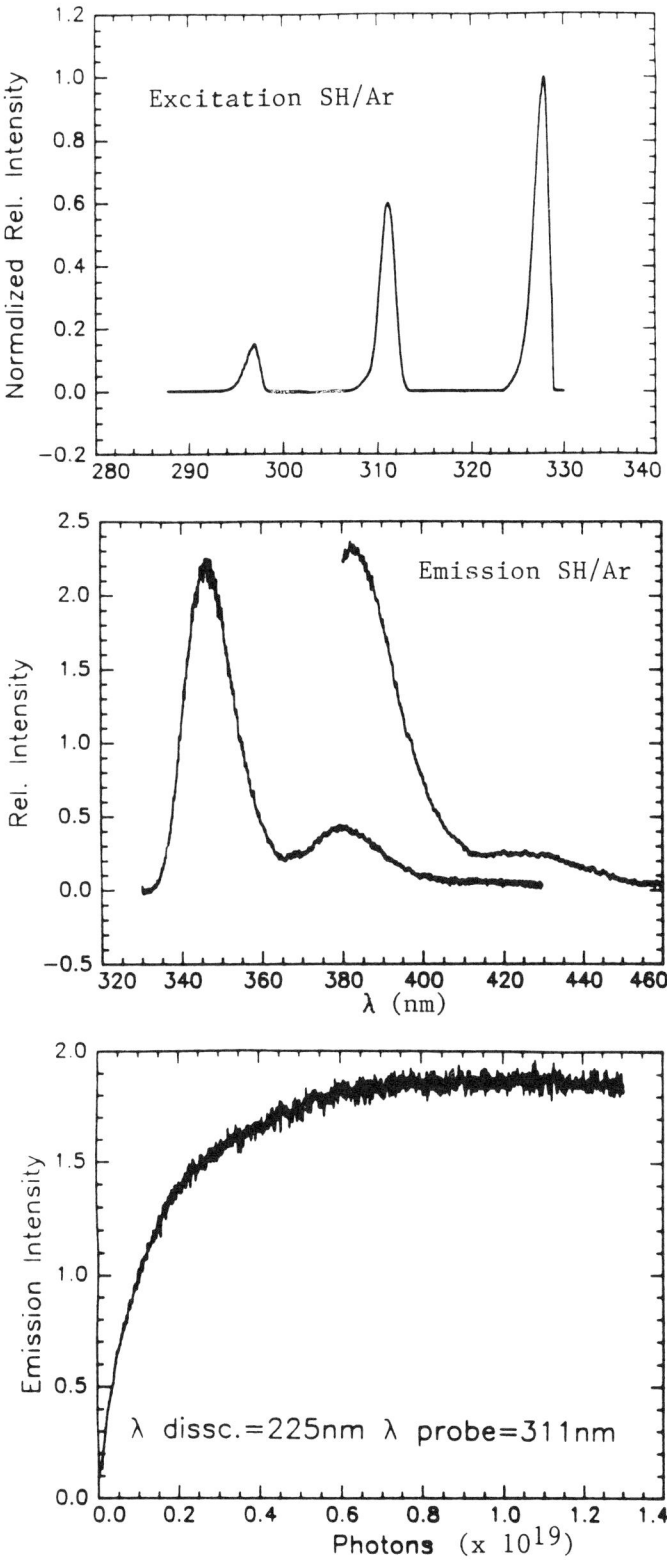

Figure 4

Structural Relaxation in Liquids and Glasses by Transient Hole Burning

Mark Berg, Jongwan Yu and Patrick Earvolino

Department of Chemistry and Biochemistry, University of Texas, Austin, Texas 78712

ABSTRACT

Ultrafast transient and permanent hole burning and fluorescence line narrowing are used to study the dynamic interaction between the electronic states of a nonpolar solute (dimethyl-s-tetrazine) and its solvent (glycerol) from low temperature glass to room temperature liquid. A model for phonon-modulated interactions not only describes the glass phase results, but also extrapolates to describe the short time dynamics in the liquid phase. However, this mechanism does not account for all of the ultrafast interaction seen at room temperature. A second mechanism connected to the structural coordinates of the liquid is identified. It has a strongly temperature dependent relaxation rate, which becomes subpicosecond at room temperature. This mechanism's rate is significantly faster than rates measured by other structural relaxation experiments.

INTRODUCTION

The time dependent perturbation of a solute's electronic states by the surrounding solvent is an important means for the solvent to exert an influence on the chemistry of the solute. In this paper, several spectroscopies commonly used for line narrowing in solid state systems are adapted to the study of electronic state perturbation dynamics in solution. These techniques are combined with a temperature variation which brings the solvent continuously from a solid glass to a room temperature liquid, allowing a direct connection between solid and liquid state dynamics. The system studied, dimethyl-s-tetrazine (DMST) in glycerol, does not have strong dipolar solute-solvent interactions. Instead, other less well studied interactions dominate. The ultrafast dynamics which were previously observed for DMST in room temperature solution[1] are shown to consist of two distinct components: one governed by phonon-like motion similar to the dynamics found in solids, and one connected with the structural dynamics which freeze out at the glass transition.

Recently, evidence for ultrafast electronic state perturbations has come from a number of sources including time-resolved Stokes' shifts,[2-4] photon echoes,[5-7] resonance Raman spectroscopy,[8] and theory.[9] Our previous hole burning measurements have shown that the nonpolar solute DMST experiences subpicosecond perturbations even in very viscous solutions.[1] We hypothesized that on a short time scale, the liquid could be viewed as a disordered solid and the perturbations treated like a phonon-modulated process.

Here temperature variation in a glass forming solvent is used to explore the connection between solid and liquid state dynamics.[10] A solid-state model of short time electronic state perturbations is found to be valid well into the liquid phase. However, this mechanism is not strong enough to account for the entire perturbation strength seen. A second mechanism is involved which is connected to the structural relaxation of the fluid. The rate of this mechanism is strongly temperature dependent and becomes subpicosecond by room temperature.

Because of the very short interaction distance, these experiments give a uniquely microscopic view of the structural dynamics. A number of other experiments have all shown a single structural relaxation time in glycerol.[11] The current experiments find a much faster relaxation time and must be measuring a different aspect of the complete structural relaxation process.

EXPERIMENTAL METHODS

In solution, the electronic transition frequencies of individual solute molecules are perturbed by the local solvent configuration. In transient hole burning (THB), the molecules with a particular transition frequency are bleached by an initial laser pulse. At a later time, the spectrum of the absorption bleach is measured by a second pulse. Initially, the bleach is confined to the excitation frequency. With increasing delay time, the solvent coordinates responsible for the electronic state perturbations relax, causing the bleach to spread over a wider spectral region. By measuring the spectral width of the bleach as a function of the delay time, the extent of solvent relaxation can be determined.

Transient hole burning was performed with a synchronously-pumped dye laser amplified by a Q-switched Nd:YAG-pumped chain of dye amplifiers. The amplified pulse had a width of 0.6 ps, a bandwidth of 13 Å, and a power of 0.7 mJ. A portion of this pulse traversed a variable delay line before being focussed onto the sample to create a transient bleach.

The remainder of the amplified pulse was used to create a white light continuum. This continuum was focussed within the bleached region of the sample before being dispersed in a monochromator. This pulse measured the spectrum of the bleach produced by the first pulse. The probe spectrum and a reference spectrum of the continuum light were detected by a dual diode array. The array allowed data collection with sufficient speed to avoid significant photochemical decomposition of the DMST during collection of a spectrum.

Permanent hole burning (PHB) is similar to THB except for the time between bleaching and measuring the spectrum. The PHB experiment is performed over a span of several minutes, so it is affected by all solvent relaxations except those which are effectively frozen into static configurations.

To do permanent hole burning, the photochemical decomposition of DMST is used to advantage. A three-plate birefringent filter was placed in the synchronously-pumped dye laser to narrow the spectrum and to broaden the pulse. The unamplified beam was focussed onto the sample for several minutes to create a permanent, photochemical bleach. Light from an arc lamp was then focussed through the bleached area to measure the spectrum of the bleach.

Fluorescence line narrowing (FLN) is used to measure the solvent relaxations occurring on a time scale of a few nanoseconds or less. As with hole burning, solute molecules within a narrow frequency region are excited. Their subsequent transition frequencies are monitored by measuring the frequency spectrum of the spontaneous fluorescence. Since most of the fluorescence occurs in a time region near the fluorescence lifetime, FLN is affected by perturbations that relax on that time scale.

Performing FLN differs from ordinary fluorescence spectroscopy only in using an excitation source and detection monochromator of sufficient resolution and in exciting the sample in a region where only a single vibronic transition is excited. An argon-ion pumped dye laser tuned to 565 nm excited the DMST samples. Fluorescence light was detected at the magic angle to eliminate rotational effects.

PHONON-MODULATED INTERACTIONS

In crystals, the interactions of an impurity electronic state with the surrounding medium is modulated by the phonon-induced motion of the neighboring solvent molecules. This process is well described by a configuration coordinate model.[12,13] The entire band of phonons is modelled by a single effective coordinate. In a lowest order approximation, the energy of the ground state is taken to be quadratic in the displacement of this coordinate. The excited state is taken to have a minimum at a displaced position, which generates a lowest order, linear coupling between the coordinate and the electronic transition energy. This coupling is parameterized by the equilibrium phonon Stokes' shift Σ_p, the difference between the transition frequency at the ground and excited state minima.

The standard deviation of the phonon-induced linewidth σ_p versus temperature T is given by[12]

$$\sigma_p^2 = \frac{\hbar\omega}{2} \Sigma_p \coth\left(\frac{\hbar\omega}{2kT}\right), \qquad (1)$$

which approaches a $\sigma_P^2 \propto T$ dependence at high temperatures. The frequency of the phonon mode is given by ω. In the solid glass, the phonon-modulated perturbations coexist with static perturbations. The absorption spectral width σ_A reflects the extent of both the rapid phonon perturbations and static perturbations. Hole burning spectral widths σ_H are only affected by rapid interactions. By combining these two experiments, the intrinsic phonon-induced width can be found,

$$\sigma_p^2 = \sigma_A^2 \{1 - [1 - (\sigma_H^2/\sigma_A^2)]^{1/2}\}. \qquad (2)$$

In the glass, phonons are the only source of picosecond modulation and their effects can be measured by transient hole burning. At sufficiently low temperatures, phonons are the only significant source of dynamics motions on any time scale and their effects can be measured by either transient or permanent hole burning. Results of these experiments in the glass, as well as in the liquid phase, are shown in Fig. 1 along with a fit to Eq. 1 for $\Sigma_p = 100$ cm^{-1}.

As Fig. 1 shows, the configuration coordinate model not only describes the rapid interactions in the solid phase, but extends continuously into the liquid phase. Thus, part of the original hypothesis, that the picosecond dynamics of a liquid can be described by a solid state picture, is confirmed.

ULTRAFAST STRUCTURAL RELAXATION

However, phonon dynamics cannot explain all of the perturbations seen by the solute. This is illustrated in Fig. 2, which shows all the widths from PHB, THB at 1.5 ps and 200 ps, and absorption spectra. The amount of broadening from phonon modulation cannot account for the total amount of perturbation indicated by the absorption width, even at room temperature.

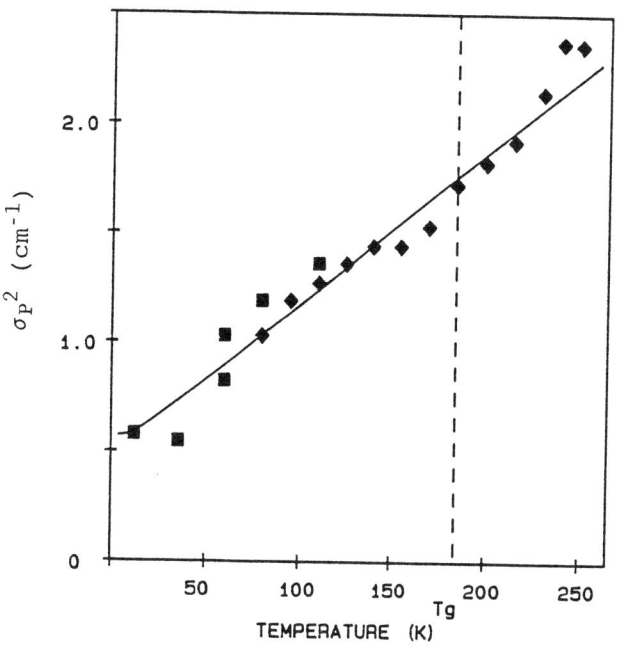

Fig. 1. Widths of the phonon-modulated perturbations σ_P derived from permanent (■) and 1.5 ps transient hole burning spectra (◆). The line is a fit to the configuration coordinate model (Eq. 1). The fit extends above the glass transition temperature T_g into the liquid phase. A constant of 4.7×10^3 cm^{-2} has been added to Eq. 1.

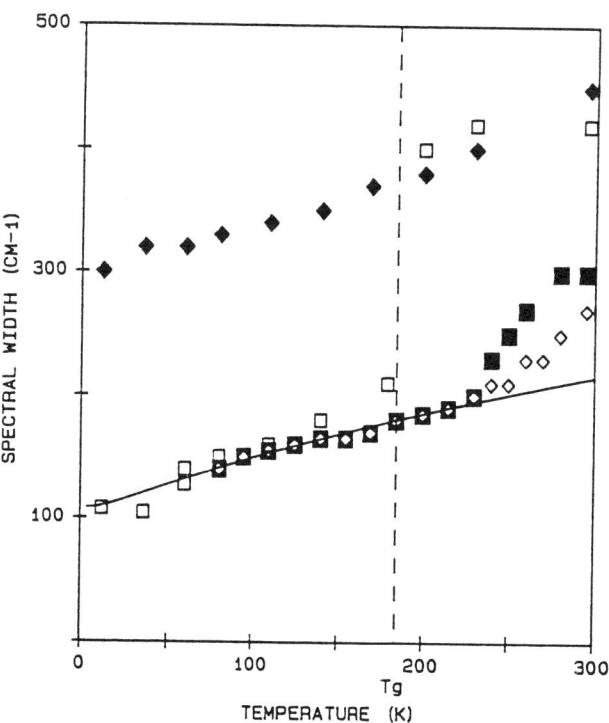

Fig. 2. Absorption (◆), permanent hole burning (□), 200 ps transient hole burning (■) and 1.5 ps transient hole burning (◇) widths versus temperature. The solid line represents the contribution of phonon-induced broadening (see Fig. 1). The glass transition temperature T_g is marked.

The extra broadening becomes evident in the various hole burning experiments at different temperatures. In PHB, which is sensitive to any modulation which occurs within the few minutes between burning and reading the hole, the extra broadening appears right at the glass transition. This indicates that the extra broadening is connected to perturbations by the structural coordinates, which are frozen in the glass, but attain a finite relaxation time above the glass transition. These are the coordinates responsible for the macroscopic fluidity of the solvent.

In the low temperature liquid region, the structural relaxation remains too slow to affect the picosecond hole burning experiments. With increasing temperature, the relaxation time increases rapidly. By 240 K, the structural relaxation time is fast enough to affect the 200 ps THB measurements, and the corresponding widths broaden beyond the phonon-induced curve. Initially, the 1.5 ps measurements are unaffected, but with slightly higher temperatures and faster relaxation rates, these widths also broaden above the phonon-induced curve.

These measurements show that although the phonon-induced perturbations are an important component of the ultrafast solvent interactions seen at room temperature, there is also a significant contribution from structural relaxation. The phonon interactions relax in a time comparable to the inverse of an acoustic phonon frequency (~100 fs) at all temperatures. On the other hand, the structural relaxation rate varies over 14 orders of magnitude and is only subpicosecond at higher temperatures.

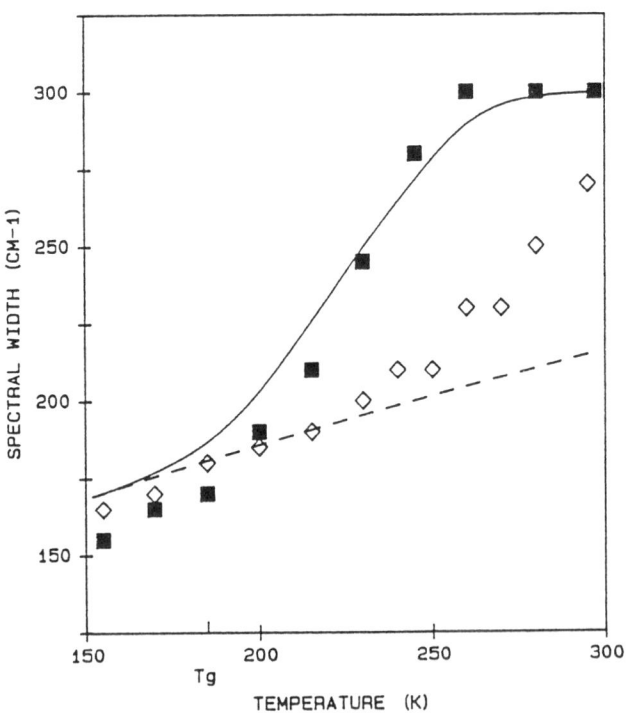

Fig. 3. Widths of fluorescence line narrowing spectra (■). Widths of 1.5 ps transient hole burning are shown for comparison (◊). The dashed line give the phonon-modulated contribution (Fig. 1). The solid line is the predicted value based on the peak shifts (Fig. 4)

FLUORESCENCE LINE NARROWING RESULTS

By using fluorescence line narrowing, the hole burning measurements can be extended in time scale to give a better picture of the structural relaxation.

Because of the higher signal-to-noise ratio and lack of interfering processes, FLN also gives somewhat more accurate results than hole burning.

The qualitative results can be seen in Fig. 3. As expected, the low temperature FLN widths match the phonon-modulated widths found from the hole burning experiments. Due to the nanosecond time scale of FLN, the FLN spectra begin to show additional broadening from structural relaxation at lower temperatures (slower rates) than the THB measurements. It is particularly clear from the combination of hole burning and FLN measurements that the deviation from the phonon-modulation model is due to an additional process becoming increasingly rapid, rather than from a break down of the phonon model at high temperature.

In addition to varying widths, FLN spectra also have temperature dependent frequency positions. Similar shifts also occur in the ordinary fluorescence spectra. These shifts are plotted versus temperature in Fig. 4a.

In order to quantify and interrelate these various shifts and widths, a model for the structural coordinates is needed. As with the phonon coordinates, a configuration coordinate model will be used. A single coordinate is used to describe the structural coordinates, and its coupling to the transition frequency is given by the equilibrium structural Stokes' shift Σ_S. The relaxation of the structural coordinate is assumed to follow Gaussian dynamics.[14]

$$K_S^e(\omega, \omega^\circ; t) = \exp\left(- \frac{\{\omega - Z(t)\omega^\circ - \Sigma_S[1-Z(t)]\}^2}{2\,\sigma_S^2[1-Z^2(t)]} \right), \tag{3}$$

$$Z(t; T) = \langle[\omega_S(t)-\Sigma_S][\omega_S(0)-\Sigma_S]\rangle / \langle[\omega_S(0)-\Sigma_S]^2\rangle,$$

where $K_S^e(\omega, \omega^\circ; t)$ is the conditional probability that the transition frequency will be ω at time t, if it is ω° at time zero. All frequencies are relative to the equilibium frequency in the ground state. The relaxation is described in terms of the autocorrelation function $Z(t; T)$ of the frequency perturbations due to the structural coordinate ω_S. Since the frequency perturbation is proportional to the displacement in the structural coordinate, $Z(t; T)$ is also the autocorrelation function for the coordinate. This function can be extracted from either the frequency shifts

$$Z(t; T) = \frac{\Omega(t; T) - (\Sigma_P + \Sigma_S)}{A\omega_e - \Sigma_S} = \frac{\Omega(t; T) - \Omega(\infty)}{\Omega(0) - \Omega(\infty)} \tag{4}$$

$$A = \sigma_S^2 / (\sigma_S^2 + \sigma_P^2)$$

or widths

$$Z^2(t; T) = \frac{\sigma_{FLN}^2(t; T) - (\sigma_P^2 + \sigma_S^2)}{A\sigma_P^2 - \sigma_S^2} = \frac{\sigma_{FLN}^2(\infty) - \sigma_{FLN}^2(t; T)}{\sigma_{FLN}^2(\infty) - \sigma_{FLN}^2(0)}. \tag{5}$$

The excitation frequency for the FLN is given by ω_e. The position and width of the FLN spectrum are given by Ω and σ respectively. The FLN experiment effectively occurs at $t = \tau_{fl}$, the fluorescence lifetime.

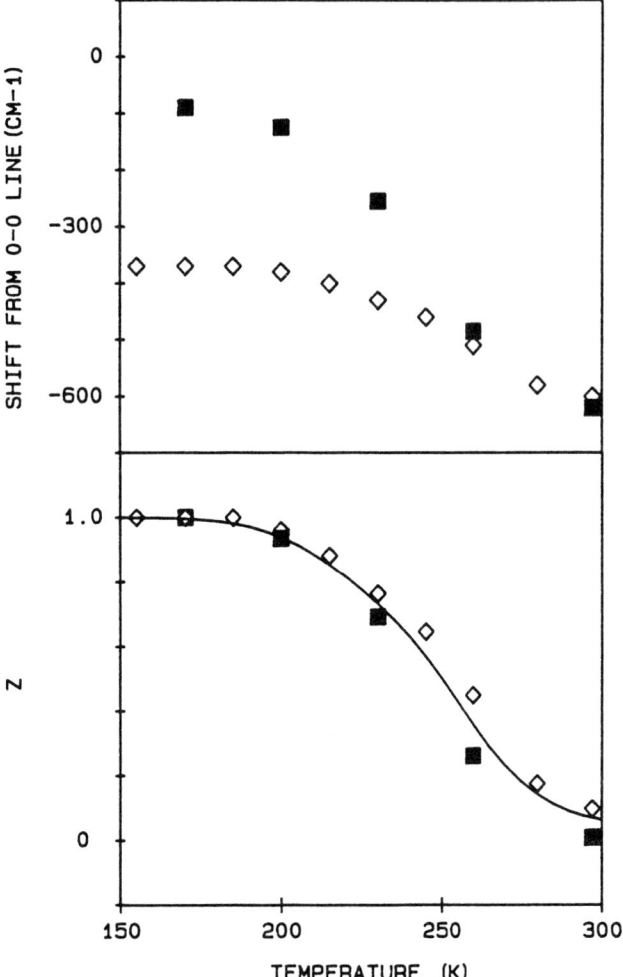

Fig. 4. Positions of the 0-0 emission line in ordinary fluorescence (■) and in FLN (◇). A) Absolute position relative to the 0-0 absorption line. B) Converted to the frequency autocorrelation function Z (Eq. 4). The solid curve is an empirical interpolation.

In ordinary fluorescence spectra, all values of the structural coordinate are excited equally, so $\omega_e = 0$. Using Eq. 4 to account for the difference in ω_e for the FLN and fluorescence spectra, the peak positions for both experiments can be converted to $Z(\tau_{fl}; T)$ (Fig. 4b). The two experiments have compatible peak shifts with temperature.

Using the $Z(\tau_{fl}; T)$ obtained from the frequency shifts, the FLN width can be predicted from Eq. 5. The prediction is compared to the data in Fig. 3. The shifts and widths are consistent. Note that the widths initially appear to reach their maximum value at lower temperatures than the shifts. This apparent difference is accounted for by the different dependence of the widths and shifts on Z predicted by the model.

This model is successful in interrelating three different measurements; fluorescence peak position, FLN peak position and FLN width. These measurements in turn are related to a more fundamental quantity, the relaxation function of the structural coordinate Z.

NATURE OF THE STRUCTURAL RELAXATION

A great deal of experimental data exists on the structural relaxation of glasses in general and glycerol in particular. Jeong, et al.[11] have shown that the data from dielectric relaxation,[15] digital correlation spectroscopy,[16] specific heat spectroscopy,[17] ultrasonic absorption, Brillouin scattering,[18] and shear relaxation all give the same structural relaxation time for glycerol

$$\tau_S(T) = (7.96 \times 10^{-16} \text{ s}) \exp[2310 \text{ K} / (T - 129 \text{ K})], \tag{6}$$

where the relaxation function is $\phi(t; T) = \exp\{-[t/\tau_S(T)]^{0.6}\}$.

These measurements can be compared to the FLN measurements by setting $Z(\tau_{fl}; T) = \phi(t; T)$ and predicting the FLN widths. This prediction is compared to the data in Fig. 5. The predicted widths broaden at much higher temperatures than the real data, i.e. at any given temperature, the relaxation is faster than predicted. In fact, the relaxation rates in Eq. 6 must be increased by a factor of ~150 before they approach the FLN data.

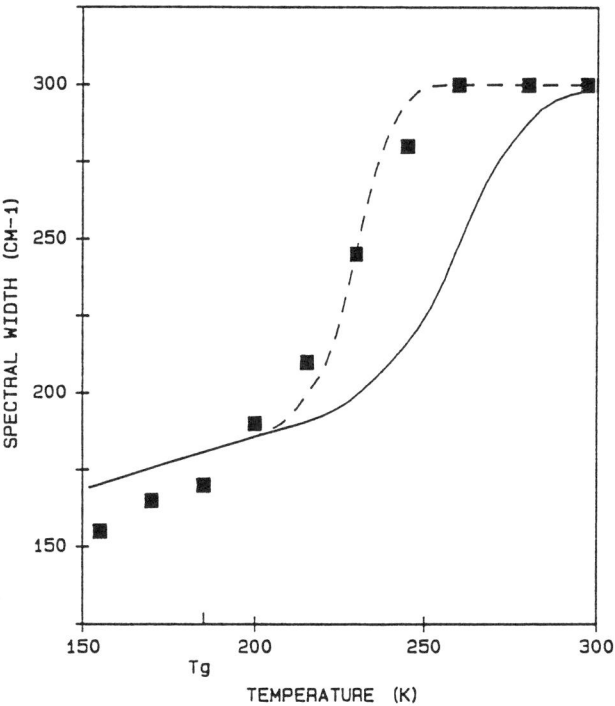

Fig. 5. Comparison of the fluorescence line narrowing widths (■) with the predictions based on the structural relaxation times of glycerol found in several other experiments (solid curve).[11] The relaxation times must be increased by a factor of 150 to match the rates seen in FLN (dashed curve).

The previous data definitely connected the extra broadening seen in FLN with structural relaxation. Clearly FLN is measuring a different aspect of the structural relaxation process than most other experiments. An interesting speculation arises in connection with hierarchical models of the glass transition.[15] These models posit that extensive relaxation of certain coordinates must precede a single step in the relaxation of other coordinates. Coordinates are arranged in a hierarchy of relaxation times ending in a maximum or ergodic time. Perhaps electronic state perturbations are strongly coupled to rapidly relaxing coordinates lower in the hierarchy, whereas other experiments sense the maximum relaxation time at the top of the hierarchy.

CONCLUSION

These experiments have shown that the subpicosecond electronic state perturbations seen in room temperature solutions are due to two distinct mechanisms. One mechanism is common to both solid and liquid states, one is unique to the liquid. At some temperature the structural relaxation time must merge with the phonon relaxation time, leading to a fundamentally different description of the liquid. Further studies linking the dynamics of solids and liquids through temperature variation across the glass transition will be very helpful in understanding the nature of relaxation in room temperature liquids.

Hole burning also has potential for the study of the dynamics leading to the glass transition. Most techniques for measuring structural relaxation can only measure relaxation over a few orders of magnitude in relaxation time, whereas the dynamics may vary over 14 orders of magnitude. In principle, hole burning can follow the entire range of structural relaxation times encountered as the liquid approaches the glass transition, from 10^{-12} to 100 s.

ACKNOWLEDGEMENTS

This research was supported by the National Science Foundation (CHE-8806381) with additional support from the Robert Welch Foundation. MB is a Presidential Young Investigator.

REFERENCES

1. T.J. Kang, J. Yu and M. Berg, Chem. Phys. Lett. **174**, 476 (1990); J. Chem. Phys. **94**, 2413 (1991).
2. P.F. Barbara and W. Jarzeba, Adv. Photochem. **15**, 1 (1990).
3. J.D. Simon, Acc. Chem. Res. **21**, 128 (1988).
4. E.W. Castner, B. Bagchi, M. Maroncelli, S.P. Webb, A.J. Ruggiero, and G.R. Fleming, Ber. Bunsenges. Phys. Chem. **92**, 363 (1988).
5. P.C. Becker, H.L. Fragnito, J.Y. Bigot, C.H. Brito Cruz, R.L. Fork and C.V. Shank, Phys. Rev. Lett. **63**, 505 (1989).
6. E.T.J. Nibbering, D.A. Wiersma and K. Duppen, Phys. Rev. Lett. **66**, 2464 (1991).
7. J.-Y. Bigot, M. Portella, R.W. Schoenlein, C.J. Bardeen, A. Migus and C.V. Shank, Phys. Rev. Lett. **66**, 1138 (1991).
8. A.B. Myers and B. Li, J. Chem. Phys. **92**, 3310 (1990).
9. A.M. Walsh and R.F. Loring, Chem. Phys. Lett. **186**, 77 (1991).
10. J. Yu, P. Earvolino and M. Berg, J. Chem. Phys., submitted.
11. Y.H. Jeong, S.R. Nagel and S. Bhattacharya, Phys. Rev. A **34**, 602 (1986); Y.H. Jeong, Phys. Rev. A **36**, 766 (1987).
12. M. Lax, J. Chem. Phys. **20**, 1752 (1952).
13. T.H. Keil, Phys. Rev. A **140**, 601 (1965).
14. R.F. Loring, Y.J. Yan and S. Mukamel, J. Chem. Phys. **87**, 5840 (1987).
15. D.W. Davidson and R.H. Cole, J. Chem. Phys. **19**, 1484 (1951).
16. C. Demoulin, C.J. Montrose and N. Ostrowsky, Phys. Rev. A **9**, 1740 (1974).
17. N.O. Birge and S.R. Nagel, Phys. Rev. Lett. **54**, 2674 (1985).
18. D.A. Pinnow, S.J. Candau, J.T. La Macchia and T.A. Litovitz, J. Acoust. Soc. Am. **43**, 131 (1968).
19. R.G. Palmer, D.L. Stein, E. Abrahams and P.W. Anderson, Phys. Rev. Lett. **53**, 958 (1984).

Ultrafast studies on electron transfer in the betaines: evidence for local heating

Nancy E. Levinger, Alan E. Johnson, Gilbert C. Walker,[a] Eva Åkesson[b] and Paul F. Barbara

Department of Chemistry, University of Minnesota, Minneapolis, MN 55455

Abstract

We have made ultrafast time resolved pump probe measurements on the intramolecular electron transfer (ET) of the betaines, specifically betaine30 and penta-t-butyl betaine. The data have been analyzed to determine the ET rate in a range of solvent environments, at various temperatures and at a variety of pump and probe wavelengths. In all cases, the observed ET rate is fast, often faster than predicted by common ET theories. As a result, we have extended some common theories to successfully predict the measured ET rates. In addition to the ET dynamics, the data also display evidence for local heat deposition in the immediate vicinity of the betaine molecule which our extended model qualitatively predicts.

1. Introduction

Current electron transfer (ET) theories predict rates which depend on static and dynamic properties of both the solute and the solvent. As such, they have had remarkable success at predicting rates for many experimentally observed ET reactions.[1-11] The success of these theories demonstrates the importance of solvation dynamics and intramolecular vibrations in promoting ET. The theories differ in the emphasis they place on various different degrees of freedom. For example, in barrierless ET reactions, several theories predict that solvent friction should limit the rate.[12-14] In inverted regime reactions, intramolecular vibrational modes can accelerate the reaction by reducing the effective barrier and decreasing the energy mismatch along classical coordinates.[9-11,15-17]

In this work and previously[15-17] we have investigated the ET dynamics of two related betaine molecules, betaine 30 (2,6-diphenyl-4-(2,4,6-triphenyl-1-pyridinio)phenolate) and penta-t-butyl-betaine (2,6-di-tertbutylphenyl-4-(2,4,6-tri-tertbutylphenyl-1-pyridinio)phenolate). Valence bond descriptions for the structures are shown below.

Here optical excitation results in a direct ET.[18,19] Hence the ET reaction we have studied is the reverse ET or ground state recovery. This electron transfer occurs in the inverted regime. Therefore, we expect that intramolecular vibrational modes will be important in ET dynamics.

We have used ultrafast pump probe spectroscopy to investigate the ET kinetics of the lowest electronic transition in the betaines, the transition corresponding to direct photoinduced ET. We have studied the dynamics of the reverse ET in many solvents, polymer films and as a function of temperature.[15-17] The major findings of our work is that the ET rate in the betaines is always fast, often faster than solvation times of the solvents used and faster than current theories predict. As a result, we have developed a hybrid model capable of accurately predicting the ET rates. The dynamics of the ET reaction were also found to be highly insensitive to changes in the bulk environment including temperature and polarity of the solvent despite the extreme sensitivity of the static absorption band to these parameters.

At probe wavelengths removed from the peak of the ET absorption band, the time resolved measurements display additional dynamics on a second time scale. We have used our hybrid model to predict these dynamics on the basis of local heating in the immediate vicinity of the betaine molecule. In this paper, we discuss the results of time resolved measurements in a variety of solvents, at various temperatures and for a variety of probe wavelengths revealing the effect of local heating. We also briefly discuss the failure of current, simple ET models and the development of our hybrid model for predicting ET.

2. Experimental Methods

All the time resolved pump probe measurements have been made using an ultrafast laser system which will be described in detail elsewhere.[23] Briefly, a mode locked Nd:YAG laser synchronously pumps a home built, linear cavity dye laser to yield 150fs (FWHM) pulses at ~800nm. The dye laser pulses are amplified to ~10µJ/pulse using the doubled output of a home built Nd:YAG regenerative amplifier to pump a three stage dye amplifier. For variable color experiments, a white light continuum was generated by focussing 80-100% of the amplified dye light into a 1cm static cell of H_2O. The light is selected using a 10nm bandpass filter and further amplified in one or two stages using ~20% of the doubled light from the regenerative amplifier. The instrument response function of the apparatus is <300fs as measured by the optical Kerr effect in H_2O.

Betaine30 dye was obtained from Aldrich and was used without further purification. Penta-t-butyl betaine was a generous gift from Professor C. Reichardt and was used without further purification. All solvents used were spectral or HPLC grade and were used without further purification. In some cases, solvents were dried over activated alumina. Static spectra were collected with a Cary 17 or Shimadzu UV160 spectrophotometer.

3. Experimental Results

Both betaine30 and penta-t-butyl betaine exhibit a photoinduced ET absorption band. These absorption bands are nearly identical for the two molecules. The static absorption spectra of betaine30 in acetone and benzene and penta-t-butyl betaine in benzene are shown in Fig. 1. The lower energy bands shown correspond to direct photoinduced ET. They are extremely sensitive to the solvent polarity shifting to higher energy with increasing solvent polarity.[18,19] Indeed, the peak of this absorption band in betaine30 is the basis for the

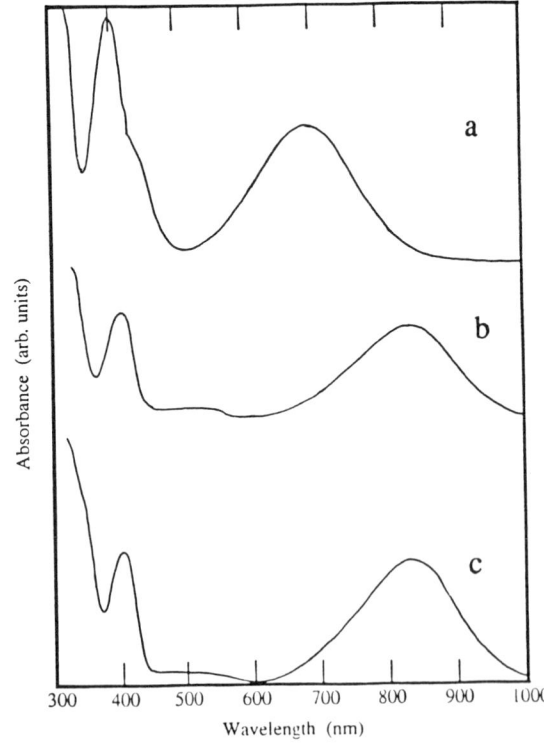

Fig. 1 Static absorption spectra for betaine30 in a) acetone, b) benzene and penta-t-butyl betaine in c) benzene.

$E_T(30)$ solvent polarity scale.[19]

We have made ultrafast time resolved pump-probe measurements of the reverse ET for betaine30 and penta-t-butyl betaine in a variety of solvent environments. Representative absorption transients for betaine30 in glycerol triacetate (GTA) and acetone are shown in Fig. 2. These solvents have similar polarities; their absorption spectra both peak near to 700nm. However, they display very different solvation dynamics. Acetone is a rapidly relaxing solvent. Its measured solvation time or the time scale for solvent fluctuations, τ_s, is approximately 0.8ps. In contrast, GTA is quite slowly relaxing, $\tau_s > 100$ps. In comparison, the ET times measured for acetone, Fig. 2b, where $\tau_{ET} \sim 1.3$ps, and GTA, Fig. 2a, $\tau_{ET} \sim 7$ps, are much closer than we would predict based on the solvation dynamics. Indeed, a comparison of the ET time versus the solvation time for a battery of solvents, as shown in Fig. 3, reveals two different kinetic regimes. For rapidly relaxing solvents, the ET time appears to be directly proportional to the solvation dynamics. But in slowly relaxing solvents, the ET times display only a mild dependence on τ_s. We have interpreted this as a manifestation of a competition between two ET mechanisms, one dominated by vibrational dynamics and one dominated by solvation dynamics. A prediction of this behavior based on our hybrid model is given below.

In addition to varying the molecule's environment, we have also made ultrafast time resolved pump-probe measurements with 700nm and 800nm excitation and 640, 700, and 800nm probe wavelengths. The measured dynamics do not appear to depend strongly on the excitation wavelength. However, we have observed a second component to the dynamics for some probe wavelengths in addition to the dynamics corresponding to the reverse ET. This component varies as a function of the probe wavelength. This effect is easily seen in Fig. 4. When we probe near the peak of the ET absorption band at 700nm, Fig. 4b, we are least sensitive to band shifting effects due to local heating hence the measured dynamics should most accurately reflect the reverse ET. However, at a wavelength longer than the

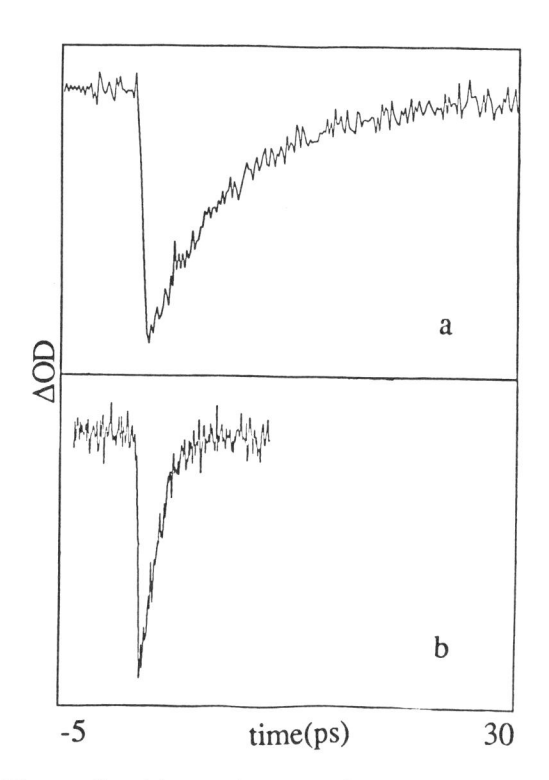

Figure 2. Absorption transients for betaine30 in a) GTA and b) acetone.

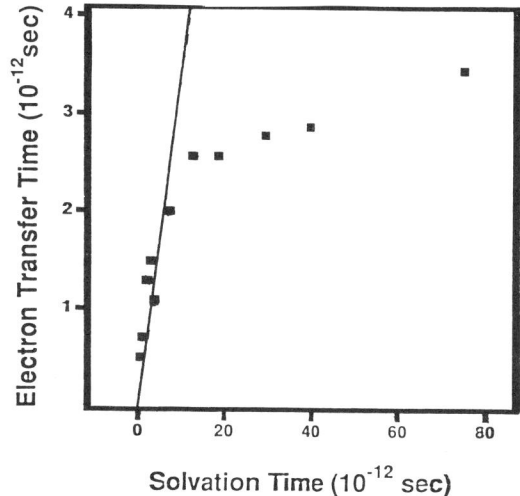

Figure 3. Electron transfer times for betaine30 in a variety of different solvents as a function of solvation time; boxes are experimental points, line indicates $\tau_{ET} = \tau_s$.

absorption peak, 800nm, Fig. 4c, we observe a fast recovery of the ground state followed by an increased absorption which decays more slowly than the initial bleach. At a wavelength shorter than the absorption peak, 640nm, Fig. 4a, multiexponential dynamics are apparent with the second time scale similar to that observed for the increased absorption at 800nm. We interpret this as evidence of local heating of the betaine molecule environment. Because the betaine absorption band is a strong function of temperature, the absorption band would be expected to shift its peak to longer wavelength with increasing temperature. The reverse ET deposits a large amount of energy into the molecule's environment thereby raising the effective local temperature. This effect would manifest itself in the measured dynamics as an increased absorption on the red edge of the spectrum and an additional bleach on the blue edge, which is what we observe. The recovery of the increased absorption and the additional bleach then correspond to local cooling of the molecule's environment.

4. Theoretical treatment

ET theories often treat the solvent polarization as the primary, inhomogeneous coordinate. For ET reactions occurring in the inverted regime, this is not always appropriate. We have used two current theories to try to model dynamics we observe for ET in the betaines. Both theories include a solvent mode and an intramolecular vibrational mode for promotion of the ET. Because the ET in the betaines is directly optically induced, all the parameters necessary to predict the dynamics can be directly obtained from the absorption spectrum. We expect these theories to be applicable to the ET of the betaines. However, they have often been invoked without testing whether they are appropriate for describing the experimental processes. As a result, we have tried to test them for the present experimental case.

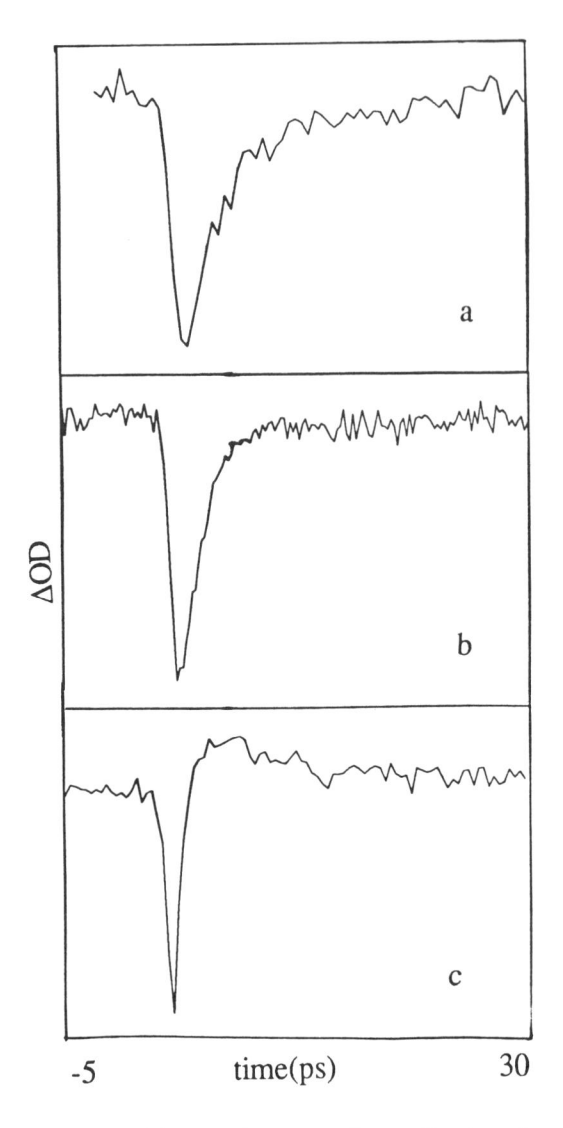

Figure 4. Absorption transients for betaine30 in acetone; 800nm excitation and a) 800nm probe, b) 700nm probe and c) 640nm probe.

Sumi and Marcus[20] presented a model that includes a classical solvent mode and an average classical intramolecular vibration as the reaction coordinate. In this model, the ET reaction occurs along vibrational coordinates with a rate constant that depends on the polarity of the local environment. Their model treats the solvent with a classical diffusion equation and vibrational coordinates are treated as a single, average classical coordinate. This theory predicts a barrier to the ET reaction. As a result, it underestimates the betaine ET rate by as much as six orders of magnitude. In fact, the theory by Sumi and Marcus always predicts rates that are orders of magnitude slower than the observed dynamics.

In their theory, Jortner and Bixon[21] use a classical solvent coordinate but add an average, quantized vibration which allows vibronic transitions. The frequency of the quantized mode is an average of the modes coupled to the ET. This model works well for quickly relaxing solvents because the addition of vibrational levels can significantly reduce the barrier to reaction. The rate then becomes limited by the solvent fluctuations. However, the Jortner/Bixon theory[18] still significantly underestimates the observed ET rates in slowly relaxing solvents. This occurs because each vibronic transition is similar to a delta function sink, so the rate is still limited by the solvent fluctuations.

Both the Sumi/Marcus[20] and Jortner/Bixon[21] theories are applicable to this inverted regime ET, but there are discrepancies between rates predicted by these theories and the experimentally measured rates. As a result, we have developed a hybrid model[17] that is an extension of the Sumi/Marcus theory which includes a quantized vibration similar to the Jortner/Bixon theory. The hybrid model accurately predicts the ET rates for the betaines because it allows the reaction to proceed via vibronic transitions similar to Jortner and Bixon's model but fluctuations in the classical vibration can also drive the reaction. Hence the reaction is not limited by the time scale of the solvent fluctuations for solvents with very slow solvation dynamics. A comparison of the predictions from each of these three models is shown in Fig. 5 along with the observed ET rates for betaine30.

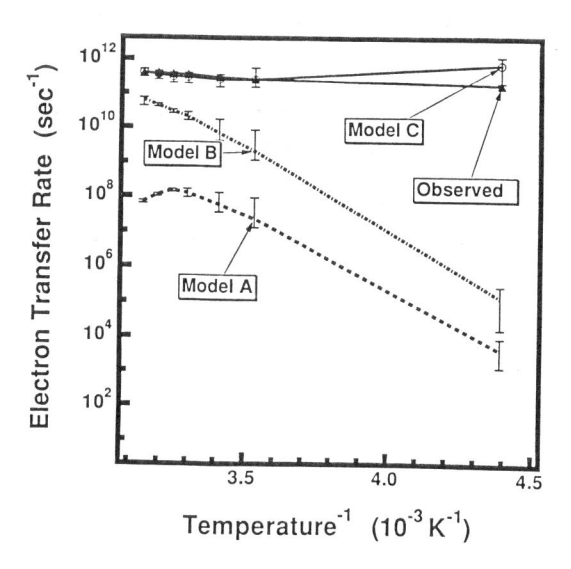

Figure 5. Electron transfer rate as a function of inverse temperature for betaine30 in GTA. Model A is the Sumi/Marcus model, Model B is the Jortner/Bixon model and Model C is the hybrid model described in the text.

To obtain the parameters for the hybrid model, we fit the absorption lineshape to a model that includes the two nuclear degrees of freedom. From the fit, we obtain values for the quantized reorganization energy, λ_{QM}, the classical reorganization energy, λ_{CL}, the driving force for the reaction, $\Delta G°$, and the quantized vibrational frequency, ν_{QM}. The classical reorganization energy is partitioned into solvent and vibrational contributions by assuming that there is no contribution to λ_{CL} in a nonpolar solvent, cyclohexane. The solvent contribution, λ_{SOLV}, is then the difference between the total λ_{CL} and the contribution from the nonpolar solvent case. The electronic matrix element, V_{EL} is estimated from the Hush relationship.[22] From the fit to the absorption spectrum, we can construct free energy curves on which to predict the dynamics.

We extended our hybrid model to predict time dependent absorption spectra.[23] This requires modelling the dynamics of the population on both the ground and the excited state surfaces. At any time, population missing from the ground state contributes to a bleach of the absorption spectrum. Population on the excited state can contribute to the signal by absorption to a higher excited state or through stimulated emission. In our analysis, we consider only the contribution from stimulated emission. All the parameters necessary for these dynamical predictions are obtained from the static absorption spectrum hence they provide a good measure of the accuracy of our hybrid model.

Figure 6 displays the dynamics of the ground and excited state populations for betaine30 in acetone. The residual population in the ground state has been subtracted for clarity. At t=0 there is a Gaussian distribution in the excited state centered at X=1, the equilibrium position in the ground state and no population in the ground state. The population in the excited state moves along the solvent coordinate towards lower energy and reacts depositing population back in the ground state. However, initially the population does not return to the ground state equilibrium position. The population in the ground state appears near X=0, the equilibrium position in the excited state and moves back to X=1. At intermediate times there is more population near X=0 in the ground state than the equilibrium population there. This effect is responsible for an increased absorption on the low energy side of the absorption spectrum. The transient absorption spectrum is calculated by convoluting the time dependent excited and ground state populations with a vibrational shape function.

5. Comparison of Predictions and Experimental Measurements

Because we can independently predict and measure the dynamics of the betaines in solution for various excitation and probe wavelengths, we are able to make direct comparisons of the predictions and the experimental measurements. Comparisons of the predicted and observed absorption transients for 700nm pump and probe light are shown for betaine30 in GTA in Fig. 7. We feel that the agreement between the predicted and observed data are remarkable especially considering that there are no adjustable parameters for the predictions. We made comparisons of the observed and predicted dynamics at other wavelengths. The general trends in the predicted and observed spectra are similar, including an increased absorption at 800nm. However, the predicted spectra do not quantitatively match the time scales for the corresponding measured spectra. This may be due to our assumption throughout this model that vibrational relaxation is instantaneous.

6. Summary

We have made measurements and predictions of the dynamics accompanying the ET of the betaines in solution. In rapidly relaxing solvents, the ET rate is dominated by solvation dynamics while in slowly relaxing solvents the rate is dominated by intramolecular vibrational dynamics. In addition to the ET dynamics, we have observed evidence for local heating of the betaine molecule environment following the ET. We have extended our hybrid model[17] to include ground and excited state population dynamics thereby allowing prediction of the spectral dynamics.[23] The model predicts the ET dynamics and some local heating due to nonequilibrium population distributions following the ET reaction. The model accurately predicts the dynamics near the peak of the absorption spectrum.

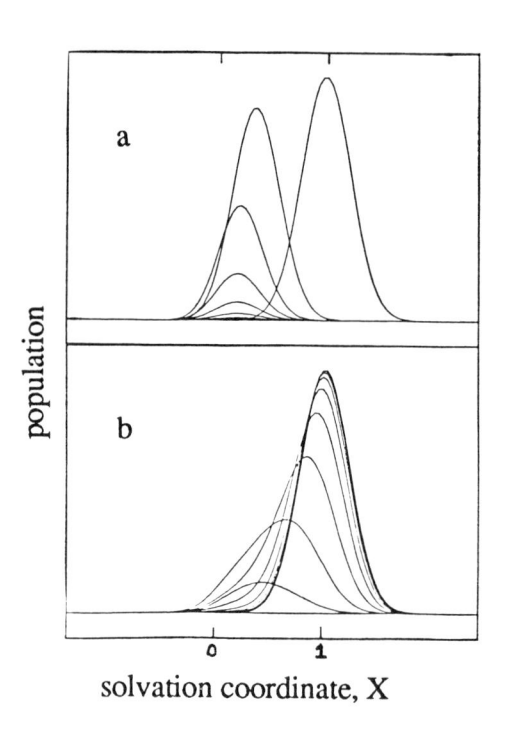

Figure 6. Time dependent a) excited state and b) ground state populations for betaine30 in acetone.

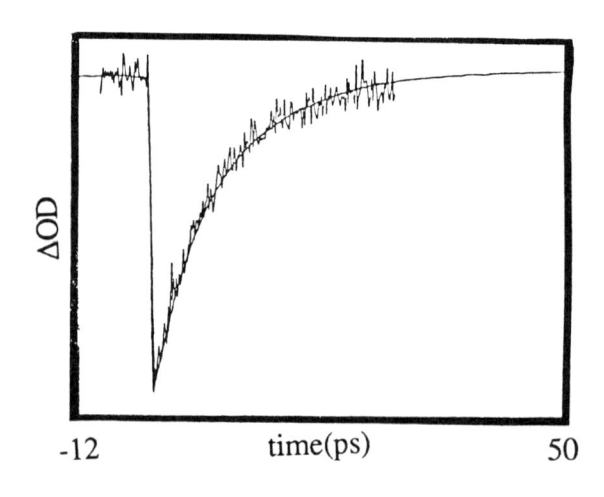

Figure 7. Observed and predicted absorption transients for betaine30 in GTA.

6. Acknowledgements

This research was supported by the National Science Foundation and the Office of Naval Research. NEL gratefully acknowledges the NSF for a postdoctoral fellowship.

7. References

a. present address: Department of Chemistry, University of Pennsylvania, Philadelphia, PA, 19104

b. present address: Department of Physical Chemistry, University of Umeå, 901 87 Umeå, Sweden.

1. M. D. Newton and N. Sutin, "Electron transfer reactions in condensed phases," *Ann. Rev. Phys. Chem.* vol. 35, pp. 437-80, 1984.

2. R. A. Marcus and N. Sutin, "Electron transfers in chemistry and biology," *Biochem. Phys. Acta,* vol. 811, pp. 265-322, 1985.

3. E. M. Kosower and D. Huppert, "Excited state electron and proton transfers," *Ann. Rev. Phys. Chem.* vol. 37, pp. 127-56, 1986.

4. J. D. Simon, "Time-resolved studies of solvation in polar media," *Acc. Chem. Res.* vol. 21, pp. 128-34, 1988.

5. M. Maroncelli, J. MacInnis and G. Fleming, "Polar solvent dynamics and electron-transfer reactions," *Science* vol. 243, pp. 1674-81, 1989.

6. M. J. Weaver and G. E. McManis, III, "Dynamical solvent effects of electron-transfer processes: recent progress and perspectives," *Acc. Chem. Res.* vol. 23, pp. 294-300, 1990.

7. P. F. Barbara and W. Jarzeba, "Ultrafast photochemical intramolecular charge and excited state solvation," *Adv. Photochem.* vol. 15, pp. 1-68, 1990.

8. P. F. Barbara, G. C. Walker and T. P. Smith, "Vibrational modes and the dynamic solvent effect in electron and proton transfer," *Science* , in press.

9. G. L. Closs and J. R. Miller, "Intramolecular long-distance electron-transfer in organic molecules," *Science* vol. 240, pp. 440-7, 1988.

10. N. Liang, J. R. Miller and G. L. Closs, "Temperature-independent long-range electron transfer in the Marcus inverted regime," *J. Am. Chem. Soc.* vol. 112, pp. 5353-4, 1990.

11. I. R. Gould, R. H. Young, R. D. Moody and S. Farid, "Contact and solvent-separated geminate radical ion pairs in electron-transfer photochemistry,"*J. Phys. Chem.* vol. 95, pp. 2068-80, 1991.

12. T. J. Kang, W. Jarzeba, P. F. Barbara and T. Fonseca, "A comprehensive photodynamical model for the excited state electron transfer of bianthryl and related molecules," *Chem. Phys.* vol. 149, pp. 81-95, 1990.

13. a) K. Tominaga, G. C. Walker, W. Jarzeba and P. F. Barbara, "Ultrafast charge separation in ADMA: experiment, simulation and theoretical issues," *J. Phys. Chem.* vol. 95, pp. 10475-85, 1991; b) K. Tominaga, G. C. Walker, T. J. Kang, P. F. Barbara and T. Fonseca,

"Reaction rates in the phenomenological adiabatic excited-state electron-transfer theory," *J. Phys. Chem.* vol. 95, pp. 10485-92, 1991.

14. J. D. Simon and S.-G. Su, "Picosecond Stokes shift studies of solvent friction: experimental measurements of time-dependent band shape and integrated intensity," *Chem. Phys.* vol. 152, pp. 143-53, 1991.

15. E. Åkesson, G. C. Walker and P. F. Barbara, "Dynamic solvent effects on electron transfer rates in the inverted regime: ultrafast studies on the betaines," *J. Chem. Phys.* vol. 95, pp. 4188-94, 1991.

16. E. Åkesson, A. E. Johnson, N. E. Levinger, G. C. Walker, T. P. DuBruil and P. F. Barbara, "Temperature dependence of the inverted regime electron transfer kinetics of betaine30 and the role of molecular modes," *J. Chem. Phys.*, in press.

17. G. C. Walker, E. Åkesson, A. E. Johnson, N. E. Levinger and P. F. Barbara, "The interplay of solvent motion and vibrational excitation in electron transfer kinetics: experiment and theory," *J. Phys. Chem.* in press.

18. A. M. Kjaer, J. Ulstrup, "Solvent bandwidth dependence and band asymmetry features of charge-transfer transitions in N-pyridinium phenolates," *J. Am. Chem. Soc.* vol. 109, pp.1934-42, 1987.

19. C. Reichardt, "Empirical parameters of solvent polarity as linear free-energy relationships," *Angew. Chem. Int. Ed. Eng.* vol. 18, pp. 98-110, 1979.

20. H. Sumi, R. A. Marcus, "Dynamical effects in electron transfer reactions," *J. Chem. Phys.* vol. 84, pp. 4894-914, 1986.

21. J. Jortner, M. Bixon, "Intramolecular vibrational excitations accompanying solvent-controlled electron transfer reactions," *J. Chem. Phys.* vol. 88, pp. 167-70, 1988.

22. N. S. Hush, "Intervalence-transfer absorption. part 2. Theoretical considerations and spectroscopic data," *Prog. Inorg. Chem.* vol. 8, pp. 391-444, 1967.

23. A. E. Johnson, N. E. Levinger, G. C. Walker, E. Åkesson, P. F. Barbara, "Spectral predictions of a diffusive model for electron transfer and direct comparisons to experiment," *J. Phys. Chem.* in preparation.

Solution Photochemistry of OClO: Excited State Dissociation and Isomerization

Robert C. Dunn, Bret N. Flanders, and John D. Simon

Department of Chemistry
University of California at San Diego
9500 Gilman Drive
La Jolla, California 92093-0341

ABSTRACT

The photodissociation dynamics of OClO in room temperature water solution were investigated using picosecond transient absorption spectroscopy. The time dependent data are consistent with the conclusion that following excitation at 355 nm, 90% of the OClO molecules photodissociate forming ClO and O and 10% isomerize forming ClOO. The photochemically generated ClOO thermally decomposes into Cl and O_2 with a rate constant of ~6.7 x 10^9 s^{-1}.

1. INTRODUCTION

Recent concern over the appearance of an ozone hole in the stratosphere of the Antarctic has initiated widespread research on the photochemical reactivity of many oxychlorinated molecules that are thought to contribute to ozone loss. One molecule that is currently receiving attention is chlorine dioxide (OClO). Stratospheric measurements in the Antarctic show that the concentration of OClO anticorrelates with that of ozone, implying that their respective chemistries are linked[1].

The photochemistry of OClO was originally thought to lead exclusively to ClO and O. Such chemistry does not lead to a net ozone loss. However, recent gas phase studies of OClO reported by Richard and Vaida show that in addition to bond cleavage, electronically excited OClO can isomerize to form ClOO[2,3]. Compared to OClO, ClOO is thermodynamically more stable by ~3 kcal/mole and kinetically less stable, rapidly dissociating into Cl and O_2[4-6]. The formation of Cl from electronically excited OClO provides a channel by which ozone can be catalytically decomposed[7].

While a number of gas phase and matrix studies on OClO have appeared, little is known about the photochemistry of OClO in solution. In this report, a summary of results of picosecond time resolved absorption experiments of room temperature OClO in water are given. A detailed account will appear shortly[8].

2. EXPERIMENTAL

The experimental details of the laser system used to collect the transient absorption dynamics have been presented elsewhere[9].

OClO was synthesized using the method reported by Bray[10]. Briefly, 30 g of oxalic acid was mixed with 8 g of potassium chlorate and placed in a 50 - 60°C water bath. Approximately 4 mL of water was added to the mixture, initiating the following reaction.

$$2ClO_3^- + H_2C_2O_4 \rightarrow 2OH^- + 2CO_2 + 2OClO \qquad (1)$$

Nitrogen was passed over the reaction mixture to collect the product gasses, OClO and CO_2. The OClO and CO_2 were subsequently bubbled through a water sample resulting in aqueous OClO concentrations of up to 2M.

3. RESULTS AND DISCUSSION

The remainder of the paper is organized in the following manner. In sections 3.1. and 3.2. the results of experimental studies of OClO in gas phase and matrices will be reviewed. In section 3.3., the solution phase properties will be discussed. This will include a summary of the steady state spectra of OClO and related molecules, the results of the time resolved absorption experiments and, a kinetic model that accounts for the absorption signals observed in the time resolved experiments.

3.1. Isolated OClO in the gas phase

Richard and Vaida have shown that excitation of gas phase OClO in the near UV produces both vibrationally excited ClO (and atomic oxygen) and the isomer ClOO[2,3]. The isomerized molecule, ClOO, subsequently dissociates into atomic chlorine and molecular oxygen[5,6]. These processes are shown in equation 2.

$$\begin{aligned} OClO + h\nu &\rightarrow OCl + O \\ &\rightarrow ClOO \rightarrow Cl + O_2 \end{aligned} \qquad (2)$$

Three closely spaced excited states of OClO are thought to give rise to this reactivity[4]. Excitation of OClO in the near UV populates the bound $^2A_2$ excited state ($^2A_2 \leftarrow {}^2B_1$). This state is proposed to couple to two excited states ($^2B_2$ and $^2A_1$) which lie nearby in energy[2-4]. Coupling to the $^2B_2$ state leads to the formation of ClO and O[2,3]. Coupling to the $^2A_1$ state forms Cl and O_2[2,3]. Accurate quantum yield measurements for these two pathways are complicated by the extreme reactivity of the ClOO isomer. Donaldson and coworkers have determined that following excitation at 362 nm, 15% of the excited OClO decay to form Cl and O_2, presumably via formation of ClOO[6].

3.2. Low temperature matrix photolysis

In contrast to the competing photochemical pathways observed in gas phase samples, excitation near the maximum of the near UV transition of OClO in matrices of nitrogen[11], argon[11,12], and sulfuric acid[13] results in exclusive formation of ClOO.

3.3. Solution photolysis

Before considering the results of the time-resolved studies of OCIO in water, the steady-state absorption spectra of several of the chemical species shown in Eq. 2 will be examined. As atomic and molecular oxygen absorb in spectral regions that are not accessible with the current laser system, the steady-state spectra for these species will not be considered.

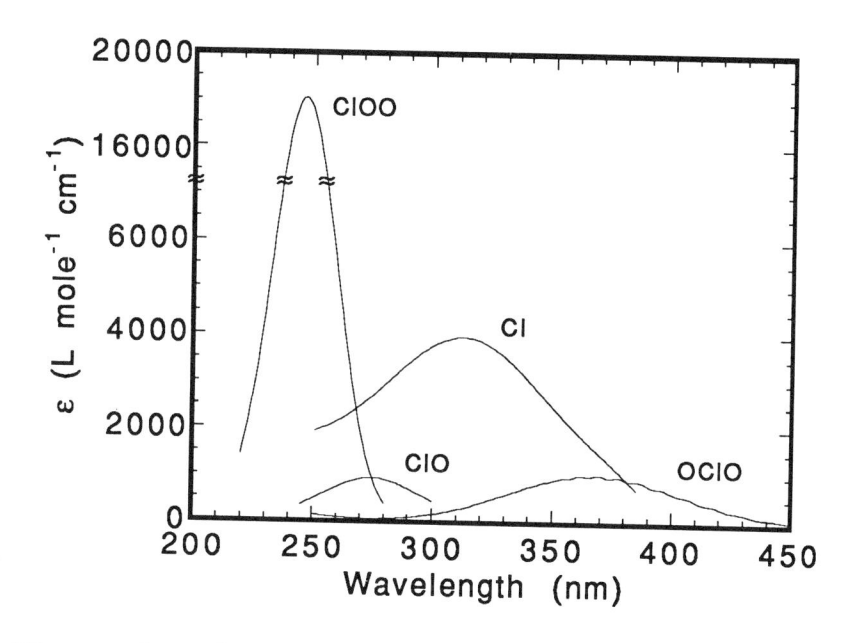

Figure 1: Steady state absorption spectra of aqueous OCIO[14], ClO[16], Cl[16], and gas phase ClOO[17,18].

We recently reported the solution phase absorption spectra of OCIO in a number of polar solvents[14]. In water, OCIO exhibits a broad, structured absorption band from 280 to 470 nm with a maximum located at 370 nm ($\varepsilon \sim 1,050$ L mole^{-1} cm^{-1}), see figure 1[15].

Comparison with the gas phase spectrum (not shown) indicates that solvation has a small effect on the shape and position of the absorption band[14]. The absorption spectra of aqueous ClO[16], Cl[16] and gas phase ClOO[17,18] are also shown in figure 1. Due to the extreme reactivity of ClOO, the solution phase spectrum has not been characterized.

Figure 2 shows the transient absorption dynamics observed for several probe wavelengths following the 355 nm photolysis of aqueous OCIO. Probing at 266 nm and 345 nm (figures 2a and 2b), an initial absorption rise within the instrument response time (80 ps) is seen following photolysis. At 266 nm, this rise is followed by a decay in signal within the first 200 ps, which levels off and remains constant for several nanoseconds. In contrast, the absorption dynamics at 345 nm exhibit a rise in signal with the same time constant as that observed for the decay at 266 nm. Probing at both 385 nm and 420 nm (figures 2c and 2d), an initial bleach in the signal is observed, which is also within the instrument response. At 385 nm, this bleach is followed by a recovery that exhibits two components. The fast component occurs within a couple of hundred picoseconds, while the slower component persists into the nanosecond time domain. The signal probed at 420 nm, however, shows only the nanosecond component. These spectra were chosen to demonstrate the range of dynamics observed. In all, 16 probe wavelengths were studied.

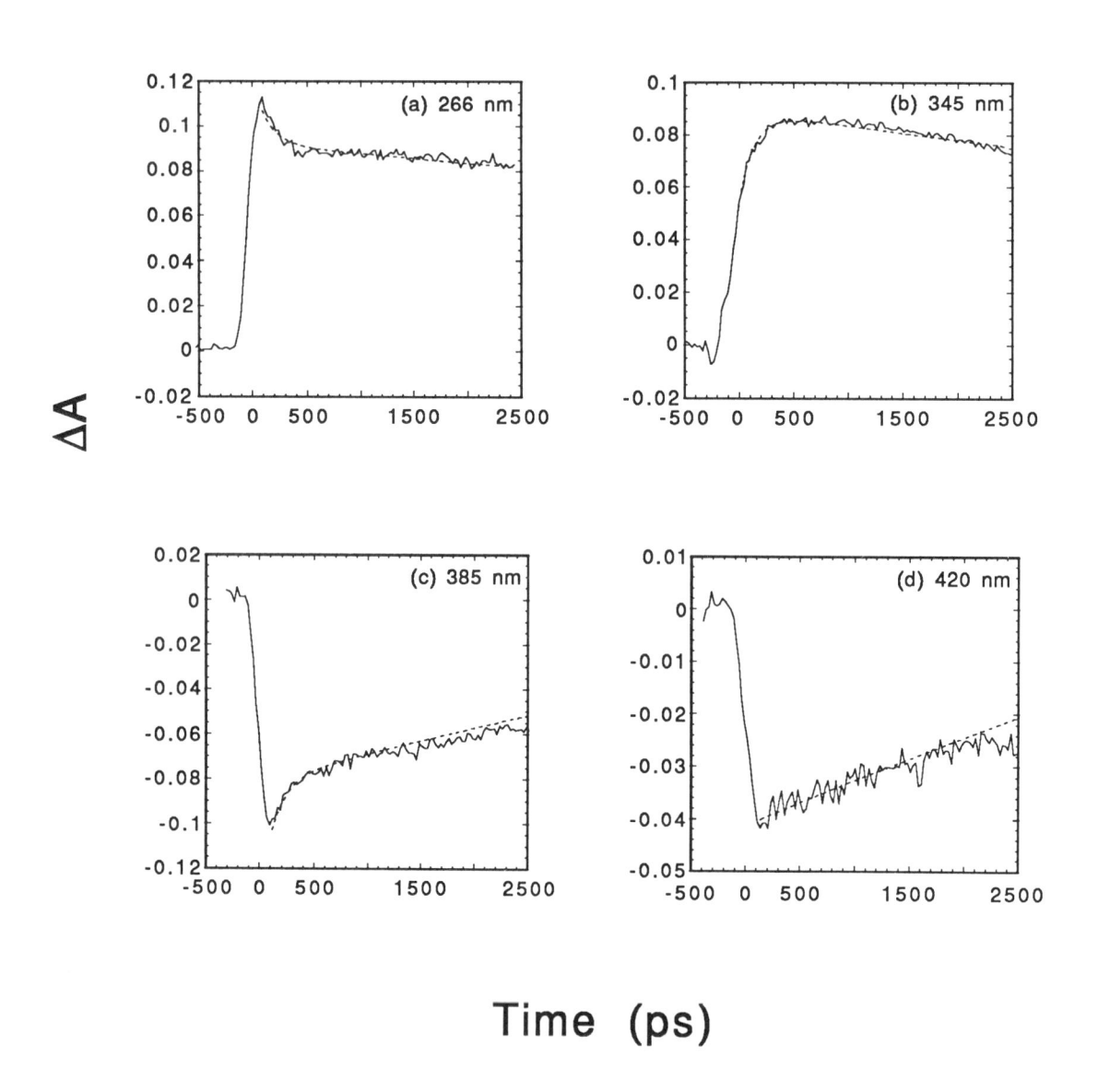

Figure 2: Transient absorption dynamics following the 355 nm photolysis of aqueous OClO. Probe wavelengths: (a) 266 nm, (b) 345 nm, (c) 385 nm, and (d) 420 nm. Dashed lines are fits to the experimental data (solid line) using the kinetic model discussed in the text.

It is reasonable to assume that the photoreactions of OCIO in solution are similar to those observed in the gas phase (i.e. Eq. 2)[8], it is possible to simulate the observed dynamics using the spectra shown in figure 1. In simulating transient absorption signals, the following reactions need to be considered in detail.

$$ClOO \xrightarrow{k_d} Cl + O_2 \qquad (3)$$

$$ClO + O \xrightarrow{k_{rec}} OClO \qquad (4)$$

It is assumed that once electronically excited, OClO partitions between the two decay channels shown in Eq. 2 on a time scale much shorter than the experimental time resolution. This is based on both the gas phase linewidth measurements, which estimate the excited state lifetime of OClO to be on the order of 1 picosecond[2,3], and the pulse limited rises observed in the solution studies. Equation 4 represents a recombination process that competes with the formation of free ClO and O. The time dependent concentrations of the various intermediates formed in the photolysis of OClO can be calculated using equations (5) - (9).

$$[ClOO](t) = [ClOO]_0 \exp(-k_d t) \qquad (5)$$

$$[Cl](t) = [ClOO]_0 \{1 - \exp(-k_d t)\} \qquad (6)$$

$$[ClO](t) = [ClO]_{rec}(t) + [ClO]_{sep} \qquad (7)$$

$$[OClO](t) = [ClO]_{rec}(0) - [ClO]_{rec}(t) \qquad (8)$$

where

$$[ClO]_{rec}(t) = \frac{[ClO]_{rec(0)}}{1 + 2k_{rec}t[ClO]_{rec(0)}} \qquad (9)$$

Once the time dependent concentrations are determined, the transient absorption signals are calculated using equation (10). The summation in this expression runs over the different chemical species which contribute at the wavelength in question (e.g. ClOO, ClO, Cl, and OClO).

$$A(l,t) = \sum_i e(l)[i](t) \qquad (10)$$

There are four adjustable parameters in this kinetic model: the partitioning between dissociation and isomerization, the thermal decomposition rate constant of ClOO (k_d), the partitioning between free ClO and recombination ($[ClO]_{rec}$ and $[ClO]_{sep}$, respectively), and the bimolecular rate constant of ClO and O recombination (k_{rec}).

The dashed lines in figure 2 show the best global fits obtained for a single set of parameters. Excellent agreement between experimental and simulated results are observable for all probe wavelengths. The optimized parameters suggest that upon photoexcitation, 10% of the excited OClO molecules undergo isomerization forming ClOO, while the remaining 90% dissociating into ClO and O. Following dissociation, 70% of the ClO, O products undergo recombination on the nanosecond time scale. Finally, the rate constant for thermal decomposition of ClOO to form atomic chlorine and molecular oxygen at room temperature is ~6.7 × 10^{9}s^{-1}.

4. CONCLUSIONS

The room temperature photochemistry of aqueous OClO was studied using picosecond absorption spectroscopy. Following excitation at 355 nm, OClO reacts by two competing pathways, consistent with recent gas phase studies[2,3,5-7]. By fitting the experimental data to a simple kinetic model, it is shown that 90% of the excited OClO molecules photodissociate to form ClO and O. The remaining 10% isomerize to form the ClOO isomer, which subsequently dissociates into Cl and O$_2$ with a rate constant of ~6.7 × 10^{9}s^{-1}.

5. ACKNOWLEDGEMENTS

This work is supported by the National Science Foundation, Division of Experimental Physical Chemistry and the MMFEL Program administered by the Office of Naval Research. We thank Professor Veronica Vaida for many stimulating and informative discussions.

6. REFERENCES

(1) S. Solomon, G. H. Mount, R. W. Sanders, and A. L. Schmeltekopf, "Visible spectroscopy at McMurdo station, Antarctica 2. Observations of OClO," *J. Geophys. Res.* vol. 92, pp. 8329-8338, 1987.

(2) E. C. Richard and V. Vaida, "The direct near ultraviolet absorption spectrum of the $\tilde{A}^2A_2 \leftarrow \tilde{X}^2B_1$ transition of jet-cooled chlorine dioxide," *J. Chem. Phys.* vol. 94, pp. 153-162, 1991.

(3) E. C. Richard and V. Vaida, "The photochemical dynamics of the $\tilde{A}^2A_2$ state of chlorine dioxide," *J. Chem. Phys.* vol. 94, pp. 163-171, 1991.

(4) J. L. Gole, "Photochemical isomerization of ClO$_2$ and the low-lying electronic state of the asymmetric isomer. Possible implications for matrix isolation spectroscopy," *J. Phys. Chem.* vol. 84, pp. 1333-1340, 1980.

(5) (a) E. Ruhl, A. Jefferson and V. Vaida, "Photodissociation of OClO: REMPI study of primary photofragments," J. Phys. Chem. vol. 94, pp. 2990-2994, 1990; (b) A. J. Colussi, "Formation and decay of (^{3}P$_j$)O atoms in the laser flash photolysis of chlorine dioxide (OClO) at 308 nm," *J. Chem. Phys.* vol. 94, pp. 8922-8926, 1990.

(6) E. Bishenden, J. Hancock and D. J. Donaldson, "Resonance-enhanced multiphoton ionization measurement of Cl (^{2}P$_{3/2}$ and ^{2}P$_{1/2}$) produced in the photolysis of OClO form 355 to 370 nm," *J. Phys. Chem.* vol. 95, pp. 2113-2115, 1991.

(7) V. Vaida, S. Solomon, E. C. Richard, E. Ruhl and A. Jefferson, "Photoisomerization of OClO: a possible mechanism for polar ozone depletion," *Nature* vol. 342, pp. 405-408, 1989.

(8) R. C. Dunn and J. D. Simon, "Excited state photoreactions of chlorine dioxide in water," *J. Am. Chem. Soc.* in press.

(9) X. Xie and J. D. Simon, "High energy and tunable picosecond laser pulses at 1 kHz: synchronously pumping a dye laser with a mode-locked, Q-switched and cavity dumped Nd:YAG laser system," *Opt. Commun.* vol. 69, pp. 303-307, 1989.

(10) (a) W. Z. Bray, *Phys. Chem.* vol. 54, pp. 569, 1906; (b) L. M. Babcock, T. Pentecost and W. H. Koppenol, "Electron affinity of chlorine dioxide,"*J. Phys. Chem.* vol. 93, pp. 8126-8127, 1989.

(11) A. Arkell and F. Schwager, "Matrix infrared study of the ClOO radical," *J. Am. Chem. Soc.* vol. 89, pp. 5999-6006, 1967.

(12) M. M. Rochkind and G. C. Pimentel, "Photolysis of matrix-isolated dichlorine monoxide: infrared spectra of ClClO and $(ClO)_2$," *J. Chem. Phys.* vol. 46, pp. 4481-4490, 1967.

(13) F. J. Adrian, J. Bohandy and B. F. Kim, "ESR investigation of the photoisomerization of OClO to ClOO in an H_2SO_4 glass: magnetophotoselection and kinetics," *J. Chem. Phys.* vol. 85, pp. 2692-2698, 1986.

(14) R. C. Dunn, B. N. Flanders, V. Vaida and J. D. Simon, "The spectroscopy of OClO in polar liquids," *Spectrochim. Acta*, in press.

(15) F. Stitt, S. Friedlander, H. J. Lewis and F. E. Young, "Photometric determination of glucose in presence of fructose," *Anal. Chem.* vol. 26, pp. 1478-1484, 1954.

(16) U. K. Klaning and T. Wolff, "Laser flash photolysis of HClO, ClO-, HBrO, and BrO- in aqueous solution. Reactions of Cl- and Br- atoms," *Ber. Bunsenges. Phys. Chem.* vol. 89, pp. 243-245,1985.

(17) (a) E. D. Morris and H. S. Johnston, "Ultraviolet spectrum of the ClOO radical," *J. Am. Chem. Soc.* vol. 90, pp. 1918-1920, 1968. (b) H. S. Johnston, E. D. Morris and J. Van den Bogaerde, "Molecular modulation kinetic spectrometry. ClOO and ClO_2 radicals in the photolysis of chlorine in oxygen," *J. Am. Chem. Soc.* vol. 91, pp. 7712-7727, 1969.

(18) A. R. Ravishankara, *private communication*.

Structure-reactivity effects in the picosecond dynamics of isolated clusters

Jack A. Syage and Jhobe Steadman

The Aerospace Corporation
P.O. Box 92957/M5-754
Los Angeles, California 90009

ABSTRACT

The link between gas-phase and condensed-phase chemical dynamics is being sought by the study of reactivity in small molecular clusters. We present measurements of reaction rates as a function of cluster size and composition to elucidate the role of solvent on a single-molecule basis. In previous work we showed that reactivity in clusters is strongly correlated to the stepwise binding energies of individual solvent molecules. Here, we show that solvent structure also plays a large role in determining chemical reactivity. This paper focuses on the excited state proton transfer reaction of phenol (PhOH) in a cluster of solvent-like molecules (i.e., NH_3 and CH_3OH). Three cases of structure-reactivity effects are reported: (1) rate inhibition by the addition of a CH_3OH molecule to an $(NH_3)_n$ solvent cluster (60 ps vs 500 ps), (2) very different reaction rates for the inequivalent phenol molecules in phenol dimer solvated by $(NH_3)_n$ (50 ps vs 500 ps), and (3) solvent reorganization following proton transfer that occurs on the time scale of 0.3 ns.

1. INTRODUCTION

Molecular clusters are proving to be an important medium for probing the correspondence between gas-phase and condensed-phase properties.[1-3] In recent work we have exploited the cluster environment using picosecond molecular beam spectroscopy to study chemistry under single-molecule solvent interactions.[4-9] The most common method of producing clusters is by supersonic expansion in that it provides molecular cooling and low energy collisions that favor cluster growth.[10] If the internal energy of the clusters is low relative to the photoexcitation energy then the excited-state clusters are nearly isoenergetic and constitute a microcanonical ensemble (i.e., an ensemble of noninteraction subsystems at constant energy). Thermal averaging effects are minimized in photochemical cluster studies.

The focus of this paper is on the correlation of solvent structure and reactivity. In previous work we showed that reactivity is strongly correlated with the stepwise binding energies of individual solvent molecules, but also showed instances where solvent structure played a major role. The results presented here are for the excited state proton transfer (ESPT) reaction of phenol in a cluster of solvent-like molecules (i.e., NH_3 and CH_3OH). We have published a series of papers on phenol/cluster proton transfer covering the subjects of solvent basicity and cluster size effects for excited state proton transfer (paper I),[6] dynamics and energetics of phenol cation (paper II),[7] and solvent structure and ground state ion-pair formation (paper III).[8] In this paper we discuss three examples of structure-reactivity effects. First we show that the addition of a single CH_3OH to an otherwise pure $(NH_3)_n$ solvent core causes a disruption of the hydrogen bonding network resulting in a reduction of reaction rate from $(\sim 60\ ps)^{-1}$ to $(>750\ ps)^{-1}$. Next, we discuss phenol dimer solvated by $(NH_3)_n$ where the inequivalent hydroxy binding sites account for a large difference in reactivity (50 ps vs 500 ps). Finally, we present measurements of solvent reorganization following proton transfer. These geometry changes are measurable because they give rise to time-dependent Franck-Condon factors for ionization.

2. EXPERIMENTAL METHOD AND BACKGROUND

The picosecond molecular beam apparatus has been described before,[4-6,11] hence, we give only the briefest details. The molecular beam employs a temperature-controlled pulsed supersonic nozzle which serves as the cluster source. The free-jet expansion passes through a skimmer to form a 5-mm-diam beam at the ionization region of a time-of-flight mass spectrometer. Phenol (Aldrich, 99+%, used as is) is contained in a sample holder in the pulsed nozzle, which is heated to about 50°C. For ammonia solvation, a 2-liter high-pressure cylinder was filled with a sample contain-

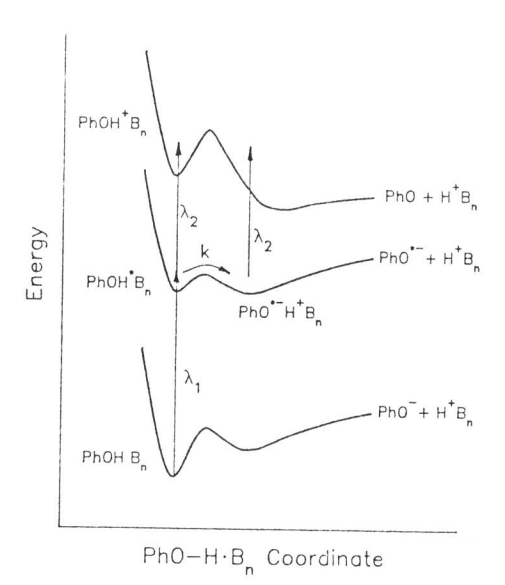

Figure 1. Energy diagram for PhOH·(NH$_3$)$_n$ consistent with a cluster size of $n \sim 5$. ESPT produces an excited state ion-pair product that is ionized to the weakly-bound outer potential well of the cluster ion.

ing 10% NH$_3$ in helium. For the NH$_3$/CH$_3$OH experiments, the NH$_3$/He sample was passed through a bubbler containing CH$_3$OH. Each of these solvent/He samples was flowed over the phenol contained in the pulsed valve. The flow mixture, typically at 35 psi, was expanded through a nozzle with a 750-μm diam aperture and 60° conical throat.

The picosecond laser is based on a pulsed active/passive mode-locked Nd:YAG laser, which produces output at 532 nm, 355 nm, and 266 nm. Time-resolved spectra were recorded by delaying a probe pulse λ_2 relative to a pump pulse λ_1 (266 nm) using a scanning optical delay line.[4-8] Calculated curves were fitted to the data using a convolution that assumed a Gaussian-shaped instrument function (typically 25–30 ps, FWHM).

The picosecond pump-probe excitation/detection scheme is given in Fig. 1 and shows approximate potential curves for the relevant ground, excited, and ionic states of the cluster. The potential curves are strongly dependent on solvent cluster size and basicity and are discussed in detail in paper I.[6] The illustration given in Fig. 1 approximates the case for phenol-(NH$_3$)$_5$. The phenol (PhOH) excited-state proton transfer (ESPT) reaction

$$\text{PhOH}^* \cdot B_n \xrightarrow{k} \text{PhO}^{*-}\text{H}^+ B_n$$

corresponds to a conversion from the locally excited S_1 state of the acid phenol to an excited ion-pair state (B$_n$ refers to the solvent cluster consisting of n molecules of base B). Reactant and product are distinguishable by mass spectrometry because each species fragments differently upon ionization. The energetics and dynamics of the solvent-clustered phenol cation are reported in paper II.[7] Briefly, a double-well minima with a large barrier exists for the cluster ions. As illustrated in Fig. 1, ionization of the reactant cluster PhOH*·B$_n$ to the inner ion potential well produces strongly bound PhOH$^+$·B$_n$ ions. The product ion-pair PhO^{*-}H$^+$B$_n$ is ionized to the outer ion well, which is weakly bound and therefore readily dissociates to form H$^+$B$_n$ ions. It is important to remember that the detected ion signals are probes of the dynamics along the neutral excited S_1/ion-pair reactive surface.

3. STRUCTURE-REACTIVITY EFFECTS OF THE SOLVENT CLUSTER

The sensitivity of chemical reactivity to solvent structure is very pronounced in small clusters. We discuss a few studies illustrating significant structural sensitivity. To begin, we present measurements of ESPT rates in pure solvent clusters representing idealized structures, thus serving as the control case for the structure-reactivity results that follow.

3.1. Reaction dynamics - pure solvent

Aromatic alcohols in solution are known to be more acidic in their S_1 state than in S_0 (i.e., the so-called pH jump).[12] In the present picosecond cluster experiment, the $\lambda_1 = 266$ nm pump pulse excites the PhOH$^+$·B$_n$ clusters to S_1 with excess energy ranging from about 1900 cm^{-1} ($n = 1$) to about 2250–2800 cm^{-1} (estimated for $n = 3 - 7$).[13,14] The S_1 origin of uncomplexed phenol is at 274.7 nm.[14] A set of reactant decay measurements are presented in Fig. 2 ($\lambda_2 = 532$ nm) as a function of solvent cluster size n for the solvents NH$_3$ and CH$_3$OH. The NH$_3$ solvent clusters induce ESPT for solvent sizes $n > 4$. No evidence of reaction was observed for CH$_3$OH (up to $n = 11$, Fig. 2). Results for other solvents are given in paper I.[6]

The sharp reaction threshold at $n = 4$ for the (NH$_3$)$_n$ solvent can be explained almost entirely by energetic arguments involving the ion-pair Coulombic potential and stepwise proton affinities for successive NH$_3$ molecules. This analysis was presented in paper I and assumes a solvent structure based on the well established geometry of ammoniated ammonium ion.[15] The assumed geometry for PhOH·(NH$_3$)$_4$, representing filling of the first solvation shell, is drawn in Fig. 3.

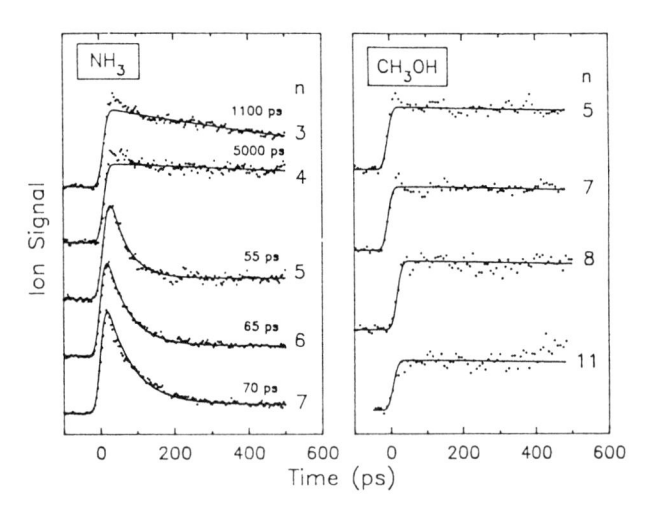

Figure 2. Observed lifetime of the reactant PhOH*·B_n (as measured by PhOH$^+$·B_n detection) in solvent clusters $(NH_3)_n$ and $(CH_3OH)_n$. The calculated curves are described in paper I. Calculated curves for the $(CH_3OH)_n$ data assume a 10 ns lifetime.

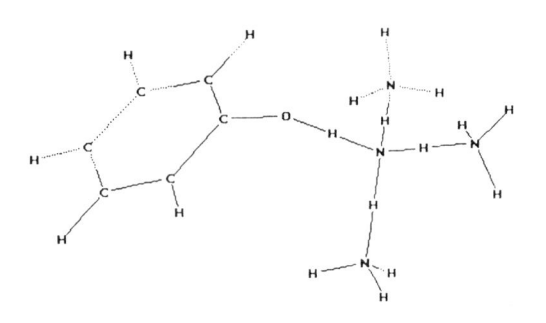

Figure 3. Assumed structure for PhOH·$(NH_3)_4$.

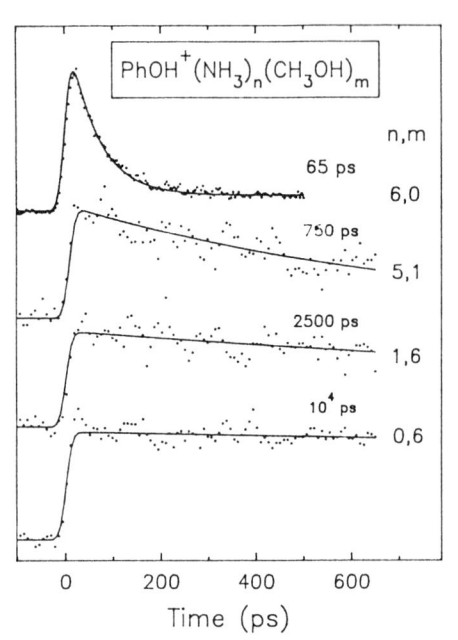

Figure 4. Measured rates of ESPT for the mixed solvent clusters PhOH*$(NH_3)_n(CH_3OH)_m$ using 266-nm pump and 532-nm probe excitation.

3.2. Disruption of the solvent shell - mixed solvents

Reaction rates for mixed solvent distributions PhOH·$(NH_3)_n(CH_3OH)_m$ of similar cluster size ($n+m$ = 6 or 7) were measured and are presented in Fig. 4. The substitution of a single CH_3OH molecule for an NH_3 molecule causes a substantial deceleration in reaction rate. Excited phenol in neat NH_3 solvent cluster $(n,m) = (6,0)$ reacts in 65 ps whereas the (5,1) cluster reacts in 750 ps. Because the (5,0) cluster reacts in 60 ps (Fig. 2), the (5,1) result indicates that the mere *addition* of a CH_3OH molecule effectively "poisons" the reactivity of the (5,0) cluster. This is quite surprising because one would not expect the proton affinity of the solvent to decrease by the adding CH_3OH. One explanation for the slowed reaction is that CH_3OH acts as an intruder that disrupts the hydrogen bonding network of the neat $(NH_3)_n$ solvent structure that is crucial to solvating and stabilizing the dissociated proton.[16,17] Although NH_3 forms the stronger bond to PhOH, CH_3OH forms the stronger bond to PhOH·NH_3 and may, therefore, displace an NH_3 from the first solvation shell.[8] Also CH_3OH can form a stable hydrogen bond to the phenol oxygen which diminishes excited state acidity, whereas NH_3 avoids that site.[19,20]

The concept of critical solvent structures in ESPT reactions has been investigated in the solution phase by Robinson and coworkers.[16] Their picosecond measurements and theoretical analyses on aromatic acids (e.g., 1- and 2-naphthol) in H_2O and alcohol (CH_3OH and C_2H_5OH) support the notion that a critical solvent cluster core is necessary to act as an efficient proton acceptor. For naphthol in H_2O, the critical solvent-core size is about four.[16] Regarding the present results, it is known from gas-phase cluster studies that the closed-shell structure $NH_4^+(NH_3)_4$ is very favorable for stabilizing a proton.[15,21] The presence of CH_3OH can disrupt this structure because the NH_3–CH_3OH bond is stronger than the NH_3–NH_3 bond.[22,23] Hence, CH_3OH can displace NH_3 from the first solvation shell and prevent a pure NH_3 closed-shell structure from forming. The structure of the solvent about a reactive solute molecule has never been directly measured. Studies of cluster distributions for mixed solvent systems are beginning to yield information on the bonding structure

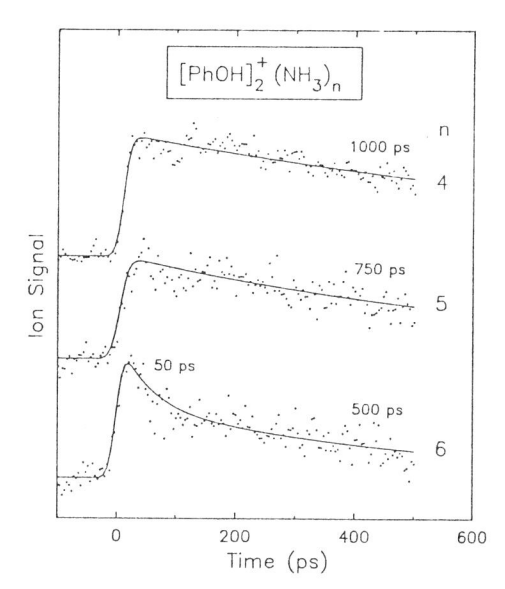

Figure 5. Time dependence for the excited state dimer cluster $[PhOH]_2^*(NH_3)_n$. Single exponential curves are calculated for $n = 4$ and 5. The sum of two exponential curves is calculated for $n = 6$.

of the solvent.[8,24] These investigations should provide new insight into the influence of solvent structure on chemical reactivity.

3.3. Different proton binding sites - phenol dimer

The phenol molecule (like many other aromatic acids and bases, e.g., carboxylic acids and certain amines) forms a stable dimer in solution.[25] The reaction rate of excited state (S_1) phenol dimer in $(NH_3)_n$ solvent clusters was measured by detecting the $[PhOH]_2^+(NH_3)_n$ ion signal (Fig. 5). Whereas phenol monomer undergoes an abrupt change in reaction rate at $n = 5$ (Fig. 2), similar behavior for the dimer acid is not evident. The decay traces in Fig. 5 show progressively faster rates with increasing n; however, it cannot be established that this is due to ESPT. There is some indication of a ~ 50 ps response for $n = 6$, followed by a slower decay of about 500 ps.

The apparent inhibition of proton transfer in phenol dimer relative to phenol monomer in comparable solvent clusters can be explained by the nature of the dimer bond. Spectroscopic evidence[26,27] indicates that the individual phenol molecules in the dimer are inequivalent as shown in Fig. 6. A linear O–H–O hydrogen bond forms in which the hydroxy proton of one phenol (type-I) is bound not to the solvent but to the oxygen of the other phenol. This other phenol (type-II) is similar to monomer phenol in that it has a proton that binds directly to the solvent. Type-I phenol has

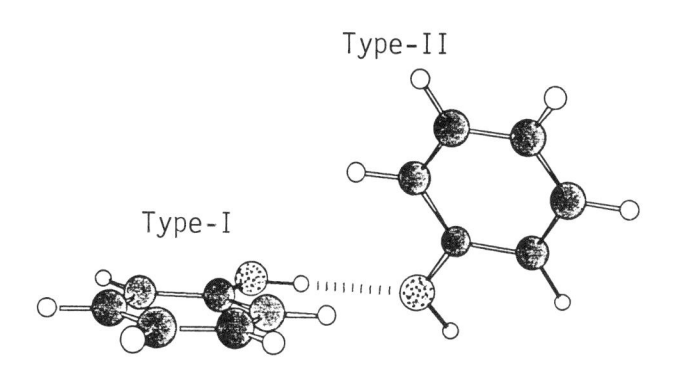

Figure 6. Structure of the phenol dimer deduced spectroscopically (Figure adapted from Ref. 27).

a restricted proton, which should quench ESPT. The fast 50 ps decay component in Fig. 5 for $n = 6$ may be due to ESPT from excitation of type-II phenol and the slow component to excitation of type-I phenol.

3.4. Solvent reorganization in clusters

Product formation kinetics, as measured by the ion-pair dissociative ionization signal $H^+(NH_3)_n$, are given in Fig. 7. This signal arises by ion-pair ionization to the weakly bound outer potential well of the cluster ion followed by cluster ion dissociation (Fig. 1).[6,7] Formation kinetics are observed for values of n less than the reaction threshold of $n = 5$ because the exothermicity of the cluster ion dissociation causes some evaporation (typically one or two molecules depending on the photoionization energy).[9] The formation curves in Fig. 7 have two time dependences; a fast one ($\simeq 65$ ps) which matches the decay of reactant in Fig. 2, and a slow one (0.3 ns), which we attribute to solvent reorganization following the proton transfer event. Other explanations seem less plausible. The slow component is not due to a structural isomer, because this would also appear in the reactant decay curves. Evaporation can be ruled out for the following reasons: (1) Evaporation in the neutral ion-pair state should show decay, not formation kinetics.[28] (2) Dissociation from the cluster ion outer-well does cause one or two solvent molecules to evaporate;[9] however, this process occurs after the λ_2 pulse and, therefore, the effect on intensity is independent of the pump-probe delay time.

The dual time dependence in Fig. 7 can be explained by fast ESPT followed by a slower reorganization of the solvent hydrogen-bond network in response to the electric-dipole forces imposed by the ion-pair structure. ESPT is expected to be fast because it in-

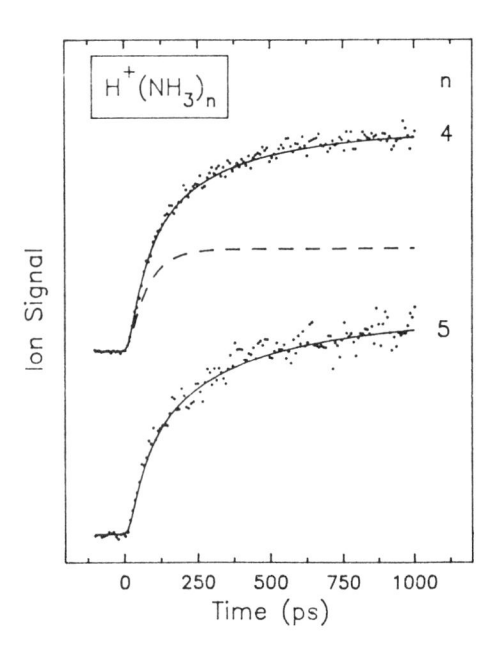

Figure 7. Observed PhO*⁻H⁺B_n ion-pair formation kinetics (as measured by H⁺B_n detection) for NH₃ solvation. Traces were recorded using 266-nm pump and 532-nm probe excitation. The data were fitted according to paper I for $k^{-1} = 65$ ps, $k_s^{-1} = 350$ ps, $i_c/i_b = 2.4$ (for $n = 4$); and $k^{-1} = 65$ ps, $k_s^{-1} = 300$ ps, $i_c/i_b = 2.0$ (for $n = 5$)

volves an electron transfer and a geometry change along a single coordinate O–H. Solvent relaxation, on the other hand, should be slower because it involves many coordinates. Time-dependent ionization efficiencies are analogous to time-dependent Stokes shifts, which are used to measure structural reorganization in solution.[29] Both phenomena are due to changing Franck-Condon factors brought about by molecular motion.

How the potential energy along the O–H coordinate might change with time due to solvent reorganization is illustrated qualitatively in Fig. 8. The equilibrium bond distance for the cluster ion PhO·H⁺B_n should be greater than for the neutral solvated ion-pair PhO*⁻H⁺B_n because of the loss of Coulombic attraction upon ionization.[6] The equilibrium solvent structure prior to proton transfer is a nonequilibrium structure for the newly formed ion pair. The solvent network must reorganize to achieve a structure that best accomodates the proton. This reorganization or "solvation of the proton" causes delocalization of charge and lengthening of the O–H bond (Fig. 8a). The relaxing ion-pair moves closer to a geometry representing the cluster-ion equilibrium configuration. This improves the Franck-Condon factors for ionization and leads to an increased ionization efficiency with time.

Calculated time dependences are plotted in Fig.

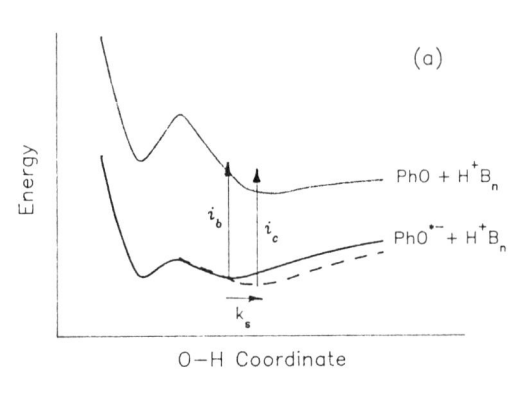

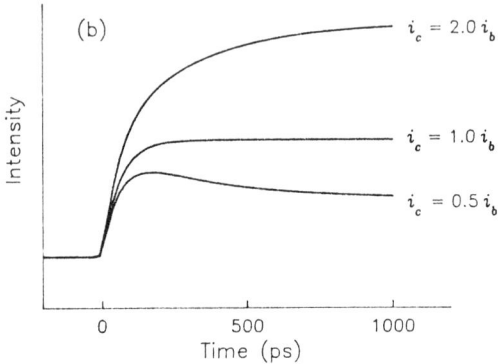

Figure 8. (a) Schematic representation of the evolving solvent-relaxed ion-pair potential. (b) Calculated ion time dependence according to paper I, for $k^{-1} = 65$ ps, $k_s^{-1} = 300$ ps and different relative values of the unrelaxed (i_b) and relaxed (i_c) ionization efficiencies.

8b for the H⁺(NH₃)_n ion signal where we define a solvent relaxation rate constant k_s and vary the relative ionization effiency from the unrelaxed (i_b) and relaxed form (i_c). The details of this simple kinetic model are given in paper I.[6] For simplicity, we assume here that ionization from the reactant excited state does not dissociate to form H⁺(NH₃)_n and that ionization from the unrelaxed and relaxed ion pair leads to complete dissociation to form H⁺(NH₃)_n. The calculations that best fit the data in Fig. 7 are those for which the ionization efficiency increases with time. These results are consistent with Franck-Condon factors for ionization that improve as a result of solvent reorganization.

It is difficult to judge whether 0.3 ns is a reasonable time scale for solvent reorganization in (NH₃)_n solvent clusters because we do not know the cluster temperatures. Most models of solvent reorientation about a dipole include more than one time constant, reflecting many types of molecular motion.[29-31] The fastest motions are due to bond rotations, whereas the slowest reorientations involve breaking and reforming of intermolecular solvent bonds. These would be particularly slow for hydrogen-bonded solvents. Extensive measurements of solvent relaxation for each of the reorientational degrees of freedom have been reported by many groups over a wide range of temperatures.[29-31] For the variety of solvents, molecular motions, and temperatures considered, relaxation times ranged from the experimentally detectable limit of about 100 fs to greater than a nanosecond. The intermolecular motional time constant for ethanol at 253 K, for example, was 112 ps (responding to a dipole excitation in a dye molecule).[30] This is an activated process because of hydrogen-bond energy and, therefore, the rates vary significantly with temperature. We have not found any measurements of NH_3 solvent relaxation rates; however, it is certainly reasonably to expect the reorientation of hydrogen bonds at low temperatures (≤ 100 K) to be on the nanosecond time scale in accord with our experimental results.

4. SUMMARY AND CONCLUSIONS

Studies of reactivity in isolated molecular clusters are providing valuable information on the interactions of individual solvent or lattice molecules on chemical dynamics. In this paper we have focused on the role of structure particularly regarding the surrounding solvent. Though stepwise solvent binding energies are a strong driving force to reactivity, simple additivity arguments are insufficient for describing the role of first- and second-shell solvent molecules. The ability of the solvent to form a optimum hydrogen bonding network for proton stabilization is a cooperative effect. Disruption of energetically favorable solvent structures have been observed to reduce reaction rates by over an order of magnitude.

5. ACKNOWLEDGEMENTS

We thank P. M. Felker for the use of Fig. 6. This work was supported by the Aerospace Sponsored Research program.

6. REFERENCES

1. J. J. Breen, W. -B. Tzeng, K. Kilgore, R. G. Keesee, and A. W. Castleman, Jr., "Intracluster reactions in phenylacetylene ammonia clusters initiated through resonant enhanced ionization" *J. Chem. Phys.* **90**, 19 (1989); O. Echt, P. D. Dao, S. Morgan, and A. W. Castleman, Jr., "Multiphoton ionization of ammonia clusters and the dissociation dynamics of protonated cluster ions," *J. Chem. Phys.* **82**, 4076 (1985).

2. O. Cheshnovsky, and S. Leutwyler, "Proton transfer in neutral gas-phase clusters: α-Naphthol·$(NH_3)_n$," *J. Chem. Phys.* **88**, 4127 (1988); T. Droz, R. Knochenmuss, and S. Leutwyler, "Excited-state proton transfer in gas-phase clusters: 2-Naphthol·$(NH_3)_n$," *J. Chem. Phys.* **93**, 4520 (1990). S. K. Kim, S. Li, and E. R. Bernstein, "Excited state intermolecular proton transfer in isolated clusters: 1-naphthol/ammonia and water," *J. Chem. Phys.* **95**, 3119 (1991).

3. J. F. Garvey and R. B. Bernstein, "Observation of intramolecular ion-molecule reactions within ionized clusters: Hetero systems ...," *J. Am. Chem. Soc.* **109**, 1921 (1987); J. F. Garvey and R. B. Bernstein, "Observation of intramolecular ion-molecule reactions within ionized clusters: The methyl halide systems," *J. Phys. Chem.* **90**, 3577 (1986).

4. J. A. Syage and J. Steadman, "Picosecond mass-selective measurements of molecular cluster reactions: $(CH_3I)_n$ $\tilde{A}$ state excitation," *Chem. Phys. Lett.* **166**, 159 (1990).

5. J. Steadman and J. A. Syage, "Picosecond mass-selective measurements of Phenol-$(NH_3)_n$ acid-base chemistry in clusters," *J. Chem. Phys.* **92**, 1630 (1990).

6. J. A. Syage and J. Steadman, "Picosecond measurements of phenol excited state proton transfer in clusters. I. Solvent basicity and cluster size effects," *J. Chem. Phys.* **95**, 2497 (1991).

7. J. Steadman and J. A. Syage, "Picosecond studies of proton transfer in clusters. 2. Dynamics and energetics of solvated phenol cation," *J. Am. Chem. Soc.* **113**, 6786 (1991).

8. J. Steadman and J. A. Syage, "Time-resolved measurements of phenol proton transfer in clusters. 3. Solvent structure and ion-pair formation," *J. Phys. Chem.* **95**, in press (1991).

9. J. Steadman, E. W. Fournier, and J. A. Syage, "Detection of neutral and ionic reaction mechanisms in molecular clusters," *Appl. Opt.* **29**, 4962 (1990).

10. M. Kappes and S. Leutwyler, "Molecular Beams of Clusters," in *Atomic and Molecular Beam Methods,* Vol. I, edited by G. Scoles, (Oxford University Press, New York, 1988), p. 380.

11. J. A. Syage, "Photodissociation and metastable decay of solvated cluster ions," *J. Chem. Phys.* **92**, 1804 (1990); J. A. Syage, "Reactivity of aniline cation to stepwise solvation in hydrogen-bonded molecular clusters," *J. Phys. Chem.* **93**, 107

(1989).

12. E. M. Kosower and D. Huppert, "Excited state electron and proton transfers," *Annu. Rev. Phys. Chem.* **37**, 127 (1986).

13. D. Solgadi, C. Jouvet, and A. Tramer, "Resonance enhanced multiphoton ionization spectra and ionization thresholds of Phenol-$(NH_3)_n$ clusters," *J. Phys. Chem.* **92**, 3313 (1988); C. Jouvet, C. Lardeux-Dedonder, M. Richard-Viard, D. Solgadi, and A. Tramer, "Reactivity of molecular clusters in the gas phase. Proton-transfer reaction in neutral phenol-$(NH_3)_n$ and phenol-$(C_2H_5NH_2)_n$," *J. Phys. Chem.* **94**, 5041 (1990).

14. N. Gonohe, H. Abe, N. Mikami, and M. Ito, "Two-color photoionization of van der Waals complexes of fluorobenzene and hydrogen-bonded complexes of phenol in supersonic jets," *J. Phys. Chem.* **89**, 3642 (1985).

15. J. M. Price, M. W. Crofton, and Y. T. Lee, "Vibrational spectroscopy of the ammoniated ammonium ions $NH_4^+(NH_3)_n$ (n = 1–10)," *J. Phys. Chem.* **95**, 2182 (1991).

16. J. Lee, R. D. Griffin, and G. W. Robinson, "2-Naphthol: A simple example of proton transfer effected by water structure," *J. Chem. Phys.* **82**, 4920 (1985); J. Lee, G. W. Robinson, S. P. Webb, L. A. Philips, and J. H. Clark, "Hydration dynamics of protons from photon initiated acids," *J. Am. Chem. Soc.* **108**, 6538 (1986)

17. D. H. Huppert, A. Jayaraman, R. G. Maines, Sr., D. W. Steyert, and P. M. Rentzepis, "Effect of pressure on proton transfer rate in aqueous solutions: A picosecond study," *J. Chem. Phys.* **81**, 5596 (1984).

18. The $S_1 \leftarrow S_0$ transition of phenol is red-shifted by the addition of one CH_3OH molecule and blueshifted by the addition of a second molecule. These observations indicate that CH_3OH bonds first to the hydroxy and then to the oxygen of phenol (Ref. 19). No blue-shift is observed for any successive addition of NH_3 to phenol, indicating that it does not bond to the oxygen. This is consistent with NH_3 being a good proton acceptor, but a poor proton donor (Ref. 20).

19. A. Oikawa, H. Abe, N. Mikami, and M. Ito, "Solvated phenol studied by supersonic jet spectroscopy," *J. Phys. Chem.* **87**, 5083 (1983).

20. D. D. Nelson Jr., G. T. Fraser, and W. Klemperer, "Does ammonia hydrogen bond?," *Science* **238**, 1670 (1987).

21. P. Kebarle, "Ion thermochemistry and solvation from gas phase ion equilibria," *Annu. Rev. Phys. Chem.* **28**, 445 (1977); M. J. Mautner and C. V. Speller, "Filling of solvent shells about ions. 1. Thermochemical criteria and the effects of isomeric clusters," *J. Phys. Chem.* **90**, 6616 (1986).

22. G. C. Pimentel and A. L. McClellan, *The Hydrogen Bond*, (W. H. Freeman, San Francisco, 1960).

23. C. E. Dykstra, "Weak interaction potentials of large clusters developed from small cluster information," *J. Phys. Chem.* **94**, 6948 (1990).

24. S. Wei, Z. Shi, and A. W. Castleman, Jr. "Mixed cluster ions as a structure probe: Experimental evidence for clathrate structure of $(H_2O)_{20}H^+$ and $(H_2O)_{21}H^+$," *J. Chem. Phys.* **94**, 3268 (1991).

25. B. H. Robinson, "Hydrogen-bonding and proton-transfer reactions in aprotic solvents," in *Proton-Transfer Reactions*, edited by C. Caldin and V. Gold, (Wiley, New York, 1975) p. 121.

26. K. Fuke and K. Kaya, "Electronic absorption spectra of phenol-$(H_2O)_n$ and (phenol)$_n$," *Chem. Phys. Lett.* **94**, 97 (1983).

27. L. L. Connell, S. M. Ohline, P. W. Joireman, T. C. Corcoran, and P. M. Felker, *J. Chem. Phys.*, in press.

28. We also employed a decreasing cluster size distribution to minimize contamination of signal due to evaporation from larger clusters.

29. M. Maroncelli, J. MacInnis, and G. R. Fleming, "Polar solvent dynamics and electron-transfer reactions," *Science* **243**, 1674 (1989).

30. J. D. Simon, "Time-resolved studies of solvation in polar media," *Acc. Chem. Res.* **21**, 128 (1988).; S-G. Su and J. D. Simon, "The importance of vibrational motion and solvent diffusional motion in excited state intramolecular electron transfer reactions," *J. Chem. Phys.* **89**, 908 (1988).

31. W. Jarzeba, G. C. Walker, A. E. Johnson, M. A. Kahlow, and P. F. Barbara, "Femtosecond microscopic solvation dynamics of aqueous solutions," *J. Phys. Chem.* **92**, 7039 (1988); M. Kozik, N. Sutin, and J. R. Winkler, "Energetics and dynamics of solvent reorganization in charge-transfer excited states," *Coord. Chem. Rev.* **97**, 23 (1990).

OPTICAL METHODS FOR
TIME- AND STATE-RESOLVED CHEMISTRY

Volume 1638

SESSION 2

Spectroscopy and Reaction Dynamics of Clusters

Chair
Robert L. Whetten
University of California/Los Angeles

Size- and state-selected cluster chemistry

Yohji Achiba, Taro Moriwaki and Haruo Shiromaru

Department of Chemistry, Tokyo Metropolitan University,
Hachioji, Tokyo 192-03, Japan

ABSTRACT

New aspects of size- and state-specific properties of carbon cluster are described on the basis of experimental evidences obtained by recently developed cluster-beam apparatus. The present paper consists of mainly two topics. One of which is spin states for small carbon neutral clusters. Focusing effects induced by a hexapole magnetic field were examined for the carbon clusters from C_1 to C_6. The second is the study of cluster-solid surface interactions. Electron emission processes induced by collision of negative cluster ions with MoS_2 single crystal were investigated in detail, paying special attention on the electron emission efficiency as functions of both size and incident kinetic energy. A new collision induced phenomenon has been found, which may be closely related with an electron detachment of negative ions induced by a surface potential of the solid target.

1. INTRODUCTION

During the last ten years, physical and chemical properties of carbon clusters have been extensively studied on account of their importance in the processes such as interstellar opacity and combustion[1]. Furthermore, the recent great success in the macroscopic quantities of C_{60} production has brought new aspects in the research of carbon clusters, leading them to the field of material science[2]. Since new discovery of fullerene productions by Kraetschmer and Huffman[2], anomalous number of papers on C_{60} and related compounds have appeared, showing interesting and unique properties closely associated with a peculiar molecular form, a followclosed cage structure. However, for rather smaller carbon clusters, there still remain many interesting and important problems to be understood. For example, understanding relatively small carbon clusters provides an important clue to clarify molecular evolutions in universe[3,4]. Furthermore, even from view points of fullerene chemistry at the present stage, more detailed information has been severely required for fully understanding formation and stability of such cluster network systems. For this purpose, understanding structures and reactivities of small carbon clusters is of particularly important.

So far, three forms have been proposed for the structures of carbon clusters as shown in Fig. 1(a)-(c), where (a) is a linear chain, (b) a monocyclic, and (c) a followclosed cage (fullerene). The preferential form among these three candidates for the carbon cluster with a specific size has been known to much dependent on their size as well as their internal status such that whether they are "hot or cold" clusters. It should be noteworthy that such a

Fig.1
Three forms of carbon clusters

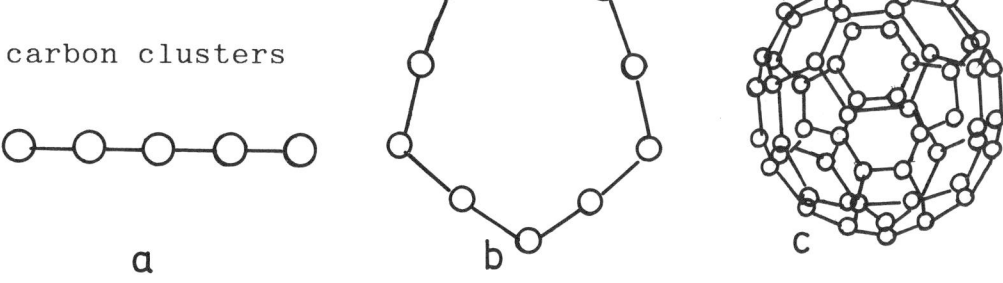

a   b   c

variation of structural change is a characteristic feature of carbon cluster system, and is directly related with many interesting size-dependent phenomena found in the studies of carbon cluster chemistry so far. In the present paper, we describe two topics: One of which is the spin state of small carbon clusters which has been verified from the observation of a focusing effect in a hexapole magnetic field. In the case of small carbon clusters, fortunately, to clarify the spin states for the size specified clusters directly leads to understanding the geometrical structure of the carbon clusters. The second topic is the study of solid surface-cluster interactions. Size selected carbon negative cluster ions (n = 5-22, 60 and 70) were examined through a phenomenon of collision induced electron emissions at solid surface as a function of incident kinetic energy (0-1 KeV).

2. SPIN STATES OF SMALL CARBON CLUSTERS[5]

2-1. STRUCTURE OF EVEN-NUMBERED CARBON CLUSTERS IN GAS PHASE

Concerning carbon clusters with odd numbered sizes up to n = 9, it has been well established that they have preferentially a linear chain structure with a singlet spin character in the ground electronic state. On the other hand, for n-even carbon clusters, the recent ab initio calculations have suggested that the ground electronic states are triplet if they exist as a linear form, whereas those of rhombic or monocyclic ring are singlet. Thus, the determination of the spin states of the carbon clusters directly leads to understanding their geomerical structures. For n-even clusters, so far two different conclusions have been deduced from the gas phase experiments. The high values of the electron affinities for n-even clusters implied that these clusters up to n = 8 are linear chain possessing triplet character[6], whereas the prefered rhombic structure of C_4 was suggested by the results of a Coulomb explosion work[7]. By means of high resolution spectroscopy, Heath and Saykally[8] have shown the existence of a linear form of C_4. In the present study, the spin states of carbon clusters with n = 4 and 6 were successfully investigated in the gas phase by measuring the degree of deflection of the carbon cluster beam in an inhomogeneous magnetic field.

The inhomogeneous magnetic field provided by a hexapole focusing lens[9] was used to identify the magnetic characters of carbon neutral clusters. The existence of the clusters possessing triplet in character was distinguished by measuring the enhancement of ion signals for the clusters of a specific size. The magnetic moments of the clusters of specific size can be evaluated from the magnetic field strength (that is the focal length of the lens) required to obtain the best focusing conditions.

2-2. APPARATUS FOR SPIN-STATE SELECTION

Figure 2 shows a schematic drawing of the apparatus employed in this work. Carbon clusters were produced by a laser vaporization method[10]. The 2nd harmonic from a Nd-YAG laser was focused onto a rotating graphite rod. A pulsed flow of He gas was used for cooling and clustering.

The neutral carbon clusters in He gas were introduced into the hexapole magnetic field through a skimmer. The hexapole magnet (16 cm long) was actuated by a water cooled coil of copper tube. After passing through the magnetic field and a field-free drift tube, neutral clusters were introduced into the detection chamber. Two apertures of 3 mm in diameter were placed at the entrance of the magnet and the exit of the drift tube so that only the beam which passes through the exit slit can be ionized in the detection chamber. Here only small part of the singlet species in the beam can enter into the detection region, because the singlet components diverge after traveling through an entrance slit, independent of the magnetic field. The neutral clusters in the detection chamber were ionized by an electron impact and analyzed by a quadrupole mass filter. The raw signals were measured using a storage oscilloscope and the signals synchronized with the pulsed flow were averaged and recorded. The intensity of magnetic field was determined by a Hall generator.

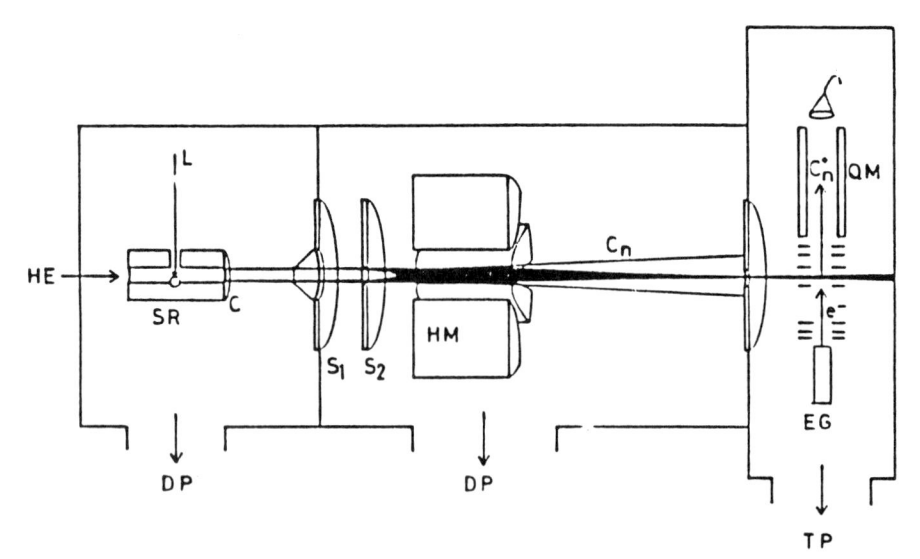

Fig.2 Schematic drawing of the apparatus. SR:sample rod, L:laser, C:reaction channel, S1:skimmer, S2:collimator, M:hexapole magnet, S3:slit, EG:electron gun, QM:Q-pole mass filter, EM:electron multiplier, DP:diffusion pump, TP:turbo-molecular pump.

2-3. EVIDENCES OF LINEAR-FORM C_4 AND C_6

The averaged time profile of ion signals without hexapole magnetic field is shown in Fig. 3a. The ion signal triggered by the gas injector opening pulse was obtained by tuning the ion lens system and quadrupole mass filter with optimized for m/e = 36 (C_3^+). It should be noted in this figure that the time profile of the peak corresponds to not only the flight time of neutral clusters but also the time profile of the action of pulsed injector of He gas. The flight time after ionization is negligible.

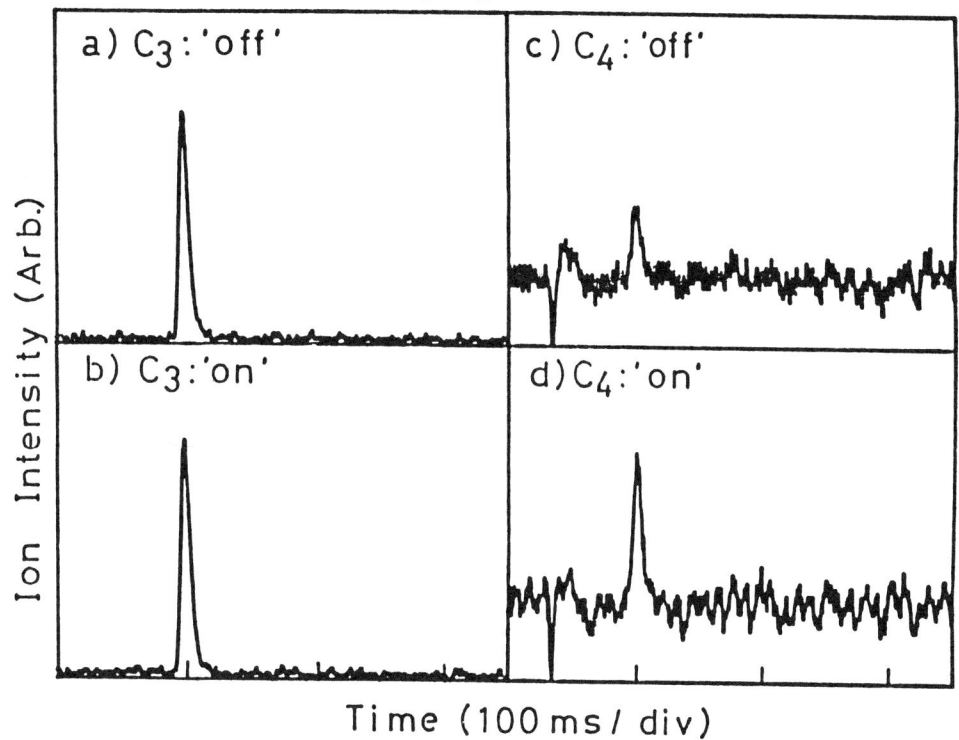

Fig.3
Ion signal intensity.(a) C_3, magnet"off",(b) C_3, magnet"on",(c) C_4, magnet"off",(d) C_4, magnet"on".

The effect of magnetic deflection was measured by changing magnetic field intensity while other parameters were kept constant. Fig. 3b shows the ion signal obtained with magnetic lens "on" optimized for C_3^+. As seen in this figure, no signal enhancement was observed in the case of C_3. In contrast to this, the signal intensities of C_4, which are shown in Fig. 2c and 2d, clearly showan enhancement by the magnetic field intensity.

The plots of the beam intensities as a function of the magnetic field strength are shown in Figs. 4(a)-(f) for C_1-C_6 carbon clusters, respectively. The ground electronic state of the carbon atom is $^3P_0$ with no magnetic moment. However, as seen in Fig. 4a, an enhancement was observed. This fact can be explained as follows: Under the present beam conditions, considerable amounts of $^3P_1$ carbon atom are populated during vaporization, since it lies only 16 cm^{-1} above $^3P_0$ state. On the basis of the present data, the velocity of the beam was estimated by using the value of magnetic moment for $^3P_1$ state as well as the magnetic field, giving an optimum focusing condition for the C_1 beam. As a result, the velocity of 1800 m/s was deduced and its value can be accepted as reasonable as a helium seeded beam.

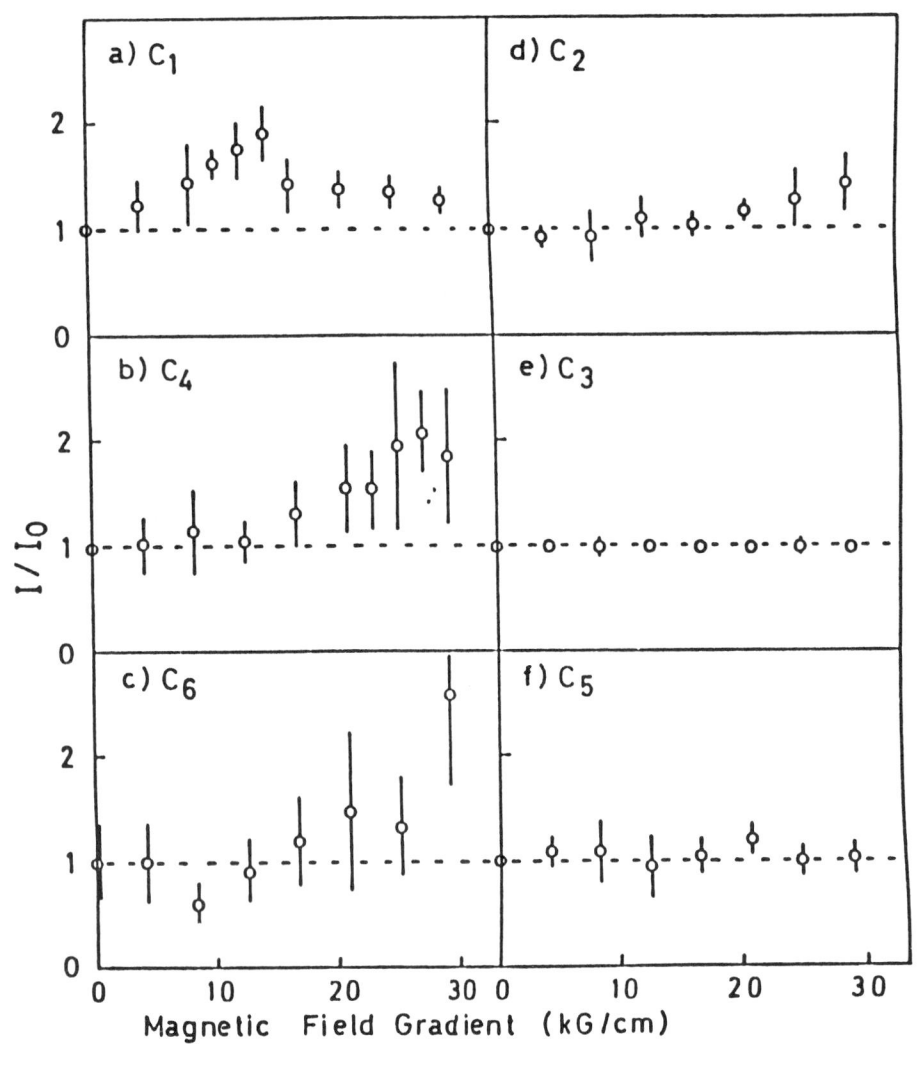

Fig.4

The plots of ion intensity of cluster beam versus magnetic field gradient.

The ground electronic states of C_2, C_3, and C_5 have so far been well explained by a linear chain with a singlet character. As expected, the beam intensities of these clusters show no dependence on the magnetic field. However, a very small enhancement is seen for C_2 at higher magnetic fields. The origin of this slight magnetic-field dependence of C_2 may be due to the fragmentation of the higher carbon clusters with triplet character.

For the clusters of C_4 and C_6, on the other hand, a large enhancement was clearly observed as shown in Figs. 4(b) and (c). This fact undoubtedly indicates that at least a considerable portion of C_4 and C_6 are of linear form with triplet character in the ground state.

Recent theoretical studies on the electronic states of C_4 have suggested that the singlet rhombus form and the triplet linear form are almost isoenergetic[11]. From the present experiment, the percentage of paramagnetic portion can be evaluated by comparing the degree of enhancement for the clusters with that of standard sample. In this sense, O_2 seems to be suitable as a standard sample. If the degree of enhancement is the same for C_n (n is specific size) and

O_2, for example, all the portion of C_n in the beam would be considered to exist as triplet species. On the other hand, if the degree of enhancement for C_n is reduced, singlet species would be simultaneously formed in the C_n beam. As a result, from the present data, we can not ruled out the possibility of the existence of the singlet rhombic C_4. The degree of the enhancement on C_4 was found to be smaller than that shown for O_2[12]. This evidence rather support that the singlet rhombic form partly exists in the beam. In order to clarify this point more qualitatively, more detailed experiments are now in progress.

3. SIZE-SPECIFIC PHENOMENA INDUCED BY CLUSTER-SURFACE COLLISION[13]

3-1. COLLISIONAL ELECTRON DETACHMENT OF SIZE SELECTED CARBON NEGATIVE CLUSTER IONS

Cluster-surface interactions are very interesting subjects from view points of not only understanding nature of clusters but also applications of clusters to new materials. Of particular interest is that whether such interactions are of size specific or not. However, so far little have been known on the size specific subjects even for energy region below 1 KeV[14,15]. Very recently Beck et al.[16] have first demonstrated the size specific phenomena induced by collision of clusters with solid surface, where impact induced cleaving and melting of alkali halide clusters were investigated. More recently Beck et al.[17] have also shown the experimental evidences of highly efficient inelastic scattering of C_{60}^+ without fragmentations[17].

We have studied collision induced processes of mass selected carbon negative ions with MoS_2 single crystal, paying a special attention on the process of electron emission as a function of the cluster size and the collision energy. The reason why we measure here electrons is that the detection of electrons is relatively easy and highly sensitive comparing with other particles. In addition to such a technical advantage, the emitted electrons induced by collision of negative ions would have unique and useful microscopic information on the cluster-surface interactions.

We found the following interesting phenomena on the electron emission processes. 1) Upon collision of carbon negative cluster ions with MoS_2 surface, there appears two different electron emission processes in the energy region lower than 1 KeV. 2) One of which dominantly occurs at rather higher collision energy (>300 eV), and is well characterized by the fact that an electron emission efficiency is totally dependent on the velocity of the clusters. No size specific phenomena have been observed in this energy range. 3) The other one occurs at low collision energy with narrow energy window. The electron emission efficiency was found to be of size specific, being dependent on rather microscopic properties of the incident clusters such as a geometric structure and an electron affinity. The plots of electron emission efficiency as a function of collision energy show maxima at a certain collision energy.

3-2. APPARATUS FOR CLUSTER-SURFACE COLLISION EXPERIMENTS

Figure 5 shows the experimental setup used in the present work. Carbon clusters were produced by an usual laser vaporization combined with a pulsed He flow[10]. Negative ion clusters were accelerated and mass selected by a TOF method[18]. The target of MoS_2 single crystal was placed on the bottom of a coaxial reflectron with which the incident kinetic energy of the clusters was controlled by applying negative voltage on a bottom of the plate. Emitted electrons were detected by a multichannel plate with a hole (o 8.0 mm) at center, through which the negative ion beam passes into the reflectron. The coaxial reflectron has an advantage in comparison with an angled reflectron[16] such that the incident angle of the incoming cluster beam is kept constant while the incident kinetic energy is changed. Therefore, the interaction area on the target does not change.

The sample of fullerene such as C_{60} and C_{70} was prepared by arc heating of graphite and isolated by a preparative HPLC described elsewhere[19]. Pure C_{60} and C_{70} powder was first dissolved in benzene and dropped onto a Cu rod for the laser vaporization.

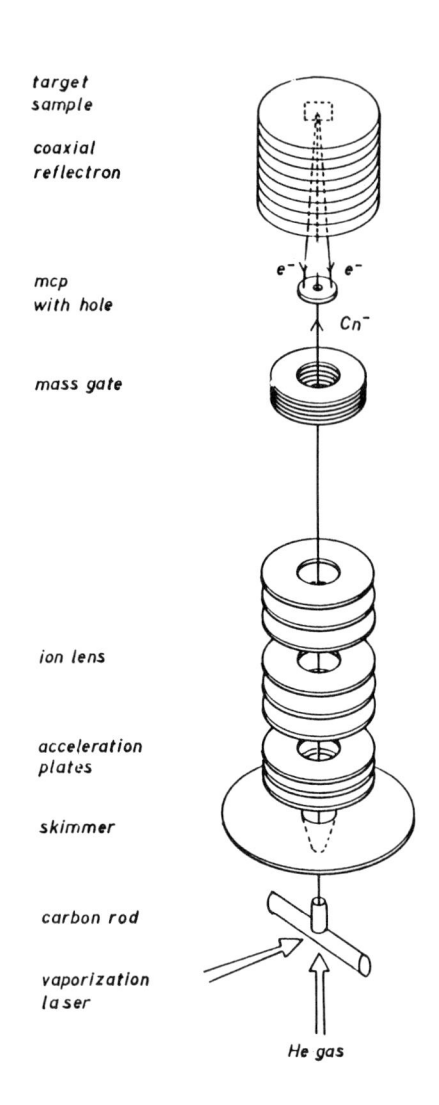

Fig.5 A schematic drawing of the apparatus.

3-3. SIZE DEPENDENCE OF COLLISIONAL ELECTRON EMISSION EFFICIENCY

According to the previous works[18], carbon negative cluster ions have three forms, i.e., a linear chain, a monocyclic ring and a hollowclosed cage structure, and a preferential form for the carbon clusters with an particular size much depends on their sizes and internal energies. Concerning the carbon clusters smaller than n = 30, both linear chain (for n < 10) and monocyclic ring (for 10 < n < 30) forms are dominant[20], while the carbon clusters of C_{60} and C_{70} are certainly hollowclosed cage[12]. Furthermore, as described in our previous paper[18], under certain laser vaporization condition (highly vibrationally hot condition), negative carbon cluster ions exist as a linear chain up to n = 15-20. Note here that the cold carbon negative ions larger than n = 10 are monocyclic. These experimental facts suggest that we can perform the cluster collision experiments

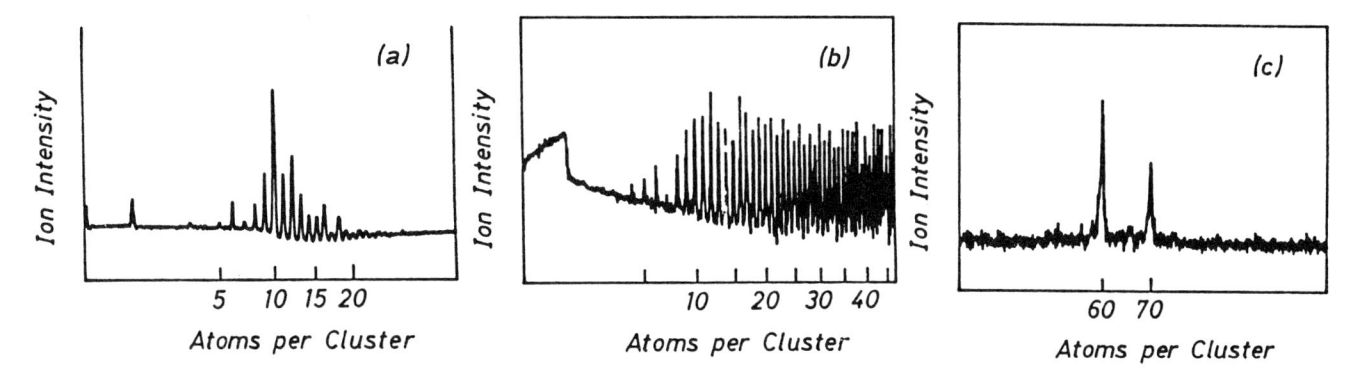

Fig.6 TOF mass spectra of carbon negative cluster ions. The condition of the laser was optimized for (a) ring-rich, and (b) chain-rich clusters. Chemically isolated C_{60} and C_{70} was used for (c).

not only as functions of incident energy and size, but also a geometrical structure.

Figures 6(a)-(c) display typical mass distributions of carbon negative ions revealing the presence of three carbon forms which were produced by the "controlled" laser vaporization. The spectrum (a) is well characterized by the presence of several strong magic numbers, reflecting that the carbon clusters larger than n = 10 exist as a monocyclic ring[20]. On the other hand, the spectrum (b) reflects the "chain dominant" mass distributions for which we can see no magic like behaviors[18]. Mass abundance smoothly distributes over whole range. Figure 6(c) is a mass spectrum of fullerenes, for which chemically isolated C_{60} and c70 powder were used.

Figure 7(a)-(c) show the plots of electron emission efficiencies for the carbon negative clusters of n = 10 (ring), n = 9 (chain) and n = 60 (hollowclosed cage), respectively, as functions of incident kinetic energies. Here, the electrons emitted by collision of negative cluster ions with the target were detected after traveling

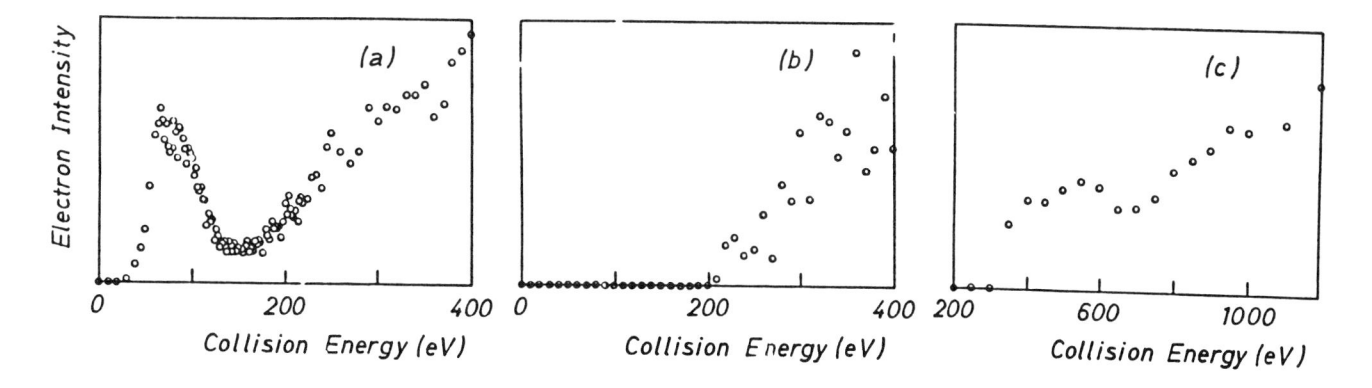

Fig.7 Variations of electron emission intensities versus the collision energy for (a) C_{10}^-, (b) C_9^-, and (c) C_{60}^-.

through the electric field of the reflectron. The electric field applied to the reflectron acts as a decceleration for the incident negative cluster ions, whereas it works as an acceleration for the electrons at this time. Most of the the electrons emitted at the target go back to the same way with a small divergence, because the intrinsic kinetic energies of the electrons initially generated at the surface by collision are negligibly small (< 3 eV)[15].

As can be seen from these curves, the general features commonly observed for all of these carbon forms are that electron emission efficiencies gradually increase by increasing kinetic energy in the higher energy region (> 300 eV). However, it is also certainly distinct that for the ring (C_{10}) and the hollowclosed-cage form (C_{60}), the curves have a peak or a hump at a certain kinetic energy lower than that mentioned above. From these experimental evidences, it is readily suggested that there are two different mechanisms in electron emission processes.

3-4. COLLISION INDUCED ELECTRON DETACHMENT

The plots of the electron emission efficiency show a peak at around 50-200 eV kinetic energy for all the ring carbon clusters examined here. The peak intensity varies within the same order from n = 10 to 18. The peak positions and shapes of electron emission against incident kinetic energies for n = 10, 12, 16 and 18 are shown in Fig. 8. Here electron emission intensities obtained for these ring carbon clusters are normalized at their peaks. The threshold energy for giving electron emission peaks slightly changes, depending on their sizes. Roughly speaking, the threshold energy tends to shift to the higher collision energy for the case of the clusters with the higher electron affinity. For the fullerenes,

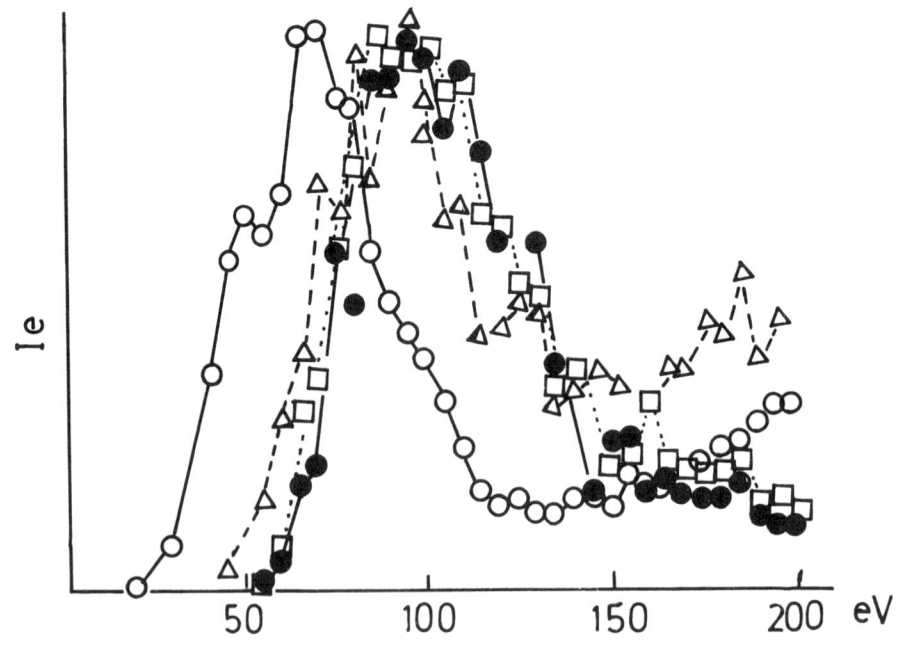

Fig.8 Collision-induced electron detachment intensities versus the collision energy for O C_{10}^-, △ C_{12}^-, □ C_{16}^- and ● C_{18}^-.

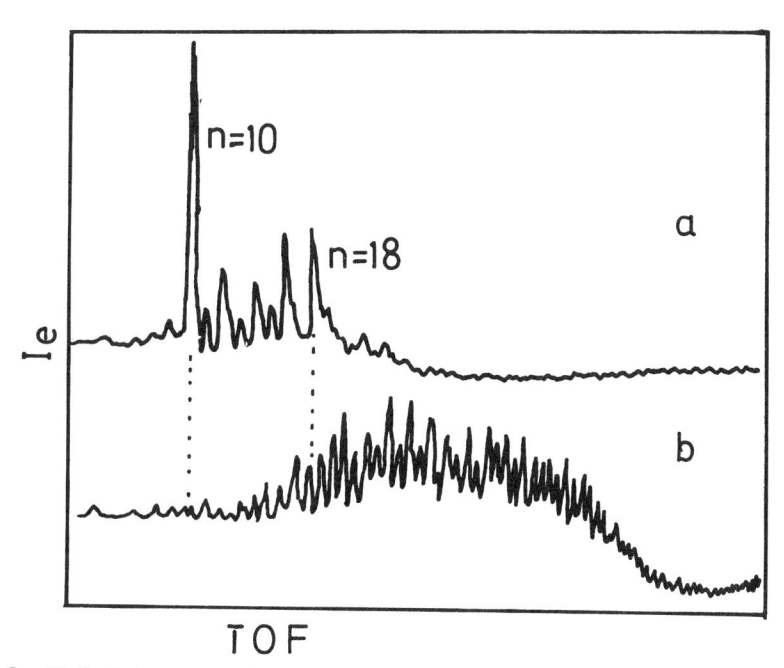

Fig.9 time-of-flight profile of electrons for the carbon beams produced under (a) "ring-rich" and (b) "chain-rich" condition.

however, the peak positions shift to much higher kinetic energy region (at around 550 eV). However, no peak-like shape appears in the plots for the chain-form clusters (n = 5-9). The electron emission efficiencies almost monotonically increase with a threshold at around 150-200 eV for these cases.

No peak appearance for the chain-form carbon was further supported by the experimental evidences that even in the case of carbon clusters larger than n = 10, if they exist as a chain form[18], they show little trace of the electron emission in the low energy region. Two spectra in Fig. 9 display such situations, in which the time of flight profiles of electron emissions from the carbon cluster beam generated under different conditions are compared with each other. Here, in the case of (a), the laser vaporization condition was adjusted so that the ring-form carbon clusters were abundant for n > 10. The spectra were measured by fixing the collision energy at 110 eV. The electron signals attributed to n-even clusters are distinctly seen. On the other hand, in the case of (b), the condition was kept so that the chain-form carbon clusters are dominant up to n = 20. There appears no electron signals attributed to the carbon negative ios with n < 20, while the electron signals attributable to the ions with n > 20 are seen. These facts are well consistent with the facts that no peak of electron emission efficiency curve appears for n = 9 shown in Fig. 7(b).

4. ACKNOWLEDGMENTS

This work was supported by the grants from the Ministry of Education, Science and Culture of Japan. The authors wish to thank M. Mizumachi, N. Yanase, N. Kobayashi, and Y. Kaneko for helpful discussions throughout the course of the present work.

5. REFERENCES

1. W. Weltner Jr., and R. J. Van Zee, "Carbon molecules, ions and clusters", Chem. Rev., 89, 1713-1747 (1989).
2. W. Kratschmer, L. D. Lamb, K. Fostiropoulos, and D. R. Huffman, "Solid C_{60}: a new form of carbon", Nature, 347, 354-358 (1990).
3. K. H. Hinkle, J. J. Keady, and P. F. Bernath, "Detection of C_3 in the circumstellar shell of IRC + 10216", Science, 241, 1319-1322 (1988).
4. P. F. Bernath, K. H. Hinkle, and J. J. Keady, "Detection of C_5 in the circumstellar shell of ICR + 10216", Science 244, 562-564 (1989).
5. H. Shiromaru, M. Mizumachi, Y. Achiba, N. Yanase, N. Kobayashi and Y. Kaneko, "Determination of the spin state for carbon clusters: Focusing effect by hexapole magnetic field", Proceeding of International Symposium on the Physics and Chemistry of Finite Systems, Richmond, Virginia (1991), in press.
6. S. Yang, K. S. Taylor, M. J. Craycraft, J. Conceicao, C. Pittiete, O. Cheshnovsky, and R. E. Smalley, "UPS of 2-30 atom carbon clusters", Chem. Phys. Lett., 144, 431-436 (1988).
7. M. Algranati, H. Feldman, D. Kella, E. Malkin, E. Miklazky, R. Naaman, Z. Vager and Z. Zajfman, "The structure of C_4 as studied by the Coulomb explosion method", J. Chem. Phys., 90, 4617-4618 (1989).
8. J. R. Heath and R. J. Saykally, "The structure of the C_4 cluster radical", J. Chem. Phys., 94, 3271-3273 (1991).
9. R. L. Christensen and D. R. Hakilton, "Permanent magnet for atomic beam focusing", Rev. Sci. Instrum., 30, 356-358 (1959).
10. T. G. Diet, M. A. Duncan, D. E. Powers and R. E. Smalley, "Laser production of supersonic metal cluster beams", J. Chem. Phys., 74, 6511-6512 (1981).
11. Z. Slanina, "ab initio estimation of the energy difference between the linear and rhombic structure of $C_4(g)$", Chem. Phys. Lett., 173, 164-168 (1990), and references therein.
12. N. Yanase, K. Miyake, N. Kobayashi and Y. Kaneko, "Measurement of magnetic moments of $(O_2)_2$ and $(O_2)_3$", Proc. 17th International Conference of the Physics on Electronic and Atomic Collision, Brisbane, (1991).
13. T. Moriwaki, H. Shiromaru and Y. Achiba, "Dynamics of the cluster-surface interactions: Collisional detachment of size selected carbon negative cluster ions", Proc. International Symposium on the Physics and Chemistry of Finite Systems, Richmond, Viginia (1991), in press.
14. H. Hsieh and R. S. Averback, "Molecular-dynamics investigation of cluster-beam deposition", Phys. Rev. B42, 5365-5368 (1990).
15. H., Shiromaru, T. Moriwaki, C. Kittaka and Y. Achiba, "Collisional electron emission at solid surface induced by mass selected negative cluster ions", Z. Phys., D20, 141-143 (1991).
16. R. D. Beck, P. St. John, M. L. Homer and R. L. Whetten, "Impact-induced cleaving and melting of alkali-halide

nanocrystals", Science, 253, 879-882 (1991).
17. R. D. Beck, P. St. John, M. M. Alvarez, F. Diederich and R. L. Whette, "Resillence of all-carbon molecules C_{60}, C_{70} and C_{84}: A Surface-scattering time-of-flight investigation", J. Phys. Chem., 95, 8402-8409 (1991).
18. Y. Achiba, C. Kittaka, T. Moriwaki and H. Shiromaru, "Evidence of linear Chain for n >10 carbon negative ion clusters revealed by mass selected photodetachment spectroscopy", Z. Phys., D19, 427-429 (1991).
19. K. Kikuchi, N. Nakahara, M. Honda, S. Suzuki, K. Saito, H. Shiromaru, K. Yamauchi, I. Ikemoto, T. Kuramochi and S. Hino, "Separation, detection and UV/VIS absorption spectra of fullerenes; C_{76}, C_{78}, and C_{84}", Chem. Lett., 1607-1610 (1991).
20. O. Cheshnovsky, C. L. Pettiette and R. E. Smalley, "UPS of metal and semiconductor clusters", Ion and cluster ion spectroscopy and structure, Elsevier, Amsterdam, p. 373-415, (1989).

Photodissociation of mass-selected clusters: solvated metal ions

James M. Farrar and Stephen G. Donnelly

Department of Chemistry, University of Rochester
Rochester, New York 14627

ABSTRACT

We present a study of the photodissociation spectra of Sr$^+$ solvated by the polar molecules H$_2$O and NH$_3$. Mass selection allows us to control the number of solvent molecules bound to the metal center. The electronic spectra are interpreted using ligand field and charge transfer concepts. For clusters with NH$_3$ as the solvent, the spectra undergo large red shifts with increasing cluster size, with absorption maxima moving from 590 nm for n=1 to 1.4 μ for n=6. Absolute cross section measurements show that $\langle r^2 \rangle$ for the valence electron increases by a factor of nearly 20 as n increases from 1 to 5. This increase is confirmed in molecular dynamics simulations by Martyna and Klein, suggesting that the initial stages of ionization and solvation involve Rydberg state formation.

2. INTRODUCTION

Clusters have often been proposed as ideal models for studying the transition from the atomic or molecular limit to the bulk. Described as a "fifth state of matter" lying between the gas and condensed phase limits,[1,2] mass-selected clusters afford unique opportunities to probe the effect of specific numbers of monomers on observable microscopic properties of the collective system. Two of the most important issues concerning chemists, the states of matter and the nature of chemical reactivity, are intimately related to the gas to liquid/solid transition. The motivation for the work we describe here has arisen from our interest in gas phase ion chemistry and our desire to make the connection between that subject and the chemistry of ions in condensed phases. We have specifically examined the environment about a singly-charged atomic ion as a variable number of polar solvent molecules surround this center. The objectives of this work have been to use electronic photodissociation spectroscopy of mass-selected solvated metal ions to study the solvation process with a controlled number of solvent molecules. The choice of alkaline earth cations is judicious, in that they are isoelectronic with alkali atoms, with $^2S_{1/2}$ ground state terms and $^2P \leftarrow {}^2S$ transitions that lie in the visible. By solvating such species with polar molecules such as H$_2$O and NH$_3$, we create the potential to address an intrinsically bulk phenomenon, the process of spontaneous ionization and electron solvation in finite clusters. This paper summarizes our experiments to date, as well as recent theoretical approaches to the electronic spectra of size selected clusters, specifically Sr$^+$(H$_2$O)$_n$ and Sr$^+$(NH$_3$)$_n$.

3. EXPERIMENTAL

The majority of the experiments reported here have been performed with a time-of-flight(TOF)/Reflectron mass spectrometer, a schematic of which is shown in Figure 1. This instrument has the capability to produce solvated metal ion clusters over a wide mass range, to mass-select the parent ions, overlap the ion packet with a laser pulse, and mass analyze the daughter ions. Based in part on the designs of Lineberger at Boulder,[3] Johnson at Yale,[4] and more recently Schlag at Munich,[5] the spectrometer embodies a source employing the laser vaporization technique, optimized for the production of solvated metal ions. We have also performed a number of experiments with a "pickup" source, in which metal ions from a thermionic emitter are injected into the supersonic flow of solvent vapor from a pulsed valve. Data to be presented later show that the internal temperatures of ions produced by these two methods are quite different, giving us the capability to study the transition from the gas to both the liquid and solid phases.

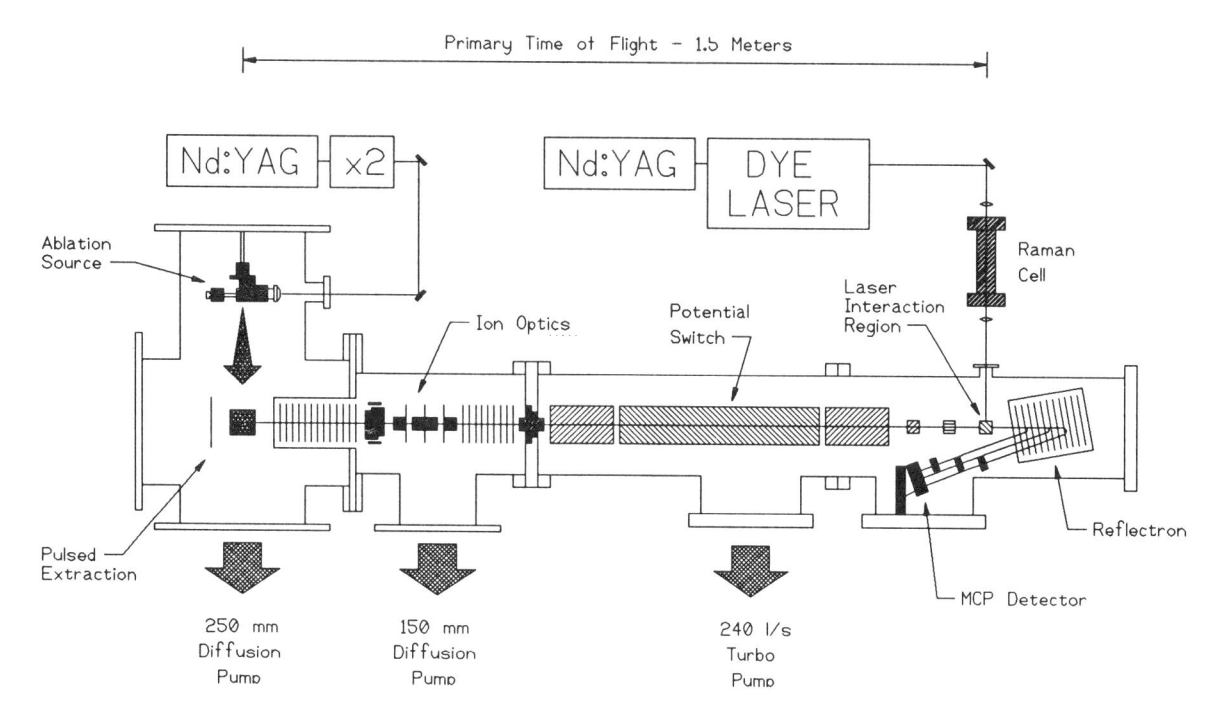

Figure 1 Schematic of TOF/reflectron instrument

4. SMALL CLUSTER SPECTRA

We measure photodissociation cross sections by collecting photofragment product signals as a function of photolysis wavelength. By careful measurements of the laser fluence and the attenuation of the ion beam, we may determine absolute photodestruction cross sections in the thin target limit from the following relation:

$$I/I_0 = \exp[-\sigma \Phi]$$

where σ is the photodestruction cross section (units of cm^2) and Φ is the laser fluence (units of photon cm^{-2}). Although most of the cross sections are reported as relative values determined by measuring signals and normalizing to parent ion beam intensity and laser power, we describe later measurements in which absolute cross sections provide important information on the cluster-size dependence of the electronic wavefunctions of these systems.

The first experiments we conducted on Sr$^+$(NH$_3$) and Sr$^+$(H$_2$O) with "warm" ions were quite informative because they violated our initial simple view that the solvent molecules would only weakly perturb the atomic ion spectra.[6] The atomic $^2P_{1/2,3/2} \leftarrow {}^2S_{1/2}$ resonance lines in Sr$^+$ occur at 407.89 and 421.67 nm (24516 and 23715 cm^{-1}).[7] Instead of fairly narrow cluster absorptions occurring in the vicinity of these resonance lines, we instead observed very broad band spectra as shown in Figure 2 for Sr$^+$(NH$_3$)$_{1,2}$ that are red-shifted from the atomic transitions by hundreds and even thousands of wavenumbers. Figure 3 shows similar spectra for Sr$^+$(H$_2$O)$_{1,2}$. The spectra are amenable to interpretation via ligand-field and molecular orbital concepts. In our published work, we have presented arguments indicating that the gross features of the spectra for the singly-solvated species are readily interpretable as metal-centered transitions based on 5s → 5p excitations. However, the strong red-shifting of the cluster transitions relative to the isolated atomic ion resonance lines is indicative of some significant ion-ligand wavefunction mixing, i.e., "tight binding." In the case of Sr$^+$(NH$_3$), *ab initio* calculations by Bauschlicher and co-workers indicate that the first excited state of this system is a singly-degenerate $^2A_1$ state.[8]

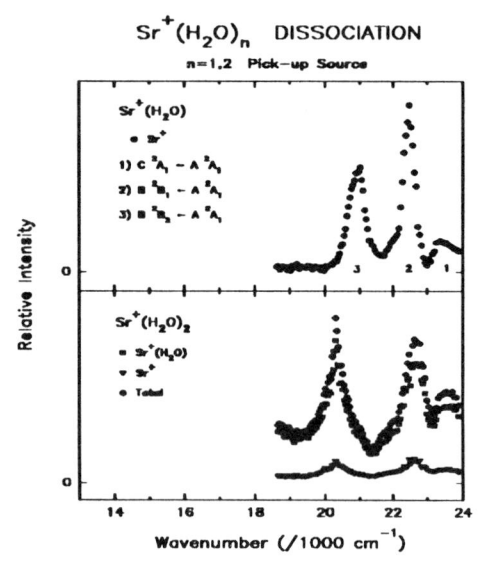

Figure 2 Sr$^+$(H$_2$O)$_{1,2}$ photodissociation spectra

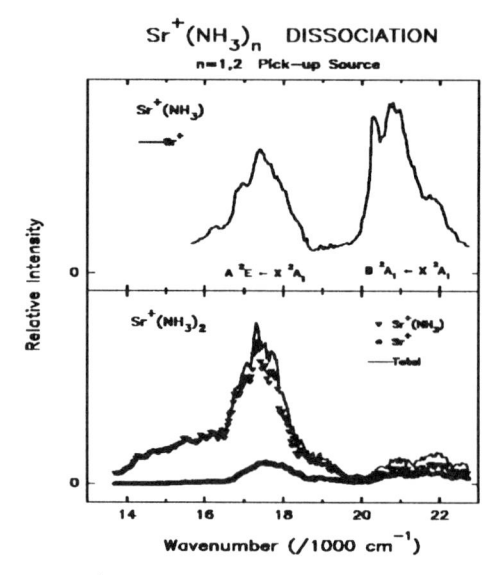

Figure 3 Sr$^+$(NH$_3$)$_{1,2}$ photodissociation spectra

The spectra shown in Figures 2 and 3 were measured for ions prepared from a "pickup" source in which low energy ions from a thermionic emitter are injected into the supersonic flow of solvent vapor from a pulsed valve. Very few collisions occur downstream from the point of injection, and thus the ions cool by the unimolecular process of evaporation. The internal energy distributions of such ions extend right up to the limit of ions that are metastable on the timescale of transit through the mass spectrometer, i.e., ions with energies just below threshold for unimolecular decay at the photolysis volume. We have performed "evaporative ensemble" calculations[9] on these ions and find that for the clusters with one and two solvent molecules, the mean internal excitation is approximately 20 kcal mol^{-1}. In contrast, we can also perform measurements on ions produced by laser vaporization, in which the clusters experience many collisions with the buffer gas in the supersonic expansion after cluster formation. The result of such measurements is a spectrum in which the internal temperatures are significantly colder. Our initial estimates, based on rotational contours, suggests a rotational temperature below 50 K. Figure 4 shows a comparison of spectra taken under both "hot" and "cold" conditions. The spectra show clear simplification, and indicate that the clusters dissociate through a predissociation mechanism. The progression shown in the cold spectrum corresponds to a spacing of 280 cm^{-1}, which we attribute to the Sr-N stretching frequency. Spectra such as those shown in Figure 4 suggest that high resolution data can be obtained for small clusters in favorable cases, but also that by employing the two different methods for preparing clusters, we can examine the approach to both the solid and liquid states.

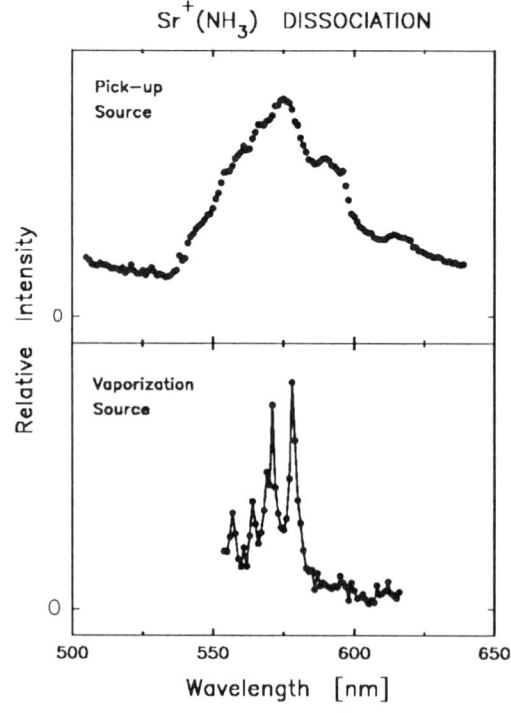

Figure 4 Hot and cold spectra for Sr$^+$(NH$_3$)

5. LARGE CLUSTER SPECTRA: TRANSITION TO THE BULK

The "holy grail" of cluster studies is to make the connection between the gas phase and the bulk. We have proposed that the alkaline earth cations might be particularly good systems to study the transition toward the intrinsically bulk phenomenon of electron solvation. Our work on $Sr^+(NH_3)_n$, even for values of n as small as 3 to 6, has shown a large size-dependent red shift that has been interpreted as electron transfer from the metal to the solvent.[10] The experimental data are shown in Figure 5. These data motivated some path integral Monte Carlo simulations on the structures of such clusters by Martyna and Klein[11], who showed that the red shift observed is correlated with increasing Rydberg state character of the ground and excited state wavefunctions of the clusters. Recent theory and experiment on $Na(H_2O)_n$ clusters[12] also suggest substantial charge transfer from the metal to the solvent in small clusters. The excitation of electronic degrees of freedom in the clusters and the resultant dissociation process are reminiscent of electronic to vibrational (E-V) energy transfer in collisions,[13] generalized in the present case to half-collision processes. Existing theoretical work[14] has shown that there is a strong connection between the E-V process and electron transfer to the collision partner undergoing vibrational excitation. We exploit this connection in interpreting the photodissociation mechanism for small clusters and generalizing to larger clusters whose behavior approximates the bulk.

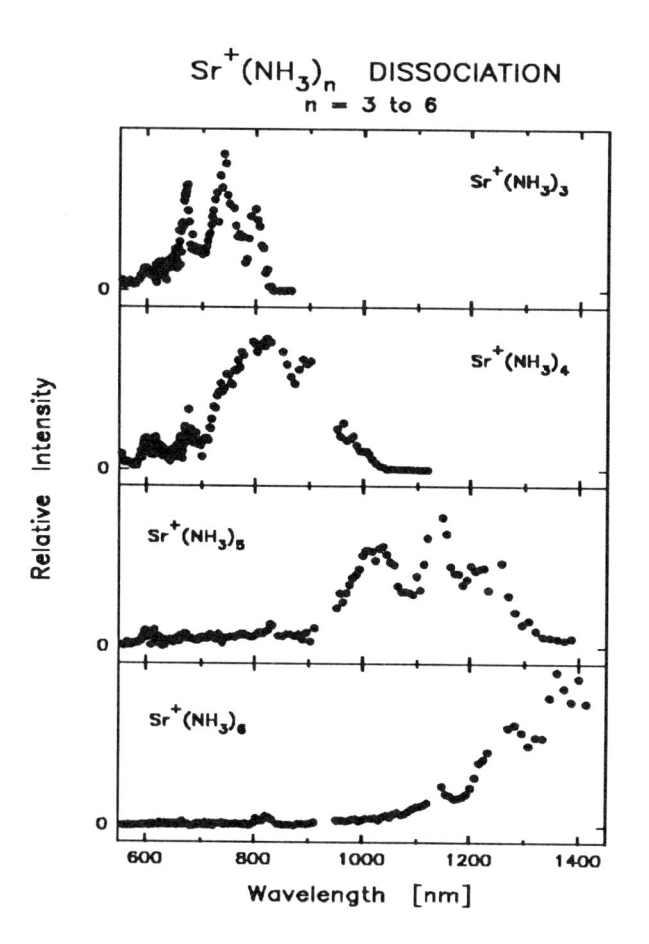

Figure 5 Absorption spectra for $Sr^+(NH_3)_{3-6}$

Our initial data and the analysis of the spectral shifts via a Franck-Platzman charge transfer to solvent (ctts)[15] model in terms of the increasing stability of a solvated ion-pair with increasing solvent number led us to suggest that we were observing transitions from a valence state cluster to an excited state having extensive ion-pair character. In contrast, Klein's molecular dynamics calculations suggest that the transitions are really s 3 p Rydberg transitions in which ion-pair character is strongly mixed into the ground and excited state wavefunctions, the mixing increasing with increasing solvent number. These two views of the red shift are not really disparate when considered in the following light. With increasing solvent number, the 5s $^2S_{1/2}$ and 5p $^2P_{1/2,3/2}$ states of Sr^+ naturally lower in energy; for the smaller clusters, this energy lowering may be 30-35 kcal mol^{-1} per additional solvent. This situation is illustrated in Figure 6. The second ionization limit for Sr, corresponding to the formation of Sr^{+2} + e$^-$ lies 8 eV above the ^2P levels in the bare ion. Solvation of this state corresponds to ion-pair formation, and the size-dependent stabilization of this state is dominated by the solvation of the doubly-charged Sr^{+2} species, stabilizing by 70-80 kcal mol^{-1} per solvent. Thus, the ion-pair state lowers in energy more rapidly than the valence states based on the solvation of ^2S and ^2P singly-charged atomic ions. In analogy with charge transfer exciton processes in crystals,[16] ion pair character begins to mix into the wavefunctions for the 5s and 5p-based valence states. This mixing is strongest for the higher-

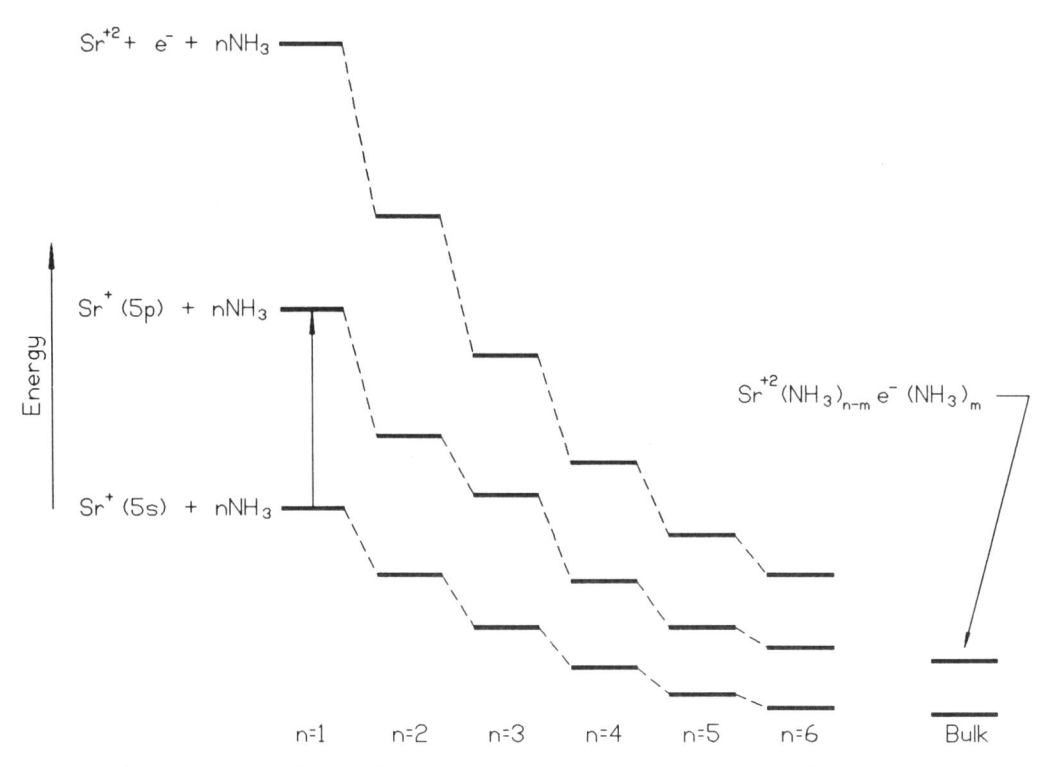

Figure 6 Schematic energy level diagram for the stepwise solvation energetics of Sr$^+$(NH$_3$)$_n$

lying 5p state, lying closer in energy to the ion-pair state. Mixing of ion-pair character into the valence states decreases the gap between the solvated atomic ion states. Klein's interpretation of our data is that we are still observing s 3 p transitions in the large clusters, with both states having significant Rydberg character, although the 5p state has significantly more ion pair character because of the increased valence and ion-pair mixing. An interesting issue then becomes understanding the point at which the ion pair states should become comparable in energy with the 5s/5p cluster states. This degeneracy might signal the onset of bulk-like behavior, or the ion-pair state may continue to lower its energy and become the ground state of the system. Such considerations are still open to controversy, as a recent article by Tuttle and Golden[17] raises the question of whether the widely-accepted cavity model or the solvated anion cluster (SAC) model[18] is a more correct description of the solvated electron ground state in the bulk.

While measuring electronic absorption band positions without regard to absolute intensities provides critical information on the size-dependent evolution of electronic structure, a singularly important piece of information can be extracted from absolute cross sections. In the case where a well-defined ground state undergoes excitation to a set of excited states such that the ground and excited states form a complete set, there exists a class of useful sum rules on intensities that have important implications for our systems. In the alkaline earth systems in particular, where the ground electronic state has significant atomic ion parentage, and where the role of solvent in mass-selected clusters is to modify significantly the ground state electron density, these sum rules have a particularly pleasing interpretation. These sum rules can be written in the following form:[19]

$$\int I(\omega) \, \omega^k \, d\omega = S(k)$$

The integration extends over all excited states, the complete absorption band for bound-free transitions. In the case of k = 0, the sum is the familiar Thomas-Reiche Kuhn sum rule, with S(0) = N, the number of valence electrons in the system. For our purposes, the critical sum rule is the one for which k = -1. Elementary evaluation of the absorption cross section for dipole radiation using completeness properties of the wavefunctions shows that S(-1) is directly proportional to $\langle r^2 \rangle$, the mean square electron radius <u>in the ground electronic state</u>. The experimental data for $Sr^+(NH_3)_{1-6}$ show a monotonic increase in cross section with increasing solvent number, and combined with the large n-dependent red shift, result in an increase in $\langle r^2 \rangle$ by a factor of 20 as n increases from 1 to 5. This result comes directly from the data and confirms the path integral Monte Carlo calculations of Martyna and Klein that show a similar increase as the valence electron becomes increasingly Rydberg-like with increasing n. This simple demonstration of the sum rule technique shows that it is a powerful method for examining the electron distribution in clusters that model the approach to the condensed phase.

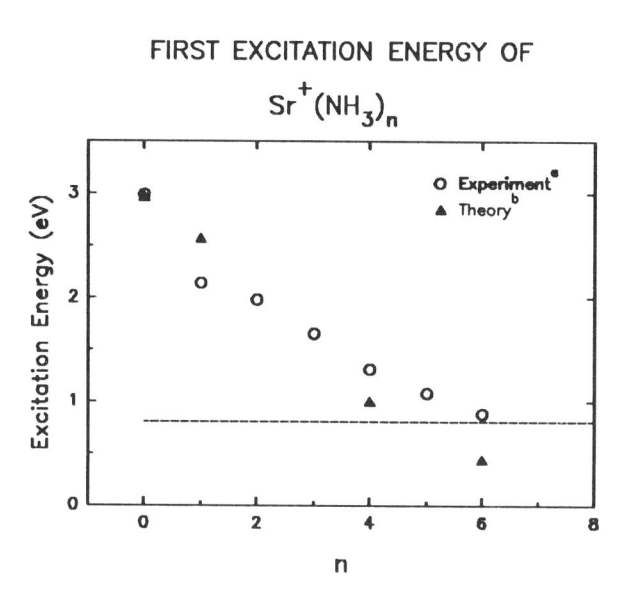

Figure 7 Electronic excitation energy vs. n for $Sr^+(NH_3)_n$, n=0-6

The issue of how closely we can approach the bulk behavior and under what circumstances has already become a stimulating issue. It is quite interesting to plot the photon energy at the absorption maxima from our data on $Sr^+(NH_3)_n$ as a function of cluster size and compare those results with the PIMC calculations of Martyna and Klein. That result is shown in Figure 7, revealing an interesting trend. The experimental and calculated points clearly lie on curves having qualitatively different slopes. Moreover, the theory clearly overshoots the bulk solvated electron limit, indicated as a dashed horizontal line. Experiment, on the other hand, approaches the bulk limit quite closely, and with a smaller slope. It is quite clear that additional data on larger clusters will be required to determine if the energy gap will continue to decrease and then finally increase to reach the bulk limit. These experiments are currently underway.

6. CONCLUSIONS

We have observed large, stepwise red shifts in the absorption spectra of the cluster ions $Sr^+(NH_3)_n$, with n=1 to 6. This systematic shift appears to be correlated with increasing transfer of electron charge from the metal ion to the bulk solvent. Both theory and experiment show that $\langle r^2 \rangle$ increases dramatically with increasing solvent number, suggesting that the early stages of spontaneous ionization in polar solvents involves the formation of Rydberg states. The data appear to provide new insights into the processes of spontaneous ionization and electron solvation in small clusters.

7. ACKNOWLEDGEMENTS

We wish to acknowledge support of this research by the National Science Foundation under grant

CHE088-07833. We also acknowledge the Donors of the Petroleum Research Fund administered by the American Chemical Society for partial support of this research.

8. REFERENCES

1. G. D. Stein, Phys. Teach. **17**, 503 (1979).
2. T. D. Märk and A. W. Castleman, Jr., Adv. At. Mol. Phys. **30**, 65 (1985); T. D. Mä2.rk, Int. J. Mass Spectrom. Ion Proc. **79**, 1 (1987).
3. M. L. Alexander, Ph.D. dissertation, University of Colorado, 1987; see also M. A. Johnson and W. C. Lineberger, in J. M. Farrar and W. H. Saunders, Editors, <u>Techniques for the Study of Ion-Molecule Reactions</u> (Wiley, New York, 1988), p. 221.
4. M. J. DeLuca, Ph.D. dissertation, Yale University, 1990; L. A. Posey, M. J. DeLuca, and M. A. Johnson, Chem. Phys. Lett. **131**, 170 (1986).
5. U. Boesl, R. Weinkauf, K. Walter, C. Weickhardt, and E. W. Schlag, J. Phys. Chem. **94**, 8567 (1990).
6. M. H. Shen and J. M. Farrar, J. Chem. Phys. **94**, 3322 (1991).
7. C. E. Moore, <u>Atomic Energy Levels</u>, Natl. Bur. Std. (U. S.) **467** (1958).
8. C. W. Bauschlicher, Jr. and H. W. Partridge, to be published.
9. C. E. Klots, J. Chem. Phys. **83**, 5854 (1985).
10. M. H. Shen and J. M. Farrar, J. Phys. Chem. **93**, 4386 (1989).
11. G. J. Martyna and M. L. Klein, **95**, 515 (1991).
12. I. V. Hertel, C. Hüglin, C. Nitsch, and C. P. Schulz, Phys. Rev. Lett. **67**, 1767 (1991).
13. I. V. Hertel, Adv. Chem. Phys. **50**, 475 (1982).
14. P. Botschwina, W. Meyer, I. V. Hertel, and W. Reiland, J. Chem. Phys. **75**, 5438 (1981).
15. R. Platzman and J. Franck, Z. Phys. **138**, 411 (1954).
16. V. M. Agranovich *et al.*, Chem. Phys. Lett. **110**, 270 (1984).
17. T. R. Tuttle, Jr. and S. Golden, J. Phys. Chem. **95**, 5725 (1991).
18. H. F. Hameka, G. W. Robinson, and J. Marsden, J. Phys. Chem. **91**, 3150 (1987).
19. See, for example, U. Fano and J. W. Cooper, Rev. Mod. Phys. **40**, 441 (1968). We thank Professor Mark Johnson for making us aware of this topic.

Stimulated Raman spectroscopy of complexes and clusters in supersonic molecular beams

B. F. Henson, G. V. Hartland, V. A. Venturo, P. M. Maxton, and P. M. Felker

Department of Chemistry and Biochemistry
University of Califonia, Los Angeles, California 90024-1569

ABSTRACT

Mass-selective, ionization-detected stimulated Raman spectroscopies and their advantages in the size-selective vibrational spectroscopy of clusters in supersonic molecular beams are described. Results from the application of such methods to benzene dimer, phenol-water, and carbazole-$(Ar)_n$ clusters are presented and discussed.

1. INTRODUCTION

The generation of atomic and molecular clusters in supersonic molecular beams has undergone considerable progress in recent years.[1] This progress in cluster synthesis has been accompanied by the development of methods, primarily spectroscopic, with which to probe the properties of these species. Work in this laboratory has focused on the development and application of stimulated Raman methods in vibrational studies of molecular clusters.[2-6] Such work has shown that mass-selective variants of ionization-detected stimulated Raman spectroscopy (IDSRS) are powerful, size-selective probes of ground-state cluster structure and dynamics. In this paper we give an overview of these methods and their useful features. In addition, we describe results obtained by the application of mass-selective IDSRS in studies of benzene dimer, phenol-water, and carbazole-$(Ar)_n$ clusters.

2. THE NATURE AND ADVANTAGES OF MASS-SELECTIVE IDSRS

2.1 Ionization-detected stimulated Raman spectroscopies

IDSRS methods are double resonance schemes that employ resonantly enhanced multiphoton ionization (REMPI) to probe the population changes induced by stimulated Raman scattering. Three general schemes of IDSRS exist. The first, depicted in the level diagram of Fig. 1a, is ionization-gain stimulated Raman spectroscopy (IGSRS).[7] In IGSRS the population gain of the final state in the stimulated Raman transition – $|b\rangle$ – is monitored by a REMPI process originating in $|b\rangle$. Ion current is measured as a function of ω_2. Raman transitions are manifest as peaks in the ion signal when $\omega_1 - \omega_2$ corresponds to a Raman resonance. The second IDSRS scheme is ionization-loss stimulated Raman spectroscopy (ILSRS).[8] In ILSRS, depicted in Fig. 1b, the population loss of the initial state in the Raman process – $|a\rangle$ – is monitored by a REMPI process originating in $|a\rangle$. Again, ω_2 is scanned while ion signal is monitored. Raman transitions are now manifest as depletions in an otherwise constant ion signal. Such depletions occur when $\omega_1 - \omega_2$ corresponds to a Raman resonance. The third IDSRS scheme is a variant of IGSRS in which ω_1 and ω_2 are fixed to a Raman resonance and ω_3 is scanned.[7] This method allows one to measure the vibronic spectrum induced by the Raman process. Features in such a spectrum may correspond to transitions from the vibrational state initially prepared by the Raman process. Or, they may correspond to transitions from states populated by a dynamical process

(IVR or predissociation) occurring subsequent to Raman excitation.

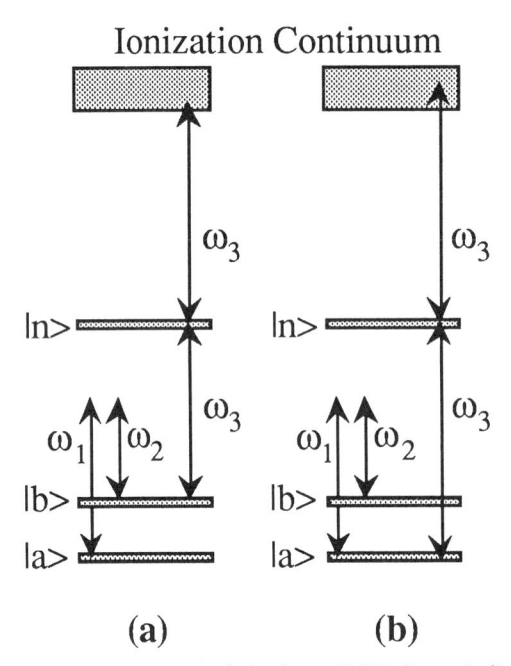

Fig. 1. Level diagrams depicting (a) the IGSRS and (b) the ILSRS processes. ω_1 and ω_2 denote the excitation pulses that drive the stimulated Raman transition from $|a\rangle$ to $|b\rangle$. ω_3 denotes the pulse that photoionizes from (a) $|b\rangle$ or (b) $|a\rangle$ via the excited intermediate state $|n\rangle$.

Both IGSRS and ILSRS may be implemented interferometrically as Fourier transform spectroscopies.[2,4,9] In such implementations ω_1 and ω_2 from relatively broad-bandwidth light sources pass through a Michelson interferometer prior to being combined with ω_3 and then proceding to the sample. Ion signal is measured as function of interferometer delay. The resulting trace is then Fourier transformed to yield that portion of the Raman spectrum of the sample that is overlapped by the $\omega_1 - \omega_2$ bandwidth of the excitation sources. The virtue of these Fourier transform schemes is that spectral resolution depends only on the interferometer delay range scanned in the experiment and not on the bandwidths of the ω_1 and ω_2 light sources employed. Results obtained with FT-IGSRS and FT-ILSRS in studies of benzene dimer will be presented below.

Because IGSRS and ILSRS methods rely on the detection of ions, both may be readily implemented with mass-selective detection by filtering the total ion signal through a mass analyzer.[3-6] In this laboratory we accomplish this by means of a time-of-flight mass spectrometer. As discussed below, mass-selective detection is very desirable in IDSRS studies of clusters.

2.2 Some valuable features of IDSRS schemes

Several features of the IDSRS schemes render them particularly valuable in cluster studies. We summarize these features in this subsection.

A. *Visible light sources drive the ground-state vibrational transitions.* There are definite advantages associated with the fact that the vibrational transitions in IDSRS can be induced with visible light sources as opposed to the infrared sources that are required for single photon-resonant vibrational transitions. Principal among these advantages are wide spectral coverage, easy generation of high-energy pulses having narrow bandwidths, and easy generation of short pulses. Because of these advantages associated with visible sources IDSRS experiments can access Raman resonances throughout the vibrational fundamental region. Reasonably high spectral resolution (0.05 cm^{-1}) can be readily achieved in frequency-domain IDSRS experiments. And, the potential for the successful implementation of picosecond-resolved IDSRS in molecular beam studies is very high.

B. *High sensitivity.* IDSRS methods are able to measure Raman spectra of minority species in the sparse medium of a seeded, skimmed, supersonic expansion. This high sensitivity derives from the very high efficiency of the REMPI probe process and subsequent ion detection, as well as from the fact that stimulated Raman processes can effect substantial transfer of vibrational population under quite reasonable experimental conditions.

C. *High species-selectivity.* The ability to measure the vibrational spectrum of a particular species without interference from other species is of paramount importance in cluster studies. Two features of IDSRS methods render them highly species-selective. First, the vibronic resonance condition associated with the probe process affords a degree of species-selectivity. Second, the mass analysis of the photoions generated in the schemes allows for further species-specificity.

D. *Minimal rotational congestion.* The Raman resonances that are best studied by IDSRS are those for which the isotropic part of the polarizability tensor dominates. Such resonances tend to be characterized, even at fairly high resolution, by rotational structure dominated by a single feature – the Q-branch.[10] This is a big advantage if one desires to measure vibrational linewidths and/or frequency shifts. Of course, it is a disadvantage if one wants to analyze rotational structure to obtain geometrical information.

E. *Vibrational state-selectivity in probing.* By virtue of the vibronic resonance condition that obtains in IGSRS, the scheme is automatically biased towards the detection of a particular vibrational state of a particular species. This feature can be exploited to characterize product vibrational state distributions subsequent to intramolecular vibrational energy redistribution (IVR) or predissociation.

F. *Access to all of the vibrational chromophores in a molecular complex or cluster.* ILSRS can be used to characterize all the Raman resonances in a weakly bound complex or cluster, even if the Raman-excited vibrational motion is localized in a moiety different than the chromophore through which the photoionization process proceeds.[4,6] For example, we show below that in the phenol-water complex the OH stretch fundamentals localized in both the phenol and the water moieties can be observed.

3. EXPERIMENTAL PROCEDURES

The mass-selective IDSRS experiments reported on here were implemented with an apparatus described in detail elsewhere.[4-6,11] Briefly, part of the frequency-doubled output of an injection-seeded Nd:YAG laser (Spectra-Physics GCR-3) pumped a grating tuned dye laser (Spectra-Physics PDL-3). The output of the dye laser (ω_2) and the remaining portion of the doubled Nd:YAG output (ω_1) served as the two excitation fields required to drive the stimulated Raman transitions. The ultraviolet REMPI probe field (ω_3) was generated as the frequency-doubled output of a second dye laser (Spectra-Physics PDL-2) which was pumped by a second Nd:YAG laser (Spectra-Physics DCR-2A). The firing of the second Nd:YAG laser was synchronized and delayed with respect to the firing of the first. The temporally overlapped ω_1 and ω_2 fields were combined collinearly on a beamsplitter, the result of which was then combined collinearly with ω_3 on a second beamsplitter. The three-color pulse train was focused into the photoionization region of a time-of-flight mass spectrometer (TOFMS) with a 25 cm fl lens. The molecular beam was formed by entraining the species of interest in a carrier gas of helium or a mixture of argon and helium. The expansion of this mixture into vacuum was controlled by a pulsed molecular beam valve. The expansion was skimmed several centimeters downstream from the expansion orifice. The skimmed beam then traveled into the ionization region of the TOFMS. Photoions of a given mass were detected with a dual microchannel plate assembly, the output of which was amplified and averaged by a boxcar integrator. The boxcar output was monitored by a computer as a function of ω_2 to yield frequency-domain IGSRS or ILSRS spectra. In cases where the vibronic spectrum originating in a Raman-excited state was desired, ω_1 and ω_2 were fixed such that $\omega_1 - \omega_2$ was resonant with the Raman transition of interest and the computer monitored the boxcar output as a function of ω_3.

The procedure employed to obtain FT-IGSRS and FT-ILSRS spectra was somewhat different from the above. Specifically, before being combined with the ω_3 beam, the ω_1 and ω_2 pulse trains were directed collinearly through a Michelson interferometer, the delay of which was calibrated by a single-mode He-Ne laser. The output of the interferometer was then combined with ω_3 and focused into the molecular beam sample. FT-IDSRS experiments were conducted by monitoring the signal of the parent ion of interest as a function of interferometer delay, with ω_1, ω_2, and ω_3 all fixed. The interferograms so obtained were Fourier-transformed to yield Raman spectra. The laser system used to obtain such spectra consisted of just the DCR-2A Nd:YAG laser pumping two PDL-2 dye lasers. ω_1 was provided by a portion of the frequency-doubled output of the Nd:YAG laser (bandwidth of about 1 cm^{-1}). ω_2 was obtained from one of the dye lasers (bandwidth of 0.3 or 3 cm^{-1} depending on the grating order used). ω_3 was obtained from the frequency-doubled output of the second dye laser and was optically delayed about 10 nsec with respect to the ω_1 and ω_2 pulses.

For some of the ILSRS and FT-ILSRS spectra presented below a normalization scheme analogous to that used in the infrared-ultraviolet double-resonance experiments of Ref. 12 was used. A suprasil flat was employed to combine the ω_3 beam with the ω_1 and ω_2 beams. Two spatially separated, approximately parallel ω_3 beams resulted from the reflection of ω_3 off of the front and back faces of the flat. Only one of these beams was combined collinearly with ω_1 and ω_2 beams. Upon being focused into the sample, the output from the flat separated into two spatially distinct regions. In one region the species in the molecular beam only encountered the ω_3 light. In the other, species encountered ω_3, and

ω_1 and ω_2. In both regions photoions were produced. However, only in the latter region could Raman processes deplete the ion signal. Owing to their spatial separation in the time-of-flight mass-spectrometer acceleration region, flight times for ions of the same mass originating in the two regions were resolvably different. The signal from ions of a given mass originating in the one-color region was averaged by one boxcar integrator, and the signal corresponding to ions of the same mass originating in the three-color region was averaged by a second boxcar integrator. The three-color signal was divided by the one-color one to give a quantity sensitive to Raman depletions but insensitive (ideally) to ion fluctuations arising from probe laser power or molecular-beam instabilities. This normalization scheme is advantageous in ILSRS (or FT-ILSRS) experiments in which the ion signals are large enough that the noise due to ion-counting statistics is negligible compared with the ion noise from other sources.

4. RESULTS AND DISCUSSION

4.1 Benzene dimer

We have done numerous experiments on benzene dimer and isotopomers thereof.[3,5,11] One aspect of this species that has been of particular interest to us is its geometry. Oddly enough, despite numerous spectroscopic studies,[13] there is no consensus on what the dimer geometry is. Indeed, there is disagreement on gross features of the geometry, such as whether or not the benzene moieties are symmetrically equivalent. Our Raman results[5,11] strongly indicate that the benzenes in benzene dimer are symmetrically *inequivalent*, in contrast to interpretations made of vibronic spectroscopic results.[13]

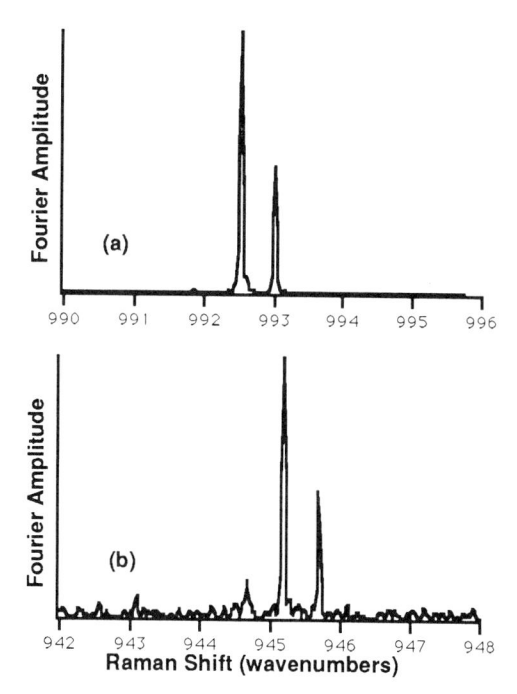

Fig. 2. FT-IDSRS spectra in the ν_1 region of (a) fully protonated benzene dimer, and (b) fully deuterated benzene dimer. The resolution in the spectra is about 0.05 cm^{-1}.

The IDSRS evidence for inequivalent benzenes in benzene dimer consists principally of the splittings and frequency shifts of the ν_1 (totally symmetric ring-breathing mode near 993 cm^{-1}) and ν_2 (totally symetric C-H stretch near 3070 cm^{-1}) fundamentals in the various isotopomers of the species. Figure 2 shows that the $h_6 - h_6$ dimer (h_6 represents fully protonated benzene) and the $d_6 - d_6$ dimer each have two Raman resonances in the ν_1 region. The ν_2 resonances of these homodimers are also split.[11] Such splittings might have been expected, even for a symmetrical dimer, given that excitation exchange interactions will mix the ν_1 (or ν_2) level of one benzene in the dimer with that of the other. However, analysis[11] shows that such mixing in a symmetrical dimer will produce two resonances with the lower frequency one carrying all of the Raman intensity and the higher frequency one having no Raman intensity at all (in the limit where the isotropic part of the Raman polarizability tensor dominates). Thus, the homodimer results in the ν_1 and ν_2 regions indicate an asymmetrical dimer.

These results are fully supported by IDSRS measurements on the $h_6 - d_6$, $h_6 - d_1$, and $d_6 - d_1$ isotopomers. For each of these dimers we find *two resonances* in the region of the ν_1 (and ν_2) fundamentals of each of the moieties.[5,11] For example, for $h_6 - d_6$ there are two resonances in the region of the h_6-localized ν_1 fundamental (at 992.66 and 992.93 cm^{-1}) and two in the region of the d_6-localized ν_1 fundamental (at 945.28 and 945.59 cm^{-1}). This is very strong evidence for inequivalent sites in the dimer. Moreover, if one looks at the magnitudes of the splittings of the fundamentals in the various dimer isotopomers, one finds further strong evidence for inequivalent benzenes.[5,11] For example, the ν_1 doublets in the $h_6 - h_6$ and $d_6 - d_6$ homodimers are split by the same amount. However, the h_6-localized ν_1 doublets in $h_6 - d_6$ and $h_6 - d_1$ are split by values smaller than in the homodimers but identical to one another. These results point to an inherent ν_1 site splitting of 0.3 cm^{-1} in all of the dimers, with an additional excitation exchange contribution in the homodimers.

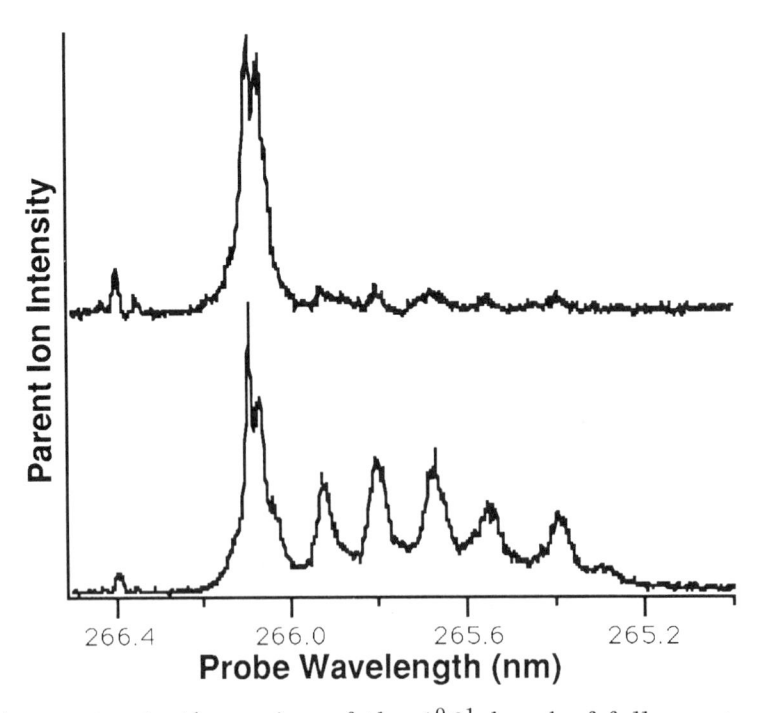

Fig. 3. Vibronic spectra in the region of the $1_1^0 6_0^1$ band of fully protonated benzene dimer obtained subsequent to excitation of each of the two components of the ν_1 Raman

doublet of the species. Top – Vibronic spectrum resulting from the excitation of the Raman band at 993.02 cm^{-1}. Bottom – Vibronic spectrum resulting from the excitation of the Raman band at 992.53 cm^{-1}.

A final set of results that bear on the geometry of the dimer were obtained by characterizing the vibronic resonances of the vibrational levels excited by the various ν_1 resonances of the dimer isotopomers. For example, Fig. 3 shows that the vibronic spectrum in the region of the $S_1 \leftrightarrow S_0$ $1_1^0 6_0^1$ band that results subsequent to excitation of the 993.02 cm^{-1} Raman resonance in $h_6 - h_6$ has a markedly different intensity distribution than that which arises subsequent to excitation of the Raman resonance at 992.53 cm^{-1} in the same species. For the mixed isotopomers the differences are even more extreme.[5,11] These results are readily interpreted in terms of vibronic excitations localized in the different moieties of an asymmetrical dimer. A further interesting implication from these double resonance vibronic spectra is that there is a large geometry change subsequent to vibronic excitation of one of the benzene moieties in the dimer, such geometry change being reflected in the prominent progressions in low frequency intermolecular modes. A full report of our IDSRS work on benzene dimer and higher benzene clusters is in preparation.[11b]

4.2 The phenol-water complex

A comprehensive IDSRS study of hydrogen-bonded complexes involving phenol is in press.[6] One result of particular interest to us from that work pertains to the O-H stretch resonances of the phenol-water complex. We reproduce these results here for two reasons. First, they provide an excellent illustration of how mass-selective ILSRS can be used to probe vibrational resonances in all of the moieties of a molecular cluster. Second, they indicate the potential of IDSRS methods for the probing of vibrational dynamics in such clusters.

From what is known about the proton-donating and proton-accepting properties of phenol and water, respectively, it is virtually certain that the most stable phenol-water complex has a hydrogen-bonded geometry in which the phenol acts as proton donor and the water acts as proton acceptor.[14] Such a geometry is borne out by the results of vibronic spectroscopy on the jet-cooled complex.[15] Given this, one expects there to be three O-H stretch fundamentals in the complex: (1) a phenol-localized fundamental significantly red-shifted from the analogous resonance in free phenol due to the involvement of the phenolic hydrogen in the hydrogen bond, (2) a water-localized symmetric stretch fundamental close in frequency to that of free water (since the water hydrogens are not involved in the hydrogen bond), and (3) a water-localized asymmetric stretch fundamental. The first two of the O-H fundamentals mentioned above have appreciable Raman strength and have been observed by us in mass-selective ILSRS experiments. Figure 4 shows the two resonances. One notes two points. First, one of the resonances is indeed significantly red-shifted (about 135 cm^{-1}) with respect to free O-H stretch frequencies. In contrast, the other is red-shifted by only a few cm^{-1}. We assign the former band to the phenol-localized O-H stretch fundamental and the latter to the water-localized symmetric O-H stretch. Second, the linewidths of the two bands are different by a factor of four, with the water-localized transition being narrower despite being at higher frequency. Translating linewidths into lifetime lower limits one finds that the complex lives for at least 17 psec upon excitation of the water-localized O-H stretch, and for at least 4 psec upon excitation of the phenolic O-H stretch. Such behavior, which is analogous to behavior in smaller hydrogen-bonded complexes,[16] implies that the proton-donating O-H stretch of phenol couples more effectively with the intermolecular modes of the complex than does the proton-accepting O-H symmetric stretch of water. An investigation of this apparent mode-selective dynamics by picosecond-resolved IDSRS would certainly be of interest. In any case, these results provide an illustration of the fact that the rotational selection rules in Raman spectroscopy can give frequency-domain IDSRS methods the power to elucidate dynamics through linewidth measurements, even when full rotational resolution is not available.

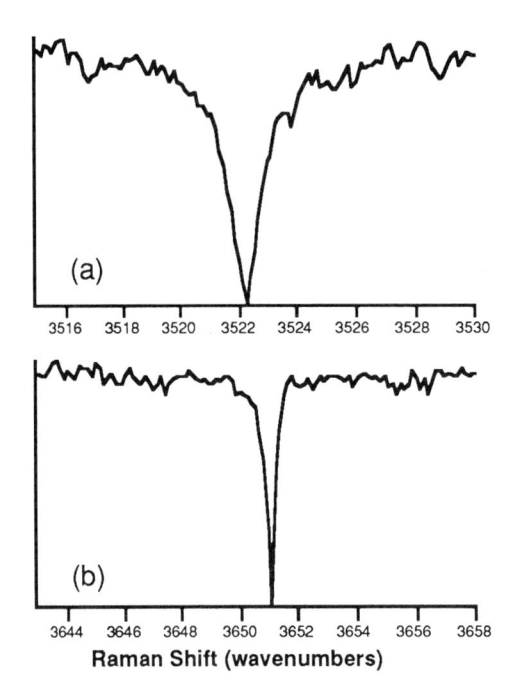

Fig. 4. Mass-selective ILSRS spectra of the phenol-water complex. (a) The phenol-localized O-H stretch fundamental. (b) The water-localized O-H symmetric stretch fundamental.

4.3 Carbazole-(Ar)$_n$ clusters

Very recently we have begun a study[17] of the vibrational spectroscopy of carbazole-(Ar)$_n$ clusters. These species have been studied in some detail both experimentally and theoretically.[18] They are of interest because they represent a convenient model system for the study of the microscopic details of solvation. Our interest in them is two-fold. First, we wonder if the vibrational spectroscopy of the species is consistent with conclusions drawn from the results of vibronic spectroscopy.[18] Second, we are interested in the vibrational dynamics of the carbazole moiety and, in particular, how the dynamics changes as a function of cluster size (i.e., degree of solvation). At this time we have progressed to the point where we are confident that reasonably high quality ILSRS spectra can be measured even for fairly large clusters. For example, Fig. 5 shows mass-selective ILSRS spectra in the 1645 cm^{-1} region of the Raman spectrum as a function of the number of Ar atoms involved in the cluster. Three points are noteworthy. First, it is clear that size-selective vibrational spectra can be measured by mass-selective ILSRS on this system. Second, there are spectral changes in this region as the cluster grows. Just what these changes mean is unclear at this point. However, their existence implies that there are vibrational signatures which may be useful as probes of solvation in this system. Third, the linewidths of some of the resonances of the larger clusters are considerably narrower than one might have expected. For example, the linewidth of the 1646 cm^{-1} resonance in carbazole-$(Ar)_{10}$ is 0.3 cm^{-1}, implying a vibrational lifetime of greater than about 17 ps. Such a lifetime seems quite long given that carbazole is a fairly large molecule and that complexation with ten argons should give rise to an enormous density of vibrational states in the cluster at this vibrational energy. Still, it is pertinent to point out that studies of the vibrational energy flow dynamics in large clusters are relatively few,[19] and one really does not yet know what typical behavior is. It is clear to us that the mass-selected ILSRS method can contribute significantly to the characterization of such behavior. We plan to continue studies in this area.

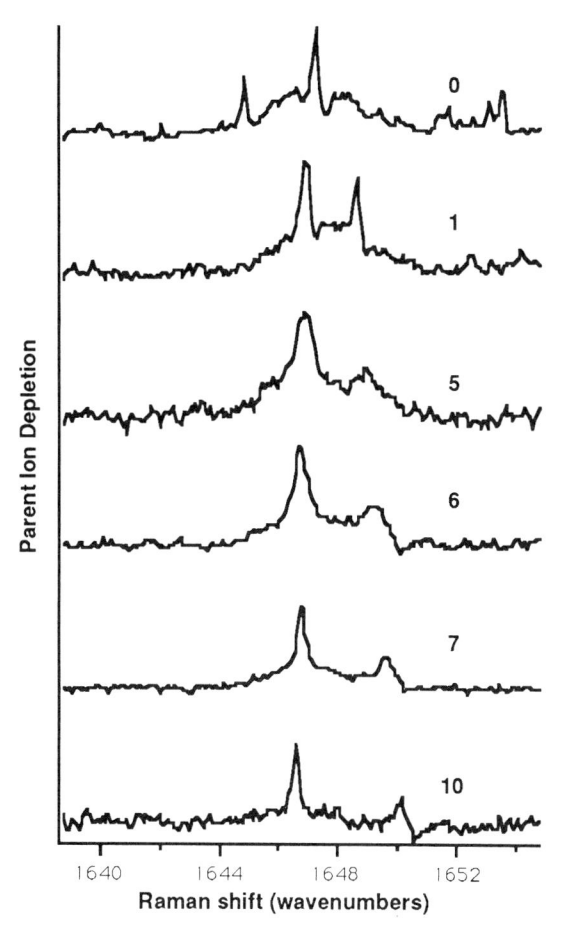

Fig. 5. Mass-selective ILSRS spectra of carbazole-$(Ar)_n$ clusters in the 1640 to 1655 cm^{-1} region. The spectra are plotted such that increasing ion depletion corresponds to the positive ordinate direction. The value of n corresponding to each spectrum is given in the figure.

5. ACKNOWLEDGEMENTS

This work was supported by the Office of Basic Energy Research of the United States Department of Energy through grant no. DE-FG03-849ER14066. Partial support of this work was also provided by the donors of the Petroleum Research Fund administered by the American Chemical Society through grant no. 22728-AC6-C.

6. REFERENCES

1. For example, see E. R. Bernstein, Editor, *Studies in Physical and Theoretical Chemistry, Vol. 68. Atomic and Molecular Clusters*, Elsevier, Amsterdam, 1990.

2. G. V. Hartland, B. F. Henson, L. L. Connell, T. C. Corcoran, and P. M. Felker, "Methods for the high resolution Raman spectroscopy of seeded molecular beams: Interferometry applied to ionization-detected stimulated Raman spectroscopy," *J. Phys. Chem.*, vol. 92, pp. 6877-6880, 1988.

3. B. F. Henson, G. V. Hartland, V. A. Venturo, and P. M. Felker, "Time-resolved Raman spectroscopy of benzene dimer: Nanosecond-plus lifetime at 992 cm^{-1} vibrational energy," *J. Chem. Phys.*, vol. 91, pp. 2751-2753, 1989.

4. G. V. Hartland, B. F. Henson, V. A. Venturo, R. A. Hertz, and P. M. Felker, "Applications of ionization-detected stimulated Raman spectroscopy in molecular beam studies," *J. Opt. Soc. Am. B*, vol. 7, pp. 1950-1959, 1990.

5. B. F. Henson, G. V. Hartland, V. A. Venturo, R. A. Hertz, and P. M. Felker, "Stimulated Raman spectroscopy in the ν_1 region of isotopically substituted benzene dimers: Evidence for symmetrically inequivalent benzene moieties," *Chem. Phys. Lett.*, vol. 176, pp. 91-98, 1991.

6. G. V. Hartland, B. F. Henson, V. A. Venturo, and P. M. Felker, "Ionization-loss stimulated Raman spectroscopy of jet-cooled hydrogen-bonded complexes containing phenols," *J. Phys. Chem.* – in press 1992.

7. P. Esherick and A. Owyoung, "Ionization-detected stimulated Raman spectroscopy," *Chem. Phys. Lett.*, vol. 103, pp. 235-240, 1983.

8. W. Bronner, P. Oesterlin, and M. Schellhorn, "Ion-dip Raman spectroscopy: A method to measure Raman spectra at 4×10^{-9} bar," *Appl. Phys. B*, vol. 34, pp. 11-15, 1984.

9. G. V. Hartland, B. F. Henson, and P. M. Felker, "The dependence of Fourier transform nonlinear spectroscopies on the temporal characteristics of the excitation fields," *J. Chem. Phys.*, vol. 91, pp. 1478-1497, 1989.

10. For example, D. A. Long, *Raman Spectroscopy*, McGraw-Hill, New York, 1977.

11. (a) B. F. Henson, *Stimulated Raman Spectroscopy Studies of Jet-Cooled van der Waals Clusters of Benzene*, Ph.D. Dissertation, Department of Chemistry and Biochemistry, University of California, Los Angeles, 1991; (b) B. F. Henson, G. V. Hartland, V. A. Venturo, and P. M. Felker – to be submitted.

12. R. H. Page, Y. R. Shen, and Y. T. Lee, "Local modes of benzene and benzene dimer, studied by infrared-ultraviolet double resonance in a supersonic beam," *J. Chem. Phys.*, vol. 88, pp. 4621-4636, 1988.

13. See (a) K. O. Börnsen, H. L. Selzle, and E. W. Schlag, "Spectra of isotopically mixed benzene dimers: Details on the interaction in the vdW bond," *J. Chem. Phys.*, vol. 85, pp. 1726-1732, 1986; (b) K. Law, M. Schauer, and E. R. Bernstein, "Dimers of aromatic molecules: (Benzene)$_2$, (toluene)$_2$, and benzene-toluene," *J. Chem. Phys.*, vol. 81, pp. 4871-4882, 1984; and references therein.

14. For example, G. C. Pimentel and A. L. McClellan, *The Hydrogen Bond*, W. H. Freeman, San Francisco, 1960.

15. (a) H. Abe, N. Mikami, and M. Ito, "Fluorescence excitation spectra of hydrogen-bonded phenols in a supersonic free jet," *J. Phys. Chem.*, vol. 86, pp. 1768-1771, 1982; (b)

H. Abe, N. Mikami, M. Ito, and Y. Udagawa, "Dispersed fluorescence spectra of hydrogen-bonded phenols in a supersonic free jet," *J. Phys. Chem.*, vol. 86, pp. 2567-2569, 1982.

16. See R. E. Miller, "Vibrationally induced dynamics in hydrogen-bonded complexes," *Acc. Chem. Res.*, vol. 23, pp. 10-16, 1990.

17. V. A. Venturo, P. M. Maxton, B. F. Henson, and P. M. Felker – to be submitted.

18. S. Leutwyler and J. Bösiger, "Rare-gas solvent clusters: Spectra, structures, and order-disorder transitions," *Chem. Rev.*, vol. 90, pp. 489-507, 1990.

19. See, for example, (a) F. G. Celii and K. C. Janda, "Vibrational spectroscopy, photochemistry, and photophysics of molecular clusters," *Chem. Rev.*, vol. 86, pp. 507-520, 1986; (b) X. J. Gu, D. J. Levandier, B. Zhang, G. Scoles, and D. Zhuang, "On the spectroscopy of SiF_4 and SF_6 in Ar clusters: Location of the solute," *J. Chem. Phys.*, vol. 93, pp. 4898-4906, 1991.

Photoinduced intra-cluster electron transfer reactions of captured intermediates in gas phase S$_N$2 reactions

D. M. Cyr, G. A. Bishea, C.-C. Han[†], L. A. Posey[‡], and M. A. Johnson[*,§]

Department of Chemistry, Yale University
225 Prospect St., New Haven, CT 06511

ABSTRACT

We describe a series of experiments designed to establish the location of the charge-transfer excited states implied by the double minimum potential energy surface for the gas phase S$_N$2 ion molecule reactions. Two distinguishable species, corresponding to the entrance and exit channel ion dipole complexes, are isolated in an ionized free jet. These complexes are interrogated by photodetachment spectroscopy in addition to the photoexcitation and collisional activation induced fragmentation patterns. Charged transfer excited states are evident in the X$^-$· CH$_{4-n}$Y$_n$ ($n \geq 2$) species in which dissociative intra-cluster electron transfer yields XY$^-$ product ions while the S$_N$2 reaction appears to be driven upon photoexcitation of the I$^-$· CH$_3$I complex.

1. INTRODUCTION

It is now well known that when the classic S$_N$2 ion-molecule reactions (e.g. Cl$^-$ + CH$_3$Br → Br$^-$ + CH$_3$Cl) are carried out in the gas phase, the long range electrostatic forces distort the familiar potential energy surface in solution to create deep minima corresponding to the entrance and exit channel ion-dipole complexes.[1-5] These minima in turn provide the experimentalist with a way to trap and localize the collision complexes at two locations along the reactive surface. These species can then be spectroscopically characterized and triggered to react either by a laser photon or by a collision with a third body. At the simplest level, we can regard these complexes as two isomeric forms of the [ClCH$_3$Br]$^-$ negative ion which are interconvertable by simultaneous methyl and electron transfer. In this case we can imagine the reaction coordinate as methyl inversion and translation between two fixed halogen atoms. Deep within the minima, the electronic structure of the two isomers is well described by the charge-localized Cl$^-$· CH$_3$Br and Br$^-$· CH$_3$Cl species. It is possible to envision the S$_N$2 potential surface as being constructed from these two diabatic electronic surfaces (corresponding to the electron localized on one halogen) while the methyl is transferred.[6] These curves are strongly repulsive along this reaction coordinate, and the adiabatic ground state surface can then be recovered by mixing these curves with an electron transfer-type matrix element which serves to split the diabatic states at the avoided crossing to reveal the transition state as illustrated below:

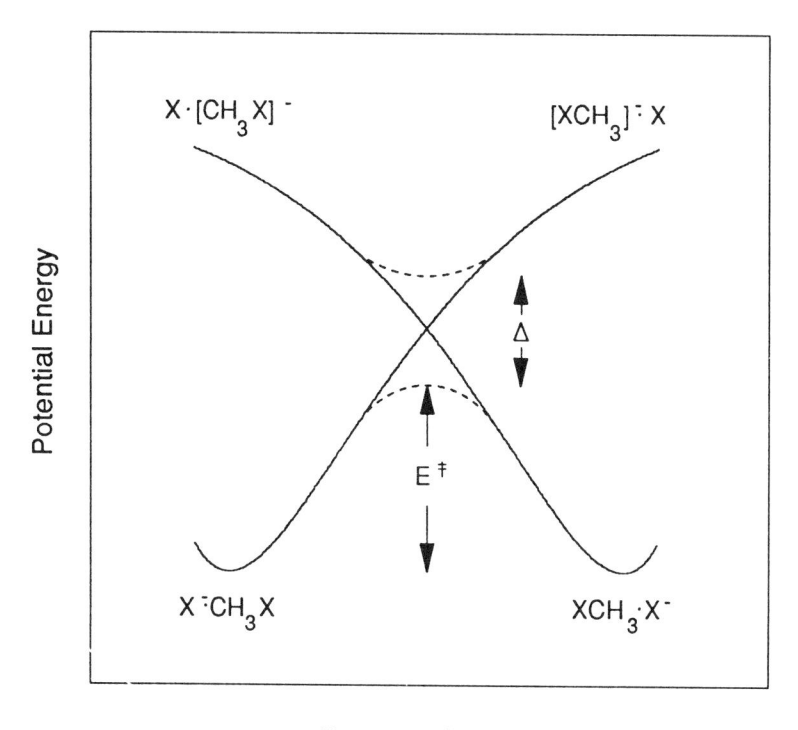

Figure 1. Correlation diagram along the S_N2 reaction coordinate.

When viewed in this fashion, there is a similarity between the S_N2 reaction intermediates and the familiar tri-halide ions, such as I_3^-. In fact, the stable trihalide ions have been used to model solvent effects in the unstable S_N2 transition states.[7,8] Mulliken pointed out that we may similarly regard the I_3^- species as derived from atom transfer between $I_2 \cdot I^-$ and $I^- \cdot I_2$; however, unlike the S_N2 complexes, the charge transfer (CT) interaction is sufficiently strong in I_3^- to overpower the potential maximum in the diabatic states to form a symmetrical minimum corresponding a "symmetrized" charge transfer complex.[9] The symmetrical geometry is not particularly stable, however, and the I_3^- ion is often found in an asymmetric linear configuration in the presence of various cations.[10,11] In this context, the near UV absorption band of the I_3^- anion is seen to correlate to the $I^- \cdot I_2 \rightarrow I \cdot I_2^-$ CT transition as the molecule distorts along the antisymmetric stretch. These remarks are relevant to the S_N2 charge transfer complexes in which the symmetrical $[X \cdot CH_3 \cdot X]^-$ configuration lies at a local maximum in the potential and corresponds to the transition state for reaction. The stable minima are therefore displaced along the antisymmetric stretch coordinate, which raises the possibility of locating the upper diabatic state in Fig. 1 by photoexcitation of the charge transfer band. In this report, we summarize our recent efforts to drive the ground state S_N2 reactions starting from each of the two minima, and present the first data on the possible identification of the charge transfer electronically excited states.

2. RESULTS AND DISCUSSION

2.1 Identification of the entrance and exit channel complexes.

Since our methodology for the preparation of the complexes involves mass analysis of the ionic species created in an ionized free jet, we are obliged to establish our ability to cleanly prepare and isolate each of the ion-dipole isomers. The synthetic strategy involves ionization and subsequent electron attachment to mixtures such as CH_2Cl_2/CH_3Br, for example, in the case of preparation of the entrance channel $Cl^- \cdot CH_3Br$ complex.[12] Since the electron binding energy of the entrance and exit complexes differs primarily in the electron affinity of the halide, we expect that the vertical detachment energy (VDE) should be different for each complex. This energy is given by:

$$\Delta VDE = \Delta EA(X) + \Delta D_0 + \beta$$

where $\Delta EA(X)$ is the difference in electron affinities of the halides, ΔD_0 is the difference in binding energies of the entrance and exit channel complexes and β accounts for the repulsive energy on the neutral surface over the geometry of the complex. The binding energies of the complexes have been measured experimentally or estimated theoretically;[6,13] however, there is little information regarding the energetics on the neutral surface. Since the neutrals do not benefit from the long range electrostatic interaction, we expect the shape of the neutral S_N2 reaction to more nearly resemble that of the ionic reaction in solution. A schematic of the probable potential curves are shown below:

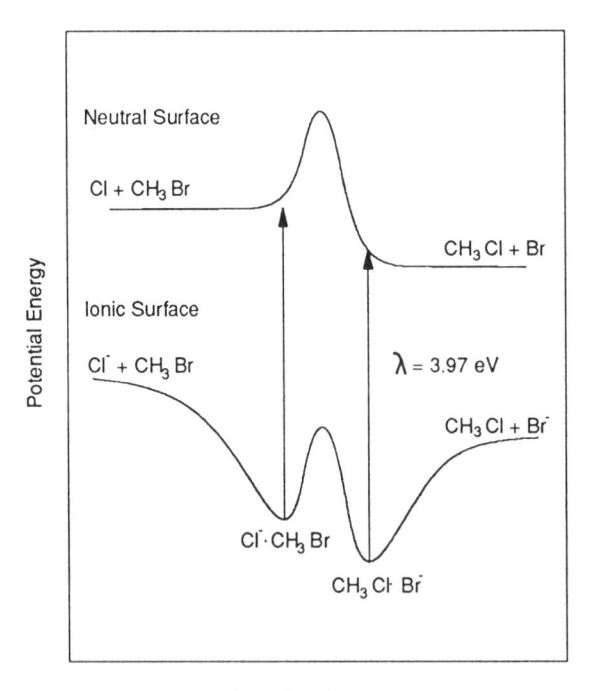

Figure 2. Distinguishing the isomers by photodetachment.

where the arrows indicate the strategy of the photodetachment experiment in the case where the repulsive energy on the neutral surface is similar for both channels. We anticipate that the exit channel will have a significantly lower vertical detachment threshold than the entrance channel. Since we expect continuum absorption once the photon energy is above the vertical detachment energy, our approach was to monitor the photodetachment cross sections for each cluster as a function of decreasing excitation energy until the absorption of one of the complexes drops dramatically. The extent to which the complexes are spectroscopically distinguishable establishes the extent of contamination of one with the other. The photodetachment data for the entrance and exit complexes (Cl$^-\cdot$ CH$_3$Br and Br$^-\cdot$ CH$_3$Cl, respectively) are displayed below:

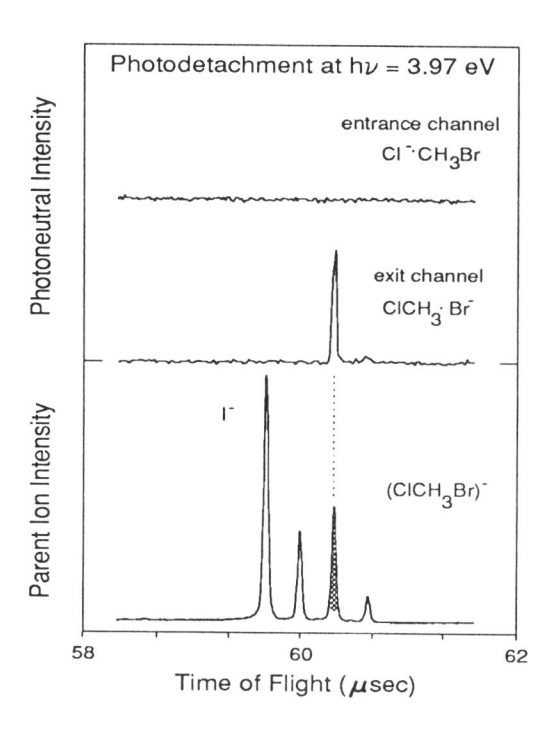

Figure 3. Photodetachment results for both entrance and exit channel complexes from a mass selected isotopomer.

The lower trace in Fig. 3 displays a typical parent mass spectrum (identical for each complex) and the upper traces record the photoneutral signal from excitation of the most intense isotopomer ([^{79}BrCH$_3$^{37}Cl]$^-$ and [^{81}BrCH$_3$^{35}Cl]$^-$) at 3.97 eV (312 nm). The photostability of the complex prepared by the (nominally) "entrance channel" synthesis compared to the exit channel provides clear evidence that the syntheses create distinguishable species. Consistent with our expectation, the exit channel complex has a lower electron affinity, and that there is essentially no contamination of the entrance channel with Br$^-\cdot$ CH$_3$Cl since there is no observable absorption at this energy. Electron photodetachment is the only experimentally observed photoproduct observed (so far) upon photoexcitation of the Cl$^-\cdot$ CH$_3$Br complexes, indicating that we cannot as yet drive the S$_N$2 reaction by excitation of the CT excited state indicated in Fig. 1. We will review our efforts to locate these states in Section 2.3.

2.2 Collisional activation of the trapped intermediates

The exothermic gas phase S_N2 reactions (under thermal conditions) are atypical ion-molecule reactions in that the reactions proceed at a collision efficiency of 1% or less, and yet often display a *negative* Arrhenius activation parameter.[6,14] The generic explanation of this effect invokes the anticipated "tight transition state" for the inversion, which significantly reduces the density of states over the transition state relative to the dissociation of the complex back to separated reactants. Now that we have access to isolated entrance and exit channel complexes, it is of some interest to activate the complexes and monitor the branching ratio for decomposition into the reactant and product halide ions. We excited the complexes by colliding the ion beam into a second argon pulsed expansion at high energy (~2.5 KeV). Superficially, this method would appear to have little relevance to reactive trajectories near the ground state surface; however, the preponderance of work on collisional excitation at high energy[15,16] indicates that the mean energy imparted is rather low (~0.5 - 1.0 eV) and the complexes should be excited in the vicinity of the transition state ($E^{\ddagger} \approx 0.5$ eV). Note that since the complex is undoubtedly cooled in the expansion to near the bottom of the wells in Fig. 2, collisional excitation on the order of 1 eV brings the complex into the range of the transition states for reaction and dissociation. Typical data for the entrance and exit channel complexes from the reactions:

$$Br^- + CH_3I \rightarrow CH_3Br + I^- \qquad [1]$$
$$Cl^- + CH_3Br \rightarrow CH_3Cl + Br^- \qquad [2]$$

are shown below:

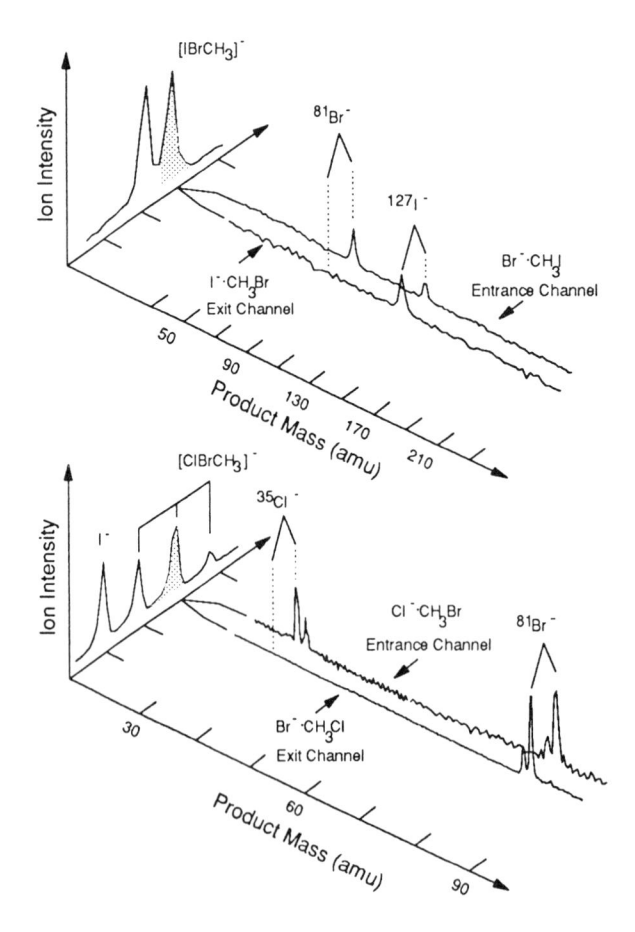

Figure 4. Collisional activation products from both entrance and exit channels of two mixed halide complexes.

The mass spectra of the parent isotopomers is displayed along the depth axis, and the two traces along the horizontal axis present the ionic products generated upon collisional activation from a single isotopomer of the entrance and exit channel complexes. In each case, the exit channel is found to break up exclusively into the exothermic product ions. The entrance channel complexes, on the other hand, are found to form the exothermic products with about the same probability as that for the complex dissociating back toward reactants. This branching is considerably higher than that observed in the corresponding thermal collision (k_{exp}/k_{ADO} =0.015 for reaction [1])[17].

It would seem that the increased efficiency of the collisional activation of the complexes is not simply a consequence of higher energy excitation, since the rate of reaction [1] decreases with increasing temperature, indicating that the collision efficiency might indeed *decrease* with increasing temperature. Thus, excitation of the complexes is conspicuously efficient at driving the reaction, perhaps due to the fact that the atoms are pre-aligned along the reaction coordinate. An alternative explanation exists, however, in which the collision preferentially excites a significant fraction of the complexes in the energy range between the barrier and the dissociation asymptote (see Figure 2). Complexes prepared in this energy range would necessarily decay to Br$^-$ products with unit efficiency since they have insufficient energy to reach the reactant asymptote. Resolution of this issue will likely require collisional excitation in the threshold region (presumably with a guided ion beam) or an alternative excitation scheme such as infrared multiphoton excitation to bring the energy close to the first potential barrier in a more controlled fashion.

We have also investigated the effects of substituents on the reaction efficiency of the entrance channel complexes in the series:

$$Cl^- \cdot CH_3Br \rightarrow CH_3Cl + Br^- \qquad [3]$$
$$Cl^- \cdot CH_3CH_2Br \rightarrow CH_3CH_2Cl + Br^- \qquad [4]$$
$$Cl^- \cdot (CH_3)_2CHBr \rightarrow (CH_3)_2CHCl + Br^- \qquad [5]$$

and the analogous Br$^- \cdot$ RI systems. In each case, the entrance and exit channel complexes could be readily prepared in the ionized free jet source. Like the X$^- \cdot$ CH$_3$Y systems, the exit channel complexes collisionally decompose exclusively into the exothermic halide product. Interestingly, the alkyl substituents have very little effect on the branching ratios for X$^-$/Y$^-$ in both the entrance channel complexes Br$^- \cdot$ RI and Cl$^- \cdot$ RBr. Thus, the Cl$^- \cdot$ (CH$_3$)$_2$CHBr combination reacts to form Br$^-$ with approximately 50% efficiency, just as in the case of the Cl$^- \cdot$ CH$_3$Br reaction (shown in the upper trace of figure 4). This is perhaps surprising since the bimolecular reaction:

$$Cl^- + (CH_3)_2CHBr \rightarrow ClCH(CH_3)_2 + Br^- \qquad [6]$$

has an immeasurably slow rate (k<1x10^{-13} cm$^3 \cdot$ molecule$^{-1} \cdot$ sec^{-1})![17] The mechanism of the enhanced reaction efficiencies of the complexes by collisional activation is as yet unclear, although similar reaction efficiencies were observed in the collisional activation of metastable entrance channel complexes.[18]

2.3 A search for excited states: Trapping the CT band

We noted above that photoexcitation of the Cl$^- \cdot$ CH$_3$Br complexes was not observed to drive the S$_N$2 reaction since photoexcitation resulted in electron detachment. One possibility is that photoexcitation *is* occurring to the CT excited state, but this transition is indistinguishable from photodetachment because the excited anionic state the crosses a neutral surface before approaching the avoided crossing in Fig. 1.[19] This is actually a likely scenario since photoexcitation is observed very close to the expected adiabatic electron detachment energy of the complex. We have briefly looked for absorption to lower energy than the electron affinity of the halide and do not find significant absorption until about 0.5 eV above the electron affinity of X^-, which is approximately the ion-dipole stabilization energy of the complex. To overcome this competition with electron ejection we seek a way to trap the CT excited state by using RI species which are known to survive in the presence of the electron continuum. Simply replacing RX by species with two or more halides is likely to provide such a trap since the dissociative electron attachment rate is known to rapidly increase with halide number in $CH_{4-n}Y_n$[20]:

$$e^- + CH_{4-n}Y_n \rightarrow Y^- + CH_{4-n}Y_{n-1} \qquad [7]$$

so that the rate for CH_2Br_2 is more than an order of magnitude faster than that of CH_3Br. Thus, there is a chance that halide ejection from the dibromo complex can compete with electron detachment to reveal the presence of the excited state by monitoring photochemical production of Br^-.

We therefore isolated the $I^- \cdot CH_2Br_2$ complex as before and verified that, like all other exit channel complexes, this species indeed collisionally fragments to the exothermic product I^-. Photoexcitation of this complex at 4.6 eV (266 nm), far above the expected VDE (3.5 eV), again only results in photodetachment of the electron; however, excitation near the electron affinity of the cluster at 3.5 eV (355 nm) results not in the expected Br^- products, but the diatomic IBr^-! These results are summarized below:

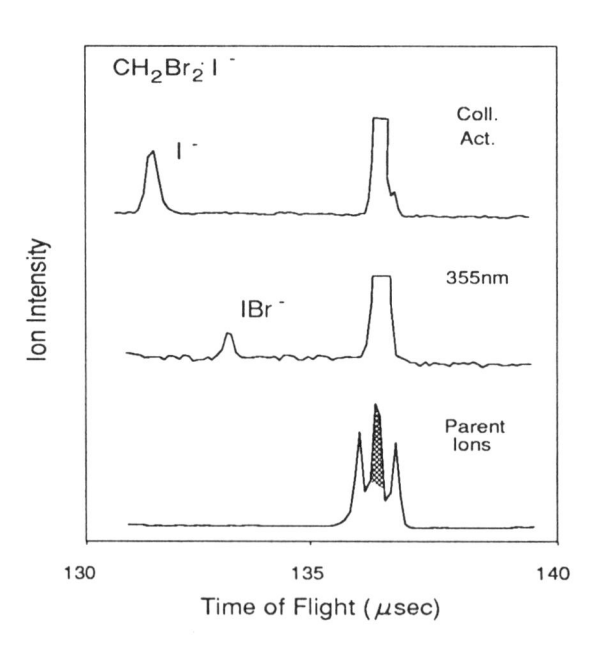

Figure 5. Comparison of collisional activation and photoexcitation of $CH_2Br_2 \cdot I^-$.

The upper trace displays the ionic products from collisional activation of the most intense isotopomer (parent mass spectrum, lower trace), while the middle trace presents the ionic products from photoexcitation near the vertical electron binding energy at 3.5 eV. Clearly, the photo- and collisional excitation yield different products. Once again, photoexcitation is not observed to drive the (endothermic) S_N2 reaction (with the expected Br⁻ product), but yields a new (previously unobserved) diatomic product.

The XY⁻ diatomic photofragment ion is a generic property of complexes with several halides excited near their electron affinities, and has now been observed in the systems:

Br⁻· CH_2Br_2 → Br_2^- + CH_2Br	(λ=298 nm)	[8]
Cl⁻· $CHCl_3$ → Cl_2^- + $CHCl_2$	(λ=280 nm)	[9]
I⁻· CH_2I_2 → I_2^- + CH_2I	(λ=355 nm)	[10]
I⁻· $CHBr_3$ → IBr⁻ + $CHBr_2$	(λ=355 nm)	[11]

At a given photoexcitation energy, the dihalide becomes increasingly important as the number of halogens on the methane is increased as shown below for the I⁻· $CH_{4-n}Br_n$ sequence:

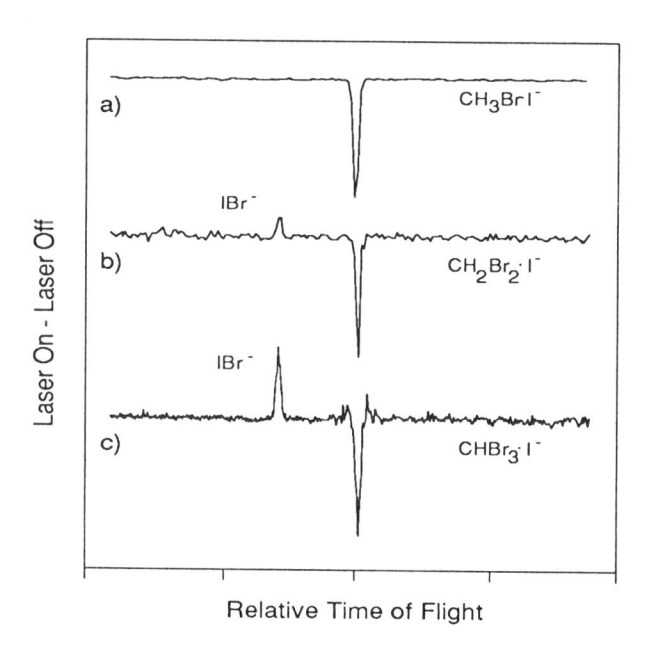

Figure 6. Photofragment IBr⁻ observed from a mass selected isotopomer for several $CH_{4-n}Br_n$· I⁻ complexes.

Interestingly, this trend parallels the electron attachment rates of the corresponding halomethanes.

2.4 Mechanism for dihalide production: anionic analogue to "harpoon" chemistry

The occurrence of photofragmentation near the vertical electron detachment energy can be rationalized upon consideration of the anticipated behavior of the I· [CH$_{4-n}$Y$_n$]$^-$ charge transfer excited state invoked in Fig. 1. The decay product of the transient [CH$_2$Br$_2$]$^-$ ions is known to be Br$^-$, but the nascent Br$^-$ anion is not free, and remains in the vicinity of the I atom even after the C-Br$^-$ bond is broken in the dissociative electron capture. We speculate on the mechanism with the sequence schematically below as:

$$I^- \cdot \underset{H}{\overset{Br}{H}}C-Br \xrightarrow{h\nu} I \cdot \underset{H}{\overset{Br^-}{H}}C-Br \longrightarrow I \cdots \underset{H}{\overset{Br^-}{H}}C-Br \longrightarrow IBr^- + CH_2Br$$

where it is clear that the mass ratio of C/Br ensures that most of the recoil energy in the C-Br$^-$ bond cleavage step in deposited in the lighter C atom, perhaps causing it to pivot about the remaining halogen atom. This leaves the I---Br$^-$ moiety bound at the outer turning point of the diatomic IBr$^-$ potential energy curve, resulting in observation of the dihalide product. This sequence amounts to an intra-cluster electron transfer process analogous to the collision complex involved to explain classic "harpoon" reactions between metal atoms and alkyl halides[21] such as:

$$K + CH_3I \rightarrow K^+ \cdots [CH_3I]^- \rightarrow K^+ \cdots I^- \cdots CH_3 \rightarrow KI + CH_3 \qquad [12]$$

in which the electron transfer step is spontaneous at the "harpoon radius." In the negative ion case, the electron transfer is photoinduced, and the recombination step is not driven by the powerful Coulomb interaction but the weaker ion-induced dipole interaction which defines the long range potential of the IBr$^-$ molecular ion. Note that in the negative ion case, photoexcitation is in the neighborhood of the VDE of the complex and ejects electrons with little out-going kinetic energy. The electron capture resonances for the processes:

$$e^- + CH_{4-n}Y_n \rightarrow Y^- + CH_{4-n}Y_{n-1} \qquad [13]$$

are rather narrow and *peak* at low electron kinetic energy, approaching zero in the case of CH$_3$I.[19] Thus, photoexcitation at high energy would create high kinetic energy electrons which are not efficiently captured by the bare CH$_{4-n}$Y$_n$ species and so we expect dihalide production to fall off upon excitation more than 0.5 eV above the VDE, which is found experimentally.

It remains to be determined whether the observed photochemistry of the X$^-$· CH$_{4-n}$Y$_n$ (n≥2) is due to the CT excited state of the S$_N$2 surface (detected due to the increased efficiency of Y$^-$ production upon electron capture), or whether it is simply an artifact of the CH$_{4-n}$Y$_n$ (n≥2) systems. We note that the dissociative electron capture rate is similar for CH$_2$Br$_2$ and CH$_3$I, and we have recently taken a closer look at the X$^-$· CH$_3$I complexes to explore the possiblity of IX$^-$ production. Interestingly, we have found photoproduction of I$^-$ from excitation of I$^-$· CH$_3$I at 3.5 eV, indicating that we can indeed find evidence of the CT excited state in the simpler X$^-$· CH$_3$Y complexes if the dissociative electron capture rate of methyl halide is sufficiently high. However, we note that the monohalomethane derivative does not yield the dihalide anion product! It would appear highly useful to attempt to drive asymmetric S$_N$2 reactions (such as Br$^-$· CH$_3$I) by photoexcitation in the vicinity of the vertical detachment energy, and such experiments are currently underway in our laboratory.

2.5 Relevance to the mechanism of collision-induced dissociation of anions.

It is often suggested that collisional excitation at high energy occurs via an *electronic* excitation.[16] It could therefore be argued that "activation" of the S_N2 reaction from the $Cl^-\cdot CH_3Br$ entrance channel complex actually proceeds by electronic excitation to the CT $[Cl\cdot(CH_3Br)^-]$ excited state (see Fig. 1) which then breaks through the avoided crossing to form the observed Br^- products. This model is weakened by the fact that photoexcitation of the $Cl^-\cdot CH_3Br$ mono-halomethanes is not observed to drive the S_N2 reaction, but results in photodetachment of the electron. The case for collision-induced electronic excitation is further called into question in this class of species by the behavior of the $I^-\cdot CH_2Br_2$ complex, where we have mapped out the photochemical pathways (in the Franck-Condon region) and find that photoexcitation of the electron and collisional excitation of the complex result in mutually exclusive products (IBr^- and/or electrons from photoexcitation and I^- from collisional activation). Therefore, it appears collisional excitation may involve vibrational excitation within the ground electronic state. Such a mechanism is clearly implied, for example, in the collisional excitation of weakly bound clusters such as $I^-\cdot (H_2O)_n$, in which the first electronically excited state correlates to the "charge transfer to solvent" (CTTS) transition in aqueous solution and lies about 4 eV above the ground electronic state. Nonetheless, collisional excitation of species such as $I^-\cdot (H_2O)_n$ and $O_2^-\cdot (H_2O)_n$ results primarily in loss of a single water monomer (which requires an excitation energy of only 0.5 eV), while relaxation of the electronically excited state should be evidenced by an "electronic metastability" channel[22] characterized by loss of 7 or 8 water molecules. Therefore, high energy collisional activation of large clusters also appears to occur *via* vibrational excitation in the ground electronic state, where vibrational excitation evident in both the large and small ionic complexes.

3. ACKNOWLEDGMENTS

M. A. J. thanks the National Science Foundation for support of this research.

4. REFFRENCES

†Current address: Institute of Atomic and Molecular Sciences, Academia Sinica, Taiwan, Republic of China.
‡Current address: Department of Chemistry, Vanderbilt University, Nashville, TN.
§NSF Presidential Young Investigator and Camille and Henry Freyfus Teacher-Scholar.

1. W. E. Farneth and J. I. Brauman, "The Isomerization of Bicyclo[2.1.0]pent-2-enes," *J. Am. Chem. Soc.* **98**(18), 5546-5552, 1970.

2. W. N. Olmstead and J. I. Brauman, "Gas Phase Nucleophilic Displacement Reactions," *J. Am. Chem. Soc.* **99**(13), 4219-4228, 1977.

3. C.-C. Han, J. A. Dodd and J.I. Brauman, "Structure and Reactivity in Ionic Reactions," *J. Phys. Chem.* **90**(3), 471-477, 1986.

4. J. A. Dodd and J. I. Brauman, "Marcus Theory Applied to Reactions with Double Minimum Potential Surfaces," *J. Phys. Chem.* **90**(16), 3559-3562, 1986.

5. G. Caldwell, T. F. Magnera, and P. Kebarle, "S_N2 Reaction in the Gas Phase. Temperature Dependence of the Rate Constants and Energies of the Transition States. Comparison with Solution," *J. Am. Chem. Soc.* **106**(4), 959-966, 1984.

6. S. S. Shaik and A. Pross, "S_N2 Reactivity of CH_3X Derivatives. A Valence Bond Approach," *J. Am. Chem. Soc.* **104**(10), 2708-2719, 1982.

7. M.-F. Ruasse, J. Aubard, B. Galland and A. Adenier, "Kinetic Study of the Fast Halogen-Trihalide Ion Equilibria in Protic Media by the Raman-Laser Temperature-Jump Technique. A Non-Diffusion-Controlled Ion-

Molecule Reaction," *J. Phys. Chem.* **90**(18), 4382-4388, 1986.

8. A. Parker, "Solvation of Ions. Part VI. Activity Coefficients of Anions in Methanol Relative to Dimethylformamide," *J. Chem. Soc. A* 220-228, 1966.

9. R. S. Mulliken and W. B. Person, Molecular Complexes, p.190 and p.327, Wiley-Interscience, New York, 1969.

10. J. C. Slater, "Note on the Interatomic Spacing in the Ions I_3^-, FHF^-," *Acta. Cryst.* **12**, 197-200, 1959.

11. R. C. L. Mooney Slater, "The Triiodide Ion in Tertaphenyl Asonium Triiodide," *Acta. Cryst.* **12**, 187-196, 1959.

12. D. M. Cyr, L. A. Posey, G. A. Bishea, C.-C. Han, and M. A. Johnson, "Collisional Activation of Captured Intermediates in the Gas Phase S_N2 Reaction $Cl^- + CH_3Br \to Br^- + CH_3Cl$," *J. Am. Chem. Soc.* **113**(25), 9697-9699, 1991.

13. K. Hirao and P. Kebarle, "S_N2 Reactions in the Gas Phase. Transition States for the Reaction: $Cl^- + RBr = ClR + Br^-$, where R = Ch_3, C_2H_5, and iso-C_3H_7, from *ab-initio* Calculations and Comparison with Experiment. Solvent Effects," *Can. J. Chem.* **67**, 1261-1267, 1989.

14. A. A. Viggiano, R. A. Morris, J. S. Paschkewitz, and J. F. Paulson, "Kinetics of the Gas Phase Reactions of Cl^- with CH_3Br and CD_3Br: Strong Experimental Evidence for Nonstatistical Behavior," preprint, submitted to *J. Am. Chem. Soc.* 1992.

15. M. S. Kim and F. W. McLafferty, "Efficiency of Collisional Activation of Gaseous Organic Ions," *J. Am. Chem. Soc.* **100**(11), 3279-3282, 1977.

16. V. H. Wysocki, H. I. Kenttämaa, and R. G. Cooks, "Energetics of Activation of $Fe(CO)_4^-$ and $Cr(CO)_5^-$ in Gaseous Collisions," *J. Phys. Chem.* **92**(22), 6465-6469, 1988.

17. C. H. Depuy, S. Gronert, A. Mullin, and V. M. Bierbaum, "Gas-Phase S_N2 and E_2 Reaction of Alkyl Halides," *J. Am. Chem. Soc.* **112**(24), 8650-8655, 1990.

18. S. T. Graul and S. T. Bowers, "The Nonstatistical Dissociation Dynamics of Cl^- (CH_3Br): Evidence for Vibrational Excitation in the Products of Gas Phase S_N2 Reactions," *J. Am. Chem. Soc.* **113**, 9696-9697, 1991.

19. W. A. Chupka, A. M. Woodward, S. D. Colson, and M. G. White, "Electron Photodetachment from Transient Negative Ions in the Multiphoton Ionization of CH_3I," *J. Chem. Phys.* **82**(11), 4880-4885, 1985.

20. D. Spence and G. J. Schulz, "Temperature Dependence of Electron Attachment at Low Energies for Polyatomic Molecules," *J. Chem. Phys.* **58**(5), 1800-1803, 1973.

21. D. R. Herschbach, "Reactive Scattering," *Faraday Disc. Chem. Soc.* **55**, 233-251, 1967.

22. T. F. Magnera, D. E. David, and J. Michl, "Metastable, Collision-Induced and Laser-Induced Decomposition of $Ar_mN_{2n}^-$ (m=0,1) Cluster Ions in the Gas Phase," *J. Chem. Soc., Faraday Trans.* **86**(13), 2427-2440, 1990.

Photo- and Collision- Induced Dissociation of Ar Cluster Ions

Tamotsu Kondow, Takashi Nagata and Shinji Nonose

Department of Chemistry, Faculty of Science,
The University of Tokyo, Bunkyo-ku, Tokyo 113
JAPAN

ABSTRACT

Photo- and collision- induced dissociation of an argon cluster ion, Ar_n^+, were investigated by use of mass spectrometry. The kinetic and angular distributions of the ionic and neutral photofragments revealed two reaction pathways; dissociation of the trimeric core ion and evaporation from its solvation shell. In the Kr and Ne collisions with Ar_n^+, the size- and collision energy- dependences of the dissociation cross sections were explained in the scheme of the charge - induced dipole, and induced dipole - induced dipole scatterings. Conversion efficiency of the collision energy into the internal energy of Ar_n^+ was found to be proportional to the internal degrees of freedom. The upper limit of the conversion efficiency was estimated to be about 60 % in the collision energy of 0.2 eV.

1. INTRODUCTION

Geometric and electronic structures of an argon cluster ion, Ar_n^+, have been studied with an emphasis on localization and delocalization of its positive charge. Several theoretical studies have predicted that Ar_n^+ is composed of a tightly bound trimeric ion core, and a weakly bound solvation shell (core ion model) [1-5]. The existence of the trimeric ion core in the small cluster ions has been examined experimentally by the photoabsorption spectra of Ar_n^+ ($n \leq 15$). The spectra have exhibited an intense broad peak in the vicinity of 520 nm [6,7], which resembles the Ar_3^+ spectrum. The thermochemical stability of Ar_3^+ [8] lends a further support of the trimeric ion core model. In addition, the ionization potentials of Ar_n have been successfully explained in terms of the trimeric ion core model with the aid of Monte-Carlo calculation in a small n-range [9,10]. As the cluster size increases more, the positive charge tends to be delocalized from the trimeric to tetrameric ion core [6,7]. Further, this specific structure reflects on the photodissociation dynamics. In fact, the angular and the kinetic energy distributions of ionic photofragments of Ar_n^+ are affected by the presence of the trimeric ion core [11]. This specific cluster structure provides us a unique opportunity to explore reaction processes which are scarcely encountered in ordinary gas phase reaction processes. In this regard, photodissociation of Ar_n^+ and dissociation of Ar_n^+ in collision with Kr and Ne atoms were investigated. In the photodissociation of Ar_n^+, two types of reaction pathways can be distinguished; pathways ascribable to the trimer ion core and those opened by the perturbation of the solvating neutral Ar atoms. On the other hand, collision-induced dissociation which depends characteristically on the collision energy and the impact parameter open different reaction pathways and provides complimentary information on the Ar_n^+ dynamics.

2. Experimental

a. Photodissociation

The experimental apparatus has been described in detail elsewhere [11]. The cluster ion, Ar_n^+, was produced by electron impact (~20 μA at 200 eV) on a neutral cluster beam generated in a free jet expansion. The ions were accelerated by a pulsed electric field up to ~1.6 keV and were irradiated by an output of a pulsed dye laser at the spatial focus located at ~120 cm downstream. The direction of the polarization vector, E, of the laser was set to be either parallel ($\theta=0°$) or perpendicular ($\theta=90°$) to the ion-beam axis (denoted as Z axis hereafter). Photofragment ions were successively mass-analyzed by a secondary reflectron mass spectrometer and detected by a Hamamatsu F2223-21S microchannel plate. Neutral photofragments were observed directly after the photodissociation.

b. collision-induced dissociation

The cluster ions produced by electron impact on neutral clusters of Ar were mass-selected by a quadrupole mass spectrometer (Extrel, 162-8), and were allowed to collide with a target atom, Rg (= Kr and Ne), in a collision region enclosed by an octapole ion guide. The product ions were mass-analyzed by a sector-magnet mass spectrometer (JEOL, JMS-D300). The ion source, the quadrupole mass spectrometer, and the collision region were floated up to 1 kV against the ground. The total cross section was measured by varying the pressure of the target gas in the range of 10^{-5} - 10^{-6} Torr. The background pressure was attained to be less than 5×10^{-7} Torr. The observed cross sections were normalized against the reported cross sections for the process, $Ar_2^+ + Rg \rightarrow Ar^+ + Ar + Rg$. [12].

3. RESULTS AND DISCUSSION

3-1. Photodissociation of Ar_n^+

The photofragment TOF spectra of Ar_n^+ ($4 \leq n \leq 6$) for the production of Ar^+ were measured at various photolysis wavelengths from 465-580nm, with the polarization vector of the photolysis laser parallel ($\theta=0°$) and perpendicular ($\theta=90°$) to Z axis. There were almost equally-separated three peaks and a single peak associated with Ar^+ and Ar_2^+, respectively in the spectra. With the polarization vector perpendicular to Z axis, the outer peaks of Ar^+ became very weak, whereas the center peak unchanged. The spectral features and θ-dependence of Ar^+ from Ar_n^+ ($4 \leq n \leq 6$) and those from Ar_3^+ lead to the conclusion that the trimeric ion core is the chromophoric core of Ar_n^+ and that Ar^+ is produced directly from the trimeric ion core as a primary product [11]. The photodissociation reaction is given as,

$$Ar_3^+(^2\Sigma_\mu^+)(Ar)_{n-3} + h\nu \rightarrow [Ar_3^+(^2\Sigma_g^+)](Ar)_{n-3}$$
$$\rightarrow Ar^+ + (n-1)Ar. \quad (4 \leq n \leq 6) \quad (1)$$

The kinetic energy distribution of Ar^+ obtained from the analysis of the photofragment spectra consists of fast and slow components; the fast component originates from the ion core with the positive charge on the outer Ar atoms and the slow component does from the ion core with the charge on the center atom. The average kinetic energy of the fast fragment was estimated to be in the range of 0.2 - 0.4 eV in the center-of-mass frame, depending on the energy of the photolysis laser. In the process associated with the fast component, about an half of the available energy turns out to be transmitted into the kinetic energy of Ar^+. The rest of the available energy is shared by its neutral counterparts. This finding further confirms that the fast Ar^+ fragment is directly produced through the dissociation of the ion core and the available excess energy is consumed as the relative kinetic energy of the separating two outer atoms; the trimeric ion core probably retains a linear configuration as predicted by theoretical calculations [3].

The TOF spectra of 532 nm photofragment neutrals from Ar_n^+ (3≤n≤24) were also obtained. Figure 1 shows the spectrum for Ar_9^+ as a typical example, in comparison with the spectrum of the photofragment ions. The energetics of the photodissociation reaction shows that the neutral fragments detected are identifiable to Ar atoms. The central peak and the two outer peaks are assigned to the Ar atom fragments having slow and fast kinetic energies, respectively. The fast fragment was produced from Ar_n^+ with n ≤ 13 but disappeared for n > 14. In the photodissociation of the larger clusters, the slow fragment was the dominant product whose kinetic energy has a 450 K thermal distribution with a high energy tail corresponding to an effective temperature of 1000 K. The branching fraction for the production of the tail part is estimated to be about 40 % at maximum. The dynamics for the production of the neutral fragments is explained as well in terms of the calculated geometric and electronic structures [2] on the basis of the reaction scheme (1). In the electronic ground state, the linear triatomic ion core is caged in the solvation shell. In a small cluster ion with 3≤n≤13, the fast atom fragment is ejected directly from the ion core as the counterparts of the Ar^+ fragments, while in a large cluster ion the major

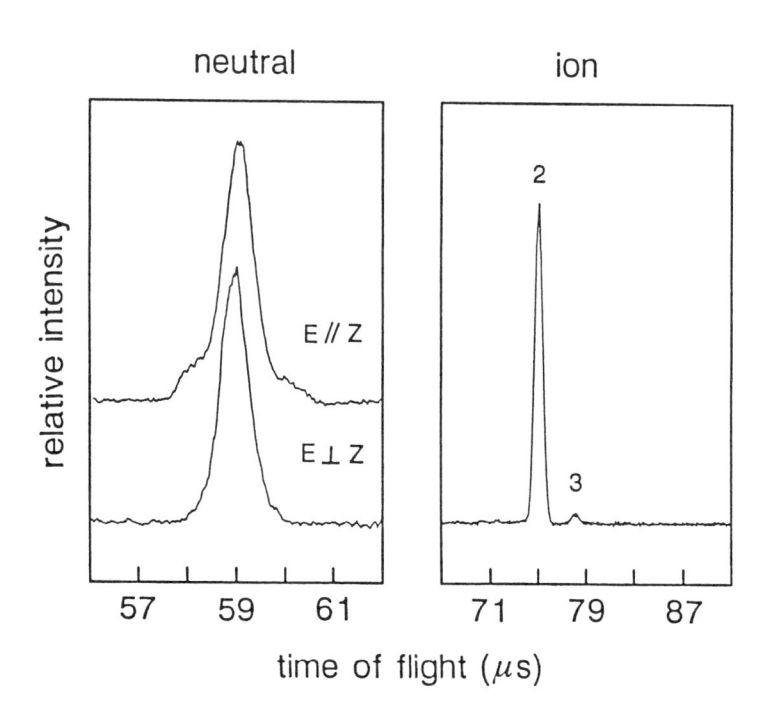

Fig. 1: Photofragment TOF spectra for Ar_9^+

portion of the photofragment atoms are not ejected directly from the cage but instead evaporated with a thermal kinetic energy. The average number of Ar atoms evaporated was estimated to be about 2 in the photolysis of Ar_{10}^+. The Ar_3^+ photolysis for the production of the neutral atoms also supports the equilibrium structure of Ar_3^+ having $D_{\infty h}$ symmetry, in agreement with the conclusion given from the Ar_3^+ photolysis for the Ar^+ production.

3-2. Collision-induced Dissociation

a. Reaction Pathway

In the collision of Ar_n^+ with Kr, the product ions, $Ar_{n'}^+$ (n'<n) and $Ar_{n'}Kr^+$ (n'=0, 1, 2 and 3) were observed, whereas in the Ne collision no charge-exchange species $Ar_{n'}Ne^+$ were detected. The collisional reactions involving Rg (= Kr, Ne) are given by the following pathways:

$$Ar_n^+ + Rg \rightarrow Ar_{n'}^+ + (n'-n)Ar + Rg \qquad \text{(evaporation)} \qquad (2)$$

$$Ar_n^+ + Rg \rightarrow Ar_{n'}Rg^+ + (n'-n)Ar. \qquad \text{(charge exchange)} \qquad (3)$$

In the Ne collision, pathway (3) does not proceed in the collision energy range studied. In the Kr collision, the branching ratio of pathway (3) decreased smoothly, and level off at n ~ 6 with the increase of n. This size-dependence suggests that pathway (3) consists of two different processes; in a small cluster ion, the charge can be transferred directly from Ar_n^+ to Kr, while in a large cluster ion, Kr can not approach the ion core without penetrating inside the cluster ion. This phenomenon may be related to the specific structure of Ar_n^+ composed of the ion core, and the solvating Ar atoms.

b. Cross Sections

Figure 2 shows the total cross section and the cross sections for pathways (2)

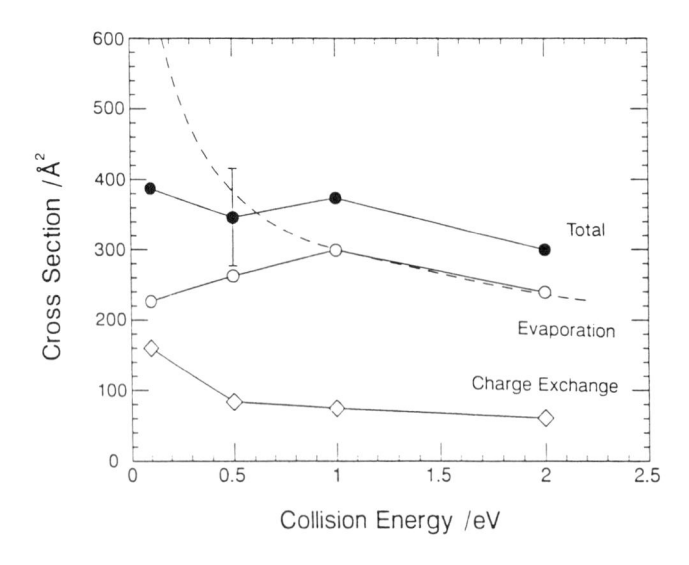

Fig. 2: Dependences of the cross sections on the collision energy for the Ar_{11}^+ + Kr system.

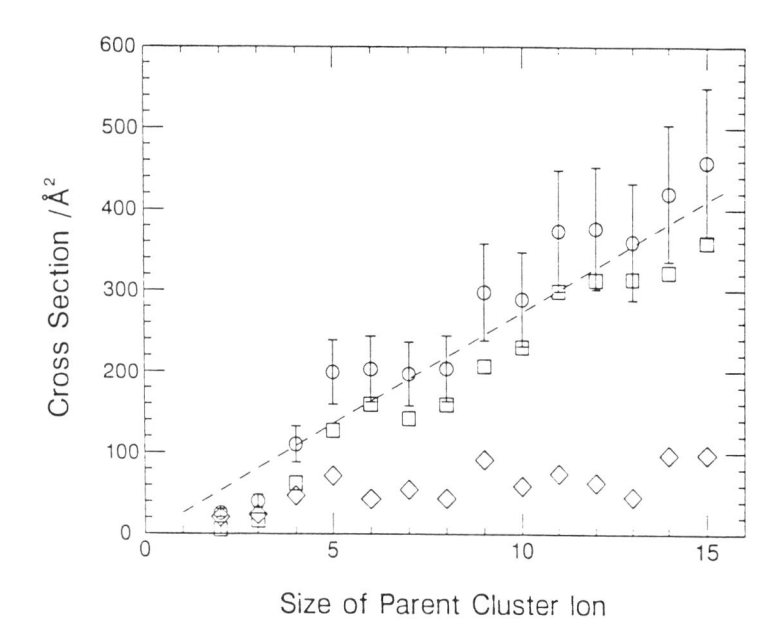

Fig. 3: The total cross section (○), the cross sections of pathway (2)(□) and pathway (3)(◇) are shown as function of the cluster size, n, in the reaction of Ar_n^+ with Kr. The dashed line shows the total cross section calculated on the basis of the model given in the text.

and (3), in the Ar_{11}^+ + Kr reaction system. As the collision energy increases, the total cross section decreases gradually but that of pathway (2) does not change appreciably. On the other hand, the cross section for pathway (3) decreases rapidly with the collision energy. This collision-energy dependence indicates that pathway (3) is an exothermic and no-activation process. In order to explain the large cross sections of these reactions, every atom in Ar_n^+ is treated as if it behaves as a scatterer of the target atom. The total cross sections are estimated, in the scheme of charge-induced dipole [7] and induced dipole-induced dipole scattering, as shown in Fig. 2. A similar argument is applicable to the Ne collision.

The size-dependences of the cross sections in the Kr collision are shown in Fig. 3. The total cross section is proportional to n; the cross section of ~ 450 Å^2 is obtained at n = 15. The cross section for pathway (2) depends similarly as the total cross section does, since the cross section for pathways (3) is much smaller than the total cross section. The size-dependence is also predicted by the model proposed as mentioned above. The linear size-dependence in the Ne collision can be interpreted by this model, as well.

c. Transmission of Collision Energy into Intracluster Modes

The average number of Ar atoms evaporated from Ar_n^+ in the 0.2 eV collision of Kr was obtained to be 0.25n with n<10 and reached a constant number, 2.5, with n>10 (see Fig. 4(a)). Almost identical result was obtained in the Ne collision with the same collision energy (see Fig. 4(b)). If this average number is proportional to the increment of the internal energy of Ar_n^+, the conversion efficiency of the collision energy into the internal energy of Ar_n^+ turns out to be proportional to the cluster size or, in other words, the internal degrees of freedom. It is inferred from this

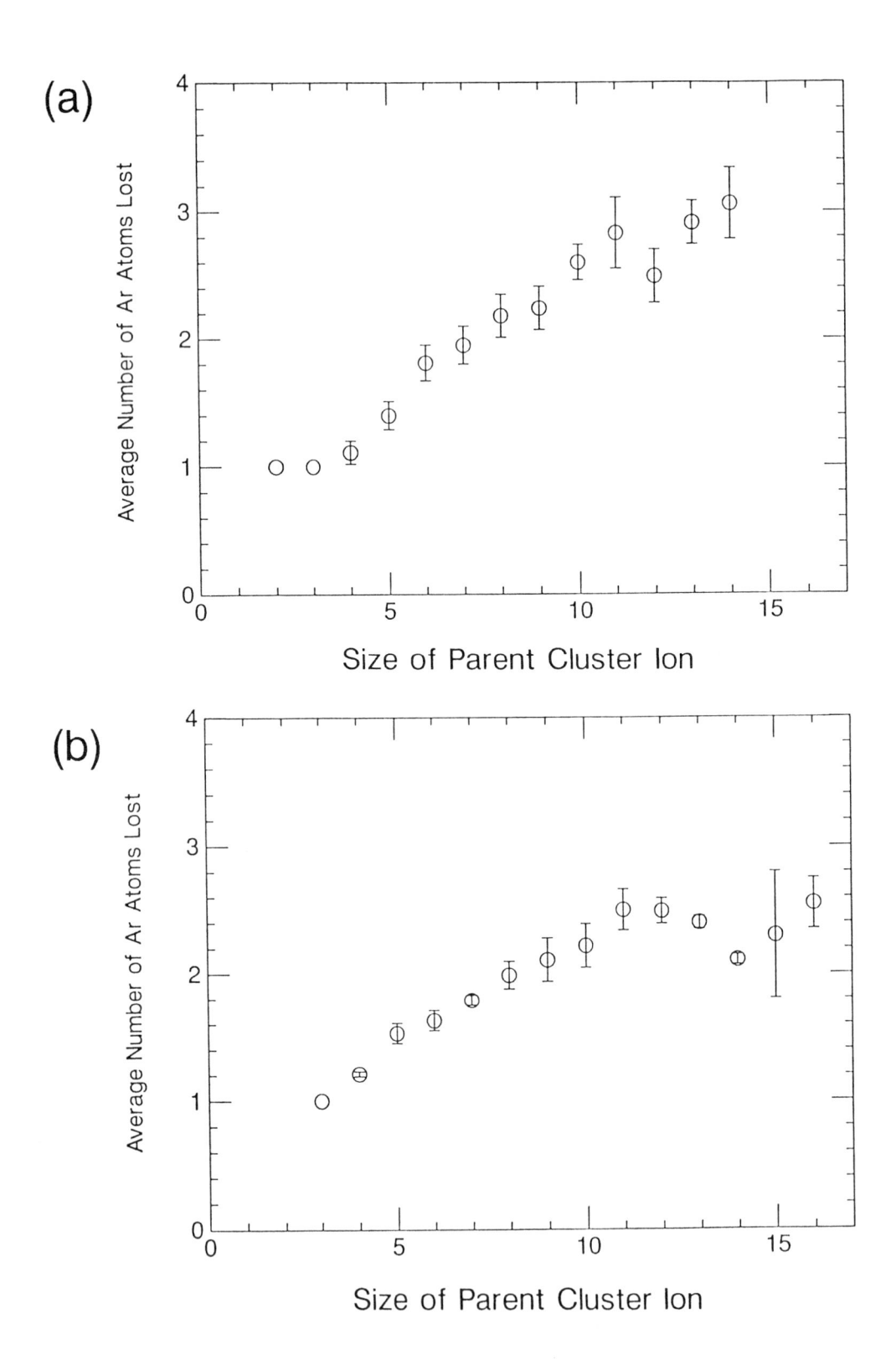

Fig. 4: Average number of At atoms evaporated in collision with Kr(panel(a)) and Ne(panel(b)).

result that the collision energy is partitioned statistically into the internal degrees of freedom. As 2.5 atoms are evaporated at maximum, the minimum energy given to the internal degrees of freedom by the collision is calculated to be 0.125 eV by considering the average evaporation energy of 0.05 eV per Ar atom [6]. This implies that about 60 % of the collision energy is converted to the internal energy, irrespective of the target atom.

ACKNOWLEDGMENTS

The present work has been supported by a Grant-in-Aid for General Scientific Research by the Ministry of Education, Science and Culture of Japan.

REFERENCES

1. P. J. Kuntz and J. Valldorf, Z. Phys. D-Atoms, Molecules and Clusters, **8**, 195 (1988).
2. I. Last and T. F. George, J. Chem. Phys., **93**, 8925 (1990).
3. H. -U. Böhmer, S. D. Peyerimhoff, Z. Phys. D-Atoms, Molecules and Clusters, **11**, 239 (1989).
4. A. Ding and J. Hesslich, Chem. Phys. Lett., **94**, 54 (1983).
5. I. A. Harris, R. S. Kidwell and J. A. Northby, Phys. Rev. Lett., **53**, 2390 (1984).
6. N. E. Levinger, D. Ray, M. L. Alexander and W. C. Lineberger, J. Chem. Phys., **89**, 5654 (1988).
7. H. Haberland, B. von Issendorff, T. Kolar, H. Kornmeier, C. Ludewigt and A. Risch, Phys. Rev. Lett., **67**, 3290 (1991).
8. K. Hiraoka and T. Mori, J. Chem. Phys., **90**, 7143 (1989).
9. W. Kamke, J. de Vries, J. Krauss, E. Kaiser, B. Kamke and I. V. Hertel, Z. Phys. D-Atoms, Molecules and Clusters, **14**, 339 (1989).
10. G. Ganteför, G. Bröker, E. Holub-Krappe and A. Ding, J. Chem. Phys., **91**, 7972 (1989).
11. T. Nagata, J. Hirokawa and T. Kondow, Chem. Phys. Lett., **176**, 526 (1990).
12. W. B. Maier II, J. Chem. Phys., **62**, 4615 (1975).

Vibrational spectroscopy of size-selected metal ion-solvent clusters

James M. Lisy

University of Illinois at Urbana-Champaign, Department of Chemistry
School of Chemical Sciences, Urbana, Illinois 61801

ABSTRACT

Vibrational spectra of metal ion-solvent clusters have been obtained size-selectively, using mass spectrometric techniques. Metal ions (Cs^+ and Na^+) have been solvated by as many as 25 molecules with solvents such as methanol and ammonia. Structural information such as solvent shell sizes and solvent orientation have been inferred from the variation in solvent vibrational frequencies as a function of cluster size. Independent Monte Carlo simulations of the solvated ions, based on established ion-solvent, and solvent-solvent interaction potentials, are in good agreement with the experimental observations. The simulations in turn have indicated that hydrogen bonding may play a significant role in determining the solvent shell size for large ions. An informative picture of microscopic ion solvation is the result of the interplay between experiment and simulation.

1. INTRODUCTION

Interactions between ions and neutral molecules play a fundamental role in atmospheric, interstellar, solution and biophysical chemistry. Yet in comparison to the spectroscopy of neutral clusters,[1] there have been very few investigations of ion-neutral clusters. The information from these detailed neutral studies has been used to formulate intermolecular interaction potentials,[2] which in turn can help answer basic questions dealing with the condensed phase. The strong electrostatic interactions between ions and neutrals in the condensed phase lead to significant disruption of the solvent structure by overwhelming the weaker neutral-neutral interactions. The study of ion clusters, where a single ion is complexed (i.e. solvated) to one or more neutral clusters, allows the competition between the ion-neutral and neutral-neutral interactions to be followed. Using mass spectrometric techniques, sequentially larger cluster ions can be studied in a size-selective manner. This approach also eliminates the complications that can arise from the presence of counter-ions.

Although high resolution spectroscopic investigations have been done for a number of protonated molecules such as H_3O^+ and NH_4^+ in the infrared,[3] there have only been a few studies of true ion clusters or solvated ions and at much lower resolution. Some of the more notable research includes the electronic spectroscopy of Sr^+ complexed by H_2O and NH_3 from Farrar et. al.[4] In the 3 μm region of the infrared, Lee and coworkers[5] have investigated a number of ion clusters formed about H_3O^+ and NH_4^+, and have resolved rotational structure in the $NH_4^+(NH_3)_N$ system. Our studies have focussed on metal ions, Na^+ and Cs^+, complexed to common solvents, methanol and ammonia, in the 9 to 10 μm region of the infrared.[6]

Theoretical investigations of ion-neutral interactions have taken a variety of approaches. One of the most common involves establishing pair-wise interaction potentials between sites on the neutral molecule and a second neutral molecule or an ion. Molecular Dynamics (MD) or Monte Carlo (MC) simulations, based on these interaction potentials, are then performed to examine the structure and/or energetics of the solvated ion.[7] While the potentials may be based on ab initio interactions, they are typically empirical, averaging in multibody interactions. As a result, simulations on the pure liquids yield properties which are in good agreement with known experimental values.

The remainder of this paper is divided into five sections. We will first discuss the preparation of the solvated ions, characterization of the internal energy distribution within the cluster ion and the procedures for obtaining vibrational spectra of mass-selected species. The MC simulation methods will be presented along with the intermolecular interaction potentials. Vibrational spectra for $Na^+(CH_3OH)_N$, $Cs^+(CH_3OH)_N$, $Na^+(NH_3)_M$ will be compared and contrasted to illustrate the dependence on ion and solvent type. The structural results from the MC simulations will then be compared with the experimental observations. The combined experimental and simulation results will then be discussed in regards to the competition between ion-neutral and neutral-neutral interactions.

2. EXPERIMENTAL

Details of the experimental apparatus may be found in our previous publications.[6] Briefly, metal ions are generated from a thermionic emitter that is positioned from 5 to 10 mm downstream from a supersonic molecular beam nozzle. The ions are guided into the expanded molecular jet, where inelastic collisions with the neutral clusters lead to the formation of ion clusters. A series of electrostatic lenses are used to guide and focus the ion cluster beam through a skimmer and into a second vacuum chamber. A second set of electrostatic lenses injects the ion cluster beam into a quadrupole mass filter where a specific ion is selected. After exiting the quadrupole, the ion cluster is accelerated into a conversion dynode and is detected by an electron multiplier. A schematic of the experimental apparatus is shown in Figure 1.

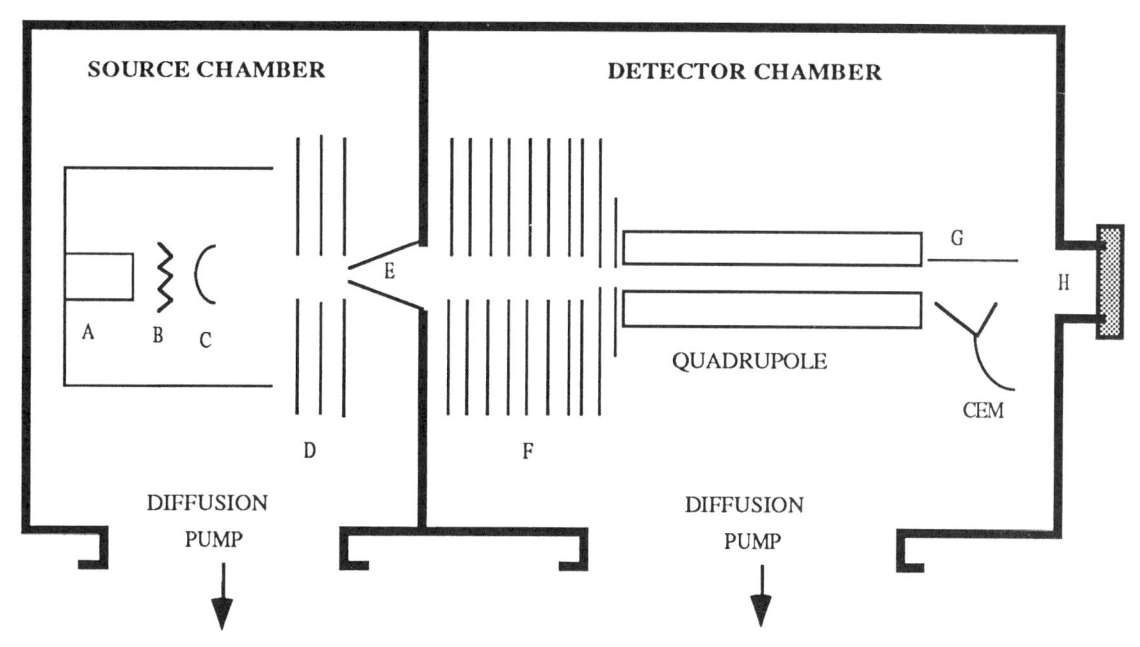

Figure 1. Schematic of ion cluster apparatus. A - supersonic nozzle, B - ion filament, C - ion deflector, D - source ion lenses, E - skimmer, F - detector ion lenses, G - conversion dynode, H - ZnSe window, CEM - channeltron electron multiplier.

When generating clusters in a molecular beam, one typically expects the internal energy or "temperature" of the clusters to be quite low. This is not the case for ion clusters generated by the method given above. After the ion collides with a cluster, the neutral molecules rearrange to "solvate" the ion. The collisional and solvation processes impart a significant amount of internal energy (300 to 400 kJ/mol) to the cluster ion. One method of energy dissipation involves the evaporation of solvent molecules from the cluster ion. After the loss of a few (5 to 10) solvents, the dissociation lifetime (the inverse of the rate of unimolecular decomposition) of the cluster ion becomes comparable to the flight time through the apparatus. The internal energy of the cluster ion can be estimated with the formalism of the evaporative ensemble as developed by Klotz.[8] The ensemble requires a relationship for the rate of unimolecular dissociation as a function of energy, k(E), and that each cluster has undergone at least one dissociative event. For our systems,[6] we have used a simple RRK model for the unimolecular dissociation rate, with an asymptotic binding energy determined from the bulk enthalpy of vaporization of the pure solvent.

$$k_N(E) = A \left(\frac{E - V_N}{E} \right)^{L-1} \tag{1}$$

By equating the average internal energy per vibrational degree of freedom with $k_B T$, we can determine a cluster "temperature." For $Cs^+(CH_3OH)_N$ and $Na^+(CH_3OH)_N$, $N \geq 7$, these temperatures decrease from 320 K (N = 7) to about 215 K (N = 25), well within the range of liquid methanol.

The vibrational spectra of size-selected ion clusters are obtained by inducing vibrational predissociation of cluster ions inside the quadrupole mass filter. Since the solvated ions have a substantial amount of internal energy, a small additional increment in energy will lead to unimolecular dissociation while the cluster is within the quadrupole. This energy is supplied by exciting a vibrational mode of the solvent using a tunable infrared laser. This vibrational energy rapidly spreads throughout the cluster by intermolecular vibrational relaxation. Following fragmentation, the daughter ion has the wrong m/e value and is ejected from the quadrupole. By modulating the laser on/off and measuring the laser intensity, the depletion can be used to determine the photodissociation cross-section, given below.

$$\sigma_D(\tilde{v}) = \frac{-\ln[1-D(\tilde{v})]}{It/hc\tilde{v}} \qquad (2)$$

where $\sigma_D(\tilde{v})$ is the photodissociation cross-section at frequency $\tilde{v}(cm^{-1})$, I is the laser intensity, t is the cluster ion flight time through the quadrupole and $D(\tilde{v})$ is the fractional depletion of the ion cluster intensity.

Since the flight times through the quadrupole were measured in the 20 to 60 μsec range, molecules with large absorption cross-sections and fairly powerful tunable lasers are required to observe substantial depletion. When considering duty cycle, only the CO_2 laser satisfies all of the requirements. As a result, suitable solvent molecules must have vibrational transitions in the 9 to 10 μm tuning range of the CO_2 laser. Furthermore, these vibrations must be sensitive to bonding environment. Both methanol and ammonia possess suitable vibrational modes: the C-O stretch for methanol and the v_2 "umbrella" mode in ammonia. The influence of bonding environment for ammonia is shown in Figure 2.

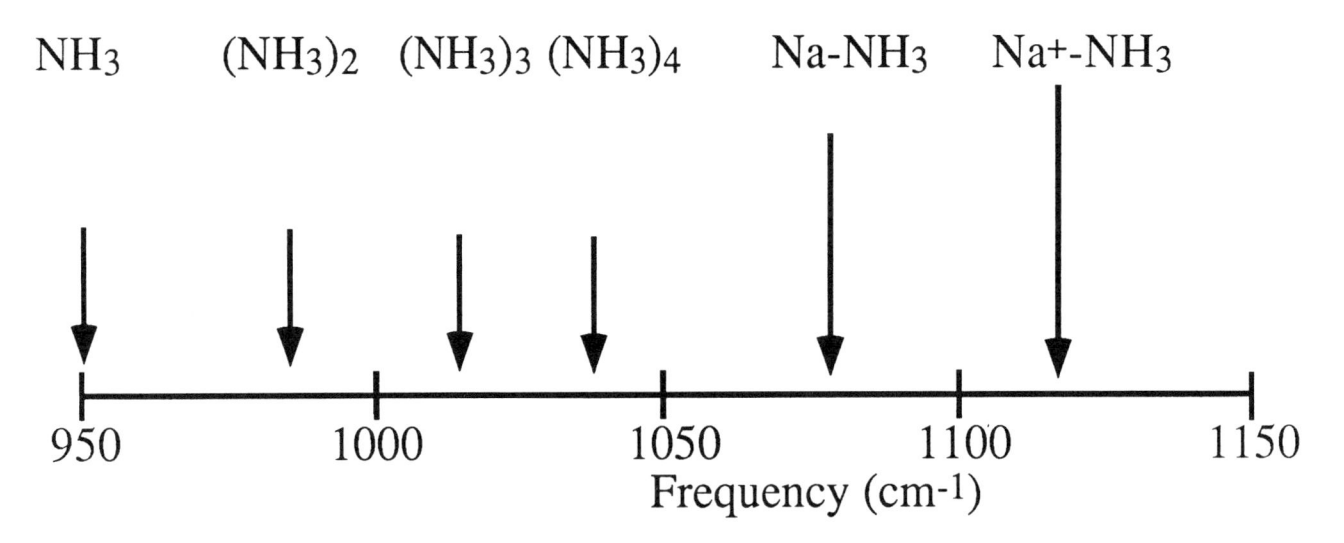

Figure 2. NH_3 v_2 umbrella mode sensitivity to bonding environment.

3. SIMULATIONS

The Monte Carlo (MC) method used in these studies follows the algorithm of Metropolis.[9] The calculations were carried out at the University of Illinois, School of Chemical Sciences computation facilities using either a MIPS Computer System M-120, a VAX 3500 or a 3200 workstation. Details concerning initial configurations, step sizes, annealing schedules and acceptance ratios may be found elsewhere.[6] Simulations were run on sequentially larger cluster ions, starting with 5 solvent molecules about the metal ion, and finishing with 22 to 25 solvents. The temperature, at which statistics were collected, was determined from the analysis of the evaporative ensemble. For clusters containing methanol, statistics were collected at 250, 225 and 200 K, while for ammonia, statistics were collected from 200 to 160 K depending on cluster size.

The methanol-methanol interactions were based on the transferable intermolecular potential functions (TIPS) developed by Jorgenson.[10] The potential functions treat methanol as a three-site system: oxygen, hydrogen and a methyl group:

$$V_{SS} = \sum_{N=1}^{3} \sum_{K=1}^{3} \frac{Q_N Q_K}{R_K} - \frac{C_N C_K}{R_K^6} + \frac{A_N A_K}{R_K^{12}} \quad (3)$$

which is a standard Lennard-Jones 6-12 potential plus a coulombic term. The ion-methanol interactions were based on a modified Rittner potential:

$$V_{IS}(R) = \sum_{K=1}^{3} \frac{Q_{ion} Q_K}{R_K} - \frac{\alpha_K Q_{ion}^2}{2R_K^4} - \frac{C_{ion} C_K}{R_K^6} + A_K e^{-R_K/P_K} \quad (4)$$

An induced dipole term and an exponential repulsive term were added for a more realistic ion-neutral interaction. Atomic polarizabilities were used for O and H, while one-half of the molecular polarizability of ethane was used for CH_3. Parameters used in these functions are given in Table I. Similar potentials were used for modelling the Na^+-ammonia interactions, but for ammonia a five site model is used: one for each atom and a fifth for a negative charge center displaced 0.156 Å from the nitrogen along the symmetry axis towards the hydrogens.

TABLE I

Pairwise Potential Parameters

Methanol-Methanol

Site	Q(e)	C_N(kJ/mol)$^{1/2}$ Å^3	A_N(kJ/mol)$^{1/2}$ Å^6
O	-0.685	50.1	1467.9
H	0.400	0.0	0.0
CH_3	0.285	100.2	5767.4

Ion - Methanol

Site	α_K (Å^3)	$10^{-2} C_{ion} C_K$ (Å^6 kJ/mol)	$10^{-3} A_K$ (kJ/mol)	P_K (Å)
O	0.800	45.47	232.0	0.2738
H	0.395	0.0	0.0	
CH_3	2.24	8.01	232.0	0.2738

4. RESULTS

Vibrational predissociation spectra have been collected for $Cs^+(CH_3OH)_N$, N = 4-25, $Na^+(CH_3OH)_N$, N = 4-22 and $Na^+(NH_3)_M$, M = 6-9. For $Cs^+(CH_3OH)_N$, three distinct spectral regions were observed and are displayed in Figure 3. The first spectral region about 1031 cm[-1] is present for clusters in the size range from N = 4 to 10. The C-O stretching frequency is shifted to lower value with respect to the gas-phase monomer transition frequency of 1034 cm[-1]. The shift is in the same direction as for the proton acceptor methanol (at 1026.5 cm[-1]) in the neutral methanol dimer.[11] Since the oxygen portion of the molecule is expected to be directed toward the Cs^+, this is consistent with the dimer results. The second spectral region about 1037 cm[-1] develops for N = 11 to 18, while the first region stays constant. We interpret this new feature to be due to methanols in a different bonding environment than the first 10, and conclude that 10 methanols fill the first solvent shell about the Cs^+. The third spectral region at 1044 cm[-1] develops for cluster sizes of N ≥ 18. This is also the same region where large neutral methanol clusters absorb,[12] or in other words the gas-phase equivalent of "bulk" methanol.

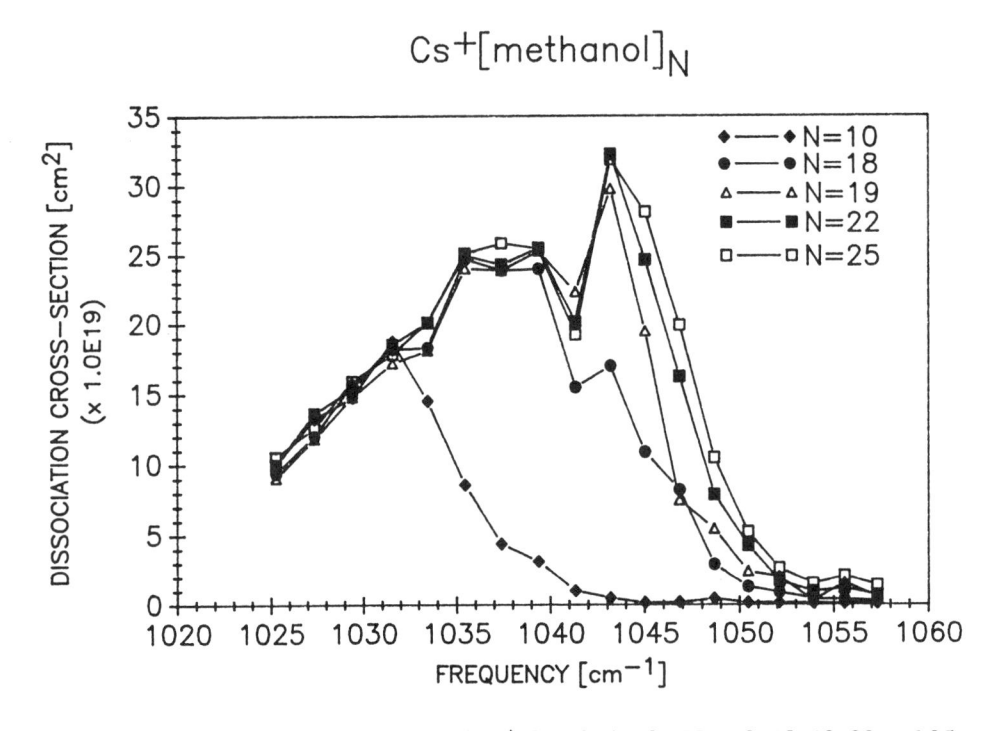

Figure 3. Photodissociation spectra of $Cs^+(CH_3OH)_N$ for N = 10, 18, 19, 22 and 25.

With the vibrational predissociation spectra of $Na^+(CH_3OH)_N$, comparison to the $Cs^+(CH_3OH)_N$ spectra can be made to examine the effects of ion size or charge density. Four distinct spectral regions were observed for $Na^+(CH_3OH)_N$ and are shown in Figure 4. We observe an initial spectral feature for N ≤ 6 at the limit of our CO_2 laser tuning range, i.e. ≤ 1025 cm[-1]. A second feature gains intensity near 1034 cm[-1] for 7 ≤ N ≤ 14, without significant change in the first region. We interpret this observation as an indication that 6 methanols fill the first solvent shell of Na^+. A third feature about 1039 cm[-1] appears for 14 ≤ N ≤ 20, once again with little effect on the first two spectra regions. For N ≥ 20, we finally observe a "bulk-like" feature about 1046 cm[-1].

For $Na(NH_3)_M^+$, we only observe vibrational predissociation spectra for M ≥ 7. In Figure 5, we show the spectra for M = 6- 9. There is an abrupt onset in the dissociation cross-section at M = 7. The feature at 1085 cm[-1] is intermediate between large neutral NH_3 clusters at 1040 cm[-1] and $Na^+ \cdots NH_3$ at 1115 cm[-1]. We interpret this feature to be consistent with ammonia molecules outside the first solvation shell of Na^+. Ammonias within the first solvent shell will absorb at frequencies closer to 1115 cm[-1] which is outside the tuning range of our CO_2 laser. From the abrupt onset, sharpness and frequency of the feature at 1085 cm[-1], it would appear that six ammonias form the first solvation shell about the Na^+ ion.

**Na+[MeOH]N, N=6-22
Vibrational Spectra**

Figure 4. Photodissociation spectra of $Na^+(CH_3OH)_N$ for N - 4, 10, 14, 17, 19, 21 and 22.

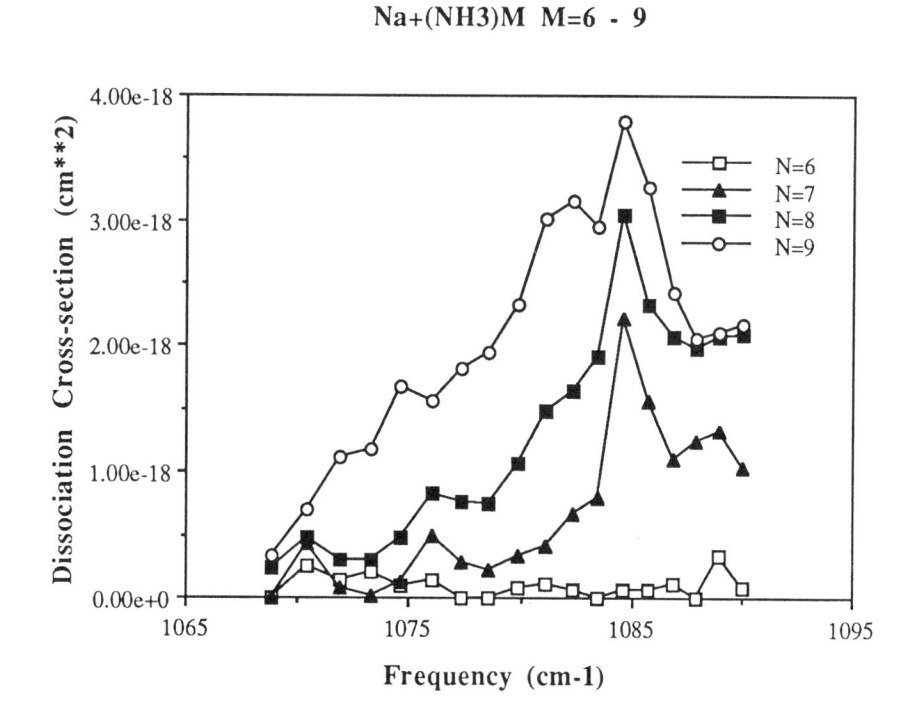

Na+(NH3)M M=6 - 9

Figure 5. Photodissociation spectra of $Na^+(NH_3)_M$ for M = 6, 7 and 9.

5. MONTE CARLO RESULTS

Radial distribution functions were calculated from the MC simulations for the metal ion-oxygen, hydrogen and methyl groups for both Na^+ and Cs^+. For Na^+ with ammonia the sodium ion-nitrogen and hydrogen radial distribution functions were determined. Additionally, the radial distribution function between oxygen and hydrogen atoms on different methanol molecules was also determined. The radial distribution functions can then be integrated to obtain solvent numbers and numbers of hydrogen bonds. We have also taken representative configurations from the MC simulations to observe some coarse structural features.

A typical series of radial distribution functions for the $Na^+ \cdots O$ distance in $Na^+(CH_3OH)_N$ is shown in Figure 6. As the number of methanols increase, new features in the radial distribution function can be observed at 4.4 Å for $N \geq 7$ and 5.8 Å for $N \geq 12$. These new features correspond to the new solvent shells as they form for larger and larger cluster ions. A summary of the ion-oxygen radial distribution functions is given in Table II.

Less extensive simulations were performed for $Na^+(NH_3)_M$. From both studies done in our laboratory and others,[13] it appears that six ammonias form the first solvent shell about the sodium ion.

Analysis of the hydrogen bonding in $Na^+(CH_3OH)_N$ and $Cs^+(CH_3OH)_N$ yielded a distinct difference between the two ions. As the number of methanols increased from $N = 8$ to $N = 10$, there was a sharp increase in the number of hydrogen bonds in $Cs^+(CH_3OH)_N^+$ from 1.0 to 7.3. The hydrogen bond criteria used specifies an $O \cdots H$ distance of 1.7 to 2.5 Å and an $O \cdots H-O)$ angle of 150 to 180°. This increase in the number of hydrogen bonds occurs if the first solvent shell about the cesium ion is filled. In contrast, the number of hydrogen bonds in $Na^+(CH_3OH)_N$, $N = 6$ and 8, are 0 and 1.5 respectively. These results from the MC simulations indicate extensive hydrogen bonding between the methanols in the first solvent shell of Cs^+, while there is no hydrogen bonding between like methanols in Na^+.

TABLE II

Comparison of Ion-O RDFs for $Na^+(CH_3OH)_N$ and $Cs^+(CH_3OH)_N$

		Ion-O
$Na^\pm$	Cluster Size	Distance (Å)
1st site	N = 4-6	2.3
2nd site	N = 7-14	4.3
3rd site	N > 14	5.7
$Cs^\pm$		
1st site	N = 4-10	2.9
2nd site	N = 11-18	4.8
3rd site	N > 18	5.8

Na+ - Oxygen Radial Distribution Functions

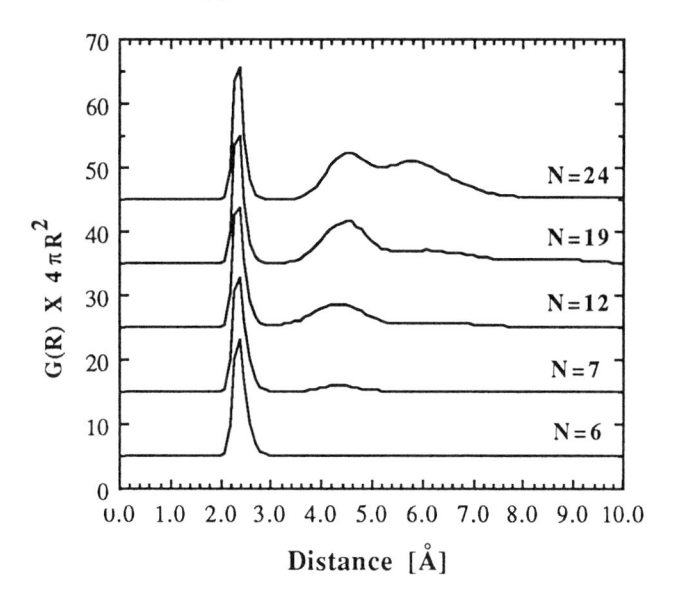

Figure 6. Na$^+$ - oxygen radial distribution function for Na$^+$(CH$_3$OH)$_N$ with N = 6, 7, 12, 19 and 24. The baselines are offset for clarity.

6. DISCUSSION

It is perhaps most useful to first establish the consistency between the experimental observations and the MC simulations. The infrared spectroscopic results gave clear indications of size-dependent behavior as did the independent simulations. The shifts in the vibrational frequencies associated with these new spectral features appear to be due to a weakening of the electrostatic influence on the solvent by the ion as the distance between the two increases. New peaks in the radial distribution functions also indicated solvent molecules taking up positions outside of existing solvent shells. The close agreement between the experiments and simulations can be seen in Table III for Na$^+$(CH$_3$OH)$_N$. The onset of new

TABLE III

Solvation Shell Occupation Number for Na$^+$(NH$_3$)$_N$

N	Experimental 1st	2nd	3rd	Monte Carlo 1st	2nd	3rd
6	6	-	-	6.0(1)	0.0	-
7	6	1	-	6.0(2)	0.9(3)	-
10	6	4	-	6.2(4)	3.2(8)	0.6
12	6	6	-	6.1(3)	4.8(9)	1.1
15	6	~8	1	6.2(6)	6.1(1.2)	2.7
19	6	~8	5	6.0(1)	8.0(1.0)	5.0

solvent shells occurs at approximately the same size and the size of the solvent shells is also in good agreement. A comparison of Table II and Figure 3 shows a similar degree of consistency for $Cs^+(CH_3OH)_N$. The less extensive measurements for $Na^+(NH_3)_N$ also were in agreement on a first solvent shell size of 6 ammonias. Thus in each case there is good agreement between the structural results from the MC simulations and the spectroscopic features from the vibrational predissociation spectra.

A comparison of the spectroscopic results for $Na^+(NH_3)_N$ and $Cs^+(CH_3OH)_N$ indicates the role of ion size on the solvent structure and perturbation of the solvent vibrational mode. The smaller Na^+ has a stronger influence on the surrounding solvent molecules, as can be seen from Table IV. First, the C–O vibrations of solvents in the first solvation shell undergo a greater shift from the monomer value than methanols in the first shell about the cesium ion. Na^+ appears to have two additional solvent binding environments which are different from the "bulk-like" phase, where Cs^+ has only one. The stronger electrostatic influence of the sodium ion also leads to smaller and more compact structures as seen from Table II, as well as Table IV.

It is also tempting to consider the ramifications of the MC results on hydrogen bonding in these systems. A solvent shell size of ten methanols is quite large for $Cs^+(CH_3OH)_N$, and appears to get some of its stability from the extensive hydrogen bonding within the first solvent shell. While the smaller sodium ion completely disrupts the hydrogen bonding of methanol within its first solvent shell, the impact of the larger cesium ion is a less extensive perturbation. It may be not that incorrect to imply that the Cs^+-methanol electrostatic interaction is comparable to the hydrogen bond strength within the first shell. The cesium ion orients the methanols so that the oxygen atom coordinates the ion, but does not break the hydrogen bond network. Sodium ions are much more disruptive, destroying the hydrogen bonds between the first shell methanols. This disruption penetrates farther away from the sodium ion into the solvent as seen by the larger frequency shifts and number of distinct bonding environments in comparison to the cesium ion. We are currently modifying our apparatus to study the O–H stretch in order to more directly probe the effects on hydrogen bonding.

TABLE IV

Comparison of Infrared Bands in $Na^+(CH_3OH)_N$ and $Cs^+(CH_3OH)_N$

Na^+	Cluster Size	Peak Position (cm^{-1})	Shift From Monomer (cm^{-1})
1st site	N = 4-6	≤ 1025	≥ -9
2nd site	N = 7-14	1033	-1
3rd site	N = 15-20	1040	+6
4th site	N > 20	1046-1048	+12 to +14
Cs^+			
1st site	N = 4-10	1025-1031	-9 to -3
2nd site	N = 11-18	1038	+4
3rd site	N > 18	1044	+10

7. ACKNOWLEDGEMENTS

The author would like to gratefully acknowledge the efforts of Drs. W.-L. Liu, J. A. Draves, Z. Luthey-Schulten, Mr. T. J. Selegue and Mr. N. Moe in this research. This work has been supported in part by the National Science Foundation (Grants No. CHE-8714735 and CHE-9111930) and the University of Illinois Research Board.

8. REFERENCES

1. R. E. Miller, "The Vibrational Spectroscopy and Dynamics of Weakly Bound Neutral Complexes," Science *240*, 447-453, 1988.

2. R. C. Cohen and R. J. Saykally, "Extending the Collocation Method to Multidimensional Molecular Dynamics," J. Phys. Chem. *94*, 7991-8000, 1990. J. M. Hutson, "The Intermolecular Potential of Ne-HCl. Determination from High-Resolution Spectroscopy," J. Chem. Phys. *91*(8), 4448-4454, 1989.

3. R. J. Saykally, "Infrared Laser Spectroscopy of Molecular Ions," Science *239*, 157-161, 1988.

4. M. H. Shen and J. M. Farrar, "Absorption Spectra of Size-Selected Solvated Metal Cations: Electronic States, Symmetries and Orbitals in $Sr^+(NH_3)_{1,2}$ and $Sr^+(H_2O)_{1,2}$" J. Chem. Phys. *94*(5), 3322-3331, 1991.

5. M. Okumura, L. I. Yeh, J. D. Myers and Y. T. Lee, "Infrared Spectra of the Solvated Hydronium Ion," J. Phys. Chem. *94*, 3416-3427, 1990. J. M. Price, M. W. Crofton and Y. T. Lee, "Vibrational Spectroscopy of the Ammoniated Ammonium Ions $NH_4^+(NH_3)_n$ (n=1–10)," J. Phys. Chem. *95*, 2182-2195, 1991.

6. J. A. Draves, Z. Luthey-Schulten, W.-L. Liu and J. M. Lisy, "Gas-Phase Methanol Solvation of Cs^+: Vibrational Spectroscopy and Monte Carlo Simulation," J. Chem. Phys. *93*, 4589-4602, 1990. T. J. Selegue, N. Moe, J. A. Draves and J. M. Lisy, "Gas-Phase Solvation of Na^+ with Methanol," J. Chem. Phys., submitted. T. J. Selegue and J. M. Lisy, "Vibrational Spectroscopy of Ammoniated Sodium Ions $Na^+(NH_3)_M$, M = 6-12," J. Phys. Chem., submitted.

7. J. K. Buckner and W. L. Jorgensen, "Energetics and Hydration of the Constituent Ion Pairs of Tetramethylammonium Chloride," J. Am. Chem. Soc. *111*, 2507-2516, 1989. M. Sprik, M. L. Klein and K. Wantanabe, "Solvent Polarization and Hydration of the Chloride Anion," J. Phys. Chem. *94*, 6483-6488, 1990.

8. C. E. Klots, "The Evaporative Ensemble," Z. Phys. D. *38*, 83-89, 1987.

9. N. Metropolis, A. W. Rosenbluth, M. N. Rosenbluth, A. H. Teller and E. Teller, "Equation of State Calculations for Fast Computing Machines," J. Chem. Phys. *21*(6), 1087-1092, 1953.

10. W. L. Jorgensen, "Transferable Intermolecular Potential Functions. Application to Liquid Methanol Including Internal Rotation," J. Am. Chem. Soc. *103*, 341-345, 1981.

11. J. P. LaCosse and J. M. Lisy, "Vibrational Predissociation Spectroscopy of $(CH_3OD)_2$ and $(CH_3OH)(CH_3OD)$ in the 9.6 μm Region," J. Phys. Chem. *94*, 4398-4400, 1990.

12. F. Huisken and M. Stemmler, "Infrared Photodissociation of Small Methanol Clusters," Chem. Phys. Letters *144*(4), 391-5, 1988.

13. M. Marchi, M. Sprik and M. L. Klein, "Solvation of Electrons, Atoms and Ions in Liquid Ammonia," Faraday Discuss. Chem. Soc. *85*, 373-389, 1988. T. J. Selegue, N. Moe and J. M. Lisy, to be published.

LASER DESORPTION JET COOLING SPECTROSCOPY OF ORGANIC CLUSTERS.

Mattanjah S. de Vries, Heinrich E. Hunziker,
Gerard Meijer, and H. Russell Wendt

IBM Research Division, Almaden Research Center, San Jose, CA 95120-6099

ABSTRACT

While laser ablation is often used to create clusters, laser desorption at lower fluences can be used to probe clusters already present on a surface or van der Waals clusters formed between desorbed species and gas phase atoms.

- As an example of the former we discuss a study of para amino benzoic acid dimers. The dimers are formed by laser desorbing the monomer into a supersonic expansion.

- With the technique of laser desorption jet cooling it is also possible to form van der Waals clusters and species in the drive gas. As an example we discuss the spectroscopy of triphenylamine/argon clusters.

- Thirdly, the same technique allows the study of species on the surface, rather than clusters formed by the volatilization itself. A dramatic example is the formation of carbon "clusters" by laser ablation in a separate step and their subsequent study by laser desorption.

1. Experimental.

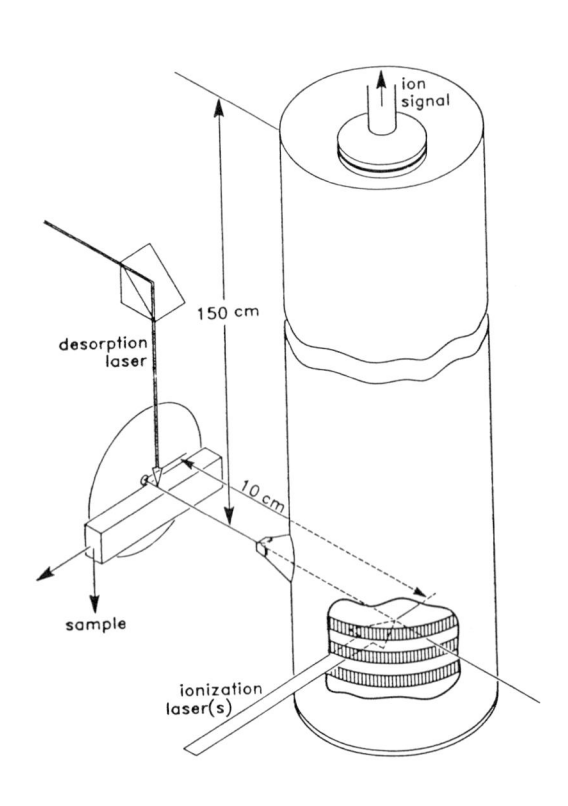

Figure 1 schematically shows the technique of laser desorption jet cooling spectroscopy, which we have described in detail elsewhere [1]. In brief, molecules, adsorbed on a surface, are laser desorbed on the vacuum side of a pulsed supersonic expansion of Ar. The desorbed molecules are entrained in the drive gas, cooled in the expansion, and intercepted downstream by the detection laser(s). One or two color resonant multiphoton ionization (REMPI) is followed by linear time of flight (TOF) mass spectrometric detection.

FIGURE 1: Schematic view of laser desorption jet cooling spectrometer.

2. Para amino benzoic acid (PABA) clusters.

Figure 2 shows spectra of PABA dimers, formed in the expansion from laser desorbed monomers. Part of the sample on the surface was labeled by replacement of two hydrogen atoms on the ring by deuterium. As a result three dimer peaks can be observed in the mass spectrum: unlabeled dimers (HH) at mass 274, mixed dimers consisting of one labeled and one unlabeled unit (HD) at mass 276, and finally, dimers in which both units are deuterated (DD) at mass 278. The dependence of these three signals on excitation laser wavelength can be measured simultaneously and results in the three spectra shown in figure 2. The top panel represents the spectrum of the HH species and the middle panel gives the spectrum of the DD species. The spectrum is shifted because of the slight change in zero point energy, but otherwise virtually identical. The bottom panel is the spectrum of the mixed dimer, the HD species, and appears as a composite of the top two. The implication is that in the case of the mixed dimer the excitation wavelength can be chosen such as to have the photon absorbed selectively either by the deuterated or by the undeuterated ring. This implies that the interaction energy in this hydrogen bonded cluster is very small. In fact, from a more detailed analysis of the spectra (ref) we can derive an interaction energy of $0.4 +/- 0.3$ cm^{-12}.

The fact that the excitation in the PABA dimer is so strongly localized also has an interesting consequence for clusters formed by the dimer and Ar atoms from the drive gas. Figure 3 shows the origin of the REMPI spectrum of such a complex. The higher frequency peak is practically unshifted compared to the bare dimer and corresponds to excitation of the ring, that does not have the Ar atom associated with it. On the other hand, the lower frequency peak results from excitation of the ring with Ar atom. In this case there is a 38 cm^{-1} redshift, identical to that of the PABA-Ar complex. Laser desorption jet cooling makes it possible to see these effects, although the monomer units have a vapor pressure of about 10^{-8} torr, and therefore don't normally occur in the gas phase with sufficient densities to form clusters.

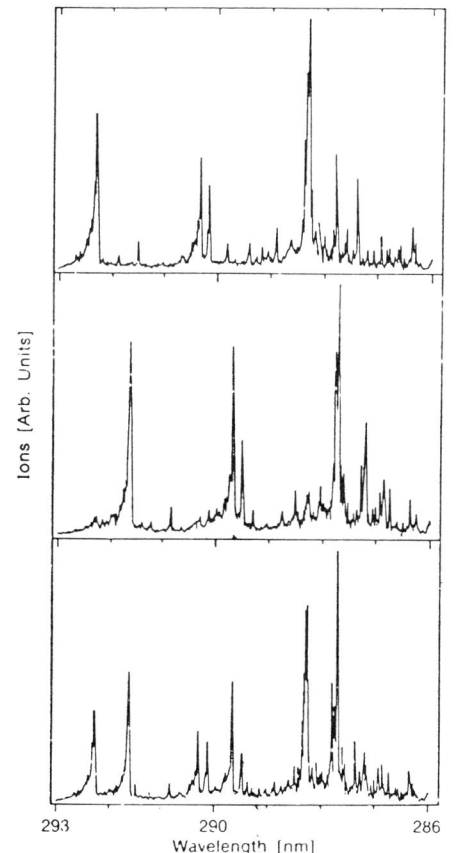

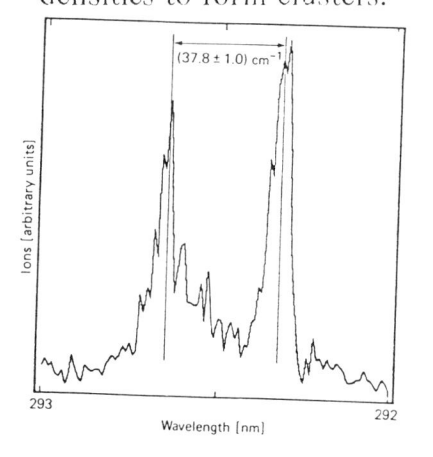

FIGURE 3: Part of the (1+1)-REMPI spectrum of the (PABA dimer)-Ar van der Waals complex, showing the double origin. Ions are recorded at the parent mass of 314 amu.

FIGURE 2: (1+1)-REMPI spectra of PABA dimers (mass 274; upper), deuterated PABA dimers (mass 276; middle), and mixed deuterated-undeuterated PABA complexes (mass 276; lower). All three spectra are recorded in the same laser scan.

3. Triphenyl amine (TPA) clusters with Ar.

An interesting example of clusters formed between desorbed and gas phase molecules can be observed with triphenyl amine [3]. Figure 4 shows REMPI spectra for TPA as well as for its clusters with Ar and Kr. Figure 5 shows the spectrum of the cluster formed between TPA and two Ar atoms. The REMPI spectrum of TPA itself shows a vibrational progression in the symmetric torsion mode (114 cm^{-1}) as well as in the symmetric C-N stretching mode (280 cm^{-1}). The cluster spectra show a remarkable *blue* shift with respect to the spectrum of the free TPA molecule (211 cm^{-1} for Ar and 216 cm^{-1} for Kr). This is exceptional since generally clusters of aromatic molecules with rare gas atoms show a characteristic small red shift. In fact, just such a red shift can be observed when we compare the spectrum of TPA with two Ar atoms with that of TPA-Ar. Analysis of high resolution spectra shows that both TPA and the blue shifted TPA-Ar complex are symmetric top molecules. It is concluded that the rare gas atom is located on the C_3 symmetry axis, either above or below the umbrella. Once this special position is occupied, a second rare gas atom is "forced" into another position, namely above one of the phenyl rings. This is the "normal" position, which produces the usual red shifting the spectrum.

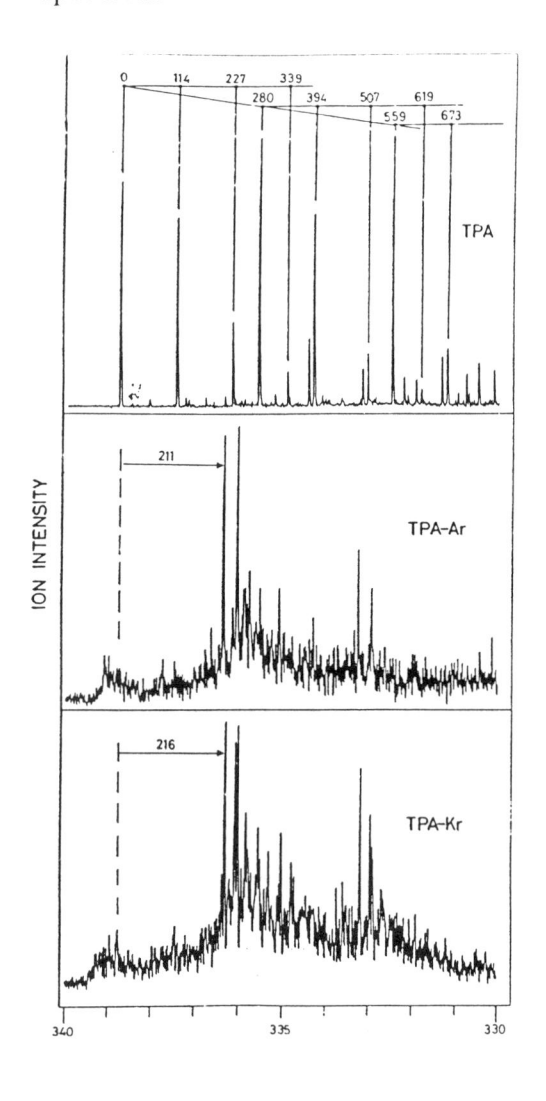

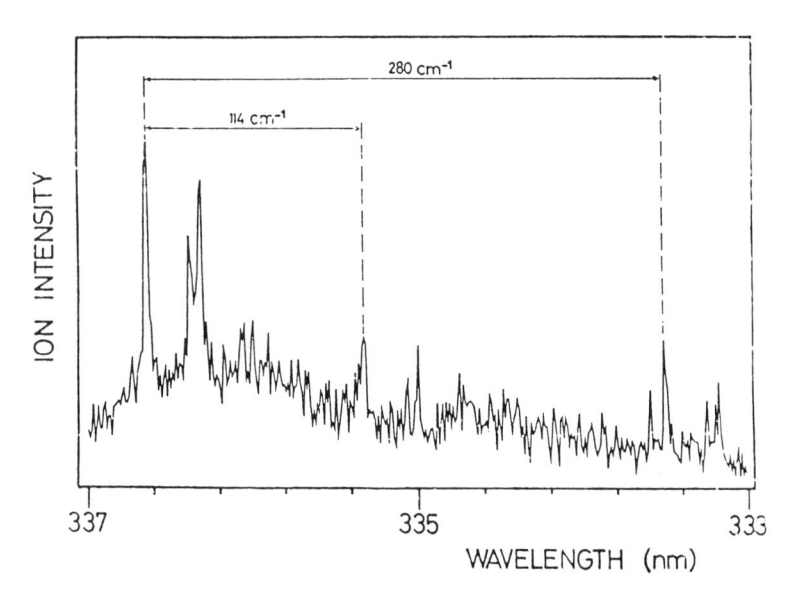

FIGURE 5: (1+1)-REMPI spectrum of the TPA-Ar$_2$ complex. The indicated origin is 187 cm^{-1} blueshifted relative to the origin of free TPA.

FIGURE 4: Vibrationally resolved (1+1)-REMPI spectra of TPA, TPA-Ar, and TPA-Kr. In all cases the parent ion is mass selectively detected. In the lower panels the shift in cm^{-1} is indicated of the van der Waals complex relative to free TPA.

4. Carbon clusters.

Fullerenes were first observed in carbon cluster research, but really should be classified as molecules, rather than as clusters. Nevertheless it is important, when using laser desorption and mass spectrometry to detect fullerenes, to demonstrate that the species are not produced in the laser step. This is especially relevant since, in fact, under certain conditions, fullerenes *can* be produced by laser ablation. Figure 6 shows the laser laser ionization TOF mass spectrum of a mixture of two carbon soots extracts. Each was produced in a carbon arc. In one case (a) La metal vapor was also present in the discharge, while in the other case (b) the carbon rods contained 13C enriched graphite. Both products were further treated by toluene extraction. Clearly both soot (a) and soot (b) produce C60 as well as C70, however only soot (a) produces a metallofullerene, LaC82. If this species would have been produced by desorption or jet cooling, then we would have also observed the 13C enriched metallofullerene, which is clearly not the case. This demonstrates that in fact the metallofullerene is present on the surface, prior to analysis, and that laser desorption jet cooling can also be used for analysis of complexes that are already present at the surface.

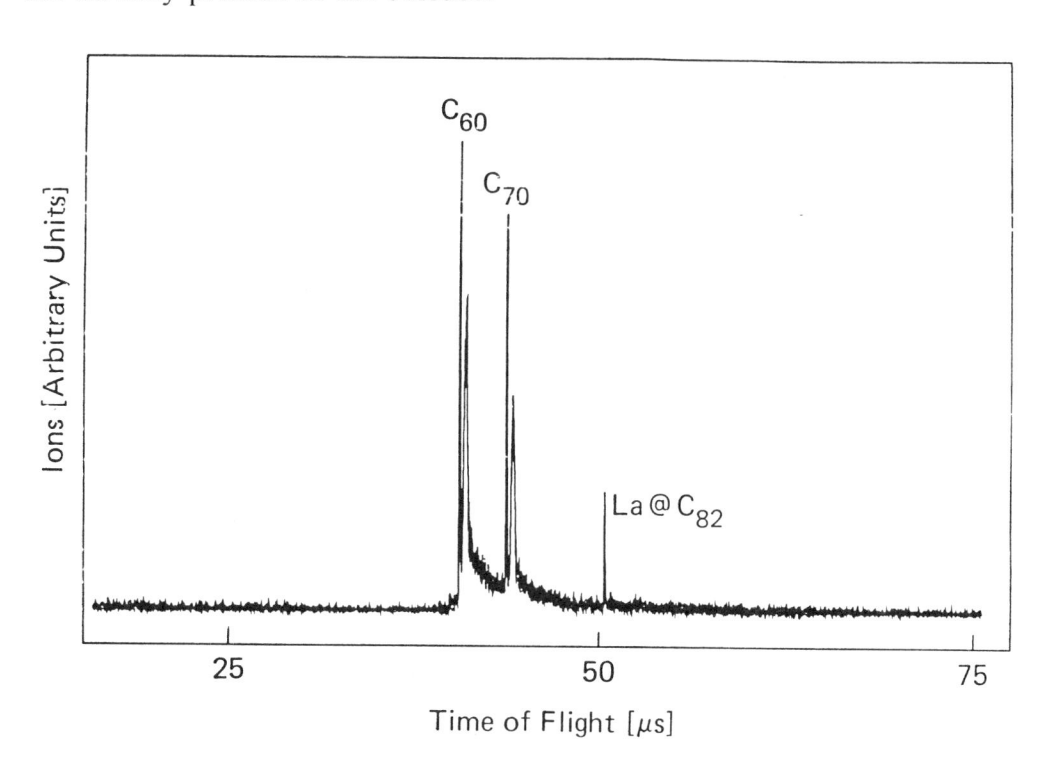

FIGURE 6: Laser desorption laser ionization time of flight mass spectrum of material containing a mixture of C_{60}, C_{70}, LaC_{82} 13C enriched C_{60}, and 13C enriched C_{70}. The latter two species show up as (unlabeled) peaks to the high mass side of the (labeled) C_{60} and C_{70} peaks with natural abundance.

5. REFERENCES

[1] G. Meijer, M. de Vries, H. Wendt, and H. Hunziker, Appl. Phys. **B 51**, 1871 (1990).
[2] G. Meijer, M. de Vries, H. Wendt, and H. Hunziker, J. Chem. Phys. **92**, (1990).
[3] G. Meijer, M. S. de Vries, H. R. Wendt, and H. E. Hunziker, Spectroscopy of Triphenylamine and its van der Waals Clusters. To be published

A gas phase study of the group VI transition metal tricarbonyl complexes by negative ion photoelectron spectroscopy[*]

A. A. Bengali, S. M. Casey, C.-L. Cheng, J. P. Dick, P. T. Fenn, P. W. Villalta, and D. G. Leopold

Department of Chemistry, University of Minnesota, Minneapolis MN 55455

ABSTRACT

Photoelectron spectra are reported for Cr(CO)$_3^-$, Mo(CO)$_3^-$ and W(CO)$_3^-$ anions prepared from the corresponding metal hexacarbonyls in a flowing afterglow ion source. The 488.0 nm spectra were obtained at an electron kinetic energy resolution of 5 meV using a newly constructed apparatus. The spectra exhibit transitions between the ground electronic states of the anions and the neutral molecules, and they show weak activity in the symmetric CO stretching, MC stretching, MCO bending and CMC bending vibrational modes. The observed vibrational structure indicates that the anions, like the neutral molecules, have C$_{3v}$ equilibrium geometries. Force constants estimated for the neutral M(CO)$_3$ molecules from the fundamental vibrational frequencies measured here are consistent with stronger metal-ligand bonding in the coordinatively unsaturated complexes than in the corresponding hexacarbonyls. Franck-Condon analyses of the spectra indicate only small differences between the equilibrium bond lengths and bond angles of the anions and the corresponding neutral molecules. The electron affinity pattern observed among the three group VI metal tricarbonyls is compared with characteristic trends within triads of transition metal atoms, and within the coinage metal dimer series. This comparison, combined with the results of previously reported theoretical calculations, suggests that the extra electron in the M(CO)$_3^-$ anions occupies an *sp* hybrid orbital. Electron affinities of 1.349 eV, 1.337 eV, and 1.859 eV (all ±0.006 eV) are obtained for Cr(CO)$_3$, Mo(CO)$_3$, and W(CO)$_3$, respectively.

1. INTRODUCTION

Coordinatively unsaturated metal carbonyls are of interest as active species in catalytic reactions,[1-4] as the building blocks of stable organometallic complexes,[5] and as computationally tractable benchmarks against which to test theoretical models of metal-carbonyl bonding.[6] Although unsaturated metal carbonyls have many features in common with their fully ligated counterparts, qualitative differences in their electronic structure, reactivity, and bonding are also expected. For example, among the neutral Fe(CO)$_n$ molecules, sequential metal-ligand bond strengths differ by an order of magnitude,[7,8] and reactivity toward CO varies by almost three orders of magnitude.[9,10]

We report here a study of gas phase Cr(CO)$_3$, Mo(CO)$_3$ and W(CO)$_3$ and the corresponding anions by negative ion photoelectron spectroscopy. These systems provide an opportunity to compare the bonding for open *d*-shell metal carbonyls of the first, second and third transition series at a degree of coordinative unsaturation intermediate between that of the bare metal atom and the saturated complex.

Previous spectroscopic studies of the Group VI metal tricarbonyls have been limited to the measurement of CO stretching vibrational frequencies by infrared spectroscopy of the matrix-isolated[11] and gas phase[10,12-14] molecules. Perutz and Turner[11] found that matrix-isolated Cr(CO)$_3$, Mo(CO)$_3$ and W(CO)$_3$ each exhibits two IR-active CO stretching modes, ruling out a planar D$_{3h}$ structure. Intensity ratios observed for Mo(CO)$_3$ and its isotopic derivatives indicate a C$_{3v}$ structure with a C-Mo-C bond angle of 105°.[11] The symmetric CO stretching vibrations of the three tricarbonyls in CH$_4$ matrices fall in the range 1975-1981 cm^{-1}, substantially reduced from the 2112-2117 cm^{-1} values observed[15] for the stable hexacarbonyls in solution. Gas phase time-resolved infrared absorption studies of Cr(CO)$_3$ and Mo(CO)$_3$ by the Weitz,[10,12] Rosenfeld[13] and Rayner[14] groups identified the more intense CO stretch of E symmetry at the frequency expected based on the matrix study, providing evidence that these molecules adopt a C$_{3v}$ structure in the gas phase as well. These studies also found that bimolecular rate constants for CO addition to Cr(CO)$_n$ (n=2-5) and Mo(CO)$_n$ (n=3-5) are within an order of magnitude of the gas kinetic collision rate.[10,12-14] Since the M(CO)$_6$ complexes are known to have singlet ground states, and reactions requiring a change in spin multiplicity are expected to be relatively slow, these data suggest that Cr(CO)$_3$ and Mo(CO)$_3$ also have singlet ground states.

In other gas phase experimental studies, bond strengths of coordinatively unsaturated Group VI metal carbonyls have been obtained by techniques involving excimer laser photolysis of the parent hexacarbonyls.[16-18] Two of these studies[16,18] found the $M(CO)_n$-CO (n=3-5) bond dissociation energies each to be within 0.3 eV of the average energy required to dissociate the first three CO ligands from the corresponding metal hexacarbonyl complex. These results were interpreted as being consistent with the view that the metal center in $M(CO)_3$, like that in $M(CO)_6$, has a low spin d^6 configuration.[16]

Theoretical studies of the Group VI tricarbonyls include several semi-empirical calculations;[19-22] to our knowledge, no *ab initio* calculations have been reported to date. Although early calculations by Kettle[19] suggested planar D_{3h} geometries for all $M(CO)_3$ fragments in accord with the VSEPR rules, subsequent molecular orbital calculations by Burdett[20] and by Elian, Hoffmann and coworkers[21] predicted low spin d^6 $M(CO)_3$ molecules to adopt a pyramidal C_{3v} structure. Elian and Hoffmann predicted the ground state of $Cr(CO)_3$, for example, to have a C-Cr-C bond angle of 93°, and a singlet d^6 configuration in which the six valence electrons occupy a set of three closely-spaced a_1+e orbitals reminiscent of the low-lying t_{2g} set in the octahedral hexacarbonyl complex.[21] The driving force for this nonplanarity is the rapid stabilization of the $a_1(d_{z^2})$ orbital (oriented along the C_{3v} symmetry axis) as the molecule bends, due to decreased metal-ligand σ repulsion and increased bonding with the carbonyl π^* orbital.[20,21] Burdett also found the low spin d^6 tricarbonyls to adopt a C_{3v} structure, but predicted planar or Jahn-Teller distorted geometries for states with high or intermediate spin.[20] These computational results, combined with the experimentally determined[11] C_{3v} geometry for matrix-isolated $Mo(CO)_3$, suggest that the neutral Group VI tricarbonyls have singlet ground states, consistent with the conclusions of the kinetics studies[10,12-14] described above. An INDO calculation has predicted the electron affinity of $Cr(CO)_3$ to be 1.19 eV, with the extra electron in the anion occupying one of a set of doubly degenerate $d_{xz,yz}$ orbitals.[22]

Negative ion photoelectron spectroscopy[23] has previously been profitably used to study the $Ni(CO)_n$[24] and $Fe(CO)_n$[7] complexes. This technique has several unique advantages as a method for studying neutral and anionic coordinatively unsaturated organometallic species. Since the selection rules for vibrational transitions are determined by Franck-Condon factors, it is often possible to observe vibrational states that are forbidden or weak in infrared spectroscopy. In particular, it is in principle possible to observe the bending and M-C stretching modes of the neutral molecule, which have thus far escaped detection for neutral binary (i.e., unsubstituted) metal tricarbonyls by other spectroscopic techniques. Negative ion photoelectron spectroscopy also allows the measurement of the electron affinity of the neutral molecule, a value that may be useful in understanding its reactivity toward electron-rich reagents. In addition, intensity ratios observed in the photoelectron spectra yield quantitative information concerning the change in molecular structure induced by addition of the extra electron.

2. EXPERIMENTAL SECTION

Spectra were obtained using a new negative ion photoelectron spectrometer, described in detail elsewhere.[25] Briefly, ions are prepared and thermalized in a flowing afterglow ion-molecule reactor. Negative ions sampled from the flowing afterglow are focused into a beam, accelerated, and mass analyzed by a 90° sector magnet. The mass-selected ion beam is then decelerated, and crossed by the intracavity radiation of a cw argon ion laser operated at 488.0 or 514.5 nm. A small solid angle of the resulting photoelectrons is energy analyzed by a hemispherical analyzer and detected by a position sensitive detector.

The best resolution (FWHM) that we have obtained on this apparatus thus far is 3.0 meV, as measured by Lorentzian fits to peaks observed in the spectrum of W⁻. The spectra reported here were obtained at 5 meV (40 cm⁻¹) resolution. Absolute electron kinetic energies are calibrated with respect to O⁻, whose electron detachment energy is accurately known.[26] The relative electron kinetic energy scale is calibrated against fine structure splittings[27] observed in the W⁻ photoelectron spectrum, typically revealing a small linear energy scale compression factor of 0.2% or less. In addition to these two corrections, obtaining accurate electron kinetic energies requires converting the measured laboratory energies to the center-of-mass frame. With these procedures, splittings measured in the photoelectron spectra of Cr⁻, Mo⁻ and other atomic anions are found to agree with the known values[27] to ±5 cm⁻¹ for electron kinetic energies above 0.3 eV, the region of interest in the present study.

3. RESULTS

Figure 1 shows photoelectron spectra obtained at 488.0 nm for the $Cr(CO)_3^-$, $Mo(CO)_3^-$ and $W(CO)_3^-$ anions. The spectra of the three Group VI tricarbonyls are quite similar in general appearance, but are shifted in absolute energy. Each spectrum exhibits several peaks arising from photodetachment transitions from the ground electronic state of the anion to different vibrational levels of the ground electronic state of the neutral $M(CO)_3$ molecule. Scans over the entire electron kinetic energy (eKE) range of 0-2.540 eV revealed no additional features due to photodetachment from $M(CO)_3^-$ anions, although very weak features attributed to anionic photofragments were observed, as discussed below.

The most intense peak in each spectrum, which appears at high eKE (low ion-to-neutral transition energy), can readily be assigned as the origin band, corresponding to the "0-0 transition" between the zero point vibrational levels of the anion and the neutral molecule. Lorentzian fits to the spectra indicate origin band positions of 1.191 eV for $Cr(CO)_3^-$, 1.203 eV for $Mo(CO)_3^-$ and 0.681 eV for $W(CO)_3^-$. Calculations based on estimates for the neutral and anion geometries[25] indicate that unresolved rotational structure will not shift the observed origin band by more than 1 meV for these room temperature anions. Adiabatic electron affinities for the neutral $M(CO)_3$ molecules can thus be simply obtained by subtracting the observed origin band positions from the 2.540 eV photon energy. The resulting electron affinities are 1.349±0.006 eV for $Cr(CO)_3$, 1.337±0.006 eV for $Mo(CO)_3$, and 1.859±0.006 eV for $W(CO)_3$.

Each spectrum also exhibits transitions from the ground vibrational level of the anion to excited vibrational levels of the neutral molecule, which produce peaks to the left (low eKE) side of the origin band. The spacings between these peaks and the origin band provide measurements of vibrational frequencies in the neutral molecule. The relative intensities of these vibrational transitions are determined by the Franck-Condon overlap of the vibrational wavefunctions of the anion and the neutral molecule. As a result, a vibrational mode which corresponds to a change in equilibrium geometry between the anion and neutral will appear active in the spectrum, and the extent of this activity provides a quantitative signature of the magnitude of the geometry change.

It is evident from a casual inspection of Figure 1 that the $M(CO)_3^-$ photoelectron spectra show only weak vibrational activity, an indication that the equilibrium structures of the anions are very similar to those of the corresponding neutral molecules. The neutral C_{3v} molecules have four totally symmetric (A_1) vibrational modes, and their expected frequencies correspond to those of the four vibrational modes observed to be active in the $M(CO)_3$ spectra, as is described below. Therefore, it appears reasonable to assign the active modes as A_1 vibrations. This assignment implies that the Group VI $M(CO)_3^-$ negative ions also have C_{3v} equilibrium geometries.

As shown in Figure 1, each $M(CO)_3^-$ spectrum displays a prominent peak 2000±10 cm^{-1} above the 0-0 transition. This interval can readily be assigned as the symmetric CO stretching frequency in the neutral molecule. The $Cr(CO)_3^-$ and $Mo(CO)_3^-$ spectra also exhibit weaker peaks at 3990 cm^{-1} due to transitions to v=2 of this mode. The corresponding overtone would appear near 0.185 eV in the $W(CO)_3^-$ spectrum, but is not detected due, in part, to the reduced sensitivity of our instrument at low electron energies.

The symmetric CO stretch was previously observed in CH_4 matrices by Perutz and Turner[11] at 1979 cm^{-1} for $Cr(CO)_3$, 1981 cm^{-1} for $Mo(CO)_3$ and 1975 cm^{-1} for $W(CO)_3$. For $Cr(CO)_3$ and $Mo(CO)_3$, the frequency of the CO stretch of E symmetry is increased by 21-29 cm^{-1} on going from the CH_4 matrix to the gas phase.[10,12-14] Thus, the 2000±10 cm^{-1} frequencies observed here for the A_1 modes fall in the range expected based on the previously reported matrix values and matrix-to-gas phase shifts. This excellent agreement provides further support for the conclusion that the neutral molecules adopt a C_{3v} structure in the gas phase as well as in matrices.

The second active vibrational mode in the photoelectron spectra gives rise to prominent peaks 680±10 cm^{-1}, 660±10 cm^{-1} and 620±10 cm^{-1} above the 0-0 transition for $Cr(CO)_3$, $Mo(CO)_3$ and $W(CO)_3$, respectively. These intervals, which again correspond to fundamental vibrational frequencies in the neutral molecules, are in the region expected for MCO bending and MC stretching vibrations.[28] The v=2 overtone, and combination bands involving 1 or 2 quanta of excitation in this mode and 1 or 2 quanta of the CO stretch, are also observed in each spectrum.

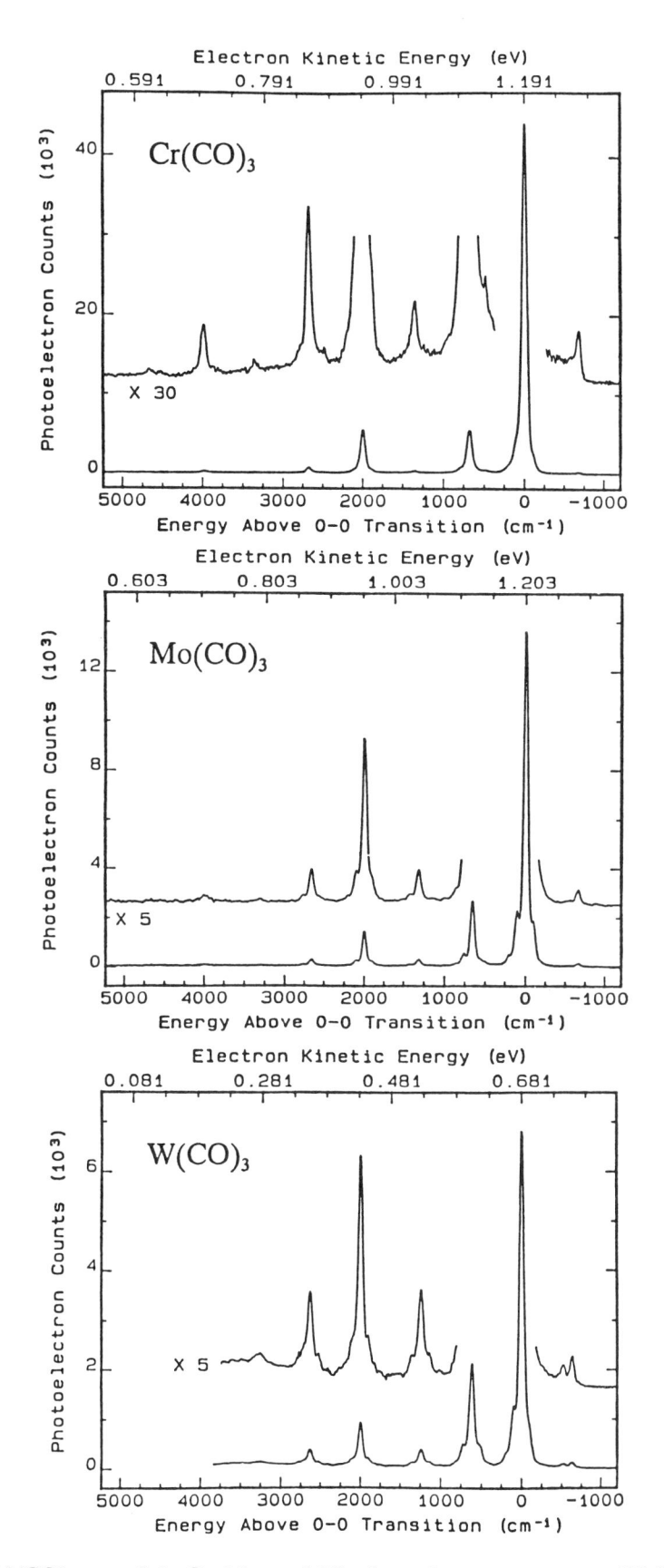

Figure 1. $M(CO)_3^- \rightarrow M(CO)_3 + e^-$ (M=Cr, Mo, and W) photoelectron spectra at 488.0 nm (2.540 eV).

These overtone and combination bands exhibit no detectable anharmonicity at our resolution, and have approximately the intensities expected based on harmonic Franck-Condon fits to the relative intensities of the fundamental and origin bands. The fundamental, overtone and combination bands involving these two modes account for all of the well-resolved peaks observed above the 0-0 transitions in the three spectra.

In the Cr(CO)$_3^-$ spectrum, a third vibrational mode is observed as a weak shoulder to the right of the 680 cm^{-1} peak, 480±20 cm^{-1} above the 0-0 transition. We assign this shoulder to a third vibrational mode with a fundamental frequency of 480±20 cm^{-1} in neutral Cr(CO)$_3$. This frequency, as well, falls in the range expected for MCO bending and MC stretching vibrations.[28] This mode also appears in combination with the CO stretch, producing the weak peak seen 2480±20 cm^{-1} above the 0-0 transition in Figure 1.

The fourth and last active vibrational mode observed in this series of molecules appears as a weak shoulder 100±20 cm^{-1} from the origin bands. These shoulders are most pronounced in the Mo(CO)$_3^-$ and W(CO)$_3^-$ spectra, which also display weak v=2 overtones. This low-frequency mode is in the region expected for CMC bending vibrations. Shoulders observed 100±20 cm^{-1} to the right of the origin bands are due to transitions from excited CMC bending levels in the anions. Combination bands involving the CMC bend in the neutral molecule, and sequence bands due to transitions from the excited 100±20 cm^{-1} anion level, are also observed as shoulders on the other strong peaks in each spectrum.

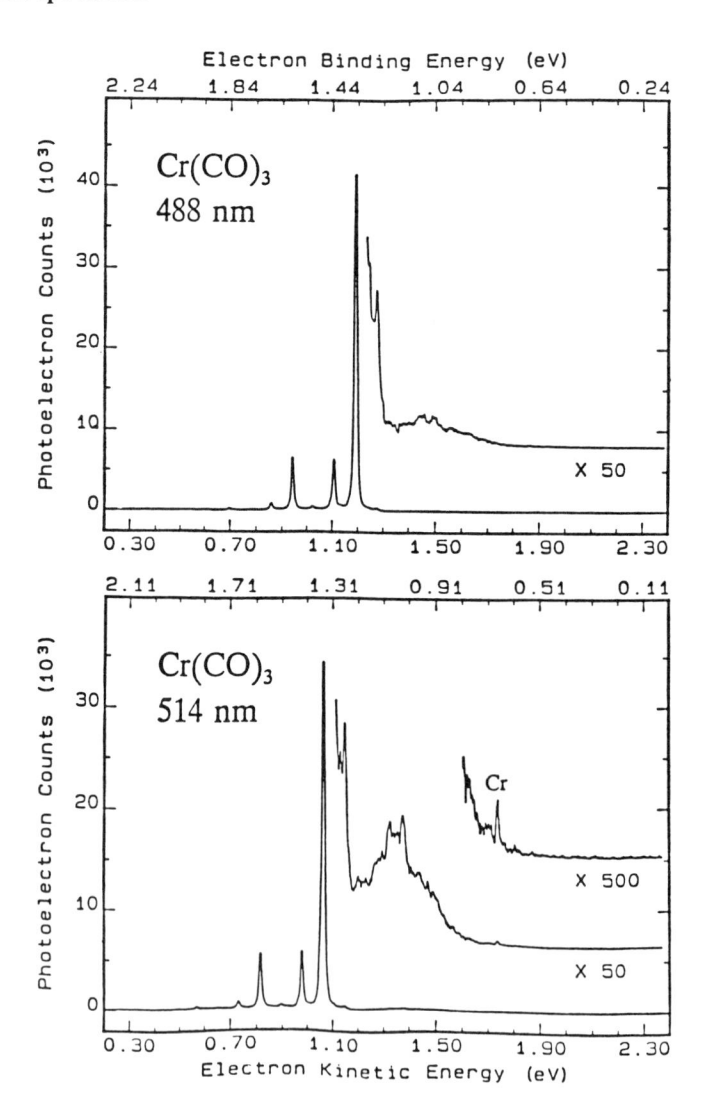

Figure 2. Cr(CO)$_3^-$ → Cr(CO)$_3$ + e$^-$ photoelectron spectra at 488.0 nm (2.540 eV) and 514.5 nm (2.409 eV).

Additional features due to transitions from vibrationally excited anions are observed to the right of the origin bands in the x30 and x5 expansions shown in Figure 1. The positions of these hot bands give ion vibrational frequencies of 680 ± 10 cm^{-1} for Cr(CO)$_3^-$, 660 ± 10 cm^{-1} for Mo(CO)$_3^-$ and 630 ± 10 cm^{-1} for W(CO)$_3^-$. These values are equal to those in the corresponding neutral molecules within the quoted experimental uncertainties. Franck-Condon analyses of the spectra indicate that these hot bands have approximately the intensities expected for room temperature anions.

Evidence for photofragmentation is found in the Cr(CO)$_3^-$ spectrum. Figure 2 compares photoelectron spectra obtained at 488.0 and 514.5 nm at a circulating laser power of about 40 W, recorded at an instrumental resolution of 9 meV. The spectra are plotted as a function of electron kinetic energy between 0.2 and 2.4 eV, and thus are shifted by the 0.131 eV difference in photon energy at these two wavelengths. At 488.0 nm, the x50 magnification reveals a weak, broad feature near 1.5 eV. When the laser power is increased from 25 to 65 W, the intensity of this feature approximately doubles relative to that of the other peaks in the spectrum. This behavior suggests that this weak feature is due to a two-photon process involving photodissociation of Cr(CO)$_3^-$ followed by photodetachment of a fragment anion. At 514.5 nm, this feature appears with a greatly enhanced relative intensity, which does not, however, increase significantly when the laser power is doubled. In addition, spectra obtained for a mass-selected Cr(CO)$_2^-$ ion beam at 488.0 and 514.5 nm also show broad features in this region, whose intensities do not vary significantly with laser wavelength. These observations suggest that the broad features observed at high eKE in the spectra of the mass-selected Cr(CO)$_3^-$ ions are actually due to photodetachment from Cr(CO)$_2^-$ photofragments. In addition, the Cr(CO)$_3^-$ → Cr(CO)$_2^-$ + CO photodissociation cross section is evidently much greater at 514.5 nm than at 488.0 nm. An increase in the Cr(CO)$_3^-$ photodissociation cross section with increasing wavelength in this region was also observed in the photodisappearance ICR studies of Dunbar and Hutchinson.[29] Photofragmentation of Cr(CO)$_3^-$ to Cr$^-$ also occurs at 514.5 nm, as is evident from the weak peak at 1.74 eV, which is due to photodetachment from ground state Cr$^-$.

Table 1 summarizes our measurements of the electron affinities of the neutral molecules, and the fundamental frequencies of the symmetric vibrational modes of the neutral molecules and anions. Based on the results of potential energy distribution calculations[25], v_2 is labelled as the symmetric MCO bend and v_3 as the symmetric M-C stretch, although in each case the other internal coordinate makes a minor contribution to the normal mode.

	Cr(CO)$_3$	Mo(CO)$_3$	W(CO)$_3$
Electron Affinity (eV)	1.349 ± 0.006	1.337 ± 0.006	1.859 ± 0.006
Fundamental Frequencies (cm^{-1}) of A$_1$ modes[a]: Neutral Molecule			
v_1 (CO stretch)	2000 ± 10	2000 ± 10	2000 ± 10
v_2 (MCO bend)	680 ± 10	660 ± 10	620 ± 10
v_3 (MC stretch)	480 ± 20	---	---
v_4 (CMC bend)	---	100 ± 20	100 ± 20
Negative Ion			
v_2 (MCO bend)	680 ± 10	660 ± 10	630 ± 10
v_4 (CMC bend)	---	100 ± 20	100 ± 20

(a) The internal symmetry coordinate that makes the largest contribution to each normal mode, based on potential energy distribution calculations described in Ref. 25, is also indicated.

The difference between the equilibrium geometry of the neutral Cr(CO)$_3$ molecule and that of the Cr(CO)$_3^-$ negative ion can be deduced from a Franck-Condon analysis of relative peak intensities observed in the photoelectron spectrum. Since the peak spacings reveal no detectable anharmonicity at our instrumental resolution, the vibrational modes were modelled as independent harmonic oscillators. Franck-Condon factors were computed using the recursion formula method of Hutchisson[30] as implemented by the least-squares spectral fitting program PESCAL.[31] The resulting normal mode displacements are given in Table 2, below.

	Cr(CO)$_3$	Mo(CO)$_3$	W(CO)$_3$
Principal force constants for the neutral molecules[a]			
$F_{CO\ Stretch}$ (mdyn/Å)	15.1 ± 0.4	15.2 ± 0.4	15.2 ± 0.5
$F_{MC\ Stretch}$ (mdyn/Å)	2.6 ± 0.3	----	----
$F_{MCO\ Bend}$ (mdyn Å/rad^2)	1.10 ± 0.20	1.15 ± 0.25	1.02 ± 0.25
$F_{CMC\ Bend}$ (mdyn Å/rad^2)	----	0.5 ± 0.4	0.6 ± 0.5
Normal mode displacements (amu^{½} Å) between anion and neutral molecule[b]			
$\Delta Q(v_1$, CO stretch)	.070 ± .015	.065 ± .013	.080 ± .015
$\Delta Q(v_3$, MC stretch)	.025 ± .013	<.03	<.03
$\Delta Q(v_2$, MCO bend)	.11 ± .02	.14 ± .03	.20 ± .04
$\Delta Q(v_4$, CMC bend)	<.2	.21 ± .04	.23 ± .05
Internal coordinate displacements between anion and neutral molecule[a]			
Δr_{CO} (Å)	.015 ± .004	.014 ± .004	.017 ± .004
Δr_{MC} (Å)	.01 ± .01	.01 ± .01	.01 ± .01
$\Delta \sphericalangle_{MCO}$ (degrees)	1.7 ± .7	2.1 ± .7	2.8 ± 1.0
$\Delta \sphericalangle_{CMC}$ (degrees)	≤ 1	1 ± 1	1 ± 1

(a) The normal mode and Franck-Condon calculations used to obtain the principal force constants and internal coordinate displacements are described in more detail in ref. 25.
(b) The internal symmetry coordinate that makes the largest contribution to each normal mode is also indicated.

To express these results in terms of changes in bond lengths and bond angles, the **L** matrix, calculated through a standard normal coordinate analysis[32] using an approximate force field[25], was used to convert the normal mode displacements, ΔQ, into internal symmetry coordinate displacements, ΔS:

$$\Delta S = L\ \Delta Q. \qquad (1)$$

Results indicating the absolute magnitude of the change in equilibrium geometry on detachment of an electron from the anion are given above in Table 2 for each of the species studied. Although the harmonic Franck-Condon analysis cannot itself disclose the directions of these geometry changes, some insight into this matter can be obtained from a consideration of the type of orbital associated with the extra electron, as is discussed in Section 4.2.

4. DISCUSSION

4.1. Anion electronic configuration

The very similar appearance of the Cr(CO)$_3$⁻, Mo(CO)$_3$⁻ and W(CO)$_3$⁻ photoelectron spectra, which reflects similar structural displacements on electron detachment, suggests that the "extra" electron occupies the same type of molecular orbital in all three anions. One clue as to whether this orbital can be characterized as primarily a *d* or an *s* metal orbital is provided by a comparison of the electron affinity pattern within this series to that typically observed for triads of transition metal atoms.

For the purpose of this comparison, it is useful to distinguish between an atomic "*s*-electron affinity"[33] for the process M($s^1 d^{n+1}$) + e⁻ → M⁻($s^2 d^{n+1}$) and a "*d*-electron affinity" for M($s^2 d^n$) + e⁻ → M⁻($s^2 d^{n+1}$). The true electron affinity, which always refers to the ground states of the atom and the anion, may correspond to one of these values, depending on the ground state electron configurations. For example, Cr and Mo, which have $s^1 d^5$ atomic and s^2d^5 anionic ground states, the true electron affinity is the s-electron affinity, from which the d-electron affinity can be obtained by adding the atomic $s^1d^5 \rightarrow s^2d^4$ promotion energy.[27]

Accurate electron affinities have been measured spectroscopically for five complete transition metal triads.[26] Although a quick perusal of these values reveals no consistent pattern within each triad, a clear pattern emerges if we distinguish between s- and d-electron affinities. For the s- electron affinities, all five triads exhibit a small (0.04-0.11 eV) increase from the first to the second transition series atom, followed by a large (0.44-1.01 eV) increase for the third transition series atom:

s-Electron Affinities (eV)

I		II		III	
V	0.787	Nb	0.893	Ta	1.532
Cr	0.675	Mo	0.747	W	1.183
Co	1.094	Rh	1.137	Ir	1.916
Ni	1.181	Pd	1.238	Pt	2.128
Cu	1.235	Ag	1.302	Au	2.309

In contrast, for the d-electron affinities, all five triads show a large (0.47-2.43 eV) increase from the first to the second transition series atom, followed by a large (0.71-1.61 eV) <u>decrease</u> from the second to the third transition series atom:

d-Electron Affinities (eV)

I		II		III	
V	0.525	Nb	1.035	Ta	0.322
Cr	1.636	Mo	2.107	W	0.817
Co	0.662	Rh	2.714	Ir	1.565
Ni	1.156	Pd	3.536	Pt	2.230
Cu	2.624	Ag	5.052	Au	3.444

Within each triad, the third transition series atom has, by far, the largest s-electron affinity, whereas the second transition series atom has, by far, the largest d-electron affinity.

These trends can be understood in part from a consideration of relativistic effects,[34] which are greatest for atoms of the third transition series. This phenomenon results in a contraction of the s (and, to a lesser extent, the p) orbitals, which have the greatest electron densities near the nucleus and thus the highest electron velocities. The smaller s and p orbitals more completely shield the d (and f) electrons from the nuclear charge, producing an expansion of those orbitals. Thus, relativistic effects result in more strongly bound s and p electrons, and more weakly bound d and f electrons, in metals of the third transition series.

To determine whether these atomic electron affinity patterns can be extended to molecular transition metal species, it is useful to consider the metal dimers Cu_2, Ag_2 and Au_2.[35] Like the $M(CO)_3^-$ systems studied here, the coinage metal dimers share a common ground state valence electron configuration, and the extra electron occupies the same type of molecular orbital in all three anions. Since this is an sp hybrid orbital,[35] it may be expected that the electron affinities of these molecules will follow the characteristic s-electron affinity pattern, with the third transition series dimer exhibiting a substantially higher electron affinity than the second. An examination of the measured electron affinities[35] of Cu_2 (0.836±0.007 eV), Ag_2 (1.023±0.007 eV) and Au_2 (1.938±0.007 eV) confirms that this is indeed the case.

For the Group VI tricarbonyls, we measure electron affinities of 1.349 eV for $Cr(CO)_3$, 1.337 eV for $Mo(CO)_3$ and 1.859 eV for $W(CO)_3$. Thus, the electron affinities are similar for the first and second transition series complexes, and increase substantially for the third transition series complex. This pattern resembles that observed for the atomic s-electron affinities, as well as for the coinage metal dimers, whose extra electrons occupy sp hybrid orbitals. Based upon these considerations, it appears likely that the extra electron in the Group VI $M(CO)_3^-$ anions occupies a molecular orbital that is primarily a metal s or sp hybrid orbital, rather than a d orbital.

Previously reported semi-empirical calculations indicate that the lowest unfilled orbitals in d^6 metal tricarbonyl fragments are a doubly-degenerate e set derived mainly from the metal d orbitals,[20,21] and an a_1 hybrid orbital which involves a mixture of the valence p, s, and d_{z^2} metal orbitals (in order of decreasing contribution) with some delocalization to the carbonyl groups.[21] Of these two possibilities, our experimental results suggest that the a_1 hybrid orbital provides a better description of the orbital occupied by the extra electron in the anion. Since single occupation of the doubly degenerate e set in a low-spin d^7 tricarbonyl complex is predicted[20] to yield a Jahn-Teller distorted geometry, the view that the extra electron resides in the a_1 orbital is also consistent with the lack of spectral evidence for a Jahn-Teller distortion in the anions. If we assume, based on the computational and indirect experimental evidence summarized in the Introduction, that the neutral Group VI tricarbonyls have low-spin $^1A_1$ ground states, then the addition of an electron to the sp hybrid orbital yields $^2A_1$ ground states for the $M(CO)_3^-$ anions.

4.2. Geometry displacements

The view that the extra electron in the $M(CO)_3^-$ anions occupies the sp hybrid orbital is also consistent with the observed structural changes on electron attachment. Since most of the electron density of this orbital is calculated[21] to point away from the $M(CO)_3$ fragment, the addition of an electron is expected to result in only small displacements between the equilibrium geometries of the anions and the corresponding neutral molecules, in agreement with our experimental results, given in Section 3.

The photoelectron data indicate a slight (0.01-0.02 Å) difference between the equilibrium CO bond lengths of the $M(CO)_3^-$ anions and the corresponding neutral molecules. Although the harmonic Franck-Condon analysis does not indicate the direction of this displacement, several considerations suggest that the CO bonds are longer in the anions. The extended Hückel calculations predict the hybrid a_1 orbital to be somewhat delocalized into the π^* orbitals of the carbonyl groups,[21] suggesting that the addition of an electron to this orbital will slightly weaken the CO bond. For NiCO$^-$, as well, calculations[36] indicate that the extra electron enters a vacant sp hybrid orbital pointed away from the metal carbonyl fragment, resulting in a slightly increased anion CO bond length. Singly-charged metal carbonyl anions studied by matrix IR spectroscopy exhibit CO stretching frequencies 110-160 cm^{-1} lower than those of the corresponding neutral molecules,[37] a result also suggestive of weaker, longer CO bonds in the anions.

The activity of the symmetric MCO bending mode in the photoelectron spectra indicates a small (1-4°) change in the equilibrium MCO bond angle in the mirror plane on electron detachment. This result implies that this angle deviates slightly from linearity in the neutral molecules and/or in the anions. X-ray and neutron diffraction studies of $(C_6H_6)Cr(CO)_3$ also show significant (with respect to the quoted experimental uncertainties) deviations of 1.5-2.0° from linear MCO bonds.[38] Similar results have been obtained for other stable organometallics incorporating pyramidal $M(CO)_3$ groups, and a possible explanation for these results based on σ-π^* electron repulsion has been proposed by Kettle.[19] Bent MCO bonds have also been observed for VCO and $V(CO)_2$ in low-temperature matrix studies,[39,40] although *ab initio* calculations[41] suggest that these systems have linear equilibrium geometries with very flat bending potentials.

4.3. Force constants

The principal CO stretching force constants for the neutral tricarbonyls, obtained from normal mode analyses[25], are given in Table 2 for each of the species studied. These results can be compared with the corresponding values[15] of 17.24±0.07 mdyn/Å for $Cr(CO)_6$, 17.33±0.06 mdyn/Å for $Mo(CO)_6$, and 17.22±0.04 mdyn/Å for $W(CO)_6$. The principal CO stretching force constants are essentially the same for the three metal complexes within each series, but are substantially lower in the tricarbonyl than the hexacarbonyl complexes. Consistent with this result, a monotonic reduction in CO stretching force constants with decreasing coordination number has been reported for a variety of unsaturated metal carbonyls studied by matrix infrared spectroscopy.[42,43] This reduction can be understood in terms of a greater metal electron density available for π backbonding to the antibonding π^* orbitals of each ligand in the coordinatively unsaturated complexes, resulting in weaker C-O bonds.

Force constants for the low-frequency vibrational modes are also consistent with stronger metal-ligand bonding in the coordinatively unsaturated complexes. The MC stretching force constant measured here for neutral $Cr(CO)_3$, the only system for which this vibration was observed, is 2.6±0.3 mdyn/Å, significantly higher than in $Cr(CO)_6$

26. H. Hotop and W. C. Lineberger, "Binding Energies in Atomic Negative Ions," *J. Phys. Chem. Ref. Data*, vol. 14, pp. 731-750, 1985. Electron affinities given in Section 4.1 include more recent measurements for Fe and Co (D. G. Leopold and W. C. Lineberger, "A Study of the Low-Lying Electronic States of Fe_2 and Co_2 by Negative Ion Photoelectron Spectroscopy," *J. Chem. Phys.*, vol. 85, pp. 51-55, July 1986), Cu (Ref. 35), Pd (J. Ho, K. M. Ervin, M. L. Polak, M. K. Gilles, and W. C. Lineberger, "A Study of the Electronic Structures of Pd_2^- and Pd_2 by Photoelectron Spectroscopy," *J. Chem. Phys.*, vol. 95, pp. 4845-4853, October 1991), Cr, Mo, and W (Ref. 25).

27. C. E. Moore, <u>Atomic Energy Levels</u>, Vols. I-III, National Bureau of Standards No. 467, Washington, DC, 1958.

28. P. S. Braterman, <u>Metal Carbonyl Spectra</u>, Academic Press, London, 1975.

29. R. C. Dunbar and B. B. Hutchinson, "Photodecomposition of Gas-Phase Transition Metal Carbonyl Anions," *J. Am. Chem. Soc.*, vol. 96, pp. 3816-3820, June 1974.

30. E. Hutchisson, "Band Spectra Intensities for Symmetrical Diatomic Molecules," *Phys. Rev.*, vol. 36, pp. 410-420, August 1930; "Band Spectra Intensities for Symmetrical Diatomic Molecules, II," vol. 37, pp. 45-50, January 1931.

31. K. M. Ervin and W. C. Lineberger, unpublished FORTRAN program.

32. E. B. Wilson, J. C. Decius, and P. C. Cross, <u>Molecular Vibrations</u>, Dover, New York, 1955.

33. C. S. Feigerle, R. R. Corderman, S. V. Bobashev, and W. C. Lineberger, "Binding Energies and Structure of Transition Metal Negative Ions," *J. Chem. Phys.*, vol. 74, pp. 1580-1598, February 1981.

34. K. S. Pitzer, "Relativistic Effects on Chemical Properties," *Acc. Chem. Res.*, vol. 12, pp. 271-276, August 1979; P. Pyykkö and J.-P. Desclaux, "Relativity and the Periodic System of Elements," *Acc. Chem. Res.*, vol. 12, pp. 276-281, August 1979; P. Pyykkö, "Relativistic Effects on Structural Chemistry," *Chem. Rev.*, vol. 88, pp. 563-594, May 1988.

35. D. G. Leopold, J. Ho, and W. C. Lineberger, "Photoelectron Spectroscopy of Mass-Selected Metal Cluster Anions," *J. Chem. Phys.*, vol. 86, pp. 1715-1726, February 1987; J. Ho, K. M. Ervin, and W. C. Lineberger, "Photoelectron Spectroscopy of Metal Cluster Anions," *J. Chem. Phys.*, vol. 93, pp. 6987-7002, November 1990.

36. M. Blomberg, U. Brandemark, J. Johansson, P. Siegbahn, and J. Wennerberg, "The Vibrational Frequencies, the Dissociation Energy, and the Electron Affinity of Nickel Carbonyl," *J. Chem. Phys.*, vol. 88, pp. 4324-4333, April 1988.

37. E. W. Abel, R. A. McLean, S. P. Tyfield, P. S. Braterman, A. P. Walker, and P. J. Hendra, "Vibrational and Electronic Spectra and Bonding in Ionic Transition Metal Hexacarbonyls," *J. Mol. Spec.*, vol. 30, pp. 29-50, April 1969; P. A. Breeze, J. K. Burdett, and J. J. Turner, "Charged Carbonyls in Matrices," *Inorg. Chem.*, vol. 20, pp. 3369-3378, October 1981.

38. B. Rees and P. Coppens, "Electronic Structure of Benzene Chromium Tricarbonyl by X-ray and Neutron Diffraction at 78 K," *Acta Cryst.*, vol. B29, pp. 2515-2528, November 1973.

39. L. Hanlan, H. Huber, and G. A. Ozin, "Direct Synthesis Using Vanadium Atoms," *Inorg. Chem.*, vol. 15, pp. 2592-2597, November 1976.

40. R. J. Van Zee, S. B. H. Bach, and W. Weltner, "ESR of $V(CO)_n$ (n=1 to 3) Molecules in Rare-Gas Matrices," *J. Phys. Chem.*, vol. 90, pp. 583-588, February 1986.

41. L. A. Barnes and C. W. Bauschlicher, "Theoretical Studies of the Transition Metal-Carbonyl Systems MCO and $M(CO)_2$, M=Ti,Sc, and V," *J. Chem. Phys.*, vol. 91, pp. 314-330, July 1989.

42. R. L. DeKock, "Preparation and Identification of Intermediate Carbonyls of Nickel and Tantalum by Matrix Isolation," *Inorg. Chem.*, vol. 10, pp. 1205-1211, June 1971; E. P. Kündig, D. McIntosh, M. Moskovits, and G. A. Ozin, "Binary Carbonyls of Platinum, $Pt(CO)_n$ (Where n=1-4)," *J. Am. Chem. Soc.*, vol. 95, pp. 7234-7241, October 1973; L. A. Hanlan, H. Huber, E. P. Kündig, B. R. McGarvey, and G. A. Ozin, "Chemical Synthesis Using Metal Atoms," *J. Am. Chem. Soc.*, vol. 97, pp. 7054-7068, November 1975; G. A. Ozin and L. A. Hanlan, "Rhodium and Iridium Atom Chemistry," *Inorg. Chem.*, vol. 18, pp. 2091-2101, August 1979.

43. C. H. F. Peden, S. F. Parker, P. H. Barrett, and R. G. Pearson, "Mössbauer and Infrared Studies of Matrix-Isolated Iron-Carbonyl Complexes," *J. Phys. Chem.*, vol. 87, pp. 2329-2336, June 1983.

44. A. M. English, K. R. Plowman, and I. S. Butler, "Vibrational Spectra and Potential Constants of the (η^6-Benzene)chromium(0) Chalcocarbonyl Complexes (η^6-C_6H_6)$Cr(CO)_2(CX)$ (X=O,S,Se)," *Inorg. Chem.*, vol. 21, pp. 338-347, January 1982.

45. S. Pignataro, A. Foffani, F. Grasso, and B. Cantone, "Negative Ions from Metal Carbonyls by Electron Impact," *Z. Phys. Chem. (Munich)*, vol. 47, pp. 106-113, October 1965.

Resonant MPI spectrum of allyl radicals

David W. Minsek, Joel A. Blush, and Peter Chen*[1]

Mallinckrodt Chemical Laboratory
Harvard University, Cambridge MA 02138

ABSTRACT

A mass-selected, partially rotationally-resolved, resonant multiphoton ionization spectrum of the allyl radical, C_3H_5, is reported. Photoelectron spectroscopy, isotopic labelling, and rotational analysis establish that the band system corresponds to the $2^2B_1 \leftarrow 1^2A_2$ transition, with an origin band at 248.15 nm. Spectral simulation indicates that the equilibrium CCC bond angle of the radical decreases from 124.6° in the ground state to 117.5° in the excited state.

1. INTRODUCTION

We wish to report a partially rotationally-resolved, mass-selected 1+1 resonant multiphoton ionization spectrum of the allyl radical, C_3H_5. This is the first electronic spectrum for this simplest of all conjugated π-radicals to show any rotational structure. We assign the upper state of the transition to the 2^2B_1 state with origin at 40306 cm^{-1}, which, while previously observed[2] in absorption at low resolution, had neither been assigned nor analyzed. A preliminary rotational analysis yields a decrease in the CCC bond angle from 124.6°, reported for the ground state[3], to approximately 117.5° in the excited state. Comparison of photoelectron spectra taken with 10.49 eV one-photon ionization, and 1+1 MPI via the 2^2B_1 state indicate that, while the CCC bond angle of allyl radical decreases from 124.6° to 117.5° upon excitation, there is little or no change in bond angle going from the 2^2B_1 excited state of allyl radical to allyl cation. This structural similarity constitutes the only experimental information on the geometry of the allyl cation.

2. EXPERIMENTAL

Allyl radicals were produced in the nozzle of a supersonic jet expansion by the pyrolysis of allyl iodide. Both the molecular beam time-of-flight mass spectrometer[4,5,6,7], and the magnetic-focusing time-of-flight photoelectron spectrometer[5,6,7] have been previously described. The nozzle consisted of an electrically-heated 1.0 mm ID silicon carbide tube with a heated zone of 15.0 mm extending to the sonic orifice, as detailed elsewhere[7]. Allyl iodide (1 torr partial pressure) was seeded into 2 atm helium and expanded via a pulsed valve (General Valves Series 9) at 20 Hz through the hot nozzle into the source region of the mass spectrometer. The jet was skimmed, photoionized, and mass-analyzed by time-of-flight. One-photon vacuum-UV photoionization (118.2 nm, 10.49 eV) of the pyrolysate showed exclusively two peaks: m/e=41 for C_3H_5, and m/e=127 for I•.

An injection-seeded Nd^{3+}-YAG laser (Spectra-Physics GCR-3, Model 6300 seeder) was used to pump a dye laser (Spectra-Physics PDL-3, DCM dye) at 20 Hz. The dye laser output was frequency-doubled and mixed with the YAG fundamental (Spectra-Physics WEX-1) to produce tunable far-UV in the 238-250 nm wavelength region with a pulse energy of approximately 1 mJ. The far-UV laser radiation was isolated from the other frequencies using an Inrad Four Prism Filter. The far-UV spectral linewidth is estimated to be 0.1 cm^{-1}.

The far-UV radiation was focused into the ionization region of the spectrometer using a 350 mm f.l. fused silica spherical lens. Spectra were obtained by scanning the dye laser and monitoring the integrated area of the m/e=41 signal as a function of wavelength. The light was slightly defocused in order to avoid fragmentation of the radicals or radical cations; under tightly-focused conditions, C_xH_y (x=1-3) fragments with m/e<41 were observed.

Allyl iodide-d_5 was prepared from allyl alcohol-d_6 (MSD Isotopes) by the method of Landauer and Rydon[8]. The 118.2 nm photoionization mass spectrum of pyrolyzed allyl iodide-d_5 showed only two peaks: m/e=46 for C_3D_5, and m/e=127 for I•.

3. RESULTS AND DISCUSSION

Figure 1 shows the 1+1 resonant MPI spectrum of allyl radical in the 238-250 nm region. The origin band of the transition, with a maximum at 248.15 nm, is detailed in Figure 2. To the blue of 238 nm, we observe additional bands that are progressively weaker and broader. We assign the 248.15 nm band to the transition from the vibrationless ground state to a vibrationless excited state based on three lines of argument: (*i*) *No large bands are observed to the red* of the 248.15 nm band. We expect cooling of the radical by supersonic expansion subsequent to pyrolysis to have suppressed hot bands in the spectrum. (*ii*) We find that the 248.15 nm band maximum exhibits only a *small spectral shift* of 14 cm^{-1} to the red upon complete deuteration of the radical. Other bands in Figure 1 shift to the red by much larger amounts. (*iii*) A time-of-flight photoelectron spectrum of allyl radical obtained in our lab by 1+1 resonant MPI through the 248.15 nm band shows only a *single, sharp peak* corresponding to the adiabatic ionization potential of the radical. We note that the one-photon vacuum-UV photoelectron spectrum of allyl radical reported by Houle *et al*[9], (HeI) or recorded in our laboratory (118.2 nm) show vertical and adiabatic ionization potentials differing by 50 meV, indicative of a geometry change upon ionization. Our current 1+1 MPI photoelectron results, and the relative intensities among the bands in Figure 1, suggest that the equilibrium geometry of the excited state resembles that of allyl cation.

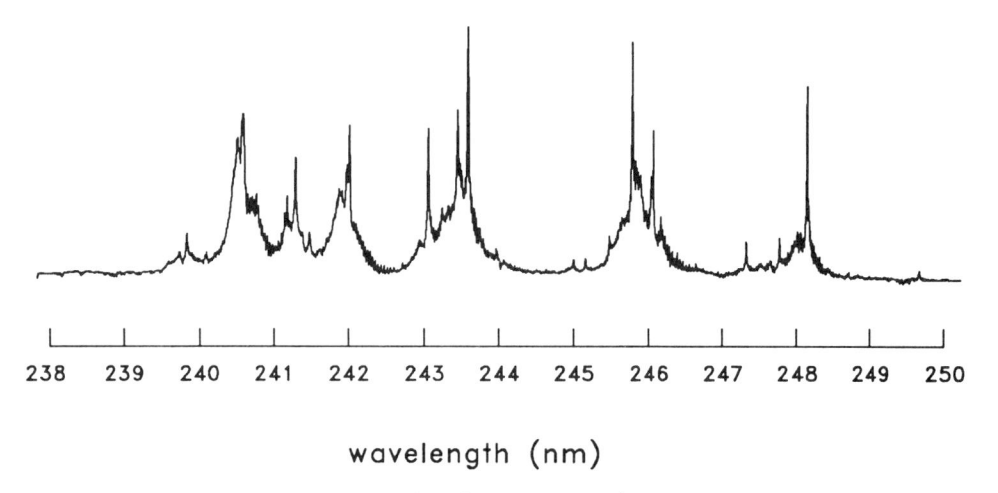

Figure 1. Survey scan of the 238-250 nm region by mass-resolved, 1+1 resonant multiphoton ionization for m/e=41. Allyl radicals, C_3H_5, are produced by supersonic jet flash pyrolysis of allyl iodide.

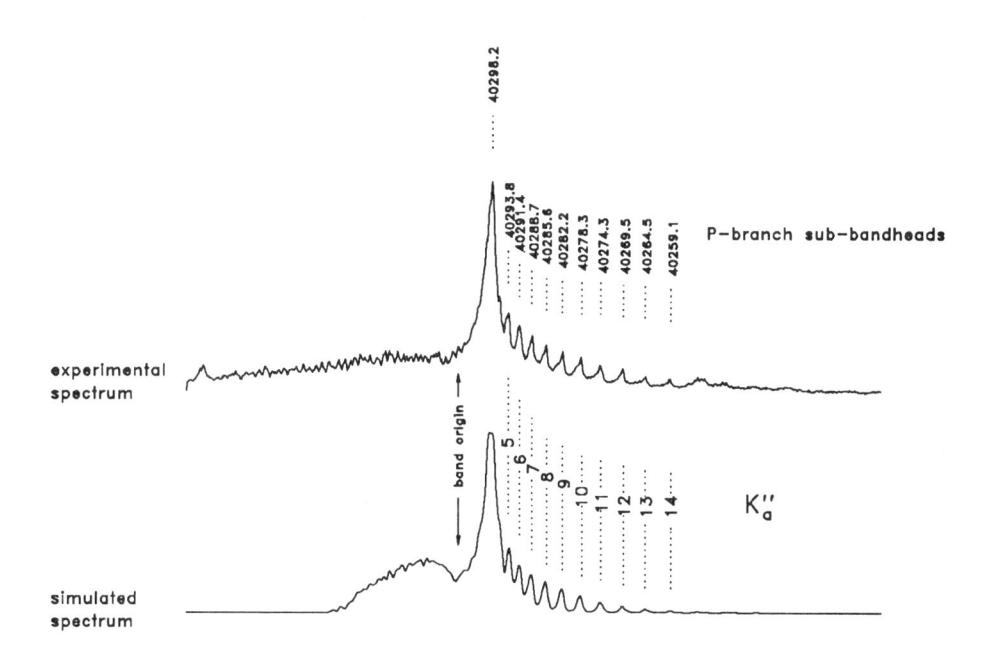

Figure 2. The 248.15 nm origin band of the $2^2B_1 \leftarrow 1^2A_2$ transition, scanned at a laser bandwidth of 0.1 cm^{-1} from 40363.7 to 40214.9 cm^{-1}. Each "line" is an unresolved P-branch sub-bandhead. The R-branches, to the blue of the origin, while less intense than in the simulation, are nevertheless present in the spectrum, and can be seen readily in the more compressed spectrum of Fig. 1.

Having established that the 248.15 nm band corresponds to the origin for the observed transition, we assigned the upper-state electronic symmetry by simulation[10] of the partially resolved structure in that band, in conjunction with simple molecular orbital arguments. One-electron excitations from the principal configuration of the ground state, $(core)(6a_1)^2(4b_2)^2(1b_1)^2(1a_2)^1$ [1^2A_2], lead to low-lying excited states with principal electronic configurations and symmetries of $(core)(6a_1)^2(4b_2)^2(1b_1)^1(1a_2)^2$ [1^2B_1], $(core)(6a_1)^2(4b_2)^2(1b_1)^2(2b_1)^1$ [2^2B_1], $(core)(6a_1)^2(4b_2)^2(5b_2)^1(1b_1)^2$ [1^2B_2], and $(core)(6a_1)^2(7a_1)^1(4b_2)^2(1b_1)^2$ [1^2A_1]. There should also be a second $^2A_1$ state, the *3s* Rydberg state, in the same energy region, which would mix strongly with the 1^2A_1 state. *Ab initio* calculations by Ha *et al*[11] at the CI level give <u>vertical</u> energies for those excited states below 6 eV (207 nm) of 3.13 eV [1^2B_1], 5.26 eV [1^2B_2], 5.33 eV [1^2A_1], 5.52 eV [2^2B_1], and 5.86 eV [2^2A_1]. The 1^2B_1 state has been observed by Currie and Ramsay[12] in a cryogenic matrix at 404 nm (3.07 eV) as a broad absorption. The 1^2A_1 state is forbidden in one-photon absorption from the ground state, but has been seen as a two-photon resonance by Hudgens *et al*[13]. Sappey and Weisshaar[14] recorded the same spectrum with much higher resolution, identifying several vibronic bands and setting the origin of the 1^2A_1 state at 4.97 eV, only 249 cm^{-1} to the red of our 248.15 nm band origin. The upper state of the 248.15 nm (5.00 eV) band in the present work can therefore be plausibly assigned to either the 1^2B_2 or 2^2B_1 excited states on energetic grounds. Inspection of a C_{2v} character table indicates that a $B_2 \leftarrow A_2$ transition would have its transition moment aligned along the inertial axis perpendicular to the molecular plane, giving a Type C band profile. Similarly, a $B_1 \leftarrow A_2$ transition would have its transition moment aligned along the long axis of the molecule, giving a Type A band profile. Allyl radical is a near-prolate asymmetric top ($\kappa = -0.9$) for which the Type C and Type A bands should resemble perpendicular and parallel bands, respectively, in the prolate symmetric top limit. Simulation as a Type A band is most consistent with the observed structure of the 248.15 nm band. We accordingly assign that band to the origin of the $2^2B_1 \leftarrow 1^2A_2$ transition.

The simulated band is shown below the observed spectrum in Figure 2. While the spectrum is insufficiently resolved for a detailed fit of the band contour to a full set of rotational constants, a reasonable model can be used for the simulation. Because the gross rotational structure in the observed spectrum is most sensitive to changes in the CCC bending angle, we take, as a model, an excited state radical for which all bond lengths and angles, except the CCC bending angle, are fixed at those in the experimental geometry of the ground state allyl radical[3]. An upper state angle of 117.5°, rotational temperature of 150 K, and linewidth of 1.0 cm^{-1} yields a close match with the rotational structure in Figure 2. The rotational constants used for the simulation are A"=1.803, B"=0.328, C"=0.278, A'=1.619, B'=0.351, and C'=0.288 cm^{-1}. Because allyl is so close to the prolate limit, only $\Delta K_a = 0$ sub-bands were included in the simulation. We note that a CCC bond angle of 117.5° is quite close to the 118° predicted for allyl cation by *ab initio* calculations[15], and is therefore consistent with our photoelectron results above. The large decrease in A displaces each successive sub-band origin to the red. In addition, the increase in both B and C causes each sub-band to shade to the blue. Each of the features to the red of the band maximum is the unresolved P-branch head for a sub-band with a single value for K_a. The band maximum itself consists of several unresolved low K_a P-branch sub-bandheads. Because of the large change in A, the sub-band Q-branches do not overlap to produce a central maximum as is typical for asymmetric rotor Type A bands. The R-branches produce a broad feature to the blue of the band maximum which, while much less prominent in the spectrum than would be expected from the simulation, is nevertheless present in all scans over the 248.15 nm band. We can speculate that the relative weakness of the R-branches with respect to the P-branches may be due to an electronic transition moment that is not constant over the band. Such behavior would not be surprising for a state that is heavily mixed with other nearby excited states.

The vibrational structure of the $2^2B_1 \leftarrow 1^2A_2$ transition evident in Figure 1 affords no simple interpretation. The gross spacings of ≈ 390 cm^{-1} can be attributed to excitation of the totally symmetric ν_7 CCC bending vibration in the excited state[16]. A long progression in that mode is consistent with the decrease in equilibrium CCC bond angle, upon electronic excitation, that comes out of the simulation of the rotational structure of the 248.15 nm band. At a more detailed level of analysis, the richness of the spectrum in Figure 1 is puzzling, especially because the absence of hot bands suggests that sequence bands should also be weak. We are currently considering two possibilities: (*i*) nonplanarity in the excited state would introduce more allowed bands, and further complicate the spectrum with a double-well potential problem, or (*ii*) vibronic coupling may allow levels of the 1^2A_1 state, the one-photon forbidden *3s* Rydberg state whose origin is only 249 cm^{-1} to the red of the $2^2B_1 \leftarrow 1^2A_2$ origin, to borrow intensity from allowed transitions to levels in 2^2B_1 and appear as extra bands. The coupling could be promoted by one quantum of the ν_{12} CH$_2$ symmetric twisting mode of b$_1$ symmetry.

Callear and Lee[2] reported a strongly broadened series of absorptions centered at 224.9 nm which they assigned to the $2^2B_1 \leftarrow 1^2A_2$ transition. None of the bands were rotationally resolved, and no vibronic assignments of individual bands were given. Interestingly, most of the bands in Figure 1 were reported in that work as "weak bands, almost line-like features," but not as part of the "main system." No regularities were reported in the spacings. We suggest a tentative interpretation for the far-UV spectrum of allyl radical. The $2^2B_1 \leftarrow 1^2A_2$ transition, whose origin band lies at 248.15 nm, shows an extended Franck-Condon envelope due, at least in part, to a substantial change in the equilibrium CCC bond angle upon excitation. Approximately 3500 cm^{-1} above the origin, a fast radiationless process markedly shortens the excited state lifetime, broadening the spectrum, and introducing irregularities to the band spacings. This is consistent with our observation that the bands to the blue of 238 nm are progressively weaker and broader in 1+1 resonant MPI despite their increasing intensity in the absorption measurements by Callear and Lee. While the spectroscopic studies presently favor no one radiationless process over another, *ab initio* calculations[17] have suggested that a disrotatory closure of allyl radical to cyclopropyl radical is favorable for the 2^2B_1 state. If so, time-delayed pump-probe 1+1 resonant MPI spectroscopy of allyl radical may prove to be an interesting kinetic probe of excited state dynamics.

4. ACKNOWLEDGMENTS

The authors acknowledge helpful discussions with Professors P.B. Kelly (UC Davis) and J.C. Weisshaar (U. of Wisconsin, Madison). Support from the Department of Energy and the Exxon Educational Foundation is acknowledged. Funding from the National Science Foundation for the lasers used in this work is also acknowledged.

5. REFERENCES

[1] NSF Presidential Young Investigator, David and Lucile Packard Fellow, Camille and Henry Dreyfus Teacher-Scholar, Alfred P. Sloan Research Fellow.

[2] Callear, A.B.; Lee, H.K. *Trans. Faraday Soc.* **1968**, *64*, 308.

[3] Vajda, E.; Tremmel, J.; Rozsondai, B.; Hargittai, I.; Maltsev, A.K.; Kagramanov, N.D.; Nefedov, O.M. *J. Am. Chem. Soc.* **1986**, *108*, 4352, report CC=1.428 Å, CH=1.069 Å, CCC=124.6°, and CCH=120.9° for the ground state of allyl radical by gas-phase electron diffraction.

[4] Blush, J.A.; Park, J.; Chen, P. *J. Am. Chem. Soc.* **1989**, *111*, 8951.

[5] Minsek, D.W.; Chen, P. *J. Phys. Chem.* **1990**, *94*, 8399.

[6] Clauberg, H.; Chen, P. *J. Am. Chem. Soc.* **1991**, *113*, 1445.

[7] Clauberg, H.; Minsek, D.W.; Chen, P. *J. Am. Chem. Soc.*, in press.

[8] Landauer, S.R.; Rydon, H.N. *J. Chem. Soc.* **1953**, 2224.

[9] Houle, F.A.; Beauchamp, J.L. *J. Am. Chem. Soc.* **1978**, *100*, 3290.

[10] Spectral band simulations were performed with the ASYROT-PC program: Judge, R.H. *Comp. Phys. Comm.* **1987**, *47*, 361.

[11] Ha, T.K.; Baumann, H.; Oth, J.F.M. *J. Chem. Phys.* **1986**, *85*, 1438.

[12] Currie, C.L.; Ramsay, D.A. *J. Chem. Phys.* **1966**, *45*, 488.

[13] Hudgens, J.W.; Dulcey, C.S. *J. Phys. Chem.* **1985**, *89*, 1505.

[14] Sappey, A.D.; Weisshaar, J.C. *J. Phys. Chem.* **1987**, *91*, 3731.

[15] Wiberg, K.B.; private communication. Calculations were at the HF/6-31G* level.

[16] This value for ν_7 for the 2^2B_1 excited state would be similar to that reported for the same vibration in the 1^2A_1 excited state (ref. 14) and the cation (ref. 9).

[17] Merlet, P.; Peyerimhoff, S.D.; Buenker, R.J.; Shih, S. *J. Am. Chem. Soc.* **1974**, *96*, 959; Farnell, L.; Richards, W.G. *J. Chem. Soc. Chem. Comm.* **1973**, 334.

High resolution inner valence UV photoelectron spectra of the O₂ molecule and CI calculations of $^2\Pi_u$ states between 20 and 26 eV.

P. Baltzer, B. Wannberg, L. Karlsson and M. Carlsson-Göthe

Department of Physics, Uppsala University
Box 530, S-751 21 Uppsala, Sweden

M. Larsson

Department of Physics I, Royal Institute of Technology
S-100 44 Stockholm, Sweden

ABSTRACT

High resolution HeI and HeII excited inner valence photoelectron spectra of the oxygen molecule have been recorded between 20 and 26 eV. In this range three photoelectron bands are clearly seen; they are associated with the B $^2\Sigma_g^-$, 3 $^2\Pi_u$ and c $^4\Sigma_u^-$ states of O_2^+. The state of $^2\Pi_u$ symmetry observed around 24 eV shows a long vibrational progression, contrary to earlier work, with spacings that decrease successively towards higher electron binding energies. The assignment is confirmed by *ab initio* calculations. These calculations show that if the potential curve is followed along the electron configuration rather than the adiabatic curve, the vibrational structure can be accounted for.

1. INTRODUCTION

The inner valence region in the photoelectron spectrum of the oxygen molecule has been the subject of several earlier studies.[1-4] In the range between 20 and 30 eV four comparatively strong photoelectron bands have been observed. The first band, associated with transitions to the B $^2\Sigma_g^-$ state, shows an extensive vibrational progression in the range above 20 eV. The second band centered at 24 eV is very broad, of the order of 3 eV, and has in previous studies been considered to be essentially structureless. It corresponds to a state of $^2\Pi_u$ symmetry associated with the third state arising from the $1\pi_u^3 1\pi_g^2$ final state configuration. We denote this state 3 $^2\Pi_u$ in this report and in a forthcoming publication.[5] The third band, which is found in the same energy region, is narrow, consisting essentially of two intense vibrational lines corresponding to the v=0 and v=1 components of the final state and an additional broad structure reflecting a higher vibrational state. A fourth band has been observed at about 27.3 eV.

In the present investigation we have obtained new HeI and HeII excited photoelectron spectra of bands in the energy range between 20 and 30 eV. The study has been carried out using an improved UV-source with very good characteristics, which has allowed recordings with both high resolution and intensity. Some of the recordings were carried out with monochromatized HeIIα radiation using a new monochromator[6] for the resonance radiation from the UV source.

Potential curves of O_2^+ have been calculated earlier.[7-9] These calculations completely corroborate the notion of a structureless band at 24 eV. In one paper[8] it is even explicitly statet that: "If the transition occurs vertically, the energy of $^2\Pi_u$ III (*i.e. 3 $^2\Pi_u$ in our notation*) at the R_e value of O_2 (R~1.2 Å) is much higher than the height of the hump (R≈1.9 Å) and it is natural that no vibrational structure is observed." According to these calculations the 3 $^2\Pi_u$ dissociates to the $O(^3P)+O^+(^2D)$ limit at 22.06 eV above the ground state of O_2. However, in the present study carried out at higher resolution than previously, an extensive progression is observed which converges towards a higher dissociation limit than 22.06 eV (in fact it converges to the $O(^3P)+O^+(^2P)$ limit at 23.75 eV). In the present work we have also been able to observe the second $^2\Pi_u$ state for the first time. In order to aid in the interpretation of these new results *ab initio* calculations of potential curves of O_2^+ have been carried out.

2. EXPERIMENTAL AND COMPUTATIONAL DETAILS

The measurements were performed by means of a UV-photoelectron spectrometer[10] based on an electrostatic hemispherical analyzer with a mean radius of 144 mm and a microchannel plate detector system. An electron lens focuses the photoelectrons onto the entrance slit of the analyzer. Target gas pressures of the order of 10 mTorr were used in the present investigation. By employing a newly developed pair of correction electrodes in the gas cell, spectra obtained in the present study are essentially background free. The helium radiation employed for the photoionization was produced in a VUV-source based on a microwave driven discharge. The disharge takes place at a pressure of about 50 mTorr in a very small volume between the pole pieces of a strong magnet which provides a field with the shape of a magnetic bottle fulfilling the ECR condition in the centre.

Potential curves for the 2 and 3 $^2\Pi_u$ states were calculated by means of the complete-active-space self-consistent-field (CAS SCF) and multireference CI methods.[11,12] The basis set was a contracted [*6s 4p 2d 1f*] Gaussian set. The active space during the CAS SCF calculations were the $2\sigma_g$, $2\sigma_u$, $3\sigma_g$, $3\sigma_u$, $1\pi_u$, $1\pi_g$ orbitals. The multireference CI calculation included all single and double excitations with respect to the CAS SCF reference function.

3. RESULTS AND DISCUSSION

The neutral (O_2) ground state electron configuration is

$$KK\ (2\sigma_g)^2(2\sigma_u)^2(3\sigma_g)^2(1\pi_u)^4(1\pi_g)^2$$

Figure 1 shows a HeII overview spectrum. The outer valence spectrum in the energy range below 20 eV includes transitions to the X $^2\Pi_g$, a $^4\Pi_u$, A $^2\Pi_u$ and b $^4\Sigma_g^-$ states. This part of the spectrum will not be discussed further. Figure 2 shows a detail of a HeI excited spectrum between 20.2 and 21.1 eV. The strong lines are due to the B $^2\Sigma_g^-$ state. The positions of the vibrational bands were used to derive vibrational constants: ω_e=1137.2 cm^{-1} and $\omega_e x_e$=19.4 cm^{-1}. In addition to the strong lines associated with the B $^2\Sigma_g^-$ state, a number of weaker lines can be clearly seen (these lines are also present in the HeII spectrum). The *ab initio* calculations suggest that these lines could be due to the 2 $^2\Pi_u$ state. The calculated potential curves are shown in Figure 3. The lines are indeed observed in the correct energy region. Also the agreement between

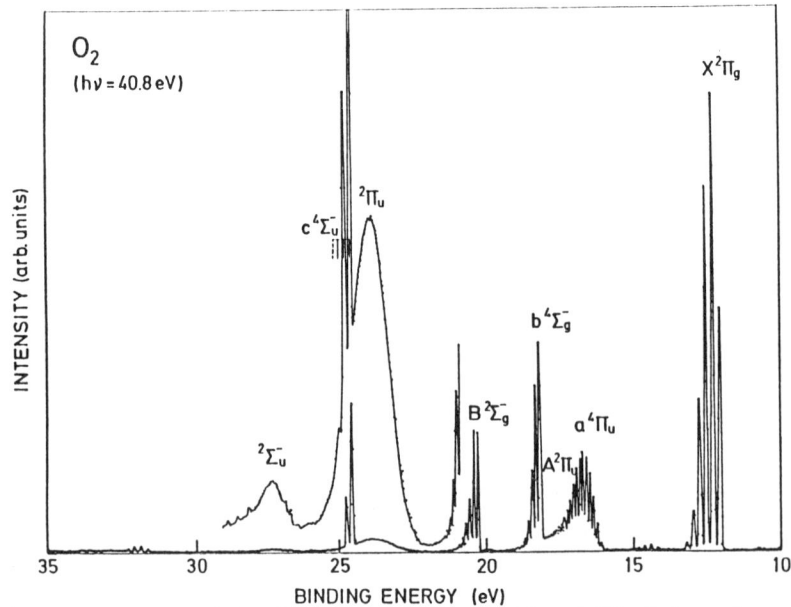

Fig. 1. A photoelectron spectrum of the O_2 molecule between 10 and 35 eV by using monochromatized HeIIα radiation at 40.8 eV. The reolution of the photoelectron lines are approximately 60 meV.

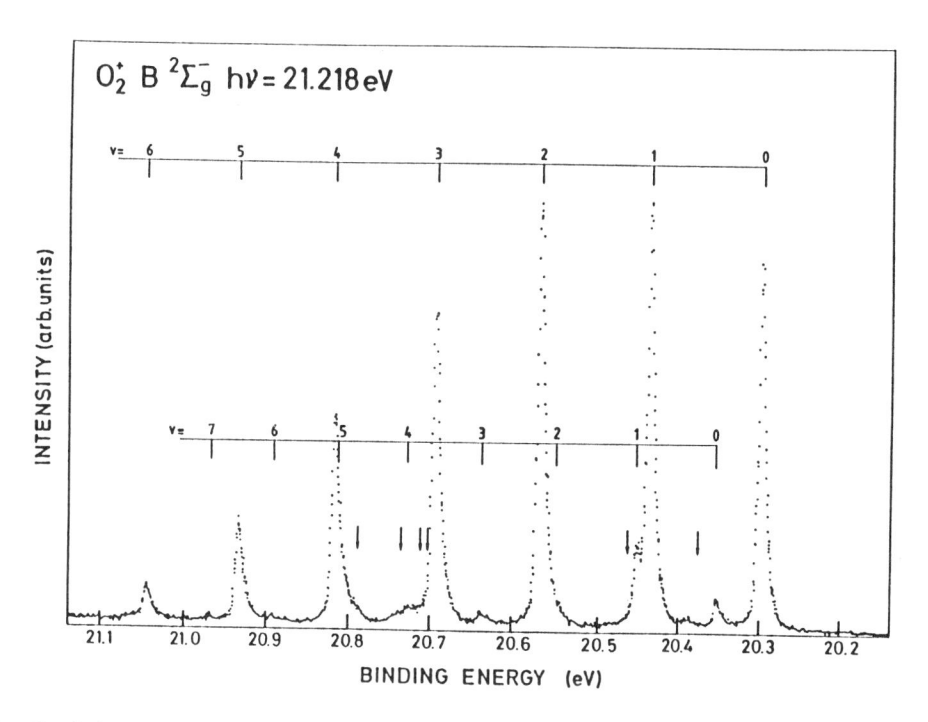

Fig. 2. A detail of the HeI excited photoelectron spectrum of the O_2 molecule between 20.2 and 21.1 eV. The channel width is 0.5 meV.

observed and calculated vibrational levels are good, however, only by assuming that the observed lines correspond to rather high vibrational levels. Thus, such an assignment would imply that a number of lines corresponding to low vibrational levels would be unobservable in the experimental spectrum; this seems unreasonable. The interpretation of this progression will therefore have to await further experimental work (which is in progress).

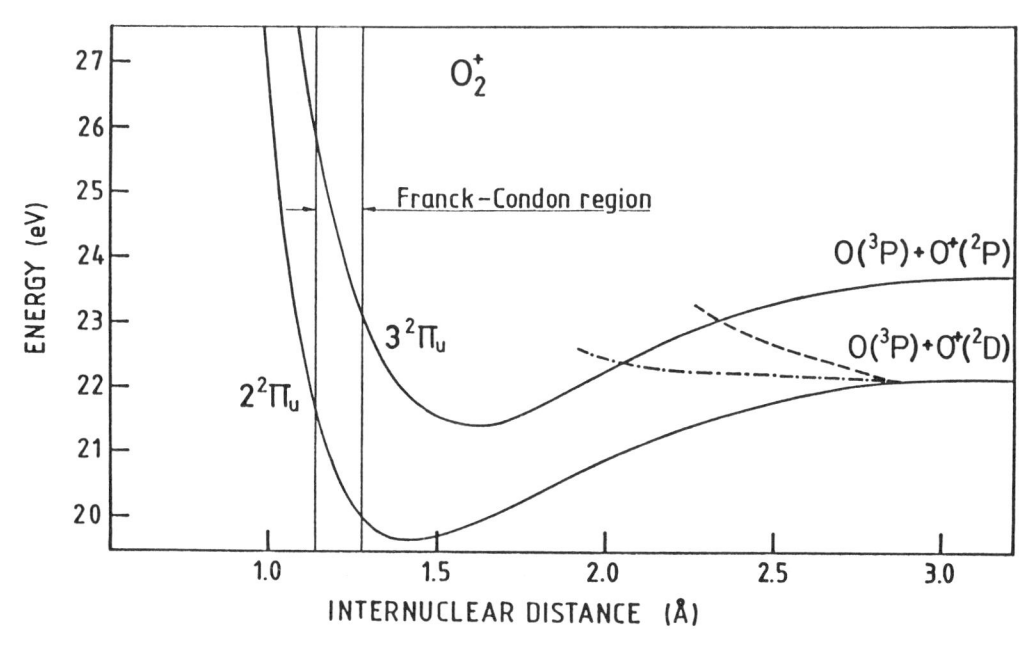

Fig. 3. Calculated potential curves for the $2\,^2\Pi_u$ and $3\,^2\Pi_u$ states (full lines) arising from the $(1\pi_u)^3(1\pi_g)^2$ configuration. Dotted lines: approximate potential curves for two other $^2\Pi_u$ states.

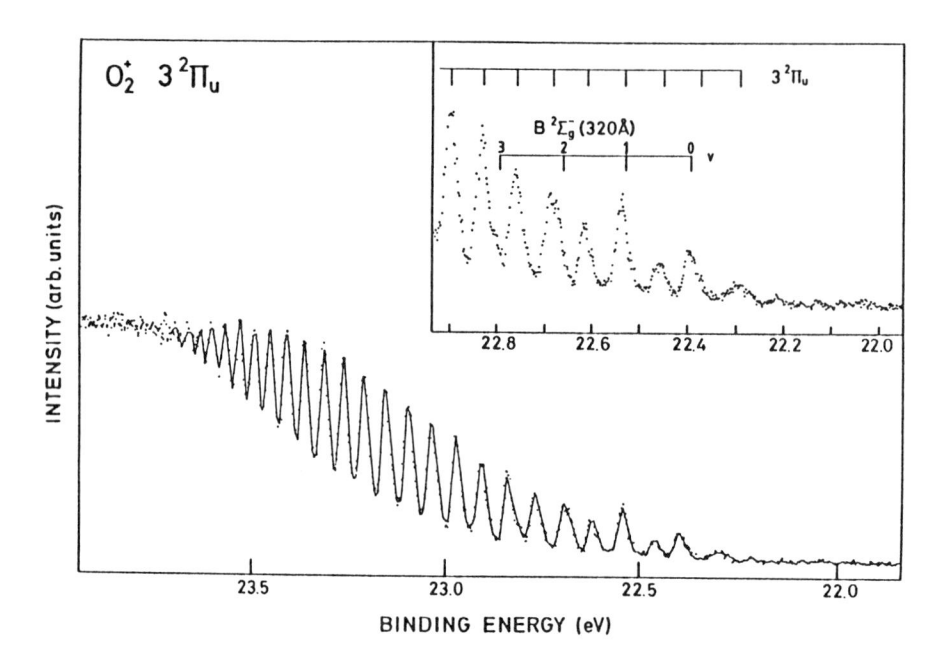

Fig. 4. A detail of the HeII excited photoelectron spectrum of the O_2 molecule between 22 and 24 eV showing the lines associated with the 3 $^2\Pi_u$ state. On the low binding energy side weak lines are seen which correspond to transitions to the B $^2\Sigma_g^-$ state.

Figure 4 shows a high resolution recording of the 22-24 eV region. The band at 24 eV, which was structureless in the earlier work[1-3] and also in the HeII spectrum (Fig.1), now have a clear vibrational structure. This structure is associated with the 3 $^2\Pi_u$ state derived from the $(1\pi_u)^3(1\pi_g)^2$ electron configuration. The spacings are in the order of 80 meV in the beginning of the band system, and decrease gradually towards the end of the progression. The convergence limit is approximately 23.8 eV. The unexpected structure in the 24 eV band can be accounted for if the 3 $^2\Pi_u$ state is allowed to dissociate to the $O(^3P)+O^+(^2P)$ dissociation limit. At the equilibrium internuclear distance the 3 $^2\Pi_u$ state is well described by the $(1\pi_u)^3(1\pi_g)^2$ configuration. If the strictly adiabatic potential is followed the 3 $^2\Pi_u$ would change character at 2.06 Å to dominating configuration $(3\sigma_g)(1\pi_u)^4(1\pi_g)(3\sigma_u)$. This configuration differs by two orbital from the $(1\pi_u)^3(1\pi_g)^2$ configuration; thus the interaction between the adiabatic potential curves in the (avoided) crossing region is very small and the diabatic picture, with a crossing, is a better representation of the situation. The curve shown in Fig. 3 for the 3 $^2\Pi_u$ has thus been obtained by following the $(1\pi_u)^3(1\pi_g)^2$ configuration, and also the second crossing has been allowed for. It is clear from a comparison between calculated and observed energy levels that there seems to be no influence from the two crossing $^2\Pi_u$ states.

4. REFERENCES

1. O. Edqvist, E. Lindholm, L.E. Selin, and L. Åsbrink, "On the photoelectron spectrum of O_2", Phys. Scripta 1, 25-30 (1970).
2. J.L. Gardner and J.A.R. Samson, "Photoion and photoelectron spectroscopy of oxygen", J. Chem. Phys. 62, 4460-4463 (1975).
3. N. Jonathan, A. Morris, M. Okuda, K.J. Ross, and D.J. Smith, J. Chem. Soc. Faraday Trans. 70, 1810 (1974).
4. H. van Lonkhuyzen and C.A. de Lange, "Modulation techniques in UV photoelectron spectroscopy of transient species; the $O_2^+(^2\Pi_u)\leftarrow O_2(^1\Delta_g)$ transition", Electron. Spectrosc. 27, 255-260 (1982).
5. P. Baltzer, B. Wannberg, L. Karlsson, M. Carlsson Göthe, and M. Larsson, "High resolution inner valence UV photoelectron spectra of the O_2 molecule and CI calculations of the $^2\Pi_u$ states between 20 and 26 eV", Phys. Rev.

A (in press).

6. P. Baltzer, M. Carlsson Göthe, B. Wannberg, and L. Karlsson, "An easy to build, high intensity monochromator for helium II radiation applied to inner valence photoelectron studies of small molecules" Rev. Sci. Instrum. **62**, 630-638 (1991).

7. N.H.F. Beebe, E.W. Thulstrup, and A. Andersen, "Configuration interaction calculations of low-lying electronic states of O_2, O_2^+, and O_2^{2+}", J. Chem. Phys. **64**, 2080-2093 (1976).

8. N. Honju, K. Tanaka, K. Ohno, and H. Taketa, "Configuration interaction calculation of the photoelectron spectra of O_2", Mol. Phys. **35**, 1569-1578 (1978).

9. C.M. Marian, R. Marian, S.D. Peyerimhoff, B.A. Hess, R.J. Buenker, and G. Seger, "*Ab initio* CI calculations of O_2^+ predissociation phenomena induced by a spin-orbit coupling mechanism", Mol. Phys. **46**, 779 (1982).

10. P. Baltzer, B. Wannberg, and M. Carlsson Göthe, "Optimization and redesign of an electron spectrometer for high-resolution gas phase UV photoelectron, Auger electron, and ion fragment spectroscopy", Rev. Sci. Instrum. **62**, 643-654 (1991).

11. P.E.M. Siegbahn, J. Almlöf, A. Heiberg, and B.O. Roos, "The complete active space SCF (CASSCF) method in a Newton-Raphson formulation with application to the HNO molecule", J. Chem. Phys. **74**, 2384-2396 (1981).

12. P.E.M. Siegbahn, Int. J. Quant. Chem. **23**, 1869 (1983).

CO(v,J) Product State Distributions from the Reaction
$O(^3P) + OCS \rightarrow CO + SO$

Scott L. Nickolaisen[a], Harry E. Cartland[b], David Veney[b], and Curt Wittig

Department of Chemistry
University of Southern California
Los Angeles, CA 90089-0482
(213)740-7760

(a) current address: Jet Propulsion Laboratory, MS 183-901, 4800 Oak Grove Drive, Pasadena, CA 91109

(b) Science Research Laboratory, U.S. Military Academy, West Point, NY 10996

ABSTRACT

The title reaction was studied by probing the CO(v,J) product state distributions. Oxygen atoms were formed by 355 nm photolysis of NO_2. Photolysis produces approximately equal populations of NO(v=0) and NO(v=1). The collision energy of oxygen atoms corresponding to NO(v=0) is 1570 cm^{-1}. This is above the O + OCS activation barrier of 1540 cm^{-1}. Oxygen atoms corresponding to NO(v=1) do not have sufficient energy to proceed over the activation barrier, thus insuring monoenergetic collisions. CO product was probed using an IR tunable diode laser. Nascent CO distributions were extracted from the transient absorption signals using an initial slope approximation. A vibrational branching ratio of [v=1]/[v=0] ≤ 0.05 was measured. CO(v≥2) was not detected. The CO(v=0) rotational Boltzmann plot was bimodal. The distribution for 0≤J≤15 had a temperature of 350 ± 20 K. For J≥15, the plot had a temperature of 4400 ± 390 K. The low J population is the result of rotational relaxation of the nascent CO distribution. The high J signals are direct measure of the nascent CO population. Surprisal analysis resulted in a parameter of $\Theta_R = 3.7 \pm 0.5$. Hence, the CO(v=0) distribution is colder than a "prior" statistical model.

1. INTRODUCTION

The reaction of atomic oxygen and carbonyl sulfide has been examined to a degree in the past. Interest in the title reaction stems from its involvement in both atmospheric and combustion chemistry of sulfur containing systems.[1] The kinetics of the reaction have been previously studied yielding an expression for the rate constant of k = 2.35 x 10^{-11} exp{-1540 cm^{-1}/kT} cm^3 molecule^{-1} s^{-1}.[2] However, the details of the potential energy surface (PES) for this system have not been determined. For example, it is unknown whether the activation barrier of 1540 cm^{-1} represents a saddle point on the potential energy surface, or if a minimum exists along the reactive pathway which corresponds to an activated intermediate.

Our examination of the O + OCS reaction is an extension of a set of experiments in which several analogous atom-molecule reactions have been studied. Other systems studied were the H + CO_2 and H + OCS reactions.[3,4] The ultimate goal of these studies is to understand the factors which determine the partitioning of energy in the reaction products. These factors include the energy of the reactive collision, the shape of the PES along the reactive pathway, and the endothermicity or exothermicity of the reaction. The product state distributions determined in these reactions can yield information about the location of the transition state, the extent of energy randomization within the products, and the possible role of reactive intermediates.

2. EXPERIMENTAL

The basic experimental approach of these studies has been outlined previously.[3,4] Atomic oxygen was generated by photolysis of NO_2 using the 355 nm output of a tripled Nd:YAG laser (Quantel, model YG581C). UV pulse energies were measured by a photodiode and were typically 150 mJ. NO_2 (Matheson) was purified by successive freeze/pump/thaw cycles in an ethanol/liquid nitrogen. OCS (Matheson) was purified from CO contamination by condensing OCS in an ethanol/liquid nitrogen slush (157 K) and pumping off the head gas. A mixture of 20% NO_2 in OCS was prepared and allowed to thoroughly mix. The gas mixture was flowed through a 2 m Pyrex cell at a total pressure of 400 mTorr.

CO(v,J) product was monitored by a tunable IR diode laser (Laser Analytics system). Because the diode laser has no internal wavelength calibration, it was necessary to use a reference cell in order to provide an absolute wavelength marker. To accomplish this, 10% of the IR beam was directed to a reference cell which consisted of a water cooled Pyrex tube mounted with CaF_2 windows and equipped with high voltage electrodes. A mixture of CO in helium was flowed through the electrical discharge cell. By adjusting the current and voltage of the discharge, CO rotational lines in vibrational levels up to $v \approx 9$ could be found. The diode laser wavelength was locked to the desired CO reference line by dithering the frequency over a very small range. The output of the reference cell detector was connected to a lock-in amplifier which was configured to take the first derivative of the incoming signal. The lock-in amplifier output was attenuated and connected to the external input of the diode laser current. The lock-in output acted as a correction signal to compensate for drift of the diode laser frequency away from line center.

The remainder of the IR beam was passed through the reaction cell collinear with the UV photolysis beam. The UV beam was separated from the IR beam by a dichroic mirror. The IR beam was passed through a 3.5 - 6.5 µm bandpass filter and focused onto a .25 mm diameter photovoltaic InSb detector (Judson, model J10D-M204) with an impedance matched preamplifier (Perry, model 730-40). The detection system had a measured response time of ~ 70 ns. The signal from the reaction cell detector was digitized by a 100 Megasample s^{-1} transient recorder (LeCroy, model 8818A). Triggering of the transient recorder was synchronized to the photolysis pulse using the trigger signal of the Nd:YAG laser. Experiments were performed at repetition rate of 0.5 Hz, and 200 transients were averaged for each recorded signal.

3. RESULTS AND DISCUSSION

The reaction of ground state atomic oxygen and carbonyl sulfide has an exothermicity of $\Delta H_{f,298}$ = -211.4 kJ mol^{-1}. The system has an activation barrier of 1540 cm^{-1} to reaction. In a study of the photolysis of NO_2 at 347 nm, it was found that NO(v=0) and NO(v=1) were produced in approximately equal amounts.[5] This is expected to hold for 355 nm photolysis as well implying that a significant portion of the oxygen atoms produced in the photolysis process will correspond to NO(v=0). For NO(v=0), the corresponding O atoms have a kinetic energy of 1570 cm^{-1} in the O+OCS center-of-mass (c.m.) system. This amount is barely sufficient to overcome the activation energy to reaction. In the case of NO(v=1), oxygen will only have 600 cm^{-1} of kinetic energy in the O+OCS c.m. Since this insufficient to overcome the activation barrier, the only collisions that will lead to products are those in which the oxygen atom corresponds to NO(v=0). Thus, the reactive collisions will be essentially monoenergetic. The energy available for partitioning among product degrees of freedom from such reactive collisions is E_{avail} = 19,240 cm^{-1}.

In order to analyze the recorded CO transient signals, it is first necessary to devise a kinetic scheme that models the relevant formation and relaxation processes occurring within the reaction cell. A simplified model is given in the following list of elementary reactions.

$$O^* + OCS \rightarrow CO(v,J) + SO \tag{1a}$$

$$O^* + OCS \rightarrow CO(v',J') + SO \tag{1b}$$

$$O^* + M \rightarrow O + M \tag{2}$$

$$CO(v,J) + M \rightarrow \sum_{v',J'} CO(v',J') + M \tag{3a}$$

$$\sum_{v',J'} CO(v',J') + M \rightarrow CO(v,J) + M \tag{3b}$$

Reaction 1a represents production of CO into the state being monitored with reaction 1b representing CO production into all other energetically accessible states. Reaction 2 represents deactivation or removal of translationally hot oxygen atoms from the probed volume without producing CO. This includes flight of oxygen out of the probed volume as well as the possible reaction with other species such as NO_2 or CO. Reactions 3a and 3b represent relaxation either out of or into the monitored CO state. Because of the summations within reactions 3a and 3b, it is not possible to explicitly solve the set of differential equations resulting from this model, but if the relaxation processes are modelled by a two state system in which the monitored state interacts with a "bath" state consisting of all other possible CO states via k_{3a} and k_{3b}, then the following expression for the time-dependent concentration of CO(v,J) may be derived:

$$[CO(v,J,t)] = \frac{[O^*]_0 [OCS] \left(C_r k_{1a} - (k_{1a}+k_{1b}) k_{3a} [M]\right)}{C_r (C_3 - C_r)} \left(\exp\{-C_r t\} - \exp\{-C_3 t\}\right)$$

$$+ \frac{[O^*]_0 [OCS] (k_{1a}+k_{1b}) k_{3a} [M])}{C_r C_3} \left(1 - \exp\{-C_3 t\}\right) \quad (4)$$

where $C_r = (k_{1a}+k_{1b})[OCS] + k_2[M]$ and $C_3 = (k_{3a}+k_{3b})[M]$. In the limit as $t \to 0$, Equation 4 reduces to a simple expression linear in time:

$$[CO(v,J,t)] \approx k_{1a} [O^*]_0 [OCS] t \quad (5)$$

Equation 5 indicates that at times shorter than the relaxation rates, the nascent CO[v,J] population is linearly proportional to the initial slope of a plot of the CO absorption signal versus time.

CO product state distributions from the O + OCS → CO + SO reaction were determined by fitting the initial 500 μs of signal to line using a linear least-squares routine. The slope of the fitted line was recorded as the raw CO(v,J) population. However, since the technique employed absorption spectroscopy which measures the population difference between the two quantum states involved in the probed transition, it is also necessary to correct the raw data for the population in the upper state. The initial ~ 3 μs of signal for several transient absorption signals at three different CO(0,J) lines is shown in Figure 1. For CO(J<15), the initial rise of the absorption signal is delayed for some time following the photolysis pulse. The magnitude of the delay was dependent upon the rotational quantum number of the line being monitored with delays up to ~ 0.6 μs for J < 5. As the rotational quantum number of the state increased, the length of the delay decreased. For J ≥ 15, the rise of the absorption signal began promptly with the photolysis pulse.

A Boltzmann plot of the CO(v=0) rotational distribution as determined above is given in Figure 2. The plot exhibits a definite bimodal behavior. The high J distribution has a rotational temperature of $T_R = 4400 \pm 390$ K. The low J portion of the distribution is fit by $T_R = 350 \pm 20$ K. The CO lines that contribute to the low J distribution are those that show a delay in the initial rise of the absorption signal. The fact that the low J distribution has a rotational temperature close to room temperature, and the rise of the signals for these lines was delayed from the photolysis pulse suggest that the low J distribution is the result of rotational relaxation of the nascent CO distribution. However, the prompt rise of the absorption signals following photolysis at higher J's indicates that the above analysis is a

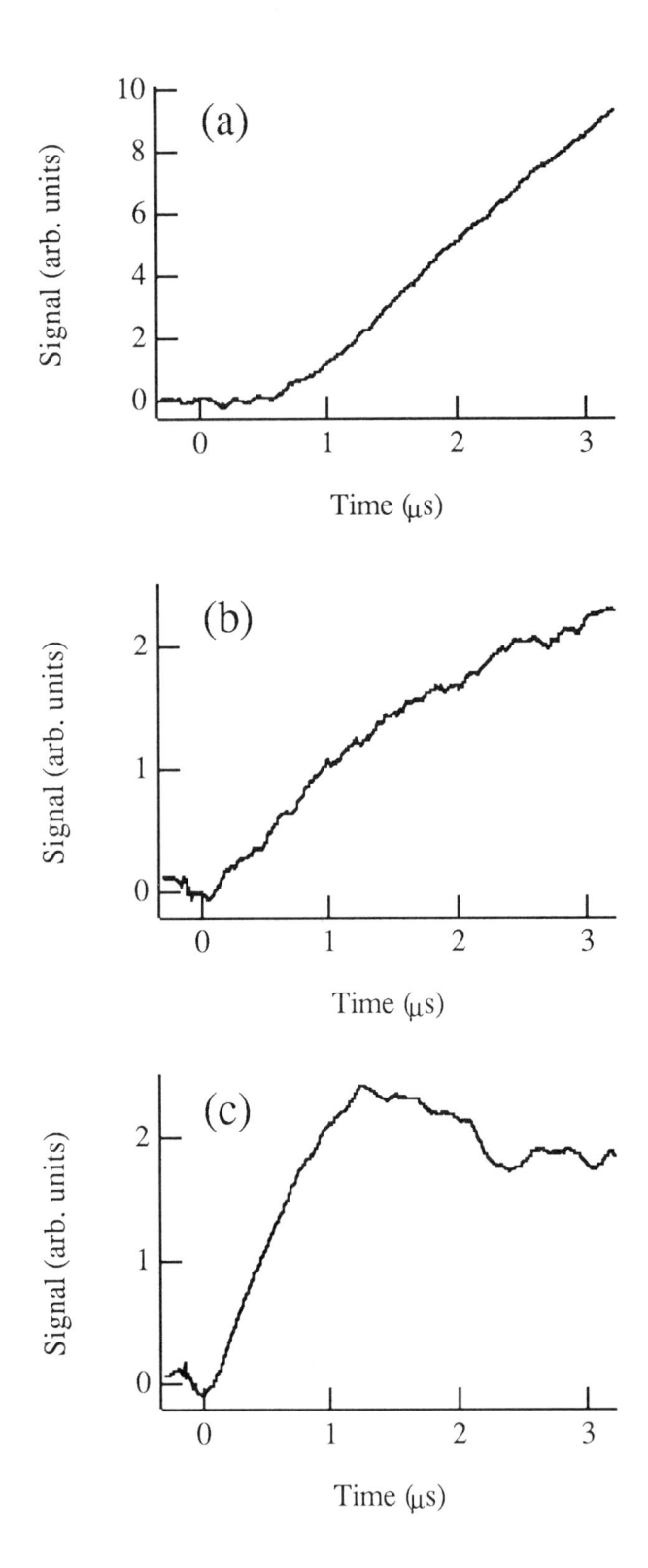

Figure 1. Transient absorption signals for (a) CO[v=0,R(5)], (b) CO[v=0,R(18)], and (c) CO[v=0,R(36)]. t = 0 indicates the photolysis pulse.

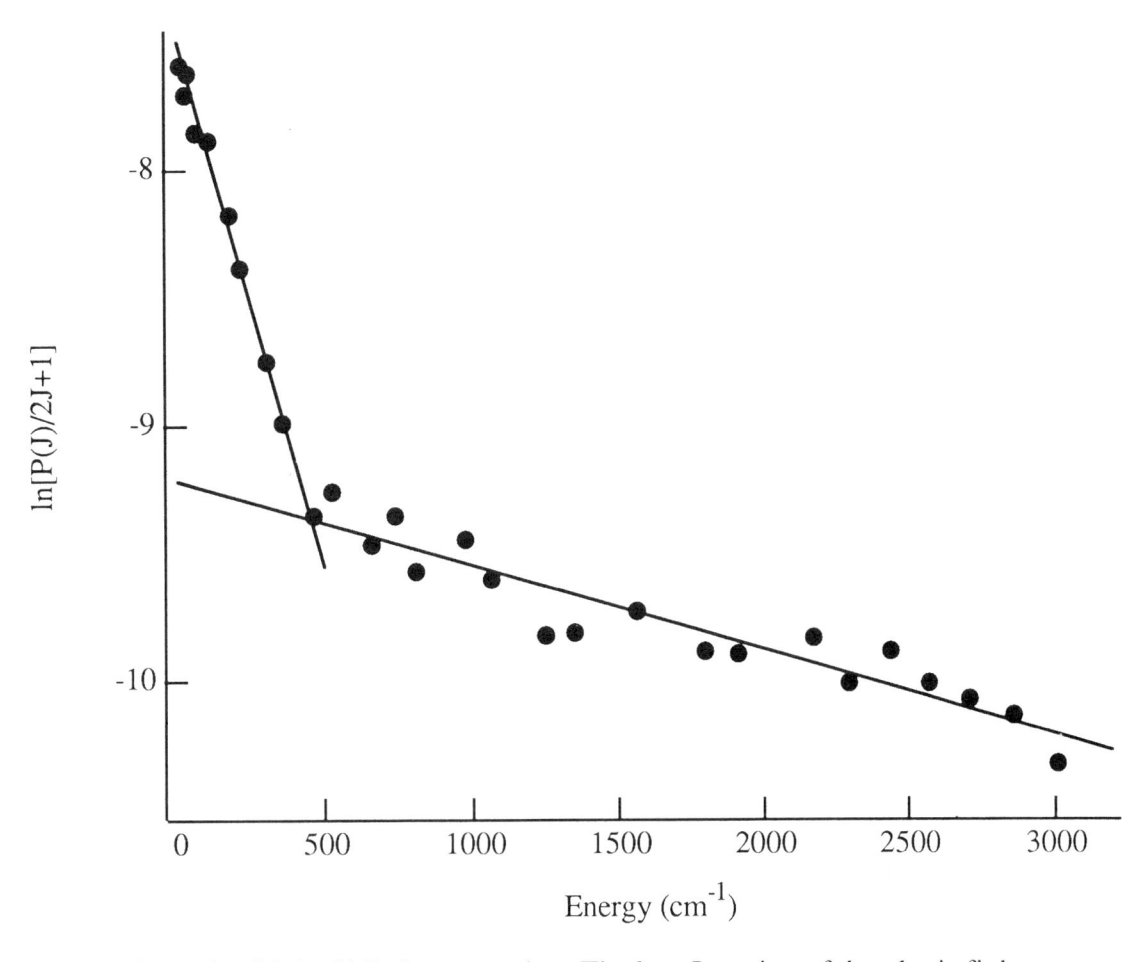

Figure 2. CO(v=0) Boltzmann plot. The low J portion of the plot is fit by a rotational temperature of $T_R = 350 \pm 20$ K, and the high J portion is fit by $T_R = 4400 \pm 390$ K. Error bounds in this and subsequent plots are expressed as $\pm \sigma$.

valid method for extracting the nascent populations from experimental data, *i.e.*, the data used to determine the nascent distributions is from the initial portion of the transient signals where relaxation processes have yet to significantly affect the populations.

The extent of statistical behavior of the CO(v=0) distribution was examined by comparing the measured distribution to a "prior" distribution in which the partitioning of energy among product degrees of freedom is constrained to maximize the entropy of the reaction. This was done using surprisal analysis in which $-\ln[P(J)/P^o(J)]$ is plotted versus a reduced energy variable, g_R, where $P(J)$ is the experimentally measured distribution, $P^o(J)$ is the "prior" distribution, and g_R is given by $E_R(J)/(E_{avail}-E_v)$.[6] The slope of such a plot, given as Θ_R, is a measure of the statistical behavior of the distribution. If $\Theta_R < 0$, there is more energy partitioned into rotations than would be statistically expected; if $\Theta_R = 0$, the distribution behaves statistically; and if $\Theta_R > 0$, the distribution is colder than expected.

A surprisal plot of the CO(v=0) distribution is shown in Figure 3. The high J distribution has a surprisal parameter of $\Theta_R = 3.7 \pm 0.5$. Thus, the nascent CO(v=0) distribution is colder than statistically expected indicating that energy is being partitioned into other product degrees of freedom at the expense of CO rotations. The extremely large surprisal parameter of $\Theta_R = 76.1$ for the low J portion of the distribution is not surprising; once collisions significantly affect the product distribution, a comparison between the nascent CO distribution and a statistical distribution with $E_{avail} = 19,240$ cm^{-1} is no longer valid as the relaxation process reduces the total energy of the CO ensemble as CO comes into thermal equilibrium with the bath gases.

Very weak transient signals were observed in the CO(v=1) manifold for $J \leq 17$. A Boltzmann plot for CO(v=1) is given in Figure 4. The resulting distribution has a rotational temperature of $T_R = 340 \pm 55$ K. Because no signals were observed at higher J in the v = 1 manifold, it was not necessary to correct the CO(v=0) distribution for population in v = 1. Using the low J distributions of each level, a vibrational branching ratio was estimated to be $[v=1]/[v=0] \leq 0.05$. From these nascent CO(v,J) distributions an average energy content of the CO product was calculated to be $\langle E_{int}(CO)\rangle = 3150$ cm^{-1} which is 16% of the available energy.

The low relative energy content of CO indicates that the processes which determine energy partitioning within the reaction products are not statistical in nature. Additionally, the internal degrees of freedom of product CO are not in thermal equilibrium, *i.e.*, a temperature of 4400 K measured for the CO(v=0) rotational distribution would produce a vibrational branching ratio of $[v=1]/[v=0] = 0.5$. This is an order of magnitude greater than the vibrational branching ratio estimated above.

There are several possible hypotheses for the reaction dynamics which might explain the observed CO energy partitioning. The first is one in which the reactants form a long-lived OCSO† complex as they traverse the PES. In the O + OCS reaction of the present study, the energy of collision is barely above the activation barrier so that a reactive

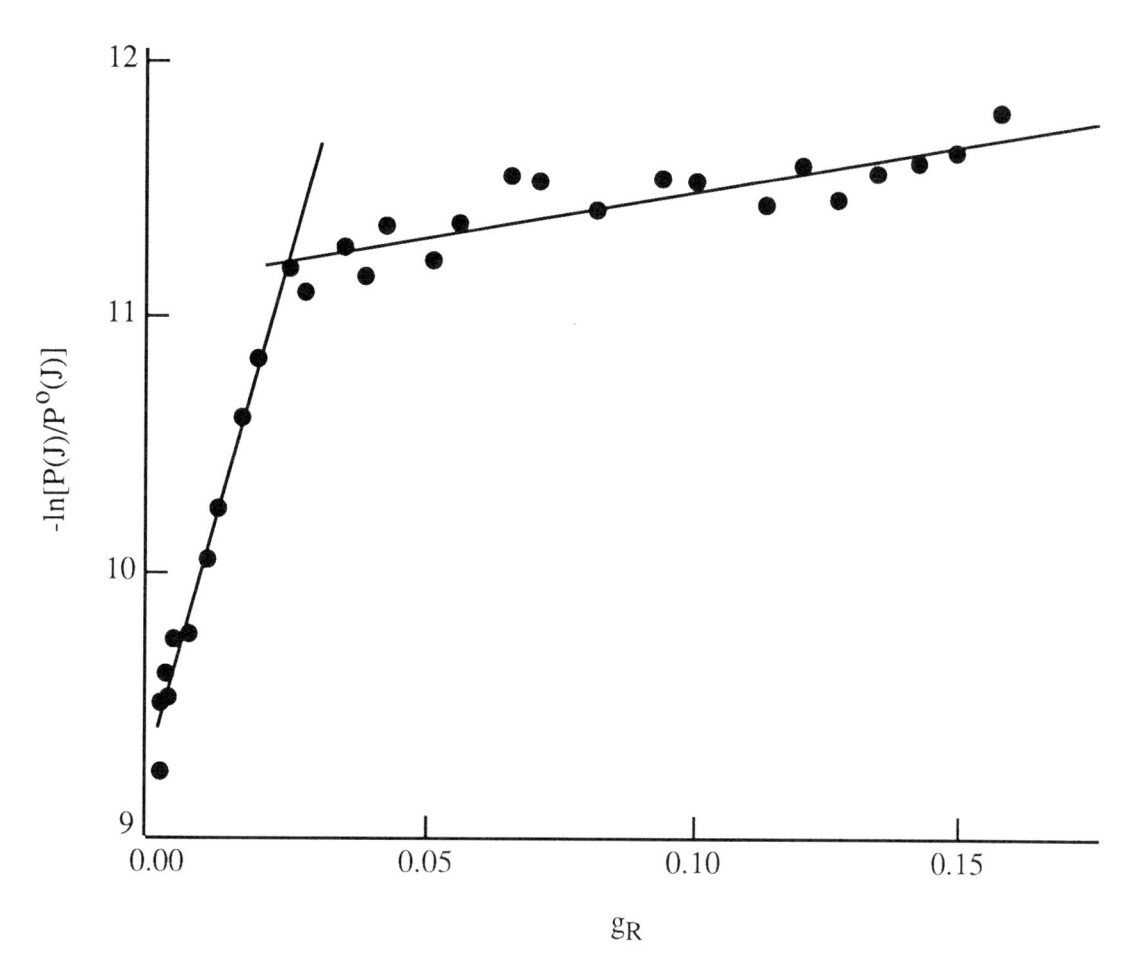

Figure 3. CO(v=0) surprisal plot. The low J portion is fit results in a surprisal parameter of $\Theta_R = 67.1$, and the high J portion is fit by $\Theta_R = 3.7 \pm 0.5$.

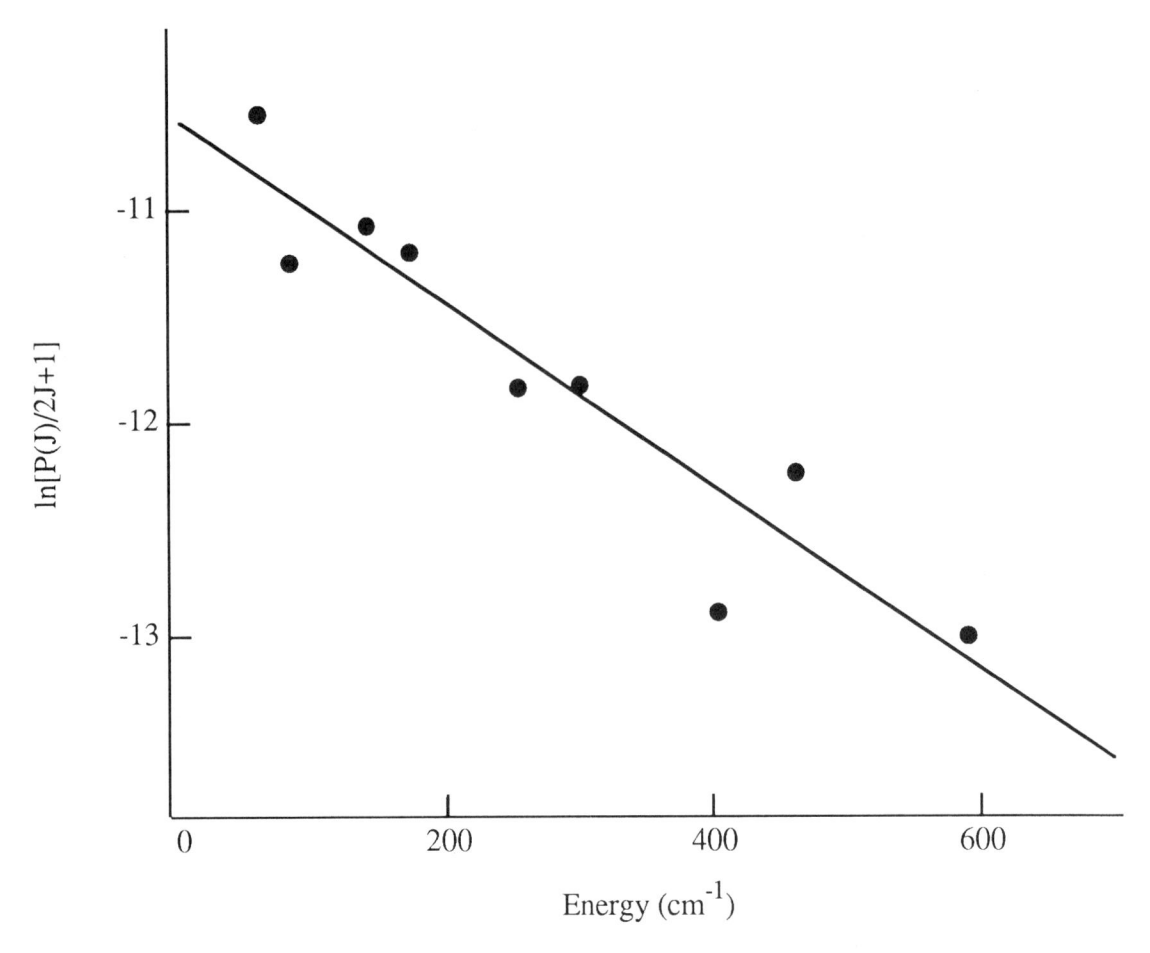

Figure 4. CO(v=0) Boltzmann plot. The plot has rotational temperature of $T_R = 340 \pm 55$ K.

intermediate could form with little energy in its internal degrees of freedom, unlike previous studies in which the energy of the initial impact was well above the activation barrier of reaction making it unlikely that a local minimum along the reaction pathway of the PES was significantly affecting the final state distributions.[3,4] In such an intermediate, randomization of the energy available at the transition state would occur before the complex dissociates to products. In this scheme, excitation of the CO product is the result of a torque acting on CO in the exit channel as energy is released.

In the second hypothesis, the reactants traverse the transition state on a time scale short enough that complete energy randomization in the OCSO† complex cannot occur, but rather energy is preferentially deposited into a degree of freedom that becomes the rotational motion of product CO. In this case, the OCSO† complex dissociates before a significant amount of energy can flow into the CO vibrational mode. The release of energy in the exit channel is governed by the repulsive forces acting between the product c.m.'s and is directed primarily into the translational motion of the separating fragments.

Further experiments on this system are necessary in order to completely elucidate the dynamics of reaction. The energy content and product state distributions of the SO product must be determined. The amount of energy partitioned into product translations must be measured. It is also essential to calculate an accurate theoretical model of the O + OCS PES in order to determine whether a minimum exists along the reaction pathway corresponding to a stable intermediate OCSO† complex as well as to examine the forces acting on the products within the exit channel. Additionally, trajectory calculations on such a model would establish the existence of a preferred collision geometry leading to reaction.

SUMMARY

1. The CO(v=0) rotational distribution was bimodal. The high J portion of the distribution had a rotational temperature of $T_R = 4400 \pm 390$ K and represent the nascent CO population. The low J distribution had a temperature of $T_R = 350 \pm 20$ K and was the result of rotational relaxation of the nascent distribution.

2. Surprisal analysis of the CO(v=0) distribution resulted in a surprisal parameter of $\Theta_R = 3.7 \pm 0.5$ indicating that the nascent CO(v=0) rotational distribution was colder than predicted by a statistical model in which the entropy in the products is maximized.

3. The CO(v=1) rotational distribution had a temperature of $T_R = 340 \pm 55$ K. Signals were not seen in this vibrational manifold for J > 17.

4. A vibrational branching ratio of [v=1]/[v=0] ≤ 0.05 was determined using the low J distributions.

5. The average energy content of the CO product was $\langle E_{int}(CO) \rangle = 3150$ cm^{-1} which is only 16% of the available to product degrees of freedom.

REFERENCES

1. (a) I. Glassman, *Combustion*, Academic Press, New York, 1977. (b) P. Warneck, *Chemistry of the Natural Atmosphere*, Academic Press, San Diego, 1988.

2. (a) R. Atkinson, D.L. Baulch, R.A. Cox, R.F. Hampson Jr., J.A. Kerr, and J. Troe, *J. Phys. Chem. Ref. Data*, p. 881, Vol. 18, 1989. (b) W.B. DeMore, D.M. Golden, R.F. Hampson, C.J. Howard, M.J. Kurylo, M.J. Molina, A.R. Ravishankara, and S.P. Sander, *Chemical Kinetic and Photochemical Data for Use in Stratospheric Modeling, Evaluation Number 8*, JPL Publication 87-41, 1987.

3. S.L. Nickolaisen, H.E. Cartland, and C. Wittig, "CO Internal Excitation from the Reaction: $H + CO_2 \rightarrow CO + OH$," *J. Chem. Phys.*, submitted.

4. S.L. Nickolaisen, *State Resolved Product Dynamics in Atom-Molecule Interactions*, Doctoral dissertation, University of Southern California, 1991.

5. G.E. Busch and K.R. Wilson, "Triatomic Photofragment Spectra. I. Energy Partitioning in NO_2 Photodissociation," *J. Chem. Phys.*, pp. 3626-3638, Vol. 56(7), 1972.

6. R.D. Levine and R.B. Bernstein, *Molecular Reaction Dynamics and Chemical Reactivity*, Oxford University Press, New York, 1987.

OPTICAL METHODS FOR
TIME- AND STATE-RESOLVED CHEMISTRY

Volume 1638

SESSION 3

Photodissociation Dynamics

Chair
Andrew H. Kung
Lawrence Berkeley Laboratory

Translational spectroscopy of H atom photofragments

Michael N.R. Ashfold, Ian R. Lambert, David H. Mordaunt,
Gregory P. Morley and Colin M. Western

School of Chemistry, University of Bristol, Bristol BS8 1TS, U.K.

ABSTRACT

We present three examples which serve to illustrate the way in which use of the technique of H atom photofragment translational spectroscopy can provide much new insight into the photodissociation dynamics of molecular hydrides.

1. INTRODUCTION

The technique of H atom photofragment translational spectroscopy, developed by Welge and coworkers at Universitat Bielefeld,[1-6] has provided little short of a revolution in the resolution attainable when measuring the translational energy spectrum of H and D atomic photofragments.

Two variants of the method have now been demonstrated. In the original version of the technique,[1-4] the nascent neutral H($n=1$) photofragments of interest were converted into H$^+$ ions *at source* — prior to their having escaped the interaction volume — and the angular and translational energy distributions of the recoiling H atom fragments were monitored by measuring the times–of–flight (TOF) of the resulting *ions*. The H$^+$ ions were formed via the following two colour two photon threshold ionization process:

$$H(n=1) + h\nu_{VUV}(\lambda=121.6 \text{ nm}) \longrightarrow H(n=2) \qquad (1a)$$
$$H(n=2) + h\nu_{UV}(\lambda=364.6 \text{ nm}) \longrightarrow H^+ + e^- \qquad (1b)$$

The Ly–α radiation required for excitation step (1a) can be obtained by frequency tripling (in krypton gas) the output of a conventional dye laser. One convenient property of the H atom is the fact that its $n=2$ level lies at an excitation energy that is three quarters that of the ionization limit. Thus the two colour two photon resonance enhanced ionization of atomic H defined in (1) above can be brought about using a *single* dye laser, since the subsequent absorption of one of the fundamental UV dye laser photons provides exactly the energy required for threshold ionization.

Even with this 'primitive' version of the technique of H atom photofragment translational spectroscopy, the energy resolution attainable (typically $\Delta E/E \sim 1\%$) is appreciably better than is achieved for H photofragments when using conventional photofragment translational spectroscopy (i.e. where the neutral fragment of interest is detected following electron bombardment and mass spectrometric detection at the end of its flight path).[7] That the resolution is not better still is largely attributable to the effects of space charge. The nett effect of the two colour two photon ionization process (1) is to create a concentration of H$^+$ ions localized in space and time. The Coulomb repulsion between these protons inevitably causes some smearing of the nascent H atom velocity and angular distributions. Recognition of this limitation led to the more recent variant of the technique — in which the kinetic energy distribution of the H atom photofragments of interest are monitored not via the ion, but *via the Rydberg atom*.[5,6,8]

This newer form of H atom photofragment translational spectroscopy requires the introduction of an additional laser, tuned so as to excite the H(n=2) atoms to a Rydberg state of high principal quantum number, i.e.

$$H(n=2) + h\nu_{UV} (\lambda \sim 364.8 \text{ nm}) \longrightarrow H(n \sim 70-90) \qquad (1c)$$

A small d.c. field, placed around the interaction region, ensures extraction of the (now unwanted) H$^+$ ions that will inevitably still be formed via excitation process (1b). The obvious advantage of this strategy is that the H fragments fly as *neutral particles* and are ionized (by field ionization) only *after* separation by TOF and immediately prior to detection. Space charge effects are obviated, and appreciably higher kinetic energy resolution ($\Delta E/E \sim 0.3\%$) results. We note in passing that the presence of the static d.c. field confers another important benefit. ℓ ceases to be a good quantum number in the presence of the field, and the H atom Rydberg states prepared *in the presence of the field* have lifetimes at least one order of magnitude longer than those of the high n, $\ell = 0$ or 2 states that would have been prepared using this same double resonant excitation scheme but under strictly field free conditions. Even under field free conditions the lifetimes of these high n ($\ell = 0$ or 2) Rydberg states of the H atom are in the many tens of microseconds range; a further order of magnitude extension in these lifetimes as a result of the applied d.c. field ensures that radiative decay is not a particularly significant loss mechanism even for the slowest H atoms in the TOF spectra.

Though still in its infancy, this new technique has shown itself to be a very powerful tool for those researching in the field of molecular photodissociation dynamics. In what follows we present the results of three recent studies – of, respectively, the photodissociation of HCN, H$_2$S and CH$_4$, each at the Lyman–α wavelength – which serve to illustrate the potential of the method.

2. EXPERIMENTAL

A detailed description of the experimental apparatus and method, and of the procedures for data analysis, have been presented elsewhere[9].

3. RESULTS and DISCUSSION

3.1 Photodissociation of HCN(DCN) at 121.6 nm

The hydrogen cyanide molecule has many of the properties one would seek if one was designing a model system for those interested in a detailed understanding of molecular photofragmentation dynamics. It is a volatile gas, admittedly highly toxic, but also relatively easy to prepare. It has just one dissociation coordinate, involving H—CN bond cleavage, in the energy range of interest to chemists. Furthermore, its electronic absorption spectrum shows structure, thus encouraging the notion that it should be possible to study the predissociation of this molecule in a state selective fashion. It should therefore come as no surprise to learn that there have been many previous studies, both experimental and theoretical, of the vacuum ultraviolet (VUV) spectroscopy and photochemistry of this molecule. What is much more surprising is the lack of definitive conclusions from these studies, even about such fundamental issues as the branching ratios into the various electronic states of the products.

Fig. 1 shows the TOF spectrum (recorded out to ~75 μs after the photolysis pulse) of H atoms resulting from 121.6 nm photolysis of HCN molecules in the early time part of a skimmed pulsed expansion of a 1:10 HCN:Ar mixture. The interaction region – detector separation in this experiment is 425 mm. The TOF spectrum of D atoms resulting from DCN photodissociation at this wavelength has a similar appearance, though the fast peak is not so dominant. No H atom signal was detected from HCN photolysis at times ≲ 16 μs; similarly we detected no D atoms with flight times ≲ 22 μs from the 121.6 nm photolysis of DCN.

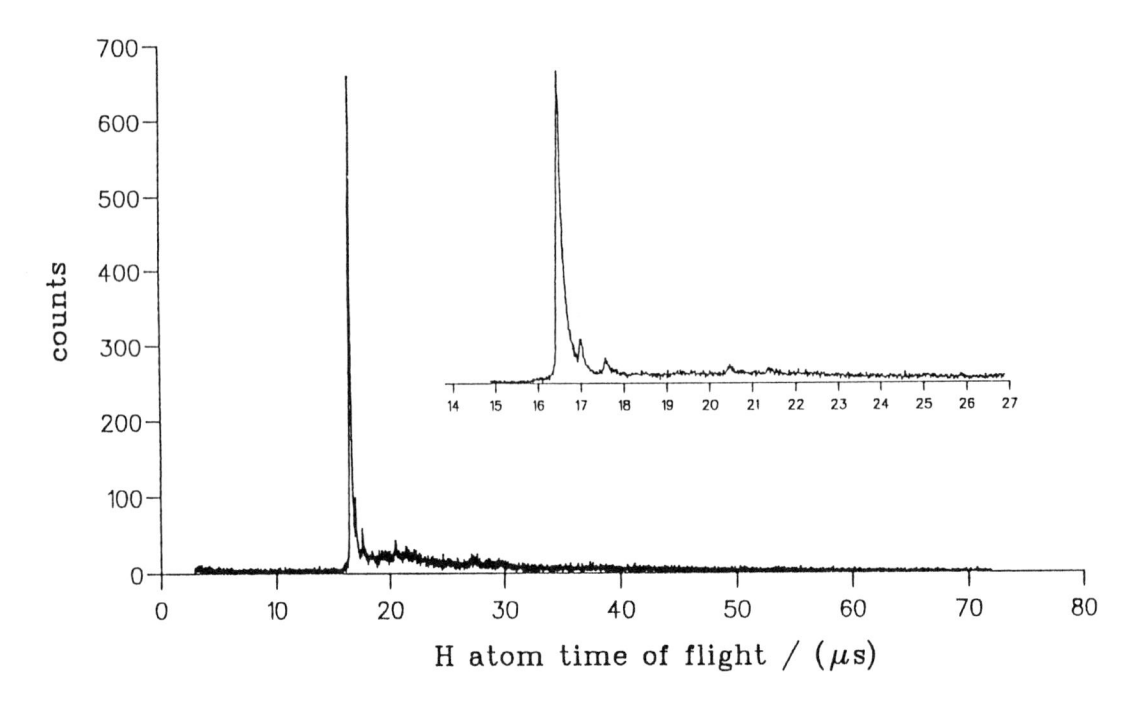

Fig. 1 H Rydberg atom TOF spectrum resulting from photolysis of jet–cooled HCN at 121.6 nm. The inset displays the early time part of this spectrum on an expanded horizontal scale.

Knowing the length of the flight path (d = 425 mm) we can transform each point in such a TOF spectrum into a spectrum of *total* fragment kinetic energy by applying the relationship

$$E_{kin,total} = E_{kin,H} + E_{kin,CN} = \frac{1}{2}m_H\left[1 + \frac{m_H}{m_{CN}}\right]\left(\frac{d}{t_H}\right)^2. \qquad (2)$$

(The corresponding equation for the total fragment kinetic energy associated with any particular D atom time–of–flight is obtained simply by replacing m_H and t_H in eq. (3) by the appropriate values for m_D and t_D.) The H/D atom recoil velocities are sufficiently large (3.6×10^4 m s^{-1} in the case of the fastest H atoms contributing to the TOF spectrum shown in fig. 1) compared with the parent beam velocity (~500 m s^{-1}) that we neglect any correction of the fragment kinetic energy spectrum to account for the lab. to centre of mass coordinate transformation. (This correction would cause a 1% change in the quoted kinetic energy for H + CN fragments recoiling with ~1000 cm^{-1} total kinetic energy; in the case of DCN dissociation we estimate an equivalent percentage error due to neglect of the lab. to centre of mass correction for D + CN fragments with ~2250 cm^{-1} recoil energy.)

However, it is necessary to rebin the data if we wish to have a sensible measure of the relative strengths of the signals associated with different fragment kinetic energies. This reflects the fact that a given time increment, δt, in the earlier part of the TOF spectrum necesarily corresponds to a larger energy spread than would the same δt at a later time in the spectrum. The rebinning procedure can be visualised as follows: The experimentally measured TOF spectrum (spanning the time range t_i to t_f) comprises some number, n, of time channels (typically 15000) of fixed width (10 or 20 ns normally). Software has been written so that the total fragment kinetic energy spectrum is partitioned into an equivalent number of energy channels, each of equal energy width ΔE, such that

$$n\Delta E = E(t_i) - E(t_f) \qquad (3)$$

with the spread of flight times corresponding to each energy increment calculated via eq. (2). The software then redistributes the measured counts in each *time* channel into the appropriate *energy* channels according to the extent of their mutual overlap.

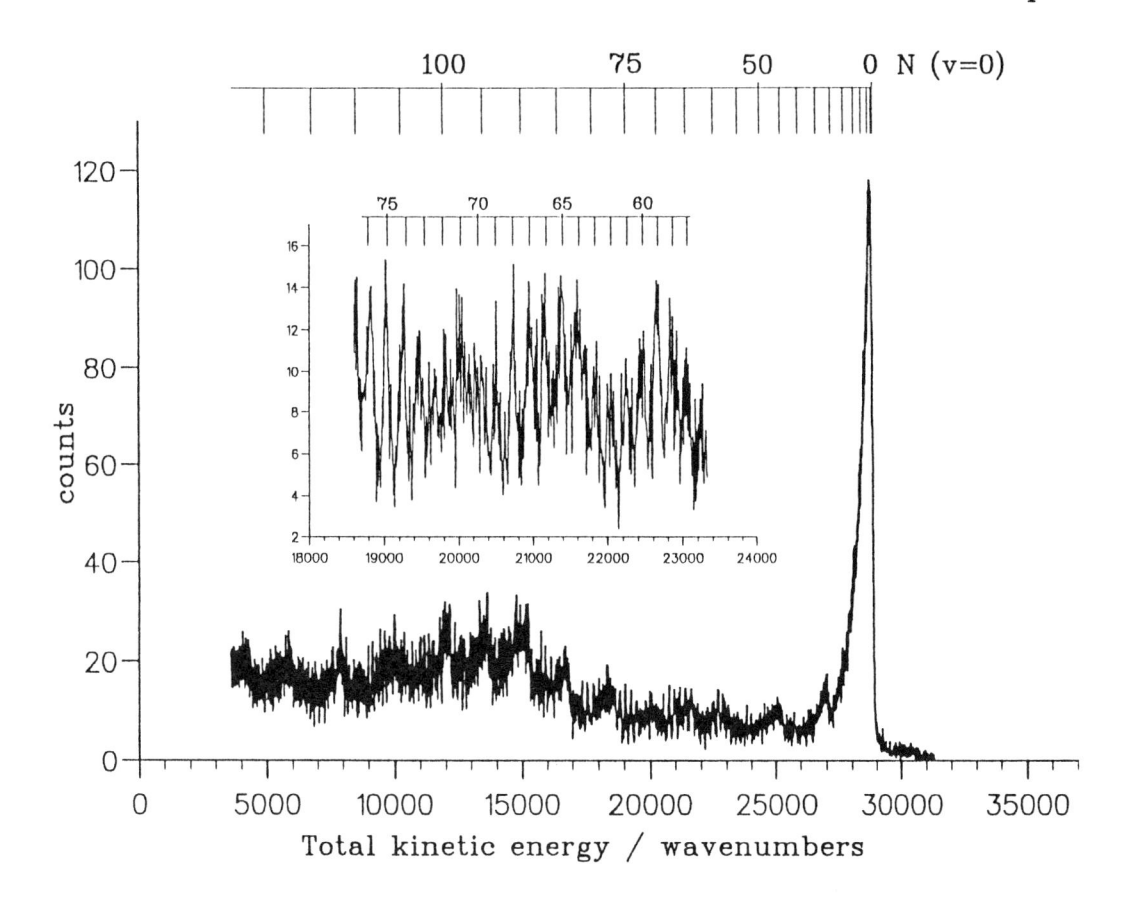

Fig. 2 Spectrum of the total kinetic energy of the D + CN fragments resulting from photolysis of DCN at 121.6 nm. The inset shows a portion of this spectrum on an expanded scale. The energies corresponding to various of the rotational levels of $CN(A)_{v=0}$ are marked above the two spectra.

Fig. 2 shows the spectrum of the *total* fragment kinetic energy spectrum arising from the 121.6 nm photolysis of the jet-cooled sample of DCN obtained after processing in the manner outlined above. We note the intense localised feature centred around 28500cm^{-1} and the reproducible fine structure observed at lower kinetic energies. The magnitudes of $D_0(H-CN)$ and $D_0(D-CN)$ have been determined previously[10,11] to sufficient accuracy for us to be able to identify the intense feature with the dissociation channel:

$$DCN \longrightarrow D + CN(A^2\Pi)_{v'=0, \text{low } N'} \tag{4}$$

The subsiduary maxima observed at kinetic energies ~ 26900 cm^{-1} and ~ 25100 cm^{-1} are similarly attributable to the fragmentation channels yielding CN(A) fragments in their v'=1 and 2 vibrational states.

We can actually use the data displayed in fig. 2 to derive a more accurate value for D_0(D–CN) and thus, in turn, for D_0(H–CN). This is possible because we are able to associate the comparatively widely spaced structure evident in the kinetic energy range 18800 – 24000 cm^{-1}, part of which is highlighted in the inset to fig. 2, with the formation of CN(A)$_{v'=0}$ fragments possessing *high rotational quantum number N'*. Given the spectroscopic constants derived through earlier analyses of the CN(A–X) emission spectrum[13] we are able derive the CN(A)$_{v=0}$ rotational quantum state numbering indicated in fig. 2. This in turn leads to a reliable estimate of the total recoil energy associated with formation of D + CN(A)$_{v'=0, N'=0}$ fragments (28820 ± 200 cm^{-1}) and thence, given the energy conserving relation

$$E_{kin}[D + CN(A)_{v'=0, N'=0}] = E_{photon} - T_{00}[CN(A-X)] - D_0(D-CN), \tag{5}$$

together with the values E_{photon} = 82280 cm^{-1} (for D Lyman–α) and T_{00}[CN(A–X)] = 9094.3 cm^{-1} for the wavenumber separation between the *lowest* rotational levels of the ground (X) and first excited (A) states of CN,[12], we obtain:

$$D_0(D-CN) = 44370 \pm 200 \text{ cm}^{-1}.$$

This value for the bond dissociation energy of DCN was used in deriving the absolute energy scale shown on the horizontal axis of the plot displaying the CN fragment internal energy distribution shown in fig. 3a. As a result we can also indicate the threshold energies for dissociation channels leading to higher vibrational states of CN(A) and for the first few D + CN(B)$_{v'}$ product channels.

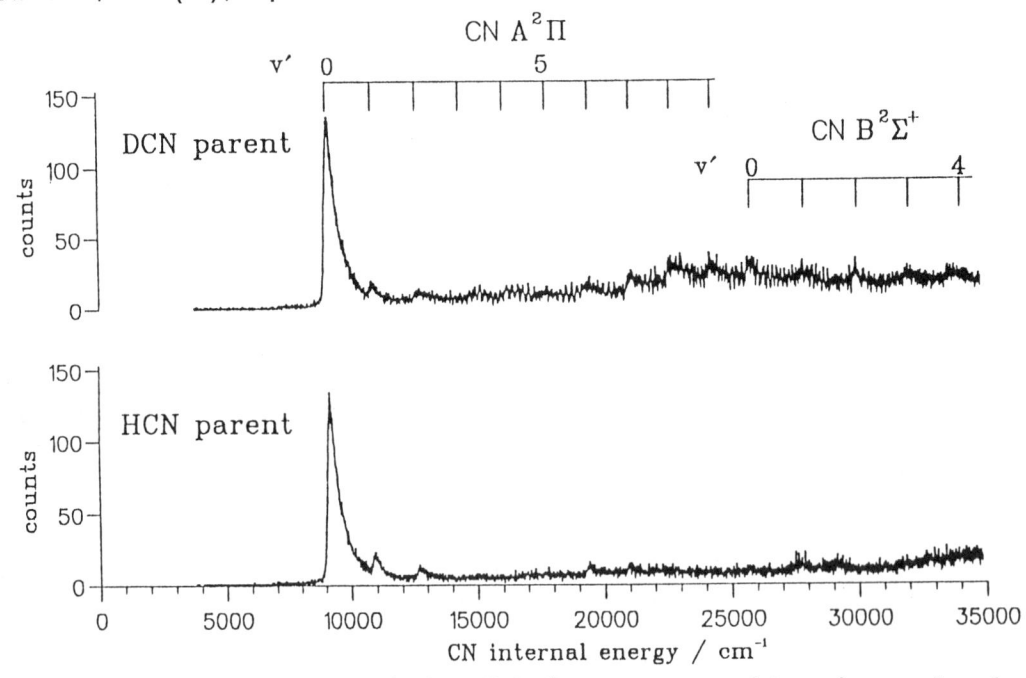

Fig. 3 Internal energy spectrum of the CN fragments resulting from photolysis of (a) DCN and (b) HCN at 121.6 nm. Thresholds associated with the various vibrational levels of the electronically excited A$^2\Pi$ and B$^2\Sigma^+$ states of CN are indicated.

The corresponding internal energy spectrum of the CN fragments resulting from 121.6 nm photolysis of HCN (fig. 3b) shows no such resolved rotational fine structure. Thus our best estimate of the H–CN bond strength:

$$D_0(H-CN) = 43800 \pm 200 \text{ cm}^{-1}$$

simply derives from the value for $D_0(D-CN)$ derived above, suitably corrected for the difference in the zero–point energies of ground state HCN and DCN.[13] This value for $D_0(H-CN)$, which serves to refine other recent estimates of this bond strength,[10,11] was used in deriving the absolute energy scale associated with fig. 3b.

The key conclusions from this set of experiments are:

(i) The dominant products arising from HCN(DCN) photolysis at 121.6 nm are H/D atoms together with CN fragments in their first excited $A^2\Pi$ electronic state. The formation of H/D atoms in conjunction with $CN(B^2\Sigma^+)$ fragments constitutes a few percent of the total dissociation yield. We see no evidence for the formation of any ground ($X^2\Sigma^+$) state CN fragments amongst the primary dissociation products. These measured branching ratios into the various electronic states of the CN product differ markedly from the values deduced through earlier synchrotron measurements of the excitation spectra for forming fluorescent CN(A) and CN(B) products. Specifically, these earlier studies[11,14] suggested a much smaller quantum yield (Φ = 0.1–0.3) for CN(A) fragment formation at this excitation wavelength. The reason for this discrepancy is easy to see, given the present finding that many of the CN(A) fragments are formed in their vibrationless level and that, as a result, their fluorescence will be centred in the (experimentally hard to detect) near infrared spectral region.

(ii) The population distribution amongst the various vibration–rotation states of the CN(A) product appears bimodal: Roughly 50% of the CN(A) fragments arising in the 121.6 nm dissociation of DCN are formed in low N' rotational levels of the vibrationless state, the remainder are formed rotationally (and to some extent vibrationally) excited. HCN dissociation at this wavelength yields a greater relative proportion of rovibrationally 'cool' CN(A) products.

Reference to the available *ab initio* calculations[15-17] of the potential energy surfaces associated with various of the excited states of HCN suggests an explanation for the observed product state branching ratios, and for the pattern of energy disposal amongst the CN(A) fragments. Both HCN and DCN show reasonably strong absorption at 121.6 nm. This absorption has been attributed[18] to the promotion of an electron from the 1π highest occupied molecular orbital (regarded as a bonding π orbital localised on the C≡N bond) to a $3s\sigma$ Rydberg orbital. Both Vazquez and Gouyet[15] and Peric *et al*[17] have reported various sections through the potential energy surface for this $2^1\Pi$ excited state of HCN. We first consider the form of this potential when the ∠HCN angle is held fixed at 180°. The minimum energy configuration for the linear $2^1\Pi$ state is found to occur at a C–N bond length somewhat greater than in the ground state. The calculations also reveal the presence of an avoided crossing at linear geometries, in the H—CN dissociation channel, between the diabatic $2^1\Pi$ surface and the surface for the $1^1\Pi$ valence state (associated with the electronic promotion $\sigma \longrightarrow \pi^*$). This diabatic $1^1\Pi$ state correlates, asymptotically, with the products H + CN(A). For linear configurations the surface for this valence state is steeply repulsive along R_{H-CN}. As a consequence, the resulting *adiabatic* $^1\Pi$ surface is predicted[15-17] to show a small barrier in the collinear H—CN dissociation channel. We now consider how our picture of this adiabatic $^1\Pi$ surface is changed when we relax the constraint that the atoms be collinear. The electronic degeneracy associated with both of the $^1\Pi$ states is lifted upon bending. Of particular relevance to the present discussion, the energy of the lower $^1A'$ component (labelled $3^1A'$) is greatly lowered. In contrast, the energy of the other ($3^1A''$) Renner–Teller component of this $^1\Pi$ state rises as the molecule moves away from linearity.[15] The photo-prepared excited state is sufficiently predissociated that individual P,Q and R transitions are not resolved in the parent absorption spectrum.[18] Thus it is reasonable to assume that we prepare roughly equal numbers of excited molecules in the two 'Λ-doublet' states (which we start by labelling $4^1A'$ and $4^1A''$, but which we also recognise are strongly coupled to the corresponding $3^1A'$ and $3^1A''$ continua). From the foregoing, it is clear that those molecules which dissociate via the $3^1A''$ state are most likely to fragment whilst retaining near linear geometries. If we assume the C–N bond length in HCN(DCN) molecules in the $^1\Pi$ Rydberg state to be similar to that in the ground state ion then, given the similarity between this bond length and that of the free CN(A) fragment, a simple Franck–Condon argument can account for the observed preponderance of $CN(A)_{v=0}$ fragments. We deduced (recall figs. 2 and 3) that the dominant fast peak in the H/D atom TOF spectrum is associated with $CN(A)_{v=0}$ fragments in low rotational levels. The population of such levels is entirely logical in the case of a near collinear dissociation, where the only significant source of fragment rotation would be the bending zero-point motion in the dissociating parent molecules.

This essentially collinear dissociation pathway would, of course, also be available for molecules dissociating via the $3^1A'$ surface, but for this component of the $^1\Pi$ state the linear geometry represents a potential *maximum* in the bending coordinate. Thus it is more probable that molecules evolving on the $3^1A'$ surface will bend *prior to* H/D—CN bond extension and ultimate dissociation to H/D + CN(A). The bending forces acting during the dissociation event will manifest themselves in the form of high levels of rotation in the eventual CN(A) fragments. In the limit that the collinear mechanism is comparatively slow (perhaps because it involves tunnelling through the potential barrier on the lower adiabatic $^1\Pi$ surface) then we can envisage that the ~50% of the molecules dissociating via the $3^1A''$ state would necessarily fragment from near collinear geometries, thereby yielding rovibrationally 'cold' CN(A) fragments, whilst almost all the molecules evolving via the $3^1A'$ surface would circumvent the potential barrier in the collinear exit channel by bending prior to dissociation and evolve to H/D atoms together with internally excited CN(A) fragments. This limiting case scenario accords well with experimental observation in the case of DCN dissociation (see fig. 3a). The relative importance of the low intensity 'slow' tail associated with highly rovibrationally excited fragments is markedly less in the case of HCN dissociation (see fig. 3b). This would be consistent with there being a smaller effective barrier to collinear dissociation in the case of HCN (i.e. the zero-point energy associated with the C–H stretching motion is larger): in this case we presume that 'direct' collinear dissociation via the $3^1A'$ excited state augments the 'fast' peak in the H atom TOF spectrum.

3.2 Photodissociation of $H_2S(D_2S)$ at 121.6 nm

The photodissociation dynamics of $H_2S(D_2S)$ molecules following excitation within their long wavelength absorption continuum has been the subject of several recent studies[4,19-21] but, until very recently, there had been no quantitative photochemical investigations at excitation wavelengths shorter than 193 nm. There are, in fact, no less than seven possible exoergic decomposition pathways when $H_2S(D_2S)$ molecules are excited at 121.6 nm.[5] Three of these involve formation of $H_2(D_2)$ fragments in conjunction with S atoms in, respectively, their ^3P ground state, and their metastable excited ^1D and ^1S states. The other four possible channels, namely:

$$H_2S \longrightarrow H + SH(X^2\Pi) \qquad E_{trans} \leq 6.29 \text{ eV} \qquad (6a)$$
$$\longrightarrow H + SH(A^2\Sigma^+) \qquad E_{trans} \leq 2.47 \text{ eV} \qquad (6b)$$
$$\longrightarrow H + H + S(^3P) \qquad E_{trans} \leq 2.58 \text{ eV} \qquad (6c)$$
$$\longrightarrow H + H + S(^1D), \qquad E_{trans} = 1.43 \text{ eV} \qquad (6d)$$

result in formation of one or more H(D) atoms and should thus be amenable to investigation by H/D atom photofragment translational spectroscopy. That such is indeed the case was first demonstrated in the study of Schnieder et al.[5] Fig. 4 shows spectra of the total H + SH and D + SD fragment kinetic energies obtained by measuring the times of flight of the H/D atom photofragments. Analyses of these spectra clearly show fragmentation channels (6b) and (6d) to be the dominant sources of H/D atoms.[5,8] The electronically excited SH(SD) fragments are seen to be formed in a number of vibrational levels; however the majority are formed in their zero–point level with a rotational state population distribution that spans *all possible bound and quasi-bound rotational levels*.[5,8]

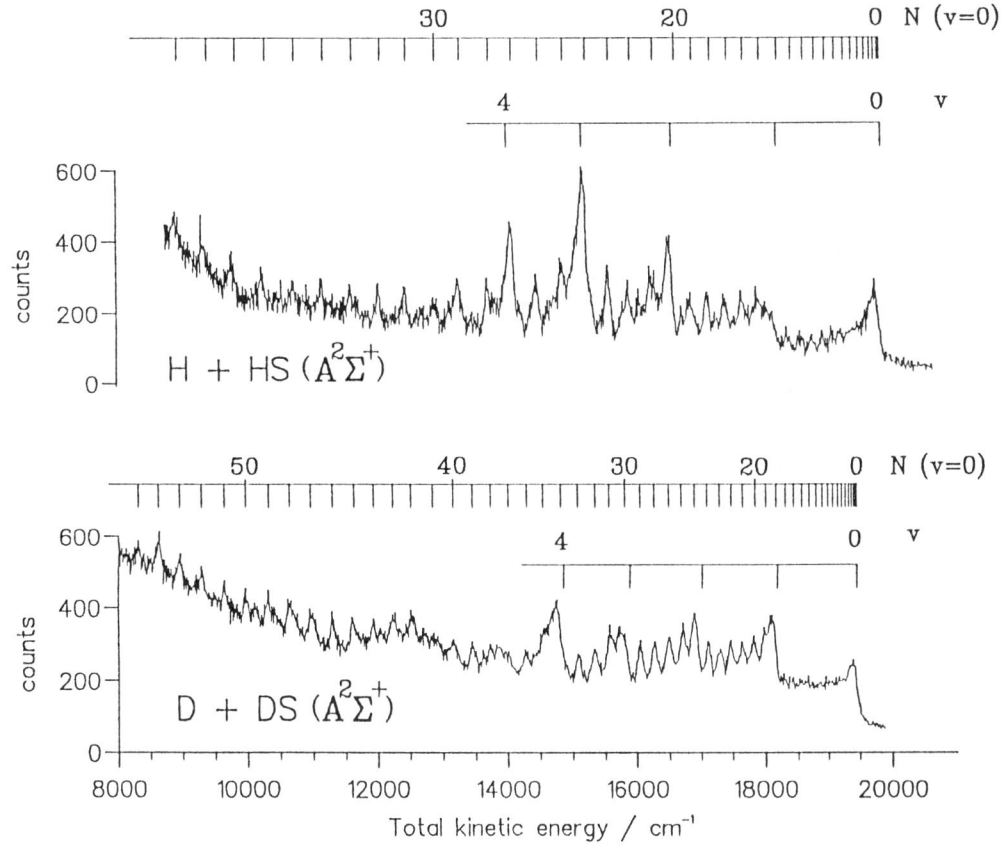

Fig. 4 Spectra of the total kinetic energy of the H + SH (D + SD) fragments resulting from photolysis of (a) H₂S and (b) D₂S at 121.6 nm. The energies corresponding to the various rovibrational levels of SH(SD) fragments in their $A^2\Sigma^+$ electronic state are indicated above the relevant spectra.

These findings, and the insignificant yield of ground state SH(SD) fragments – channel (6a) – have been rationalized[5] by assuming that:
(i) 121.6 nm photo–excitation prepares H₂S(D₂S) molecules in the first excited state of $^1A_1$ symmetry, either directly or via radiationless transfer from one or more of the Rydberg states which are apparent in this region of the respective parent absorption spectra.
(ii) the calculated topology of the potential energy surface for this $\tilde{B}^1A_1$ state of H₂S(D₂S)[22] is at least qualitatively correct.
(iii) the primary photofragmentation processes occur on this diabatic surface which, along the R_{H-SH} dissociation coordinate, correlates with the (observed) asymptotic products H + SH($A^2\Sigma^+$).

Given these assumptions, *ground* state SH(SD) fragments – channel (6a) – could arise only from those dissociating molecules that experience such large angle bending forces that they pass through linear geometries prior to substantial extension of the breaking S–H(S–D) bond, since only at near linear configurations are the non–adiabatic couplings to the lower potential energy surfaces (that correlate with ground state products) effective It has been argued[5] that the centrifugal forces associated with such a rapid opening of the bond angle in the excited state molecules would cause direct three body dissociation – channel (6d) – before such molecules have the chance to reach near linear geometries. Of course, it still remains for future studies to establish the relative importance of the H(D) atom and $H_2(D_2)$ molecular elimination channels when these two isotopomers are excited at this (or any other) vacuum ultraviolet wavelength.

3.3 Photodissociation of $CH_4(CD_4)$ at 121.6 nm

Methane, not least because of its high relative abundance, plays an important role in the atmospheres of the outer planets and their satellites where, for example, its photodissociation provides a synthetic route for higher hydrocarbons. Clearly, given the relative intensity of Lyman–α emission in the solar spectrum, 121.6 nm is *the* single most important photolysis wavelength for which we require reliable photochemical data for this molecule. The following five photodecomposition pathways are both spin conserving and energetically allowed at this excitation wavelength:

$$\begin{aligned}
CH_4 &\longrightarrow H + CH_3 & E_{trans} &\leq 5.72 \text{ eV} & (7a)\\
&\longrightarrow H_2 + {}^1CH_2 & E_{trans} &\leq 5.51 \text{ eV} & (7b)\\
&\longrightarrow H + H + {}^3CH_2 & E_{trans} &\leq 1.03 \text{ eV} & (7c)\\
&\longrightarrow H + H + {}^1CH_2 & E_{trans} &\leq 0.66 \text{ eV} & (7d)\\
&\longrightarrow H + H_2 + CH & E_{trans} &\leq 1.15 \text{ eV} & (7e)
\end{aligned}$$

Earlier flash photolysis studies at nearby excitation wavelengths established the presence of CH and CH_2 radicals amongst the dissociation products, and attempted to estimate the relative yields of H atoms and H_2 molecules.[23] The concensus view to emerge from these earlier studies was' that the decomposition pathways yielding methylene fragments in conjunction with atomic H (i.e. channels (7c) and (7d) combined) and with molecular H_2 (channel (7b)) are of roughly comparable importance, that the dissociation (7e) is a minor channel and, rather surprisingly, that the simple H elimination process (7a) has a negligible quantum yield.

We have used the Rydberg atom variant of the technique of H atom photofragment translational spectroscopy to further investigate the primary photochemistry of CH_4 (and CD_4) at 121.6 nm. The H atom TOF spectrum so obtained is displayed in fig. 5. Conversion of this TOF spectrum into a spectrum of the total fragment kinetic energy requires some caution since there is clearly more than one decomposition channel capable of yielding H atoms. However it is equally clear from energy conservation considerations that the fastest of the observed H atoms can only possibly be formed in conjunction with methyl fragments (i.e. channel (7a)). The horizontal scale of the total kinetic energy spectrum of the fragments (fig. 6) has been derived on the assumption that the observed H atoms carry 15/16 of the total recoil energy; i.e. that the partner fragment in each case is a methyl radical. The validity of this assumption must be questionable at the low energy end of fig. 6, as we pass the energetic thresholds for various of the other H atom forming decomposition pathways (i.e. channels (7c)–(7e) – see fig. 6). However, this ambiguity does not affect the principal finding from this preliminary reinvestigation of the Lyman–α photolysis of CH_4 – namely that fission of a single C–H bond is, in fact, a significant contributor to the total decomposition yield.

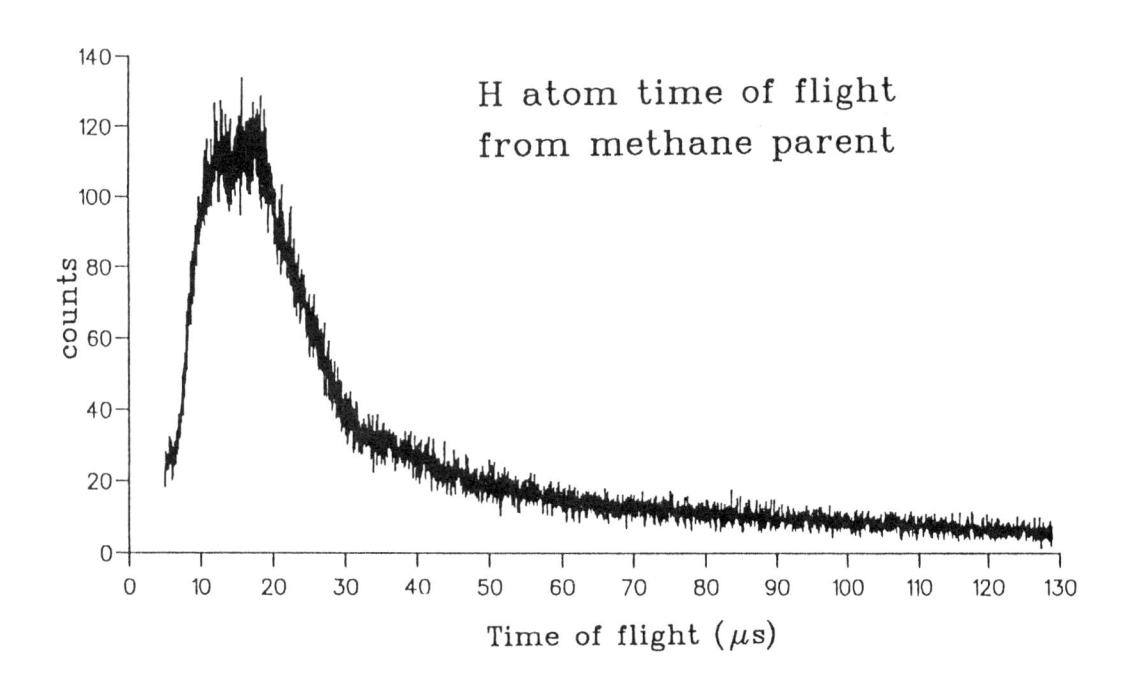

Fig. 5 H Rydberg atom TOF spectrum resulting from photolysis of jet–cooled CH_4 molecules at 121.6 nm.

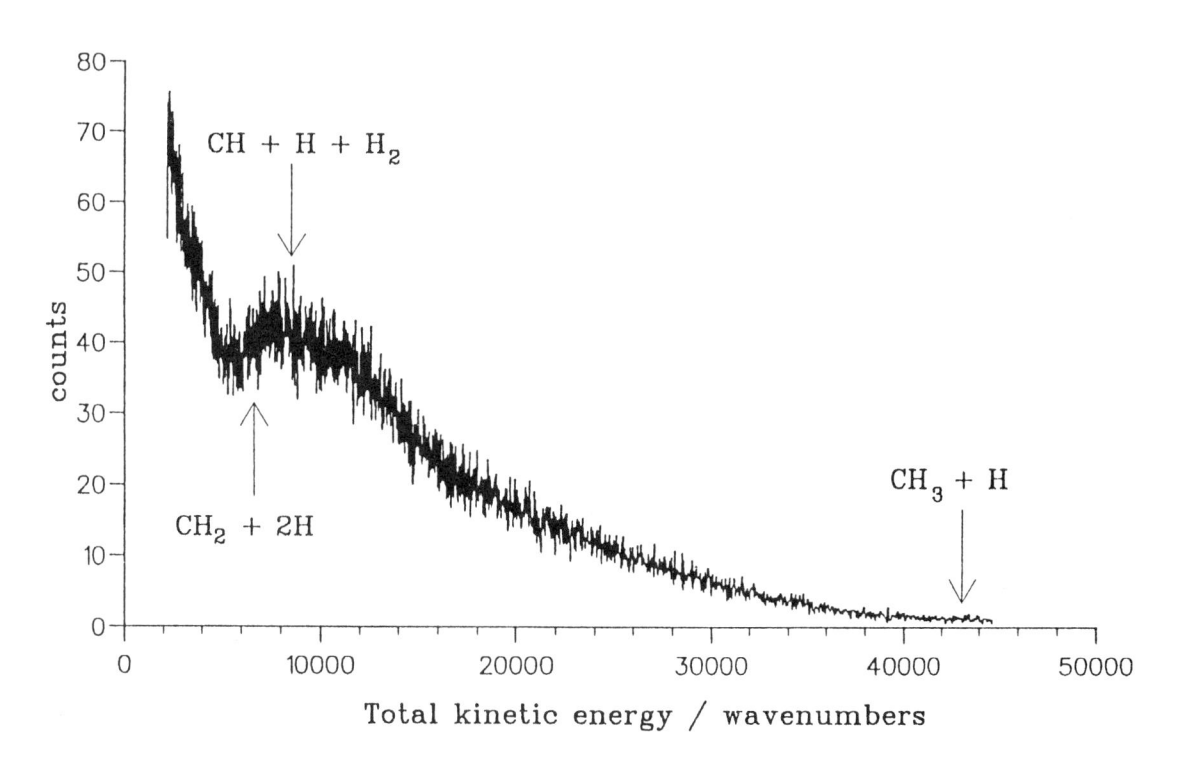

Fig. 6 Spectrum of the total kinetic energy of the fragments (assumed to be H + CH_3 – see text) arising from CH_4 photolysis at 121.6 nm. The threshold energies for the various fragmentation pathways yielding one or more H atom products are superimposed over the spectrum.

Future work must address the intriguing question of the detailed energy disposal within the partner CH_3 fragments. The total kinetic energy spectrum (fig. 6) shows no structure attributable to specific internal modes of the methyl fragment, but a crude analysis of the spectrum suggests that the mean internal energy of the CH_3 fragments, $<E_{int}>$, must be in excess of 30000 cm^{-1}. This internal excitation must be in the form of vibration–rotation excitation of CH_3 fragments in their ground electronic state, since the first excited electronic state of the radical (the $\tilde{B}^2A_1'$ state[24]) has its origin at 46205 cm^{-1}. Some excitation of the fragment out-of-plane bending mode, ν_2, is to be expected, given the fact that the CH_3 moiety in the parent CH_4 molecule has a geometry (tetrahedral) that is markedly different from the equilibrium geometry of the free radical (planar D_{3h}); the centrifugal forces associated with significant out-of-plane bending motion would inevitably promote some excitation of the ν_1 symmetric stretch mode also. However, the out-of-plane bending fundamental has a wavenumber of only $\sim$ 606 cm^{-1} and no simple Franck–Condon picture would account for the formation of methyl fragments with $\sim$ 50 quanta of such excitation. We also note, however, that Lyman–α excitation of CH_4 is likely to cause the electron promotion $t_2^5 3sa_1^1 \leftarrow t_2^6$, thereby populating the first $^1T_2$ excited state of CH_4. Analogy with the $\tilde{X}^2T_2$ ground state of CH_4^+ [25] suggests that this mixed Rydberg–valence excited state will be subject to significant Jahn–Teller distortion, which may well manifest itself as rotation and/or as excitation of the asymmetric vibrational modes of the resulting methyl photoproducts.

4. ACKNOWLEDGEMENTS

M.N.R.A. is indebted to Professor K.H. Welge and his colleagues at Universitat Bielefeld for allowing him to participate in the development and early exploitation of the technique of H atom photofragment translational spectroscopy, and to NATO for the award of the travel grant (Grant no. 85/0015) which rendered the collaboration possible. We are all very grateful to Mr. K.N. Rosser for his marvellous practical support, and to the Science and Engineering Research Council for financial support – in the form of equipment grants, a post–doctoral research fellowship (I.R.L.) and research studentships (D.H.M. and G.P.M.). C.M.W. is grateful to the Royal Society for the award of a 1983 University Research Fellowship.

5. REFERENCES

1. H.J. Krautwald, L. Schnieder, K.H. Welge and M.N.R. Ashfold, "Hydrogen atom photofragment spectroscopy: Photodissociation dynamics of H_2O in the $\tilde{B}$–$\tilde{X}$ absorption band," *Farad. Disc. Chem. Soc.*, vol. 82, pp. 99–110, 1986.
2. J. Biesner, L. Schnieder, J. Schmeer, G. Ahlers, X. Xie, K.H. Welge, M.N.R. Ashfold and R.N. Dixon, "State selective photodissociation dynamics of Ã state ammonia: I," *J. Chem. Phys.*, vol. 88, pp. 3607–3616, 1988.
3. J. Biesner, L. Schnieder, G. Ahlers, X. Xie, K.H. Welge, M.N.R. Ashfold and R.N. Dixon, "State selective photodissociation dynamics of Ã state ammonia: II," *J. Chem. Phys.*, vol. 91, pp. 2901–2911, 1989.
4. X. Xie, L. Schnieder, H. Wallmeier, R. Boettner, K.H. Welge and M.N.R. Ashfold, "Photodissociation dynamics of H_2S/D_2S following excitation within its first absorption continuum," *J. Chem. Phys.*, vol. 92, pp. 1608–1616, 1990.
5. L. Schnieder, W. Meier, K.H. Welge, M.N.R. Ashfold and C.M. Western, "Photodissociation dynamics of H_2S at 121.6 nm and a determination of the potential energy function of $SH(A^2\Sigma^+)$," *J. Chem. Phys.*, vol. 92, pp. 7027–7037, 1990.
6. M.N.R. Ashfold, R.N. Dixon, S.J. Irving, H.–M. Koeppe, W. Meier, J.R. Nightingale, L. Schnieder and K.H. Welge, "Stereochemical and angular momentum constraints in the photodissociation of ammonia," *Phil. Trans. Roy. Soc. (Lond.)*, vol. A332, pp. 375–386, 1990.

7. A.M. Wodtke and Y.T. Lee, "High resolution photofragment translational spectroscopy", in "Molecular Photodissociation Dynamics", (eds. M.N.R. Ashfold and J.E. Baggott", Royal Society of Chemistry, pp. 31–60, 1987.

8. M.N.R. Ashfold, I.R. Lambert, D.H. Mordaunt, G.P. Morley and C.M. Western, "Photofragment translational spectroscopy", J. Phys. Chem., (in press).

9. G.P. Morley, I.R. Lambert, M.N.R. Ashfold, K.N. Rosser and C.M. Western, "Dissociation dynamics of HCN(DCN) following photo-excitation at 121.6 nm", J. Chem. Phys., (submitted).

10. D.D. Davis and H. Okabe, "Determination of the bond dissociation energies in hydrogen cyanide, cyanogen and cyanogen halides by the photodissociation method", J. Chem. Phys., vol. 49, pp. 5526–5531, 1968.

11. G.A. West, Ph.D. Thesis, University of Wisconsin, 1975.

12. D. Cerny, R. Bacis, G. Guelachvili and F. Roux, "Analysis of the CN molecule", J. Mol. Spectrosc., vol.73, pp. 154–167, 1978.

13. H.C. Allen, Jr., E.D. Tidwell and E.K. Plyler, "Infrared spectrum of hydrogen cyanide and deuterium cyanide", J. Chem. Phys., vol. 25, pp. 302–307, 1960.

14. L.C. Lee, "CN($A^2\Pi \to X^2\Sigma^+$) and CN($B^2\Sigma^+ \to X^2\Sigma^+$) yields from HCN photodissociation", J. Chem. Phys., vol. 72, pp. 6414–6421, 1980.

15. G.J. Vazquez and J.F. Gouyet, "SCF CI calculation on HCN excited states", Chem. Phys. Lett., vol. 57, pp. 385–389, 1978; "SCF CI calculation on HCN: H–CN dissociation curves", Chem. Phys. Lett., vol. 65, pp. 515–522, 1979.

16. M. Peric, H. Dohmann, S.D. Peyerimhoff and R.J. Buenker, "Potential surfaces for valence-type singlet electronic states of the HCN molecule", Z. Phys. D, vol. 5, pp. 65–75, 1987.

17. M. Peric, R.J. Buenker and S.D. Peyerimhoff, "Theoretical study of the vibronic structure of the $1^1\Pi \leftarrow X^1\Sigma^+$ electronic transition in HCN and DCN", Mol. Phys., vol. 62, pp. 1323–1338, 1987; "Ab initio CI study of the vibrational structure of the $1^1\Sigma^-(1^1A'') \leftarrow X$ and $1^1\Delta(2^1A',2^1A'') \leftarrow X$ electronic transitions in HCN and DCN", Mol. Phys., vol. 64, pp. 843–864, 1988.

18. T. Nagata, T. Kondow, Y. Ozaki and K. Kuchitsu, "Absorption spectra of hydrogen cyanide and deuterium cyanide in the 130–80 nm range", Chem. Phys., vol. 57, pp. 45–53, 1981.

19. G.N.A. van Veen, K.A. Mohamed, T. Baller and A.E. de Vries, "Photofragmentation of H_2S in the first continuum", Chem. Phys., vol. 74, pp. 261–271, 1983.

20. B.R. Weiner, H.B. Levene, J.J. Valentini and A.P. Baronavski, "Ultraviolet Photodissociation Dynamics of H_2S and D_2S", J. Chem. Phys., vol. 90, pp. 1403–1414, 1989.

21. R.J. Brudzynski, R.J. Sension and B.S. Hudson, "Resonance Raman study of the first absorption band of H_2S", Chem. Phys. Lett., vol. 165, pp. 487–493, 1990.

22. F. Flouquet, "Ab initio study of the potential energy surface of the lowest A_1 symmetry excited state of H_2S", Chem. Phys., vol. 13, pp. 257–263, 1976.

23. T.G. Slanger and G. Black, "Photodissociative channels at 1216 Å for H_2O, NH_3 and CH_4", J. Chem. Phys., vol. 77, pp. 2432–2437, 1982, and references therein.

24. G. Herzberg, "The spectra and structures of free methyl and free methylene", Proc. Roy. Soc. (Lond.), vol. A262, pp. 291–317, 1961.

25. R.N. Dixon, "On the Jahn–Teller effect in CH_4^+", Mol. Phys., vol. 20, pp. 113–126, 1971.

Site-specific photochemistry

Brent Koplitz, Jeffrey L. Brum, Subhash Deshmukh,
Xiaodong Xu, Zhongrui Wang, and Yu-Fong Yen

Tulane University, Department of Chemistry
New Orleans, Louisiana 70118

ABSTRACT

We present results on "site-specific" H-atom production in photolysis experiments conducted under collisionless conditions. H and D atoms are used as labels to investigate the site(s) at which C-H (or C-D) bond cleavage occurs in a variety of haloalkane systems. Experiments using two photolysis lasers clearly indicate that photon absorption by an intermediate, presumably an alkyl radical, is important in many of the systems studied. The site(s) (e.g. α, β, or γ) at which C-H (or C-D) bond cleavage occurs is dependent not only on the nature of the molecule, but also on the photolysis wavelength. As a diagnostic tool, H- and D-atom Doppler spectroscopy allows us to gain insight into the energetics associated with the various dissociation processes. Our overall aim is to gain a further understanding of the photolysis properties of a variety of simple molecules and their associated radicals.

1. INTRODUCTION

Selective bond cleavage continues to be a topic of considerable interest in photochemical studies on small molecules.[1] Recently, attempts to understand and influence photolysis reactions that involve competing H- and D-atom channels in HOD[2-5] and HC_2D[6] have produced notable results. These interesting experiments focus on understanding and/or controlling the H versus D competition that occurs with respect to bond cleavage at a particular reactive site. In this paper, we demonstrate and discuss how H and D atoms can be used as labels to investigate photolysis involving competition between <u>chemically distinct</u> reactive sites. The question is not whether H or D is formed. Rather, H and D are used as labels to determine at which site bond cleavage occurs.

This paper reports on studies that systematically explore "site-specific" H-atom production in photolysis experiments. Note the use of the term "specific" as opposed to "selective." In our opinion, the word selective implies choice via understandable control. Earlier work on mixed haloalkanes by Butler et al.[7] or recent experiments on HOD by Crim and co-workers[2,3] would fall into the category of being "selective." Our experiments are not yet to this stage (although we hope that eventually they will be). We do, however, classify our experiments as being specific in that nature often chooses to break chemical bonds at well-defined, identifiable locations. Moreover, these locations may or may not be the sites that one expects. By measuring relative H/D ratios in selectively deuterated compounds, competition between α and β, or α, β, and γ sites can be quantitatively compared.

Our earlier work on the 248 nm photolysis of iodoethane clearly shows that C-H bond cleavage occurs primarily at the β carbon, most likely through an ethyl radical intermediate.[8] Results on selectively deuterated n-iodopropanes: $ICD_2CH_2CH_3$, $ICH_2CD_2CH_3$, and $ICH_2CH_2CD_3$ also demonstrate a preference for C-H (or C-D) bond cleavage at the β position when using 248 nm radiation.[9] In contrast, experiments involving 2-iodopropane show enhancement of C-H bond cleavage at the α, not the β site[10]. In all cases, radical photochemistry appears to play a significant role. As a diagnostic tool, H- and D-atom Doppler spectroscopy allows us to investigate the energetics associated with the various dissociation processes. Our overall aim is to more fully understand the absorption and dissociation properties of haloalkane compounds and their associated radicals. As a theoretical approach, *ab initio* calculations are used to gain additional insight into the site-specific behavior observed in the radical photochemistry. An understanding of haloalkane photochemistry is of potential environmental importance, and how such knowledge may be utilized with compounds implicated in ozone depletion is also discussed.

2. EXPERIMENTAL SECTION

As depicted in Fig. 1, the experimental arrangement for these studies is fairly standard and has been described elsewhere.[9,10] Briefly, the ionization region of a time-of-flight mass spectrometer (TOFMS) has photolysis and probe laser beams passing through it. Commercially available haloalkanes (Aldrich, MSD, or Cambridge Isotopes) are introduced into the TOFMS via a leak valve. A pulsed nozzle (General Valve) is also available for sample introduction. The output from a photolysis excimer laser (Questek 2220) is focused into the ionization region of the TOFMS with a lens (focal length (f.l.) = 1 m or 250 mm). After a short delay (typically 20 ns), the output of an excimer-pumped dye probe laser (Lambda Physik LPX 105; FL 3002) counterpropagates through the chamber. A small fraction ($\sim 10^{-6}$) of the probe radiation ($\sim$365 nm) is frequency tripled prior to entering the chamber by focusing the laser output into a cell containing $\sim$80-100 Torr of Kr. The resulting vacuum ultraviolet (VUV) light is tuned through resonance with the Lyman-α transition of atomic hydrogen (121.6 nm). Both the tripled output and the remaining fundamental radiation are re-focused by a LiF lens (f.l. = 64 mm) and overlapped with the photolysis beam inside the ionization region. Sequential absorption of 121.6 and 364.7 nm photons produces H^+ ions from the neutral H atoms. Upon formation, the ions are accelerated, pass through a flight tube, and impinge on a microchannel plate detector. The amplified signal is captured and processed by a 100 MHz digitizer (DSP Co.) controlled by a microcomputer. When a second photolysis excimer laser is used, the firing sequence is adjusted accordingly.

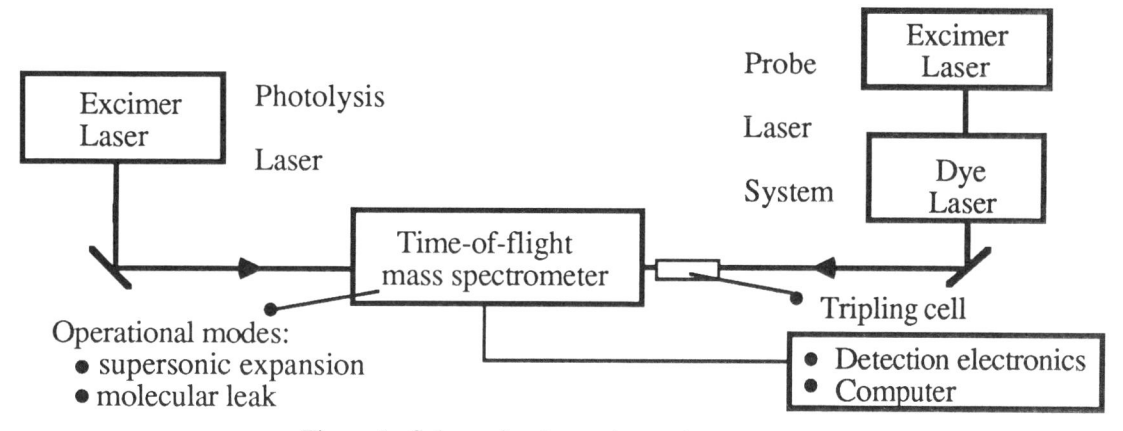

Figure 1. Schematic of experimental apparatus.

3. RESULTS AND DISCUSSION

3.1 Iodoethane Photolysis

Iodoethane has been well-characterized spectroscopically; however, photochemical experiments have naturally tended to focus primarily on processes relevant to halogen-carbon dissociation,[11] the dominant reactive pathway. Recently, we have observed H-atom formation subsequent to iodoethane photolysis at 193 and 248 nm.[8,12] The transition at each wavelength has long been thought to involve an iodine lone pair (non-bonding) electron,[13] and the predominant initial reactive event should be simple C-I bond scission. (However, at 193 nm direct C-H bond cleavage may constitute a minor, but competitive, pathway. Evidence for a small contribution ($\sim$3%) from this channel has been shown for CH_3I photolysis by Continetti et al.[14]) Assuming the initial step is cleavage of the C-I bond, an ethyl radical remains:

$$\alpha \quad {}^{H}_{H}\!\!>\!\!\overset{\bullet}{C}-\overset{H}{\underset{H}{C}}\!-H \quad \beta$$

where α and β distinguish the two different carbon atoms. Selectively deuterated iodoethanes, ICD_2CH_3 and ICH_2CD_3 (98% D-atom enrichment), are used to label the sites initially. At which carbon atom does C-H bond cleavage occur? In the case of the 193 nm dissociation of ICD_2CH_3 shown in Fig. 2a, there is no obvious evidence for dissociation occurring at a particular carbon atom.[8] The observed ratio (1.6:1) is consistent with the initial H:D ratio of 3:2 present in the parent molecule. However, when the wavelength for ICD_2CH_3 photolysis is switched to 248 nm, the situation is remarkably different as is shown in Fig. 2b. Here, the H/D ratio is ~11:1, clearly indicating a preference for bond cleavage at the β carbon.[8] Even if one chooses to adjust the ratio by a factor of 1.5 for simple statistical bias, the ratio is still 7:1. By using ICH$_2$CD$_3$, i.e. the H and D positions have been reversed, we demonstrate in Fig. 2c that the dissociation process is indeed carbon-atom specific and not due solely to some type of H-atom dynamics. As shown, *the D-atom channel now dominates*, although the D/H ratio of 3:1 (2:1 if statistically adjusted) is not quite as dramatic as the H/D ratio for ICD$_2$CH$_3$ shown in Fig. 2b. Perhaps the relative mobility of an H atom versus a D atom plays a role in this reduction. In any case, these results demonstrate unequivocally that C-H (or C-D) bond cleavage occurs predominantly on the β carbon following excitation at 248 nm, and it is indeed the carbon site that is important.

H-atom power dependence studies on C$_2$H$_5$I at 248 nm (not shown) yield a log-log plot that is clearly nonlinear, the slope being ~1.7.[12] This finding is consistent with the argument that given various energy considerations (e.g. the I-CH$_2$CH$_3$ bond energy and the I-CH$_2$CH$_3$ dissociation energetics),[7a-c,12] absorption of a single 248 nm photon by iodoethane will not sequentially break both a C-I and a C-H bond in most cases. For the majority of C-I dissociation events, there simply will not be enough internal energy deposited in the ethyl radical to overcome the barrier (~1.7 eV) to H-atom production.[12] Consequently, a logical candidate for secondary absorption is the ethyl radical, which possesses a reasonable cross section (2 x 10^{-18} cm^2)[15] at 248 nm. Additional discussion concerning radical intermediate photochemistry can be found in subsequent sections.

3.2. 1-Iodopropane Photolysis

A natural extension of the iodoethane work involves the photolysis of 1-iodopropane. With this system, we ask the following question. Is the site preference exhibited by iodoethane at 248 nm a β carbon effect, or is it a terminal carbon effect? To address this question, we studied the photolysis of three selectively deuterated 1-iodopropanes. In each case, a specific carbon site has been deuterated α, β, or γ with respect to the I atom (i.e., ICD$_2$CH$_2$CH$_3$, ICH$_2$CD$_2$CH$_3$, and ICH$_2$CH$_2$CD$_3$). As is the case with iodoethane, atomic hydrogen production resulting from the the 193 nm photolysis of 1-iodopropane displays no dominant site preference (Fig. 3a).[9] In contrast with the 193 nm data, however, 248 nm excitation clearly produces a propensity for carbon-hydrogen bond cleavage at the β position (Fig. 3b), not the terminal (γ) site. The determined ratios are 0.2, 0.7, and 0.1 for the α, β, and γ sites, respectively. Once again, a key step appears to be photon absorption by an intermediate, since power dependence studies produce nonlinear log-log plots (not shown).[9] The logical candidate for the intermediate is the *n*-propyl radical.

3.3 2-Iodopropane Photolysis

We have also conducted experiments on the excimer laser initiated photolysis of 2-iodopropane.[10] By using the selectively deuterated 2-iodopropanes, ICH(CD$_3$)$_2$ and ICD(CH$_3$)$_2$ (shown below), the "site-specific" nature of the

$$\textbf{I} \quad CD_3-\underset{H}{\overset{I}{\underset{|}{\overset{|}{C}}}}-CD_3 \qquad \textbf{II} \quad CH_3-\underset{D}{\overset{I}{\underset{|}{\overset{|}{C}}}}-CH_3$$

carbon-hydrogen bond cleavage is once again investigated. For compound **I**, photolysis at 193 nm results in an observed D/H ratio of ≥7:1. This result is shown in Fig. 4a. In contrast, 248 nm photolysis enhances carbon-hydrogen bond cleavage at the α-carbon site to the point where the D/H ratio is almost unity, as shown in Fig. 4b. The isopropyl radical is implicated as being a key intermediate, and power dependence studies at 248 nm suggest that an overall two-photon pathway is important.[10]

The observed propensity toward 248 nm photo-induced C-H bond cleavage at the β carbon site seems to occur for both the ethyl radical[8,16] and the *n*-propyl radical.[9] Thermodynamically, this result is conveniently explained by noting that the formation of a carbon-carbon double bond (i.e. H$_2$C=CH$_2$ + H or H$_2$C=CHCH$_3$ + H) could be a strong driving force. However, such a simple explanation fails to explain why it is C-H bond cleavage at the α site that apparently is enhanced by 248 nm excitation of the *isopropyl* radical.[10] In other words, our experimental results show that the likely products are CH$_3$CCH$_3$ + H, not H$_2$C=CHCH$_3$ + H. Certainly more work needs to be done in order to understand these observations.

3.4 Photolysis of Choroethane

In the previously mentioned 248 nm photolysis studies on iodoalkanes, power dependence studies suggest that photon absorption by an intermediate is occurring, and the likely candidates are the respective alkyl radicals. However, finding definitive proof is hampered by the fact that the iodo compounds absorb strongly at ~248 nm[13] as do the alkyl radicals.[15] Fortunately, using chloroalkanes as radicals precursors helps circumvent this problem, since the first absorption band for the chloro compounds is shifted further down into the UV than their iodo counterparts. In fact, 193 nm light *will* access the first electronic state of $ClCH_2CH_3$, but 248 nm radiation will not. (Note that this electronic state is directly dissociative, analogous to the *A*-state transition observed at 248 nm in ICH_2CH_3 and other iodoalkanes.)[13] If our understanding about the overall mechanism for H-atom generation in these haloethane systems is correct at 248 nm, one should be able to generate CH_2CH_3 by photolyzing $ClCH_2CH_3$ with 193 nm radiation and use a second photolysis laser operating at 248 nm to enhance H-atom production through irradiation of the CH_2CH_3 intermediate. Since CH_2CH_3 has an appreciable cross section at 248 nm as stated above, H-atom enhancement is expected. As anticipated, substantial H-atom enhancement is in fact achieved.[16] Moreover, experiments on $ClCH_2CD_3$ and $ClCD_2CH_3$ clearly show that this enhancement occurs preferentially through carbon-hydrogen bond cleavage at the β carbon site.[16] It is apparent that 248 nm photon absorption by the ethyl radical is an important step in the overall mechanism.

3.5 Calculations

In order to further understand site-specific photochemistry in haloalkanes, we have carried out *ab initio* calculations on the intermediate alkyl radicals using the GAUSSIAN 90 system of programs. Single excitation configuration interaction (CIS) computations were performed to obtain optimized geometries in the excited states. These calculations used a 6-31G basis set supplemented by diffuse functions and a set of polarization functions on the carbon atom (6-31+G*). As an example of our preliminary results, a brief discussion of our calculations on the 3s Rydberg state of the ethyl radical is presented. The optimized geometry of the ground state ethyl radical at the UHF/6-31G level was used as the starting geometry for the excited state calculation. The optimized geometry in the 3s Rydberg state shows that the C-H bond on the β-carbon site is 1.32 Å, while this bond length in the ground state is 1.08 Å. The C-C bond length reduces to 1.34 Å in the excited state from a starting length of 1.49 Å in the ground state, while the C-H bond on the α-carbon site remains practically unchanged upon excitation to the 3s Rydberg state. The input Z-matrix for these calculations did not have any bias towards any of the bonds and all symmetry constraints were removed. It appears, from these results, that the excitation of the ethyl radical to the 3s Rydberg state tends to stretch the C-H bond on the β-carbon. Similar calculations on the isopropyl radical clearly show an extension of the C-H bond at the α-carbon when the radical is excited to the 3p Rydberg state. We do not speculate, at this point, on the usefulness of these excited state geometry optimization results in explaining the observed site-specific C-H bond cleavage behavior in these systems. However, we do to point out that these results match our experimental observations very well and therefore could certainly be described as suggestive.

3.6 Application to Atmospheric Chemistry

Site-specific photochemistry is not only of interest fundamentally, but, from a practical perspective as well. For example, there exist a number of halocarbon compounds that influence global change by depleting the ozone layer, contributing to global warming, etc. As chlorofluorocarbons (CFCs) are phased out of widespread production, there are a variety of hydrochlorofluorocarbons that are being considered as replacements. The photochemistry of these compounds, especially with respect to the production of reactive H atoms, is a topic of interest as well as possible concern. Note that the photolysis of CH_3CCl_3 will produce the CH_3CCl_2 radical, which has three β hydrogen atoms. Likewise, the photolysis of CH_3CCl_2F (HCFC-141b) will produce a radical with three β hydrogen atoms. In contrast, the photolysis of CF_3CHCl_2 (HCFC-123) will produce a radical containing *no* β hydrogen atoms. Will the site-specific C-H bond cleavage described above for CH_3CH_2 (i.e. preferential bond cleavage at the β carbon site) affect H-atom generation in systems of this type? In other words, will we see *appreciably* less H-atom production when CF_3CHCl_2 is photolyzed not only because there are fewer H atoms, but also because they are in the "wrong" place?

Along these same lines, we point out that atmospheric chemistry of a compound such as CH_3CCl_3 actually produces two primary radicals. Photolysis in the stratosphere will initially break a C-Cl bond leaving the CH_3CCl_2 radical as discussed above. However, the OH radical acting on CH_3CCl_3 in the troposphere will abstract an H atom, thereby leaving CH_2CCl_3 (note the absence of β hydrogen atoms). We submit that these two radicals will have substantially different H-atom yields when photolyzed in the ultraviolet. Experiments that address issues of this sort are currently underway in our laboratory.

4. SUMMARY

Our current focus is on the experimental identification and characterization of H-atom production in a variety of molecular systems. We are beginning to identify the important photolytic pathways that are involved in this site-specific chemistry; however, a good deal more research effort (both experimental and theoretical) is needed before a clear picture develops. For example, we know that 248 nm radical photolysis constitutes an important step in producing site-specific behavior, but the mechanism behind the "probabilistic" H/D ratios for the 193 nm photolysis of selectively deuterated iodo compounds is still not obvious. Future experiments on haloalkanes involving two photolysis lasers will facilitate our ability to isolate/identify these important mechanistic steps. Ultimately, we hope to influence the photochemistry of these compounds throughout the ultraviolet region. Theoretically, the input from our *ab initio* calculations should offer insight. In general, the importance of understanding such photochemistry has practical significance in areas ranging from atmospheric ozone depletion to semiconductor fabrication.

5. ACKNOWLEDGMENTS

Acknowledgment is made to the Petroleum Research Fund, administered by the American Chemical Society, for partial support of this research. We also acknowledge support by the Department of Energy through its NIGEC program, the Louisiana Board of Regents, and the Tulane University Center for Bioenvironmental Research.

6. REFERENCES

1. Mode Selectivity in Unimolecular Reactions. Chemical Physics **193** (1989) Special Issue.
2. R.L. Vander Wal, J.L. Scott, and F.F. Crim, "Selectively Breaking the O-H Bond in HOD," Journal of Chemical Physics **92**, 803-4 (1990).
3. A. Sinha, M.C. Hsiao, and F.F. Crim, "Bond Selected Bimolecular Chemistry: H + HOD($4\nu_{OH}$) --> OD + H_2," Journal of Chemical Physics **92**, 6333-5 (1990).
4. N. Shafer, S. Satyapal, and R. Bersohn, "Isotope Effect in the Photodissociation of HDO at 157.5 nm," Journal of Chemical Physics **90**, 6807 (1989).
5. I. Bar, Y. Cohen, D. David, S. Rosenwaks, and J.J. Valentini, "Direct Observation of Preferential Bond Fission by Excitation of a Vibrational Fundamental: Photodissociation of HOD (0,0,1)," Journal of Chemical Physics **93**, 2146-8 (1990).
6. T.A. Cool, P.M. Goodwin, and C.E. Otis, "H/D Isotope Effect in the Predissociation of C_2HD," Journal of Chemical Physics **93**, 3714-5 (1990).
7. L.J. Butler, E.J. Hintsa, S.F. Shane, and Y.T. Lee, "The Electronic State-Selective Photodissociation of CH_2BrI at 248, 210 and 193 nm," Journal of Chemical Physics **86**, 2051-74 (1987).
8. J.L. Brum, S. Deshmukh, and B. Koplitz, "Iodoethane Photolysis: Which C-H Bond Leads to H-Atom Formation?," Journal of Chemical Physics **93**, 7504-5 (1990).
9. J.L. Brum, S. Deshmukh, and B. Koplitz, "Site-Specific Branching Ratios for H-atom Production Resulting from the 193 and 248 nm Photolysis of 1-Iodopropane," Journal of Physical Chemistry **95**, 8676-80 (1991).
10. J.L. Brum, S. Deshmukh, and B. Koplitz, "Site-Specific Bond Cleavage Leading to H-Atom Production in the Photolysis of 2-Iodopropane," Journal of the American Chemical Society **113**, 1432-4 (1991).
11. (a) S. Riley and K. Wilson, "Excited Fragments from Excited Molecules: Energy Partitioning in the Photodissociation of Alkyl Iodides," Discussions of the Faraday Society **53**, 132-46 (1972); (b) P. Brewer, P. Das, G. Ondrey, and R. Bersohn, "Measurement of the Relative Populations of $I(^2P_{1/2})$ and $I(^2P_{3/2})$ by Laser Induced Vacuum Ultraviolet Fluorescence," Journal of Chemical Physics **79**, 720-23 (1983); (c) H. Okabe, *Photochemistry of Small Molecules*, Wiley, New York, 1978.
12. S. Deshmukh, J.L. Brum, and B. Koplitz, "Photolysis of Iodoethane: Atomic Hydrogen Generation," Chemical Physics Letters **176**, 198-202 (1991).
13. M.B. Robin, *Higher Excited States of Polyatomic Molecules,* Vol. I, Academic Press, New York, 1974.

14. R.E. Continetti, B.A. Balko, and Y.T. Lee, "Symmetric Stretch Excitation of CH$_3$ in the 193.3 nm Photolysis of CH$_3$I," Journal of Chemical Physics **89**, 3383-4 (1988).
15. H.R. Wendt and H.E. Hunziker, "The UV Spectra of Primary, Secondary, and Tertiary Radicals," Journal of Chemical Physics **81**, 717-23 (1984).
16. J.L. Brum, S. Deshmukh, and B. Koplitz, "Conclusive Evidence for Site-Specific C-H Bond Cleavage Resulting from 248 nm Photolysis of the Ethyl Radical," Journal of Chemical Physics **95**, 2200-2 (1991).

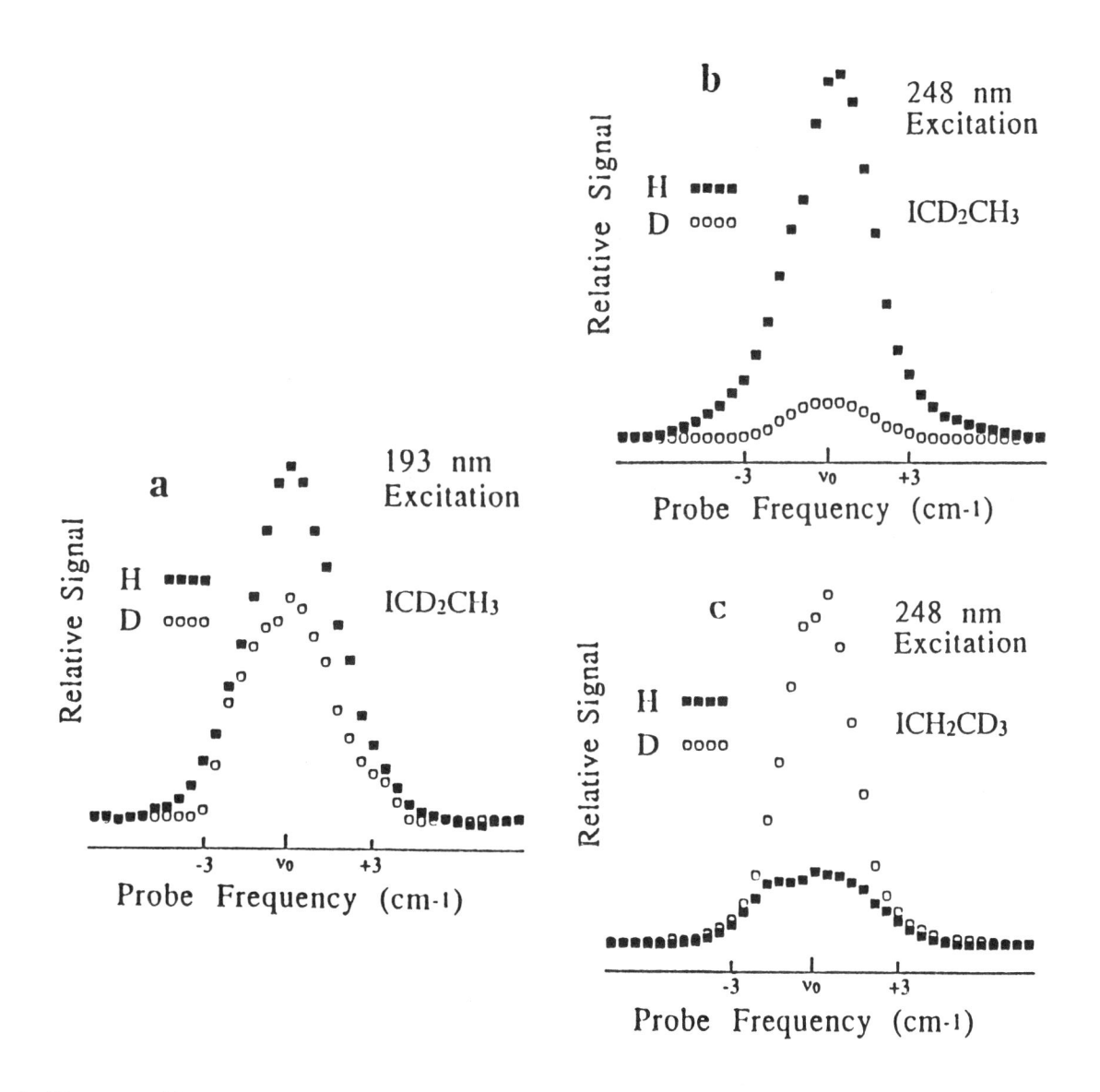

Figure 2. H-atom and D-atom Doppler profiles resulting from the 193 nm photolysis of (a) ICD$_2$CH$_3$ and the 248 nm photolysis of (b) ICD$_2$CH$_3$ and (c) ICH$_2$CD$_3$. In (a), the observed H/D ratio is 1.6:1, very close to the simple statistical ratio of 1.5:1. In (b) and (c), bond cleavage from the β carbon atom is clearly preferred. For the H-atom signal, $v_0 = 82,259.1$ cm^{-1}. For the D-atom signal, v_0 is shifted upward in energy from this value by 22 cm^{-1}.

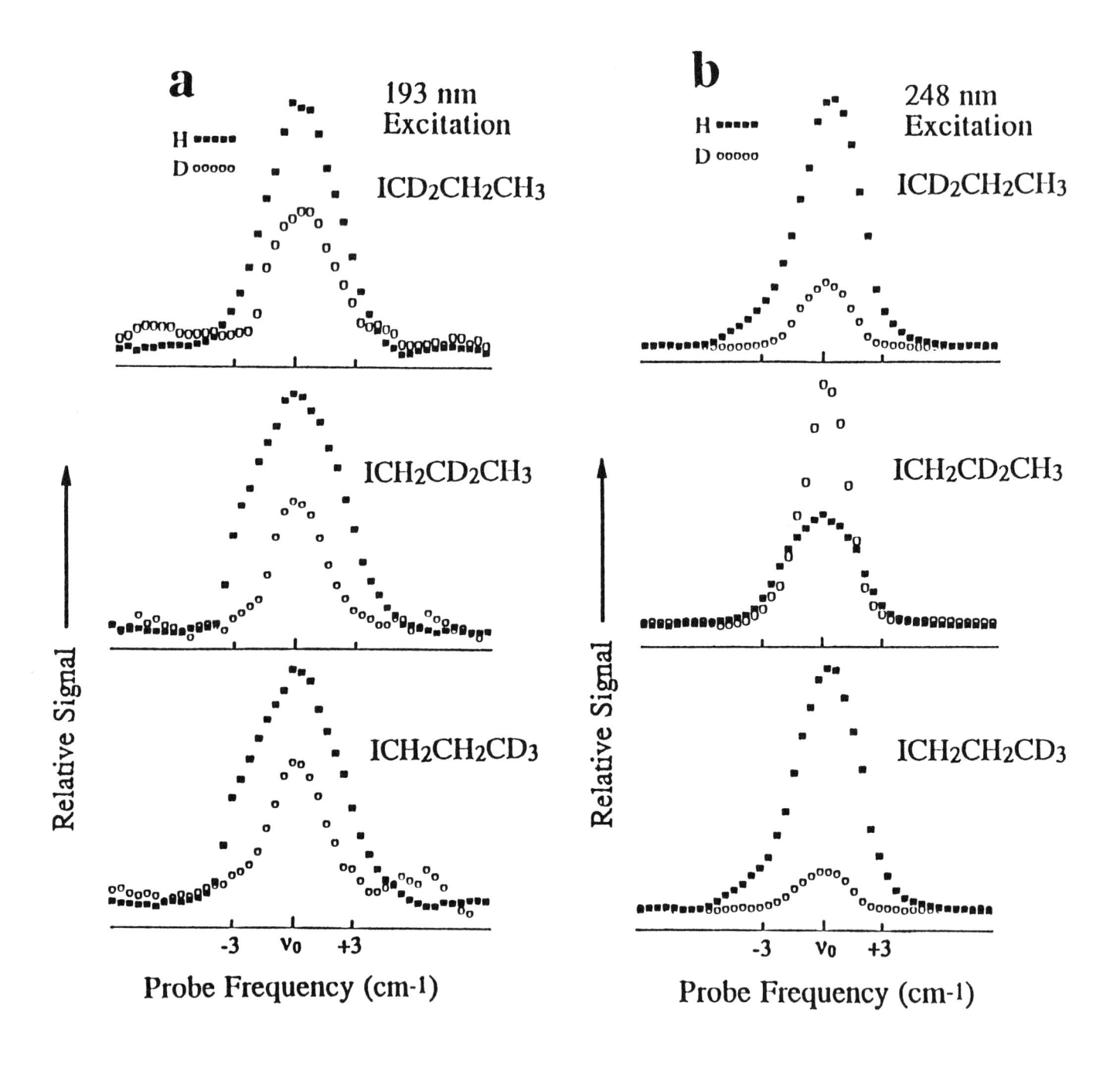

Figure 3. Doppler profiles for H and D atoms resulting from the (a) 193 nm and (b) 248 nm photolysis of selectively deuterated 1-iodopropanes. For atomic hydrogen, $\nu_0 = 82,259.1$ cm^{-1}, while for atomic deuterium this value is shifted upward in energy by 22 cm^{-1}. Each pair of profiles is normalized. The 193 nm induced profiles display no obvious site-specificity, but the 248 nm induced profiles clearly indicate a prefence toward C-H (or C-D) bond cleavage at the β carbon position.

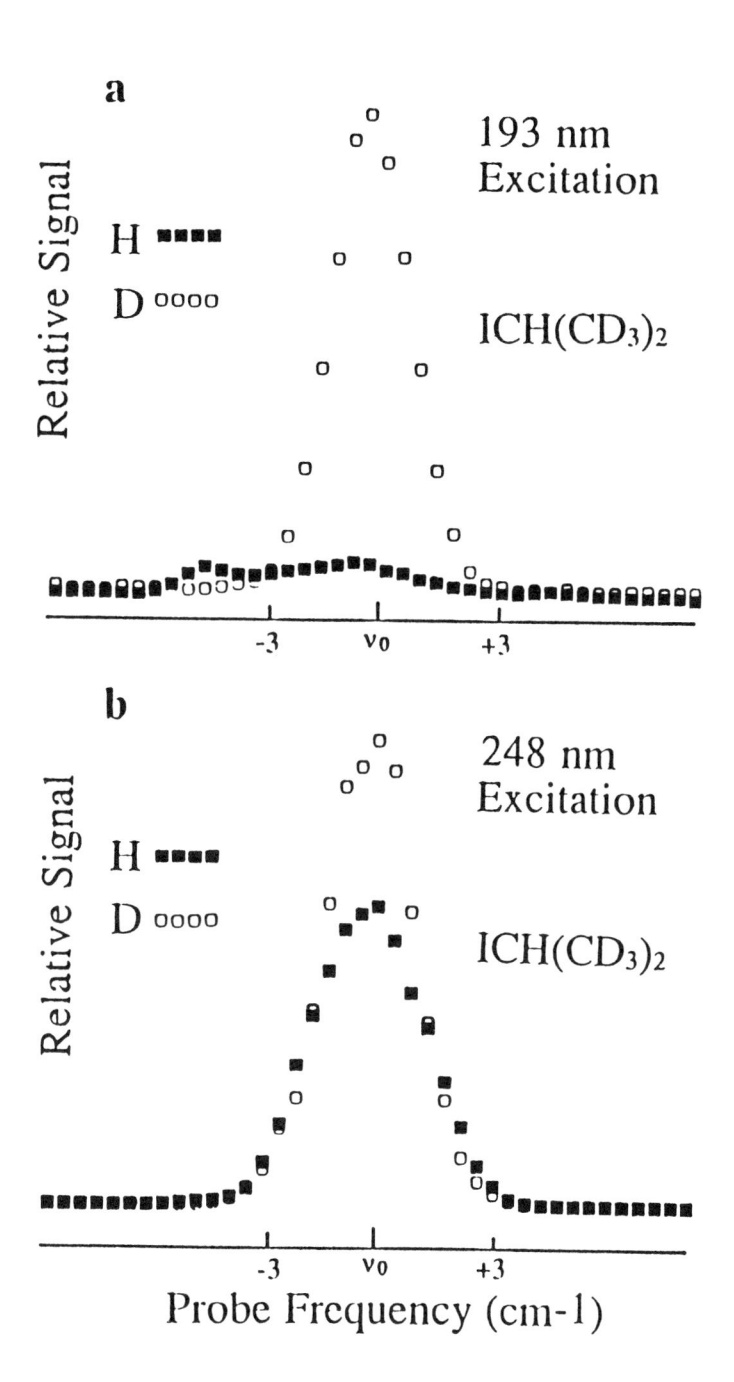

Figure 4. H-atom and D-atom Doppler profiles resulting from the photolysis of ICH(CD$_3$)$_2$ at (a) 193 nm and (b) 248 nm. Clearly, C-H bond cleavage (the α site) is enhanced at 248 nm. For the H-atom signal, ν_0 = 82,259.1 cm^{-1}. For the D-atom signal, ν_0 is shifted upward in energy from this value by 22 cm^{-1}.

C-H Dissociation rate constants of alkylbenzenes from hot molecule formed by 158 nm (F_2 laser) irradiation

Nobuaki Nakashima,[#] Tetsuya Shimada,[#] Yuichi Ojima,[$]
Yasukazu Izawa,[#] and Chiyoe Yamanaka[$]

[#]Institute of Laser Engineering, Osaka University, Yamada-Oka Suita Osaka, JAPAN 565
[$]Institute for Laser Technology, Yamada-Oka Suita Osaka, JAPAN 565

ABSTRACT

The C-H bond in the methyl group of toluene, p-xylene and mesitylene dissociates to the corresponding radicals from the hot molecules with internal energy of 8 eV. The dissociation rate constants were measured by the method of nanosecond laser photolysis. These rate constants can be predicted by a statistical theory on the basis of those measured by 193 nm irradiation. The C-C bond in paracyclophane was found to dissociate by two photons at 193 nm.

1. INTRODUCTION

We applied an F_2 laser to measure simple dissociation reaction rate constants of alkylbenzenes and studied multiphton dissociation of paracyclophane by an ArF laser.

The dissociation rate constant from hot toluene, prepared by isomerization of excited cycloheptatriene, has been measured by Hippler et al. in 1981.[1] It has been shown that hot toluene can be prepared by internal conversion by excitation of toluene and the hot toluene dissociates in 1985 by us.[2] We call this dissociation pattern as hot molecule mechanism. Now more than 20 rate constants have been reported[3,4,5] and can be explained in terms of a thermal reaction theory of SACM(Statistical Adiabatic Channel Model).[3]

In this paper, an important molecule, toluene, was irradiated at short wavelengths of F_2 laser light(158 nm). The same mechanism has found to be still operative. The constant will be shown to be explained quite well in terms of SACM, where the rate constants previously obtained at 193 nm are taken into account. The rate constants of other alkylbenzenes were similarly discussed. Two photon dissociation of paracyclophane has been studied related to the hot molecule mechanism.

2. EXPERIMENTAL

Figure 1 shows our set up for VUV nanosecond laser photolysis. An F_2 laser (Mitsubishi Heavy Industries) has a typical pulse width of 20 ns(FWHM) and the power of 50 mJ/pulse. The output power was measured by a Gentec ED 500 without absolute calibration. The detectors were in a shielded enclosure.

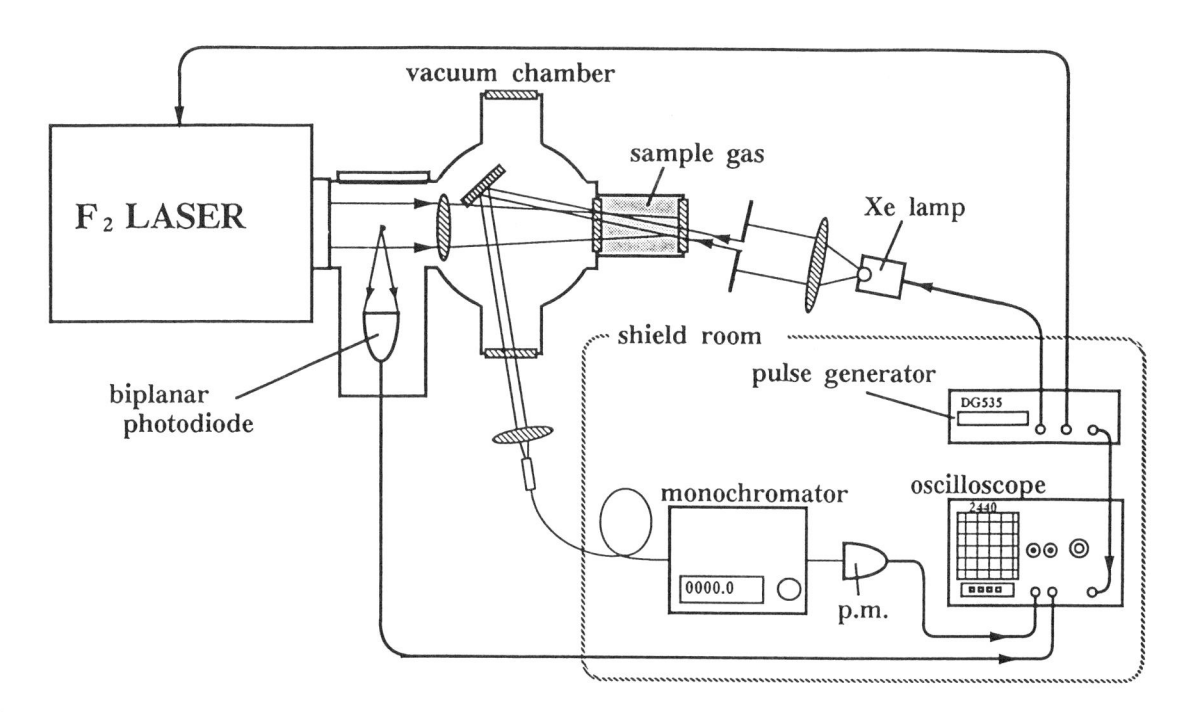

Fig. 1. Schematic diagram of F$_2$ laser photolysis.

3. RESULTS AND DISCUSSION

3.1. Absorption spectra of alkylbenzene and photochemistry

The absorption spectra of p-xylene, paracyclophane, and p-xylyl radical are plotted in the Fig. 2. For p-xylene, F$_2$ laser light(158 nm) pumps into the tail of the strong absorption band corresponding to the $^1E_{1u}$ -- $^1A_{1g}$ transition in benzene. A Rydberg band may coincide with the laser light. The spectral features of other alkylbenzenes treated here and benzyl type radicals are similar to those in the figure.

The following photochemistry is expected for p-xylene:

$$CH_3-\bigcirc-CH_3 \xrightarrow{h\nu} S_n \xrightarrow{k_{ic}} S_0^{**} \xrightarrow{k_1} CH_3-\bigcirc-\dot{C}H_2 + H$$
$$S_0 \xleftarrow{k_q[M]}$$

After rapid internal conversion (k_{ic}) the hot molecule is formed. S_0^{**} of p-xylene has an internal energy of 8.03 eV, which corresponds to a vibrational temperature of 2700 K. The major process is the C-H dissociation to xylyl radical (denoted by k_1). Collisional energy transfer leading finally to the ground state molecule in a thermal average state (denoted by $k_q[M]$) will be important under relatively high pressures.

The rate constant observed by measuring the rise curve of the benzyl type radical will be the following;

$$k_{obs} = k_1 + k_q[M] \tag{1}$$

The second term can be neglected for toluene and p-xylene, in other words, k_{obs} was great enough under low pressures (less than 0.5 Torr). For mesitylene and deuterated p-xylene($4.6 \times 10^6 s^{-1}$), the rate constants free from $k_q[M]$ were obtained by extrapolation to the zero pressure limit.

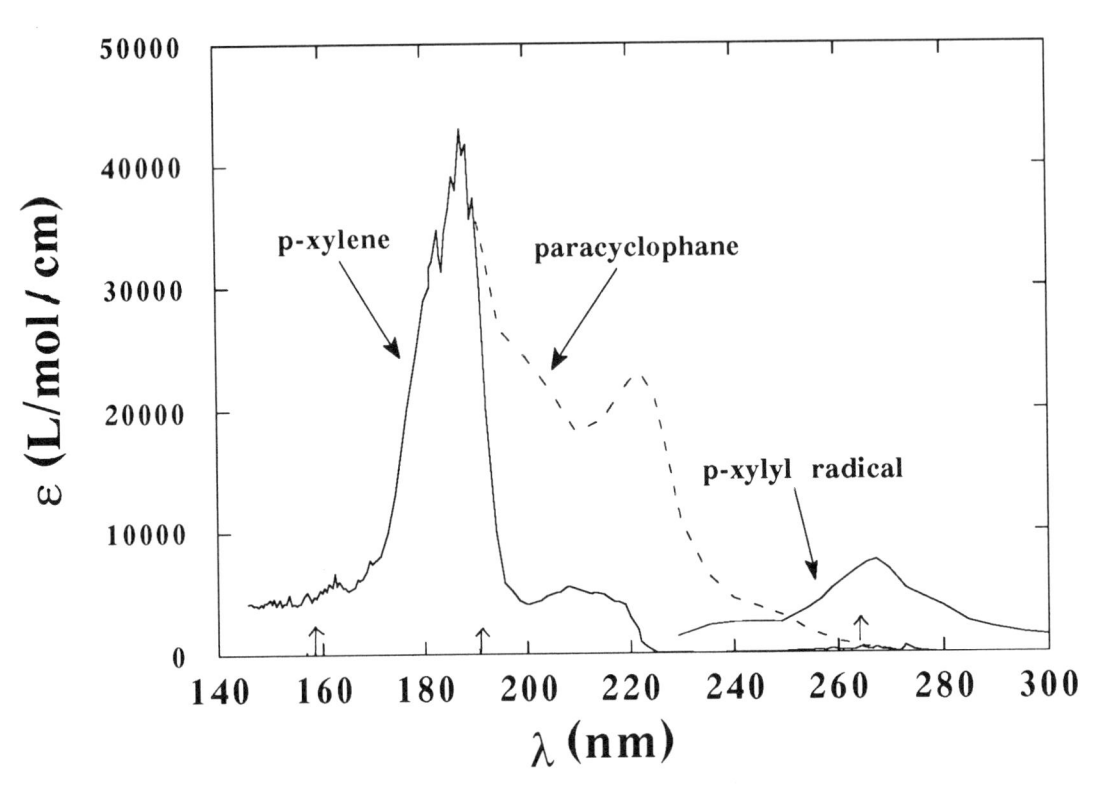

Fig. 2. Absorption spectra of p-xylene,[6] paracyclophane (420 K), and p-xylyl radical (1700K).[3] The excitation and observed wavelengths were indicated with arrows.

3.2. Time profiles of rise curve of radicals and determination the time constants

Time dependence of the concentration of radicals from three alkylbenzenes irradiated at 158 nm is shown in Fig. 3. The rise time k_{obs} was determined by analyzing a function for the time dependence of the absorbance. The function at a observing wavelength is assumed to have a single time constant, k_{obs}, and two parameters, C_1 and C_2 ;

$$f(t) = C_1(1 - \exp(-k_{obs}t)) + C_2 \tag{2}$$

We can evaluate k_{obs} and a ratio of C_2 / C_1 on the basis of a rise curve. The term C_1 is mainly due to absorption of benzyl type radical. The second term C_2 comes from absorptions of hot molecule and of products by multiphoton absorption. The observed time profiles were compared with the function convoluted with the excitation laser pulse. By repeated calculations the best fit k_{obs}'s were determined. The results are summarized in Table I with rate constants from the literature.

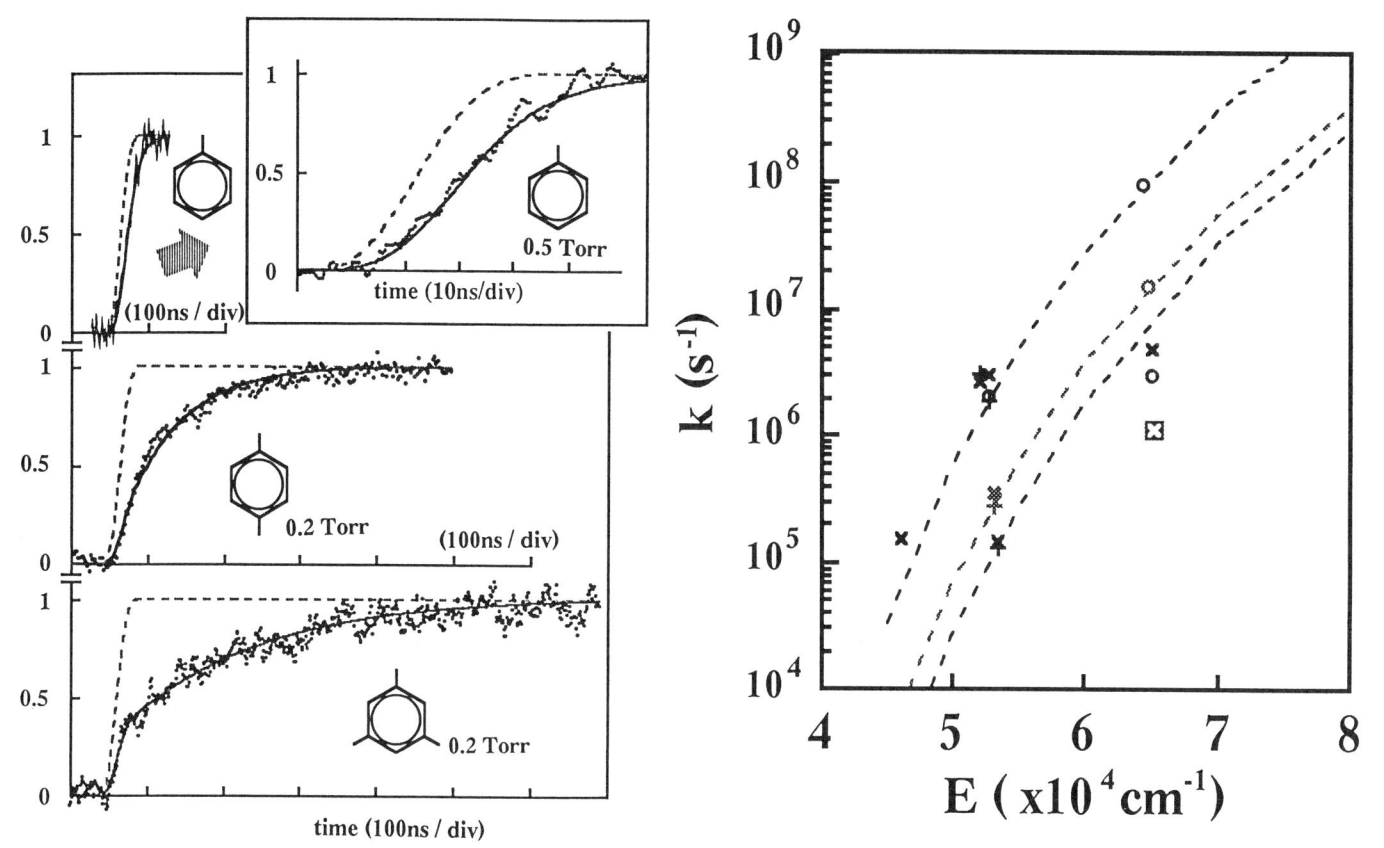

Fig. 3. Rise curves of benzyl type radicals. The dotted lines are integrated curves of the laser pulse shape, i.e., $k_{obs} = \infty$. The solid lines are the best fitted rise curves.

Fig. 4. Observed (o(by this group), +(Ref.3), x and white x(Durene)(Ref.5), and calculated curves.[3]

Table I. Dissociation rate constants (selected values)[a]

	193 nm	Ref.	158 nm	Ref.
	s^{-1}		s^{-1}	
Toluene	2.0×10^6	4	9.3×10^8	This work
	1.9×10^6	3		
(222 nm)	1.5×10^5	5		
p-Xylene	2.7×10^6	3	1.5×10^7	This work
Mesitylene	1.2×10^5	5[b]	4.8×10^6	5
	1.3×10^5	3	3.0×10^6	This work
Durene			1.1×10^6	5

a. Error values are omitted (see the text). b. From Fig.6 in Ref. 5.

3.3. Comparison and explanation of dissociation rate constants including previous studies

In conclusion, dissociation rate constants for 193 as well as 158 nm irradiation have been found to be theoretically predictable. But there are some deviations from theory and scatters in experimental values as follows.

For the cases of toluene and p-xylene, predictions using the statistical adiabatic channel calculations (SACM) by Brand et al. explain well the present results. They fitted SACM calculations to dissociation rate constants obtained using 193 nm laser irradiation.[3] The rate constant of toluene for 222 nm deviated from the calculation, but it is still in the experimental errors. Two values of p-xylene for 193 and 158 nm are found to be fitted very well with the SACM calculation, as seen in Fig. 4.

For mesitylene, experimental data have scatters. The mean value of rate constants obtained from the present work was about 2/3 of that determined by the LIF technique at 158 nm. Error values in the present experiments ($3.0 \pm 0.9 \times 10^6$) were not small, therefore, two values were fairly in agreement. Still the calculated values based on the rate constant at 193 nm were a little higher than the observations.

If other reactions, for example, the C-C dissociation forming phenyl type radical, compete with the C-H dissociation, the observed constants are the sum of the processes. For the case of toluene, Luther et al. have measured the formation rate constant of methyl radical and discussed the C-C dissociation.[7] We compared experimentally determined k_{obs} with calculated the C-H dissociaton constants. Contribution from other reaction channels than the C-H dissociation will be greater in the higher energy region, therefore, we may need to re-evaluate the calculated constants.

3.5. Photochemistry of paracyclophane

Laser chemical vapor deposition (LCVD) is an important technique of laser applications. But there are not many reports starting from organic molecules. One of the reasons will be that many organic molecules dissociate to neutral radicals by UV irradiation harder than we may expect.

Pracyclophane is known to be one of parent molecules of thermal CVD and to form a parylene film.[8] There has been no report UVU laser chemistry as far as we know. We expect the following dissociation mechanism.

We observed transient spectra which shifted by about 10 nm compared to that of p-xylyl radical and was ascribable to the biradical. Pracyclophane in the gas phase at ca. 420 K was irradiated by an ArF laser in the presence of 600 Torr of nitrogen. Preliminary results of the laser fluence dependence (Fig. 5) indicate that the biradical forms by two photon processes for 193 nm irradiation.

The mechanism is suggested to be the same as the case of two photon dissociation of toluene to benzyl radical.[9,10] Hot molecule forms by the first photon and the next photon induces dissociation to the biradical presumably via hot molecule (S_0^{***}) with the internal energy of the two photons.

The biradical formed under low laser fluence of a few mJ/cm^2. These results suggest that the above multiphoton dissociation is very efficient and is applicable to LCVD.

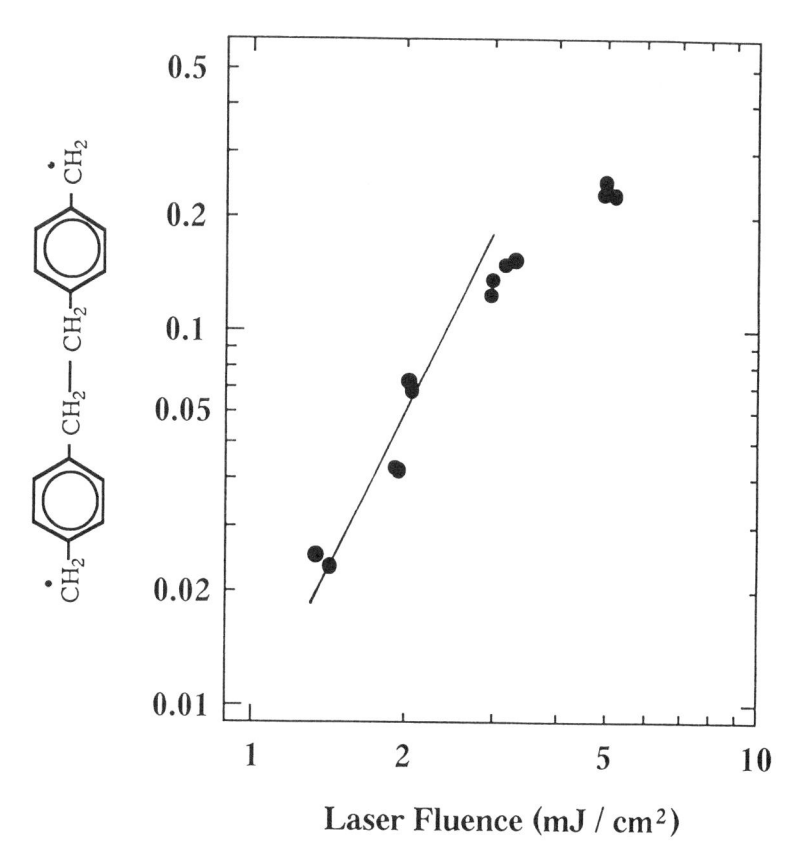

Fig. 5. Two photon dissociation of paracyclophan using ArF laser irradiation. The biradical concentration was measured as a transient absorbance at 260 nm. The straight line has a slope of 2.0.

4. REFERENCES

1. H. Hippler, V. Schubert, J. Troe, H.J. Wendelken, " Direct observation of unimolecular bond fission in toluene", Chem. Phys. Lett. 84, 253(1981).
2. N. Ikeda, N. Nakashima, and K. Yoshihara, "Photochemistry of toluene vapor at 193 nm. Direct measurements of formation of hot toluene and the dissociation rate to benzyl radical", J. Chem. Phys. 82, 5285(1985).
3. U. Brand, H. Hippler, L. Lindemann, and J. Troe, "C-C and C-H bond splits of laser-excited aromatic molecules. 1. Specific and thermally averaged rate constants", J. Chem. Phys. 94, 6305(1990).
4. Y. Kajii, K. Obi, I. Tanaka, N. Ikeda, N. Nakashima, and K. Yoshihara, "Deuterium isotope effects on photodecomposition of alkylbenzenes", J. Chem. Phys. 86, 6115(1987).
5. J. Park, R. Bersohn, and I. Oref, "Unimolecular decomposition of methylsubstituted benzenes into benzyl radicals and hydrogen atoms", J. Chem. Phys. 93, 5700(1990).
6. A. Bolovinos, J. Philis, E. Pantos, P. Tsekeris, and G. Andritsopoulos, "The methylbenzenes vis-a-ais benzene. Comparison of their spectra in the valence-shell transition region", J. Mol. Spectrosc., 94, 55(1982).
7. K. Luther, J. Troe, and K. -M. Weitzel, "C-C and C-H bond splits of laser-excited aromatic molecules. 2. In situ measurements of branching ratios", J. Chem. Phys. 94, 6316(1990).
8. S. Iwastuki, "Polymerization of quinodimethane compounds", Adv. Polym. Sci. 58, 93(1984).
9. N. Nakashima, N. Ikeda, and K. Yoshihara,"Hot toluene as an intermediate of UV-multiphoton dissociation", J. Phys. Chem. 92, 4389(1988).
10. H. Hippler, Ch. Riehn, J. Troe, and K. -M. Weitzel, "C-C and C-H bond splits of laser-excited aromatic molecules. 3. UV multiphoton excitation studies", J. Chem. Phys. 94, 6321(1990).

How to photograph a chemical reaction

K.R. Leopold, S.W. Reeve, M.A. Dvorak, W.A. Burns

Department of Chemistry
University of Minnesota
Minneapolis, MN 55455,

R.S. Ford

Scott Community College
500 Belmont Rd.
Bettendorf, IA 52722-6804,

and

F.J. Lovas and R.D. Suenram

Molecular Physics Division
National Institute of Standards and Technology
Gaithersburg, MD 20899

ABSTRACT

We report the structural characterization of the gas phase adducts HCN and CH_3CN with BF_3. Both have symmetric top structures with the nitrogen end of the R-CN toward the boron, reminiscent of the well known dative bond chemistry of BF_3 with nitrogen donors. The B-N bond lengths and N-B-F angles, however, are intermediate between those expected for van der Waals or covalent interactions. Moreover, in CH_3CN-BF_3, where comparison with X-ray crystallographic studies is possible, the gas phase adduct shows a markedly longer bond length and smaller N-B-F angle. We show that in a series of related BF_3 and BH_3 adducts, the bond length and bond angle can, in fact, be tuned almost continuously between the covalent and van der Waals limits. By analogy with classic crystallographic work by Bürgi and Dunitz and coworkers, we discuss how members of such a series can be interpreted as snapshots along a generalized reaction path for the formation of the dative bond. Finally, in the context of such a path, we examine the evolution of other (non-structural) properties of the BF_3 adducts as the donor-acceptor bond formation proceeds.

1. INTRODUCTION

Molecular spectroscopy has long played a vital role in the study of chemical bonding [1]. Likewise, modern high resolution techniques applied to the study of weakly bound molecular complexes [2] have brought our present understanding of intermolecular interactions far beyond that of early treatments [3]. Curiously, however, the intermediate regime *between* van der Waals and covalent interactions has remained relatively unexplored by gas phase spectroscopic methods. Indeed, neither the large body of elegant work on weakly bound systems, nor the even larger body of work on small, covalently bonded molecules has found much opportunity to examine the transition from van der Waals to covalent interactions.

In this report, we present the results of our recent studies of gas phase BF_3 adducts with HCN and CH_3CN. Using microwave spectroscopy, we have determined symmetric top geometries for the adducts, reminiscent of the well known dative bond chemistry of BF_3 with nitrogen donors [4]. We find, however, that the B-N bond lengths and N-B-F angles are intermediate between those expected for van der Waals and covalent bonds, making these systems excellent probes of the transition between these two limiting types of interactions. In interpreting these results, we draw analogy with classic work of Bürgi and Dunitz and coworkers on crystallographic structure correlations [5], and interpret the structures of these species as snapshots along the generalized reaction path for the formation of the boron - nitrogen dative bond. With this interpretation, we proceed to examine the evolution of other properties of the complexes (e.g. force constants, energetics, and electronic structure) as the systems pass from the van der Waals to the covalent limits.

2. EXPERIMENTAL

Spectra were observed using a Balle-Flygare type pulsed-nozzle Fourier transform microwave spectrometer at NIST in Gaithersburg [6]. For the CH_3CN-BF_3, initial spectral searches were conducted by entraining the vapor of a heated sample of the crystalline material in a flowing stream of argon, or by on-line mixing of 1% mixtures of CH_3CN and BF_3 in argon immediately prior to the expansion. The latter method turned out to be successful and was used throughout the studies of both the CH_3CN and HCN complexes. Time domain signals were averaged for 200 to 5000 pulses (depending on transition strength), and were Fourier transformed into the frequency domain to obtain line centers. In most cases, 256 or 512 data points were taken per pulse, providing resolution of 8 or 4 kHz, respectively.

Initial spectral searches for both species were guided by recent *ab initio* calculations of Jurgens and Almlöf [7], which predicted values of the B-N bond length, R(B-N) and the N-B-F angle, α(NBF), of 2.17 Å and 98°, respectively, for the CH_3CN adduct and 2.56 Å and 93°, respectively, for the HCN complex. Spectra were identified as arising from the 1:1 adducts by both the expected nuclear hyperfine structure, and by the observation of rotational progressions for several isotopically substituted forms of each species.

3. RESULTS

The spectral assignments for both CH_3CN-BF_3 and $HCN-BF_3$ are relatively straightforward, as both species are observed to display symmetric top spectra. Details are provided elsewhere [8,9], as are the spectroscopic constants determined for the CH_3CN adduct. Spectroscopic constants for $HCN-BF_3$ are given in Table 1.

Table 1

Spectroscopic Constants for Isotopic Derivatives of $HCN-BF_3$[a]

	$HC^{15}N-^{10}BF_3$	$HC^{15}N-^{11}BF_3$
B	2035.162(1)	2028.103(1)
D_{JK}	0.0146(1)	0.0173(1)
D_J	0.00464(2)	0.00315(2)
eqQ(B)	5.91(1)	2.78(1)
	$HC^{14}N-^{10}BF_3$	$HC^{14}N-^{11}BF_3$
B	2055.832(1)	2048.884(1)
D_{JK}	0.0144(1)	0.0170(4)
D_J	0.00457(4)	0.00306(9)
eqQ(B)	5.89(2)	2.78(1)
eqQ(N)	-3.99(1)	-3.96(1)

(a) All symbols have their usual definitions [1], and all values are given in MHz.

The spectroscopic constants for both species may be used to determine the structures of the complexes, and a detailed analysis for the CH₃CN adduct has been given previously [8]. The essence of the treatment is that the rotational constants for the ¹⁰B and ¹¹B derivatives with a given isotopic form of CH₃CN may be used to determine two structural parameters for the complex, and spectra of several isotopic forms of the CH₃CN provide redundant determinations which serve as a check. In reference [8], we argued that the B-N bond length, R(BN), and the N-B-F angle, α(NBF), must be determined from spectroscopic data, while the remaining intramolecular bond lengths could be reasonably constrained to their monomer values without incurring appreciable error. This followed from both chemical intuition, as well as a detailed consideration of the bond lengths and bond angles observed in the free, gas phase monomers, and the corresponding crystal structure of the CH₃CN-BF₃ solid. The results of careful analysis yielded R(BN) = 2.011 ± 0.007 Å and α(NBF) = 95.6° ± 0.6° for the gas phase species CH₃CN-BF₃. It is interesting to note that this structure is markedly different from that observed in the crystal, in which R(BN) = 1.630(4) Å and α(NBF) = 105.6(6)° [10].

A similar analysis may be applied to the rotational constants given in Table 1 for the HCN-BF₃ adduct. Detailed arguments will be presented elsewhere [9]. We note here, however, that the constraint of intramolecular structural parameters to those appropriate for the free monomers is an approximation routinely used in the analysis of van der Waals molecule spectra, and, as evident from the bond length and NBF angle determined below, the complex HCN-BF₃ is actually somewhat more like a van der Waals species than is CH₃CN-BF₃. Thus, similar constraints to those noted above will apply, and values of the boron - nitrogen bond length and N-B-F angles are determinable from measured rotational constants. The results of our preliminary analysis give R(BN) = 2.54(6) Å and α(NBF) = 88.6° ± 3.3°. Here, the full range of the error bars on α(NBF) are unphysical, and we reject NBF angles of less than 90° as being chemically unreasonable. The data, therefore, determine the angle to be between 90° and 91.9°, and we thus take the final structure of the complex to be R(BN) = 2.54(6) Å and α(NBF) = 91.0° ± 1°.

4. DISCUSSION

In reference [8], we showed that for a series of BF₃ adducts with nitrogen donors, there exists a clear correlation between the values of α(NBF) and R(BN). In particular, species with long B-N bond lengths have the BF₃ moiety essentially undistorted from its planar free-monomer structure, while progression to shorter B-N bond lengths is accompanied by an increasing distortion toward a tetrahedral geometry at the boron. By analogy with well known crystallographic work of Bürgi and Dunitz and coworkers [4], we interpreted a plot of R(BN) vs. α(NBF) as a representation of the reaction path for the formation of the boron - nitrogen dative bond. This plot is reproduced here in Figure 1. With this interpretation, Figure 1 represents the evolution of structural changes which take place as the dative bond formation proceeds. Note that the point corresponding to N_2-BF_3, with R(BN) = 2.78 Å and α(NBF) = 90.5(5)° represents the true "van der Waals limit", while the crystal structures depicted on the right side of the plot, with R(BN) = 1.60 Å and a near tetrahedral geometry at the boron represent the an ordinary chemical bond. The plot thus graphically shows the intermediate character of the HCN and CH_3CN adducts. Moreover, it may be inferred from the plot that by the time the nitrogen atom has reached half way between the van der Waals and covalent bond lengths, the angular deformation of the BF_3 is only about 5°, meaning that the bulk of the structural changes in the BF_3 which take place upon formation of the dative bond do so in the last half of the approach (i.e., the last 0.5 Å).

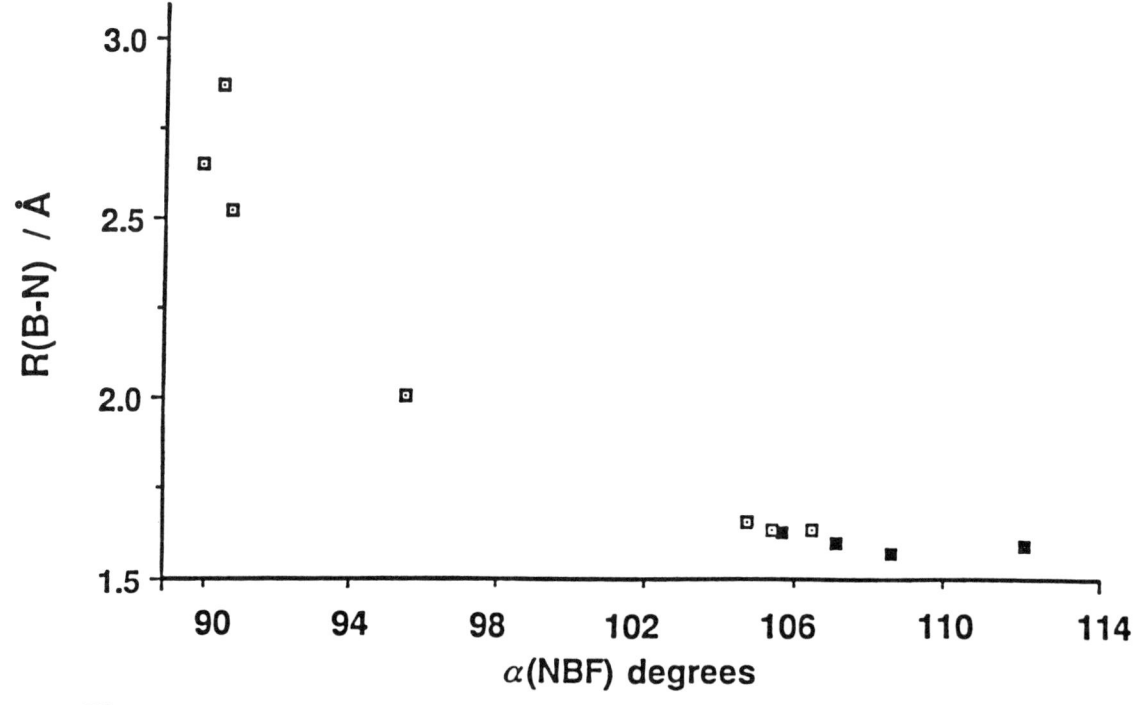

Figure 1. A plot of R(BN) vs α(NBF) for a series of BX_3 adducts with nitrogen donors. Open squares are gas phase structures and closed squares are crystal structures. The complexes are, from upper left to lower right, N_2-BF_3, NCCN-BF_3, HCN-BF_3, CH_3CN-BF_3, H_3N-BH_3, $(CH_3)_3N$-BH_3, CH_3CN-BF_3, $(CH_3)_3N$-BF_3, H_3N-BF_3, CH_3H_2N-BF_3, and $(CH_3)_3N$-BF_3. See reference [8] for additional details.

An important feature of Figure 1 is that the gas phase points span a very significant range of the bond lengths and bond angles represented. Thus, the observation of such a structure correlation across a series of gas phase compounds suggests the possibility of being able to use high resolution gas phase spectroscopy to follow the evolution of other physical properties of the system as the bond formation proceeds. Table 2 lists a number of BF_3 adducts, in order of increasing B-N bond length, and gives values for R(BN), α(NBF), k_s (stretching force constant for the B-N bond, estimated from the purely J-dependent centrifugal distortion), eqQ(B) (the nuclear quadrupole coupling constant of the ^{11}B nucleus, and ΔE (the binding energy determined from *ab initio* calculation). We consider each of these separately below. Ar-BF_3 is included in the table for comparison.

Table 2

Structures, Force Constants, Quadrupole Coupling Constants, and Binding Energies for Selected Adducts of BF_3[a]

Donor[b]	R(BN) (Å)	α(NBF) (deg)	k_s (N/m)	eqQ(^{11}B) (MHz)	ΔE (kJ/mol)	Reference
Ar	-	-	3	2.70(5)	-	11
N_2	2.78	90.5(5)	-	-	-	11
NCCN	2.64	90-100	-	-	15	12
HCN	2.54	91.0(10)	10.	2.78(1)	22	9,15
CH_3CN	2.01	95.6(6)	9.4	2.377(9)	24	8
NH_3	1.80[c]	-	-	-	75	7,15
	1.60[d]	107(2)	-	-		16

(a) A dash indicates that the quantity is either not applicable, or not known. (b) All species are symmetric tops, with the axis of the donor coincident with the C_3 axis of BF_3; (c) *ab initio* result; (d) X-ray crystal structure

First, regarding the B-N stretching force constants, it is apparent that the values of k_s for HCN-BF_3 and CH_3CN-BF_3 are larger than that for the Ar complex. Although this is not particularly surprising in light of the shortness of the B-N bonds compared with that for a true boron - nitrogen van der Waals interaction, we note that a typical stretching force constant for single bonds between first row elements is of the order of *a few hundred* N/m [1]. Thus, although the structure of the adducts are rather intermediate between van der Waals and covalent type interactions (especially for the CH_3CN species), the stretching force constants are still markedly *closer* to the van der Waals limit.

Similar behavior is observed in the calculated binding energies for the adducts. The 24 kJ/mole (5.8 kcal/mole) binding energy, for example, calculated for the CH_3CN-BF_3 adduct is somewhat large for a van der Waals interaction (c.f. the 15 kJ/mole (3.6 kcal/mole) value for NCCN-BF_3), yet it is still significantly smaller than the bond energy for a typical single bond between first row elements. Thus, although the acetonitrile adduct is *structurally* intermediate between the van der Waals and covalent limits, it is *energetically* much more like a van der Waals molecule.

The ^{11}B nuclear quadrupole coupling constants provide a somewhat different view of the complexes. Although the values measured in the weakly bound systems are certainly a convolution of projective and electronic effects, a comparison across a series can still provide some insight as to the changes taking place as the bond formation proceeds. It is readily seen in Table 2, for example, that the value of eqQ(B) for CH_3CN-BF_3 is less than either that of HCN-BF_3 or Ar-BF_3. Moreover, since an increasing angular rigidity of the complex can only *increase* the projective contribution to eqQ, the distinct *decrease* for the CH_3CN adduct may be interpreted as a reflection of changes in the electronic environment at the boron. Although a quantitative interpretation is difficult without a detailed knowledge of the angular potential for motion of the BF_3 within the complex, it is interesting to note that the eqQ(B) observed in CH_3CN-BF_3 is almost identically the average of that found in the complexes Ar-BF_3 (a van der Waals complex) and $(CH_3)_3N$-BH_3 (a stable, chemically bonded molecule for which eqQ(B) = 2.064 MHz [17]). It is unfortunate that we are, so far, unaware of any measurements of a boron quadrupole coupling constant in a stable BF_3 adduct, since this would provide a better indicator of the limiting quadrupole coupling constant for a BF_3 adduct with a nitrogen donor.

From the above comparisons, we may infer that different properties of the BF_3-Donor systems evolve at different rates as the dative bond formation proceeds. While the force constant and energetics appear to evolve slowly, the structure of the complex appears to evolve more rapidly. Similarly rapid changes also may be characteristic of the electronic environment at the boron, though a better measure of the quadrupole coupling constant in the chemically bonded limit is needed. The wide range of behavior displayed by this series of BF_3 systems suggests exciting new opportunities to examine the details of bond formation in a variety of other systems as well.

5. REFERENCES

1. K.P. Huber and G. Herzberg, <u>Molecular Spectra and Molecular Structure</u>, van Nostrand Reinhold, New York, 1979

2. see, for example <u>Dynamics of Polyatomic van der Waals Complexes</u>, edited by N. Halberstadt and K.C. Janda, NATO ASI Series, Physics Series B, vol. **227**, Plenum, London, New York, 1990

3. J.O. Hirschfelder, C.F. Curtiss, and R.B. Bird, <u>Molecular Theory of Gases and Liquids</u>, Wiley, New York, 1954

4. E.L. Muetterties, <u>The Chemistry of Boron and its Compounds</u>, Wiley, New York, 1967

5. H.G. Bürgi and J.D. Dunitz, Acc. Chem. Res., **16**, 153 (1983) and references therein.

6. (a) F.J. Lovas and R.D. Suenram, J. Chem. Phys., **87**, 2010 (1987); (b) F.J. Lovas, R.D. Suenram, G.T. Fraser, C.W. Gillies, and J. Zozom, J. Chem. Phys., **88**, 722 (1988)

7. R. Jurgens and J. Almlöf, Chem. Phys. Lett., **176**, 263 (1991)

8. M.A. Dvorak, R.S. Ford, R.D. Suenram, F.J. Lovas, and K.R. Leopold, J. Am. Chem. Soc., in press

9. S.W. Reeve, W.A. Burns, F.J. Lovas, R.D. Suenram, and K.R. Leopold, manuscript in preparation.

10. (a) J.L Hoard, T.B. Owen, A. Buzzell and O.N. Salmon, Acta Cryst., **3**, 130 (1950); (b) B. Swanson, D.F. Shriver and J.A. Ibers, Inorg. Chem., **8**, 2183 (1969)

11. K.C. Janda, LS. Bernstein, J.S. Steed, S.E. Novick, and W. Klemperer, J. Am. Chem. Soc., **100**, 8074 (1978)

12. K.R. Leopold, G.T. Fraser, and W. Klemperer, J. Am. Chem. Soc., **106**, 897 (1984)

13. J.M. LoBue, J.K. Rice, T.A. Blake, and S.E. Novick, J. Chem. Phys., **85**, 4261 (1986)

14. (a) P. Cassoux, R.L. Kuczkowski, and A. Serafini, Inorg. Chem., **16**, 3005 (1977); (b) P.S. Brian and R.L. Kuczkowski, Inorg. Chem. **10**, 200 (1971)

15. R. Jurgens-Lutovsky and J. Almlöf, private communication

16. J.L. Hoard, S. Geller and W.M. Cashin, Acta. Cryst., **4**, 396 (1951)

17. W. Kasten, H. Dreizler and R.L. Kuczkowski, Z Naturforsch, **40a**, 1262 (1985)

Primary Processes Involved in the Photodissociation of Saturated Hydrocarbons at 157 nm

Kimberly A. Prather and Yuan T. Lee

Chemical Sciences Division
Lawrence Berkeley Laboratory
and
Department of Chemistry
University of California
Berkeley, California 94720 USA

BACKGROUND

Much speculation exists regarding the primary processes involved in the unimolecular dissociation of saturated hydrocarbons dating back to the early 1960s. Previous experiments typically involved photolysis of a hydrocarbon with subsequent detection of products by mass spectrometry.[1-9] The conditions of the experiments (i.e. high pressure) led to uncertainty as to which of the detected products were primary products resulting from the unimolecular decomposition of the hydrocarbon of interest. Using the technique of photofragmentation translational spectroscopy (PTS), we have determined the primary and secondary photodissociation channels of cyclopropane, n-propane, n-butane, and isobutane at 157 nm. The PTS technique involves the use of a molecular beam which inherently gives collision free conditions, ideal for the study of unimolecular photodissociation processes.

EXPERIMENTAL

The experiment involves expansion of a 10 percent hydrocarbon/He mixture through a heated nozzle to form a molecular beam which is crossed with a pulsed excimer laser at 157 nm. The molecule decomposes to yield various products which recoil from the beam with a particular recoil velocity. Ions are created by electron impact ionization and mass selected with a quadrupole mass spectrometer. The ion arrival times are recorded on a multichannel scaler to generate time of flight spectra for all detectable masses. The experiment involves use of a rotating source machine which allows one to vary the molecular beam to detector angle and thus obtain angularly resolved time of flight spectra. Using the time of flight data, the center of mass translational energy probability distribution [P(E)] is calculated for each reaction channel.

RESULTS AND DISCUSSION

Signal from the photodissociation of cyclopropane was observed at mass to charge ratios of 36-41, 24-28, and 12-14. The dominant channel is molecular elimination to form methylene and ethylene. The time of flight spectra taken at 20 degrees of these fragments are shown in Figure 1 (below). The observed lab velocities of the two molecules are related through conservation of linear momentum, an indication that these fragments are "partners" from the same reaction channel.

CYCLOPROPANE

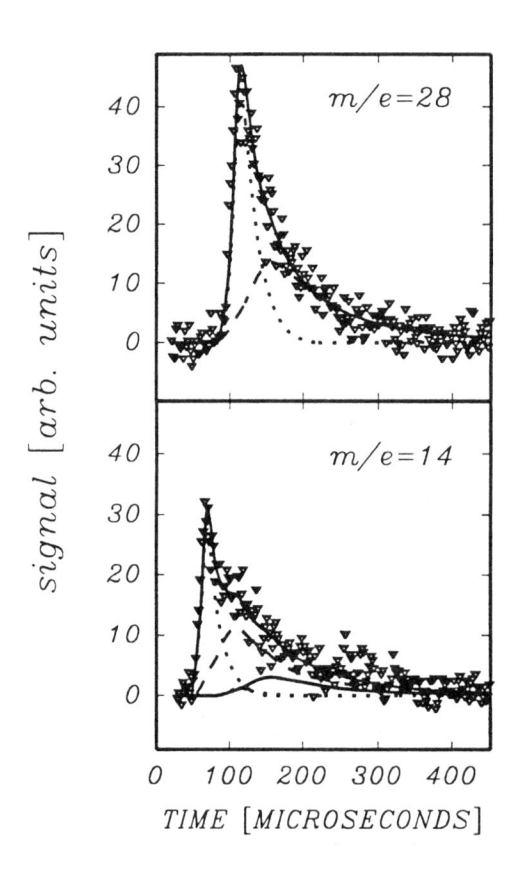

Figure 1. Time of flight spectra of m/e 28 and 14 at 20 degrees

The P(E) for this channel is shown in Figure 2. The shape of the P(E) (i.e. peaked away from zero) is indicative of a molecular elimination process involving some barrier. How far the P(E) peaks away from zero serves as a qualitative indicator of the size of the potential energy barrier for the dissociation process. Hydrogen atom elimination and methyl loss are also observed as primary decomposition channels in the photolysis of cyclopropane.

Cyclopropane

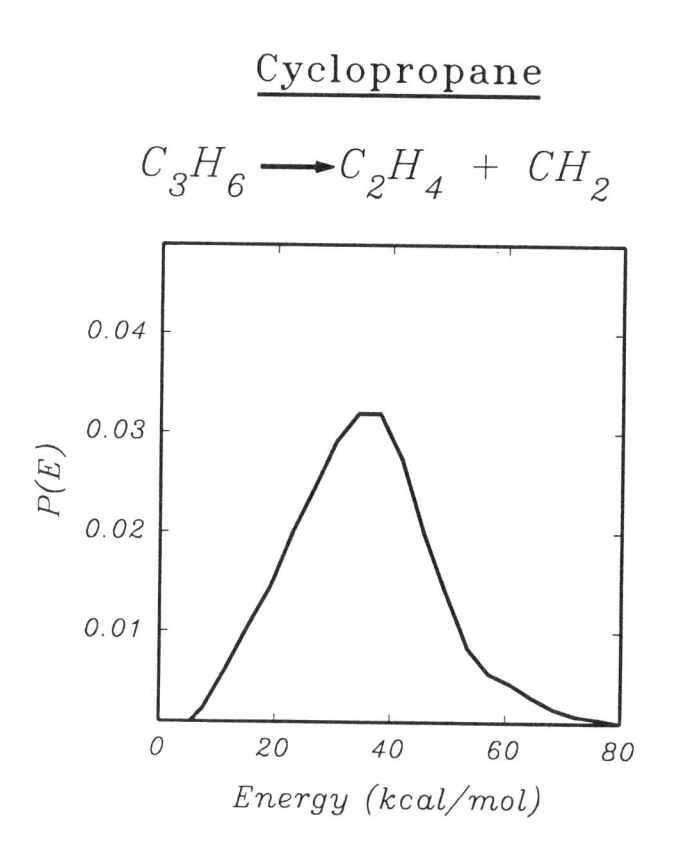

$$C_3H_6 \longrightarrow C_2H_4 + CH_2$$

Figure 2. Center of mass translational energy probability distribution for the methylene plus ethylene elimination channel

The dominant primary elimination channels observed in the photolysis of n-propane and n-butane are H and H_2 loss. At high laser powers, primary products absorb a second photon leading to secondary H and H_2 loss as well as carbon-carbon bond rupture fragments.

In the photodissociation of isobutane, H atom, H_2, and methyl plus isopropyl loss are observed as the primary elimination channels. The time of flight spectra for the methyl loss channel are given in Figure 3. The P(E) is given in Figure 4. It is peaked at zero translational energy release as expected for a simple bond rupture dissociation process.

ISOBUTANE

$$C_4H_{10} \rightarrow C_3H_7 + CH_3$$

Figure 3. Time of flight spectra of m/e 43 and 15 at 20 degrees.

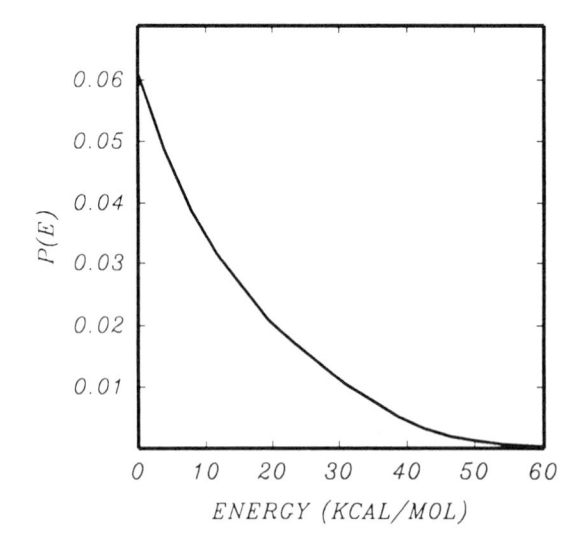

Figure 4. Translational energy probability distribution.

In summary, we now have a clear understanding of the primary processes involved in the photodissociation of a series of saturated hydrocarbons. Photofragmentation translational spectroscopy allows differentiation of the primary dissociation processes from other fragmentation channels (i.e secondary processes, ionizer fragments). The results of these experiments demonstrate the unmatched power of PTS as a tool for sorting out complicated processes in a series of related molecules.

ACKNOWLEDGEMENTS

This work was supported by the Director, Office of Energy Research, Office of Basic Energy Sciences, Chemical Sciences Division of the U.S. Department of Energy under Contract No. DE-AC03-76SF00098.

REFERENCES

1. P. Ausloos, S. G. Lias, and I. B. Sandoval, "Gas Phase Radiolysis of Propane," Discuss. Faraday Soc., vol. 36, pp. 66-74, 1963.

2. A. K. Dhingra and R. D. Koob, "Methylene Produced by Vacuum Ultraviolet Photolysis. II. Propane and Cyclopropane," J. Phys. Chem., vol. 74(26), pp. 4490-4496, 1970.

3. H. Okabe and D. A. Becker, "Vacuum Ultraviolet Photochemistry. VII. Photolysis of n-Butane," J. Chem. Phys., vol. 39(10), pp. 2549-2555, 1963.

4. R. D. Koob, "Methylene Produced in the Vacuum-Ultraviolet Photolysis of Propane," J. Phys. Chem., vol. 73(9), pp. 3168-3170, 1969.

5. P. Ausloos and S. Lias, "Gas Phase Radiolysis of Propane," J. Chem. Phys., vol. 36(12), pp. 3163-3170, 1962.

6. J. H. Vorachek and R. D. Koob, "Allyl Radicals in the Vacuum Ultraviolet Photolysis of Propane," J. Phys. Chem., vol. 74(25), pp. 4455-4456, 1970.

7. R. D. Koob, "Methylene Produced by Vacuum-Ultraviolet Photolysis, IV. Energy Distribution for the Reaction $C_3H_8 + h\nu$ (123.6) = $CH_2 + C_2H_6$," J. Phys. Chem., vol. 76(1), pp. 9-14, 1972.

8. A. K. Dhingra, J. H. Vorachek, and R. D. Koob, "Methylene Produced by Vacuum Ultraviolet Photolysis. Energy of the Methylene," Chem. Phys. Let., vol. 9(1), pp. 17-18, 1971.

9. K. Tonokura, Y. Matsumi, M. Kawasiki, and K. Kasatani, "Doppler Spectroscopy of Hydrogen Atoms from the Photodissociation of Saturated Hydrocarbons and Methyl Halides at 157 nm," J. Chem. Phys., vol. 95(7), pp. 5065-5071, 1991.

Photodissociation Dynamics of Ethylene Sulfide and Allene at 193 nm Studied by Time-of-Flight Mass Spectroscopy

Kiyohiko Tabayashi and Kosuke Shobatake

Institute for Molecular Science, Myodaiji, Okazaki 444, Japan

ABSTRACT

Photodissociation dynamics of ethylene sulfide (C_2H_4S, thiirane) and allene (C_3H_4) at 193 nm is studied by time-of-flight (TOF) photofragment mass spectroscopy. In the photodissociation of C_2H_4S the two fragments, S and SH, from two dissociation processes, $C_2H_4S + h\nu \rightarrow S(^1D) + C_2H_4$ (1) and $\rightarrow SH + C_2H_3$ (2) were detected. Translational energy distributions of the $S(^1D)$ fragment from reaction (1) exhibit two components, a low and a high energy component, indicating the contribution of two different dissociation channels; a direct dissociation and a long-lived near statistical dissociation producing a vibrationally excited counter fragment C_2H_4. This interpretation was confirmed by the observed center-of-mass (CM) angular distributions which exhibit two $S(^1D)$ components with different lifetimes. Translational energy and CM angular distributions of SH from reaction (2) are found to indicate that the excited molecule is long-lived and near statistical, and is consistent with the dissociation process of CH_2-CH-SH intermediate formed after ring opening and subsequent 1,2-hydrogen migration. The energy flow in the excited ethylene sulfide by 193 nm photon impact which results in forming S atom and SH fragments is discussed.

Photodissociation dynamics of allene (C_3H_4) has been also studied using the same technique. The primary dissociation product channel detected at m/e = 39 is C_3H_3 + H and the minor channel $C_3H_2 + H_2$ has been also detected at mass m/e = 38.

1. INTRODUCTION

Ethylene sulfide has a three-membered cyclic structure with a C_{2v} symmetry in the ground electronic manifold,[1]

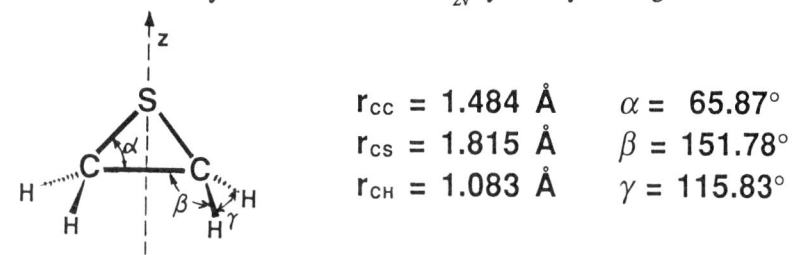

r_{CC} = 1.484 Å α = 65.87°
r_{CS} = 1.815 Å β = 151.78°
r_{CH} = 1.083 Å γ = 115.83°

The excited electronic states of C_2H_4S studied by optical UV and VUV[2] absorption, and circular dichroism[3] spectroscopies. The broad and diffuse bands appearing in the low energy region 230 - 260 nm are assigned as (n,$\sigma \rightarrow \sigma^*$) valence transitions, whereas the sharp and structured spectrum in the energy region below 210 nm has been assigned as (n $\rightarrow$ 4s,p,3d) Rydberg transitions. The 193 nm excitation of ethylene sulfide corresponds to (n $\rightarrow$ 4p, A_1-A_1) Rydberg transition whose dipole moment is parallel with respect to the major symmetry axis.

From their 193 nm photodissociation studies of ethylene sulfide Bersohn and coworkers[4,5] have shown that a sulfur atom is produced in the electronically excited $S(^1D)$ state:

$$C_2H_4S + h\nu \rightarrow C_2H_4 + \dot{S}(^1D). \quad (1)$$

Using a laser induced fluorescence technique, they have measured Doppler broadened fluorescence[5] spectra of sulfur atom $S(^1D)$ against the polarization angle between the dissociating light and the probe direction, and its angular dependent velocity distribution has been obtained. The average kinetic energy of the fragmentation was found to be $\bar{E}_t$ = 20.5 kcal/mol; the lifetime of the excited molecule was also estimated to be ~300 fs. With the anisotropy parameters, in addition, they have concluded that the dissociation is direct and symmetrical one.

Presently, in addition to the $S(^1D)$ formation (1), we have observed new photodissociative process (2) leading to SH,[6]

$$C_2H_4S \rightarrow C_2H_4 + S(^1D), \quad \Delta H = 84.4 \text{ kcal/mol} \quad (1)$$
$$\rightarrow C_2H_3 + SH, \quad \Delta H = 81.4 \text{ kcal/mol} \quad (2)$$

using TOF photofragment spectroscopy on a crossed molecular beam apparatus with a rotatable mass detector. The translational energy and angular distribution of $S(^1D)$ from the process (1) are found to have two components, indicating that another dissociation channel via a long-lived intermediate is open along with the direct one that Bersohn et al.[5] have proposed. It has been also found that the process (2) exhibit a long-lived and near statistical distribution, which is consistent with the dissociation of C_2H_4S after the ring opening followed by migration of a hydrogen atom.

Photodissociation dynamics of allene (C_3H_4, $CH_2=C=CH_2$) has been studied[7] since the rotatable detector apparatus can detect the heavier photofragments because of kinematic reasons, that is, 1) photofragmentation of allene was expected to form very light/heavy fragments pairs such as $H + C_3H_3$ and $H_2 + C_3H_2$, 2) as the result of photodissociation kinematically the heavy fragments gain small momenta relative to the initial molecular beam momentum, 3) thus the laboratory velocity vector of heavier fragments is directed within the acute angle cone with respect to the molecular beam velocity, and 4) in order to detect the heavier fragments the detector should be able to move closer to the molecular beam direction. Recently the photolysis of allene has been studied by Jackson et al.[8] at Lee's laboratory using essentially the same technique. They measured the light species such as H and H_2 fragments as well as the heavy fragments. In the present paper our results will be a only briefly mentioned.

2. EXPERIMENTAL

Experiments were performed using a crossed molecular beam apparatus with a mass spectrometer detector rotatable in the horizontal plane defined by two perpendicularly crossed beams.[9,10] For the present photodissociation experiments, one of the molecular beam is replaced by an excimer laser beam of 193 nm. The schematics of the apparatus is shown in Fig. 1. The flight length from the interaction zone to the ionizer of the mass detector, L, was 23.3 cm, laboratory angular resolution defined by the detector apertures was estimated to be no greater than 1.0 degree.

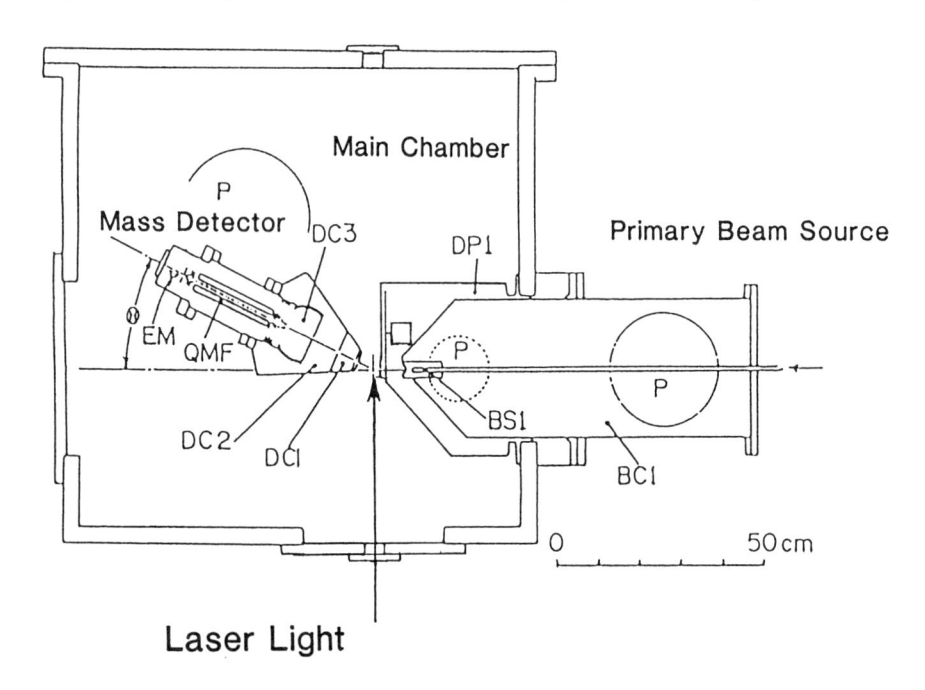

Fig. 1. Cross-sectional top view of the crossed laser-molecular beam apparatus. MC: main scattering chamber, BC1: primary beam source chamber, BS1: fuel injector, DP1: differential pumping region, DC3: mass detector chamber, DC1 and DC2: 1st and 2nd differential pumping region of the mass spectrometer chamber, QMF: quadrupole mass filter, EM: electron multiplier, P: pumping port.

The pulsed beam of ethylene sulfide was generated by expansion of pure gas at 160 Torr from a fuel injector, the nozzle of which was 0.82 mm in diameter. The pulsed beam was further gated by a mechanical chopper, placed in a differential pumping region, with FWHM of 18 μsec to minimize possible background molecules in the main scattering chamber. The timings of the injector and the laser pulse were carefully adjusted and were operated at a repetition rate of about 25 Hz in order to synchronize the three pulses. A TOF measurement of the ethylene sulfide beam shows a distribution with a peak velocity of 6.0×10^4 cm/sec and a FWHM of 25 %. Excimer laser (Lambda-Physik EMG 103 MSC) was driven with ArF at pulse power range from 30 to 80 mJ. The laser power is found to have no discernible effect on the TOF spectra of the fragments. The output laser beam was weakly focused at the interaction zone through a LiF focal lens. TOF fragment spectra were recorded on a multichannel scaler at a minimum channel width of 2 μsec. The ion drift time to travel from the ionizer to the ion detector was estimated from those for representative ions directly observed by measuring the time of flight spectra triggered by electron beam pulses with 1 μsec width. The actual flight time of the neutral fragments from the laser-molecule interaction region to the ionizer was calculated in the data processing.

In order to reduce the background level due to residual gaseous molecules going into the detector, a cooled Cu wall attached to the end of the sample holder of Cryo Kelvin (Daikin Model V202L) was placed so that the detector sees mostly the cold surface except for the apertures through which the molecular beam and the laser beam go before entering the interaction region. The background was reduced by a factor of about ten by placing the cold wall cooled down to about 20K.

In the TOF fragment-mass spectroscopy in which number density is measured like the present, the observed TOF signals $N(t, \Theta)$ at laboratory angle Θ and arrival time t (= L/V: where L is the neutral flight length of the fragment) are related to CM flux intensities $I_{CM}(E_t, \theta)$ in terms of CM translational energy E_t and angle θ by

$$I_{CM}(E_t, \theta) \propto t^3 u N(t, \Theta) \qquad (3)$$

where u is the CM velocity of the fragment detected and is given by the formula:

$$u = (v_0^2 + V^2 - 2v_0 V \cos\Theta)^{1/2}. \qquad (4)$$

The CM flux distributions in the CM energy space E_t and θ presented here were calculated by the transformation of Eq. (3).

3. RESULTS AND DISCUSSION

3.1. Ethylene Sulfide Photodissociation

In the laser photolysis of C_2H_4S at 193 nm, the TOF signals at mass m/e = 26, 32, and 33 were detected with high S/N ratios. From the comparisons of these spectra we find that the profiles of m/e = 26 have nearly the same features as those of m/e = 32. Therefore it is concluded that the signals at mass m/e = 26 originate mainly from dissociative ionization of C_2H_4 product by electron impact in the ionizer. The possible contribution of dissociative ionization of SH to the S signal has also been checked by comparing the spectra of SH and S under the same detection conditions. Taking into account the difference of the ion drift time, they are essentially different in features (represented by the peak position and TOF profiles) even at longer flight times. The observations lead us to conclude that dissociative ionization affects little upon the present TOF measurements of the S fragment. The relative intensity ratios for these signals at masses m/e = 32, 26, and 33 are approximately determined as 8: 5: 1, respectively.

The energetics of the fragmentation processes can be also examined from the initial rise of the TOF signals where fast fragments are expected to appear with the maximum translational energy available. The bond dissociation energy (ΔH) basically corresponds to the difference between the photon energy and the maximum CM translational energy released to the product pair. Table I lists the observed maximum E_{th} values as well as the literature values of heat of dissociation ΔH with which fragmentation proceeds. From Table I, S atom is reasonably considered to appear exclusively in the excited electronic (^1D, ^1S) states. Since the formation of $S(^1S)$ was previously denied[5], one could then attribute signals at mass m/e = 32 and 26 (28) to the process (1) $C_2H_4S + h\nu \longrightarrow S(^1D) + C_2H_4$, and those at mass 33 to the process (2) $C_2H_4S + h\nu \longrightarrow SH + C_2H_3$.

Table I. The maximum CM translational energies available, E_{th}, to and the dissociation energy measured, D, for the product pairs from the appearance threshold times in the observed TOF spectra (the unit of energy: kcal/mol).

Photofragments	Detected Mass	E_{th}	D(hν - E_{th})	ΔH_0^0
$S(^1D) + C_2H_4$	S^+	64.7 ± 2.0	83.3 ± 2.0	84.4
	$C_2H_2^+$	65.2 ± 4.6	82.8 ± 4.6	84.4
$SH + C_2H_3$	SH^+	66.8 ± 2.0	81.2 ± 2.0	81.4

3.1.A. S Atom Formation Process

The TOF spectra of S (m/e=32) were observed at laboratory angles Θ = 7.5°, 15°, 25°, 35°, and 50°. The typical spectra recorded at the lowest (7.5°) and highest (50°) laboratory angles are shown in Fig. 2. Two peaks can be observed at the most probable CM recoil energies of ~30 and ~7 kcal/mol. While the intensities are comparable at lower laboratory angles Θ's, the earlier peak becomes faint at higher angles Θ's.

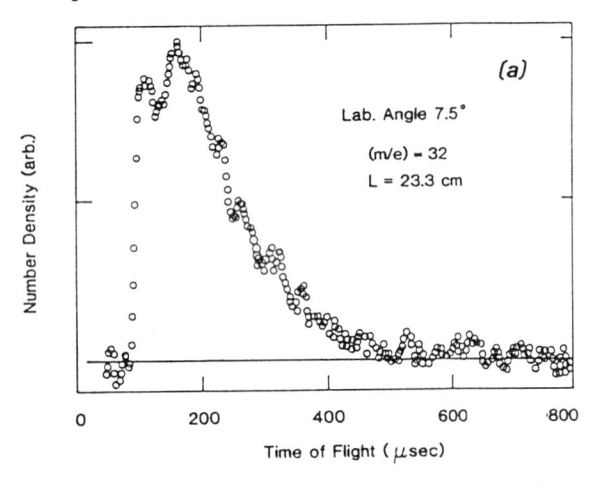

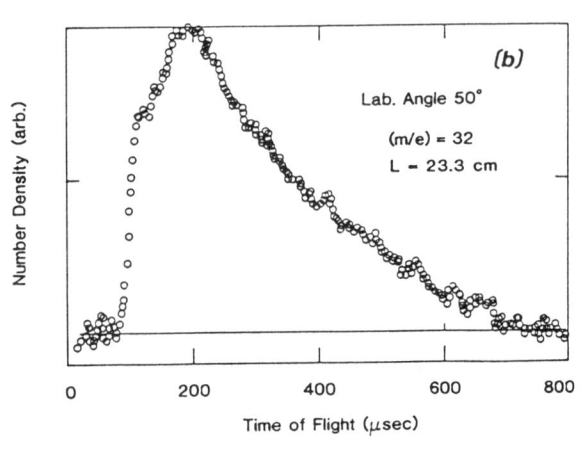

Fig. 2. Time-of-flight spectra observed for m/e = 32 at (a) laboratory angle Θ = 7.5°, and (b) Θ = 50°.

Fig. 3 shows the velocity vector diagram for 193 nm photodissociation of C_2H_4S forming $S + C_2H_4$ when S atom is detected. The E_t distributions derived from the TOF spectra (Fig. 2) of the S atom at Θ = 7.5° and 50° are compared in Fig. 4. The variation of the CM angle θ is also shown as a function of the CM recoil energy E_t. Here, the distributions have been normalized to unity at the energy of ~ 5 kcal/mol, corresponding to the late TOF peak. They are found to have a similar profile except that the relative intensities for the low and high energy components, which correspond to the late and early TOF peaks, respectively, are different depending on the CM angle θ. This observation leads us to conclude that the process (1) consists of two different sub channels.

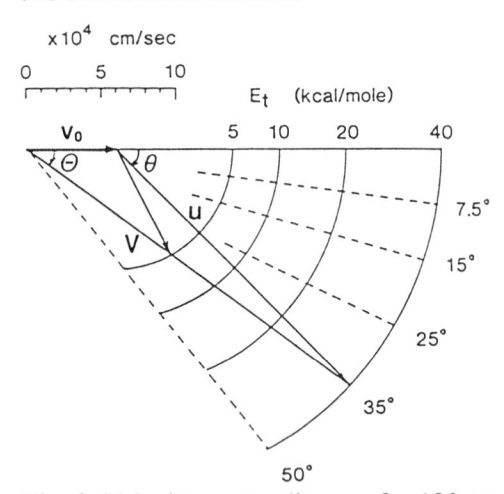

Fig. 3. Velocity vector diagram for 193 nm photodissociation of C_2H_4S forming $S + C_2H_4$ when S atom is detected. The bold arrow represents the molecular beam velocity in the laboratory frame. The laboratory angles Θ's shown are where data were taken. Velocity circles of S atom, corresponding to the CM recoil energies, are also shown.

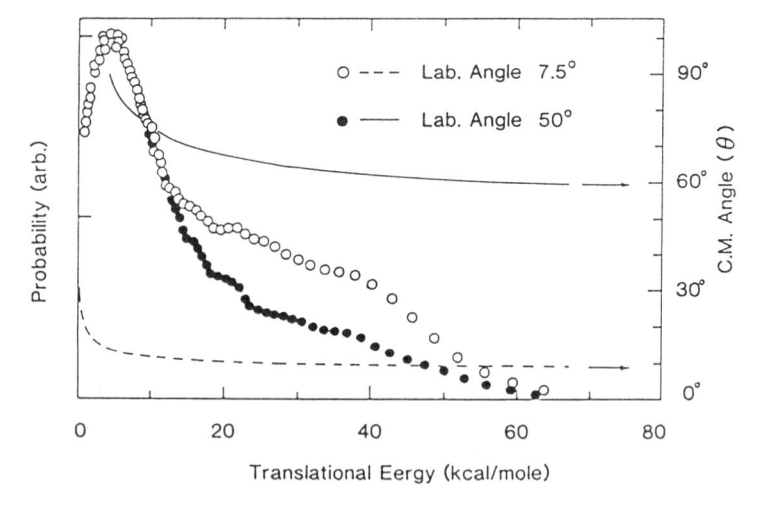

Fig. 4. CM flux (probability) distribution vs. translational energy E_t, obtained by conversion of TOF distributions in Fig. 2. Variation of the CM angle θ at each energy is also shown.

The low energy component of the distribution can be fitted to an RRK-type function with parameters α and β as

$$f(E_t) = (E_t - E_0)^\alpha (E_{max} - E_t)^\beta \tag{5}$$

where E_{max} is a maximum available CM translational energy of the fragment S and E_0 is the barrier height, if any, in the exit channel. Using α and β as adjustable parameters, a best fitted distribution for the low energy component is expressed by

$$f_{obs}(E_t) = E_t^{0.7} (64.7 - E_t)^{9.7} \tag{6}$$

The distribution of the high energy component is then obtained by subtracting Eq. (6) from the total E_t distributions. Typical results for the resolution of the high energy distribution are shown in Fig. 5. The kinetic energies are found to be relatively widely distributed with an average value $\bar{E}_t = 32.3$ kcal/mol.

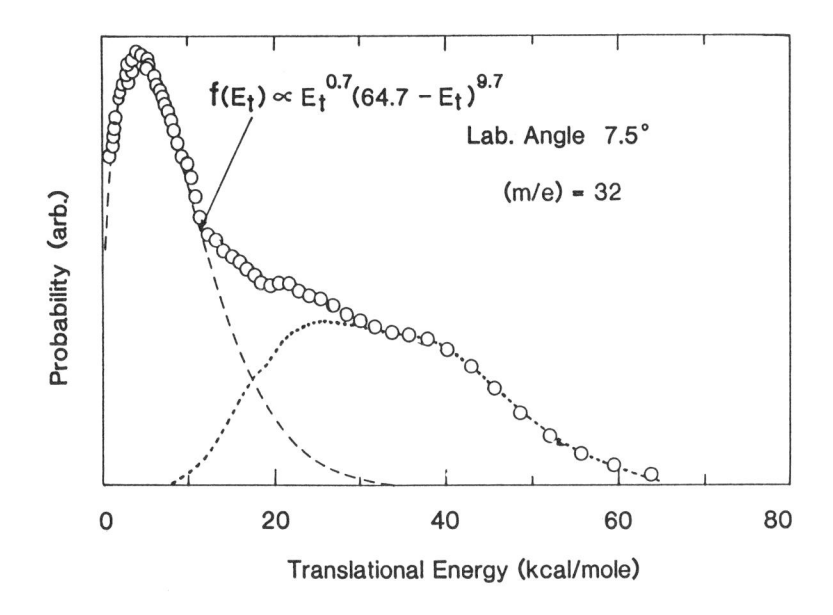

Fig. 5. Decomposition of the translational energy distribution, obtained by conversion from Fig. 2(a). Dashed curve represents parametric fits of RRK-type distribution to the low energy component; dotted curve shows the high energy component obtained by subtracting the low energy component calculated for Eq.(6) from the original S distributions.

Since the distributions for these individual components are resolved thus far, it is possible to examine its angular dependence by sampling the signals for each component out of the least blended E_t region. The CM angular distribution $W(\phi)$ is generally expressed by the following formula as

$$W(\phi) = (1/4\pi) [1 + 2 B P_2(\cos\phi)] \tag{7}$$

where $(1/4\pi)$ is a normalization factor, B a reduced anisotropy factor, P_2 the second-degree Legendre Polynomial in $\cos\phi$, and ϕ is the angle between the CM recoil velocity vector u and the electric vector E. If the laser electric vector is polarized along the molecular beam velocity vector, v_0, then the angle ϕ equals the CM angle θ, that is, $\phi = \theta$. In the following the analysis of angular distribution is made assuming that the laser electric vector is polarized along the molecular velocity vector. For the low energy component of S fragments, the flux intensities between $E_t = 5$ and 10 kcal/mol are integrated, normalized to the accumulated laser excitation power, and then plotted against the average CM angle θ over the same energy region. For the high energy component, the intensities between $E_t = 35$ and 40 kcal/mol are processed as mentioned above. The angular distribution results for these two components are shown in Fig. 6. The least-squares treatment of the data fitted to Eq. (7) for $\phi = \theta$ gives the smooth curves as indicated, with B values of 0.55 ± 0.04 and 0.24 ± 0.01, for the high and low energy components, respectively.

The angular distributions for the high and low energy components were fitted to the conventional form of Eq. (7). However the laser polarization direction is not along the molecular beam velocity, since an unpolarizated laser was used in the present experiment. We now derive the angular distribution of the fragment when unpolarized laser is used. The coordinates are chosen as shown in Fig. 7, in which the propagation vector of the laser is along the y axis, and the direction of molecular beam is z direction. Then the direction of laser polarization is in the x-z plane, and electric vector E can be expressed as

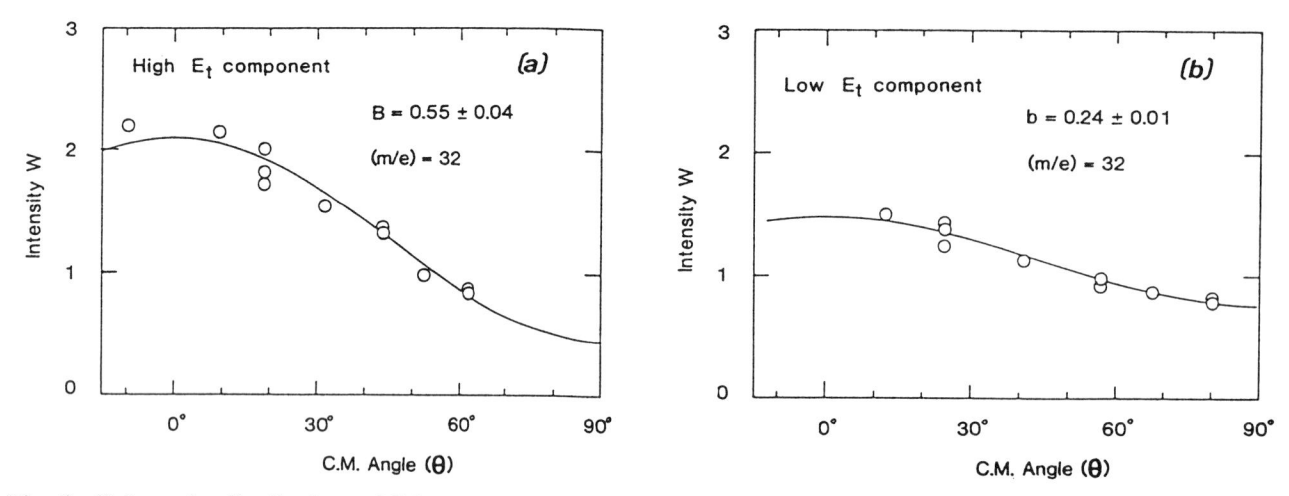

Fig. 6. CM angular distributions of S fragments. (a) Flux intensities integrated between 5 and 10 kcal/mol for the low energy component. Average θ value for these energies (5 ~ 10 kcal/mol) was adopted. (b) Flux intensities integrated between 35 and 40 kcal/mol for the high energy component. The average CM angle θ value for these energies was adopted.

$$\boldsymbol{E} = E\,(\boldsymbol{i} \sin \alpha + \boldsymbol{k} \cos \alpha) \tag{8}$$

where the unit vectors for x-, y- and z-directions are defined as $\boldsymbol{i}, \boldsymbol{j}$, and $\boldsymbol{k}$, respectively. The CM velocity of S fragment is expressed as

$$\boldsymbol{u} = u\,(\boldsymbol{k} \cos \theta + \boldsymbol{j} \sin \theta) \tag{9}$$

The cosine of the angle between the polarization vector $\boldsymbol{E}$ and the CM velocity vector $\boldsymbol{u}$, $\cos \phi$, is obtained by

$$\cos \phi = \cos \theta \cos \alpha. \tag{10}$$

Since the angular distribution of the S fragment for a given electric vector $\boldsymbol{E}$ in Eq. (8) and CM velocity vector $\boldsymbol{u}$ in Eq. (9) is expressed by formula (7), the observed angular distribution is obtained by averaging the intensity over all possible angles, α's which define the electric vector as

$$I(\theta) = (1/4\pi)(1/\pi) \int_0^\pi d\alpha\, [1 + 2 B_0 (3 \cos^2 \alpha \cos^2 \theta - 1)/2]. \tag{11}$$

in which B_0 is the true anisotropy factor dependent upon the CM recoil energy E_t. The integration of the right-hand side of Eq. (11) gives us an integrated form:

$$I(\theta) = (1/4\pi)\,[(2-B_0)/2 + B_0\, P_2(\cos \theta)]. \tag{12}$$

The standardized angular distribution for this formula is reduced to be

$$I(\theta) = [(2-B_0)/2]\,(1/4\pi)\{1 + 2\,[B_0/(2-B_0)]\, P_2(\cos \theta)\}. \tag{13}$$

The integrated distribution given by Eq. (13) is flatter than the angular distribution given by the original form expressed in terms of the "true" anisotropy factor B_0, which is the very angular distribution obtained when the laser polarization vector is parallel to the molecular velocity vector $\boldsymbol{v}_0$:

$$I(\theta) = (1/4\pi) [1 + 2B_0 P_2(\cos\theta)]. \tag{13}$$

Since the observed B value can be expressed in terms of the "true" anisotropy factor B_0, B_0 can be solved in terms of B as:

$$B_0 = 2B/(B+1). \tag{14}$$

One thus obtains the "true" anisotropy factors $B_0 = 0.71 \pm 0.036$ for the fast component (B = 0.55 ± 0.04) and $B_0 = 0.387 \pm 0.013$ for the slow component (B = 0.24 ± 0.01). These values are summarized in Table II along with other photofragmentation parameters.

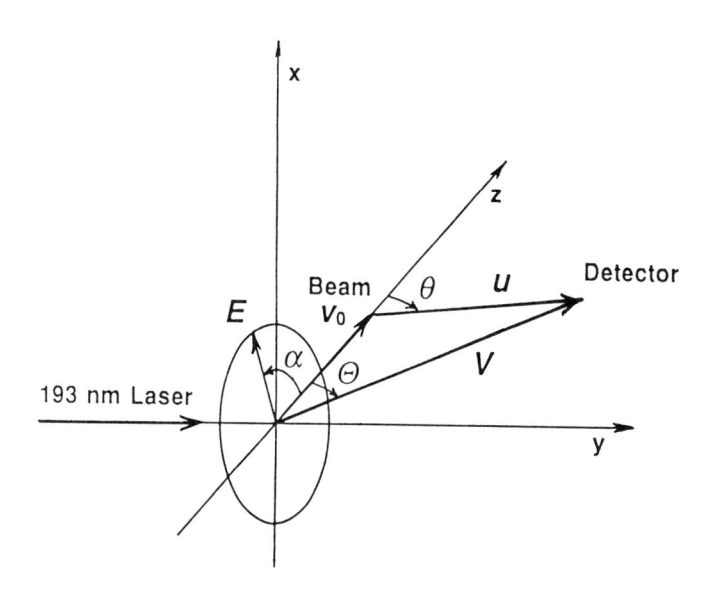

Fig. 7. Schematic vector diagram for electric polarization E, CM vector of the fragment detected u, and the molecular beam velocity vector v_0.

The B_0 values determined here should be compared with the twice of B_0 value, $2 B_0 = 1.03 \pm 0.19$ obtained by Kim et al.[5] The anisotropy factor determined for the high energy component, $B_0 = 0.71 \pm 0.036$, is close to the maximum available value (which is unity), which indicates the fragmentation occurs via a direct process. Even that for the low energy component, $B_0 = 0.387 \pm 0.013$, is comparable to the value obtained by Kim et al, although somewhat smaller than $B_0 = 0.52 \pm 0.10$. Since the complete analyses of our data have not been done, at present it is hard to say that the naive model that the fragment recoil distribution can be decomposed into two components with two different anisotropy parameters and CM recoil energy distributions, since an anisotropy factor can in principle vary with recoil energy. However one can conclude that i) the S atom fragment has two components, fast (high energy) and low (low energy), ii) the fast component undergoes the direct

Table II. Summary of photofragmentation parameters for the photolysis of ethylene sulfide at 193 nm. E_{av}: Maximum energy available in CM translational energy, E_t; $\bar{E}_t$: The average CM translational energy of the fragment pair; E_{peak}: E_t at the peak CM translational energy distribution; RRK-type parameters: α and β in Eq. (5); B: Experimentally determined best-fit anisotropy parameter; B_0: "True" anisotropy parameter

Photo-fragments	E_{av} (kcal/mol)	$\bar{E}_t$ (kcal/mol)	E_{peak} (kcal/mol)	RRK-parameters α	β	B	B_0
$S(^1D) + C_2H_4$ [a]	64.7	32.3	~ 26	...	...	0.55 ± 0.04	0.71 ± 0.036
$S(^1D) + C_2H_4$ [b]	64.7	8.9	4.5	0.7	9.7	0.24 ± 0.01	0.387 ± 0.013
$SH + C_2H_3$	66.5	9.0	< 2	0.1	6.0	0.10 ± 0.02	0.18 ± 0.03

a) High E_t energy component
b) Low E_t energy component

dissociation with a short life-time in the excited state and the slow component via relatively long-lived excited intermediate. Since the S atom is found to be formed in the electronically excited ^1D state,[5] the Doppler shift spectra of $S(^1D)$ fragment should be explained in terms of the B_0 values and the recoil energy distributions obtained from the present study.

3.1.B. SH Formation Process

The TOF signals of SH(m/e = 33) were obtained at the same angles Θ as were recorded for the S(m/e = 32) runs. Fig. 8 shows a representative TOF spectrum. Each spectrum exhibits a single peak and shows a similar distribution profile irrespective of laboratory angle Θ. The $\bar{E}_t$ distribution derived for the SH fragment is shown in Fig. 9. The distributions are found to be narrow, and hence were also fitted to an RRK-type formulae as

$$f_{SH}(E_t) = E_t^{0.1} (66.5 - E_t)^{6.0}$$

The average kinetic energy for the observed distribution was obtained as $\bar{E}_t = 9.0$ kcal/mol as shown in Table II.

The angular distribution are plotted in Fig. 10 for the flux intensities integrated between $E_t = 5$ and 10 kcal/mol; an anisotropy parameter of $B = 0.10 \pm 0.02$ is obtained. The true "anisotropy" factor $B_0 = 0.18 \pm 0.03$. The value together with the results of E_t distribution indicates that the dissociation forming SH fragments proceeds via an excited intermediate state(s) with a relatively long lifetime.

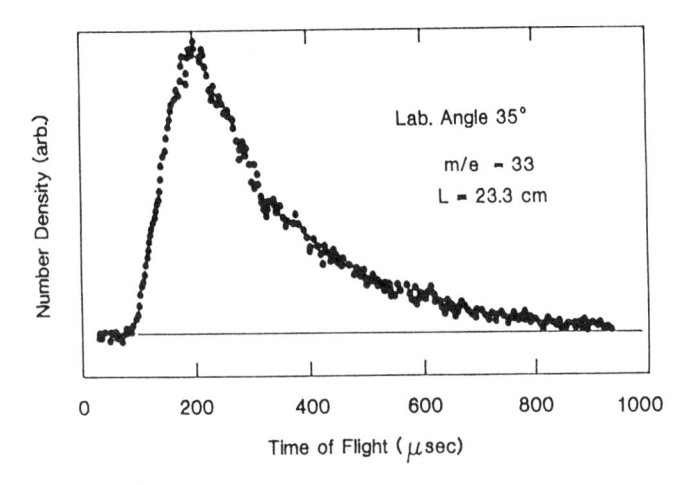

Fig. 8. A typical TOF spectrum of photofragment measured at m/e = 33 (SH⁺). Laboratory angle $\Theta = 35°$.

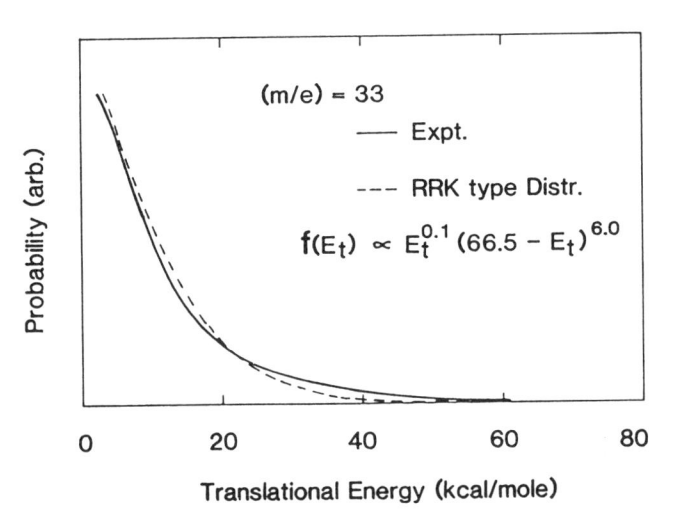

Fig. 9. CM flux (probability) distribution vs. CM recoil energy E_t, obtained from photofragment TOF spectrum of SH. Dashed curve represents a parametric fit of RRK-type distribution to the observed (solid curve) profile.

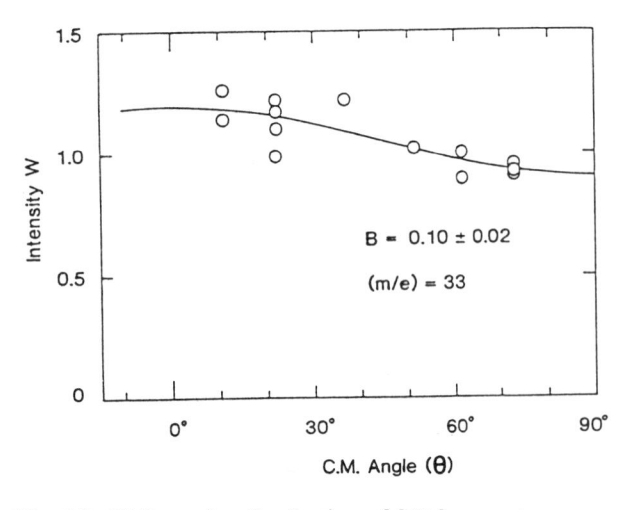

Fig. 10. CM angular distribution of SH fragment.

3.1.C. Discussion on Photodissociation of Ethylene Sulfide

The excited state of ethylene sulfide and correlation to the relevant dissociation products which are based on the experimental observations and theoretical calculations are shown in Fig. 11. Excitation of ethylene sulfide by 193 nm radiation is assigned to the transition of nonbonding ($3p_y$) electron on S to ($4p$) Rydberg orbital, and the molecular structure is considered to be almost the same as that of the ground electronic state. Adjacent to the lowest Rydberg states, it is expected that there are S_1 and S_2 dissociative levels originating from (n, σ*) and/or (σ, σ*) transition. Strausz et al.[11] predicted in their MO calculation that S_1 has a stable configuration in the ring-distorted structure with CCS angle of around 100°, while S_2 was found to be repulsive both in bent and symmetric geometry. These predissociative states are considered to carry relatively high excess energy as the translational motion in a time shorter than the rotational period. Internal conversion to vibrationally excited levels of the ground electronic state will show a long lifetimes and turn to form relaxed products $S(^1D) + C_2H_4$. Isomerization to vinylthiol is also expected to occur in the ground electronic state[12] via C-S ring-opening and followed by intramolecular H-migration after effective energy transfer. Dissociation of SH from vinylthiol is almost isoenergetic with $S(^1D) + C_2H_4$ formation. The distribution parameters thus determined for both fragmentation processes (1) and (2) are summarized in Table II.

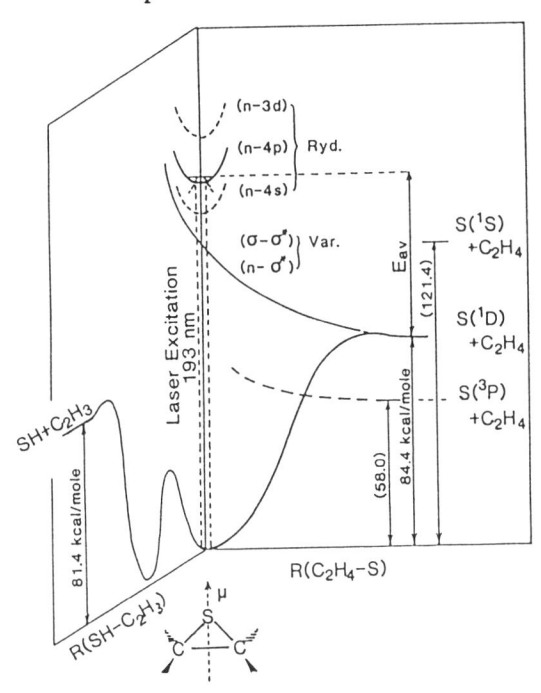

Fig. 11. Schematic potential energy surface diagram for the photofragmentation process upon laser excitation to Rydberg (n -> 4p) state.

S(^1D) Formation Channel

High Energy Component   The recoil energy distribution obtained from the TOF spectra reveals the presence of different dissociation channels in atomic S(^{1}D) formation. The average E_t value of the high energy component is determined to be 32 kcal/mol, indicating some 30 kcal/mol of maximum available energy (~ 65 kcal/mol) is left in internal modes of the ethylene fragment. The "true" anisotropy factor B_0 for the high energy component is determined as 0.71 ± 0.036, the photodissociation should proceed via a very direct process. That is, the dissociation should proceed on a repulsive, dissociative potential surface of valence bond character.

Low Energy Component   If the S dissociation proceeds retaining its C_{2v} symmetry, a significant structural change in C_2H_4 moiety is expected in the course of dissociation; the C-C bond distance should contract from 1.48 Å to 1.32 Å and the hydrogen atoms in the CHH planes which are off by 28 ° from the C-C bond should move back to the ethylene plane. Since the structural change is large, a considerable fraction of the available energy is expected to be partitioned to internal (vibrational) excitation of ethylene fragment, if the excited intermediate is deeply bound. If ethylene sulfide is excited to a state of Rydberg character, the intermediate state has a longer life-time and the energy available to the relative translational motion would be small. This may be the case of the low energy component. However the "true" anisotropy factor for this component which has been estimated as 0.387 ± 0.013, is still fairly large, and thus the dissociation proceeds via a fairly direct process.

From the best-fit parameters obtained for the CM translational energy distributions of low energy component, $f_{obs}(E_t) = E_t^{0.7}(64.7 - E_t)^{9.7}$, the exponent $\beta = 9.7$ means that the energy randomization of the available energy among the vibrational modes of the intermediate state is almost complete (it is well known that the exponent β determined from the experiment is about half of the available degrees of freedom when the RRK distribution is used. Since the high energy end of the translational energy distribution overlaps with the high energy component, the discussion on the active number of modes involved in the energy partitioning in the excited intermediate in terms of the β parameter may not be justified.

SH Formation Channel

We have obtained the following two experimental results: i) the translational energy E_t distribution is peaking toward zero and the best-fit β parameter $\beta = 6.0$ is about half of the number of degrees of freedom available in the complex s = 15, which means that the energy randomization is almost complete in the excited intermediate, and ii) the "true" anisotropy factor obtained for SH formation channel, $B_0 = 0.18 \pm 0.03$, is close to zero, that is the fragments are almost isotropically distributed. They indicate that the SH production proceeds via dissociation of vinylthiol ($CH_2=CH-SH$) formed by ring opening of the cyclic ring followed by H atom migration. It is highly probable that the intersystem crossing occurs from the electronically excited state to the ground state and the dissociation of the hot intermediate forming SH + C_2H_3.

If one simply extends RRKM treatment[13] with $E_{av} = 65$ kcal/mol, the isomerization rates via a bicyclic complex are expected to be as fast as ~ 10^{-13} sec, while unimolecular dissociation rates for -> $S(^1D) + C_2H_4$ and SH + C_2H_3 are predicted ~ 10^{-10} sec from the hot ground-state molecules.

3.2. Allene Photodissociation

The TOF photofragment spectra were measured at masses m/e = 39, 38, 26, 25, and 13. The TOF spectra for mass m/e = 39 were measured at $\Theta = 7.5, 12, 18$, and $25°$ whose velocity vector diagram is shown in Fig. 12. Typical TOF spectra measured at mass m/e = 39 and at laboratory angles $\Theta = 7.5$ and $18°$ are illustrated in Fig. 13. They are due to the reaction process:

$$C_3H_4 + h\nu(193nm) \longrightarrow C_3H_3 + H \tag{15}$$

From the appearance threshold of the TOF spectra the upper bound of the dissociation energy has been determined as 100.6 $\pm$ 5.6 kcal/mol, which is compared with the value <111.7 kcal/mol determined by Jackson et al.[8] Both of the values determined are considerably higher than the best bond dissociation energy, $\Delta H_0^0 = 85.6$ kcal/mol, estimated from the heats of formation of H, C_3H_3, and C_3H_4.[14] The TOF spectra obtained by us compare well with those of Jackson et al.,[8] except that their experiment was done with allene accelerated in helium carrier gas. Deconvolution of the TOF spectra gave us CM translational energy E_t distribution, as is shown in Fig. 14. Since we have not measured the TOF spectra for H atom fragments, the flux distribution in the low E_t energy region was not determined. The translational energy distribution obtained here is some-

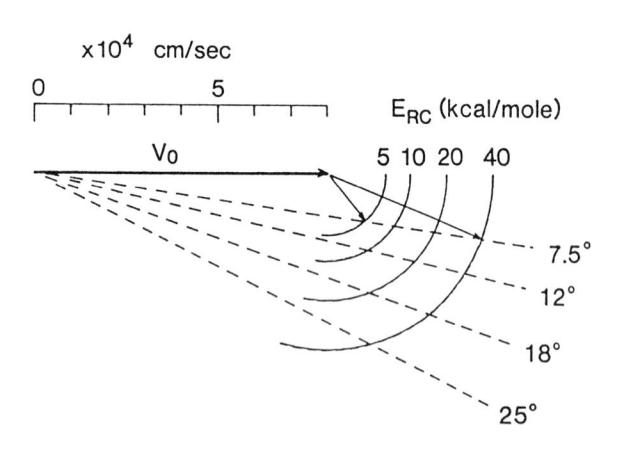

Fig. 12. Velocity vector diagram for the photodissociation of allene (C_3H_4) at 193 nm forming H + C_3H_3, when mass m/e = 39 is detected.

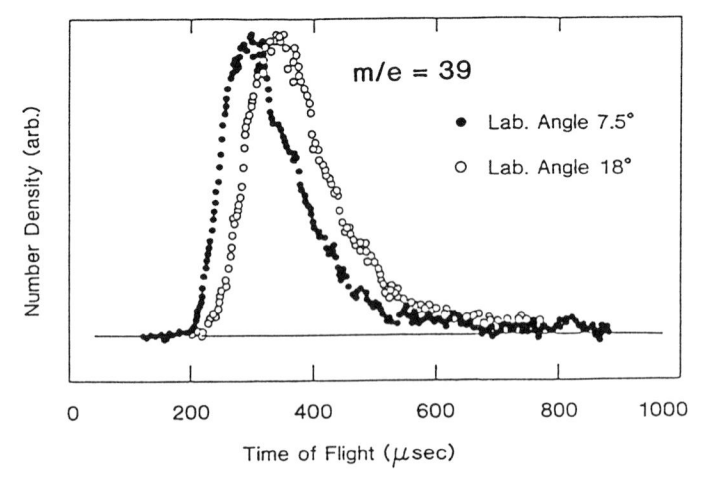

Fig. 13. TOF spectra observed for the fragment mass m/e = 39 at two laboratory angles, $\Theta = 7.5°$ and $18°$.

what broader than the one determined by Jackson et al.[8] since in our analysis the velocity distribution of the parent beam was not taken into account in transforming the TOF data to the CM translational energy flux distribution.

Since the translational energy distribution exhibits a typical one observed for the unimolecular dissociation via a long-lived complex with a low potential barrier, if any. Considering the facts that i) the maximum energy available in the CM relative translational motion is about 15 kcal/mol less than the calculated value from the best dissociation energy $\Delta H_0^0 = 85.6$ kcal/mol of Melius,[14] and ii) average recoil energy is fairly low (as high as ~ 10 kcal/mol), very large fraction of energy remains as vibrational energy of C_3H_3 fragment.

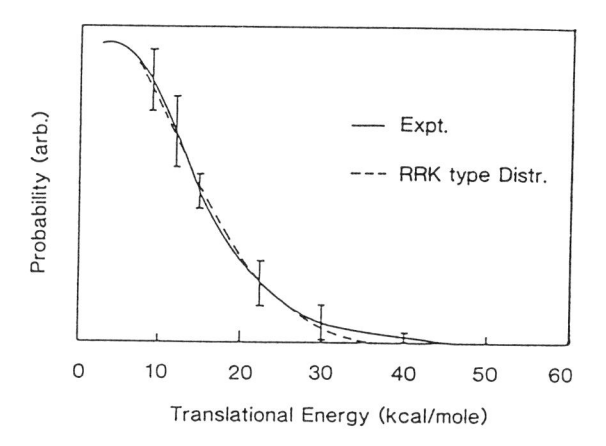

Fig. 14. CM translational energy distribution determined by transformation of TOF spectra for allene photofragmentation at 193 nm forming H + C_3H_3 measured at mass m/e = 39. The dashed curve represents a best-fit RRK-type distribution to the observed one.

The TOF spectra for reaction forming $H_2 + C_3H_2$:

$$C_3H_4 + h\nu(193\ nm) \rightarrow C_3H_2 + H_2 \qquad (16)$$

were also measured at mass m/e = 38. A typical spectrum observed at laboratory angle $\Theta = 25°$ is shown in Fig. 15. The strong band peaking at a flight time of 364 μsec has been assigned to that of the daughter ion $C_3H_2^+$ (m/e = 38) of the fragment C_3H_3 formed from the H formation reaction (15), because the profile is exactly the same as that obtained for mass m/e = 39

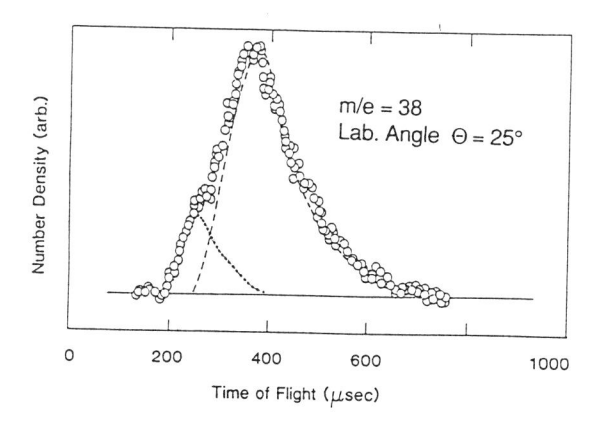

Fig. 15. TOF spectrum measured for mass m/e = 38 ($C_3H_2^+$) at a laboratory angle $\Theta = 25°$. The dashed curve is the contribution originating from the reaction product C_3H_3 + H. The dotted curve is that from reaction (16) forming $C_3H_2 + H_2$.

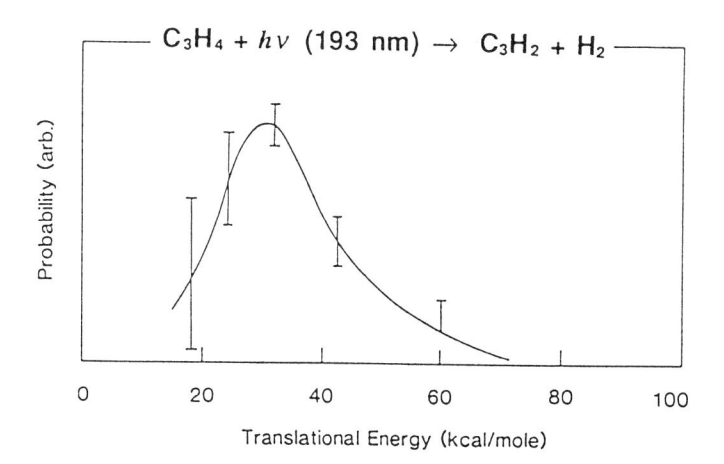

Fig. 16. The CM translational energy distribution for the product $C_3H_2 + H_2$ formed from the photodissociation of allene at 193 nm obtained by the transformation of the dotted distribution shown in Fig. 15.

at the same angle. The contribution of the H formation reaction (15) as shown by the dashed line has been subtracted from the raw TOF spectrum and that for the H_2 formation reaction has been obtained by the dotted line. The transformation of the TOF spectrum for the H_2 formation reaction has given us a translational energy distribution as illustrated in Fig. 16. The relative intensity is not so reliable, especially in the lower energy region. It seems however that the product translational energy distribution is peaking at $E_t = \sim 30$ kcal/mol which is considerably higher than the zero recoil energy and has a FWHM of about 21 kcal/mol. The translational energy distribution for H_2 formation reaction of Jackson et al.[8] is peaking at about 20 kcal/mol and has a FWHM of 15 kcal/mol. The distribution obtained by Jackson et al.[8] is expected to be more reliable than ours, since they determined it from the TOF spectrum for H_2 products detected at mass m/e = 2, which considerably reduces the kinematic uncertainties inherent in the detection of the heavier products. The high product translational energy is qualitatively consistent with decomposition of an activated complex after overcoming a high barrier which is expected for a three-centered transition state.

We have also observed TOF spectra for other masses than 39 and 38 mentioned above, i.e. at masses m/e = 26 ($C_2H_2^+$), 25 (C_2H^+), and 13 (CH^+). They would be mostly from the photofragments from the secondary photodissociation processes of the primary photofragments, C_3H_3 and C_3H_2.

4. ACKNOWLEDGMENT

K.T. would like to thank Prof. S. Nagase of Yokohama National Univ. for valuable discussions.

5. REFERENCES

1. K. Okiye, C. Hirose, D. G. Lister, and J. Sheridan, *Chem. Phys. Lett.* **24**, 111 (1974).
2. a) M. B. Robin, *Higher Excited States of Polyatomic Molecules*, vols. 1,3 (Academic, New York, 1974, 1985); b) I. Tokue, A. Hiraya, and K. Shobatake, *J. Chem. Phys.* **91**, 2808 (1989).
3. D. D. Altenloh and B. R. Russell, *J. Phys. Lett.* **77**, 217 (1981).
4. R. Bersohn, G. Ondrey, S. Kanfer, P. Brewer, and S. Yang, *J. Photochem.* **17**, 257 (1981).
5. H. L. Kim, S. Satyapal, P. Brewer, and R. Bersohn, *J. Chem. Phys.* **91**, 1047 (1989).
6. K. Tabayashi and K. Shobatake, "Photodissociation Dynamics of Ethylene sulfide at 193 nm Studied by Photofragment TOF-mass Spectroscopy", *Annual Review of IMS*, **1989**, p. 103.
7. K. Tabayashi and K. Shobatake, "Crossed Laser-molecular Beam Study of Allene Photofragmentation at 193 nm," *Annual Review of IMS*, **1989** p. 104.
8. W. M. Jackson, D. S. Anex, R. E. Continetti, B. A. Balko, and Y. T. Lee, *J. Chem. Phys.*, **95**, 7327 (1991).
9. Y. T. Lee, J. D. McDonald, P. R. LeBreton, and D. R. Herschbach, *Rev. Sci. Instrum.* **40**, 1402 (1969).
10. a) K. Shobatake, S. Ohshima, K. Tabayashi, M. Faubel, T. Horigome, M. Suzui, N. Mizutani, H. Yoshida, and K. Hayakawa, "Construction of a Crossed Molecular Beam Apparatus with an Electron Bombardment Ionization Mass Spectrometer (MBC-I)," *Annual Review of IMS*, **1983**, p. 109, and b) J. R. Grover, Y. wen, Y. T. Lee, and K. Shobatake, *J. Chem. Phys.* **89**, 938 (1988).
11. O. P. Strausz, H. E. Gunning, A. S. Denes, I. G. Csizmadia, *J. Am. Chem. Soc.* **94**, 8317 (1972).
12. M. L. McKee, *J. Am. Chem. Soc.* **108**, 5059 (1986).
13. A. G. Sherwood, I. Safarik, B. Verkoczy, G. Almadi, H. A. Wiebe, and O. P. Strausz, *J. Am Chem. Soc.* **101**, 3000 (1979).
14. The bond dissociation energy $\Delta H_0^0 = 85.6$ kcal/mol adopted by Jackson et al.[8]

157 nm Photodissociation dynamics of CO_2 via Photofragment-Translational Spectroscopy

Albert Stolow & Yuan T. Lee

Department of Chemistry, University of California, Berkeley &
Division of Chemical Sciences, Lawrence Berkeley Laboratory
Berkeley, California 94720 U.S.A.

ABSTRACT

The photodissociation of CO_2 at 157nm was studied by the photofragment-translational spectroscopy technique. Product time-of-flight spectra were recorded and center-of-mass translational energy probability distributions were determined. Two electronic channels were observed - one forming $O(^1D)$ with 92% probabilty and the other $O(^3P)$ with 8% probabilty.

INTRODUCTION

The vacuum-ultraviolet (VUV) photochemistry of CO_2 is an intriguing and important problem. Not only is CO_2 an important atmospheric constituent, but its relative simplicity should allow for detailed study and comparison with theory. For wavelengths in the range 140-170nm, CO_2 dissociation has two open channels:

(1) $\qquad CO_2(^1\Sigma_g^+) \rightarrow CO(^1\Sigma^+) + O(^1D)$

(2) $\qquad CO_2(^1\Sigma_g^+) \rightarrow CO(^1\Sigma^+) + O(^3P)$

The latter channel, reaction (2), represents a spin-forbidden process.

Early work[1] in a gas bulb indicated that the quantum yield for reaction (1) is unity at 131nm and 147nm, consistent with the expectation that a spin-forbidden process should be very unfavourable in a triatomic molecule consisting of carbon and oxygen atoms (*i.e.* relatively small spin-orbit coupling). Although, O(^{3}P) was observed, it could be completely accounted for via:

(3) $\qquad$ O(^{1}D) + CO$_2$($^1\Sigma_g^+$) → CO$_2$($^1\Sigma_g^+$) + O(^{3}P)

It is well know that O(^{1}D) is quenched by atmospheric molecules[2] with near gas kinetic efficiency. The weak spin-orbit coupling, however, suggests that such processes should be inefficient in simple encounters: the observed high quenching efficiency arises from the formation of a long-lived complex (*i.e.* several vibrational periods), increasing the probability of spin transition via multiple crossings of the singlet-triplet intersection region.

Recently, the photolysis of CO$_2$ at 157 nm was investigated using a chemical scavenging technique[3]. It was suggested that the primary photoprocess producing O(^{3}P) (reaction 2) contributes about 6% to the quantum yield. A subsequent study of the O(^{3}P$_j$, j=2,1,0) state distribution and Doppler profiles in a molecular beam experiment[4] confirmed the primary character of this channel. Interestingly, the Doppler profiles were analyzed to give an anisotropy parameter of β=2. A β-value of two is usually expected for a very direct dissociation wherein the recoil velocity vector is parallel to the electronic transition moment vector.

Most recently, the vibrational and rotational distributions of the CO($^1\Sigma^+$) product were measured via VUV laser induced fluorescence[5]. The CO product was found to have a highly excited rotational distribution, terminating abruptly at the energetic limit. This is indicative of the dissociation occuring from a bent excited state. Consistent with this was a measurement an anisotropy parameter of β=0, corroborating the suggestion that the excited state is bent. If the molecule bends strongly as it dissociates, the recoil velocity vector will be at a large angle relative to the transition moment vector and, hence, the β parameter will be small.

The excited states of CO$_2$ are very complicated[6]. There are two weak bands between the CO$_2$($^1\Sigma_g^+$) ground state and the group of {$^1\Pi_g$, $^1\Sigma_u^-$, $^1\Delta_u$, and $^1\Sigma_g^+$} excited states that occur in the range 120 - 200 nm and are electric dipole forbidden (D$_{\infty h}$ symmetry). The electronic transition, however, can be vibronically induced by a bending vibration. The first band, beginning around 6eV, with a maximum near 8.4 eV, is irregular and diffuse and no assignments have been made. Excitation at 157 nm corresponds to this transition. At higher energy, the second transition is sharper and more regular, peaking near 9.3 eV. The first optically allowed transition appears around 11.1 eV. High accuracy electronic structure calculations[7] for CO_2 have been performed in the Franck-Condon region (*i.e.* linear $D_{\infty h}$ structures) for the lowest singlet excited states: $^1\Pi_g$, $^1\Sigma_u^-$ and $^1\Delta_u$. It was shown that in the Franck-Condon region between 120 and 170 nm, the nearly overlapping $^1\Delta_u$ and $^1\Sigma_u^-$ states are involved in conical intersections with the $^1\Pi_g$ state. Thus, the photophysics of CO_2 is very complex and cannot be described within the Born-Oppenheimer approximation. Along the bending co-ordinate, the electronically degenerate $^1\Pi_g$ and $^1\Delta_u$ states split into Renner-Teller pairs of $^1A_2$ and $^1B_2$ symmetry (C_{2v}). These are strongly stabilized by bending, as is the non-degenerate $^1\Sigma_u^-$ state. The bent $^1B_2$ component of the $^1\Delta_u$ state has been analyzed (*i.e.* the carbon monoxide flame bands)[8] and the OCO bond angle was found to be 122° - very strongly bent.

In this study, we employ the technique of photofragment-translational spectroscopy to study the photodissociation dynamics of CO_2 at 157 nm in a molecular beam. The excited states very complicated and the dynamics cannot be described as occuring on a single potential energy surface. The existence of the $O(^3P)$ channel further suggests that triplet surfaces must also be involved in the dissociation dynamics. The excited states are calculated to be strongly bent and this feature should be revealed in the product state distributions.

Experimental

The high resolution rotating source photofragment-translation spectrometer has been described previously in detail[9]. A brief description follows; a schematic drawing of the apparatus is shown in Figure 1. A molecular beam is formed by passing gas through a heatable nozzle, (1) into a source chamber (2), where is was skimmed, passed into a differential chamber (3), skimmed again and finally passed into the main interaction chamber (4). A pulsed laser crosses the molecular beam at (5), which is also the viewing axis of the triply differentially pumped quadruple mass spectrometer. Time-of-flight spectra are obtained by recording the distribution of arrival times of photodissociation product molecules at the detector. Angular distributions are obtained by rotating the molecular beam about the point (5).

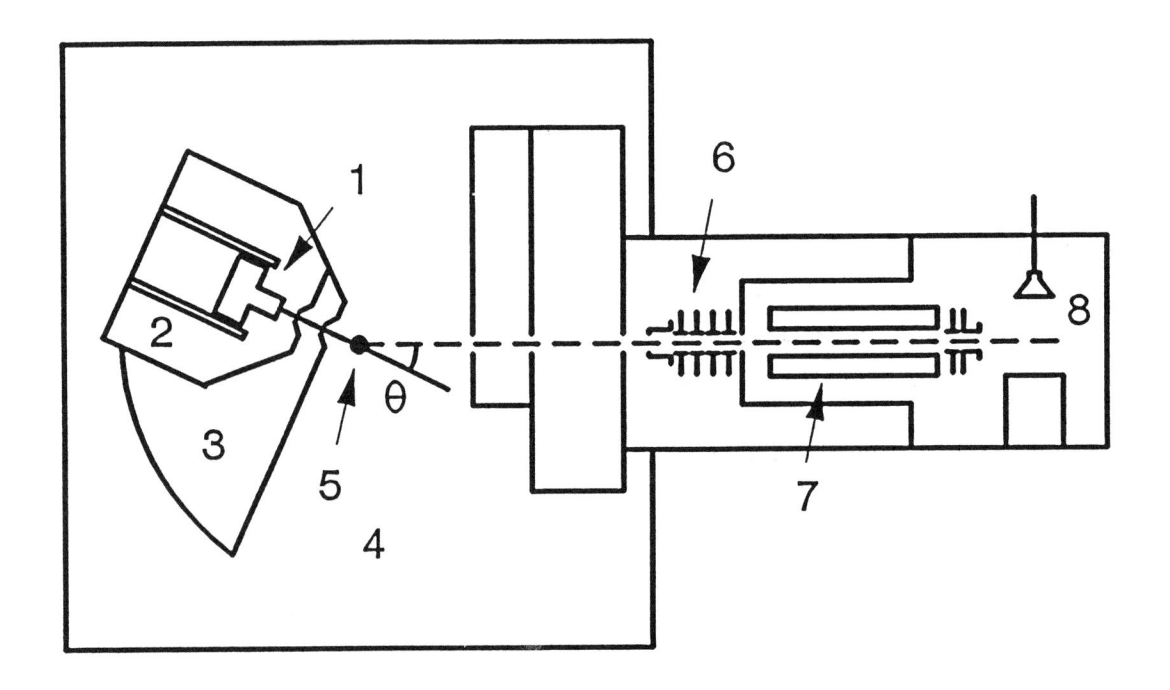

Figure 1: A schematic drawing of the photofragment spectrometer. (1) molecular beam nozzle, (2) source chamber, (3) differential chamber, (4) main chamber, (5) laser interaction point, (6) ionizer, (7) quadrupole mass spectrometer, (8) detector.

Seeded mixtures of isotopically substituted carbon dioxide (5% $^{13}CO_2$, 95% He) with a stagnation pressure of 200 torr were expanded through a 175μm diameter nozzle, which was heated to 115° C. The expansions were typically characterized by a lab velocity of 1.6×10^5 cm/s and a speed ratio of 11. The heated nozzle ensured that no clusters were present. This was checked by looking for parent $^{13}CO_2$ molecules recoiling from the molecular beam at small angles. No evidence of cluster formation was found. The bending frequency of $^{13}CO_2$ in the ground state is rather low ($v_2=654$ cm^{-1}) and doubly degenerate. Therefore, at room temperature $CO_2(v_2 = 1)$ constitutes around 8% of the ground state population whereas at 115° C, it constitutes about17%. There may, however, have been some relaxation of this mode during the supersonic expansion.

The 157 nm laser used in this experiment was a Lambda-Physik VUV excimer LPF-205. Specifically optimized for operation at 157 nm, it was capable of producing over 100 mJ/pulse at 50 Hz with a gas lifetime of 300,000 shots. These features proved important in a small signal experiment such as photofragment-translational spectroscopy. When necessary, the laser was attenuated with an O_2 gas cell neutral density filter. The laser power was continuously monitored during the experiments.

PARENT	PRODUCT	ENERGY AVAILABLE
$^{13}CO_2(v_2=0)$	$^{13}CO(v=0) + O(^3P)$	55.458 kcal/mol
	$^{13}CO(v=1) + O(^3P)$	49.467 kcal/mol
	$^{13}CO(v=2) + O(^3P)$	43.548 kcal/mol
	$^{13}CO(v=3) + O(^3P)$	37.701 kcal/mol
	$^{13}CO(v=4) + O(^3P)$	31.928 kcal/mol
$^{13}CO_2(v_2=0)$	$^{13}CO(v=0) + O(^1D)$	10.142 kcal/mol
	$^{13}CO(v=1) + O(^1D)$	4.151 kcal/mol
$^{13}CO_2(v_2=1)$	$^{13}CO(v=0) + O(^1D)$	12.011 kcal/mol
	$^{13}CO(v=1) + O(^1D)$	6.020 kcal/mol
	$^{13}CO(v=2) + O(^1D)$	0.101 kcal/mol

Table 1: The available translational energies of the fragmentation channels for CO_2 dissociation at 157nm. The $O(^3P)$ and $O(^1D)$ channels are given, as well as the $O(^1D)$ channel for dissociation of the 'hot' molecule.

Time-of-flight spectra consisting of 300,000-600,000 co-added shots were recorded by a minicomputer and transferred to another computer for analysis. Product center-of-mass translational energy probability distributions are extracted using the forward convolution technique[9]. The use of $^{13}CO_2$ significantly improved the signal-to-noise ratio in this experiment due to the reduction in background at m/e = 29 ($^{13}CO^+$) relative to m/e = 28 ($^{12}CO^+$) and m/e = 16 (O^+). The instrumental response function was obtained by photodissociating a seeded oxygen beam (5% O_2, 95% He), which at 157 nm produces only $O(^1D) + O(^3P)$ and, therefore, a single translational energy. This served to determine the delta function response of the instrument which was used in the subsequent deconvolution of the experimental data.

The 157 nm excimer laser operated at two narrow VUV lines (approx. 10 cm^{-1} bandwidth). The main line (85%) lases at 157.63 nm (64,440 cm^{-1} or 181.34 kcal/mol). $D_0(O^{13}C\cdots O)$ was taken to be 171.20 kcal/mol for the $O(^1D)$ channel and 125.88 kcal/mol for the $O(^3P)$ channel (*i.e.* corrected for ^{13}C isotopic shifts)[10]. The available energies for the various fragmentation channels are given in Table 1. The Newton diagram illustrating these center-of-mass recoil energies is shown in Figure 2. The lab angles of 10° and 30° were used most frequently in recording the TOF spectra as these two angles were most sensitive to the slower and faster components, respectively, of the $O(^1D)$ translational energy distributions. The $O(^3P)$ channel is better resolved at larger angles (30°).

Results and Discussion

Time-of-flight spectra for the ^{13}CO photofragment at lab angles of 10° and 30° are shown in Figures 3 and 4, respectively. Definitive evidence for the $O(^3P)$ primary channel (reaction 2) is seen in the small peak at early times, shown in greater detail in Figure 5. The leading edge of this peak is consistent with the formation of CO(v=0,J=0) + $O(^3P)$. The dashed and solid lines show the fits to the $O(^3P)$ and $O(^1D)$ channels, respectively.

CO(v=0) + O(3P)

CO(v=3) + O(3P)

CO(v=0) + O(1D)

CO(v=1) + O(1D)

1 0

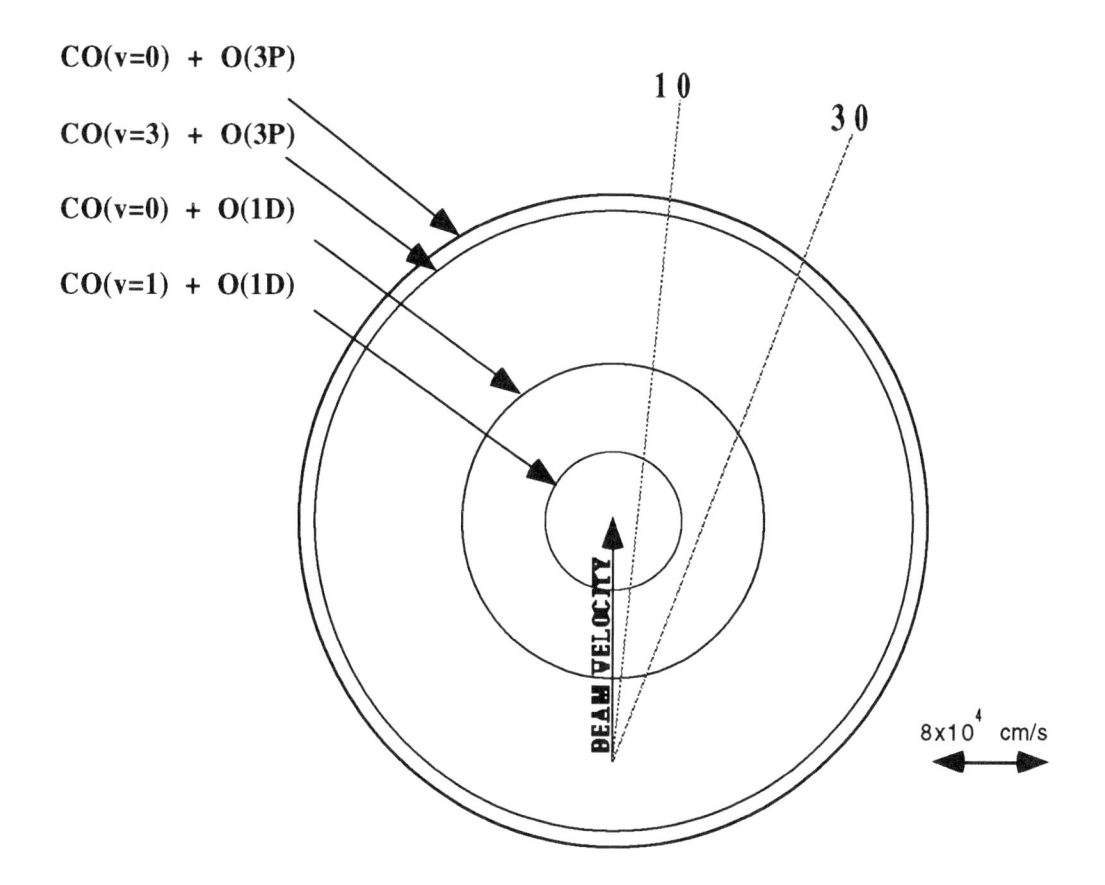

3 0

BEAM VELOCITY

8×10^4 cm/s

Figure 2: Newton diagram showing the relationship between lab frame and center-of-mass frame velocities for $CO_2 \rightarrow CO(v) + O(^3P, ^1D)$ at 157nm. The lab velocity of the beam is given by the arrow, as indicated. Product CO recoil velocities are shown as circles centered at the tip of the arrow.

The probability distribution, P(E), of translational energies for the $O(^3P)$ channel is shown in Figure 6: it is clear that there is a broad range of translational energies, consistent with the CO molecular product being formed in a range of vibrational states. In the analysis of the $O(^3P)$ Doppler profiles[4], it was assumed, for the sake of simplicity, that all recoiling atoms had the same translational energy. This allowed for a simple fit to the Doppler lineshape using a single velocity and a single anisotropy parameter, suggested to be β = 2. Due our poor signal-to-noise ratio for the small peak corresponding to the $O(^3P)$ channel, the P(E) obtained is not definitive. We can, however, compare our result with the assumption of reference 4 that all O(^{3}P) product appears as if recoiling from CO(v=4). When the shape of the P(E) was adjusted so as to have a maximum at

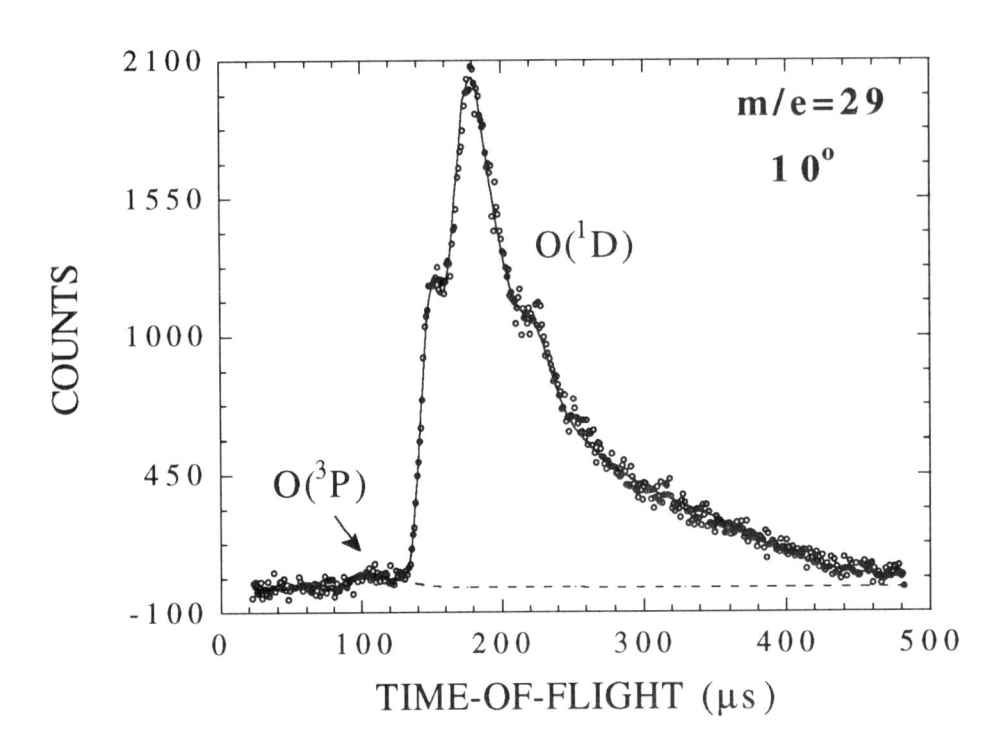

Figure 3: A time-of-flight distribution at a lab angle of 10° for the ^{13}CO fragment from ^{13}CO$_2$ photolysis at 157nm. The two electronic channels, O(^{3}P) and O(^{1}D), are indicated. The solid line shows the fit to the O(^{1}D) channel whereas the dashed line shows the fit to the O(^{3}P) channel.

CO(v=4), it was found that the fit did not match the leading edge in the experimental data. It was necessary to have to P(E) peak near CO(v=0) and have a significant breadth (30kcal/mol) in order to best fit the data. The assumption of a single recoil energy is not a good one. Since both the translational energy distribution and the anisotropy contribute to the Doppler lineshape, it is not possible to unambiguously determine the anisotropy parameter,β, from a measurement in a collinear pump-probe configuration alone. We suggest that the anisotropy parameter for the O(^{3}P) channel is close to $\beta = 0$, as is the O(^{1}D) channel, The reasons for this are as follows. The excited states of CO$_2$, both singlet and triplet, are strongly bent. Therefore, the recoil velocity vector should *not* be parallel to the transition moment vector and β should be significantly reduced from the limiting value of 2. This is the explanation offered for the β = 0 result of the O(^{1}D) channel[5]. Furthermore, the existence of a long-lived intermediate was invoked[3] to explain the observation of the spin-forbidden O(^{3}P) channel. This should further reduce the anisotropy parameter towards β = 0. With the assumption of β = 0, we obtain an electronic branching ratio of 8% O(^{3}P), close to that obtained previously by the chemical scavenging technique[3]. For reference, an assumption of β = 2 would yield a branching ratio of 6% O(^{3}P).

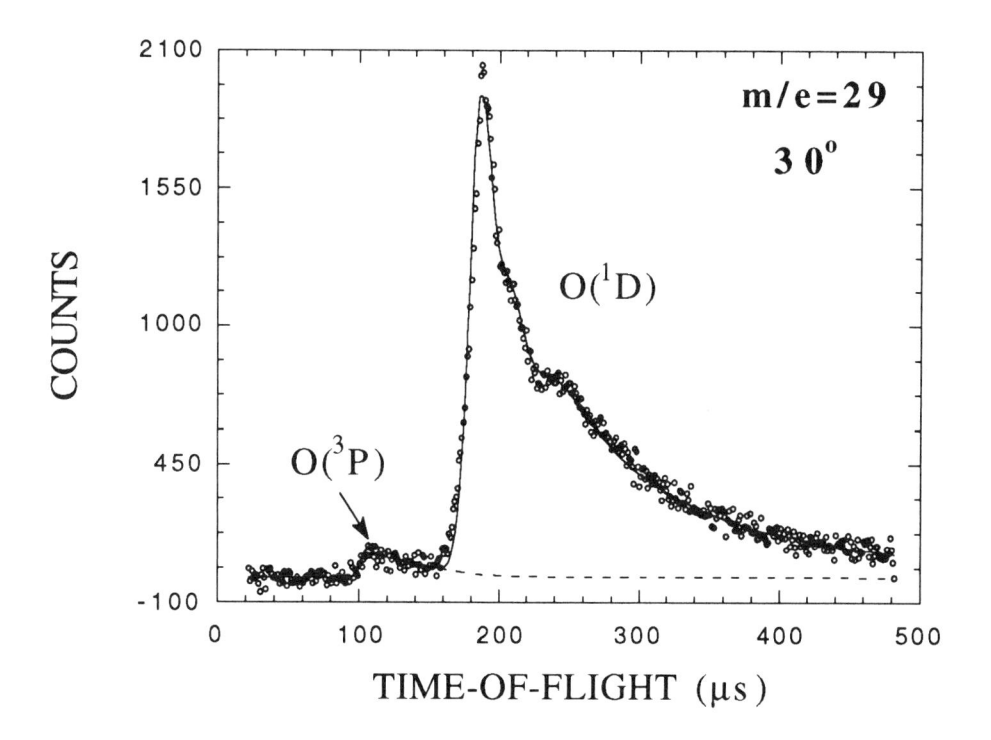

Figure 4: A time-of-flight distribution at a lab angle of 30° for the ^{13}CO fragment from ^{13}CO$_2$ photolysis at 157nm. The two electronic channels, O(^{3}P) and O(^{1}D), are indicated. The solid line shows the fit to the O(^{1}D) channel whereas the dashed line shows the fit to the O(^{3}P) channel.

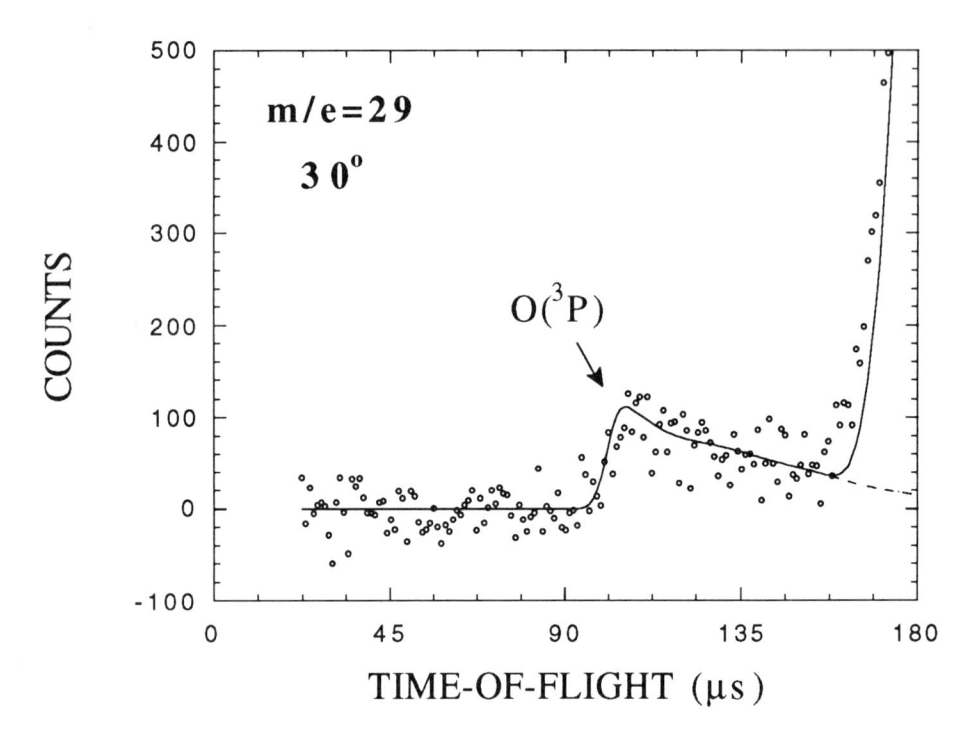

Figure 5: A time-of-flight distribution at a lab angle of 30° for the ^{13}CO fragment from $^{13}CO_2$ photolysis at 157nm showing in detail the electronic channel $O(^3P)$. The dashed line shows the fit to the $O(^3P)$ channel.

The structure in the large $O(^1D)$ channel peak (Figures 3 and 4) is related to rovibrational distributions in the CO product formed in reaction (1). The leading edge of the $O(^1D)$ channel shows that the formation of CO(v=0,J=0) from the 'hot' molecule $^{13}CO_2(\nu_2=1)$ is a small contribution. We first discuss the results at 10° (Figure 3). Based upon the Newton diagram of Figure 2, we expect to see both CO(v=0) and CO(v=1) at this angle. The first shoulder on the large $O(^1D)$ channel peak at 10° is due to the formation of CO(v=0) whereas the second, larger, peak is due to CO(v=1) - both from 'cold' $^{13}CO_2(\nu_2=0)$. The third feature, another shoulder, is due to the rotational envelope of CO(v=1,J). This indicates the significant degree of rotational excitation in the CO product, consistent with the general results of reference 5. At 30°, Figure 4, only CO(v=0) reached the detector (see Figure 2). The large peak is due to CO(v=0) and the first shoulder is related, in this case, to the rotational envelope of CO(v=0,J) - again corroborrating the results of reference 5. The second shoulder is due to the formation of CO(v=1) from the 'hot' molecule $^{13}CO_2(\nu_2=1)$. These translational distributions will be discussed in detail in a forthcoming publication[11].

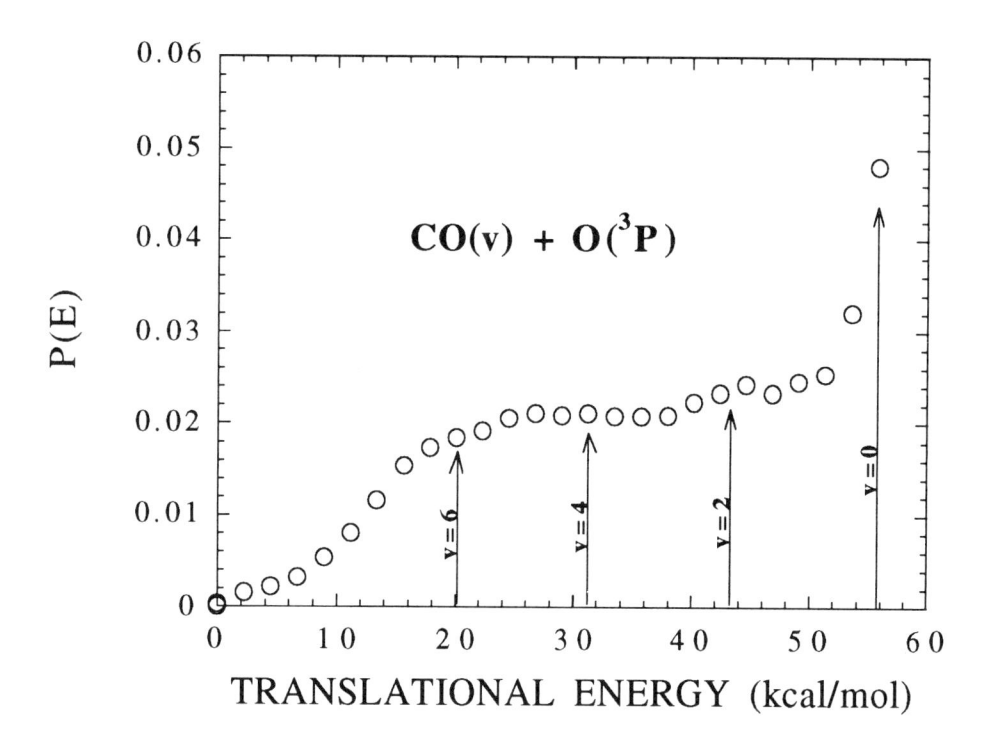

Figure 6: A translational energy distribution for $^{13}CO_2 \rightarrow {}^{13}CO(v) + O(^3P)$ at 157nm, corresponding to the dashed line fit in Figure 5. The energetics for forming several different ^{13}CO vibrational levels are indicated.

CONCLUSION

The photodissociation of CO_2 at 157 nm was studied by the photofragment-translational spectroscopy technique. The existence of the $O(^3P)$ channel was confirmed and a translational energy distribution, P(E), was obtained. A much larger, structured, signal, due to the CO(v=0,1) + $O(^1D)$ channel, was also observed.

The anisotropy parameter for each channel should be close to $\beta = 0$. With this assumption, an electronic branching ratio of 8% $O(^3P)$ was obtained, consistent with previous results. The translational energy distribution for the $CO(v) + O(^3P)$ channel was very broad (over 30kcal/mol) and appeared to peak near $CO(v=0)$.

ACKNOWLEDGEMENTS

We should like to thank Lambda-Physik for the generous loan of a VUV excimer laser LPF-205 for these (and other) studies. A.S. would like to acknowledge NSERC (Canada) for the receipt of a post-doctoral fellowship. This work was supported by the Division of Chemical Sciences, Office of Basic Energy Sciences, U.S. Department of Energy under Contract Number DE-AC03-76F00098.

REFERENCES

1. T.G. Slanger & G.Black, J.Chem.Phys. 54, 1889 (1971); 68, 1844 (1978)
2. J.C.Tully, *ibid* 61, 61 (1974); 62, 1893 (1975)
3. Y.F.Zhu & R.J.Gordon, *ibid* 92, 2897 (1990)
4. Y.Matsumi, N.Shafer, K.Tonokura, M.Kawasaki, Y.-L.Huang & R.J.Gordon, *ibid* 95, 7311 (1991)
5. R.L.Miller, S.H.Kable, P.L.Houston & I.Burak, *ibid* 96, 332 (1991)
6. R.W.Rabalais, J.M.McDonald, V.Scherr & S.P.McGlynn, Chem.Rev. 71, 73 (1971)
7. P.J.Knowles, P.Rosmus & H.-J.Werner, Chem.Phys.Lett. 146, 230 (1988)
8. R.N.Dixon, Proc.Roy.Soc. A275, 431 (1963)
9. A.M.Wodtke & Y.T.Lee, J.Phys.Chem. 89, 4744 (1985)
10. H.Okabe, *Photochemistry of Small Molecules*, Wiley, New York (1978)
11. A.Stolow & Y.T.Lee, *to be published*.

OPTICAL METHODS FOR
TIME- AND STATE-RESOLVED CHEMISTRY

SESSION 4

Photoionization, Photoelectron, and Ion Photodissociation Spectroscopy

Chair
John W. Hepburn
University of Waterloo (Canada)

Laser Spectroscopy of Radicals and Carbenes

Horst Clauberg, David W. Minsek, and Peter Chen[1]

Mallinckrodt Chemical Laboratories
Harvard University
Cambridge, Massachusetts

ABSTRACT

Photoelectron spectroscopy of isomeric C_3H_2 and C_3H_3 molecules, produced by flash pyrolysis in a supersonic jet nozzle, is used to test a valence-bond picture for radical bond strengths and carbene singlet-triplet gaps. The good agreement suggests that ionization potential measurements can be used to determine these hard-to-obtain numbers.

1. INTRODUCTION

Among the many criteria for mechanistic plausibility in organic transformations, the most fundamental is thermochemical. The enthalpy or free energy change in each step of a proposed mechanism must conform to reasonable expectations for that mechanism to be accepted for a given reaction. Accordingly, estimation schemes for the heats of formation of reactive intermediates, for which experimental thermochemical data are sparse, are widely used. These schemes, such as Benson's group equivalents[2], are usually based on an additivity assumption. Even the correction for systematic errors in *ab initio* heats of formation[3] by calculation of a homodesmotic reaction relies upon an implicit assumption of additive increments to ΔH_f by component parts of a molecule. Used properly, group or bond additivity schemes adequately predict ΔH_f for closed-shell molecules, and even for monoradicals[4], to within a few kcal/mol. For carbenes and biradicals though, additivity has a mixed record of predictions, of which the best one can say is that the predictions rest upon no solid thermochemical database, and are of uncertain reliability. The uncertainty in ΔH_f[carbene or biradical] is equivalent to saying that, while we can predict the bond dissociation energy in a molecule to give a radical, we cannot predict with certainty the second bond dissociation energy which takes that radical to a carbene or biradical.

We now propose, and verify by experiment, a general, and surprisingly simple, solution to the problem of second bond dissociation energies and carbene heats of formation. We now describe the experimental verification[5,6], for geometrically-constrained carbenes that: (1) simple bond additivity calculations correctly estimate ΔH_f for triplet ground-state carbenes, but, for singlet ground-state carbenes, additivity estimates of ΔH_f should be adjusted down by the carbene singlet-triplet gap, and (2) the singlet-triplet gap for a singlet carbene, Δ_{ST}[singlet carbene], can be approximated by

ΔIP, where ΔIP is the difference in ionization potentials between the carbene and the radical related by a C-H scission to that carbene. We therefore suggest that the equation below is a good description of carbene thermochemistry.

$$\Delta H_f[C_nH_{m-2}] \approx \Delta H_f[C_nH_m] + (2 \times BDE[C_nH_m]) - (2 \times \Delta H_f[H\bullet]) - \Delta_{ST}[C_nH_{m-2}]$$

The equation above refers a singlet carbene's ΔH_f to that of the closed-shell hydrocarbon with two more hydrogens. Alternatively, reference can be made to the monoradical that differs by one hydrogen if good thermochemical data are available for that radical. Furthermore, the singlet-triplet gap, which is the correction from additivity, can be approximated as ΔIP.

$$\Delta_{ST}[C_nH_{m-2}] \approx \Delta IP = IP[C_nH_{m-2}] - IP[C_nH_{m-1}]$$

We have performed mass and photoelectron spectroscopic measurements on isomeric C_3H_2 carbenes and C_3H_3 radicals to test this Δ_{ST}-corrected additivity estimate of singlet carbene ΔH_f values.

2. THEORETICAL MODEL

This picture of carbene thermochemistry, which at first sight appears dangerously oversimplified, is justified by: (1) straightforward application of VB ideas, and (2) detailed thermochemical measurements on benchmark model systems. Recent *ab initio* GVB calculations of the energetics of cleaving two bonds (C=C σ and π) in olefins[7], as well as sequential cleavage of C-H bonds from ethylene and acetylene, studied experimentally[8] and computationally[9], suggest that a "diabatic" bond scission proceeds to give a formal "valence state[10]" of the remaining fragment, which differs from the actual ground state of the molecule by a valence promotion energy. For geometrically-constrained carbenes, the valence state is closely approximated by the triplet state of the molecule. Therefore, for ground-state triplet carbenes, the VB promotion energy ought to be small, while for those with singlet ground states, the VB promotion energy is approximately Δ_{ST}. Similar approaches have been used for organometallic bond energies[11]. The conceptual basis (but not the quantitative application) for the VB promotion energy model for homolytic bond dissociation energies harkens back to the Heitler-London spin-valence theory. Specific examples[12] of anomalously weak bonds in some diatomic hydrides, e.g. BeH, were explained in this way as early as 1934.

3. RESULTS

We prepare reactive intermediates in a supersonic free jet expansion by high-temperature, short contact time pyrolysis in a heated ceramic nozzle. Details of the nozzle design, the time-of-flight mass and photoelectron spectrometers, and the 118.2 nm (10.49 eV) laser photoionization source can be found in our earlier publications. For the purposes of this discussion, we simply note that we produced each of the three isomers of C_3H_2, and one isomer of C_3H_3 in quantitative yield, without cross-contamination by other isomeric species, by flash pyrolysis of appropriately synthesized precursors. The precursors are described in references 5, 6, and 14. Cleanliness and specificity of the pyrolysis for each molecule was checked by photoionization mass spectroscopy. Photoelectron spectra of C_3H_2 produced in this manner established adiabatic ionization potentials, that together with ion thermochemical data available from the literature, allows us to determine the heat of formation of the reactive intermediate. Proof that each pyrolysis produced the correct structure was done by comparison of the observed photoelectron spectra with simulated Franck-Condon envelopes (*ab initio* geometries and force constants) for each isomer. We determined IP[propargyl radical] = 8.67 ± 0.02 eV, IP[cyclopropenylidene] = 9.15 ± 0.03 eV and IP[propadienylidene] = 10.43 ± 0.02 eV. Spectra were in each case consistent with the simulations.

4. DISCUSSION

We justify the quantitative application of the simple valence bond picture of bond dissociation energies by detailed thermochemical measurements on benchmark model systems. Mass and photoelectron spectroscopy of isomeric C_3H_2 carbenes established that additivity estimates for ΔH_f[carbene], corrected by the Δ_{ST}[singlet carbene], agree with experiment or the best available *ab initio* heats of formation.

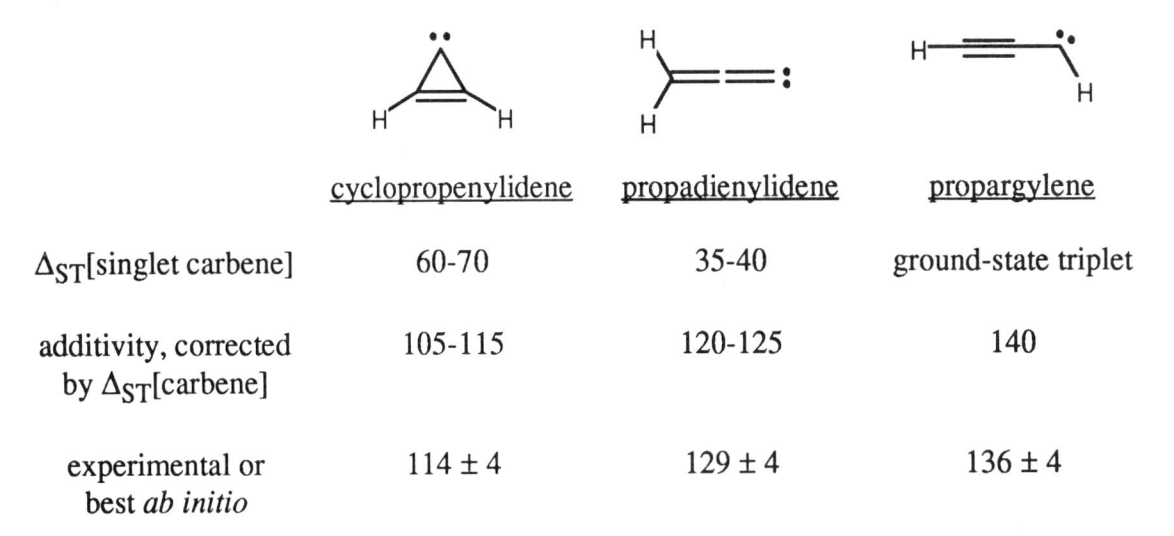

	cyclopropenylidene	propadienylidene	propargylene
Δ_{ST}[singlet carbene]	60-70	35-40	ground-state triplet
additivity, corrected by Δ_{ST}[carbene]	105-115	120-125	140
experimental or best *ab initio*	114 ± 4	129 ± 4	136 ± 4

The heats of formation, and singlet-triplet gaps, are in kcal/mol units, and are discussed in ref. 6. That reference also details the auxiliary ion thermochemical data, which along with our measured ionization potentials, produced the $\Delta H_f[C_3H_2]$. The uncertainties in the Δ_{ST}-corrected additivity estimates come primarily from uncertainties in the singlet-triplet gaps. The level of agreement nevertheless demonstrates that deviations of singlet carbene heats of formation from strict additivity can be quantified by the carbene singlet-triplet gap to within about 5 kcal/mol.

The further prediction that $\Delta_{ST} \approx \Delta IP$ is demonstrated by comparison of the ionization potentials of C_3H_2 carbenes and C_3H_3 radicals. Both propadienylidene and propargylene, for example, are derived from propargyl radical by a C-H bond scission. We predict that the acetylenic C-H bond dissociation energy (BDE) of propargyl radical should be weakened by a VB promotion energy, which we approximate as Δ_{ST}[propadienylidene] = 40 kcal/mol[13]. By contrast, the methylenic C-H bond of propargyl radical ought to be an typical sp^2 hybridized bond. Further examination of the electronic configurations of the cations finds no analogous weakening for any C-H bond in $C_3H_3^+$, i.e. diabatic C-H bond scission in the ground state of propargyl cation gives the ground state of either radical cation. Therefore, if we assume no other large bond strength effects in $C_3H_3^+$, simple arithmetic gives:

$$BDE[C_3H_3^+] - BDE[C_3H_3] \approx \Delta_{ST}[C_3H_2] \approx IP[C_3H_2] - IP[C_3H_3]$$

with the relationships in the equation depicted in the web of structures below. We emphasize that, while the picture is drawn, and the equation written, for C_3H_2, the treatment is general, applying to any singlet carbene and its corresponding radical.

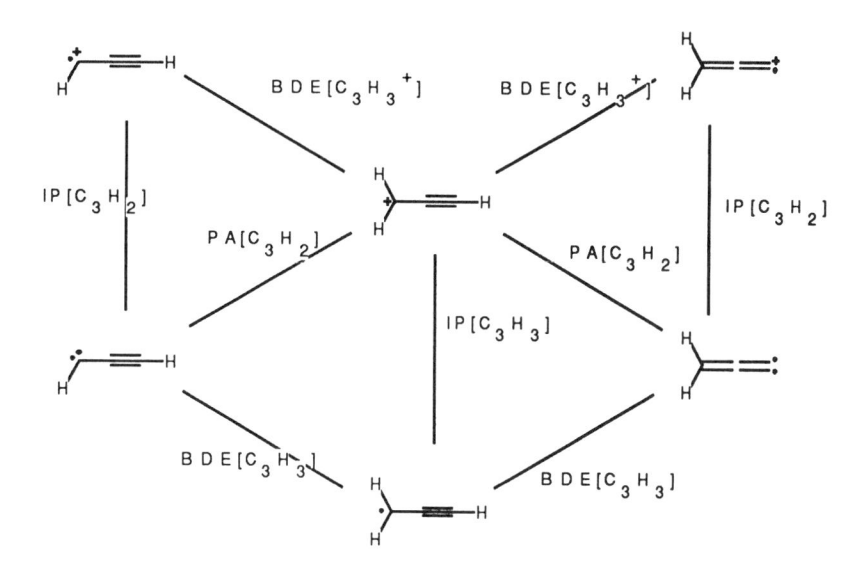

We measured IP[propadienylidene] = 10.43 ± 0.02[6], and IP[propargyl radical] = 8.67 ± 0.02[14] eV which give an estimated singlet-triplet gap of Δ_{ST}[propadienylidene] ≈ ΔIP = 40.6 ± 0.7 kcal/mol, in excellent agreement with the best available theoretical values[13]. The same picture suggests that, because propargylene is a ground-state triplet carbene, IP[propargylene] ≈ IP[propargyl radical]. We estimate IP[propargylene] by reference to our experimental IP[cyclopropenylidene]. The validity of the procedure can be checked by comparing IP[propadienylidene], estimated by this procedure, with the measured value. Relative energies of the C_3H_2 and $C_3H_2^+$ isomers for this exercise have been calculated[13,15] by *ab initio* methods. Propargylene and propargylene radical cation are calculated to be 0.95 and 0.56 eV higher in energy than cyclopropenylidene and cyclopropenylidene radical cation, respectively. Using our experimentally determined

IP[cyclopropenylidene] = 9.15 ± 0.03 eV, we derive IP[propargylene] ≈ 8.76 eV. This is nearly identical to our experimental IP[propargyl radical] = 8.67 ± 0.02 eV, as predicted by the valence bond picture above.

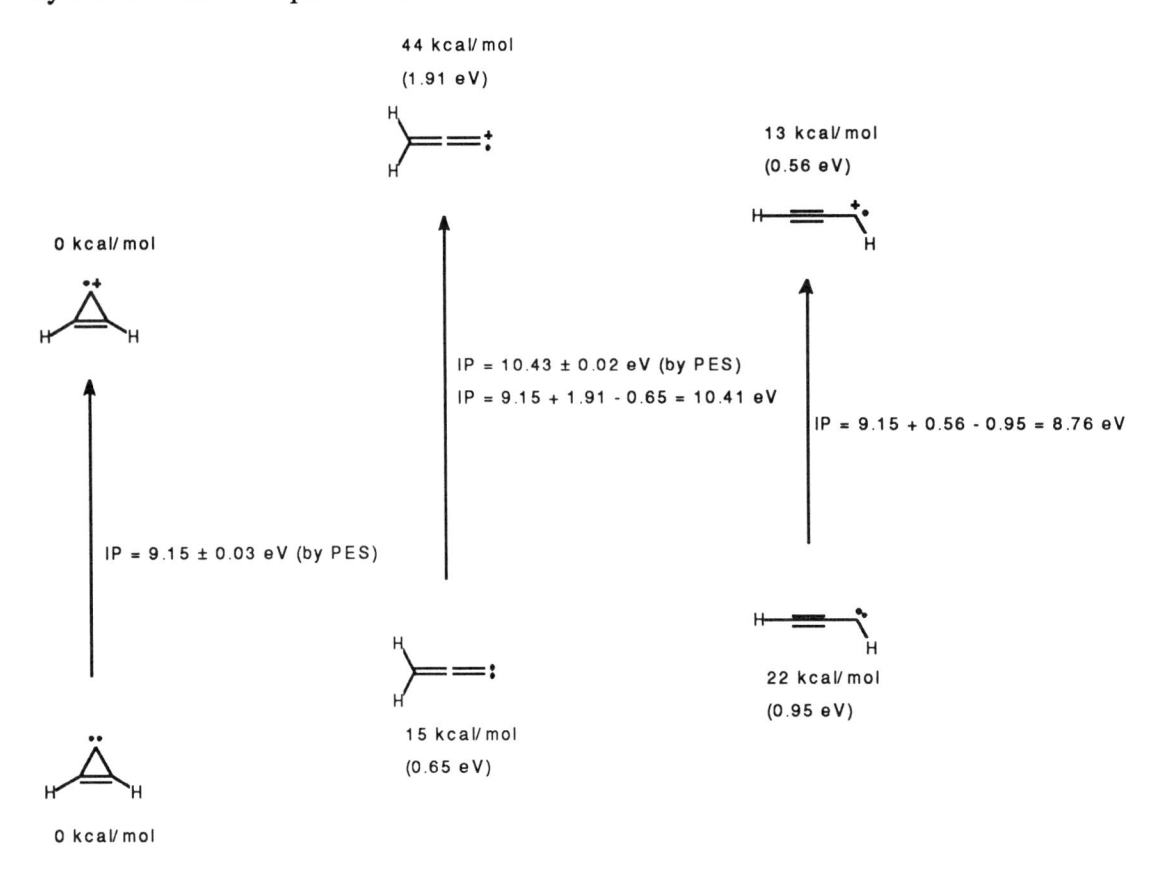

The surprisingly good agreement for propadienylidene and propargylene suggests that an analogous thermochemical cycle for Δ_{ST} can be written for any carbene and its corresponding monoradical. This reduces the measurement of Δ_{ST}[carbene], for which there is no generally applicable method, to a pair of IP or BDE measurements, for which there are a variety of methods.

The two reductions of ionization potential data available by photoionization of pyrolyically produced radicals and carbenes validate the two proposed thermochemical cycles: one for ΔH_f[carbene], and one for Δ_{ST}[carbene], at least within the C_3H_2 family of structures. Further work to broaden the range of benchmark test molecules is underway.

5. ACKNOWLEGEMENTS

We acknowledge support from the National Science Foundation for the purchase of the laser equipment used in this work. Funding from the Department of Energy (HC), the donors of the Petroleum Research Fund (DWM), the Exxon Educational Foundation, and the David and Lucile Packard Foundation is gratefully acknowledged.

6. REFERENCES

1. NSF Presidential Young Investigator, David and Lucile Packard Fellow, Camille and Henry Dreyfus Teacher-Scholar, Alfred P. Sloan Research Fellow.

2. Benson, S.W.; Cruickshank, F.R.; Golden, D.M.; Haugen, G.R.; O'Neal, H.E.; Rodgers, A.S.; Shaw, R.; Walsh, R. *Chem. Rev.* **1969**, *69*, 279.

3. For a successful example of this approach, see: Disch, R.L.; Schulman, J.M.; Sabio, M.L.; *J. Am. Chem. Soc.* **1985**, *107*, 1904.

4. O'Neal, H.E.; Benson, S.W. in *Free Radicals*, vol. II, Kochi, J.K., ed.; John Wiley & Sons: New York, 1973; pp. 338-340.

5. Clauberg, H.; Chen, P. *J. Am. Chem. Soc.* **1991**, *113*, 1445.

6. Clauberg, H.; Minsek, D.W.; Chen, P. *J. Am. Chem. Soc.* in press.

7. Carter, E.A.; Goddard, W.A. *J. Phys. Chem.* **1986**, *90*, 998; See also: Simons, J.P. *Nature* **1965**, *205*, 1308.

8. Ervin, K.M.; Gronert, S.; Barlow, S.E.; Gilles, M.K.; Harrison, A.G.; Bierbaum, V.M.; DePuy, C.H.; Lineberger, W.C.; Ellison, G.B. *J. Am. Chem. Soc.* **1990**, *112*, 5750.

9. Wu, C.J.; Carter, E.A. *J. Am. Chem. Soc.* **1990**, *112*, 5893; Wu, C.J.; Carter, E.A. *J. Phys. Chem.* **1991**, *95*, 8352.

10. A good discussion of the valence state can be found in: Murrell, J.N.; Kettle, S.F.A.; Tedder, J.M. *The Chemical Bond*, 2nd ed.; John Wiley & Sons: New York, 1985; pp. 247-249.

11. Armentrout, P.B.; Georgiadis, R. *Polyhedron* **1988**, *7*, 1573; Ohanessian, G.; Goddard, W.A. *Acc. Chem. Res.* **1990**, *23*, 386.

12. Heitler, W. *Marx's Handb. d. Radiol. VI* **1934**, *2*, 485; Herzberg, G. *Molecular Spectra and Molecular Structure I. Spectra of Diatomic Molecules*, 2nd ed., Van Nostrand Reinhold Co.: New York, 1950; pp. 357-359.

13. DeFrees, D.J.; McLean, A.D. *Astrophys. J.* **1986**, *308*, L31.

14. Minsek, D.W.; Chen, P. *J. Phys. Chem.* **1990**, *94*, 8399.

15. Wong, W.; Radom, L. *Org. Mass Spec.* **1989**, *24*, 539.

High-brightness cm^{-1}-resolution threshold photoelectron spectroscopic technique

Katsumi Kimura and Masahiko Takahashi

Institute for Molecular Science
Okazaki 444, Japan

ABSTRACT

We first describe some characteristics of our compact photoelectron analyzers capable of high-resolution (1-2 cm^{-1}) cation spectroscopy of molecules in two-color (ω_1, ω_2) laser experiments of REMPI (resonantly enhanced multiphoton ionization). Threshold ionization photoelectrons lower than a few cm^{-1} are collected as a function of ω_2 by applying a pulsed electric field at 500 ns at each laser shot, while energetic photoelectrons are removed by angular- and time-resolved discrimination. Secondly we demonstrate a rotational spectrum due to NO$^+$ obtained by this technique to examine our energy resolution. Thirdly we mention a few examples of low-frequency vibrational spectra of cation radicals; (1) rotational isomers, (2) Ar-NO van der Waals complex, and (3) aniline-Ar$_n$ complexes (n=1,2).

2. INTRODUCTION

During the last decade, 'resonantly enhanced multiphoton ionization' (REMPI) with a nanosecond UV/VIS laser system has been combined with a photoelectron spectroscopic technique, providing excited-state photoelectron spectroscopy or REMPI photoelectron spectroscopy.[1-3] Figure 1 shows a schematic drawing of two-color REMPI process producing threshold photoelectrons (e_0^-) as well as kinetic photoelectrons (e_k^-). This has been developed as a powerful tool to study highly excited states as well as ionic states of molecules.[1-3] Laser photoelectron spectra thus obtained by REMPI have been interpreted in terms of ionization transitions between the resonant excited states and the final ionic states.

In connection with two-color (ω_1, ω_2) REMPI photoelectron spectroscopy, Müller-Dethlefth et al.[4,5] have developed a ZEKE (zero kinetic energy) photoelectron technique to perform very high resolution photoelectron spectroscopy, measuring zero energy electrons as a function of the wavelength of ω_2.

In this laboratory, earlier Achiba and Kimura[6] have developed an eccentric-type threshold photoelectron analyzer useful for two-color (1+1') REMPI experiments, and obtained (1+1') threshold photoelectron spectra of benzene and aniline in a fairly high resolution (20 cm^{-1}). Recently Takahashi et al.[7-9] have designed and developed two kinds of compact analyzers capable of high-resolution (1-2 cm^{-1}) threshold photoelectron spectroscopy. One is a capillary type and the other is a deflector type, which are described in the present paper. With these analyzers, we can efficiently collect threshold photoelectrons as a function of the wavelength of the ionization laser (ω_2) in two-color (n+1') REMPI experiments.

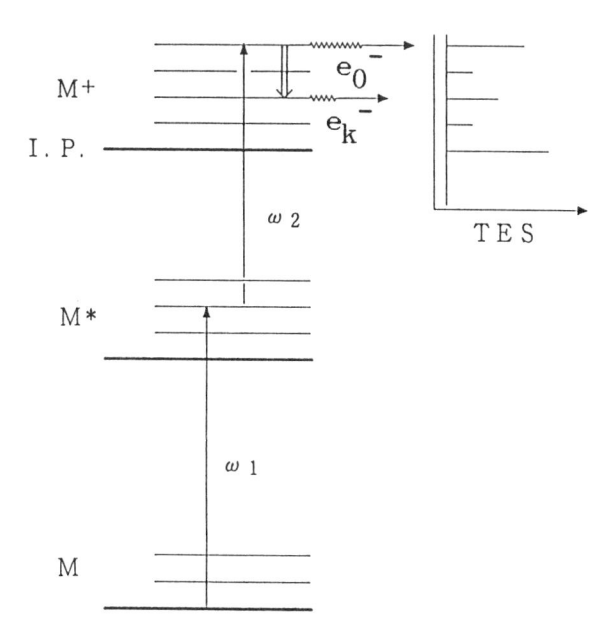

Fig. 1. Two-color (ω_1, ω_2) REMPI process producing threshold photo-electrons (e_0^-) and kinetic photoelectrons (e_k^-). The corresponding threshold photoelectron spectrum (TES) is schematically shown.

3. TWO-COLOR (n+1') IONIZATION

In laser photoelectron spectroscopy, two-color (n+1') REMPI experiments with two lasers (ω_1 and ω_2) are especially important, since the second ionization source can be employed independently of the first excitation source. In other words, it is possible to scan the wavelength of ω_1 while keeping ω_2 constant.

From a photoelectron spectroscopic point of view, it is desirable to use a single photon ionization process from the excited state. Therefore, two-color (n+1') process is most important for REMPI photoelectron spectroscopy. The advantage of the two-color ionization is that only a special molecule among a mixture of analogous molecular species can be selectively ionized to provide its cation spectroscopy.

4. COMPACT cm^{-1}-RESOLUTION THRESHOLD PHOTOELECTRON ANALYZER

Figure 2 shows two types of our threshold photoelectron analyzers, which have been developed in this laboratory; one is 'capillary type' and the other is 'deflection type'. Photoelectrons with various kinetic energies are initially produced at the ionization point (Q) at each laser shot, and then quickly disperse from the ionization region. In order to collect only threshold photoelectrons, we can use two techniques of 'time-resolved discrimination' and 'angular discrimination'. The idea of our technique is the following.

Let us consider a sphere of 10 mm in diameter surrounding the ionization point (Q). Electrons with energies lower than a few cm^{-1} should remain in the sphere for as long as 500 ns after each laser shot. In other words, only threshold photo-

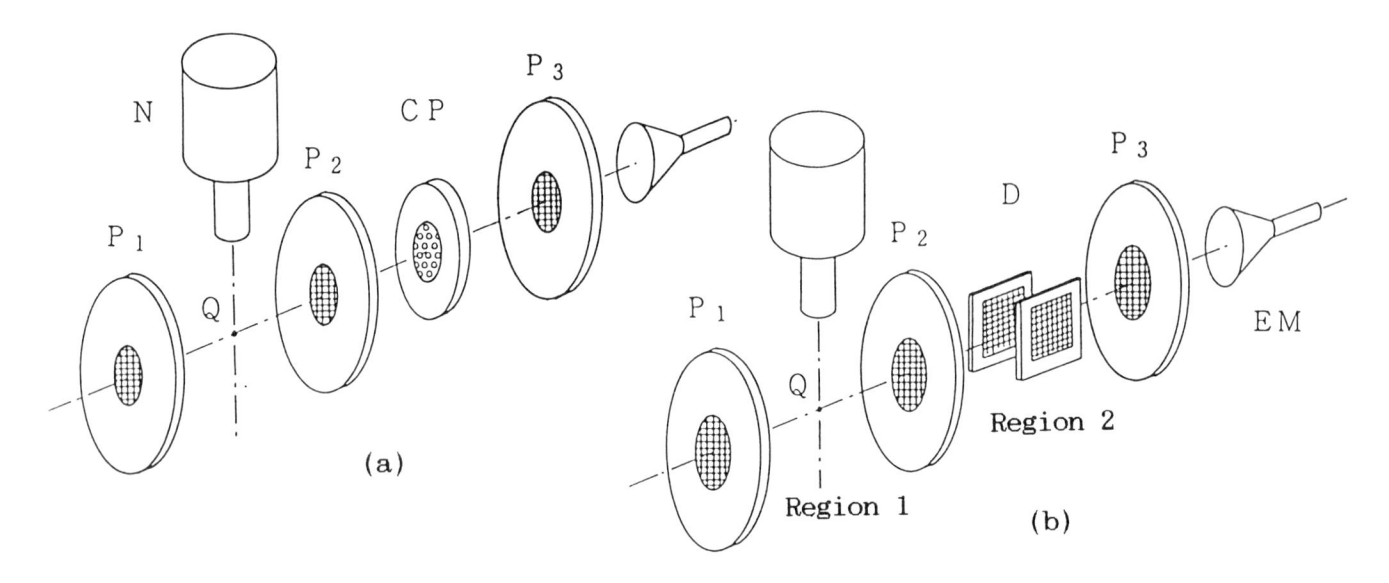

Fig. 2. Threshold photoelectron analyzers: (a) capillary type and (b) deflection type.

electrons can be collected typically at 500 ns after each laser shot. Such very low energy electrons can be therefore collected with a small-size compact analyzer by applying a pulsed electric field (a few V/cm) across P_1 and P_2. In addition to time-resolved discrimination, it is also possible to carry out angular discrimination with a capillary plate in threshold photoelectron measurements.

The capillary type (Fig. 2a) consists of a capillary plate (CP: Hamamatsu Photonics), a pair of electrodes (P_1, P_2) to extract threshold photoelectrons, an electron multiplier (EM: Murata Ceratron EMS-6081B), and a ground electrode (P_3) to cut off the electric potential of the electron multiplier (EM). A capillary plate of $\ell/d=80$ can be used, where ℓ and d are the length and diameter of the capillary holes, respectively. The capillary plate is located at 10 mm apart from the ionization point (Q). Most energetic photoelectrons are removed by the capillary plate. However, some of them penetrate through the capillary plate toward the electron detector, producing a background noise. In order to remove such energetic electrons, we apply a delayed pulsed electric field. The resolution is a few cm^{-1}.

The deflection-type analyzer (Fig. 2b) has a pair of deflection plates (D). Most energetic electrons can be removed by the deflector before reaching the detector. Figure 3 shows schematically how to apply the pulsed electric pulses in regions 1 and 2 with respect to the ns laser pulses (ω_1, ω_2).

Our deflector analyzer is very bright, very high resolution (1-2 cm^{-1}), and very compact for the following reasons. 1) All the threshold photoelectrons ejected in the whole space (4π) are collected. 2) The detector is located only at 5 cm from the ionization point. 3) The analyzer is quite simple in structure and small in size. The two-color threshold photoelectron spectra thus obtained with this analyzer is similar in intensity to ordinary one-color MPI ion-current spectra.

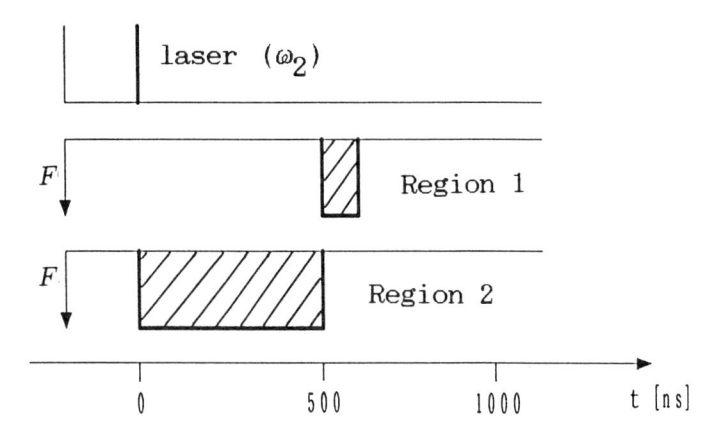

Fig. 3. Pulse profiles of the laser (ω_2) and the electric fields in Regions 1 and 2.

5. EXPERIMENTAL APPARATUS

The laser photoionization apparatus used has been described elsewhere,[7] consisting of a photoionization vacuum chamber, a pulsed-nozzle sample inlet system, and a laser system producing two kinds of tunable dye lasers. In addition to the above-mentioned threshold photoelectron analyzer, we have an ion detector and a TOF mass analyzer. The vacuum in the main chamber during photoionization experiments is kept at 2×10^{-5} Torr by an oil diffusion pump with a liquid nitrogen trap.

A home-made 'conical' nozzle is used on the top of a commercially available fuel injector to make the divergence of the free jet smaller. Argon is used as a carrier gas (2 atm) through the pulsed nozzle. An Ar-seeded molecular jet is irradiated by the two Nd-YAG pumped dye lasers (ω_1 and ω_2). Photoelectrons are initially produced (at Q in Fig. 2) at zero electric field.

Ion-current measurements to observe MPI excitation spectra are carried out as a function of the excitation laser wavelength prior to threshold photoelectron measurements. The total ion current can be converted to the voltage mode through a fast current amplifier (Keithley Model 425) and averaged with a boxcar integrator (NF Model 530-A). The resulting output signals are recorded on a chart recorder.

6. PHOTOELECTRON ROTATIONAL SPECTRUM AND BANDWIDTH

Figure 4 shows a threshold photoelectron spectrum due to NO$^+$, obtained by two-color (1+1') REMPI via the $A^2\Sigma^+$ state ($v'=0$, $N'=7$) with the deflection-type analyzer, indicating well-resolved rotational peaks ascribed to $\Delta N = N^+ - N' = 0, \pm 1, \pm 2,$ and ± 3 transitions.[7] Although the energy resolution (fwhm) depends on the delay time of the electric field as well as on the field strength, we have evaluated the energy resolution to be 2 cm^{-1} (fwhm) from the threshold photoelectron spectrum shown in Fig. 4.[7] On the basis of the adiabatic ionization potential of NO[10] and the rotational constant of the ground-state cation NO$^+$,[11] the

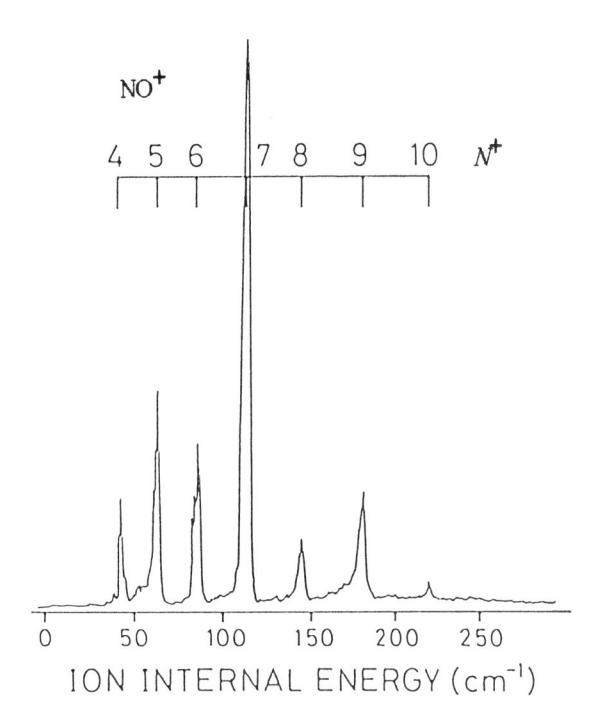

Fig. 4. The (1+1') threshold photoelectron spectrum due to NO^+, showing rotational peaks.

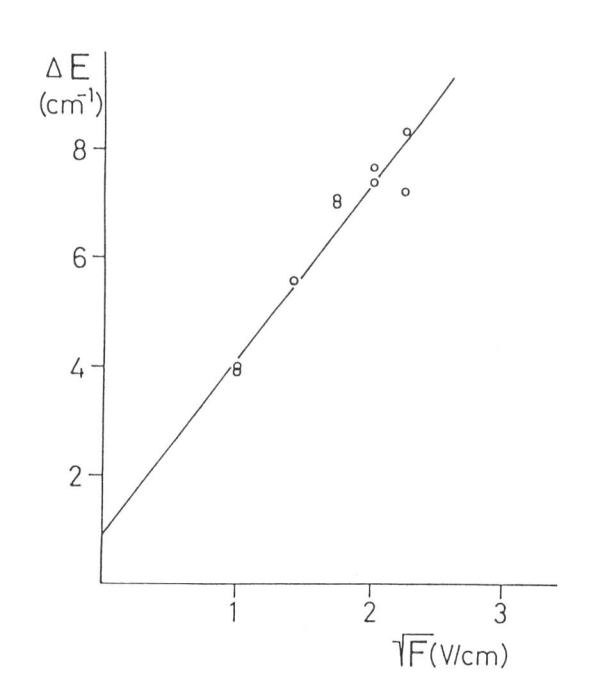

Fig. 5. The bandwidths (ΔE) are plotted against the square root of F.

seven photoelectron bands appearing in Fig. 4 have been assigned to the rotational levels (N^+=4~10) of NO^+.

Using our capillary type analyzer, we have previously obtained an energy resolution of 4 meV using a 'continuous' electric field (without using a pulsed-field technique).[8] However, since then, using a time delay technique, we have been able to drastically improve our resolution.

The bandwidth (fwhm) ΔE depends on the field strength F. This F dependence may be explained as follows. In addition to the threshold photoelectrons, field-ionization electrons of NO are also ejected from its highly excited Rydberg states in the two-color REMPI experiments. Consequently, the threshold photoelectrons should mix with the field-ionization electrons. The field ionization gives rise to a red shift in the ionization potential. According to a simple Coulomb model,[10] the energy shift is given by $\Delta E = (1/2) \times (eF/\pi\varepsilon_0)^{1/2}$. The resulting energy shift ΔE should be taken into account in very high-resolution photoelectron spectroscopy, but not significant in cation vibrational spectroscopy.

The ΔN=0 bandwidth gradually increases with F, as plotted in Fig. 5, which shows a linear relationship; ΔE (fwhm) is proportional to the square root of F. This linear relationship indicates that the field ionization of highly excited Rydberg states takes place to eject electrons. Consequently the field-ionization electrons more or less mix with the threshold photoelectrons at a delay time of 500 ns.

The intercept (0.9 cm^{-1}) of the linear line shown in Fig. 5 corresponds to the energy resolution expected in our two-color REMPI experiments under the field-free conditions. Since the effective diameter of the capillary plate (CP) in Fig. 2(a) is 10 mm, the threshold photoelectrons remaining in the effective volume after 500 ns should have kinetic energies lower than 2.3 cm^{-1}. The intercept of 0.9 cm^{-1} might correspond to the width of the ionizing laser (ω_2).

It should be mentioned that the energy resolution of our threshold photoelectron analyzer at a delay time of 500 ns may be divided into two terms governed by the laser wavelength and the field strength: Namely, $\Delta E_{1/2} = \Delta E_\ell + \Delta E_f$, where the first term ΔE_ℓ is due to the laser wavelength resolution, and the second term ΔE_f is proportional to the square root of F.[7] In order to achieve a higher resolution, it is desirable to lower F.

7. SOME EXAMPLES OF (n+1') THRESHOLD PHOTOELECTRON SPECTRA

7.1. Cations of n-propylbenzene rotational isomers

Jet-cooled n-propylbenzene exists in *trans* and *gauche* forms. Energy level diagram is shown in Fig. 6. The $S_1 \leftarrow S_0$ transition energy of the *trans* has been reported to be 49 cm^{-1} larger than the *gauche* isomer.[12] Therefore, by tuning ω_1 in two-color REMPI experiments, it is possible to completely separate the two isomers from each other in threshold photoelectron spectra. Two-color threshold photoelectron spectra thus obtained via $S_1 0^0$ are shown in Fig. 7, showing vibrational structures of the *trans* and *gauche* cations below 600 cm^{-1}.[8]

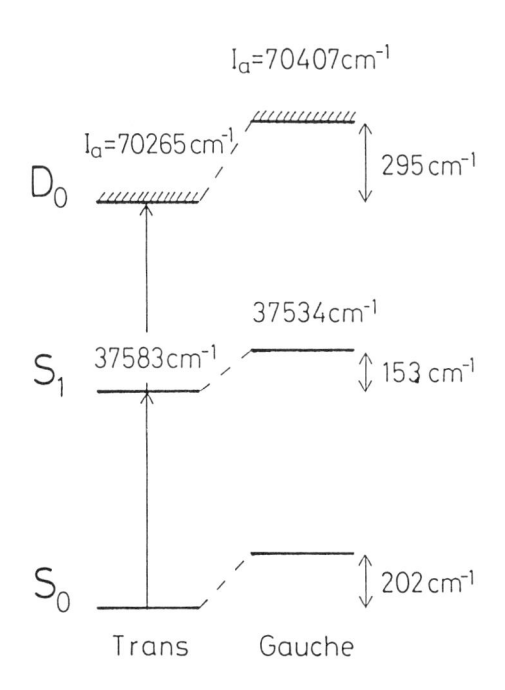

Fig. 6. Energy level diagram of the isomers of n-propylbenzene. The cation is indicated by D_0.

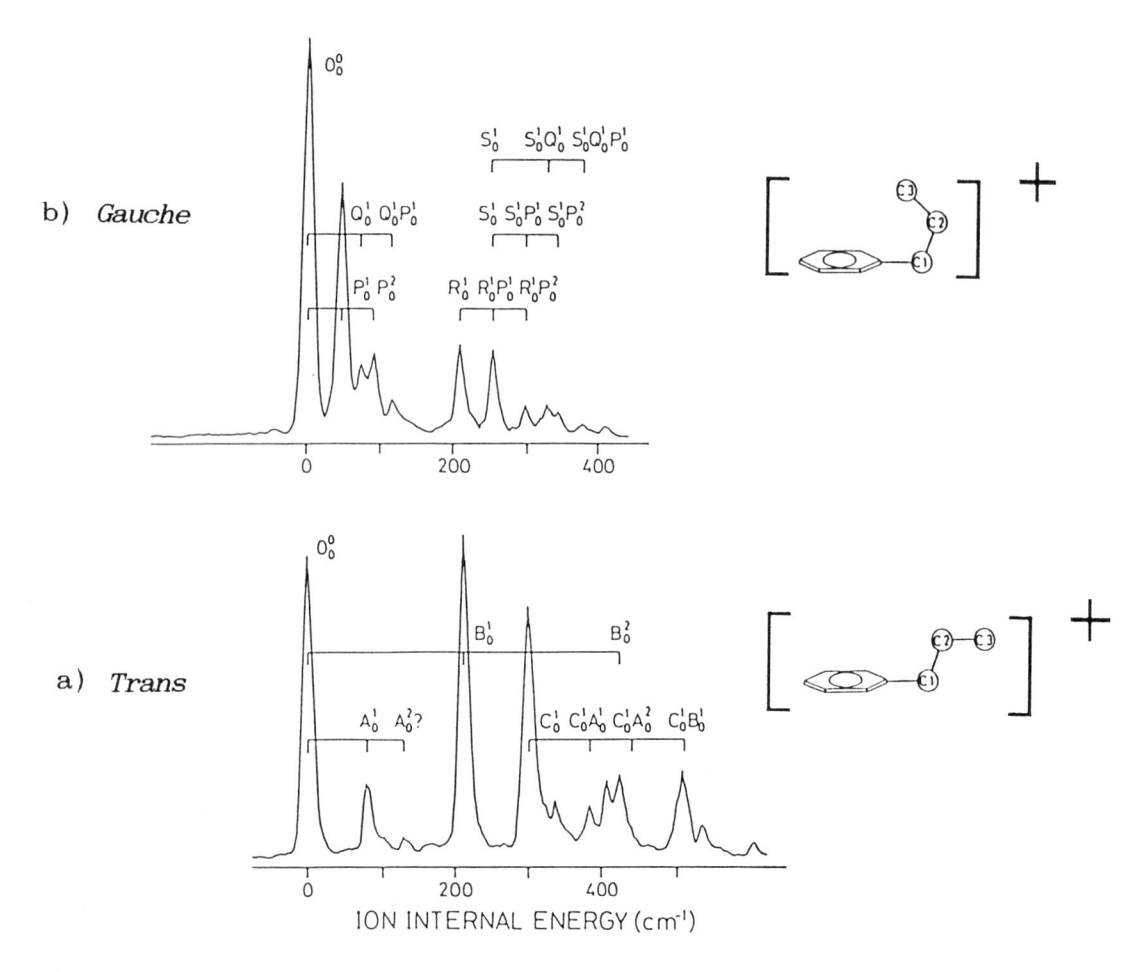

Fig. 7. The (1+1') threshold photoelectron spectra due to the [n-propylbenzene]$^+$ cations: (a) trans form and (b) gauche form.

The first threshold photoelectron bands shown in Fig. 7 correspond to the adiabatic ionization potentials 70265 ± 8 cm^{-1} (trans) and 70407 ± 8 cm^{-1} (gauche). The trans spectrum in Fig. 7 may be explained in terms of combination bands of three low-frequency modes labeled as A, B, and C whose frequencies are 82 cm^{-1}, 212 cm^{-1}, and 300 cm^{-1}, respectively. On the other hand, the gauche spectrum in Fig. 7 may be explained in terms of four low-frequency modes labeled as P, Q, R, and S of 46 cm^{-1}, 73 cm^{-1}, 207 cm^{-1}, and 252 cm^{-1}, respectively. These low-frequency modes are possibly associated mainly with the n-propyl group, suggesting that their frequencies should be sensitive to the interaction of the substituent with the benzene ring.

In order to understand these low-frequency vibrational modes, we have carried out ab initio calculations for the trans cation of n-propylbenzene at the IMS Computer Center, using the STO-2G basis sets and the GAUSSIAN 82 program. After the geometry optimization, normal mode analyses have been carried out.[8] The calculated vibrational frequencies (multiplied by a factor of 0.9) are in good agreement with the experimental frequencies.

It should be mentioned that all the Franck-Condon active vibrational modes in Fig. 7(a) should have A' symmetry. The agreement between the experimental and theoretical frequencies indicates that the spectrum (a) in Fig. 7 can be attributed to the *trans* form with C_s symmetry. Thus, the assignment earlier suggested[12] has been confirmed from our work.

7.2. Ar-NO van der Waals cation

A bound Rydberg state ($C^2\Pi$) of the Ar-NO complex has been found by Sato *et al.*[13] by using a REMPI technique. An energy diagram of Ar-NO relevant to its (2+1') REMPI process via the $C^2\Pi$ state is shown in Fig. 8. A one-color (2+1') REMPI excitation spectrum obtained with a mixture of Ar and NO in a supersonic jet is shown in Fig. 9. The progression (with spacings of 54.4, 51.0, and 46.4 cm^{-1}) appearing on the lower energy side of the free NO band is attributed to the intermolecular stretching mode of Ar-NO($C^2\Pi$).[14] The (0-0) band is red shifted by 326 cm^{-1} from the NO transition.

Figure 10 shows a threshold photoelectron spectrum due to [Ar-NO]$^+$,[14] obtained by two-color (2+1') REMPI via the intermolecular stretching vibrational level v'=0 of the $C^2\Pi$ state. The first band in Fig. 10 gives the adiabatic ionization potential of Ar-NO: $I_a = 2h\nu_1 + h\nu_2 = 73869 \pm 6$ cm^{-1} (9.159 $\pm$ 0.001 eV).

The observed threshold photoelectron vibrational progressions (Fig. 10) show fundamental frequencies of 79 and 94 cm^{-1} with strong anharmonicities. According to a study of molecular beam electric resonance microwave and radio-frequency spectroscopy,[15] the neutral ground-state Ar-NO has a 'T-shape' structure with an internuclear distance of 3.711 Å and an angle of 95.175 degrees to the NO axis.

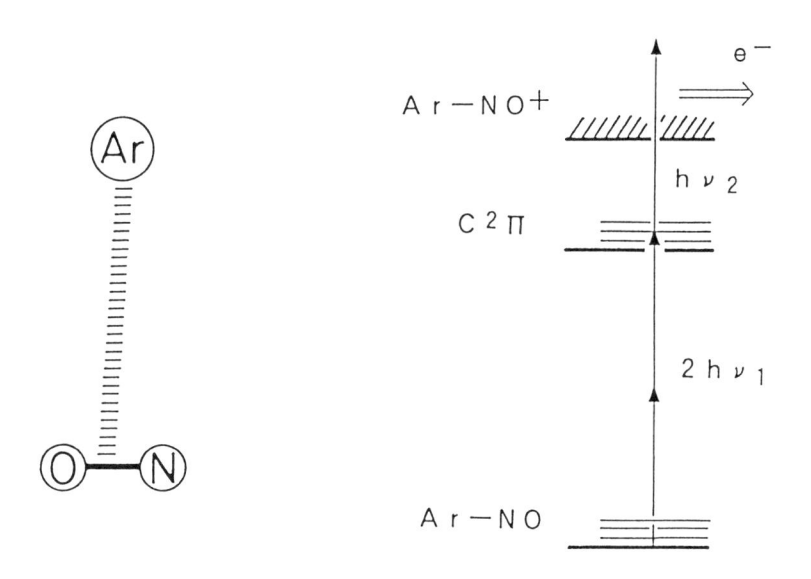

Fig. 8. Molecular geometry of the Ar-NO van der Waals complex and its energy diagram relevant to two-color (2+1') REMPI via the Rydberg $C^2\Pi$ state.

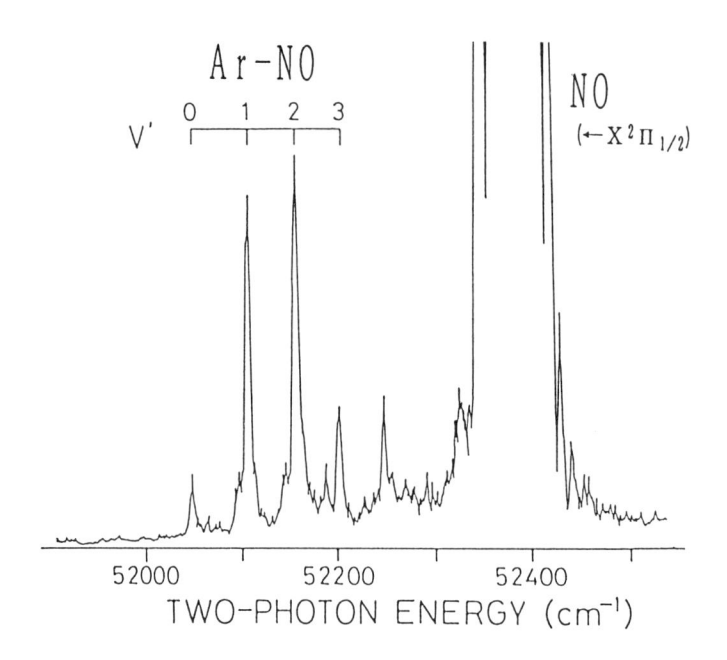

Fig. 9. One-color (2+1) REMPI excitation spectrum of Ar-NO in the $C^2\Pi$ Rydberg region.

The T-shaped triatomic [Ar-NO]$^+$ cation should have three vibrational modes, as schematically shown in Fig. 10. One of the three is almost the same as the stretching mode of the free NO$^+$ cation. Thus the two observed low-frequency vibrational structures appearing in Fig. 10 should be attributed to the intermolecular bending and stretching modes of (Ar-NO)$^+$.

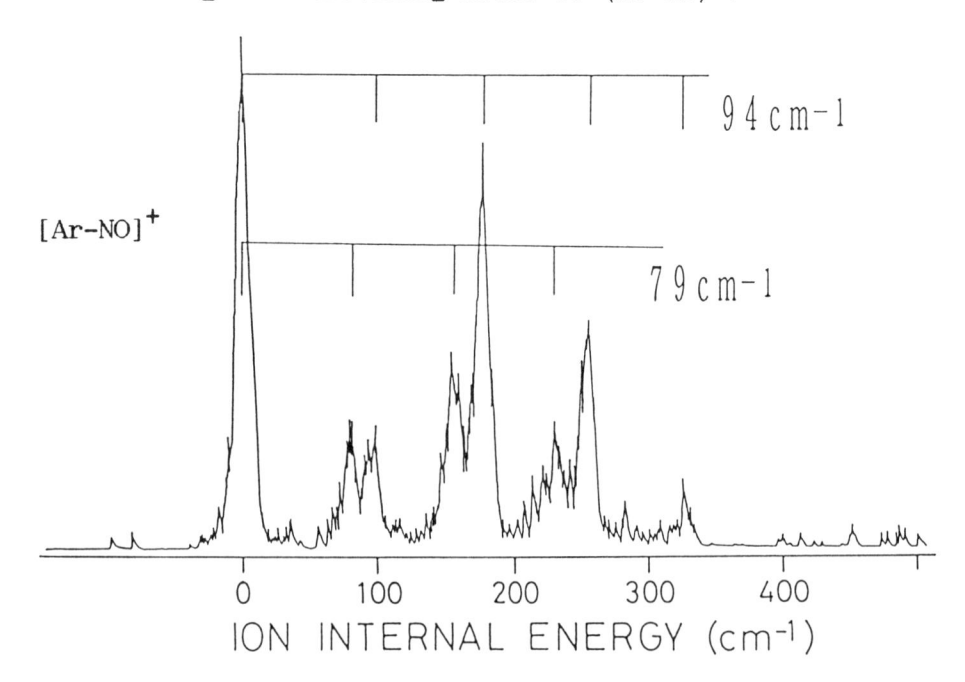

Fig. 10. The (2+1') threshold photoelectron spectrum due to the [Ar-NO]$^+$ van der Waals cation.

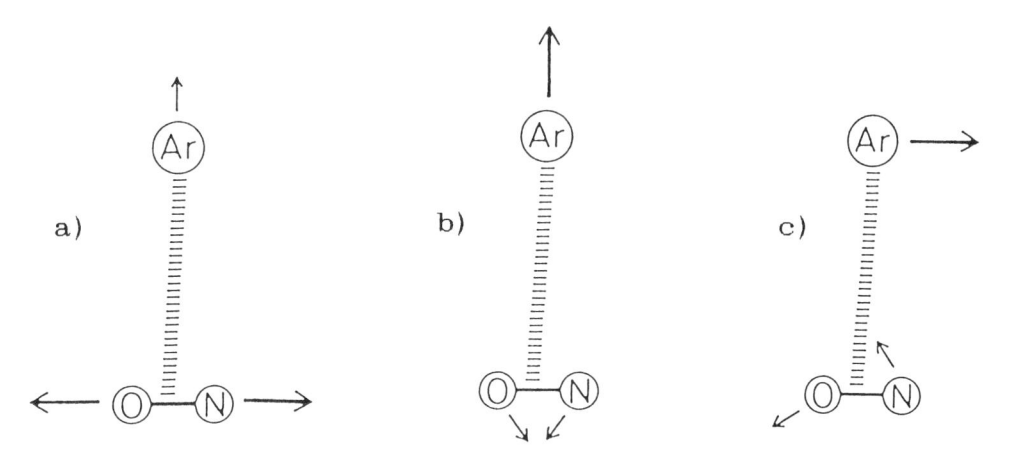

Fig. 11. Three vibrational modes of [Ar-NO]$^+$: a) the N-O stretching (2376 cm^{-1}), b) the van der Waals stretching (94 cm^{-1}), and c) the van der Waals bending (79 cm^{-1}).

7.3. Aniline-Ar$_n$ van der Waals cations (n=1,2)

The van der Waals vibrations of the aniline-Ar$_n$ complexes (n=1,2) in the S$_1$ states have been studied by Bieske et al.[16] by using a mass-selected REMPI technique. The geometrical structure of the neutral aniline-Ar complex (n=1) has been studied by a LIF rotational analysis by Yamanouchi et al.[17]

Figure 12 show threshold photoelectron spectra of aniline and its Ar complexes (n=1,2), obtained by (1+1') REMPI via S$_1$. Each cation spectrum shows several vibrational bands below 1700 cm^{-1}.[18] The threshold photoelectron spectrum of aniline in Fig. 12 is essentially the same as a TOF spectrum earlier reported by Meek et al.,[19] although the spectral resolution in the present work is much higher. The vibrational bands in Fig. 12 are correlated by broken lines. Each vibrational band of [aniline]$^+$ is split into several sub-bands in the aniline-Ar$_n$ cations (n=1,2). The vibrational frequencies of [aniline]$^+$ are little perturbed by the Ar atoms. The band splittings observed for the van der Waals cations are due to combination of the [aniline]$^+$ vibrational mode with some low-frequency van der Waals vibrational modes.

From the photoelectron origin bands (0^{+0}) in Fig. 12(a-c), the adiabatic ionization potentials ($I_a = h\nu_1 + h\nu_2$) have been determined as I_a(aniline) = 62268 ± 4 cm^{-1}, I_a(aniline-Ar) = 62157 ± 4 cm^{-1}, and I_a(aniline-Ar$_2$) = 62049 ± 4 cm^{-1}.

The I_a shift of the 1:2 complex (219 cm^{-1}) is almost twice as much as that of the 1:1 complex (111 cm^{-1}), this fact suggesting that the [aniline]$^+$ component is planar. This is also supported by the fact that the photoelectron intensity due to the [aniline-Ar]$^+$ inversion mode is very weak. The molecular geometries of the amino groups in [aniline-Ar$_n$]$^+$ are almost the same as those in the neutral S$_1$ states. The geometry of aniline in the S$_1$ state has been indicated to be planar from an analysis of the inversion potential function by Hollas et al.[20] As a result, we may conclude that the [aniline]$^+$ component has a planar geometry.

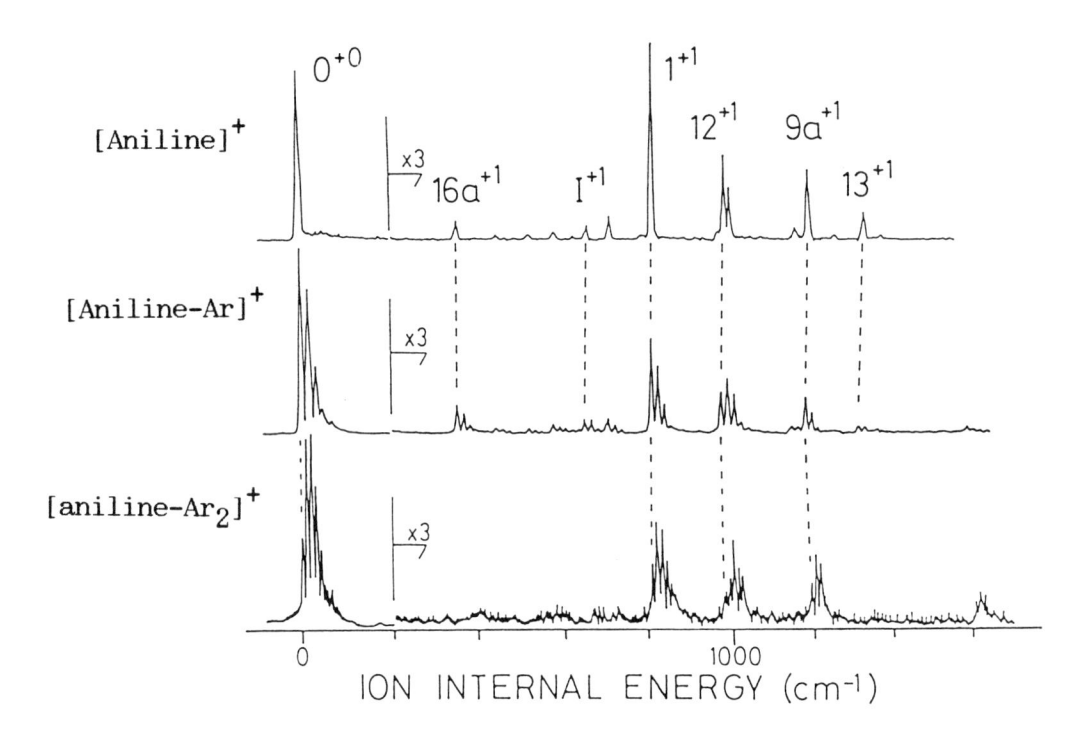

Fig. 12. The (1+1') threshold photoelectron spectra due to [aniline]$^+$ and [aniline-Ar$_n$]$^+$ (n=1,2).

From the threshold photoelectron spectra of the aniline-Ar$_n$ complexes (n=1,2), the low-frequency vibrational frequencies have been determined to be 16 cm^{-1} (n=1) and 11 cm^{-1} (n=2). From Franck-Condon calculations, these low-frequency vibrational modes have been assigned to the van der Waals 'symmetric bending' and 'in-phase bending' mode in the [aniline-Ar]$^+$ and [aniline-Ar$_2$]$^+$ cations, respectively.

8. CONCLUDING REMARKS

Our compact analyzer useful for two-color REMPI threshold photoelectron measurements with time-resolved discrimination is excellent for carrying out threshold photoelectron spectroscopy with a few cm^{-1}. Such a very high resolution photoelectron technique has enormous potential for 'cation spectroscopy' and it is applicable for any molecules and molecular complexes in supersonic jets. It should also be mentioned that individual molecular species or molecular complexes can be selectively ionized in REMPI experiments.

This compact threshold photoelectron analyzer makes it possible even to measure rotational structure of simple molecular cations such as NO$^+$. If we have an appropriate VUV laser system, it is also possible to study higher excited electronic states of molecular cations from threshold photoelectron measurements. This is an advantage of photoelectron spectroscopy.

9. REFERENCES

1. (a) K. Kimura, *Adv. Chem. Phys.*, 60, 161 (1985) 161; (b) K. Kimura, *Intern. Rev. Phys. Chem.*, 6, 195 (1987).
2. S. T. Pratt, P. M. Dehmer and J. L. Dehmer, in S. H. Lin (Ed.), *Advances in Multi-Photon Processes and Spectroscopy*, World Scientific, Singapore, 1988, Vol. 4, p 69.
3. R. N. Compton and J. C. Miller, in D. K. Evans (Ed.), *Laser Applications in Physical Chemistry*, Marcel Dekker, New York, 1989, p. 221.
4. K. Müller-Dethlefs, M. Sander, and E. W. Schlag, *Chem. Phys. Lett.*, 112, 291 (1984)
5. K. Müller-Dethlefs and E. W. Schlag, *Ann. Rev. Phys. Chem.*, 42, 109 (1991).
6. Y. Achiba and K. Kimura, "Observation of IVR by Two-Color Threshold Photoelectron Measurements", Abstract of the Symposium of Chemical Reaction held at Okazaki, the Chemical Society of Japan, 1985.
7. M. Takahashi, H. Ozeki, and K. Kimura, *Chem. Phys. Lett.*, 181, 255 (1991).
8. M. Takahashi, K. Okuyama, and K. Kimura, *J. Mol. Struct.*, 249, 47 (1991); M. Takahashi and K. Kimura, J. Chem. Phys. to be published.
9. M. Takahashi and K. Kimura, "Molecular Ion Vibrational Spectroscopy by a Time-Resolved REMPI Threshold Photoelectron Technique", Presented at the 5th International Conference on Time-Resolved Vibrational Spectroscopy held at Waseda University, Tokyo, June 1991.
10. (a) G. Reiser, W. Habenicht, K. Müller-Dethlefs and E. W. Schlag, *Chem. Phys. Lett.*, 152, 119 (1988); (b) L. A. Chewter, M. Sander, K. Müller-Dethlefs and E. W. Schlag, *J. Chem. Phys.*, 86, 4737 (1987).
11. K. P. Huber and G. Herzberg, *Constants of Diatomic Molecules*, Van Nostrand Reinhold, New York, 1977.
12. J. B. Hopkins, D. E. Powers and R. E. Smalley, *J. Chem. Phys.*, 72, 5039 (1980).
13. K. Sato, Y. Achiba, and K. Kimura, *J. Chem. Phys.*, 81, 57 (1984).
14. M. Takahashi, J. Chem. Phys. in press.
15. P. D. A. Milles, C. M. Western, and B. J. Howard, *J. Phys. Chem.*, 90, 4961 (1986).
16. (a) E. J. Bieske, M. W. Rainbird, 1. M. Atkinson, and A. E. W. Knight, *J. Chem. Phys.*, 91, 752 (1989); (b) E. J. Bieske, M. W. Rainbird, and A. E. W. Knight, *J. Chem. Phys.*, 94, 7019 (1991).
17. K. Yamanouchi, S. Isogai, S. Tsuchiya, and K. Kuchitsu, *Chem. Phys. Lett.*, 116, 123 (1987).
18. M. Takahashi, H. Ozeki and K. Kimura, *J. Chem. Phys.*, in press.
19. J. T. Meek, E. Sekreta, W. Wilson, K. S. Viswanathan, and J. P. Reilly, *J. Chem. Phys.*, 82, 1741 (1985).
20. J. M. Hollas, M. R. Howson, and T. Ridley, *Chem. Phys. Lett.*, 98, 611 (1983).

Studies of doubly-charged molecular ions by means of ion photofragment spectroscopy

Mats Larsson and Göran Sundström

Department of Physics I
Royal Institute of Technology
S-100 44 Stockholm, Sweden

Lars Broström and Sven Mannervik

Manne Siegbahn Institute of Physics
S-104 05 Stockholm, Sweden

ABSTRACT

Some recent work on doubly-charged molecular ions are described. The method of ion photofragment spectroscopy has been applied to the N_2^{2+} and NO^{2+} dications. The (8,2) band of the $A\ ^1\Pi_u - X\ ^1\Sigma_g^+$ transition was recorded at high spectral resolution. Line positions and linewidths were measured. It was found that levels of f-symmetry of the A state were more strongly predissociated than levels of e-symmetry. Attempts to observe structured photodissociation of NO^{2+}, despite aid from large scale *ab initio* calculations, failed.

1. INTRODUCTION

The standard critical compilation of data[1] concerning the states of diatomic species contains very little information on doubly charged molecular ions (dications) AB^{2+}. The last ten years have witnessed an increase of activity from both experimentalists and theorists;[2,3] however, high resolution data are still rather sparse. A typical feature of dications is that the thermodynamic limit for dissociation lies *below* the double ionization energy (if the energy of the $A^+ + B^+$ limit is lower than that of $A^{2+} + B$), due to Coulomb repulsion forces, thus making the ion unstable with respect to dissociation to singly charged products. In many cases the potential barrier that separates the quasi-bound vibrational levels from the dissociation continuum is so broad that dissociation is negligible, and the dication behaves as if it was a stable, albeit highly reactive, molecule. Since the quasi-bound vibrational levels can be separated from the dissociation limit by as much as 6-8 eV, dications possess substantial internal energy. This type of high energy metastable states could be of potential interest for generation of propulsive energy via induced unimolecular fragmentation. In order to exploit these properties of dications, structural and dynamical information is needed. The most powerful technique to provide this information is ion photofragment spectroscopy (IPS).[4]

The main problem with experimental studies of dications is the difficulty to obtain sufficient ion densities. Several techniques are suitable for extracting properties such as ionization potentials and excited electronic state energies.[2,3] In order to obtain more detailed information high spectral resolution is needed. Optical emission spectra of a few diatomic dications have been recorded at rotational resolution. Carroll[5] recorded and rotationally analyzed a singlet transition in N_2^{2+} and the same transition has more recently been studied by Cossart *et al.*[6-8] NO^{2+} completes the list of dications for which an emission spectrum is known.[9,10] The most favorable circumstance to observe an emission band occurs when the upper and lower state in the transition have nearly the same equilibrium internuclear distance. Even in those few favorable cases, needless to say, contamination by emission from neutral and singly charged molecules, dissociation, or secondary, products is a severe problem.

In 1983 Cosby, Möller and Helm[11] applied ion photofragment spectroscopy to N_2^{2+}. They merged an accelerated, mass selected beam of N_2^{2+} with a laser beam and recorded the appearance of N^+ photoproducts as a function of laser wavelength; it was found that N_2^{2+} has a very rich photodissociation spectrum between 14 900 and 19 500 cm^{-1}. Only a small fraction of the spectrum was recorded at high resolution. Several of the observed bands were assigned to the

A $^1\Pi_u$ - X $^1\Sigma_g^+$ transition; more recently Masters and Sarre[12] were able to establish the vibrational numbering for the A-X transition by investigating the photofragment spectrum of both $^{14}N_2^{2+}$ and $^{15}N_2^{2+}$. A band which was unassigned in the 1983 paper[11] has been rotationally analyzed and assigned to the $^3\Pi_g$ - $^3\Sigma_u^+$ transition in N_2^{2+}, also by means of IPS (Szaflarski et al.[13]).

The IPS work carried out so far on dications have been performed with acceleration voltages of 5 kV or less.[11-13] In the present work we have used an accelerator with five times higher acceleration voltage. This has enabled us to work with higher beam currents and, as a consequence, study one of the weaker bands observed by Cosby et al.[11] In particular we were interested to make a detailed study of a band for which the widths of the rotational lines were not Doppler limited. The lines of the (7,1) band[12] were all Doppler limited with a width of approximately 120 MHz. The few lines of the (8,1) band recorded at high resolution by Cosby et al.[11] appeared to be lifetime broadened to 1 GHz; however, it is not clear whether this line width was the Doppler limit of the experiment or not. In the present work[14] we have studied the (8,2) band at an spectral resolution comparable to that of Masters and Sarre[12] with a twofold purpose (in order of priority): *i*) To obtain predissociation rates for as many rotational levels of A $^1\Pi_u$ (v=8) as possible and *ii*) to obtain molecular constants for the X $^1\Sigma_g^+$ (v=2) and A $^1\Pi_u$ (v=8) states.

The lowest doublet states of NO^{2+} (A $^2\Pi$ and X $^2\Sigma^+$) are similar in appearance to the A and X states of N_2^{2+}. One would be inclined to believe that NO^{2+} would photodissociate via the A $^2\Pi$ - X $^2\Sigma^+$ transition. Despite large ab initio calculations of potential curves[15] and careful search[15] for discrete photodissociation of NO^{2+} only bound-free transitions have been observed.[16]

2. EXPERIMENTAL

The experimental apparatus is a 25 kV electrostatic accelerator that acts as an ion injector for a cryogenic electron beam ion source for highly charged ions. It is also designed to work in a secondary mode as an accelerator for fast ion beam laser spectroscopy. The ions are produced in a low-voltage electron impact ion source operated with N_2 or NO gas at a pressure of 10^{-4}-10^{-5} Torr and at an electron energy of 120 eV. Ions are extracted from the source by applying a 25 kV acceleration voltage and further injected into a 0.5 m radius analyzing magnet. Two quadrupole triplets and horizontal and vertical deflection plates are used to focus and steer the beam through two 2 mm diameter apertures (1.4 m apart). In order to monitor the beam after interaction with a collinear laser beam, the beam is deflected by two deflection plates separated by 50 mm into a Faraday cup, which allows a measurement of the transmitted beam current. Typical currents for the present experiments are 30 nA for NO^{2+} and 5 µA for N_2^{2+} and N^+ (the latter ion is probably the dominating). The separation and detection of photo induced N^+ ions relies on the fact that there is a large kinetic energy release when N_2^{2+} dissociates, thus making it possible to separate N^+ photoproducts from N^+ and N_2^{2+} ions created in the ion source by means of an energy analyzer. In the experiment the beam is injected into an electrostatic energy analyzer composed of two 90° cylindrical electrodes separated by 5 mm and with a 0.5 m radius. After exiting the energy analyzer the ions are detected by an electron multiplier operated in DC mode. Figure 1 shows the experimental set-up including the laser systems.

The ion beam is irradiated coaxially by a CR-699-29 c.w. tunable ring dye laser with a linewidth of 1 MHz. The laser beam is entered into the beam line through a Brewster window at the end of the straight part of the beam line, transmitted through the two 2 mm apertures and exited through another Brewster window positioned at the analyzing magnet. In order to suppress a background of collision-induced fragment ions the laser beam was mechanically chopped at a frequency of 1.2 kHz and the multiplier output current was recorded by a phase-sensitive lock-in amplifier. The CR-699-29 laser system has an internal wavemeter with an accuracy of ±200 MHz. As an additional check of the wavelength a fraction of the laser beam was deflected into an iodine cell thus providing a simultaneous recording of N_2^{2+} and I_2 lines. Since the (8,2) lines were expected to be lifetime broadened they were to not suitable for measuring the instrumental resolution, which in the present case is set by the velocity spread in the ion beam. Instead the Q(14) line of the (7,1) band was measured and the resolution of the experiment was determined to 150 MHz. The Q(14) line was also used in order to provide a measurement of the Doppler shift due to the 25 kV acceleration. The laser beam was sent antiparallel, as shown in Fig. 1, and parallel to the ion beam and the Q(14) and I_2 lines were recorded simultaneously thus providing a measure of the Doppler shift. Between recordings of N_2^{2+} lines, NO^{2+} was searched for. This procedure was undertaken in order to minimize experimental errors during the search for NO^{2+}.

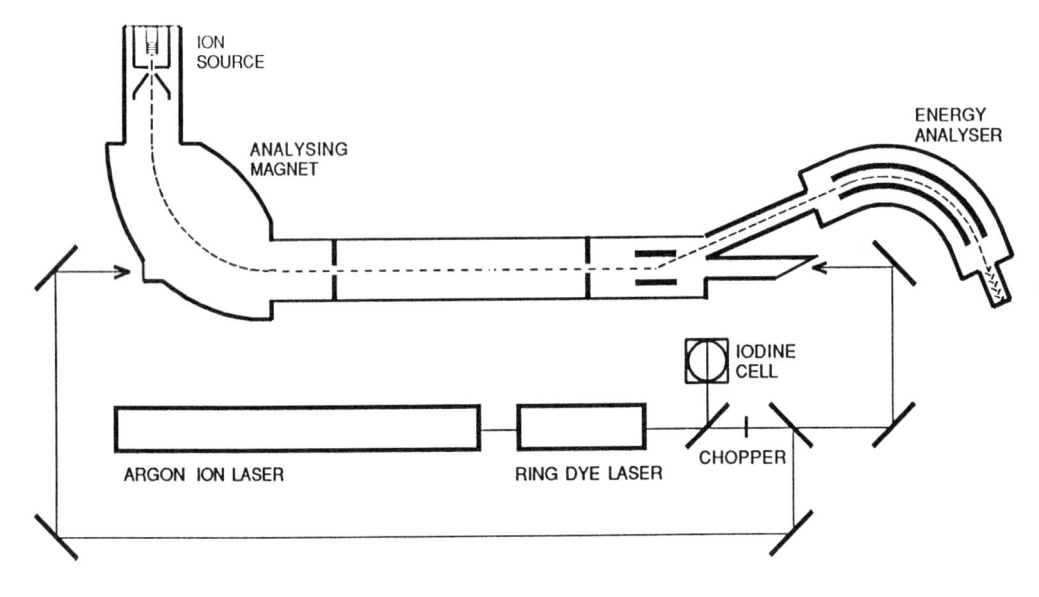

Fig.1. Experimental set-up.

The (8,2) lines were all found to be broader than the (7,1)Q(14) line and other (7,1) lines. Thus all rotational lines of the (8,2) band have predissociation lifetimes that make them broader than the Doppler limited linewidth. The measurement and interpretation of these linewidths are the main purpose of this work.

3. RESULTS AND DISCUSSION

The (8,2) band, denoted α_1 by Cosby et al.[11], consists of single P, Q and R branches. The analysis was straight forward and the spectrum displayed the expected intensity alternation for a homonuclear molecule with nuclear spin I=1. Approximately 75 lines were recorded at high resolution; line positions are given in ref. 14. In order to reduce the measured wavenumbers to molecular constants the hybrid term value method was used.[17] In this model it is assumed that the ground state is unperturbed and the determination of the ground state constants is done simultaneously with the calculation of the term values of the excited state. The validity of the assumption of an unperturbed ground state follows from the fact that the calculated molecular constants reproduces the term values obtained from a fit to the raw lines. Moreover, ab initio calculations on N_2^{2+} [18,19] strongly suggest that there are no suitable perturbers of low vibrational levels of the ground state. The energy levels of the ground state were thus directly fitted to an expression of type $BJ(J+1)-DJ^2(J+1)^2+HJ^3(J+1)^3$. Preliminary constants (in cm^{-1}; uncertainties in parentheses) for v''=2 are $B_2=1.82828(3)$, $D_2=6.48(1) \times 10^{-6}$ and $H_2=-9.1(1) \times 10^{-10}$. If the new result for B_2 is combined with the B_1-value from Masters and Sarre,[12] one obtains $B_e=1.89438$ and $\alpha_e=0.02644$.

The term values of the upper state were represented by the relation $\upsilon_0 + B[J(J+1)-1] - D[J(J+1)-1]^2 + H[J(J+1)-1]^3$. Since A $^1\Pi_u$ is a degenerate electronic state, separate values for B and D were used in order to allow for the Λ-type doubling. The average values for the e and f components were $B_8=1.24535$ and $D_8=1.490 \times 10^{-5}$. Q(22) was excluded from the fit since it deviated 0.04 cm^{-1} from its expected position.

Figure 2 shows the relevant potential curves of N_2^{2+} as calculated in ref. 19. Triplet states have not been included.

Depending on if the A $^1\Pi_u$ state is accessed via a Q-branch transition or via a P- and R-branch transition f and e components of the A state are populated, respectively. For symmetry reasons the $^1\Sigma_u^-$ state can only interact with the f components of the A $^1\Pi_u$ state. This is the reason why, for v'= 6, only Q-lines are observed in the photodissociation spectrum.[11] Thus the e-levels of v'=8 can only predissociate by tunneling through the potential barrier, i.e. the they are all shape resonances. The f-levels, on the other hand, may predissociate by tunneling and by electronic predissociation. The $^1\Pi_u$~$^1\Sigma_u^-$ interaction takes place through the L-uncoupling operator,[20] $-(1/2\mu R^2)(\mathbf{J^+L^-} + \mathbf{JL^+})$, which, owing to the presence of the $\mathbf{J^+}$ operator, makes the total matrix element proportional to $[J(J+1)]^{1/2}$.

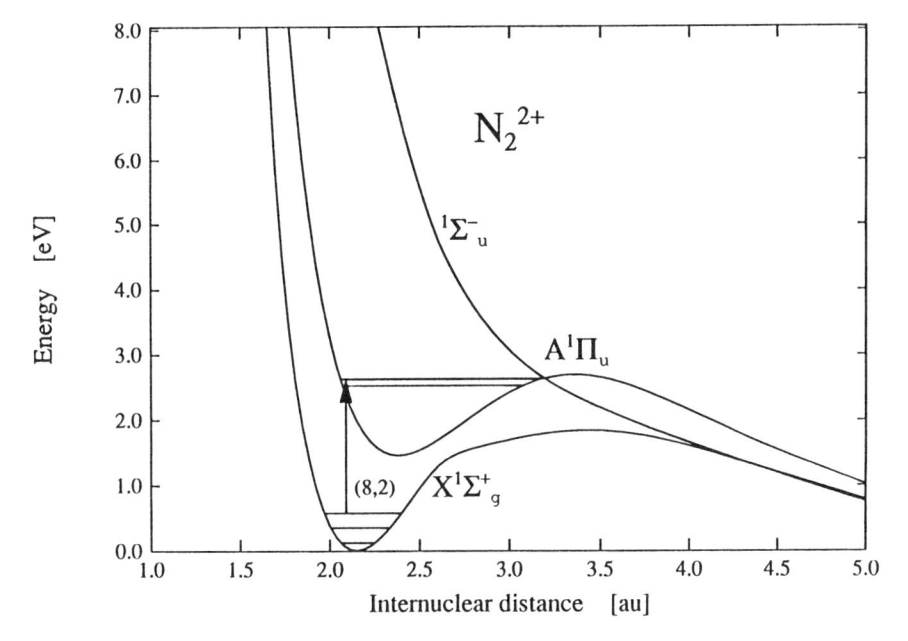

Fig. 2. Potential curves for the $^1\Sigma_u^-$, A $^1\Pi_u$ and X $^1\Sigma_g^+$ states of N_2^{2+} calculated using multireference CI wavefunctions.[19]

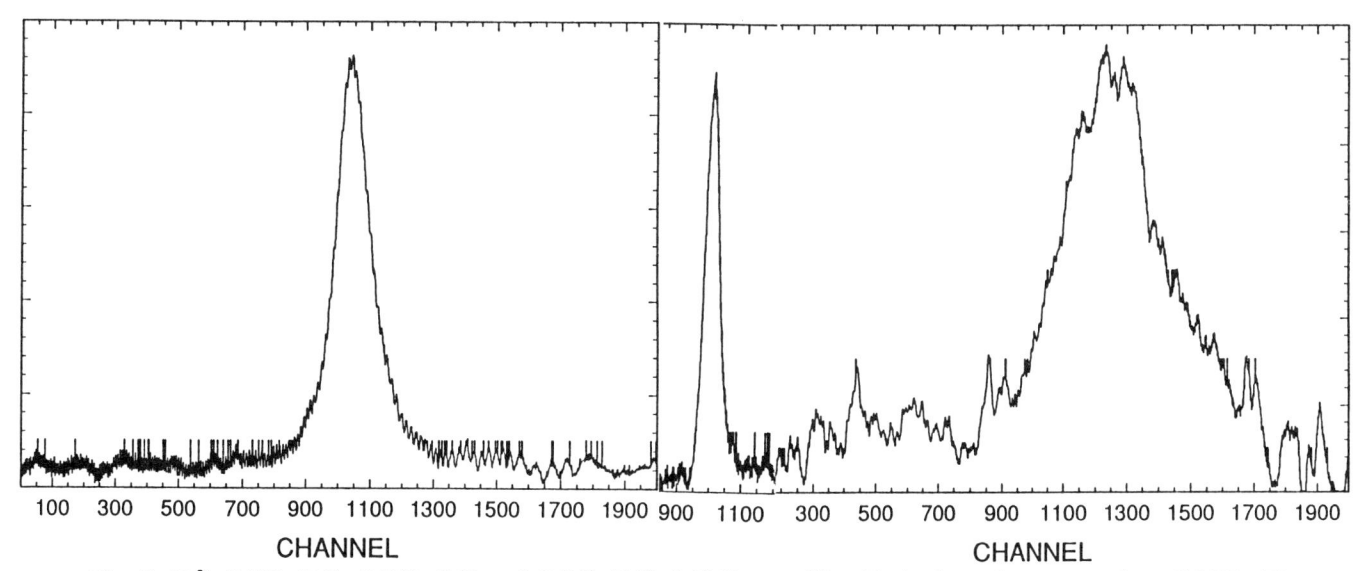

Fig. 3. N_2^{2+} Q(10) (7,1), Q(10) (8,2) and Q(23) (8,2) A-X line profiles. Each channel corresponds to 5 MHz. The intensities are not comparable.

Figure 3 shows three Q-lines of which two, Q(10) and Q(23), are from the (8,2) band and one, Q(10), is from the (7,1) band. For a predissociated rotational level, in the absence of Doppler broadening, the line shape is Lorentzian. A Doppler limited line, on the other hand, is normally well described by a Gaussian line shape. From the work by Masters and Sarre[12] it is clear that the (7,1) lines are Doppler limited. Thus the Gaussian line width was determined from the Q(10) line of the (7,1) band. The (8,2) lines were fitted using a Lorentzian component, and the Gaussian line width was kept fixed during the folding of the Lorentzian with the Gaussian. Figure 4 shows the Lorentzian line widths for v=8 levels of the A $^1\Pi_u$ state of e- (filled triangles) and f-symmetry (open squares). It is evident that the levels of f-symmetry are more strongly predissociated than levels of e-symmetry for all values of J. This shows clearly how the additional predissociation mechanism is operative for the f-levels. The measured line widths correspond to lifetimes toward predissociation between 1 and 0.1 ns. A more detailed discussion of the predissociation will be given in the forthcoming publication.

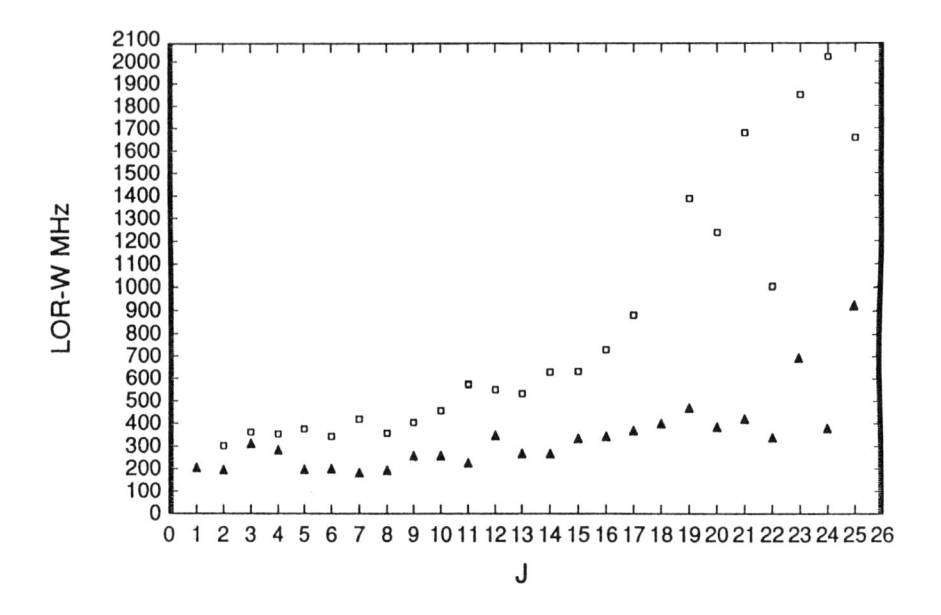

Fig. 4. Lorentzian linewidths (in MHz) for rotational levels of A $^1\Pi_u$ (v=8). The filled triangles denote levels of e-symmetry and the open squares levels of f-symmetry.

It would appear from the discussion above that the spectroscopy of N_2^{2+} is well understood and that the knowledge about N_2^{2+} should be directly applicable to a molecule like NO^{2+}. Both these conclusions are incorrect. The bands denoted γ_1 and γ_2 bands by Cosby et al.[11] are as yet unassigned, and, even if there reasons to believe that they could be assigned to the (5,1) and (5,0) bands of the A $^1\Pi_u$ - X$^1\Sigma_g^+$ transition, no assignment of lines (except their branch identity) has been possible.[11,12] The (8,2)Q(22) line is slightly perturbed and appears to be narrower than expected. While recording the (8,2) band we found at least one line which could not be assigned to any of the bands identified by Cosby et al.[11] Since we found no trace of perturbations (apart from the predissociation) in the A and X states in that wavelength region some other explanation have to be sought for. It should in this context be noticed that, owing to the mass selectivity and energy filtering of the fragment ions, a line appearing in our spectrum necessarily must be due to parent N_2^{2+}. Finally, all attempts to find shape resonances in the A $^2\Pi$ state of NO^{2+} have failed. This latter point is somewhat puzzling. In the first place, the A $^2\Pi$ and X $^2\Sigma^+$ states have potential curves similar to those shown in figure 1. Secondly, the ground states of NO and NO^{2+} are expected to have very similar equilibrium internuclear geometries (just as N_2 and N_2^{2+}), which would imply an efficient population of low vibrational levels in the ion source. Thirdly, A $^2\Pi$(free)←X $^2\Sigma^+$(bound) transitions have been observed by replacing the ring dye laser in Fig. 1 by the 514 and 488 nm output from an Ar-ion laser.[16] It should in principle be possible to lower the photon energy and, with tunable radiation, access the shape resonances below the barrier. In order to aid in the search for these resonances, *ab initio* calculations of A $^2\Pi$ and X $^2\Sigma^+$ potential curves were performed.[15]

The calculations were carried out with the complete active space SCF (CASSCF) followed by multireference CI (MRCI) calculations which included all CAS configurations as reference states. The basis set included g-functions on both nitrogen and oxygen. The expected line positions of various A-X bands were calculated and used to direct the experiment. The absence of a structured photodissociation spectrum of NO^{2+} is at present not understood.

4. ACKNOWLEDGMENTS

The authors wish to thank Jörg Senekowitsch for communicating data prior to publication and for valuable discussions.

5. REFERENCES

1. K.P. Huber and G. Herzberg, Spectroscopic Constants of Diatomic Molecules (Van Nostrand, Princeton, 1979).
2. W. Koch and H. Schwarz, "Experimental and theoretical studies of small organic dications, molecules with highly remarkable properties", Structure/Reactivity and Thermochemistry of Ions, edited by S.G. Lias and P.A. Ausloos (Reidel, Dordrecht, 1987), p. 413.
3. D. Mathur, "Multiply charged molecular ions", Electronic and Atomic Collisions, edited by H.B. Gilbody, W.R. Newell, F.H. Read and A.C.H. Smith (North Holland, Amsterdam, 1988), p. 623.
4. J.T. Moseley, "Ion photofragment spectroscopy", Advan. Chem. Phys. **40**, 245-298 (1985).
5. P.K. Carroll, "A new transition in molecular nitrogen", Can. J. Phys. **36**, 1585-1587 (1958).
6. D. Cossart, F. Launay, J.M. Robbe, G. Gandara, "The optical spectrum of the doubly charged molecular nitrogen ion", J. Mol. Spectrosc. **113**, 142-158 (1985).
7. D. Cossart and F. Launay, "The vacuum UV emission spectrum of the $^{15}N_2^{2+}$ molecular ion, J. Mol. Spectrosc. **113**, 159-166 (1985).
8. D. Cossart, C. Cossart-Magos, and F. Launay, "Intensity distribution of the rotational lines in the D $^1\Sigma_u^+$ - X $^1\Sigma_g^+$ (0,0) band of the N_2^{2+} ion", J. Chem. Soc. Faraday Trans. **87**, 2525 (1991).
9. D. Cossart, M. Bonneau, and J.M. Robbe, "Optical emission spectrum of the NO^{2+} dication", J. Mol. Spectrosc. **125**, 413-427 (1987).
10. D. Cossart and C. Cossart-Magos, "Optical spectrum of the $^{15}NO^{2+}$ dication", J. Mol. Spectrosc. **147**, 471 (1991).
11. P.C. Cosby, R. Möller, and H. Helm, "Photofragment spectroscopy of N_2^{2+}", Phys. Rev A **28**, 766-772 (1983).
12. T.E. Masters and P.J. Sarre, "High-resolution laser photofragment spectrum of N_2^{2+} ($^1\Pi_u$ - $^1\Sigma_g^+$)", J. Chem. Soc. Faraday Trans. **86**, 2005-2008 (1990); T.E. Masters, Thesis, University of Nottingham, 1990.
13. D.M. Szaflarski, A.S. Mullin, K. Yokoyama, M.N.F. Ashfold, and W.C. Lineberger, "Characterization of triplet states in doubly-charged positive ions: assignment of the $^3\Pi_g$ - $^3\Sigma_u^+$ electronic transition in N_2^{2+}", J. Phys. Chem. **95**, 2122 (1991).
14. M. Larsson, G. Sundström, L. Broström, and S. Mannervik, "Photofragment spectroscopy of resonances in the A $^1\Pi_u$ state of N_2^{2+}", (submitted to J. Chem. Phys.).
15. L.G.M. Pettersson, P.E.M. Siegbahn, L. Broström, S. Mannervik, and M. Larsson, "Theoretical potential curves for the A $^2\Pi$ and X $^2\Sigma^+$ states and a search for the A-X transition of NO^{2+}", (submitted to Chem. Phys. Lett.).
16. L.G.M. Pettersson, L.Karlsson, M.P. Keane, A. Naves de Brito, N. Correia, M. Larsson, L. Broström, S. Mannervik, and S. Svensson, "The X-ray excited Auger electron spectrum of NO and potential curves and photodissociation of the NO^{2+} ion", J. Chem. Phys. (in press).
17. N. Åslund, "A numerical method for the simultaneous determination of term values and molecular constants", J. Mol. Spectrosc. **50**, 424-434 (1974).
18. P.T. Taylor and H. Partridge, "Theoretical determination of the ground state of N_2^{2+}", J. Phys. Chem. **91**, 6148-6151 (1987).
19. J. Senekowitsch, S. ONeil, and W. Meyer, "On the bonding in doubly charged diatomics", Theoret. Chim. Acta (in press).
20. H. Lefebvre-Brion and R.W. Field, Perturbations in the Spectra of Diatomic Molecules, (Academic Press, London, 1986).

Intracavity laser photoelectron spectroscopy of Cr_2^-, Cr_2H^- and Cr_2D^-

Sean M. Casey and Doreen G. Leopold

Department of Chemistry, University of Minnesota,
Minneapolis MN 55455

ABSTRACT[1]

The photoelectron spectrum of Cr_2^- shows vibrational levels in the $^1\Sigma_g^+$ ground state of the neutral molecule up to 7300 cm^{-1} above its zero point level. These data, obtained at an instrumental resolution of 5 meV (40 cm^{-1}), reveal a panoramic view of the controversial ground state potential curve of Cr_2. Low-lying vibrational levels are found to fit a Morse potential with $\omega_e = 479 \pm 2$ cm^{-1} and $\omega_e x_e = 13.5 \pm 1.0$ cm^{-1}. This unusually large anharmonicity extrapolates to a dissociation asymptote of only 0.5 eV, considerably lower than the true 1.44 eV value. Between 4875 and 7320 cm^{-1} above the zero point level, we observe twenty peaks at 130 ± 15 cm^{-1} intervals, which we assign as transitions from the ground electronic and vibrational state of the anion to high vibrational levels of the Cr_2 ground state. Using an RKR inversion procedure, we have obtained a potential curve that fits all of the observed vibrational levels to within our experimental uncertainty. This potential curve is compared with the predictions of Goodgame and Goddard's modified GVB calculation. Transitions to highly excited vibrational levels of the Cr_2 ground state are far more intense than would be expected for a direct photodetachment process, and are also strongly wavelength dependent. These non-Franck-Condon intensities are attributed to a resonance of the laser with one or more metastable states of the negative ion far above the electron detachment threshold. The electron affinity of Cr_2 is measured to be 0.505 ± 0.005 eV. An excited electronic state of Cr_2 with a vibrational frequency of 580 ± 20 cm^{-1} is observed $14,240 \pm 30$ cm^{-1} above the ground state. For Cr_2^-, we obtain $\omega_e = 470 \pm 25$ cm^{-1}, $\omega_e x_e = 20 \pm 10$ cm^{-1}, and $r_e = 1.71 \pm 0.01$ Å. Tentative state assignments of $^1\Sigma_u^+$ or $^3\Sigma_u^+$ for the excited Cr_2 state, and $^2\Sigma_u^+$ for the anion, are discussed. Preliminary results for Cr_2H^- and Cr_2D^- are also presented. The photoelectron spectra of these anions reveal the Cr-Cr and Cr-H stretching frequencies in the neutral molecules, and exhibit partially resolved rotational structure.

1. Cr_2

1.1. Introduction

Multiple metal-metal bonding[2] can be examined in its clearest form in the unligated transition metal dimers. Chromium dimer,[3] in particular, has been the subject of numerous experimental[3-8] and theoretical[9-14] studies. Since the ground state of the Cr atom has a high spin $3d^5 4s^1$ configuration, a closed shell configuration for Cr_2 would lead to a formal bond order of six. Consistent with this picture, Cr_2 is known[3,5,6] to have a $^1\Sigma_g^+$ ground state and a bond length of only 1.6788 Å, even shorter than the very short "quadruple" Cr-Cr bonds in $Cr_2(DMP)_4$ [1.847(1) Å] and $Cr_2(2\text{-MeO-5-MeC}_6H_3)_4$ [1.828(2) Å].[15]

However, other results belie the simple picture of a hextuple bond in Cr_2. For example, the bond dissociation energy (D_o) of Cr_2 is only 1.44 ± 0.05 eV (33.2 ± 1.2 kcal/mol),[4] actually <u>lower</u> than the 2.01 ± 0.08 eV bond energy[3] of the singly 4s-4s bonded Cu_2 molecule! Theoretical studies have shown that the situation is further complicated by the 2.7-times greater size of the 4s than the 3d orbital in atomic chromium,[9a] which may cause the 4s-4s interaction in Cr_2 to be repulsive at the optimum distance for 3d-3d bonding. Modified GVB calculations by Goodgame and Goddard[11] predict that the ground state of Cr_2 has a double minimum potential energy curve, dominated by five covalent 3d-3d bonds at short internuclear distance, and by a single 4s-4s bond (with the d electrons antiferromagnetically coupled) at long distance. The "long-bond form" of Cr_2 is predicted to have a 110 cm^{-1} vibrational frequency, a 3.06 Å bond length, and a binding energy of only 0.3 eV.[10,11] Results of elegant two-color resonance Raman experiments by Moskovits and coworkers on matrix-isolated Cr_2 have been interpreted as possible evidence for this double minimum ground state potential.[7b]

We report here a gas phase study of Cr_2 by photoelectron spectroscopy of the Cr_2^- anion. Results reveal low-lying vibrational and electronic states of the neutral molecule, and provide a test of theoretical calculations of its ground state potential curve.

1.2. Experimental Section

Spectra were obtained at an instrumental resolution of 5 meV (40 cm^{-1}) using a new negative ion photoelectron spectrometer described in detail elsewhere.[16] Our instrument is similar to the Lineberger apparatus,[17] but incorporates an 8" radius 90° sector electromagnet rather than a Wien velocity filter for mass selection. The electrostatic ion optics throughout the instrument have been redesigned to achieve good mass resolution and ion transmission with this mass selector. The measured mass resolving power of our instrument is $m/\Delta m = 400$ (where Δm is the full width at half maximum intensity of a peak at mass m). This result represents a ten-fold improvement over the mass resolution typically obtained in previous instruments of this type using Wien filters.[17,18] This improved mass resolution should make a wide variety of transition metal species, including metal clusters containing hydrogen, accessible to study by this powerful spectroscopic technique. We have begun to take advantage of this new capability in our very recent studies, presented below, of the Cr_2H^- and Cr_2D^- anions.

The Cr_2^- anions were prepared in a flowing afterglow ion-molecule reactor equipped with a 2.45 GHz microwave discharge ion source. Although the ions may initially be formed in highly excited states, they are thermally equilibrated by 10^4-10^5 collisions with the helium buffer gas during their journey of several milliseconds down the field-free flow tube. In these experiments, the flow tube was operated at a helium pressure of 0.8 Torr, and $Cr(CO)_6$ was added at its room temperature vapor pressure just downstream of the microwave cavity. The mass spectrum of the resulting anions, displayed in Figure 1, contains $Cr_2(CO)_n^-$ in addition to the more abundant $Cr(CO)_n^-$ (n≤5) species. Our spectra of $Cr(CO)_3^-$ and the other Group VI tricarbonyl complexes have previously been reported.[16]

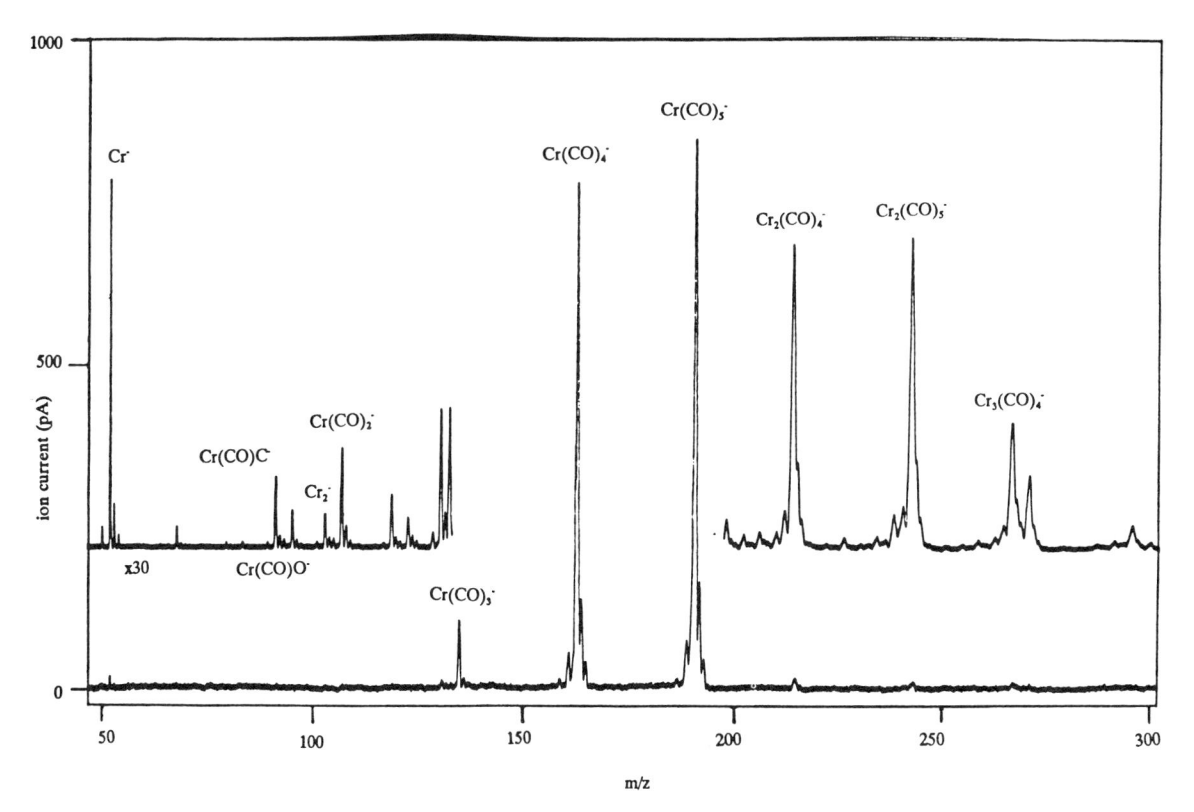

Figure 1. Mass spectrum of the $Cr(CO)_6$ precursor seeded into the flowing afterglow ion source. Each ion displays a quartet of peaks due to the four naturally occurring Cr isotopes. The mass resolving power (m/Δm, FWHM) pictured is about 140.

Presumably, the Cr_2-containing anions in Figure 1 were produced by ion-molecule reactions in the flow tube. For example, one might imagine that the abundant $Cr(CO)_5^-$ anion might react with $Cr(CO)_6$ to produce $Cr_2(CO)_{11}^-$, which could undergo partial or complete decarbonylation in the plasma. As indicated in the figure, $Cr_3(CO)_n^-$ anions were also observed. To obtain the photoelectron spectrum of Cr_2^-, the mass selector was tuned to pass only 104 amu anions into the ultrahigh vacuum photoelectron spectroscopy chamber. Photoelectron spectra of the 104 amu anion beam prepared from a ^{13}C-enriched $Cr(CO)_6$ precursor were identical to those reported below, demonstrating that all features in the spectra arise from photodetachment of Cr_2^- and not CrC_3O^-, the other 104 amu anion that could potentially be produced under these conditions.

1.3. Results and Discussion

The 488 nm (2.540 eV) photoelectron spectrum of Cr_2^- is shown in Figure 2. The strong feature at 2.035 eV electron kinetic energy corresponds to the "0-0" transition between the zero point vibrational levels of the ground electronic states of the anion and the neutral molecule. The position of this peak yields a Cr_2 electron affinity of 0.505 ± 0.005 eV. The 455 ± 15 cm^{-1} separation between the 0-0 transition and the neighboring $(v_{neutral}=1) \leftarrow (v_{ion}=0)$ transition agrees with the known $\Delta G_{\frac{1}{2}}$ value[6] of 452.34 cm^{-1} for Cr_2, confirming that we are indeed observing the $^1\Sigma_g^+$ ground state of the neutral molecule.

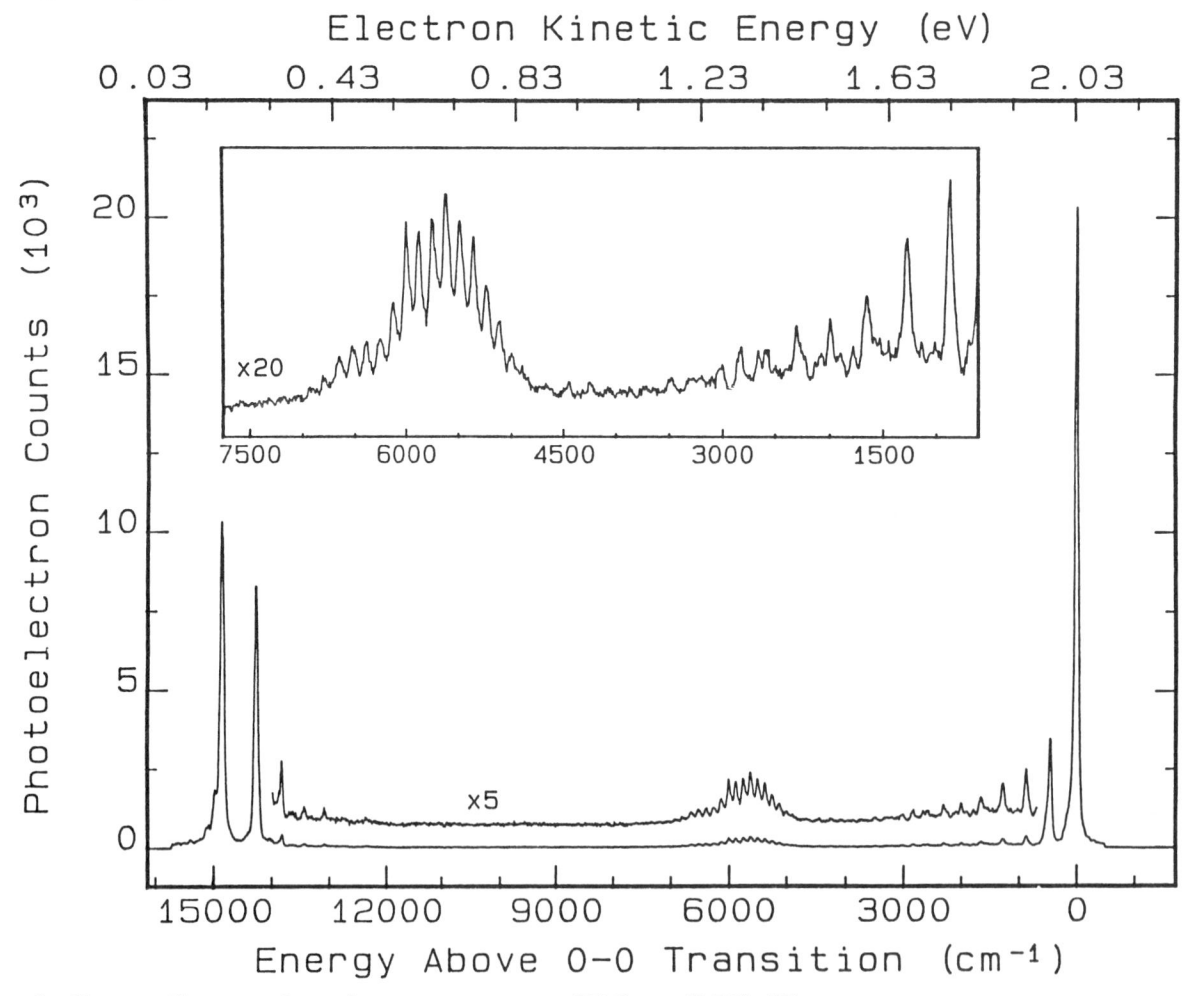

Figure 2. $Cr_2^- \rightarrow Cr_2 + e^-$ photoelectron spectrum at 488.0 nm (2.540 eV).

The two intense features at lowest electron kinetic energy reveal an excited electronic state of Cr_2 $14,240 \pm 30$ cm^{-1} above its ground state, with a vibrational frequency of 580 ± 20 cm^{-1}. The left-most peak is suppressed by about a factor of two by the reduced sensitivity of our instrument at very low electron kinetic energies. In the simplest picture of the Cr_2 ground state, the 12 valence electrons completely occupy the 6 bonding molecular orbitals.

From this viewpoint, it is rather surprising that the excited state, which must have at least one electron in an antibonding orbital, displays a considerably higher vibrational frequency (580 cm^{-1}) than the ground state (452 cm^{-1}).

This increase in vibrational frequency on electronic excitation may be due to the repulsive nature of the $4s$-$4s$ interaction at short internuclear separations, making the $4s\sigma_g$ orbital somewhat antibonding. If the excited state observed here differs from the $^1\Sigma_g^+$ ground state in that one electron has been excited out of the $4s\sigma_g$ orbital and into an essentially nonbonding orbital, then an increased vibrational frequency may result. A possible candidate for this nonbonding orbital may be a $4s\sigma_u^*$ orbital whose antibonding character may be greatly reduced by $4s\sigma$-$4p\sigma$ mixing, so as to direct the electron density away from the internuclear region. This type of highly polarized $4s\sigma_u^*$ orbital has been predicted by extensive *ab initio* calculations of Cu_2.[19] If this assignment is correct, then the excited state of Cr_2 observed here is a $^1\Sigma_u^+$ or $^3\Sigma_u^+$ state with a $(4s\sigma_g)^1(4s\sigma_u^*)^1$ s-electron configuration and the same $(3d)^{10}$ configuration as the $^1\Sigma_g^+$ ground state. Assignment to a singlet state, rather than to a triplet, is suggested by the observation in matrix studies of an absorption near 14000 cm^{-1} originating from the singlet ground state.[8c]

The photoelectron spectrum also contains information regarding Cr_2^-. Hot bands arising from ($v_{neutral}=0$) ← ($v_{ion}=n$, n=1-3) transitions to the excited state, shown at x5 magnification in Figure 2, yield $\omega_e = 470\pm25$ cm^{-1} and $\omega_e x_e = 20\pm10$ cm^{-1} for the Cr_2^- anion. If it is assumed that the Cr_2^- bond length exceeds that of Cr_2, a Franck-Condon analysis of the relative intensities of the origin and v=1 peaks of the ground state transition yields an anion bond length of 1.71±0.01 Å. Based upon a comparison of the photodetachment intensities of Cr_2^- and Cr^-, we conclude that the extra electron in Cr_2^- enters an orbital of mainly $4s$ rather than $3d$ atomic parentage. Addition of an extra electron to the $4s\sigma_u^*$ orbital of the $^1\Sigma_g^+$ state of Cr_2 gives a $^2\Sigma_u^+$ ground state for the Cr_2^- anion. The detachment of an electron from this $(4s\sigma_g)^2(4s\sigma_u^*)^1$ anion could then produce Cr_2 in its $(4s\sigma_g)^2$ ground state or in a $(4s\sigma_g)^1(4s\sigma_u^*)^1$ state, consistent with the excited state assignment proposed above.

For the neutral Cr_2 molecule, additional information concerning the ground state is obtained from a closer inspection of the high electron kinetic energy region of the spectrum, shown at x20 magnification in the insert to Figure 2. As shown on the right side of the insert, vibrational states of neutral Cr_2 from v=2 to v=6 are observed 875, 1275, 1650, 1995 and 2305 cm^{-1} (±15 cm^{-1}) above the 0-0 transition. With the v=1 level constrained to its literature value,[6] the observed energies can be fit to within our experimental uncertainty to a Morse potential with $\omega_e = 479\pm2$ cm^{-1} and $\omega_e x_e = 13.5\pm1.0$ cm^{-1}. This anharmonicity constant is unusually large: $\omega_e x_e$ values for all other neutral homonuclear first row transition metal dimers measured to date[3] fall in the range 1-4 cm^{-1}. Interestingly, the excellent Morse potential fit to vibrational levels up to v=6 (0.3 eV above the zero point level) extrapolates to a dissociation asymptote of only 0.5 eV, far below the true[4] Cr_2 bond energy (D_o) of 1.44±0.05 eV. Thus, it is clear that the Cr_2 ground state potential energy curve strongly diverges from a Morse potential at higher energies.

A weak vibrational progression of the neutral molecule comprised of more than a dozen peaks is observed from 5120 to 6660 cm^{-1} above the 0-0 transition, as shown on the left side of the x20 insert to Figure 2. Surprisingly, the peak spacings are only 130±15 cm^{-1}, and they exhibit no detectable anharmonicity. Increasing the flow tube temperature from room temperature to 250° C did not change the intensity profile of this progression, or its intensity relative to those of the v=0 through v=6 features discussed above. This result indicates that all of these transitions originate from the zero point level of the anion ground state, rather than from an excited vibrational or electronic state of the anion.

We can thus consider two possible explanations for the 130 cm^{-1} progression. Either it is associated with a low-lying excited electronic state of Cr_2, or it arises from highly excited vibrational levels of the $^1\Sigma_g^+$ ground state. We believe that the latter assignment is more likely, for the following reasons. The first member of the 130 cm^{-1} progression observed in the 488 nm spectrum occurs only 5120 cm^{-1} (0.6 eV) above the zero point level of the Cr_2 ground state. Since the $^1\Sigma_g^+$ state dissociates to ground state atoms and its binding energy is 1.44 eV, such an excited state would have a well depth of at least 0.8 eV. However, the 130 cm^{-1} vibrational frequency indicates that this state would also have a weak, long bond. In fact, Weisshaar's Badger's rule plot for first transition series diatomics,[20] which includes both ground and excited state data, suggests that this frequency would correspond to a Cr-Cr bond length of 2.9 Å. Since the frequency in the ground state is much higher (452 cm^{-1}) and its equilibrium bond length much shorter (1.6788 Å), it seems unlikely that such an excited state would have a well depth as much as 60% of the ground state value.

On the other hand, it does appear reasonable that the 130 cm^{-1} intervals may be associated with highly excited vibrational levels of the ground electronic state. As noted above, the Morse potential fit to the first six vibrational levels of the ground state converges toward an asymptote of only 0.5 eV, much lower than the experimentally determined dissociation energy of 1.44 eV. This result suggests that the ground state potential curve exhibits a "shelf" near 0.5 eV, and then continues to increase. In the intermediate region, such a potential could conceivably support the regularly, closely spaced vibrational levels observed here.

To explore this possibility, we used an RKR inversion procedure[21] to find a potential energy curve that fits both the v=1 through v=6 levels that we had previously modelled using a Morse potential, and the higher energy levels spaced by 130 cm^{-1}. We also included in the fit seven additional members of the 130 cm^{-1} progression observed at other laser wavelengths (see below), giving a total of 20 vibrational levels between 4875 and 7320 cm^{-1}, all spaced by 130 cm^{-1}. The fit also included weak peaks associated with transitions to v=7 and v=8 levels at 2580 and 2835 cm^{-1}. The width (r_{max}-r_{min}) of the RKR potential depends on the vibrational energies and quantum numbers, but the absolute values of the inner and outer turning points are a sensitive function of the rotational constants. Use of the two known rotational constants of Cr$_2$ (B_e = 0.2303 cm^{-1} and α_e = 0.0038 cm^{-1})[6] gave an RKR potential with a double-valued inner wall. To correct this problem, we assumed a Morse function for the inner wall, and adjusted the outer wall so that the new potential had the same width at each energy as the original one. Similar methods for correcting nonphysical RKR inner walls have been used by others.[22] Since we did not observe vibrational levels between v=8 and the 130 cm^{-1} progression, it was also necessary to guess the quantum number of the first member of this progression. In the fit shown here, this level was somewhat arbitrarily assigned as v = 23, and subsequent levels were numbered sequentially.

The results of the RKR fit are shown in Figure 3, where the 28 vibrational levels included in the fit are also

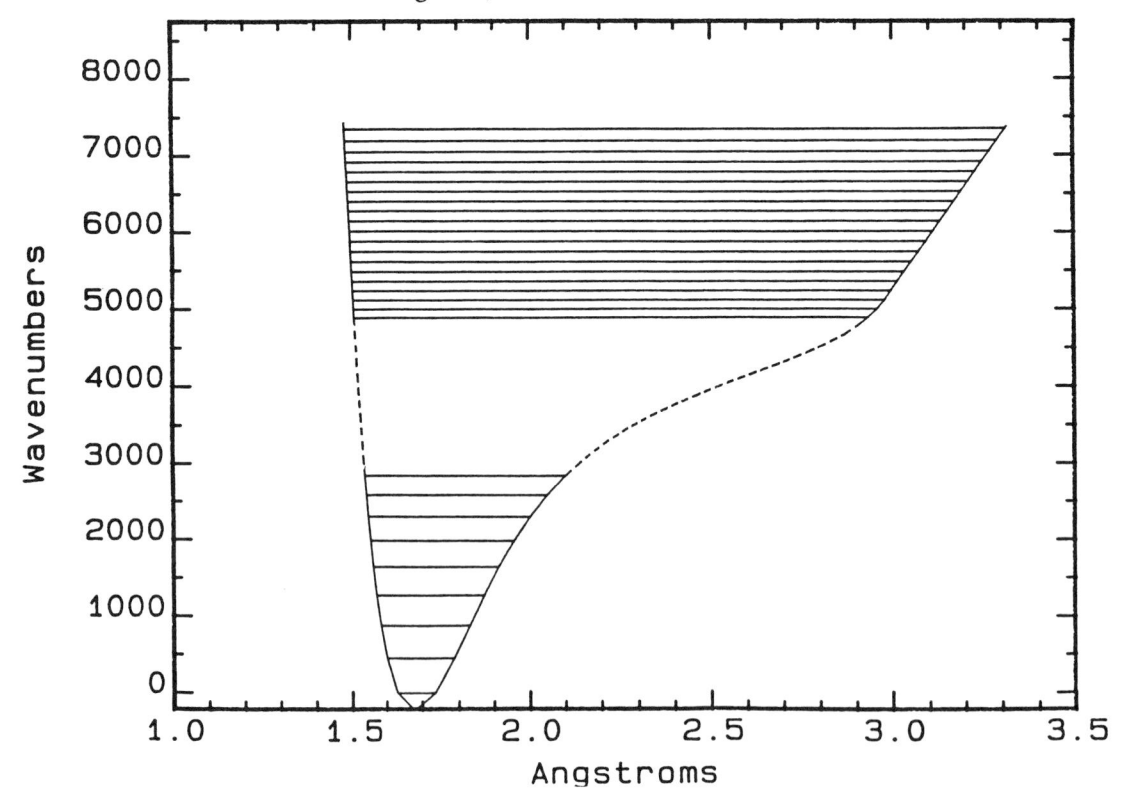

Figure 3. A plot of the inner and outer turning points against energy obtained from the RKR analysis. Solid lines represent levels observed in our spectra; dashed lines represent turning points for levels predicted by the RKR analysis, but not observed.

indicated. This curve fits all of these levels to within our experimental uncertainty of 15 cm^{-1}. Clearly, the width of the potential is not well-determined in the region indicated by dashed lines, where no vibrational levels were observed. Since the shelf of the potential occurs in this region, it could thus differ significantly from that shown in the figure. For example, a flatter shelf results if we assign higher quantum numbers to the 130 cm^{-1} progression. It is also possible that the true potential has a shallow second minimum in this region.

Despite these caveats, however, the overall shape of the fitted potential in Figure 3 provides a revealing illustration of the different contributions of *d-d* and *s-s* bonding in Cr$_2$ at different internuclear distances. Near its minimum, the curve reflects the high vibrational frequency associated with multiple *d-d* bonding. As the atoms separate, the overlap between the compact *d* orbitals is quickly lost and *s-s* bonding becomes more important, producing a large anharmonicity and a levelling out of the potential. At larger internuclear distances, the curve once again rises and gradually approaches the dissociation asymptote, reflecting the dominance of bonding between the diffuse *s* orbitals. This effect manifests itself in the lower (130 cm^{-1}) vibrational frequency observed in this region.

Thus, in a qualitative sense, the overall shape of our fitted potential curve agrees with the predictions of Goodgame and Goddard's calculations.[10,11] In fact, the 130 cm^{-1} vibrational frequency observed here is quite close to the 110 cm^{-1} value predicted for the singly 4*s*-4*s* bonded, "long-bond form" of the Cr$_2$ ground state. However, our results do not agree quantitatively with the MGVB potential. For example, the 130 cm^{-1} progression observed here appears only 0.6 eV above the 0-0 transition, and displays no detectable anharmonicity over a range of 0.30 eV. In contrast, the MGVB calculation[11] predicts the 4*s*-4*s* well to lie 1.6 eV above the 3*d*-3*d* well, and to be bound by only 0.3 eV.

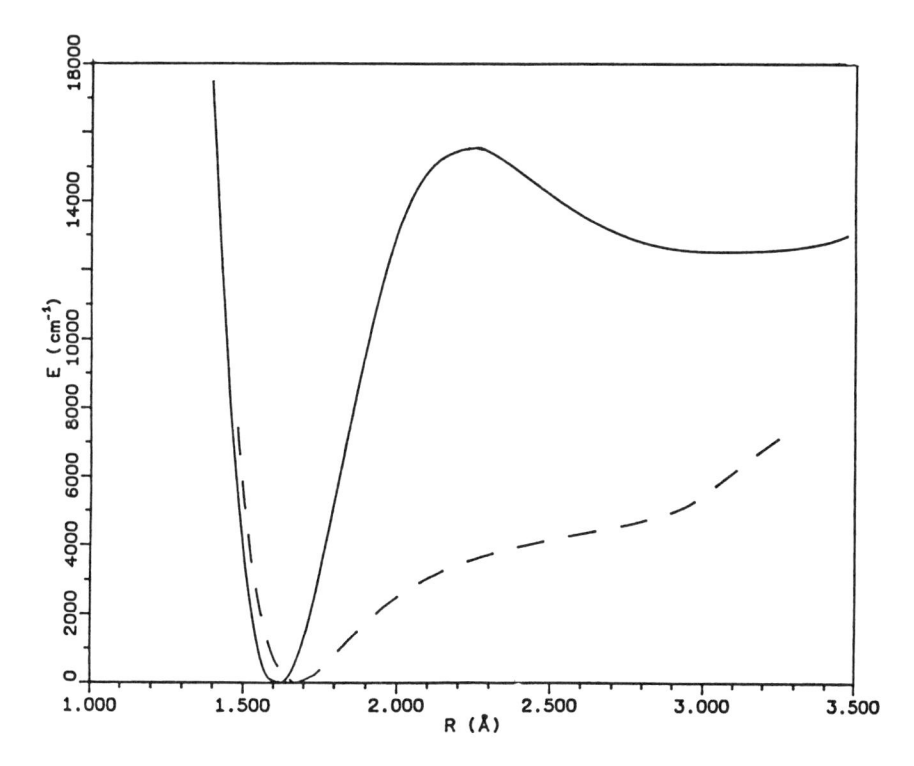

Figure 4, above, compares the predicted MGVB potential (solid line) to the RKR potential obtained here (dashed line). To clarify this comparison, the curves are plotted with their minima at the same energy. We emphasize again that our fitted potential is not unique, and that it is even possible that the true potential could contain a shallow minimum rather than a shelf in the dashed region of the potential pictured in Figure 3. (Clearly, the RKR method, which gives only two turning points, cannot produce a local minimum. However, this point is moot in our case, since the second minimum would occur in the dashed region of the potential pictured in Figure 3, where we have no vibrational data to which to fit the curve.) Nevertheless, it is clear that even in this case, the two curves would still differ dramatically. These results suggest the need for improved calculations of the ground state potential energy curve of Cr$_2$.

If the 130 cm^{-1} progression is indeed associated with highly excited vibrational levels of the ground state, one must ask why these transitions appear in the photoelectron spectrum with detectable intensity, while transitions to lower energy vibrational levels (above v=8) are not observed. As noted above, even the v=2 through v=8 peaks are substantially more intense than would be expected based upon a Franck-Condon fit to the relative intensities of the origin and v=1 peaks. The intensities observed in the spectra are also unusual in that the intensity profile of the 130 cm^{-1} progression is noticeably more irregular than that expected for the photoelectron spectrum of a diatomic species.

These observations suggest an accidental resonance of the laser with a metastable state of the anion. To test this hypothesis, we recorded spectra using the 514, 497 and 476 nm lines of the argon ion laser. For a direct photodetachment process in which the laser photon energy considerably exceeds the electron detachment energy, as is the case here, relative intensities would not vary significantly with laser wavelength. However, in the present experiment, dramatic intensity variations are observed, as is illustrated in Figure 5. In all cases, however, the spacing between each peak and the 0-0 transition does not vary with laser wavelength.

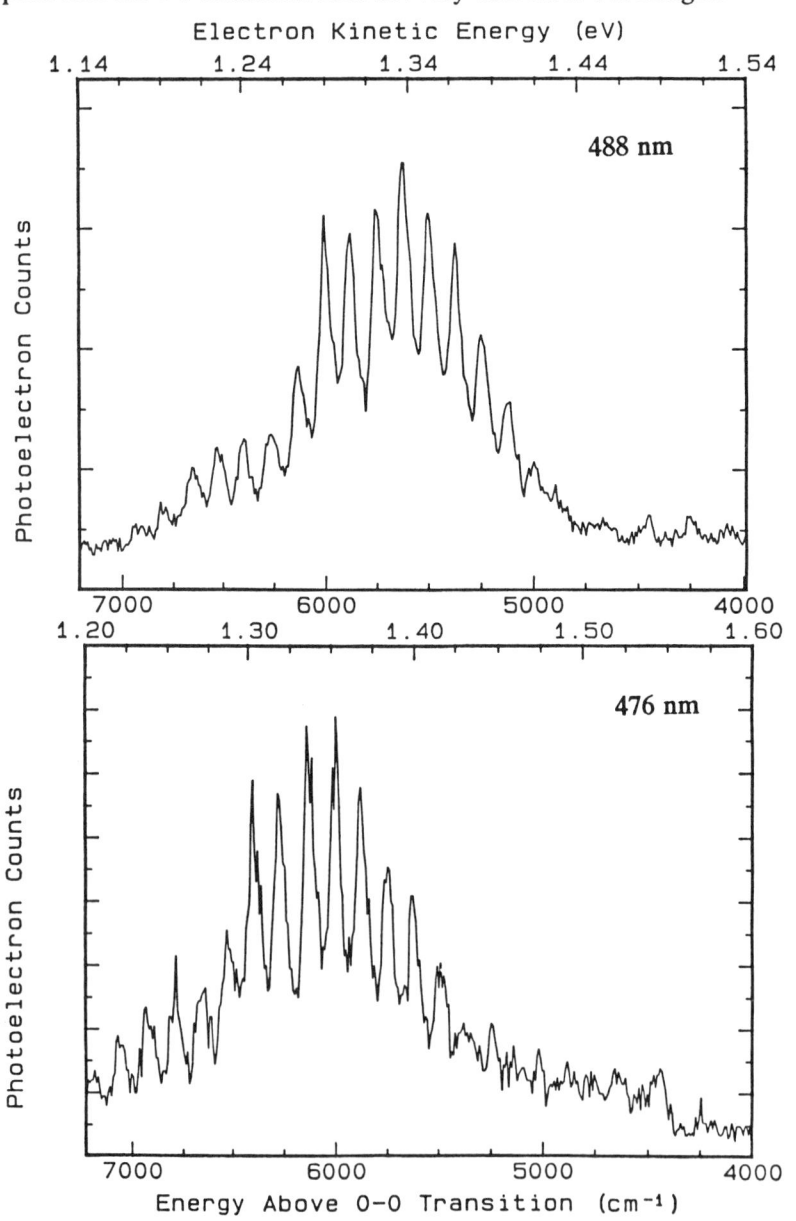

Figure 5. Comparison of the region between 4000 and 7000 cm^{-1} above the origin in the photoelectron spectrum of Cr$_2^-$ taken with the argon ion laser operating at 488.0 nm (top) and at 476.5 nm (bottom).

Based upon these observations, we conclude that the intensities observed in the spectrum result from the combined effects of the usual direct photodetachment process, and an indirect process involving autodetachment from one or more metastable states of the anion which are resonant with the laser. Since the photon energy was varied between 2.4 and 2.6 eV and the electron affinity of Cr_2 is only 0.5 eV, the resonant state(s) must be far above the electron detachment threshold. The analogous formation of intermediate autoionizing resonances is well-documented in the literature of neutral molecule photoelectron spectroscopy.[23,24] However, there are only a few examples of this type of phenomenon in the negative ion photoelectron spectra of species lacking permanent dipole moments.[25] In the present experiment, this propitious resonance allows us to probe highly excited vibrational levels of the neutral molecule potential curve that would normally be inaccessible due to negligible Franck-Condon overlap with the anion ground state.

One can speculate about the nature of the resonant anion state(s). A possible mechanism for the observed intensity variations between 476 and 514 nm involves a Feshbach resonance to an excited autodetaching anion state which decays to the ground state of the neutral molecule via a two-electron process. For example, this metastable state may correspond to the $A^1\Sigma_u^+$ state of neutral Cr_2 at 21,751 cm^{-1} (460 nm), which is thought to arise from a strongly allowed $4p\sigma_u^* \leftarrow 4s\sigma_g$ transition correlating with the $Cr(^7P_z^o) \leftarrow Cr(^7S)$ atomic transition.[6] The anion might be expected to have an analogous $(4s\sigma_g)^1(4s\sigma_u^*)^1(4p\sigma_u^*)^1\ ^2\Sigma_g^+ \leftarrow (4s\sigma_g)^2(4s\sigma_u^*)^1\ ^2\Sigma_u^+$ transition in approximately the same region. Autodetachment from high vibrational levels of this metastable anion state could prepare the neutral molecule in vibrational levels that would not be accessible via direct photodetachment from the anion ground state. This autodetachment would have to involve electronic relaxation and detachment to reach the ground $(4s\sigma_g)^2$ state of the neutral. Peaks in the photoelectron spectrum arising from this two-electron event would be expected to be weak compared to peaks arising from direct one-electron detachment. Different laser wavelengths would prepare this metastable state in different vibrational levels, which would autodetach to form a different distribution of Cr_2 vibrational levels. Alternatively, it is possible that the laser excites the anion to a state that dissociates on the timescale of electron ejection, leaving the neutral molecule in highly excited vibrational states. The latter type of mechanism has been invoked by Weisshaar and coworkers to explain resonances observed in the photoelectron spectrum of neutral Cu_2.[24h] It is also possible that both mechanisms are important in our case, and that the wavelength-dependent intensities observed are due to a resonance or near-resonance of the laser with more than one metastable state of the anion.

2. Cr_2H^- and Cr_2D^-

We have recently obtained photoelectron spectra of Cr_2H^- and Cr_2D^-, shown in Figure 6. The Cr_2H^- and Cr_2D^-

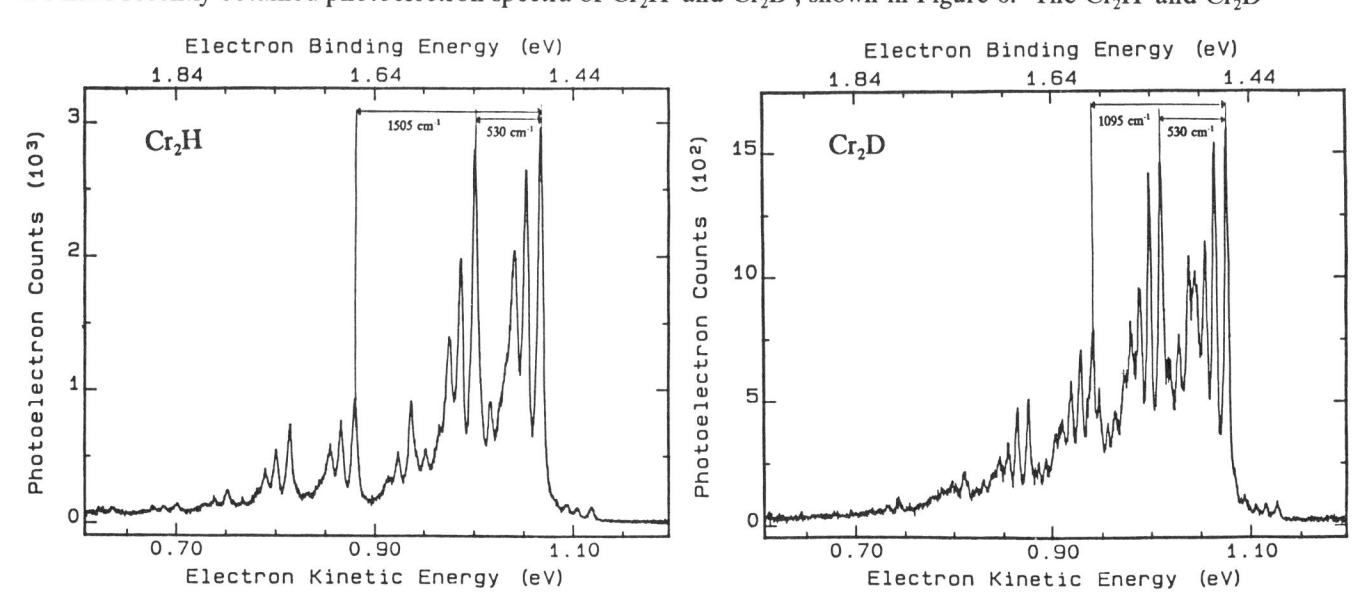

Figure 6. Cr_2H^- (left) and Cr_2D^- photoelectron spectra recorded at 488.0 nm.

anions were prepared by adding $Cr(CO)_6$ and either H_2 or D_2 to the flow tube just downstream of the microwave discharge cavity. The spectra shown here were obtained with the flow tube immersed in liquid nitrogen in order to cool the anions and simplify the spectra. The photoelectron spectra yield an electron affinity of 1.47 eV, and show vibrational activity in the Cr-Cr and the Cr-H (or Cr-D) stretching modes. For the Cr-Cr stretch, values of $\omega_e = 550$ cm^{-1} and $\omega_e x_e = 10$ cm^{-1} are obtained for both Cr_2H and Cr_2D. Surprisingly, this frequency is <u>higher</u> than that observed in bare Cr_2 ($\omega_e = 479$ cm^{-1}, $\omega_e x_e = 13.5$ cm^{-1}). The Cr-H and Cr-D stretching frequencies are 1505 cm^{-1} and 1095 cm^{-1}, respectively. The former value is quite close to that of diatomic CrH ($\Delta G_{1/2} = 1517$ cm^{-1}).[26] In addition, each vibrational transition appears as a series of closely-spaced peaks. We have qualitatively modelled these features as a rotational contour associated with a nonlinear structure, and a quantitative analysis is in progress.

3. ACKNOWLEDGMENTS

We thank Peter Villalta and Paul Deck for preparing the ^{13}C-enriched $Cr(CO)_6$ used in these studies. We are also grateful to Dr. Carl Lineberger and Dr. Joe Ho for sharing their unpublished Cr_2 data, and for valuable advice. This research was supported by the NSF under PYI Grant CHE-8858373, and by generous matching grants from the Allied Signal, Amoco, Coherent, Cray, Ford Motor, GE, IBM, and Newport Corporations. We also gratefully acknowledge support from the American Society for Mass Spectrometry, the Dreyfus Foundation, the Electric Power Research Institute, the McKnight Foundation, the Donors of the Petroleum Research Fund administered by the ACS, the Research Corporation under a Bristol-Myers Company Grant, and the University of Minnesota. One of us (S.M.C.) would like to acknowledge the financial support provided by a Fellowship from the Elmore H. Northey Endowment.

4. REFERENCES

1. This report is, in part, abstracted from: S. M. Casey, P. W. Villalta, A. A. Bengali, C.-L. Cheng, J. P. Dick, P. T. Fenn, and D. G. Leopold, "A Study of Cr_2 by Negative-Ion Photoelectron Spectroscopy," *J. Am. Chem. Soc.*, vol. 113, pp. 6688-6689, 1991. A full report on this work is in preparation for the *Journal of Chemical Physics*.

2. (a) F. A. Cotton and R. A. Walton, *Multiple Bonds Between Metal Atoms*, John Wiley & Sons, New York, 1982; (b) F. A. Cotton, "Discovering and Understanding Multiple Metal-to-Metal Bonds," *Acc. Chem. Res.*, vol. 11, pp. 225-232, 1978; (c) W. C. Trogler and H. B. Gray, "Electronic Spectra and Photochemistry of Complexes Containing Quadruple Metal-Metal Bonds," *ibid.*, vol. 11, pp. 232-239, 1978.

3. M. D. Morse, "Clusters of Transition-Metal Atoms," *Chem. Rev.*, vol. 86, pp. 1049-1109, 1986.

4. K. Hilpert and K. Ruthardt, "Determination of the Dissociation Energy of the Cr_2 Molecule," *Ber. Bunsenges. Phys. Chem.*, vol. 91, pp. 724-731, 1987.

5. (a) Yu. M. Efremov, A. N. Samoilova, and L. V. Gurvich, "The $\lambda=4600$-Å band in a spectrum produced by pulsed photolysis of chromium carbonyl," *Opt. Spectrosc.*, vol. 36, pp. 381-382, 1974; (b) D. L. Michalopoulos, M. E. Geusic, S. G. Hansen, D. E. Powers, and R. E. Smalley, "The Bond Length of Cr_2," *J. Phys. Chem.*, vol. 86, pp. 3914-3916, 1982; (c) S. J. Riley, E. K. Parks, L. G. Pobo, and S. Wexler, "The A ← X transition in Cr_2," *J. Chem. Phys.*, vol. 79, pp. 2577-2582, 1983.

6. V. E. Bondybey and J. H. English, "Electronic Structure and Vibrational Frequency of Cr_2," *Chem. Phys. Lett.*, vol. 94, pp. 443-447, 1983.

7. (a) D. P. DiLella, W. Limm, R. H. Lipson, M. Moskovits, and K. V. Taylor, "Dichromium and trichromium," *J. Chem. Phys.*, vol. 77, pp. 5263-5266, 1982; (b) M. Moskovits, W. Limm, and T. Mejean, "A weakly bound metastable state of Cr_2," *ibid.*, vol. 82, pp. 4875-4879, 1985; (c) M. Moskovits, W. Limm, and T. Mejean, "Dichromium Revisited," *J. Phys. Chem.*, vol. 89, pp. 3886-3890, 1985.

8. (a) M. J. Pellin and D. M. Gruen, "Emission, ground, and excited state absorption spectroscopy of Cr_2 isolated in Ar and Kr matrices," *J. Chem. Phys.*, vol. 79, pp. 5887-5893, 1983; (b) P. A. Montano, H. Purdum, G. K. Shenoy, T. I. Morrison, and W. Shulze, "X-Ray Absorption Fine Structure Study of Small Metal Clusters Isolated in Rare-Gas Solids," *Surf. Sci.*, vol. 156, pp. 228-233, 1985; (c) M. P. Andrews and G. A. Ozin, "Divanadium, the $\delta \rightarrow \delta^*$ Transition," *J. Phys. Chem.*, vol. 90, pp. 2852-2859, 1986.

9. (a) S. P. Walch, and C. W. Bauschlicher, *Comparison of Ab Initio Quantum Chemistry with Experiment for Small Molecules*, R. J. Bartlett, Ed., Reidel, Dordrecht, 1985, pp. 17-51; (b) S. R. Langhoff and C. W. Bauschlicher, "Ab Initio Studies of Transition Metal Systems," *Ann. Rev. Phys. Chem.*, vol. 39, pp. 181-212, 1988;

(c) S. P. Walch, C. W. Bauschlicher, B. O. Roos, and C. J. Nelin, "Theoretical Evidence for Multiple 3d Bonding in the V_2 and Cr_2 Molecules," *Chem. Phys. Lett.*, vol. 103, pp. 175-179, 1983.

10. M. M. Goodgame and W. A. Goddard, "The 'Sextuple' Bond of Cr_2," *J. Phys. Chem.*, vol. 85, pp. 215-217, 1981; M. M. Goodgame and W. A. Goddard, "Nature of Mo-Mo and Cr-Cr Multiple Bonds," *Phys. Rev. Lett.*, vol. 48, pp. 135-138, 1982.

11. M. M. Goodgame and W. A. Goddard, "Modified Generalized Valence-Bond Method," *Phys. Rev. Lett.*, vol. 54, pp. 661-664, 1985.

12. C. Wood, M. Doran, I. H. Hillier, and M. F. Guest, "Theoretical Study of the Electronic Structure of the Transition Metal Dimers, Sc_2, Cr_2, Mo_2, and Ni_2," *Faraday Symp. Chem. Soc.*, vol. 14, pp. 159-169, 1980; A. Wolf and H.-H. Schmidtke, "Nonempirical Calculations on Diatomic Transition Metals," *Int. J. Quantum Chem.*, vol. 18, pp. 1187-1205, 1980; P. M. Atha and I. H. Hillier, "Correlation effects and the bonding in Mo_2 and Cr_2," *Mol. Phys.*, vol. 45, pp. 285-293, 1982; R. A. Kok and M. B. Hall, "A Theoretical Investigation of the Bond Length of Dichromium," *J. Phys. Chem.*, vol. 87, pp. 715-717, 1983; A. D. McLean and B. Liu, "f-Type Functions in the Orbital Basis for Calculating Molecular Interactions Involving d Electrons," *Chem. Phys. Lett.*, vol. 101, pp. 144-148, 1983; G. P. Das and R. L. Jaffe, "Theoretical Study of Electron Correlation Effects in Transition Metal Dimers," *ibid.*, vol. 109, pp. 206-211, 1984; W. von Niessen, "Ionization energies of the transition metal diatomics Cu_2, Ag_2, Cr_2, and Mo_2," *J. Chem. Phys.*, vol. 85, pp. 337-345, 1986; K. W. Richman and E. A. McCullough, "Numerical Hartree-Fock calculations on diatomic chromium," *ibid.*, vol. 87, pp. 5050-5051, 1987; H.-J. Werner and P. J. Knowles, "An efficient internally contracted multiconfiguration-reference configuration interaction method," *ibid.*, vol. 89, pp. 5803-5814, 1988; Y. Takahara, K. Yamaguchi, and T. Fueno, "Potential Energy Curves for Transition Metal Dimers and Complexes Calculated by the Approximately Projected Unrestricted Hartree-Fock and Møller-Plesset Perturbation (APUMP) Methods," *Chem. Phys. Lett.*, vol. 158, pp. 95-101, 1989; G. E. Scuseria and H. F. Schaefer, "Diatomic Chromium (Cr_2)," *ibid.*, vol. 174, pp. 501-503, 1990; G. E. Scuseria, "Analytic evaluation of energy gradients for the singles and doubles coupled cluster method including perturbative triple excitations," *J. Chem. Phys.*, vol. 94, pp. 442-447, 1991.

13. (a) W. Klotzbücher, G. A. Ozin, J. G. Norman, and H. J. Kolari, "Bimetal Atom Chemistry," *Inorg. Chem.*, vol. 16, pp. 2871-2877, 1977; (b) J. Harris and R. O. Jones, "Density functional theory and molecular bonding," *J. Chem. Phys.*, vol. 70, pp. 830-841, 1979; (c) B. I. Dunlap, "Xα, Cr_2, and the symmetry dilemma," *Phys. Rev. A*, vol. 27, pp. 2217-2219, 1983; (d) B. Delley, A. J. Freeman, and D. E. Ellis, "Metal-Metal Bonding in Cr-Cr and Mo-Mo Dimers," *Phys. Rev. Lett.*, vol. 50, pp. 488-491, 1983; (e) J. Bernholc and N. A. W. Holzwarth, "Local Spin-Density Description of Multiple Metal-Metal Bonding," *ibid.*, vol. 50, pp. 1451-1454, 1983; (f) N. A. Baykara, B. N. McMaster, and D. R. Salahub, "LCAO local-spin-density and Xα calculations for Cr_2 and Mo_2," *Mol. Phys.*, vol. 52, pp. 891-905, 1984; (g) G. S. Painter, "Density Functional Description of Molecular Bonding within the Local Spin Density Approximation," *J. Phys. Chem.*, vol. 90, pp. 5530-5535, 1986.

14. W. F. Cooper, G. A. Clarke, and C. R. Hare, "Molecular Orbital Theory of the Diatomic Molecules of the First Row Transition Metals," *J. Phys. Chem.*, vol. 76, pp. 2268-2273, 1972; A. B. Anderson, "Structures, binding energies, and charge distributions for two to six atom Ti, Cr, Fe, and Ni clusters and their relationship to nucleation and cluster catalysis," *J. Chem. Phys.*, vol. 64, pp. 4046-4055, 1976; W. Klotzbücher and G. A. Ozin, "Diniobium, Nb_2, and Dimolybdenum, Mo_2," *Inorg. Chem.*, vol. 16, pp. 984-987, 1977.

15. Ref. 2(a), pp. 150-182.

16. A. A. Bengali, S. M. Casey, C.-L. Cheng, J. P. Dick, P. T. Fenn, P. W. Villalta, and D. G. Leopold, "Negative Ion Photoelectron Spectroscopy of Coordinatively Unsaturated Group VI Metal Carbonyls," *J. Am. Chem. Soc.*, in press.

17. D. G. Leopold, K. K. Murray, A. E. Stevens Miller, and W. C. Lineberger, "Methylene," *J. Chem. Phys.*, vol. 83, pp. 4849-4865, 1985; C. S. Feigerle, Ph.D. Thesis; University of Colorado; 1983.

18. H. B. Ellis and G. B. Ellison, "Photoelectron spectroscopy of HNO^- and DNO^-," *J. Chem. Phys.*, vol. 78, pp. 6541-6558, 1983; J. V. Coe, J. T. Snodgrass, C. B. Freidhoff, K. M. McHugh, and K. H. Bowen, "Photoelectron spectroscopy of the negative ion SeO^-," *J. Chem. Phys.*, vol. 84, pp. 618-625, 1986.

19. C. W. Bauschlicher, S. R. Langhoff, and P. R. Taylor, "Theoretical study of the electron affinities of Cu, Cu_2, and Cu_3," *J. Chem. Phys.*, vol. 88, pp. 1041-1045, 1988.

20. J. C. Weisshaar, "Application of Badger's rule to third row metal diatomics," *J. Chem. Phys.*, vol. 90, pp. 1429-1433, 1989.

21. J. Tellinghuisen, "A Fast Quadrature Method for Computing Diatomic RKR Potential Curves," *Computer Phys. Comm.*, vol. 6, pp. 221-228, 1974.

22. J. Tellinghuisen and S. D. Henderson, "The Use of Morse-RKR Curves in Diatomic Calculations," *Chem. Phys. Lett.*, vol. 91, pp. 447-451, 1982; H. Wang, T.-J. Whang, A. M. Lyyra, L. Li, and W. C. Stwalley, "Study of the 4 $^1\Sigma_g^+$ 'shelf' state of Na_2 by optical-optical double resonance spectroscopy," *J. Chem. Phys.*, vol. 94, pp. 4756-4764, 1991.

23. R. S. Berry, "Elementary Attachment and Detachment Processes," *Adv. Electronics Electron Phys.*, vol. 51, pp. 137-182, 1980; R. S. Berry and S. Leach, "Elementary Attachment and Detachment Processes," *ibid.*, vol. 57, pp. 1-144, 1981; J. H. D. Eland, *Photoelectron Spectroscopy*, Butterworths, London, 1984, pp. 72-82; T. Baer, "Vacuum UV Photophysics and Photoionization Spectroscopy," *Annu. Rev. Phys. Chem.*, vol. 40, pp. 637-669, 1989.

24. Examples include: (a) G. Caprace, J. Delwiche, P. Natalis, and J. E. Collin, "Preionization Effects in Photoelectron Spectra," *Chem. Phys.*, vol. 13, pp. 43-49, 1976; (b) J. H. D. Eland, "Vibrational level populations in the autoionization of oxygen," *J. Chem. Phys.*, vol. 72, pp. 6015-6019, 1980; (c) P. M. Guyon, T. Baer, and I. Nenner, "Interactions between neutral dissociation and ionization continua in N_2O," *ibid.*, vol. 78, pp. 3665-3672, 1983; (d) J. B. Pallix and S. D. Colson, "Direct Observation of Photodissociation Pathways in Ammonia by Multiphoton Ionization Photoelectron Spectroscopy," *J. Phys. Chem.*, vol. 90, pp. 1499-1501, 1986; (e) J. C. Miller and R. N. Compton, "Multiphoton ionization studies of ultracold nitric oxide," *J. Chem. Phys.*, vol. 84, pp. 675-683, 1986; (f) W. A. Chupka, "Photoionization of molecular Rydberg states," *ibid.*, vol. 87, pp. 1488-1498, 1987; (g) M. A. O'Halloran, S. T. Pratt, P. M. Dehmer, and J. L. Dehmer, "Photoionization dynamics of H_2 C $^1\Pi_u$," *ibid.*, vol. 87, pp. 3288-3298, 1987; (h) A. D. Sappey, J. E. Harrington, and J. C. Weisshaar, "Resonant two-photon ionization-photoelectron spectroscopy of Cu_2," *ibid.*, vol. 91, pp. 3854-3868, 1989.

25. G. F. Gantefor, D. M. Cox, and A. Kaldor, "Resonances in the photodetachment cross section of Au_2^-," *J. Chem. Phys.*, vol. 94, pp. 854-858, 1991; K. J. Taylor, C. Jin, J. Conceicao, L.-S. Wang, O. Cheshnovsky, B. R. Johnson, P. J. Nordlander, and R. E. Smalley, "Vibrational Autodetachment Spectroscopy of Au_6^-," *J. Chem. Phys.*, vol. 93, pp. 7515-7518, 1990.

26. K. P. Huber and G. Herzberg, *Molecular Spectra and Molecular Structure*, vol. IV, Van Nostrand Reinhold, New York, 1979.

Dynamics of S production in the 193 nm photodissociation of
CH$_3$SCH$_3$, CH$_3$SSCH$_3$, CH$_3$SH and H$_2$S

C.-L. Liao, C.-W. Hsu, C. Y. Ng

Ames Laboratory,[a] U. S. Department of Energy
and
Department of Chemistry,
Iowa State University
Ames, Iowa 50011

ABSTRACT

The production of S(^{3}P,^{1}D,^{1}S) in the 193 nm photodissociation of CH$_3$SCH$_3$, CH$_3$SSCH$_3$, CH$_3$SH, and H$_2$S has been studied using 2+1 REMPI techniques. The 193 nm photodissociation cross sections of the radicals CH$_3$S and HS initially formed in the photodissociation of CH$_3$SCH$_3$, CH$_3$SSCH$_3$, CH$_3$SH, and H$_2$S have been determined as 1×10^{-18} cm^2 using a rate equation scheme. The dominant product from CH$_3$S is S(^{1}D), while that from SH is S(^{3}P). The formation of S(^{3}P) in the 1-photon (193 nm) photodissociation of CH$_3$SSCH$_3$ is also observed.

1. INTRODUCTION

The study of molecular photodissociation is valuable for elucidating the detailed dynamics of chemical reactions involving a collison complex mechanism. Unlike unimolecular dissociation on the ground electronic potential energy surface, molecular dissociation in excited electronic states often involves adiabatic and/or nonadiabatic interactions with other electronic states. Translational energy and internal state distributions and final product branching ratios can be critically sensitive to the region of the potential energy surface to which the molecule is excited. The experimental techniques applied to the study of molecular photodissociation include crossed laser-molecular beam analysis of product translational energy,[1] emission spectroscopy of photofragments,[2] and laser-induced fluorescence or multiphoton ionization detection of product quantum state distributions.[3]. In the past decade, tremendous progress has been made in understanding the photodissociation dynamics of triatomic molecules. However, the photodissociation study of radicals, especially polyatomic radicals, has remained essentially an unexplored research area. Detailed state-to-state photodissociation cross sections for radicals will provide not only challenges for dynamical calculations, but also for *ab initio* quantum chemical studies.

Recent photofragmentation studies on a series of organosulphur compounds performed in our laboratory have yielded valuable information on the energetics and the photodissociation dynamics of these molecules.[4,5]. Based on the kinetic energy distributions observed for S, the photodissociation studies of CH$_3$SSCH$_3$ and CH$_3$SCH$_3$ suggest that the CH$_3$S photofragment further dissociates by the absorption of a second 193 nm photon to produce S predominantly in the ^{1}D state. In order to further examine this conclusion, we have measured directly the nascent electronic state distributions of S atoms formed in the 193 nm photodissociation of CH$_3$SCH$_3$, CH$_3$SSCH$_3$, CH$_3$SH and H$_2$S using the 2+1 resonance-enhanced multiphoton ionization (REMPI)

---

[a]Operated for the U.S. Department of Energy by Iowa State University under Contract No. W-7405-Eng-82. This work was supported by the Director for Energy Research, Office of Basic Energy Sciences.

method. By calibrating the experimental results to known photodissociation cross sections at 193 nm for the formation of S(^{3}P,^{1}D) from CS$_2$, we have obtained estimates for the photodissociation cross sections for CH$_3$S and HS at 193 nm.

2. EXPERIMENTAL

The experiment basically involved the delayed REMPI detection of S(^{3}P,^{1}D,^{1}S) produced in 193 nm laser photodissociation of selected organosulfur compounds. The schematic of the experimental setup is shown in Fig. 1(a). A home-built TOF mass spectrometer of the two-stage Wiley-McLaren design, which is shown in Fig. 1(b), was used to detect S$^+$ ion. A pulsed beam of CH$_3$SSCH$_3$, CH$_3$SCH$_3$, CH$_3$SH, or H$_2$S was

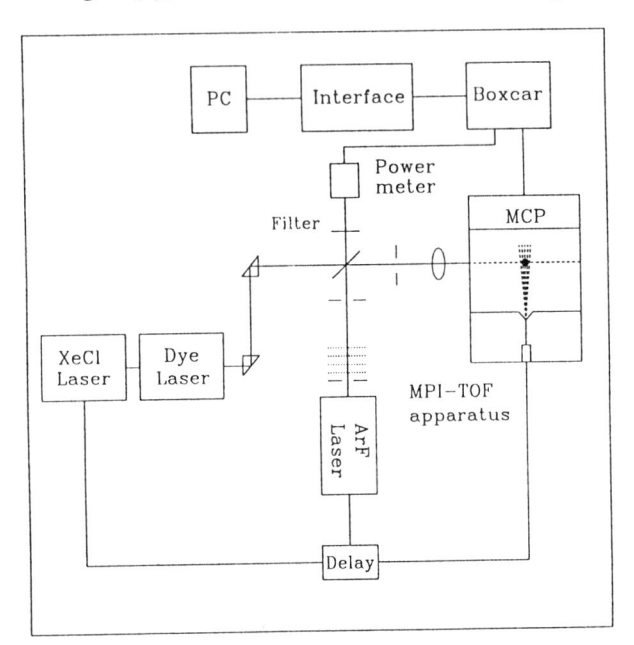

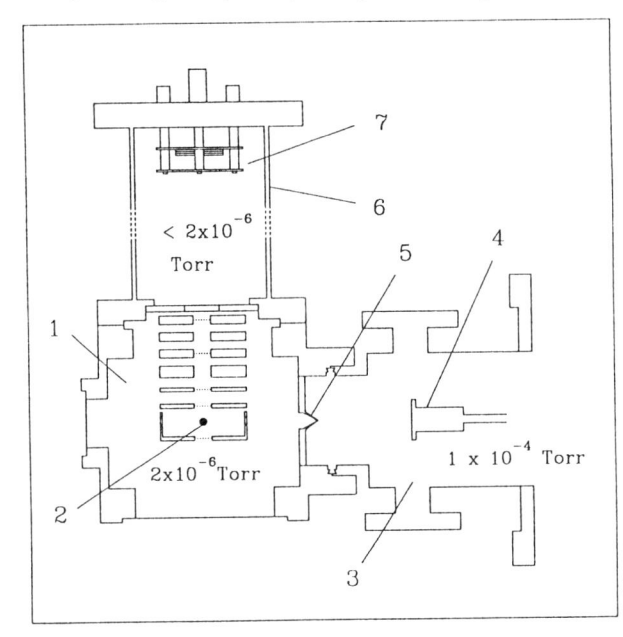

Fig. 1 (a) Schematic diagram of the experimental setup; (b) Cross section view of the TOF mass spectrometer; 1) photodissociation chamber; 2) photodissociation & ionization region; 3) beam source chamber; 4) pulsed valve; 5) skimmer; 6) TOF tube; 7) MCP

produced by supersonic expansion through a commercial pulsed valve (General Valve No. 9, nozzle diameter = 0.5 mm, temperature ≈ 298 K, stagnation pressure ≤ 150 Torr). The molecular beam was skimmed (7 cm from the nozzle) by a conical skimmer (1 mm dia.) and intersected with both the dissociation and ionization lasers 8.5 cm downstream from the skimmer. The molecular beam source chamber was pumped by a liquid-nitrogen trapped 6 in. diffusion pump (pumping speed ≈ 2,000 ℓ/s), while the photodissociation and TOF tube were evacuated by a 50 ℓ/s turbomolecular pump. For a pulsed valve of repetition rate 17 Hz, the beam source and photodissociation chambers were maintained at pressures of ≈ 1 × 10^{-4} and 2 × 10^{-6} Torr, respectively. The ArF photodissociation laser (Questek 2460) was operated in the constant pulse energy mode. The laser beam was attenuated by layers of stainless steel wire mesh and spatially filtered by two irises before being focused by a fused-silica lens (200 mm f.l.) to a spot of ≈ 2x2 mm^2 at the intersection region. The energies used were in the range of 160 to 1000 μJ/pulse.

The ionization detection of S(^{3}P,^{1}D,^{1}S) was accomplished with an excimer laser (Lambda Physik EMG 201 MSC) pumped dye laser (FL 3002) system. Rhodamine 6G and Kiton Red were used to produce the fundamental in the 570-630 nm region. The UV second harmonic was generated using a temperature stabilized (FL-37-1) angle-tuned BBO crystal. The pulse energy, typically 0.2 mJ/pulse, was monitored with a pyroelectric detector (Molectron J3-05). The dye laser beam propagated coaxially with the ArF laser beam into the reaction chamber and intersected with the molecular beam at 90°. The firing of the excimer laser was delayed by 570-800 μs with respect to the triggering pulse for the opening of the pulsed valve. The delay between the two lasers was varied from 30 to 100 ns. Within this delay range, the S ion intensity remained nearly constant. Therefore, a delay of 50 ns between the two lasers was set in all the experiments. Probing the S atomic states was accomplished by 2-photon absorption [$3\ ^3P_J \to 4\ ^3P_{J'}$ (308-311 nm) and $3\ ^1D_2 \to 4\ ^1F_3$ (288.19 nm)] followed by absorption of a third photon to produce S^+ in the $^4S°$ and $^2D°$ states.

The firing sequence of the pulsed valve and the two lasers was controlled by two digital delay units (Stanford Research, Model DG 535). The ion signal from the microchannel plate and the dye laser signal from pyroelectric detector were fed into two identical boxcar integrators (Stanford Research, SR-250), which were interfaced to an IBM AT computer. All chemical samples were obtained from Aldrich. The liquid samples, CH_3SCH_3, CH_3SSCH_3 and CS_2 (99% pure), were degassed by a series of freeze- pump-thaw cycles. The gaseous samples, CH_3SH and H_2S (99.9% pure), were used without further purification.

3. RESULTS AND DISCUSSION

The fine-structure distributions of $S(^3P_J)$ were measured by summation over the upperstate fine-structure levels ($4^3P_{J'}$).[6] The fine-structure distributions of $S(^3P_J)$ generated by the 193 nm photodissociation of CS_2 were determined to be $^3P_2 : ^3P_1 : ^3P_0 = 0.71 \pm 0.02 : 0.20 \pm 0.02 : 0.09 \pm 0.04$, in excellent agreement with the result obtained by Waller and Hepburn[7] using the vacuum ultraviolet (VUV) laser-induced fluorescence technique. Thus, the overall branching ratio $S(^3P)/S(^1D) = 2.78$ for CS_2, measured previously by VUV laser-induced fluorescence[7] and time-of-flight (TOF) mass spectrometric[4a] methods, was used as a standard to determine the $^3P/^1D$ ratio for the individual sulfur compounds in this study. In the analysis of our experiments, a rate equation scheme[8] was used to interpret the excimer laser (ArF) power dependence observed for $S(^3P, ^1D)$.

3.1 CH_3SCH_3

Upon the absorption of 193 nm photons, CH_3SCH_3 may dissociate to produce S according to processes (1) and (2):

$$CH_3SCH_3 + h\nu\ (193\ nm) \xrightarrow{\sigma_1} CH_3 + SCH_3 \qquad (1)$$

$$SCH_3 + h\nu\ (193\ nm) \xrightarrow{\sigma_2} S + CH_3 \qquad (2)$$

where σ_i are the photodissociation cross sections at 193 nm. The TOF mass spectrometric study on this molecule supports this stepwise mechanism for the production of S atoms.

The fine-structure distribution has been determined to be $^3P_2 : ^3P_1 : ^3P_0 = 0.59 : 0.32 : 0.09$, which is close to the statistical distribution $5/9 : 3/9 : 1/9$. The S^+ intensities corresponding to the formation of $S(^3P_2)$ and $S(^1D)$ are plotted as a function of the photodissociation laser power in Figs. 2(a) and 2(b). The similarity of the power dependences for $S(^3P)$ and $S(^1D)$ is consistent with the conclusion that both $S(^3P)$ and $S(^1D)$ originate from the same precursor molecule, i.e. CH_3S. The overall branching ratio $^3P/^1D$ is 0.15/0.85. Using this branching ratio and $\sigma_1 = 1 \times 10^{-17}\ cm^2$ for the primary photodissociation cross sections[9], the best-fitted photodissociation cross section for process (2) is determined to be $\sigma_2 = 1 \times 10^{-18}\ cm^2$. The dominance of the production of $S(^1D)$ from the photodissociation of CH_3S suggests that the dissociation of CH_3S proceeds on the $\widetilde{D}\,^2A'$ potential surface.[4b] The previous photodissociation study of ethylene sulfide, $(CH_2)_2S$, has shown that S atoms are produced exclusively in the ^1D state.[10]

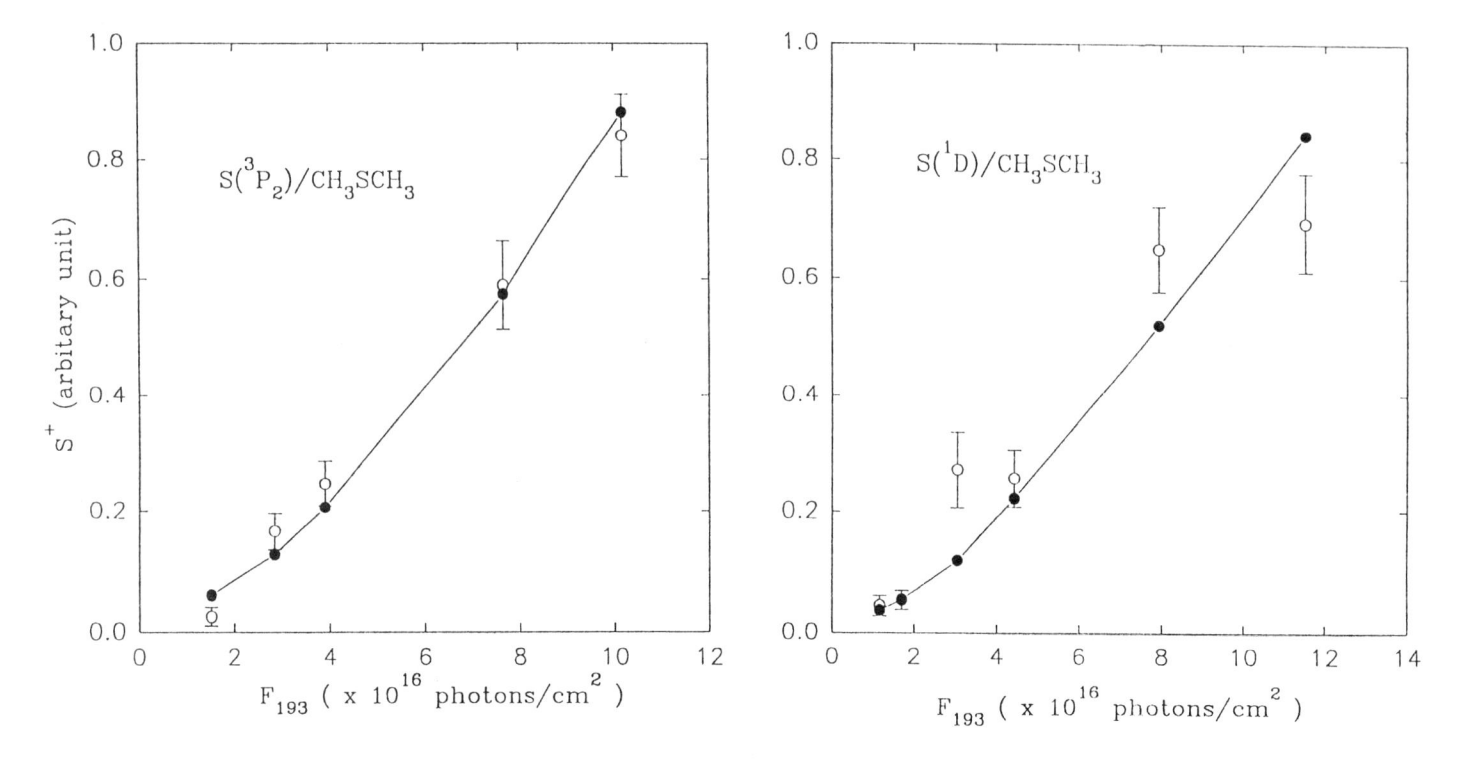

Fig. 2 Dissociation laser power dependence of (a) $S(^3P_2)$ and (b) $S(^1D)$ from CH_3SCH_3.
(○) Experimental data; (●) best-fitted values (total S^+ signal).

3.2 CH$_3$SSCH$_3$

The dissociation scheme considered for the production of S upon the absorption of 193 nm photons is:

$$CH_3SSCH_3 + h\nu\,(193\text{ nm}) \xrightarrow{\sigma 3} CH_3S + SCH_3, \quad (3)$$

$$CH_3S + h\nu\,(193\text{ nm}) \xrightarrow{\sigma 4} CH_3 + S, \quad (4)$$

$$CH_3SSCH_3 + h\nu\,(193\text{ nm}) \xrightarrow{\sigma 5} CH_3 + SSCH_3, \quad (5)$$

$$CH_3SS + h\nu\,(193\text{ nm}) \xrightarrow{\sigma 6} CH_3S + S, \quad (6)$$

$$CH_3SSCH_3 + h\nu\,(193\text{ nm}) \xrightarrow{\sigma 7} CH_3 + S + SCH_3. \quad (7)$$

Sulfur atoms produced through processes (3) and (4) or (5) and (6) involve a two-photon stepwise mechanism, while S atoms generated by process (7) involve a one-photon process.

The fine-structure distribution of $S(^3P_J)$ formed in the 193 nm photodissociation of CH_3SSCH_3 has been measured to be $^3P_2 : {}^3P_1 : {}^3P_0 = 0.77 : 0.17 : 0.06$, which is significantly different from that of $S(^3P_J)$

generated from CH$_3$SCH$_3$. We expect the fine-structure distribution of S(^{3}P$_J$) from CH$_3$SSCH$_3$ to be similar to the case of CH$_3$SCH$_3$ if S atoms were only produced via processes (3) and (4). The overall ^{3}P/^{1}D branching ratio has been found to depend on the dissociation laser power. The fraction of S(^{3}P) decreases monotonically from 0.97 to 0.86 as the photon flux of the 193 nm laser increases from 1x10^{16} to 1.2x10^{17} photons/cm^2. This observation suggests that at low photon flux, process (7) is an important channel for the production of S(^{3}P). We note that, due to energy constraint, S formed by process (7) can only be in the ground ^{3}P state. Formation of SSCH$_3$ by process (5) was observed in the previous TOF mass spectrometric study.[4b] Further photodissociation of SSCH$_3$ to form S and SCH$_3$ [process (6)] is a viable pathway.

Using the literature[10] total absorption cross section of 4.8x10^{-17} cm^2 for CH$_3$SSCH$_3$ at 193 nm (i.e., σ_3 + σ_5 + σ_7) and the dissociation cross section of CH$_3$S σ_4 = 1x10^{-18} cm^2, determined in the photodissociation experiment of CH$_3$SCH$_3$, we obtain the set of best-fitted σ-values to the rate equation scheme as σ_3 = 8x10^{-18} cm^2, σ_5 = 2.2x10^{-17} cm^2, σ_6 = 3.2x10^{-17} cm^2 and σ_7 = 1.8x10^{-17} cm^2. The contributions of S from processes (4), (6) and (7) are indicated in Fig. 3(a) and 3(b). At low photon flux, the S formation is dominated by the 1-photon dissociation [process (7)] yielding sulfur atoms solely in the ground S(^{3}P) state. At higher photon flux, process (6) also contributes significantly to the formation of S(^{3}P). These results indicate that S(^{1}D) atoms are produced mainly by processes (4) and (6). At low photon flux, the formation of S(^{1}D) is dominated by process (6), but the production of S(^{1}D) by process (4) becomes more important than that by process (6) at high photon flux. The rate equation scheme also predicts the S(^{3}P)/S(^{1}D) branching ratio for the dissociation of SSCH$_3$ radical [process (6)] to be 0.9/0.1.

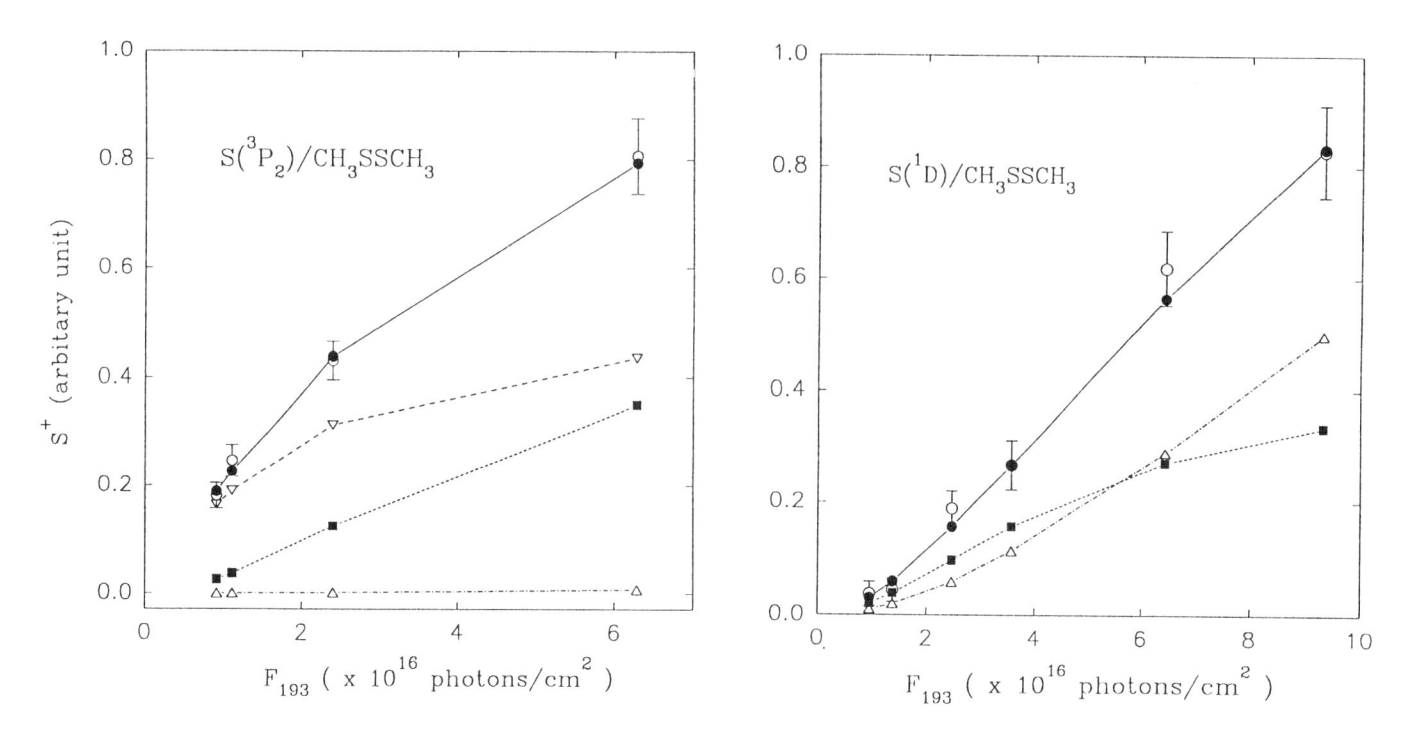

Fig. 3 Dissociation laser power dependence of (a) S(^{3}P$_2$) and (b) S(^{1}D) from CH$_3$SSCH$_3$. (○) Experimental data; (●) best-fitted values (total S$^+$ signal); (△) process (4); (■) process (6); (▽) process (7).

Our observation that the photodissociation of CH$_3$S produces S predominantly in the ^{1}D state is consistent with the conclusion of the previous TOF mass spectrometric study[4(b),4(c)] based on the kinetic energy release measurement of S.

3.3. H₂S

As a model triatomic molecular system for detailed understanding of photodissociation dynamics, the UV photodissociation of H$_2$S has been the subject of considerable study in recent years. The photodissociation of H$_2$S at 193 nm produces HS radicals with up to six vibrational quanta. Secondary photodissociation of HS radicals to produce S(^{3}P) and S(^{1}D) has been observed by H-atom photofragment-translational spectroscopy.[11,12] The MPI study of Steadman and Baer[13] indicates that H$_2$S may dissociate to form S + H$_2$ by two-photon absorption in the wavelength range of 285-297 nm. Thus we consider the scheme for the production of S from the 193 nm photodissociation of H$_2$S to involve processes (8)-(10).

$$H_2S + h\nu \text{ (193 nm)} \xrightarrow{\sigma_8} H + HS \quad (8)$$

$$HS + h\nu \text{ (193 nm)} \xrightarrow{\sigma_9} H + S \quad (9)$$

$$H_2S + h\nu \text{ (193 nm)} \xrightarrow{\sigma_{10}} H_2 + S \quad (10)$$

The fine-structure distribution of S(^{3}P$_J$) has been determined to be $^3P_2 : {}^3P_1 : {}^3P_0 = 0.67 : 0.25 : 0.08$. The power dependences of S(^{3}P$_2$) and S(^{1}D) are shown in Fig. 4. The slope in S(^{3}P$_2$) increases gradually as the

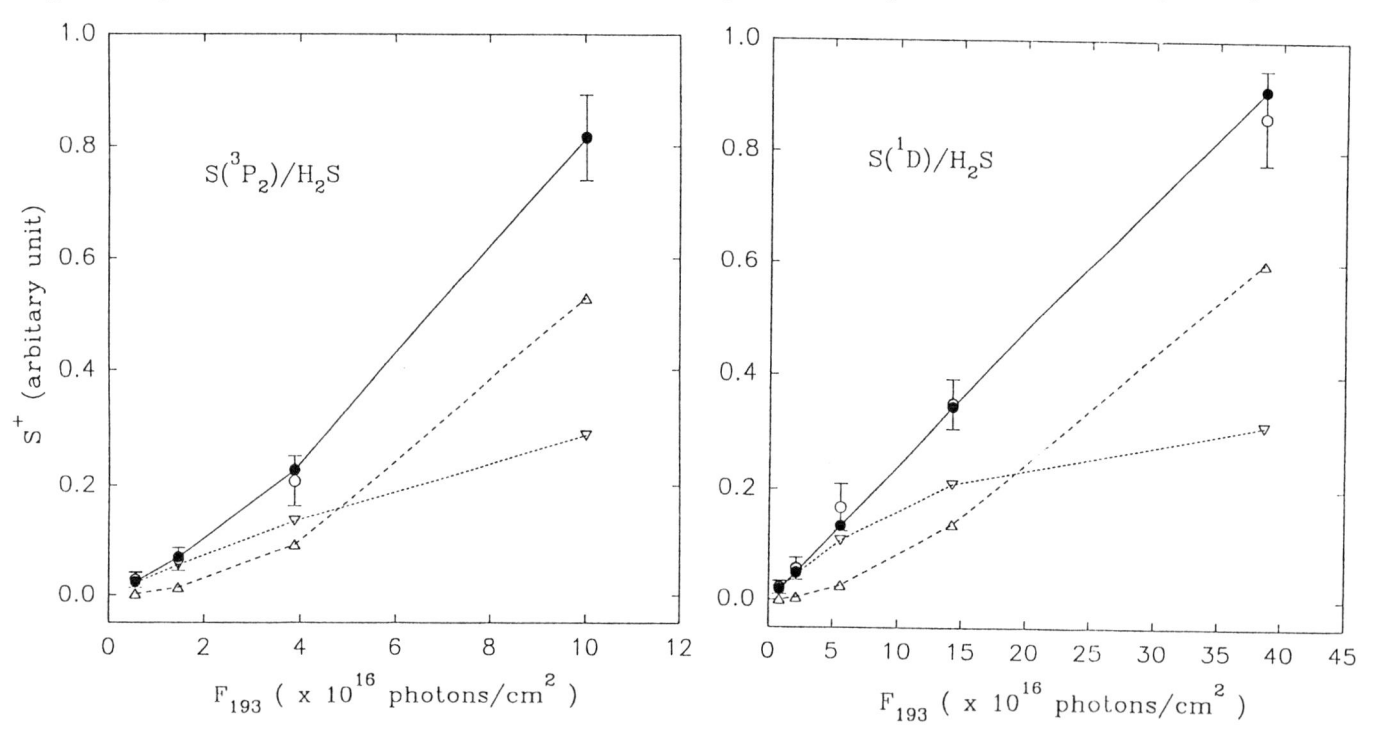

Fig. 4 Dissociation laser power dependence of (a) S(^{3}P$_2$) and (b) S(^{1}D) from H$_2$S. (○) Experimental data; (●) best-fitted values (total S$^+$ signal); (△) Process (9); (▽) process (10).

photon flux increases. In contrast, the S(^{1}D) data show the opposite trend. The fraction of S(^{3}P) increases from 0.6 to 0.8 as the photon flux is increased in the range 0.5-40×10^{16} photons/cm^2. The best fit of the data yields the following cross sections: $\sigma_8 = 6.5 \times 10^{-18}$ cm^2, $\sigma_9 = 1.1 \times 10^{-18}$ cm^2, and $\sigma_{10} = 3 \times 10^{-19}$ cm^2. The ^{3}P/^{1}D branching ratio for the process (9) is 0.86/0.14, in excellent agreement with the result observed in the H-atom translational spectroscopic study.[11] The branching ratio ^{3}P/^{1}D = 0.6/0.4 is found for process (10).

In order to gain insight about the photodissociation of HS [process (9)], the potential curves accessible by dipole-allowed transitions are shown in Fig. 5. If dissociations of HS radicals via the excited $^2\Sigma^-$ and $^2\Delta$ states are direct, the branching ratio $^3P/^1D$ should reflect the vibrational distributions of the ground state $HS(X^2\Pi_i)$ initially formed by process (8). The energy of a 193 nm photon can pump $HS(X^2\Pi_i, v=0,1)$ to the $^2\Sigma^-$ excited state which dissociates asymptotically to $S(^3P) + H(^2S)$. However the radicals $HS(X^2\Pi_i, v=2,3,4)$ will be excited to the $^2\Delta$ state, which leads to the formation of $S(^1D) + H(^2S)$. The ratio of the population in $HS(X,v=0,1)$ to that in $HS(X,v=2,3,4)$ is observed to be 0.784/0.128[11,12], in good accord with the $^3P/^1D$ branching ratio determined here for process (9).

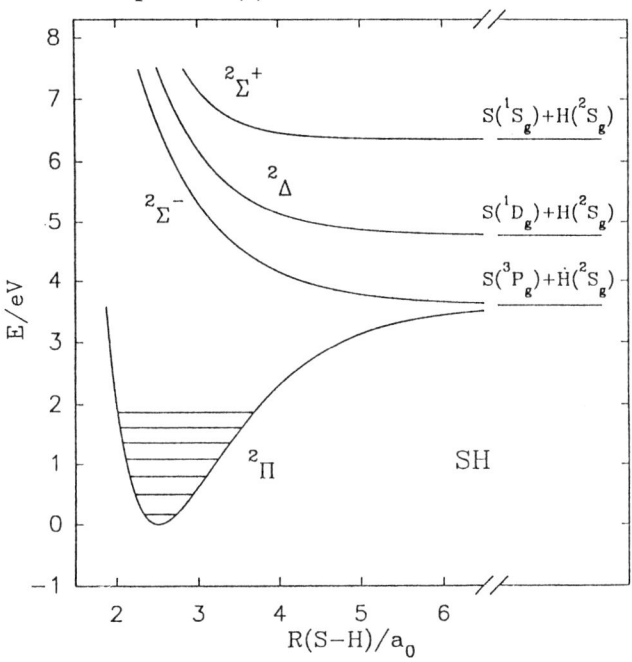

Fig. 5 Potential energy states involved in the 193 nm photodissociation of HS.

3.4 CH₃SH

The production of S atoms by 193 nm photodissociation of CH₃SH is considered to follow a scheme which includes processes (11), (12), (13), (14) and (15).

$$CH_3SH + h\nu\ (193\ nm) \xrightarrow{\sigma_{11}} CH_3S + H \qquad (11)$$

$$CH_3S + h\nu\ (193\ nm) \xrightarrow{\sigma_{12}} CH_3 + S \qquad (12)$$

$$CH_3SH + h\nu\ (193\ nm) \xrightarrow{\sigma_{13}} CH_3 + SH \qquad (13)$$

$$SH + h\nu\ (193\ nm) \xrightarrow{\sigma_{14}} H + S \qquad (14)$$

$$CH_3SH + h\nu\ (193\ nm) \xrightarrow{\sigma_{15}} CH_4 + S \qquad (15)$$

Process (15) represents the possible production of S by one-photon absorption. The quantum yield for process (11) (or (13)) has been reported to be 0.79 (0.21) and 0.93 (0.07) at 185 and 254 nm, respectively.[14] Thus the ratio $\sigma_{11}/\sigma_{13} = 4/1$ is used in the model fitting of CH₃SH photodissociation at 193 nm.

The fine-structure distribution for S(^3P_J) has been measured to be $^3P_2 : {}^3P_1 : {}^3P_0 = 0.66 : 0.24 : 0.10$ at high photon flux. Figures 6(a) and 6(b) depict the signals for S(^3P) and S(^1D) formed in the photodissociation of CH_3SH as a function of the ArF laser power. Using the cross sections derived here for

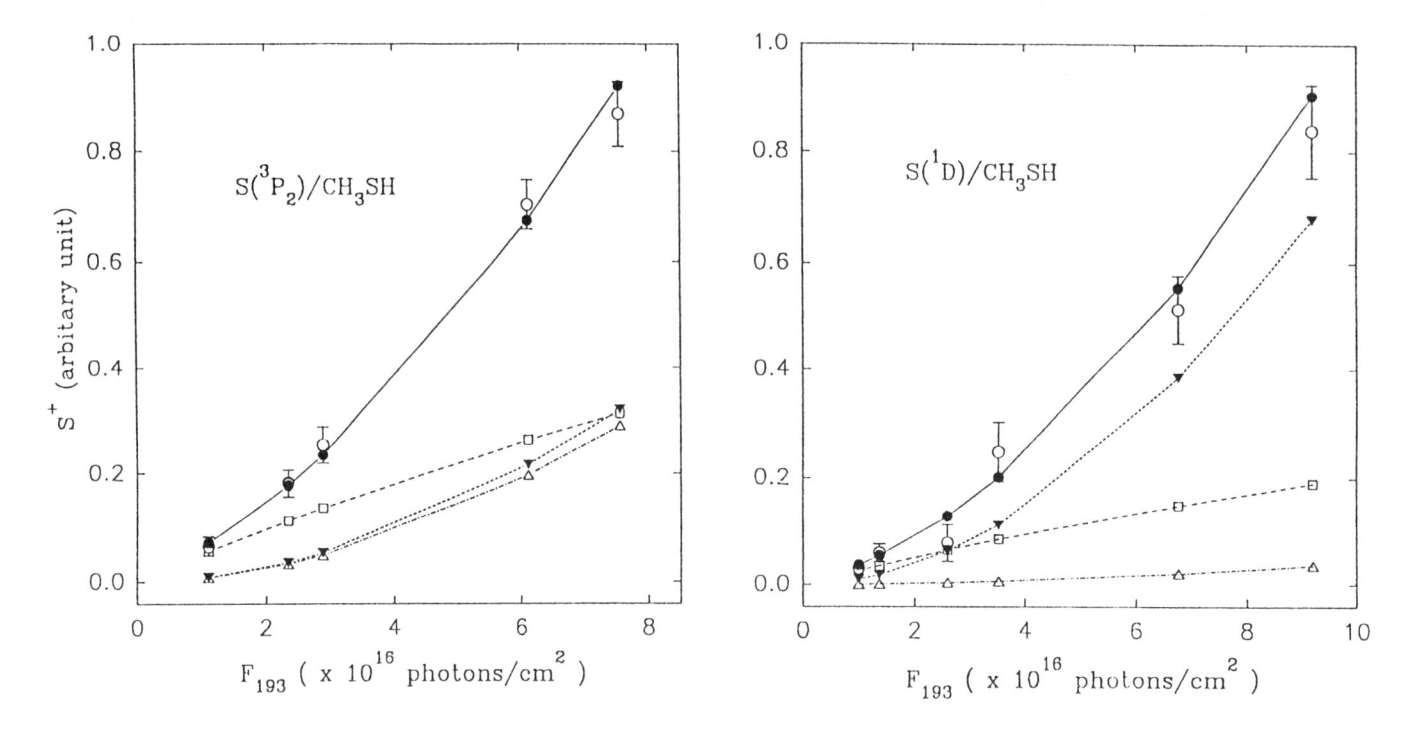

Fig. 6 Dissociation laser power dependence of (a) S($^3P_2$) and (b) S(^1D) from CH_3SH. (○) Experimental data; (●) best-fitted values (total S$^+$ signal); (△) Process (14); (▼) process (12); (□) process (15)

the photodissociation of CH_3S and HS radicals, the set of best-fitted cross sections are: $\sigma_{11} = 3.5 \times 10^{-18}$ cm^2, $\sigma_{13} = 8.8 \times 10^{-19}$ cm^2, and $\sigma_{15} = 6.8 \times 10^{-20}$ cm^2. The branching ratios $^3P/^1D$ due to the secondary processes (12) and (14) are 0.26/0.74 and 0.86/0.14 respectively. As shown in Fig. 6(b), the photodissociation process $CH_3S \rightarrow S(^1D) + CH_3$ completely dominates the S(^1D) production at high laser fluence. It is interesting to note that processes (12), (14) and (15) contribute almost equally to the formation of S(^3P) at high laser fluence. We also observe the formation of S(^1S) atoms at laser wavelength 299.575 nm. The intensity for S(^1S) relative to those for S($^3P, {}^1D$) cannot be quantified because S(^1S) is not produced in the calibration reaction $CS_2 \rightarrow S + CS$. The formation of S(^1S) is likely if HS($^2\Pi, v \geq 5$) radicals are formed in process (13). The previous TOF photofragmentation studies[5,14] of CH_3SH and H_2S indicate that about 30% and 5% of the available energy appears as product internal energy. The formation of S(^1S) may result from the population of the excited HS($^2\Sigma^+$) state.

$$HS(^2\Pi_i, \geq 5) + h\nu \text{ (193 nm)} \rightarrow HS(^2\Sigma^+) \tag{16}$$

$$HS(^2\Sigma^+) \rightarrow S(^1S) + H \tag{17}$$

Therefore the detection of S(^1S) is consistent with the conclusion that HS(X,v $\geq$ 5) radicals are formed in the 193 nm photodissociation of CH_3SH.

4. CONCLUSIONS

We have measured the fine-structure distributions of $S(^3P_J)$ atoms produced by the photodissociation of CH_3SCH_3, CH_3SSCH_3, H_2S and CH_3SH at 193 nm. The branching ratio $^3P/^1D$ for each molecule has been determined by calibrating to that for CS_2. The dominant channels for the photodissociation of CH_3S and HS are

$$CH_3S + h\nu(193\ nm) \rightarrow S(^1D) + CH_3,$$

$$HS + h\nu(193\ nm) \rightarrow S(^3P) + H.$$

A rate equation scheme has been used to analyze the formation of $S(^3P, ^1D)$ as a function of the ArF laser power. The analysis yields the absolute photodissociation cross sections for CH_3S and HS at 193 nm.

5. REFERENCES

1. G. E. Busch, J. F. Cornelius, R. T. Mahoney, R. I. Morse, D. W. Schlosser, and K. R. Wilson, *Rev. Sci. Instrum.*, Vol. 41, 1066 (1970); G. E. Busch and K. R. Wilson, *J. Chem. Phys.*, Vol. 56, 3626, 3638, 3655 (1972); R. K. Sander and K. R. Wilson, *J. Chem. Phys.*, Vol 63, p. 4252, 1975; R. J. Oldman, R. K. Sander, and K. R. Wilson, *J. Chem. Phys.*, Vol. 63, p. 4252, 1975.

2. M. D. Person, K. Q. Lao, B. J. Eckholm, and L. J. Butler, *J. Chem. Phys.*, Vol. 91, p. 812, 1989.

3. See, for example, extensive references cited in "Dynamics of Molecular photofragmentation" *Faraday Discussions of the Chemical Society.* No. 82, 1986.

4a. W.-B. Tzeng, H.-M. Yin, W.-Y. Leung, J.-Y. Luo, S. Nourbakhsh, G. D. Flesch, and C. Y. Ng, *J. Chem. Phys.*, Vol. 88, p. 1658, 1988.

  b. S. Nourbaksh, C.-L. Liao, and C. Y. Ng, *J. Chem. Phys.*, Vol. 92, p. 6587, 1989.

  c. S. Nourbaksh, K. Norwood, H.-M. Yin, C.-L. Liao and C. Y. Ng, *J. Chem. Phys.*, Vol. 95, p. 5014, 1991.

5. S. Nourbaksh, K. Norwood, H.-M. Yin, C.-L. Liao and C. Y. Ng, *J. Chem. Phys.*, Vol. 95, p. 946, 1991.

6. D. J. Bamford, M. J. Dye, and W. K. Bishel, *Phys. Rev.*, Vol. A36, p. 3497, 1987.

7. I. M. Waller and J. W. Hepburn, *J. Chem. Phys.*, Vol. 87, p. 3261, 1987.

8. C.-W. Hsu, C.-L. Liao, and C. Y. Ng, to be published.

9. J. G. Calvert and J. N. Pitts, Jr., *Photochemistry*, John Wiley and Sons, Inc., 1966.

10. H. L. Kim, S. Satyapal, P. Brewer, and R. Bersohn, *J. Chem. Phys.*, Vol. 91, p. 1047, 1989.

11. R. E. Continett, B. A. Balko, and Y. T. Lee, *Chem. Phys. Lett.*, Vol. 182, p. 400, 1991.

12a. G. N. A. Van Veen, K. A. Mohamed, T. Baller, and A. E. DeVries, *Chem. Phys.*, Vol. 74, p. 261, 1983.

  b. Z. Xu, B. Koplitz and C. Wittig, *J. Chem. Phys.*, Vol. 87, p. 1062, 1987.

  c. X. Xie, L. Schnieder, H. Wallmeier, R. Boettner, K. H. Welge and M. N. R. Ashfold, *J. Chem. Phys.*, Vol. 92, p. 1608, 1990.

13a. J. Steadman and T. Baer, *J. Chem. Phys.*, Vol. 89, p. 5507, 1988.

  b. J. Steadman and T. Baer, *J. Chem. Phys.*, Vol. 91, p. 6113, 1989.

14. A. B. Callear and D. R. Dickson, *Trans. Faraday Soc.*, Vol. 66, p. 1987, 1970.

Electronic spectroscopy of ionic clusters

E.J. Bieske, A.M. Soliva, A. Friedmann and J.P. Maier

Institut für Physikalische Chemie, Universität Basel
Klingelbergstrasse 80, CH-4056 Basel, Switzerland

ABSTRACT

The electronic transitions of the ionic clusters N_2^+-He_n (n=1,2,3), N_2^+-Ne, and N_2O^+-Ne have been detected. The cluster ions, produced by electron impact of a pulsed, seeded supersonic expansion, are mass-selected and injected into an octopole where the electronic transition is induced by a tunable, pulsed laser. The electronic spectra are recorded by detecting the yield of fragment ions, produced by vibrational predissociation within the octopole, in a further quadrupole mass-selector, as a function of the laser wavelength. The spectra show rotational and/or vibrational structure, which has been analysed to provide information on the structure, binding energies and vibrational frequencies of these species.

1. INTRODUCTION

Although current literature is abundant with articles on the generation and reactivity of ionic clusters, their spectroscopic characterization is by and large outstanding. The direct spectroscopic approaches in this direction have involved infrared multiphoton absorption-dissociation studies on mass-selected ions,[1] photodissociation of transition metal/raregas dimer ions,[2] and laser-induced fluorescence[3] and photodissociation[4] measurements on fluorobenzene-rare gas ions. Information on structures has been inferred indirectly from collision-induced fragmentation patterns,[5] ion-molecule reactions[6] or photodissociation measurements with product ion energy distribution.[7] We have been developing an approach to study the electronic spectra of mass-selected elementary cluster ions and in this article the first results on N_2^+-He_n (n=1,2,3),[8,9] N_2^+-Ne,[10] and N_2O^+-Ne,[11] are briefly overviewed. Our method involves exciting discrete vibronic transitions in clusters which lead either directly to vibrational predissociation on the excited state surface or on lower potential energy surfaces after fluorescence or internal conversion. Dissociation renders facile the observation of electronic transitions of ionic complexes using mass spectroscopic techniques.

Ionic clusters have binding energies intermediate between van der Waals molecules and chemical bonds. Thus, for example, the calculated binding energy of N_2^+-He is ≈ 0.02 eV,[12] whereas N_2^+-Ar is bound by ≈ 1 eV.[13] For our initial studies we chose N_2^+-He because: (1) the system had been studied theoretically,[12] the main results being that for the weakly bound complex the He atom sits 295 pm from the N_2^+ center of mass but that there is essentially no orientational preference of the He atom, (2) due to the minor perturbation by the He the B-X electronic transition of N_2^+-He should be found close to that of N_2^+ (B $^2\Sigma_u^+$-X $^2\Sigma_g^+$), (3) post excitation vibrational predissociation to produce N_2^+ was likely and would be slow enough that appreciable spectroscopic broadening would not occur.

2. APPROACH

In the experiment itself, the first obstacle to be surmounted is the ability to produce a weakly bound ionic cluster such as N_2^+-He. This was accomplished by the construction of a special ion source,[10] shown in fig. 1, which involves a pulsed supersonic expansion of N_2 in He (≈ 1:100, backing pressure a few bar, nozzle diameter 800 µm) crossed by ≈ 200 eV electrons.

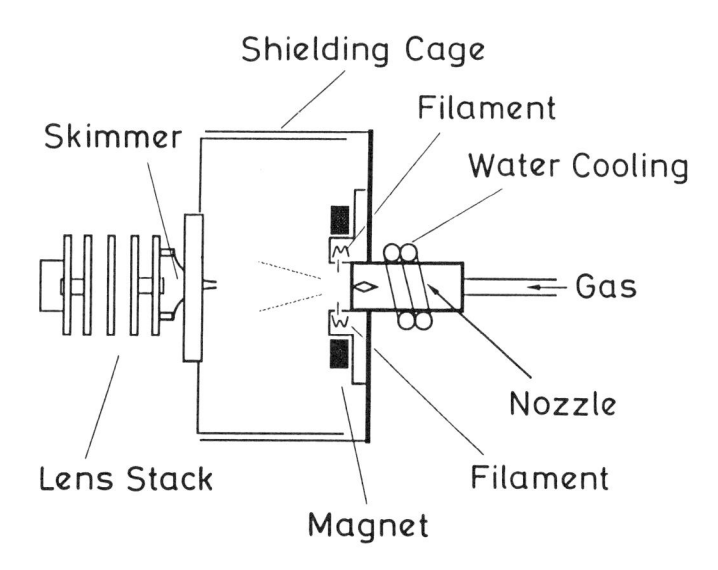

Fig. 1. Electron impact ion source used for the production of cooled ionic clusters. A pulsed supersonic expansion is crossed by 200 eV electrons to produce ions. The ions are extracted through the skimmer by a mild electric field (< 1V cm) and are focussed into the first mass selecting quadrupole by a series of lenses.

The cooled (≈20K) N_2^+-He cluster ions are produced in the early part of the expansion either through clustering of a He atom on an N_2^+ core or possibly by associative ionization of electronically excited He atoms with neutral N_2 molecules. The entire ion source is in an electrically well-shielded region from which ions are gently extracted and focused into the apparatus.

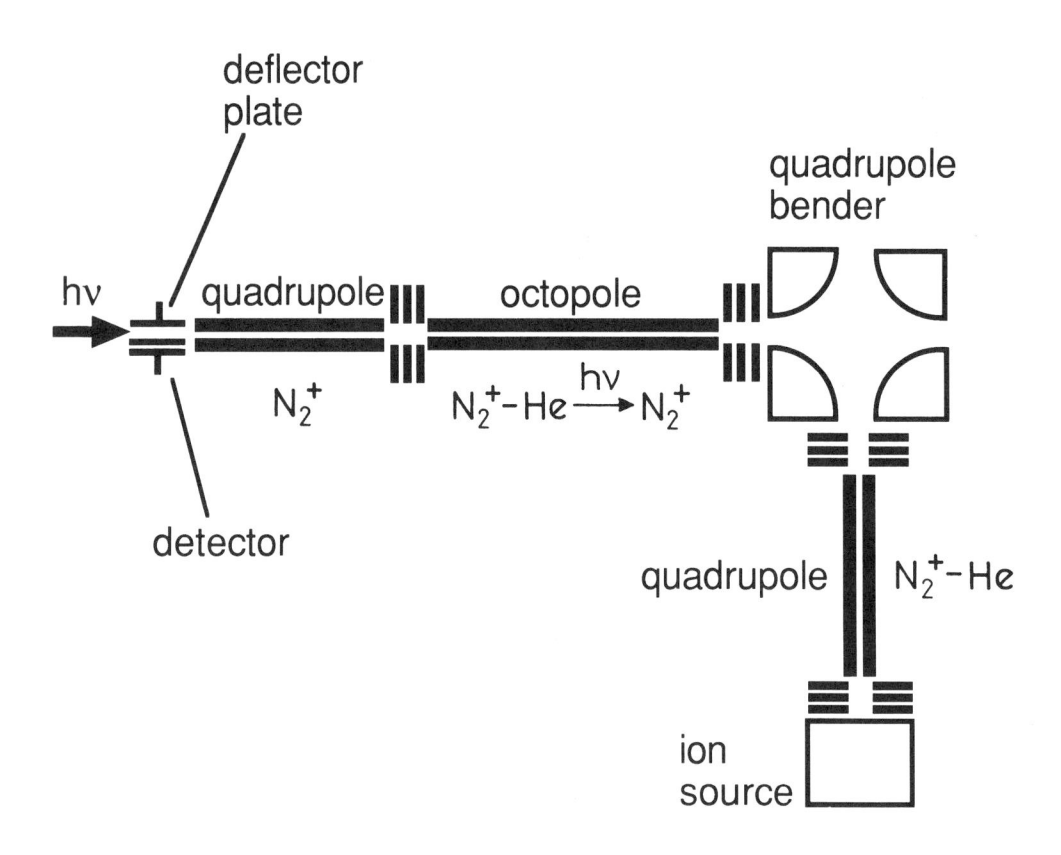

Fig. 2. Schematic of the quadrupole-octopole-quadrupole apparatus. Cluster ions produced in the source are mass selected by the first quadrupole before being deflected by the quadrupole bender into the octopole ion guide. The third quadrupole is tuned to transmit fragment N_2^+ ions which are subsequently detected by dual microchannel plates. Ion lenses are represented by broken triple bars.

In our approach, as depicted schematically in fig.2, the unambiguous assignment of the induced transition to a particular cluster species is guaranteed by mass selection of the the relevant cluster. The ions produced in the source are passed into a mass-selecting quadrupole (Q_1), which is set to transmit only N_2^+-He ions. Thus the dominant N_2^+ ions, along with other undesired cluster species, are immediately eliminated from the measurements. The N_2^+-He ions are then deflected 90^0 by a quadrupole bender and injected into an octopole ion guide. A pulsed, tunable, dye laser is propagated coaxially down the middle of the octopole, and induces the electronic transition. After some time the photoexcited N_2^+-He ions undergo vibrational predissociation; the thus generated N_2^+ ions continue through the octopole and into a second mass-selecting quadrupole (Q_2), which is set to transmit only the N_2^+ product ions. The electronic transition is consequently mapped by scanning the wavelength of the dye-laser while the intensity of the N_2^+ ions is monitored.

3. ELECTRONIC SPECTRA OF N_2^+-He$_n$

In fig.3 is reproduced the observed electronic spectrum of N_2^+-He$_n$ (n=1,2,3) along with a simulation of the B $^2\Sigma_u^+$-X $^2\Sigma_g^+$ transition of N_2^+. The resolved structure of the N_2^+-He spectrum bears a good deal of resemblance to the rotational structure of N_2^+ at a rotational temperature of 20-30K. Of particular note is the 2:1 intensity alternation, a consequence of the equivalence of the N atoms in the complex and the nuclear spin of ^{14}N (I=1). When analogous measurements are carried out on ^{15}N$_2^+$-He, the spectrum shown in fig.4 is observed, revealing instead a 1:3 intensity pattern due to the I=1/2 nuclear spin of ^{15}N$_2$. Either a 'T' shaped structure or a free internal rotor structure where the He does not have a well defined angular location, are consistent with equivalent N atoms. Simulation of the band profile excludes the possibility that the structure of the N_2^+-He cluster is T-shaped; the conclusion is that there is free internal rotation of the N_2^+ moiety in both the X and B states. This is in accord with *ab initio* and rovibrational calculations that have been undertaken for this complex in which the zero point energy was found to be above the barrier for internal rotation.[12] The band profile of N_2^+-He thus arises chiefly from $\Delta j=\pm 1$ changes in the quantum number pertaining to the internal rotation of the N_2^+ core.

The band origin of N_2^+-He is shifted by merely $\approx 1 cm^{-1}$ from that of N_2^+ indicating that the binding energies in the X and B states are almost the same, consistent with the absence of any discernible progression in the He----N_2^+ stretching vibration in the spectrum. Furthermore, from the breaking off of the internal rotational struture in the "R-branch" (at j =7), a lower estimate for the dissociation energy of N_2^+-He of 100-130 cm^{-1} may be inferred. This may be compared with the calculated value of $\approx$ 98 cm $^{-1}$.[12] How the vibrational predissociation operates is not quite clear. It appears, however, that fluorescence from the electronically excited B state back to vibrationally excited levels of the ground state X does not take place, rather an internal conversion process involving the intermediate A state occurs.[9]

For the species with n=2 and 3 the spectra (fig.3) show similar features to that of n=1, though a broadening of the structure is apparent with increasing n.

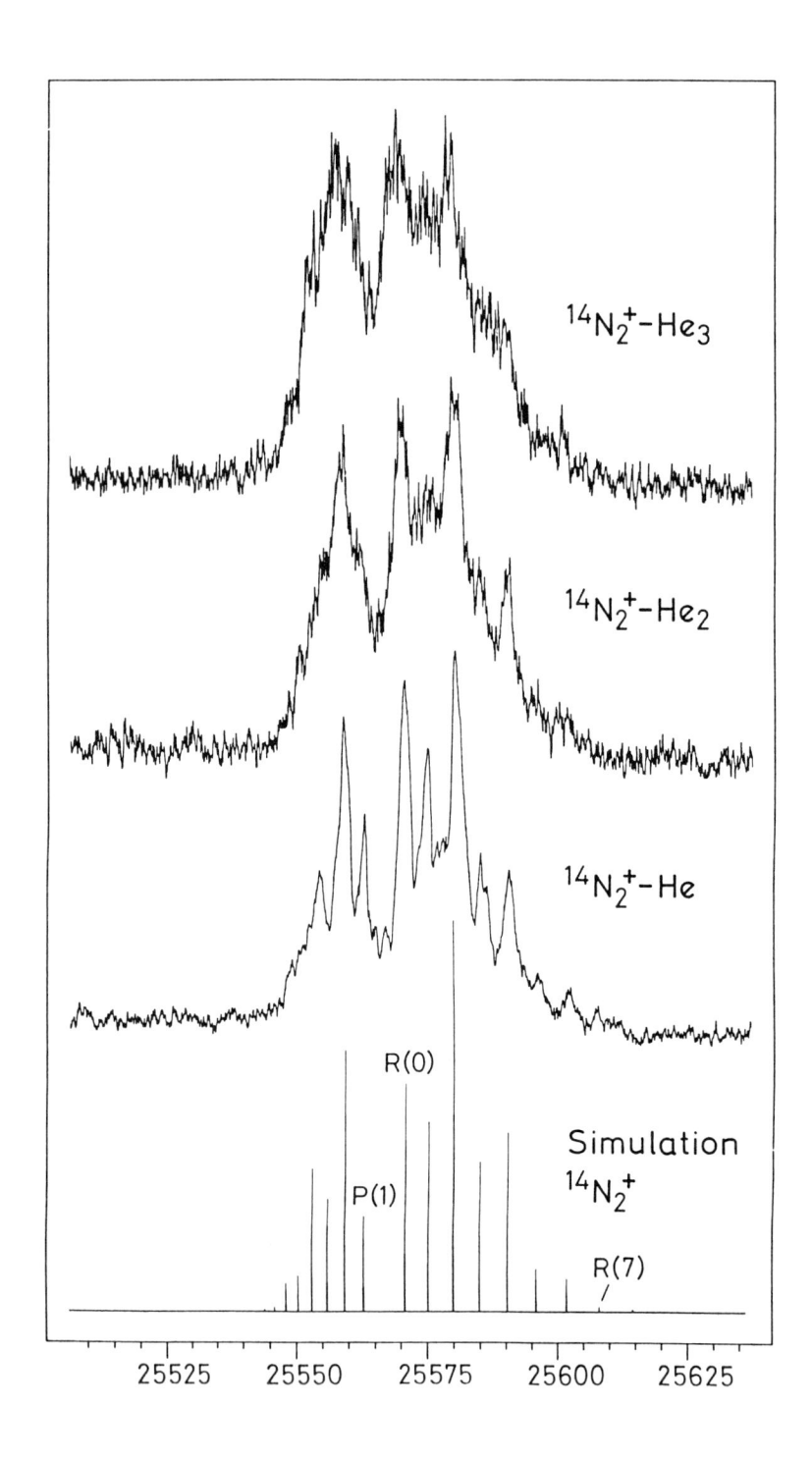

Fig. 3. $^{14}N_2^+$-He, $^{14}N_2^+$-He$_2$, and $^{14}N_2^+$-He$_3$ B←X photodissociation spectra and a simulated $^{14}N_2^+$ spectrum in the 391 nm region. Peaks labelled P(1) and R(0) correspond to the P(1.5) and R(0.5) transitions of free $^{14}N_2^+$.

This similarity supports the notion that free internal rotation persists with several helium atoms attached. In fact similar structure persists for clusters with up to six He atoms, the largest cluster for which we have recorded spectra. It also can be seen that as for the n=1 case, the 'R-branch' extends to the same amount for n≥2, suggesting similar binding energies for the additional He atoms ($\approx 100 cm^{-1}$). This seems reasonable if one considers that the binding of He atoms should be dominated by interactions with the N_2^+ core, with He-He interactions making a minor contribution to the cohesion of the clusters.

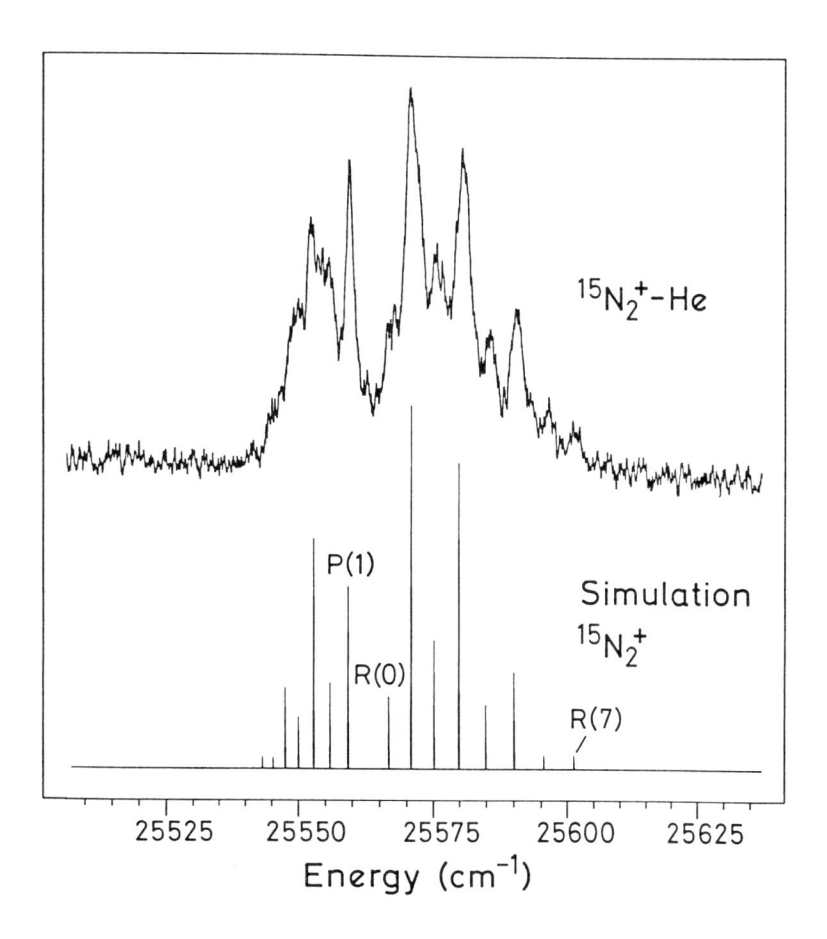

Fig. 4. $^{15}N_2^+$-He B←X photodissociation spectrum and simulated $^{15}N_2^+$ spectrum in the 391 nm region. Peaks labelled P(1) and R(0) correspond to the P(1.5) and R(0.5) transitions of free $^{15}N_2^+$.

4. ELECTRONIC SPECTRUM OF N_2^+-Ne

When the B-X electronic transition of N_2^+-Ne is excited (fig.5) the observed spectrum shows well-resolved vibrational structure at the lowest cluster temperatures (bottom trace), which disappears at higher temperatures (top trace). Fits of the rotational profile of the origin band (b in fig.5), lead to the conclusion that the N_2^+-Ne complex is probably linear, or quasi-linear, in the X and B states,

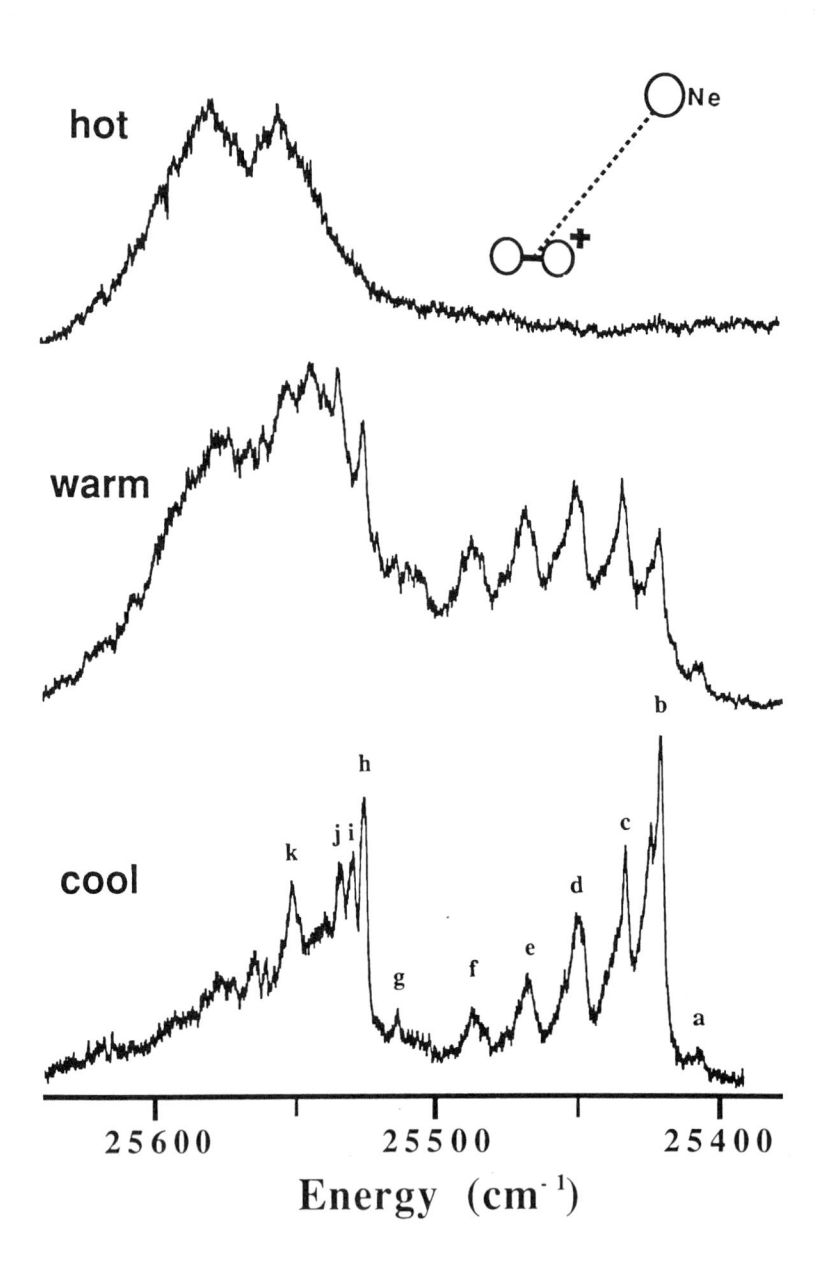

Fig. 5. B←X photodissociation spectrum of N_2^+-Ne in the 391 nm region. The three spectra were taken under different conditions and have different effective temperatures (as labelled). The lower two spectra were taken using the electron impact source shown in fig. 1 while the upper spectrum was taken using a discharge ion source. The dye laser bandwidth was 0.3 cm^{-1}.

with a ≈ 400 pm separation of the neon atom from the N_2^+ center of mass in the X state, decreasing by 10 pm upon excitation into the B state.[10] The origin band is shifted by 146.5 cm^{-1} from that of N_2^+, indicating that the bonding increases considerably upon electronic excitation. In the limit of N_2^+ isolated at 4K in a neon matrix, this shift amounts to 463 cm^{-1}.

The vibrational structure corresponds to progression, sequence and combination bands involving the low frequency van der Waals type modes. Thus, for example, the Ne----N_2^+ stretching mode has a frequency of ≈104 cm^{-1} (band b to band h spacing in fig. 5) in the B state. The bands labelled c,d,e,f and g correspond either to progressions in the low frequency bend or possibly to bending vibration hot bands. Whatever the case, it seems that the complex is much stiffer in the B than the X state. As the temperature of the complex is increased, higher lying vibrational levels in the X state are populated which are more like the free internal rotor levels observed in N_2^+-He. The profile at the highest temperature (fig.5) then comes to resemble that of N_2^+ just as it does in N_2^+-He at low internal energies. Furthermore, irrespective of the temperature, the intensity of the 'R-branch' (to the blue of the origin) drops to zero at a certain value indicating that once the internal rotation quantum number j exceeds a certain value the cluster is no longer bound. From these data one gleans an approximate dissociation energy of ≈ 300 cm^{-1} for the complex.

5. ELECTRONIC SPECTRUM OF N_2O^+- Ne

That the described approach can also be applied to polyatomic ions bound to rare gases, is illustrated for N_2O^+-Ne. The situation is complicated by the fact that the N_2O^+ ion core, in its $^2\Pi$ ground state, posseses orbital angular momentum. An off-axis perturber mixes the $^2\Pi_{3/2}$ and $^2\Pi_{1/2}$ spin-orbit components to produce substates of A´ and A´´ symmetries, the lower lying of which is expected to be preferentially populated amongst the ions produced by our source. The N_2O^+-Ne complex was formed (with an effective temperature of ≈30K) in the same manner as outlined above, and the A-X electronic spectrum was detected when the N_2O^+ product ion was monitored (fig.6).[11] The most prominent peak in the spectrum is due to transitions from the lower electronic substate whilst the smaller band shifted by 141 cm^{-1} to lower energy results from transitions from the upper substate. The splitting between the A´ and A´´ states is given by $(A_{so}^2 + \varepsilon^2)^{1/2}$ where A_{so} is the spin orbit coupling constant of X state N_2O^+, with the ordering of the A´ and A´´ states depending on the sign of the interaction parameter ε. From the observed splitting and the previously determined value for A_{so}=133.2 cm^{-1},[14] we deduce that $|\varepsilon|$=50±8 cm^{-1}, a non-zero value implying a nonlinear geometry for N_2O-Ne$^+$.

The shift of the origin band for the cluster ion is 33 cm⁻¹ to the blue from the monomer, in line with shifts in this energy direction observed for Σ-Π transitions of ions embedded in neon matrices.[15] Finally it can be noted that whereas for the zeroth and v_2=1 (NNO bend) levels of the A state of N_2O^+-Ne, only N_2O^+ fragment ions are produced, for the v_1=1 (NNO stretch) level, both N_2O^+ and a small fraction NO^+ ions are observed.

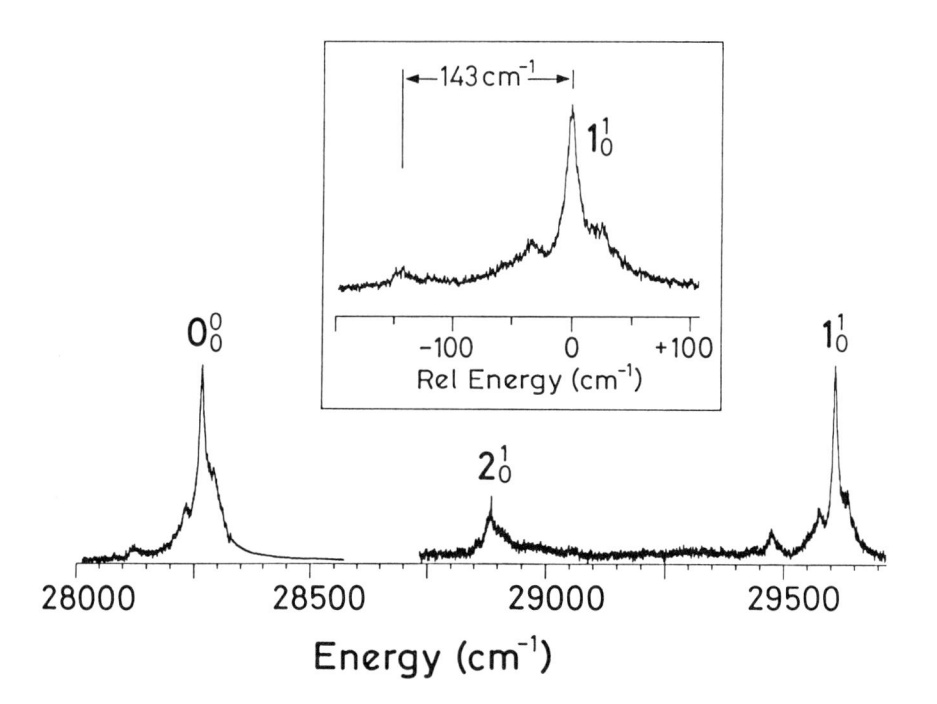

Fig. 6. A←X electronic spectrum of N_2O^+-Ne in the 357 to 336.5 nm region. Two dyes were used to scan the entire range and intensities of the various bands do not represent the proper relative absorption cross sections of the different bands. The inset shows details of the 1_0^1 band.

6. CONCLUSIONS

By measuring the photoinduced vibrational predissociation of ionic complexes we have been able to ascertain details of the structure, binding energies and vibrational frequencies of several hitherto uncharacterized ionic complexes. Some of the results are novel. N_2^+-He has been shown to have no well defined angular structure in either the X or B electronic states, and behaves as an almost free internal rotor at even the lowest internal energies. Free internal rotor structure persists for N_2^+-He$_n$ complexes with n up to six. On the other hand N_2^+-Ne appears to be linear though reasonably floppy in the X state, but becoming considerably more rigid in the B electronic state with a marked increase in binding energy also accompanying the transition. At higher internal energies the N_2^+-Ne complex also becomes essentially a free internal rotor. N_2O^+-Ne differs from the N_2^+ complexes in that it has a $^2\Pi$ ground state and displays a mixing of the spin-orbit components of the N_2O^+ monomer induced by an off-axis Ne perturber.

7. ACKNOWLEDGEMENTS

The studies at Basel are part of Project No.20-29886.90 of the " Schweizerischer Nationalfonds zur Förderung der wissenschaftlichen Forschung".

8. REFERENCES

1. M. Okumura, L. I. Yeh and Y. T. Lee, J. Chem. Phys., **83**, 3705 (1985).
2. D. Lessen and P. J. Brucat, J. Chem. Phys., **90**, 6296 (1989).
3. R. A. Kennedy and T. A. Miller, J. Chem. Phys., **85**, 2326 (1986).
4. E. J. Bieske, R. I. McKay, F. R. Bennett and A. E. W. Knight, J. Chem. Phys., **92**, 4620 (1990).
5. T. Weiske, D. K. Böhme, J. Hrusak, W. Krätschmer and H. Schwarz Angew Chem. Int. Ed. Engl., **30**, 884 (1991).
6. S. W. McElvany, B. I. Dunlap and A. O'Keefe, J. Chem. Phys., **86**, 715 (1987).
7. M. T. Bowers, in "Ion and Cluster Ion Spectroscopy and Structure ", ed. J. P. Maier, pp. 241-273, Elsevier, New York, 1989.
8. E. J. Bieske, A. Soliva, M. Welker and J. P. Maier, J. Chem. Phys., **93**, 4477 (1990).
9. E. J. Bieske, A. M. Soliva, A. Friedmann and J. P. Maier, J. Chem. Phys., 1992, in press.
10. E. J. Bieske, A. M. Soliva and J. P. Maier, J. Chem. Phys., **94**, 4749 (1991).
11. A. M. Soliva, E. J. Bieske and J. P. Maier, Chem. Phys. Lett., **179**, 247 (1991).
12. S. Miller, J. Tennyson, B. Follmeg, P. Rosmus and H. Werner, J. Chem. Phys., **89**, 2178 (1988).
13. V. Frecer, D. C. Jain and A-M. Sapse, J. Phys. Chem., **95**, 9263 (1991).
14. J. H. Callomon and F. Creuzberg, Phil. Trans. R. Soc. Lond. A, **157**, 277 (1974).
15. J. Fulara, S. Leutwyler, J. P. Maier and U. Spittel, J. Phys. Chem., **89**, 3190 (1985).

Experimental methods for probing structure and dynamics of gas-phase molecular dications

Kazushige Yokoyama, Diane M. Szaflarski,[a] Amy S. Mullin,[b] and W. C. Lineberger

Joint Institute for Laboratory Astrophysics, University of Colorado and National Institute of Standards and Technology and Department of Chemistry and Biochemistry, University of Colorado, Boulder, Colorado 80309 - 0440 USA

ABSTRACT

Coaxial ion-laser beam photofragmentation spectroscopy combined with coincidence detection is a powerful method for probing doubly charged molecular ions. The high-resolution photofragmentation spectra contain detailed information on the molecular structure and dissociation dynamics of doubly charged cations. Our study on N_2^{2+} reveals detailed intramolecular interactions with other states, and also provides direct information on previously unobserved electronic states of this dication. We discuss these intramolecular interactions and the dissociation mechanism and demonstrate the high sensitivity of this technique.

1. INTRODUCTION

The physical and chemical properties of molecular dications are of great interest both experimentally and theoretically. For some molecular dications the chemical bonding forces overcome the electrostatic Coulomb repulsion between two positively charged centers, resulting in strongly bound but metastable species. Gas-phase dications may be important players in chemical reactions which take place in low pressure, low collision environments such as in astrophysical clouds and in upper atmospheric reactions.

Photofragmentation spectroscopy has proven to be an especially powerful tool for obtaining detailed information regarding the structure and dissociation dynamics of molecular dications.[1,2,3,4] The coaxial laser-ion beam photofragment spectrometer used in our group consists of an electron impact ionization ion source, a high-resolution laser and a sensitive scheme for collecting the photofragment ions in coincidence. The coaxial beam geometry and acceleration of the ion beam provide sub-Doppler resolution.[5,6] The utility of this technique has recently been demonstrated[3,4] in the investigation of the $^3\Pi_g$ - $^3\Sigma_u^+$ transition of N_2^{2+}.

Electron impact ionization is commonly used to generate gas-phase molecular dications.

---

[a]Present address: Naval Ocean System Center, Code 553, San Diego, CA 92152-5000.
[b]Present address: Department of Chemistry, Columbia University, New York, NY 10027.

However, since the efficiency of electron impact ionization is determined by the Franck-Condon overlap of the wave functions of the neutral precursor and the dication state, dication electronic states with bond lengths similar to the neutral precursor ground state are preferentially populated. The high-resolution of the coaxial photofragment spectrometer, however, allows us to probe electronic states which are not readily populated by electron impact ionization, but are electronically coupled to the populated states and observed through perturbations or predissociation. In particular, electronic coupling of two discrete levels which results in spectral line shifts and intensity variations that can be accurately accounted for through perturbation analysis, thus provides detailed information on an otherwise inaccessible electronic state. Additionally, accurate linewidth measurements provide lifetimes of the predissociating state and insight into the nature of the electronic coupling between the discrete levels and a continuum.

2. EXPERIMENTAL

2.1. Ion formation

The electron impact ionization source used for dication production is illustrated in Fig. 1(a). An effusive beam of neutral precursor gas is introduced through a translatable nozzle with a 1.0 mm orifice; the source chamber pressure is maintained at ca. 5×10^{-5} torr. A 3 - 4 mA electron beam is generated by running a current through a coiled tungsten filament, directing the electrons downward with 120 eV of energy via a repeller plate, and collecting them in a Faraday cup. The electron beam crosses the neutral beam about 1 cm from the nozzle. Ions formed in the electron impact region drift through a 9 mm skimmer orifice into a differentially pumped region, as shown in Fig. 1(b). Typical mass-selected dication beam currents are 30 - 100 pA.

The production of doubly-charged ions competes with loss mechanisms such as charge transfer and collisionally induced dissociation. To maximize dication formation, source conditions were initially optimized using NO^{2+}, which has a mass-to-charge ratio m/q = 15 and can be readily distinguished from its N^+ and O^+ fragments. To determine the region of optimum dication production in the effusive neutral beam, the distance between the electron beam and the nozzle was varied. Figure 2 shows mass spectra at two different electron-nozzle distances and illustrates the enhanced dication production when the ionization occurs further downstream in a region of fewer collisions. Numerous dication species have been formed using this source, including NO^{2+}, N_2^{2+}, CO^{2+}, CF_2^{2+}, and CF_3^{2+}. The mechanism of formation for these dications is believed be a one-electron process.[7]

The N_2^{2+} electronic states prepared by electron impact are predominantly populated by the vertical transition from the ground state of neutral N_2, with a bond length of 1.0998 Å in the lowest vibrational level.[8] Therefore the electronic states that are most likely to be populated are the ground $^1\Sigma_g^+$ and $^3\Sigma_u^+$ states which have bond lengths of 1.1316 Å and 1.09406 Å in the lowest vibrational levels, respectively.[1,3] The rotational temperature of the ion source was characterized to be approximately 300 K by using the intensity distribution of lines appearing in the photofragment spectra of N_2^{2+}.

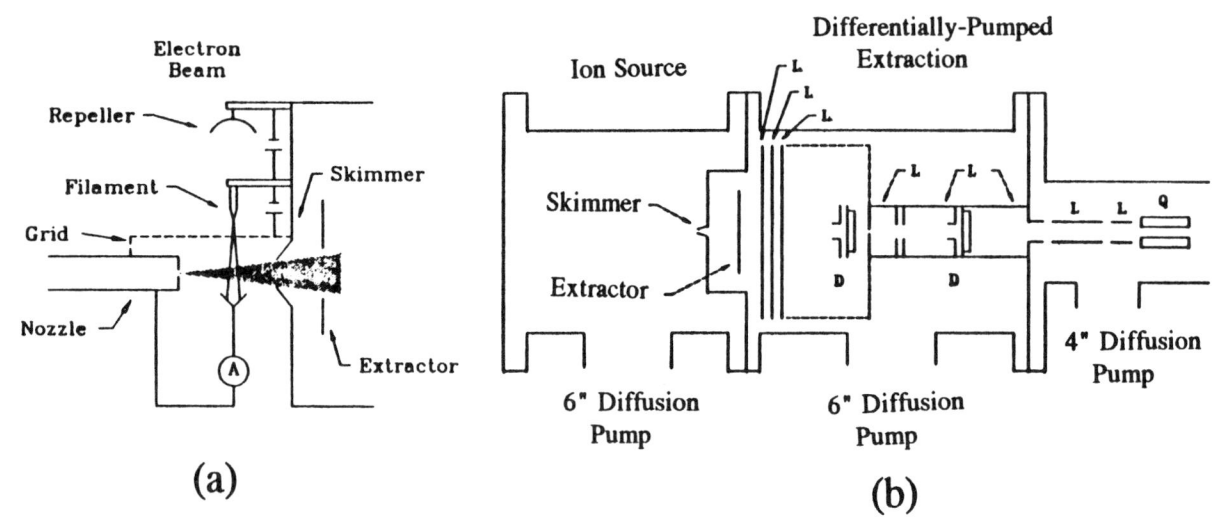

Fig. 1. (a) Doubly-charged positive ion source. An electron beam crosses an effusive beam of precursor gas. The filament is maintained at about 120 V negative with respect to the nozzle, grid, and skimmer. Positive ions are extracted with 100 - 200 V. (b) Schematic view of the ion source chamber and differentially pumped extraction region used in the dication photofragment experiments. Lenses are denoted by L, deflectors by D, apertures by A, and quadrupole ion lenses by Q.

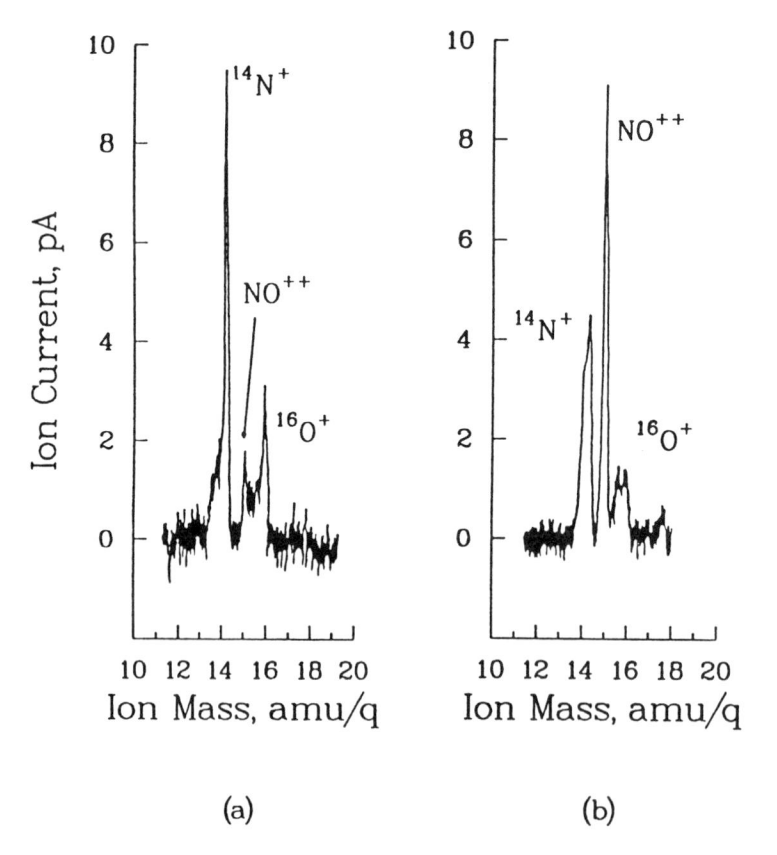

Fig. 2. Mass spectra resulting from electron impact on NO. In (a), the electron beam ionizes at the throat of nozzle. In (b), the electron beam crosses the neutral beam ~ 1 cm downstream.

2.2. Spectra collection and laser systems

Following extraction from the ion source, positive ions are mass-selected by a 90° magnetic mass sector, collimated, accelerated to 2300 eV/q, and merged coaxially with the output of a tunable CW laser. The dications are photodissociated in a 30 cm coaxial interaction region and the two singly charged photofragments are spatially separated from the parent ion beam and each other, and are collected. The detection scheme is illustrated in Fig. 3. The approach we have employed to monitor the fragmentation process is coincidence detection of the photofragments. In the case of $^{14}N_2^{2+} + h\nu \rightarrow {}^{14}N^+ + {}^{14}N^+$, this technique is quite useful. The mass selected ion with m/q = 14 includes both N^+, and N_2^{2+}. However, when the dication dissociates, the Coulomb repulsion between the two ionic fragments typically results in ~5 - 10 eV of kinetic energy release with respect to the center-of-mass. Thus, the photofragments of N_2^{2+} have a maximum energy difference of ca. 400 eV between the forward and the backward scattered N^+ fragments.[11] This energy spread is sufficient to result in significant spatial separation of the photoproducts after passing through the quadrupole energy analyzer. The spatial separation is then enhanced by the parallel plate deflectors which bend the fragments by 90°. The backward scattered fragments are deflected more sharply than the forward scattered fragments. This is illustrated in Fig. 3. The fragments are counted with microchannel plate detectors,[4] as a function of the photon energy. The N^+ photofragments are collected as coincidence events, thereby discriminating against stray ions hitting the detectors.

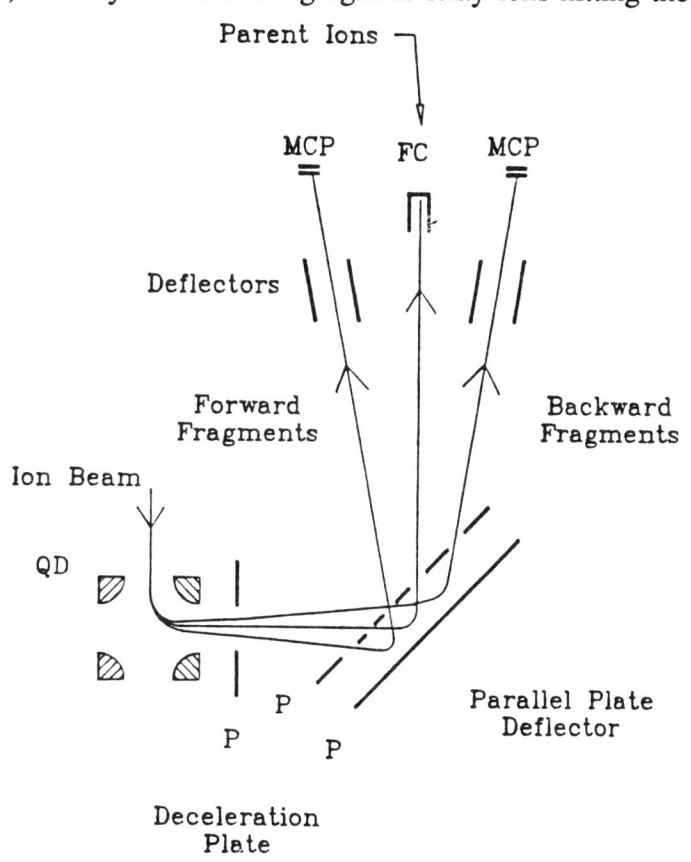

Fig. 3. Schematic view of photofragment separation and detection in the photofragment spectrometer. The quadrupole deflector is denoted by QD, microchannel plate detectors by MCP, electrostatic plates by P, and Faraday cups by FC.

The laser wavelengths available in our laboratory range from 500 nm to 1000 nm by use of either a CW home-built dye laser or a modified Ti:Sapphire laser, both of which can provide a resolution of $\sim 1 \times 10^{-4}$ cm^{-1}. Details of the dye laser system have been reported previously,[4] here we focus on the modifications and upgrades of the Ti:Sapphire laser which permit continuous single-mode frequency scanning.[9] The Ti:Sapphire laser (Schwartz Electro-Optics) is available in a ring configuration and has a three-plate birefringent tuner, unidirectional device, and fixed position thin etalon. The crystal is typically pumped with up to 7 W of an argon-ion laser all visible lines and provides 600 mW of single mode light achieving a resolution of < 2 MHz with no active stabilization. In order to obtain single mode tuning capability, a piezo-driven scanning thick etalon and dual galvo driven Brewster plates are added and electronically locked to the cavity mode.[10]

3. DETAILS OF INTRAMOLECULAR INTERACTIONS IN N_2^{2+}

The $^3\Pi_g$ - $^3\Sigma_u^+$ transition band in the photofragment spectrum of N_2^{2+} has been studied extensively in our group.[3,4] A detailed spectroscopic analysis has shown that electronic coupling with nearby states is responsible for perturbations in the spectrum, as well as for the predissociation of the upper state. *Ab initio* curves obtained by J. Senekowitsch and coworkers[12] are shown in Fig. 4. The observed perturbation arises from coupling between the $^3\Sigma_u^+$ (v=0) state and the $^3\Pi_u$ (v≈10) state. The predissociation of the $^3\Pi_g$ (v=0) state results from electronic coupling with the continuum of the $^3\Sigma_g^-$ state. By investigating details of the electronic couplings, we have obtained information about electronic states which are not

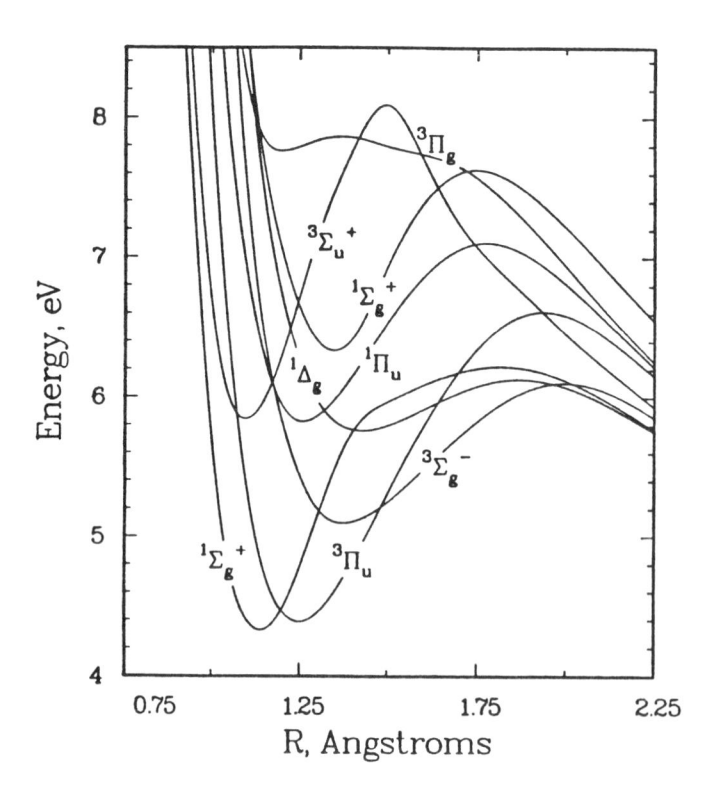

Fig. 4. Low-lying quasibound states of N_2^{2+}. These curves are given in Ref. 12. The energy scale is set to zero at the $N^+(^3P_g) + N^+(^3P_g)$ dissociation limit.

populated by electron impact ionization, the $^3\Sigma_g^-$ state and the $^3\Pi_u$ state. In this section, we focus on the details of these interactions and extend the previous analysis.

The perturbation between the $^3\Sigma_u^+$ (v=0) and the nearby $^3\Pi_u$ state has been observed for rotational levels J=14 and above and has been analyzed using a two-level interaction model.[4,11] Based on the interaction matrix elements given by Kovács,[13] the spin-orbit coupling constant ξ and the electronic-rotation coupling constant η are obtained by non-linear least-squares fitting as $\xi = 1.5$ (± 0.1) cm^{-1} and $\eta = 0.220$ (± 0.003) cm^{-1}. The rotational constant of the perturbing $^3\Pi_u$ (v $\approx$ 10) state is estimated as $B_v = 1.24$ ($\pm$ 0.05) cm^{-1}. The vibrational numbering of the perturbing state is estimated using the *ab initio* curves of Senekowitsch *et al.* In the region of the perturbation, the molecular eigenstates have both $^3\Sigma_u^+$ and $^3\Pi_u$ character and can be expressed as a linear combination of these states. As shown in Fig. 5, the perturbed region of the spectra has features which result from the $^3\Pi_g$ - $^3\Pi_u$ transition as well as the $^3\Pi_g$ - $^3\Sigma_u^+$ transition. Using our data, the mixed eigenstates are determined to be,

$F_1(N=22)$: 0.71 $|^3\Sigma_u^+\rangle$ - 0.70 $|^3\Pi_u\rangle$,
$F_2(N=22)$: 0.71 $|^3\Sigma_u^+\rangle$ + 0.70 $|^3\Pi_u\rangle$,
and $F_3(N=22)$: 0.60 $|^3\Sigma_u^+\rangle$ + 0.80 $|^3\Pi_u\rangle$.

Although the $^3\Pi_u$ state is not efficiently prepared through electron impact ionization and is not accessible directly through an optical transition, it does have significant Franck-Condon overlap with the $^3\Sigma_u^+$ state and therefore can be studied indirectly through the observed perturbation.

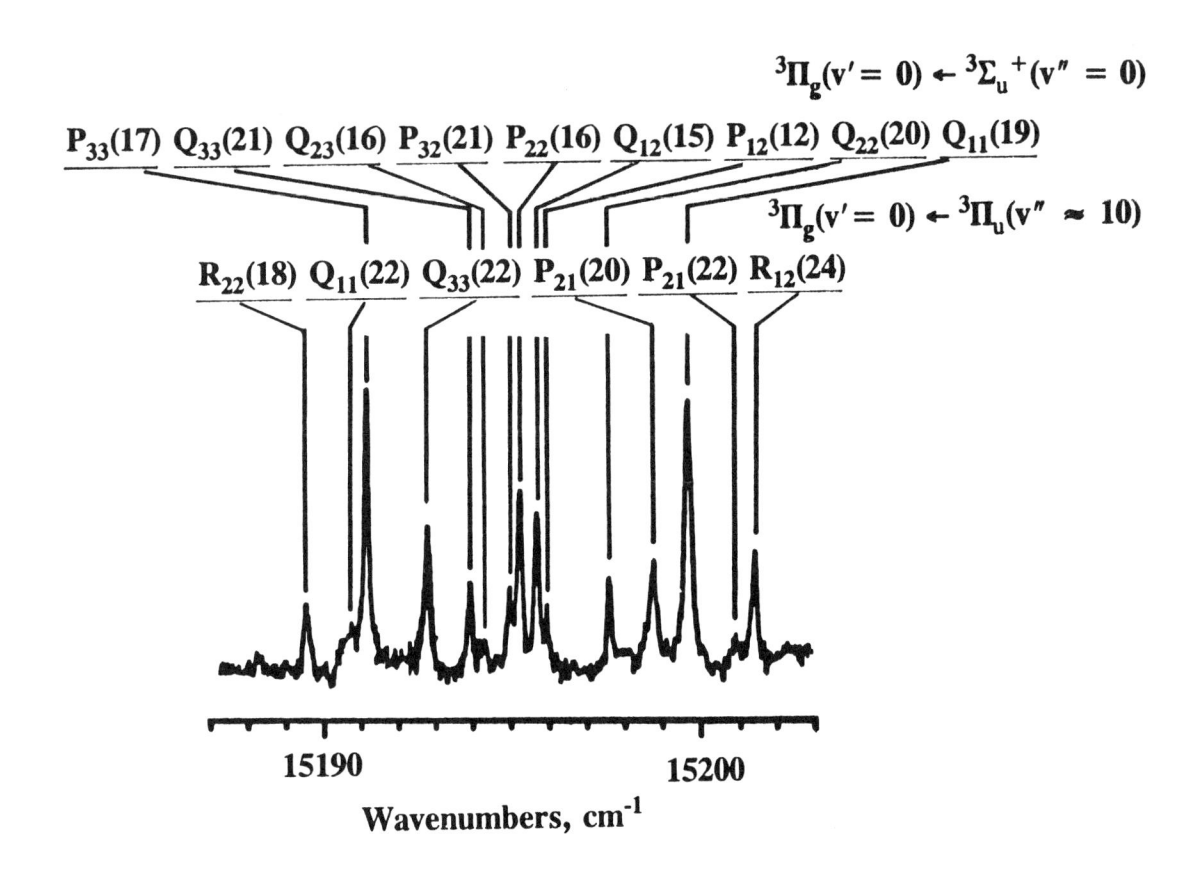

Fig. 5. The perturbed spectral regions of the $^3\Pi_g(v' = 0)$ - $^3\Sigma_u^+(v'' = 0)$ transition and $^3\Pi_g(v' = 0)$ - $^3\Pi_u(v'' = 10)$ transition as denoted in the figure.

The upper state of the $^3\Pi_g$ - $^3\Sigma_u^+$ transition is predissociated by electronic coupling with the continuum of the $^3\Sigma_g^-$ state. The observed linewidths of the $^3\Pi_g$ (v=0) state are equally broadened to ca. 3 GHz which corresponds to predissociation lifetimes ~50 ps, as presented in Table I. The $^3\Sigma_g^-$ and $^3\Pi_g$ states interact through spin-orbit and electronic-rotation coupling and predissociation linewidths Γ (s^{-1}) can be described using Fermi's Golden rule[14]

$$\Gamma = (2\pi/\hbar) \sum_i |\langle b|H|c_i\rangle|^2 \rho,$$

where $\langle b|H|c_i\rangle$ is the interaction matrix element between bound states b and continuum states c_i, ρ is the density of the states. Analytic expressions for Γ can be determined using the matrix coupling elements from Kovaćs,[13] as given in Table II. Some levels, namely F_1 and F_3 f-parity rotational levels [(1) and (5) in Table II], are dissociated only by the spin-orbit interaction. Thus, by assuming ρ is smooth around the region under consideration, ξ is uniquely determined by these states. We report $\xi = 0.21$ (± 0.05) cm^{-1}, and $\eta = 2 \times 10^{-3}$ ($\pm 1 \times 10^{-3}$) cm^{-1}. The magnitude of these constants indicates spin-orbit coupling of the $^3\Pi_g$ state, with the $^3\Sigma_g^-$ continuum being primarily responsible for the predissociation of $^3\Pi_g$ (v=0). This is consistent with the notable lack of rotational dependence in the observed linewidths. It should be pointed out that reported constants contain the overlap of the vibrational wave functions and the additional information on the potential curves will be needed to provide ρ.

TABLE I. The observed lifetime of the $^3\Pi_g$ state is in picoseconds. The lifetime is obtained by fitting observed lines to a Lorentzian function. The error associated with any individual measurement is $\sim \pm 6$ ps.

	$^3\Pi_g(F_1,f)$	$^3\Pi_g(F_1,e)$	$^3\Pi_g(F_2,f)$	$^3\Pi_g(F_2,e)$	$^3\Pi_g(F_3,f)$	$^3\Pi_g(F_3,e)$
J = 5	54		61			
6		53				
7			79	59		
8						67
9						
10		46	64	61	60	
11	62	48	75			
12						59
13						
14		50				
15						
16		51				46
17				50		
18		44				

TABLE II. Formulas (Ref. 13) for the predissociation linewidth (Γ) of the $^3\Pi_g$ state caused by interaction with the $^3\Sigma_g^-$ state.

Rotational Level	Γ
(1) $\Gamma[^3\Pi_g(F_1,\text{ f-level})]$	$(2\pi/\hbar)\rho\{(J-1)(J+1)/[2J(2J+1)]\}\xi^2$
(2) $\Gamma[^3\Pi_g(F_1,\text{ e-level})]$	$(2\pi/\hbar)\rho[(J-1)/2J][\xi-4\eta J]^2$
(3) $\Gamma[^3\Pi_g(F_2,\text{ f-level})]$	$(2\pi/\hbar)\rho[2J(J+1)]\{\xi/[2J(J+1)]+2\eta\}^2$
(4) $\Gamma[^3\Pi_g(F_2,\text{ e-level})]$	$(2\pi/\hbar)\rho[(J+1)^3+J^3]\xi^2/[2J(J+1)(2J+1)]$
(5) $\Gamma[^3\Pi_g(F_3,\text{ f-level})]$	$(2\pi/\hbar)\rho\{J(J+2)/[2(J+1)(2J+1)]\}\xi^2$
(6) $\Gamma[^3\Pi_g(F_3,\text{ e-level})]$	$(2\pi/\hbar)\rho[(J+2)/(2J+2)][\xi+4\eta(J+1)]^2$

4. CONCLUSIONS

The techniques we have described offer the possibility of obtaining detailed information about the potential energy curves of dications. In particular, detailed spectroscopic data have enhanced the current understanding of the molecular potentials, and provided insight into the mechanisms responsible for the fragmentation processes.

5. ACKNOWLEDGMENTS

We are grateful to Jan Hall, Miao Zhu, and Steve Swartz for assisting with the Ti:Sapphire laser system, and to M. N. R. Ashfold, C. M. Western, and G. Gerber for stimulating discussions on the photofragment spectra and the spectral analysis. This research was supported in part by the AFOSR - HEDM project, grants AFOSR-89-0074 and F49620-J-0071.

6. REFERENCES

1. P. C. Cosby, R. Möller, and H. Helm, "Photofragment Spectroscopy of N_2^{2+}," Phys. Rev. A, vol. 28 No. 2, pp. 766-772, 1983.
2. T. E. Masters and P. J. Sarre, "High-resolution Laser Photofragment Spectrum of N_2^{2+} ($^1\Pi_u$-$^1\Sigma_g^+$)," J. Chem. Soc. Faraday Trans., vol. 86 No. 11, pp. 2005-2008, 1990.
3. D. M. Szaflarski, A. S. Mullin, K. Yokoyama, M. N. R. Ashfold, and W. C. Lineberger, "Characterization of Triplet States in Doubly Charged Positive Ions: Assignment of the $^3\Pi_g$-$^3\Sigma_u^+$ Electronic Transition in N_2^{2+}," J. Phys. Chem., vol. 95 No. 6, pp. 2122-2124, 1991.
4. A. S. Mullin, D. M. Szaflarski, K. Yokoyama, G. Gerber, and W. C. Lineberger, "Triplet State Spectroscopy and Photofragment Dynamics of N_2^{2+}," J. Chem. Phys., in press.
5. U. Hefter, R. D. Mead, P. A. Schulz, and W. C. Lineberger, "Ultrahigh-resolution Study of Autodetachment in C_2^-," Phys. Rev. A, vol. 28 No. 3, pp. 1429-1439, 1983.
6. R. D. Mead, "Negative Ion Excited States," PhD. Thesis, University of Colorado, Boulder, 1984.
7. T. D. Märk, and G. H. Dunn eds., *Electron Impact Ionization*, p. 168, Springer-Verlag, New York, 1985.

8. P. G. Wilkinson, "High-resolution Absorption Spectra of Nitrogen in the Vacuum Ultraviolet," Astrophys. J., vol. 126, pp. 1-9, 1957.
9. P. F. Moulton, Photonics Spectra, pp. 119-122, March 1991.
   P. F. Moulton, Photonics Spectra, pp. 123-125, April 1991.
10. L. W. Hollberg, "Measurement of the Atomic Energy Level Shift Induced by Blackbody Radiation," PhD. Thesis, University of Colorado, Boulder, 1984.
11. A. S. Mullin, "Autodetachment and Photofragment Spectroscopy of Metastable Molecular Ions," PhD. Thesis, University of Colorado, Boulder, 1991.
12. J. Senekowitsch, S. V. ONeil, P. Knowles, and H-J. Werner, "The $^3\Pi_g \leftarrow {}^3\Sigma_u^+$ Transition in N_2^{2+}," J. Phys. Chem., vol. 95 No. 6, pp. 2125-2127, 1991.
13. I. Kovács, *Rotational Structure in the Spectra of Diatomic Molecules*, p. 258, American Elsevier Publishing, New York, 1969.
14. L. I. Schiff, *Quantum Mechanics,* 2nd ed., p. 283, McGraw Hill, New York, 1955.

Rotationally Resolved Threshold Photoionization of H$_2$S

Ralph T. Wiedmann and Michael G. White

Chemistry Department, Brookhaven National Laboratory, Upton, NY 11973

1. INTRODUCTION

Photoionization leading to a singly charged ion and free electron represents one the simplest molecular fragmentation processes and determination of the final states of the products (ion and photoelectron) provides a complete description of the photoejection dynamics. Quantum characterization of the molecular fragment requires resolution of the internal rovibronic state distribution which in turn reflects the exchanges of energy and angular momentum in the "scattering" of the photoexcited electron and the anisotropic molecular ion potential. With the recent development of zero kinetic energy threshold photoelectron spectroscopy (ZEKE-PES) by Müller-Dethlefs, Schlag and coworkers[1] it is now possible to take advantage of the high resolution capabilities of laser radiation and measure the rotational distributions of many small cations produced via photoionization.

We have used threshold photoelectron spectroscopy in conjunction with a VUV laser source to probe the rotational distributions of several molecular cations following one-photon VUV photoionization.[2-6] Spectra are obtained by the delayed, pulsed field ionization (PFI) method which is a variant of the ZEKE-PES technique as first demonstrated by Reiser, et al.[7] The PFI technique takes advantage of the similarity between the distribution of Rydberg levels populated 1–2 cm^{-1} below an ionization threshold and the partial wave distribution of the photoelectron with near zero kinetic energy. The narrowband ($\leq$1 cm^{-1}), tunable VUV radiation required for rotationally resolved photoionization measurements is produced by third harmonic or sum frequency generation of visible and/or near UV laser light in pulsed, free jet expansions of the rare gases (Ar, Kr, Xe) or the diatomic gases CO and N$_2$.[8,9]

In this paper, we report on the rotationally-resolved threshold photoelectron spectra of the non-linear triatomic H$_2$S. These measurements are an extension of our earlier study on H$_2$O which provided the first look at the symmetry properties of allowed rotational transitions in the photoionization of non-linear molecules.[5] For photoionization of asymmetric top systems, transitions between levels involving the rotational angular momentum projections (K_a, K_b, K_c) permit resolution of the photoelectron continua according to symmetry. For H$_2$O, both type A ($\Delta K_c = \pm 1, \Delta K_a = 0$) and type C ($\Delta K_c = 0, \Delta K_a = \pm 1$) rotational transitions were observed and these were shown to be associated with ka_2 and ka_1 photoelectron continua, respectively. The appearance of type A transitions with $\Delta K_a = 0$ is in variance with the predictions of a multichannel quantum defect theory (MQDT) analysis based on the symmetry properties of the initial state[10] in which the near atomic-like nature of the outermost $1b_1$ molecular orbital is invoked. The PFI results are also the subject of two very recent theoretical calculations which draw very different conclusions concerning the near-threshold photoionization dynamics for H$_2$O.[11,12] The new spectra reported here for the (000) and (010) states of H$_2$S$^+$ exhibit both type A and type C rotational photoionization transitions and coupled with the previous H_2O^+ results are interpreted as a clear demonstration of the *non-atomic* nature of photoionization from the $1b_1$ orbital in the H_2X (X=O, S) molecules.

2. EXPERIMENTAL

The photoelectron spectrometer and VUV source have been described in detail elsewhere.[9] Briefly, we use a small, separately pumped frequency tripling chamber attached to a standard photoelectron/photoion time–of–flight (TOF) spectrometer. Tunable VUV radiation in the range 118.6–117.0 nm required to reach the H_2S^+, $\tilde{X}^2B_1$ (000) and (010) vibrational bands was produced by $2\omega_{uv} + \omega_{vis}$ sum frequency mixing in Xe gas near the $5p \to 5d$ and $5p \to 7s$ atomic resonances, respectively.[13] As a result of poor conversion efficiency at shorter wavelengths, it was not possible to obtain the PFI spectrum for the (100) symmetric stretch. The collinear visible (593.0–583.5 nm; ~40 mj/pulse; 0.07 cm^{-1}, 20 Hz) and UV (296.5–291.8 nm; ~5 mj/pulse) laser beams were focussed by a 100 mm focal length achromatic lens (Optics for Research) inside the tripling chamber and into a pulsed, free jet expansion of Xe. The diverging fundamentals (visible and UV) and sum frequency VUV radiation are captured by a pyrex capillary tube (30 cm long; 1 mm id.) and directed into the TOF spectrometer, where it passes between two parallel plates which define the extraction field for the TOF spectrometer. The VUV intensity at the spectrometer is estimated to be $\sim 10^9$ photons per pulse, with an energy bandwidth of 0.7 cm^{-1}.

High-resolution threshold spectra were obtained by the delayed, pulsed field ionization (PFI) method as first demonstrated by Reiser, *et al.*[7] PFI takes advantage of the well known Stark shift of an ionization threshold in an external electric field (given by $\sim 6\sqrt{F}$ cm^{-1}, where F is the field strength in volts/cm). Bound Rydberg levels very near the ionization threshold which are stable in a field free environment become open ionization channels when a Stark shift larger than their binding energy is applied. For fields used in this work (0.3-0.5 V/cm) only Rydberg levels with $n \geq 150$ can be field ionized. Rydberg states with such high principal quantum numbers are very long lived due to the weak interaction of the Rydberg electron and the ion core. A small positive DC voltage (~ 0.05 V) was maintained on the repeller (lower) plate to sweep out any slow or near ZEKE photoelectrons produced directly by the laser pulse. After a delay of 700 ns, a fast, negative pulse ranging in amplitude from 0.3 − 0.5 volts was applied to the repeller plate. Electrons produced by this pulsed field were readily distinquished from photoelectrons arising from direct ionization as the arrival time of the former depend only on the pulse delay and voltage.

Rotationally cold PFI spectra for H_2S were obtained by expanding 300 Torr of a 5% mixture of H_2S in Ar into the source chamber ($\sim 5 \times 10^{-5}$ Torr) through a pulsed valve (Laser Technics) fitted with a 0.5 mm diameter nozzle. A 1 mm skimmer located 3 cm downstream of the valve collimated the beam so it passed cleanly through the TOF apparatus (background pressure with sample $\sim 2 \times 10^{-7}$). To obtain comparable signal levels for the weaker (010) vibrational band, the distance between the pulsed valve and skimmer was reduced to 1 cm. In assigning rotationally cold spectra, we assume that the populations in the nuclear spin states (ortho and para) do not redistribute during the expansion. For normal H_2S this fixes the ratio of ortho to para at 3:1, regardless of the final rotational temperature. The ortho and para nuclear spin manifolds are then treated separately, each with the same rotational temperature (relative to its lowest energy level) and then normalized to recover the overall 3(ortho):1(para) ratio.

The spectra were placed on an absolute energy scale by calibration to known transitions in H$_2$ detected via 2-photon (VUV+UV) resonant multiphoton ionization (REMPI).[9,14] and corrected for the small Stark-shift (1-2 cm^{-1}) introduced by the pulsed field. Uncertainties in the wavelength calibration and spectral peak location result in an overall uncertainty of ±2 cm in the reported ionization energies.

3. RESULTS AND DISCUSSION

3.1 SYMMETRY AND PHOTOIONIZATION OF ASYMMETRIC TOPS

For asymmetric tops, the rotational energy levels can be labeled by J_{K_a,K_c} (or $N^+_{K_a,K_c}$ for the cation) where J is the total angular momentum and K_a and K_c describe the projection of the total angular momentum on the symmetry axis for corresponding pure prolate and oblate top wavefunctions. For the ion, N^+ labels the angular momentum excluding spin, and a small spin-orbit splitting results in F$_1$ and F$_2$ levels (J=N+1/2 or N-1/2) which we generally do not resolve. We use the convention that the x, y and z axes correspond to the c, a and b molecular axes, where the molecular axes are defined such that $I_a < I_b < I_c$ ($A > B > C$). With this choice of axes, z is the C$_2$ axis and y the orthogonal in-plane axis. It should be noted that while the energy levels and rotational symmetry of an asymmetric top can be *labelled* with respect to the correlating symmetric top wavefunctions, K_a and K_c are not good quantum numbers, and the true asymmetric top wavefunction is a linear combination of symmetric top functions.

The highest occupied molecular orbital in H$_2$S is essentially a non-bonding sulfur $3p_x$ orbital pointing out of the molecular plane with b_1 symmetry. Since H$_2$S is a closed shell molecule, its ground electronic state is A_1 symmetry, and removal of one outer electron yields an ion with $^2B_1$ symmetry. Dipole selection rules require that the final state (ion core + electron) must have A_1, B_1 or B_2 electronic symmetry, corresponding to kb_1, ka_1, and ka_2 Rydberg (or continuum) electrons. Restricting the Rydberg electron to l≤ 2, the possible atomic orbitals of A_1 symmetry include s, p_z, d_{z^2} and $d_{x^2-y^2}$. Only d_{xy} orbitals have A_2 symmetry, while p_x and d_{xz} are of B_1 symmetry.

A more detailed analysis of the allowed dipole transitions must include the symmetry of the rotational wavefunctions as well. The overall selection rules were outlined in our earlier paper on H$_2$O,[5] where it was shown that for outgoing photoelectrons with ka_1 symmetry, $\Delta K_c = 0, \pm 2 \ldots$; $\Delta K_a = \pm 1, \pm 3 \ldots$, for outgoing ka_2 electrons, $\Delta K_c = \pm 1, \pm 3 \ldots$; $\Delta K_a = 0, \pm 2 \ldots$, and for outgoing kb_1 electrons, $\Delta K_c = \pm 1, \pm 3 \ldots$; $\Delta K_a = \pm 1, \pm 3 \ldots$. These transitions are called type C, type A and type B, respectively, corresponding to the axes along which the transition dipoles are located. Type B transitions were shown to be strictly forbidden since they connect ground and ionic state rotational levels of different nuclear spin symmetry.

3.2 PFI SPECTRA OF H$_2$S

The jet-cooled, PFI spectrum for the $\tilde{X}^2B_1$ (000) and (010) vibrational levels of H$_2$S$^+$ are shown in Figures 1 and 2, respectively. The rotational assignments are based on accurate spectroscopic constants for the neutral ground state[15] and the (000) and (010) vibrational levels of the $\tilde{X}^2B_1$ ionic ground state.[16] Transition energies were calculated using the parameterized Hamiltonians given by Townes and Schawlow for rotational energy levels of asymmetric top molecules.[17] Spin doubling of the H$_2$S$^+$ rotational levels is generally smaller than our overall resolution ($\sim$2 cm^{-1}) so that the final ionic levels are labeled by core angular momentum exclusive of spin, i. e. $N^+_{K_a,K_c}$. The H$_2$S$^+$ spectra are similar to those obtained previously for the (000) and (100) levels of H$_2$O$^+$,[5] in that threshold photoionization is accompanied by only small changes in core angular momentum, i. e. $\Delta N = 0, \pm 1$ and $\Delta K_{a,c} = 0, \pm 1$ and *both* type C ($\Delta K_c = 0; \Delta K_a = \pm 1$) and type A ($\Delta K_c = \pm 1, \Delta K_a = 0$) transitions are observed. As noted earlier, type A transitions were not expected to appear in the H$_2$O spectra based on the near-atomic nature of the $1b_1(2p_O) \rightarrow kd$ photoionization transition.[10] Although the type A transitions are generally weaker than type C transitions, the H$_2$S$^+$ spectra clearly contain lines which can be unambiguously assigned to type A. Our results for both H$_2$O and H$_2$S suggest that the near-atomic nature of the initial state does not exclusively determine the ionization dynamics of the H$_2$X molecules.

The lower traces in Figs. 1 and 2 are simulations of the one-photon ionization spectra of the (000) and (010) vibrational states assuming uniform line strengths, a ground state rotational temperature of 15 K and a Gaussian line shape with a width of 1 cm^{-1}(FWHM). The absolute energy scales are fixed by taking the ionization potentials to be the energies of the nuclear spin forbidden $0_{00} \leftarrow 0_{00}$ transitions at 84,434 $\pm$ 2 cm^{-1} for the (000) level and 85,592 $\pm$ 2 cm^{-1} for the excited (010) state. These ionization energies can be compared to the rotationally unresolved PES values of 84,413 $\pm$ 2 cm^{-1} and 85,550 $\pm$ 3 cm^{-1} obtained by Karlsson, *et al* using a room temperature gas sample and the He I resonance line (21.2 eV).[18] A type C to type A intensity ratio of 4:1 is found to give the best overall simulations for both vibrational levels and these can be compared to a 2:1 ratio found for H$_2$O$^+$.[5] A 2:1 type C to type A intensity ratio for H$_2$O$^+$ was rationalized by noting that two da_1 ($d_{z^2}, d_{x^2-y^2}$) final states are allowed for type C transitions while there exists only one da_2 (d_{xy}) final state of the correct symmetry for type A transitions. As will be discussed below, this correlation between type A/type C transition strengths and the number of associated d final states is only fortuitous; type A photoionization transitions in H$_2$O are actually dominated by $kp(l = 1)$ partial waves which result from torques exerted on the photoexcited electron by the molecular ion potential. Overall, the reasonable agreement between the PFI spectra and the simulations provides confidence in the line assignments as well as in the characterization of the spectra as having nearly independent type C and weaker type A contributions.

Comparison of Figs. 1 and 2 shows that the $\tilde{X}^2B_1$ (000) vibrationless ground state and the (010) bending excited state have virtually identical PFI spectra. Small intensity differences between the $0_{00} \leftarrow 1_{10}$ and $0_{00} \leftarrow 1_{01}$ lines may reflect slightly different rotational temperatures as the (010) required a smaller nozzle-to-skimmer distance (1 cm) to obtain reasonable signal levels. A similar lack of vibrational state dependence was observed in the PFI spectra for the ground (000) and (100) excited symmetric stretch levels of H$_2$O$^+$.[5] These results indicate that near threshold photoionization for the H$_2$X molecules can be treated in a Franck-Condon model in which vibrational motion does not strongly influence the molecular ion potential experienced by the escaping photoelectron.

The absence of transitions with $|\Delta N| > 1$ in the (000) and (010) H$_2$S$^+$ spectra also suggests that little angular momentum is transferred between the photoelectron and the ion core, particularly with respect to rotation. Neglecting spin, the range of ΔN is given by $l+1, \ldots, -l-1$ and since $l = 2$ partial waves are expected to dominate, transitions with $|\Delta N|$ as high as 3 are possible. Such large changes in core angular momentum occur in the threshold photoionization spectrum of O$_2$ ($|\Delta N| \leq 5$) and are attributed to the presence of a near-threshold shape resonance which enhances the $l = 3(f\sigma_u)$ partial wave.[2,3] The sensitivity of the shape resonance potential to the internulear distance also leads to a pronounced vibrational state dependence of the O_2^+ rotational state distribution. The relative simplicity of the H_2O^+ and H_2S^+ rotational spectra are consistent with the predictions of the "rotation spectator" model in which the photoelectron carries away the angular momentum of the incident photon.[19,20] This model was found to be adequate to describe the low resolution threshold photoelectron spectrum of HCl.[21]

3.3 GENERAL ASPECTS OF H_2X (X = O, S) PHOTOIONIZATION

The observation of both type A ($\Delta K_c = \pm 1, \Delta K_a = 0$) and type C ($\Delta K_c = 0; \Delta K_a = \pm 1$) rotational photoionization transitions in H_2O^+ and H_2S^+ is contrary to the original predictions of Child and Jungen[10] who used MQDT to analyze the high resolution photoionization spectrum of H_2O reported by Page et al.[22] Specifically, Child and Jungen predict only type C transitions and moreover, only those with $|\Delta K_a| = \pm 1$ and $|\Delta N| \leq 1$. The limits on the changes of core angular momenta arise from the assumption that the $1b_1$ molecular orbital can be described exclusively in terms of an atomic p_x orbital with both l and its body fixed projection, λ, equal to one. Although the authors characterize these as propensity rules, they are predicted to become stronger as the upper state approaches the Hund's case (d) limit, *i. e.* as the Rydberg levels converge to the ionization threshold. Our own symmetry analysis of the selection rules for H_2X photoionization showed that, in general, only type B transitions are strictly forbidden (due to nuclear spin considerations) and that there was no *a priori* way to exclude type A transitions without a detailed knowledge of the partial wave distribution of the continuum.[5] Our analysis assumed that the molecular point group symmetry, C_{2v} in this case, was valid for determining the overall symmetry properties of the combined molecular ion core and photoexcited electron complex. This was based on the fact that the absorption process takes place near the ion core where a Hund's case (b) coupling scheme is appropriate for both the ground and ion core–photoelectron excited states. As noted above, the asymptotic, long-range ionization channels are better characterized in a Hund's case (d) basis where the orbital angular momentum l of the photoelectron is well defined. In this basis, symmetry considerations show that type A and type C transitions are associated with continua of a given parity, specifically $l = odd$ with type A and $l = even$ with type C transitions.[10,12] Consequently, $d(l = 2)$ partial waves cannot contribute to both types of transitions and one is left with the conclusions of Child and Jungen, *i. e.*, type A transitions are not likely due to the atomic p_x nature of the outer $1b_1$ orbital and the $\Delta l = \pm 1$ dipole selection rule for atoms.

In an effort to explain the observed PFI spectra of H_2O^+, Gilbert and Child[11] attributed the appearance of type A transitions to field-induced autoionization of nominally very weak np Rydberg states via dipole coupling with strongly allowed and near-degenerate $n'd$ Rydberg levels. This mechanism has many attractive features and has been invoked to explain the observation of anomalous rotational branch intensities in the PFI spectra of N_2O and HCl.[6] However, the PFI spectra for both H_2O^+ and H_2S^+ do not exhibit features which point to field-induced autoionization. The most obvious indication would be the observation of rotational branches with large *positive* changes of core angular momentum ($\Delta N > +1$). In the cases of N_2O[4] and HCl,[6] rotational branches with $\Delta J > 2$ were found to be relatively intense, whereas the corresponding rotational branches differing only in the sign of ΔJ were often not observed. Such unusual intensity distributions are not evident the PFI spectra of H_2O^+ and H_2S^+, neither of which exhibit strong features which can be confidently assigned to $|\Delta N| > 1$. Furthermore, such field-induced autoionization processes are expected to be very sensitive to the density of Rydberg states near successive ionization thresholds.[11] These will vary considerably from H_2O to H_2S which have rotational constants which differ by nearly a factor two. The observed similarity between the H_2O^+ and H_2S^+ transition types would therefore be quite unexpected. It therefore seems unlikely that field-induced autoionization is the cause of type A photoionization transitions in the H_2X molecules.

A more straightforward explanation for the appearance of type A transitions is due to a very recent *ab initio* calculation on H_2O by Lee, et al[12] using the Schwinger variational method which has been generalized to include cation rotational distributions. In this calculation, no assumptions concerning the atomic character of the initial and final states are made and the continuum is calculated in the full anisotropic potential of the ion core. The results are surprising in that the type A photoionization transitions are predicted with intensities in nearly quantitative agreement with experiment and a partial wave analysis shows these transitions to be accompanied by nearly pure p wave continua. Furthermore, the type C transitions are not exclusively d partial waves as expected from the atomic analogies, but include significant and for some lines (*e. g.* $0_{00} \leftarrow 1_{10}$) dominant s wave character. These calculations emphasize the importance of the non-spherical nature of the molecular ion core which can "scatter" or torque the escaping photoelectron into different partial waves. For H_2S, we expect that similar dynamics are applicable since the $1b_1$ molecular orbital is primarily localized on the sulfur atom. However, the participation of the nominally unoccupied $3d$ levels centered on the heavier sulfur atom could lead to additional high l components in the continuum and modify the rotational line strengths.

ACKNOWLEDGEMENTS

The authors would like to thank M.-T. Lee, K. Wang and Professor V. McKoy (Cal Tech) for providing us with the results of their H_2O calculations prior to publication. This work was performed at Brookhaven National Laboratory and supported by the US Department of Energy, Office of Basic Energy Sciences under contract No. DE-AC02-76CH00016.

References

[1] K. Müller-Dethlefs and E. W. Schlag, *Annu. Rev. Phys. Chem.*, **42**, 109 (1991).

[2] R. G. Tonkyn, J. W. Winniczek and M. G. White, *Chem. Phys. Lett.*, **164**, 137 (1989).

[3] M. Braunstein, V. McKoy, S. N. Dixit, R. G. Tonkyn and M. G. White, *J. Chem. Phys.*, **93**, 5345 (1990).

[4] R. T. Wiedmann, E. R. Grant, R. G. Tonkyn and M. G. White, *J. Chem. Phys.*, **95**, 746 (1991).

[5] R. G. Tonkyn, R. T Wiedmann, E. R. Grant and M. G. White, *J. Chem. Phys.*, **95**, 7033 (1991).

[6] R. G. Tonkyn, R. T. Wiedmann and M. G. White, *J. Chem. Phys.*, in press.

[7] G. Reiser, W. Habenicht, K. Müller-Dethlefs and E. W. Schlag, *Chem. Phys. Lett.*, **152**, 119 (1988).

[8] R. H. Page, R. J. Larkin, A. H. Kung, Y. R. Shen and Y. T. Lee, *Rev. Sci. Instrum.*, **58**, 1616 (1987).

[9] R. G. Tonkyn and M. G. White, *Rev. Sci. Instrum.*, **60**, 1245 (1989).

[10] M. S. Child and Ch. Jungen, *J. Chem. Phys.*, **93**, 7756 (1990).

[11] R. D. Gilbert and M. S. Child, *Chem. Phys. Lett.*, in press.

[12] M.-T. Lee, K. Wang, V. McKoy, *private communication*.

[13] B. P. Stoicheff, J. R. Banic, P. Herman, W. Jamroz, P. E. LaRocque and R. H. Lipson, Laser Techniques for Extreme Ultraviolet Spectroscopy, T. J. McIlrath and R.R. Freeman, editors (American Institute of Physics, New York, 1982), p.19.

[14] I. Dabrowski and G. Herzberg, *Can. J. Phys.*, **52**, 1110 (1974).

[15] L. L. Strow, *J. Mol. Spectrosc.*, **97**, 9 (1983).

[16] G. Duxbury, M. Horani and J. Rostas, *Proc. Roy. Soc. London*, **331**, 109 (1972).

[17] C. H. Townes and A. L. Schawlow, Microwave Spectroscopy (McGraw-Hill, New York, 1955), p.86.

[18] L. Karlsson, L. Mattsson, R. Jadrny, T. Bergmark and K. Siegbahn, *Physica Scripta*, **13**, 229 (1976).

[19] R. N. Dixon, G. Duxbury, M. Horani and J. Rostas, *Mol. Phys.*, **22**, 977 (1971).

[20] G. Duxbury, Ch. Jungen and J. Rostas, *Mol. Phys.*, **48**, 719 (1983).

[21] H. Frohlich, P. M. Guyon and M. Glass-Maujean, *J. Chem. Phys.*, **94**, 1102 (1991).

[22] R. H. Page, R. J. Larkin, Y. R. Shen and Y. T. Lee, *J. Chem. Phys.*, **88**, 2249 (1988).

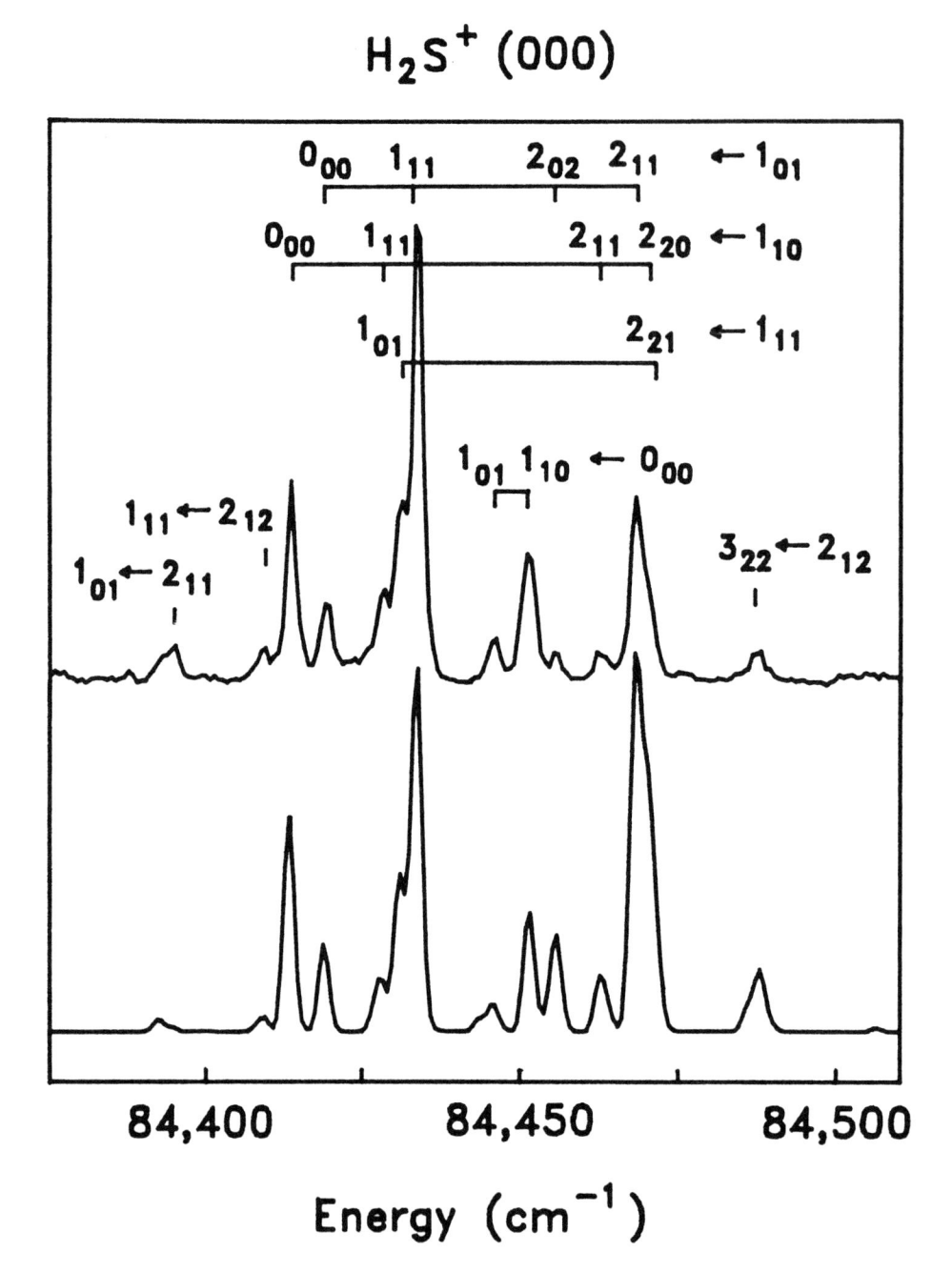

Figure 1: Upper trace: Rotationally-resolved, pulsed field ionization spectrum for one-photon ionization of H_2S to the $\tilde{X}^2B_1$ (000) level of H_2S^+. Lower trace: Simulated photoionization spectrum of H_2S at 15 K including both type A ($\Delta K_c = \pm 1$, $\Delta K_a = 0$) and type of C ($\Delta K_c = 0$; $\Delta K_a = \pm 1$) rotational transitions (1:4 ratio) and assuming ionization energy of 84,434 cm^{-1}.

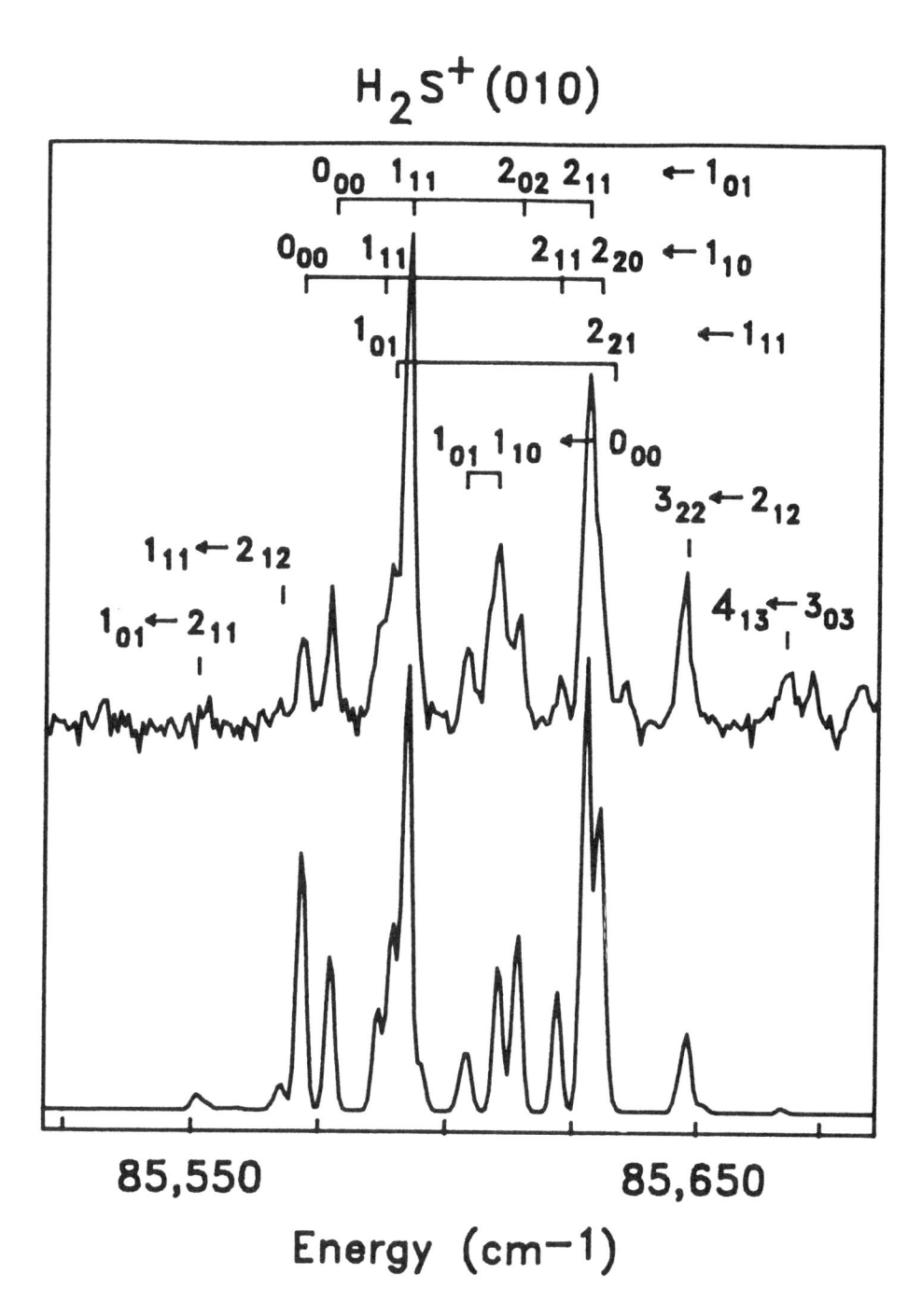

Figure 2: Upper trace: Rotationally-resolved, pulsed field ionization spectrum for one-photon ionization of H_2S to the $\tilde{X}^2B_1$ (010) excited bend level of H_2S^+. Lower trace: Simulated photoionization spectrum of H_2S at 15 K including both type A ($\Delta K_c = \pm 1$, $\Delta K_a = 0$) and type C ($\Delta K_c = 0$; $\Delta K_a = \pm 1$) rotational transitions (1:4 ratio) and assuming an adiabatic ionization energy of 85,592 cm^{-1}.

Ultrafast Raman echo experiments in the liquid phase

Mark Berg, Laura J. Muller and David Vanden Bout

Department of Chemistry and Biochemistry, University of Texas, Austin, Texas 78712

ABSTRACT

The Raman echo experiment, which can distinguish homogeneous and inhomogeneous broadening of vibrational spectra, is reported for the first time in liquids. The symmetric methyl stretch of acetonitrile, which has been extensively studied with contradictory results, is shown to be homogeneously broadened. The vibration interacts with only with rapid collisional motions of the solvent.

INTRODUCTION

The dephasing of vibrational transitions is a powerful probe of liquid phase dynamics.[1] The dephasing time is determined by both the rate and magnitude of the solvent perturbations exerted on the vibration. Dephasing times have been extensively measured from isotropic Raman line widths in the frequency domain[1-6] and time-resolved coherent Raman scattering in the time domain.[7-12] However, major questions about these measurements and about liquid dynamics can only be answered if it is known whether the dephasing process is homogeneous or inhomogeneous.

Solvent motions which are much slower than the dephasing time (typically ~1 ps) cause inhomogeneous broadening, while motions much faster than the dephasing time cause homogeneous broadening. These are limiting behaviors and intermediate cases are also possible. Interaction mechanisms with a variety of rates, such as collisions, or fluctuations of the local density, electric field, or concentration, are conceivable. Determining where a liquid lies along the homogeneous-inhomogeneous continuum gives an important piece of information about the solvent motions which couple to internal coordinates. In addition, further interpretation of vibrational dephasing experiments cannot be made, without knowing the homogeneity of the transition.

Many indirect experimental and theoretical arguments have been advanced for homogeneous or inhomogeneous broadening in particular cases, often with contradictory results.[2-11, 13-15] It has been shown that the experiments performed so far cannot give a direct measure of the homogeneity of a vibrational transition.[16,17] An unambiguous determination requires a multiple pulse experiment which can exploit a nonlinear response of the material. Specifically, the Raman echo[18] has been theoretically demonstrated to directly measure the homogeneity of vibrational dephasing.[17]

Unfortunately, the difficulty of the Raman echo has delayed its experimental execution. The effect first arises in seventh order perturbation theory and corresponds to a $\chi^{(7)}$ process. As a result, it has a low cross section and requires high light intensities. Subpicosecond time resolution is needed for most vibrations. In addition, synchronized pulses at two different and tunable frequencies are needed. At least one group has argued that the Raman echo is not experimentally possible in liquids.[19]

In this paper, the feasibility of the Raman echo is demonstrated and results on the symmetric methyl stretching vibration of acetonitrile presented.[19] This vibration has been extensively investigated by line shape analysis,[2-6] time-resolved coherent Raman scattering experiments (TRCRS),[8-11] computer simulations[15] and theory[13,14] with contradictory results on its homogeneity. Here it is conclusively shown that the transition is homogeneously broadened and therefore that the vibration interacts primarily with rapid, collision-like motions of the solvent.

THE RAMAN ECHO EXPERIMENT

Time domain coherence experiments are well known in a variety of contexts.[21,22] The simplest of these experiments is the free induction decay (FID), which consists of exciting the sample with a single pulse and then monitoring the decay of the resulting coherence as a function of time. The coherence decay is exactly the Fourier transform of the line shape of the transition in the frequency domain and contains no information on the homogeneity of the decay.

The homogeneity can be determined by an echo experiment. In an echo, a second pulse excites the sample at a time t_1 after the first pulse and the coherence is measured as a function of time t_2 after the second pulse. The second pulse induces a rephasing of any inhomogeneous processes, but cannot rephase homogeneous ones. Thus the form of the coherence decay is sensitive to the degree of homogeneity.

Loring and Mukamel pointed out the similarity of TRCRS and the FID and therefore the inability of TRCRS to determine vibrational homogeneity.[17] In TRCRS, two synchronous visible laser pulses are combined to create a single pulse of vibrational excitation by stimulated Raman scattering. The resulting coherence is monitored as a function of time by coherent Raman scattering from a third visible laser pulse.

Loring and Mukamel went on to show that the Raman echo is an analogue of other echo experiments and is sensitive to the vibrational homogeneity.[17] In the Raman echo, two pairs of visible pulses act on the vibration, each pair causing a single pulse of vibrational excitation, and the resulting coherence is monitored through coherent Raman scattering by a fifth laser pulse.

Thus the Raman echo can be interpreted in terms of the well established behavior of other echo experiments such as the spin or photon echoes. However, the unique physical properties of liquid dynamics can place the experiment in limits which are not commonly found in other situations. For example in most other contexts, the inhomogeneous broadening is much larger than the homogeneous, which leads to the simple result that the echo coherence will form a narrow peak at $t_1 = t_2$. In liquids, the amount of inhomogeneous broadening is expected to be comparable to the amount of homogeneous broadening. As a result, the coherence would be delayed and broadened, but would not be expected to form a classic "echo" (see below).

The type of inhomogeneity just discussed assumes the vibration is perturbed by two independent processes: one rapid and homogeneous, and one slow and inhomogeneous. Another potential cause of a nonhomogeneous vibrational line is a single perturbation process with a modulation rate intermediate between the homogeneous and inhomogeneous limits. Although seldom encountered in other phases, this is a plausible scenario for liquids. Again the resulting echo coherence decay would not have the classic form, but would be readily distinguishable from a homogeneous decay (see below).

PREDICTED RESULTS

Although the echo decay shapes expected are not the most well studied forms, the theory for the echo decay shape is well established for all cases. Recall that the echo signal $S_{RE}(t_1, t_2)$ is a function of two variables, the time between the first two pulse pairs, and the time from the second pulse pair to the probe pulse. Regardless of the homogeneity of the line, the signal decay in t_1 with t_2 held constant will always match the FID signal decay. This occurs because there is no rephasing during the first time interval.

The interesting dynamic information is obtained by varying t_2 with t_1 held constant. If $t_1 = 0$, the signal decay will again match the FID decay regardless of the nature of the vibrational perturbations. With no time for dephasing during the first time interval, there is nothing to rephase during the second time interval.

If the line is completely homogeneous, there will be no rephasing possible even as t_1 is increased. Thus a homogeneous line is characterized by a signal decay shape in t_2 which is independent of t_1 and always matches the FID decay shape (Fig. 1A).

If the vibrational perturbations are caused by two processes, one homogeneous with a homogeneous dephasing time of T_2, and one inhomogeneous with a Gaussian frequency distribution of standard deviation σ, the shape of the echo decay is given by[21]

$$S_{RE}(t_1, t_2) = \exp[-2(t_1 + t_2)/T_2] \exp[-\sigma^2 (t_2 - t_1)^2]$$

An example of the decays for a line with homogeneous and inhomogeneous contributions of equal width are shown in Fig. 1B. A 1.18 ps FWHM Gaussian has been convolved with the signal to represent the experimental time resolution.

In the alternative model of a single modulation process with frequency width σ and a modulation time τ, the echo signal is[23]

$$S_{RE}(t_1, t_2) = \exp\{-2(\sigma\tau)^2 [(t_2/\tau) + 2 \exp(-(t_2-t_1)/\tau) + 2 \exp(-t_1/\tau) - \exp(-t_2/\tau) - 3]\}$$

Homogeneous broadening occurs when $\sigma\tau \ll 1$; a nonhomogeneous line results from violating this limit. An example of the signal decays for a longer modulation time ($\sigma\tau = 1.9$) is shown in Fig. 1C

The results for these two specific models illustrate a general principle. Regardless of whether a classic "echo" consisting of a sharp delayed peak in the signal is seen, the form of the echo decay provides a sensitive and quantifiable measure of the dynamic of the vibrational perturbations, which cannot be obtained from TRCRS or from Raman line shapes.

Fig. 1 Calculated Raman echo decays at $t_1 = 0$ ps and $t_1 = 4$ ps for three dynamic models. In each case, the corresponding Raman line has a FWHM of 5.0 cm^{-1}. A) Homogeneous broadening ($T_2 = 2.12$ ps). B) Homogeneous and inhomogeneous components with equal FWHM's ($\sigma = 1.28$ cm^{-1}, $T_2 = 3.43$ ps). C) Single process with slow modulation, $\sigma\tau = 1.9$ ($\sigma = 2.56$ cm^{-1}, $\tau = 4$ ps).

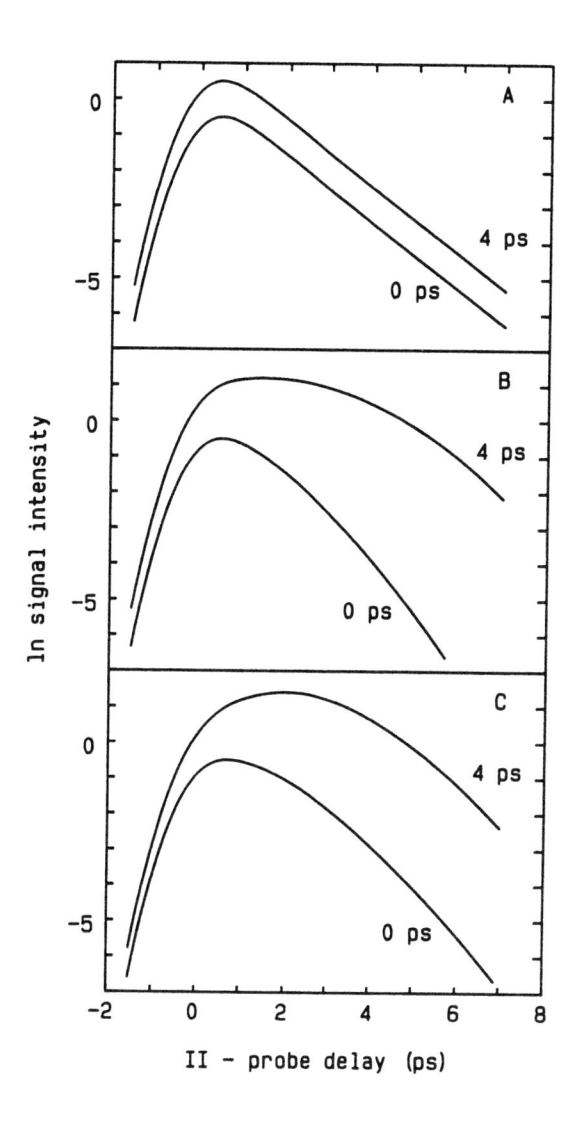

EXPERIMENTAL

To generate short pulses at two frequencies, two synchronously-pumped dye lasers tuned to 570 nm (yellow) and 683 nm (red) were used. To reduce the pulse width and decrease the timing jitter between the lasers, they were pumped by a single CW mode-locked Nd:YAG laser whose output was compressed to 2 ps before frequency doubling. The high intensities needed were obtained by amplifying each beam in an independent chain of dye amplifiers pumped by a Q-switched Nd:YAG laser.

The amplified yellow pulses typically had a pulse width of 0.45 ps (sech^2) which determined the time resolution of the experiment. The red pulses were kept longer at 0.85 ps to reduce the effects of timing jitter. Typical amplified energies were 0.55 mJ in the yellow pulse and 0.15 mJ in the red pulse.

The amplified beams were divided by beam splitters, recombined on dichroic filters and focussed onto the sample to meet the phase matching condition

$$k_e = k_{L3} + 2(k_{L2} - k_{S2}) - (k_{L1} - k_{S1}),$$

where k_e is the wave vector of the echo signal, k_{L1}, k_{L2} and k_{L3} are the wave vectors of the yellow pulses and k_{S1} and k_{S2} are the wave vectors of the red pulses. The pulses with wave vectors k_{L1} and k_{S1} were combined collinearly and temporally coincident to form the first excitation pulse pair (I). The pulses with wave vectors k_{L2} and k_{S2} were similarly combined to form the second excitation pair (II), which had an angle of 125 mrad from the first excitation pair. The probe pulse k_{L3} was brought in at an angle of 125 mrad from the second excitation pair to generate coherent anti-Stokes scattering in the k_e direction, collinear with the probe.

The Raman echo signal was isolated by a bandpass filter and a pinhole before detection on a photomultiplier tube. The Raman echo signal was identified by its appearance at the anti-Stokes' frequency, its appearance at the correct phase matching angle and by its disappearance when any of the five beams was blocked.

The major source of interference was nonphase-matched anti-Stokes scattering of the probe beam from the coherence generated by the second excitation beam. This signal appeared only 10 mrad from the echo signal. A chopper blocked the first red pulse on alternate shots and the amount of this background was measured and subtracted from the signal.

RESULTS

The first measurement is the FID or TRCRS signal needed for comparison to the echo results. This experiment has previously been done on acetonitrile[10,11] and should correspond to the Fourier transform of the isotropic Raman line.[2] Our results are shown in Fig. 2. The decay is exponential with a decay constant of 0.8 ps in agreement with the previous results.

Figure 3 shows the decay of $S_{RE}(t_1, t_2)$ with t_1 when t_2 is held constant. As predicted, this decay exactly duplicates the FID signal.

Fig. 2 Points: Natural log of the Raman free induction decay intensity versus delay between excitation and probe. Line: Decay predicted from the isotropic Raman line shape.

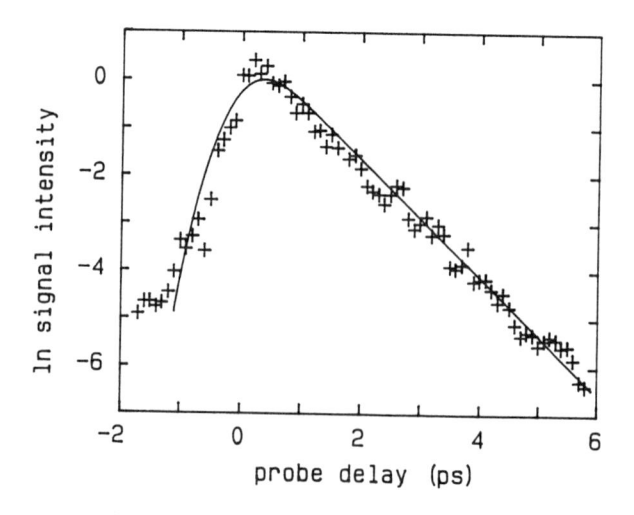

Fig. 3 Points: Natural log of the Raman echo intensity versus t₁ with t₂ = 1 ps. Line: Decay predicted from the isotropic Raman line shape.

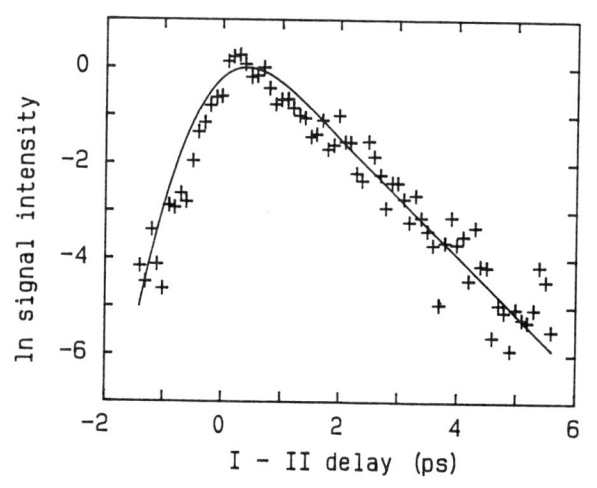

The decay of $S_{RE}(t_1, t_2)$ with t_2 for several values of t_1 is given in Fig. 4. As anticipated, the signal for $t_1 = 0$ again matches the FID decay. The critical question is whether this decay changes as t_1 is increased. The results are unambiguous that the decay is unperturbed up to $t_1 = 4$ ps. This shows that the vibration is completely homogeneous.

The sensitivity of the experiment to the presence of inhomogeneity is illustrated in Fig. 5. The experimental data are compared to predicted behavior for a second inhomogeneous process in Fig. 5A. If an inhomogeneous process as large as indicated by earlier coherent Raman experiments[9] or theoretical calculations[13,14] (5.0 cm⁻¹ FWHM) were present, the echo would have been dramatically different. Even an inhomogeneous width 35% of the total line width would be clearly detectable.

In Fig. 5B, the consequences of a single, slow process are examined. If the perturbation correlation time were as short as 0.5 ps, detectable effects would have been present. Thus, the correlation time for solvent interaction with the vibration must be less than 0.5 ps.

CONCLUSIONS

These experiments have demonstrated the feasibility of Raman echo experiments in the liquid phase. In the system examined, the symmetric methyl stretch of acetonitrile, the vibration was shown to be completely homogeneous. To be in this limit, the lifetime of the solvent perturbations must be significantly less than 0.5 ps. This is approaching the velocity autocorrelation time and so the perturbations could be appropriately described as resulting from single collisions.

In many ways this system is a good candidate for homogeneous broadening. Acetonitrile is known to be a rapidly relaxing solvent by other measures. In addition, the interaction of the symmetric methyl stretch with the solvent are not particularly strong as indicated by the relatively narrow Raman line width. It is not clear if this result will also hold for vibrations with stronger interactions or for more highly structured solvents.

Fig. 4 Points: Natural log of the Raman echo intensity versus t_2 for various values of t_1. Lines: Identical decays predicted for a homogeneous vibrational line.

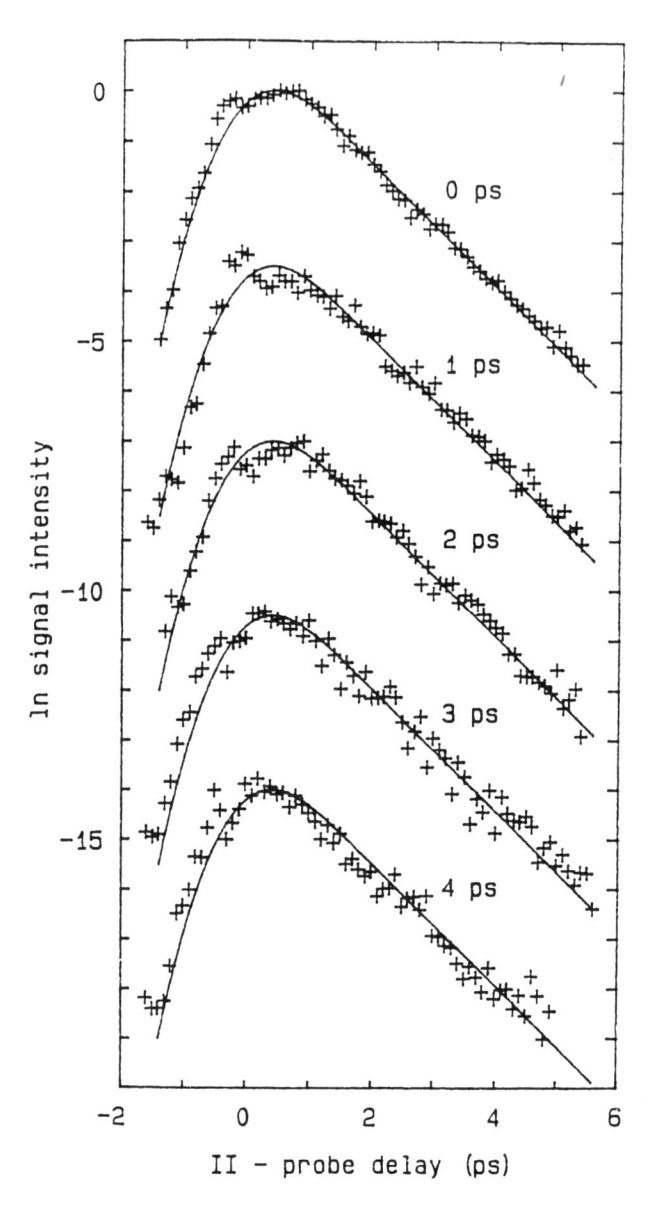

The Raman echo now promises to answer these questions about solvent dynamics. Particularly important is that the Raman echo does not simply tell whether slower coordinates exist in liquids. Rather it tells whether those coordinates interact significantly with intramolecular coordinates, and thus whether they can influence intramolecular processes of chemical interest.

ACKNOWLEDGEMENTS

This work was supported by the National Science Foundation (CHE-8806381). MB is a Presidential Young Investigator (1990-95). LJM is a National Physical Science Consortium Fellow.

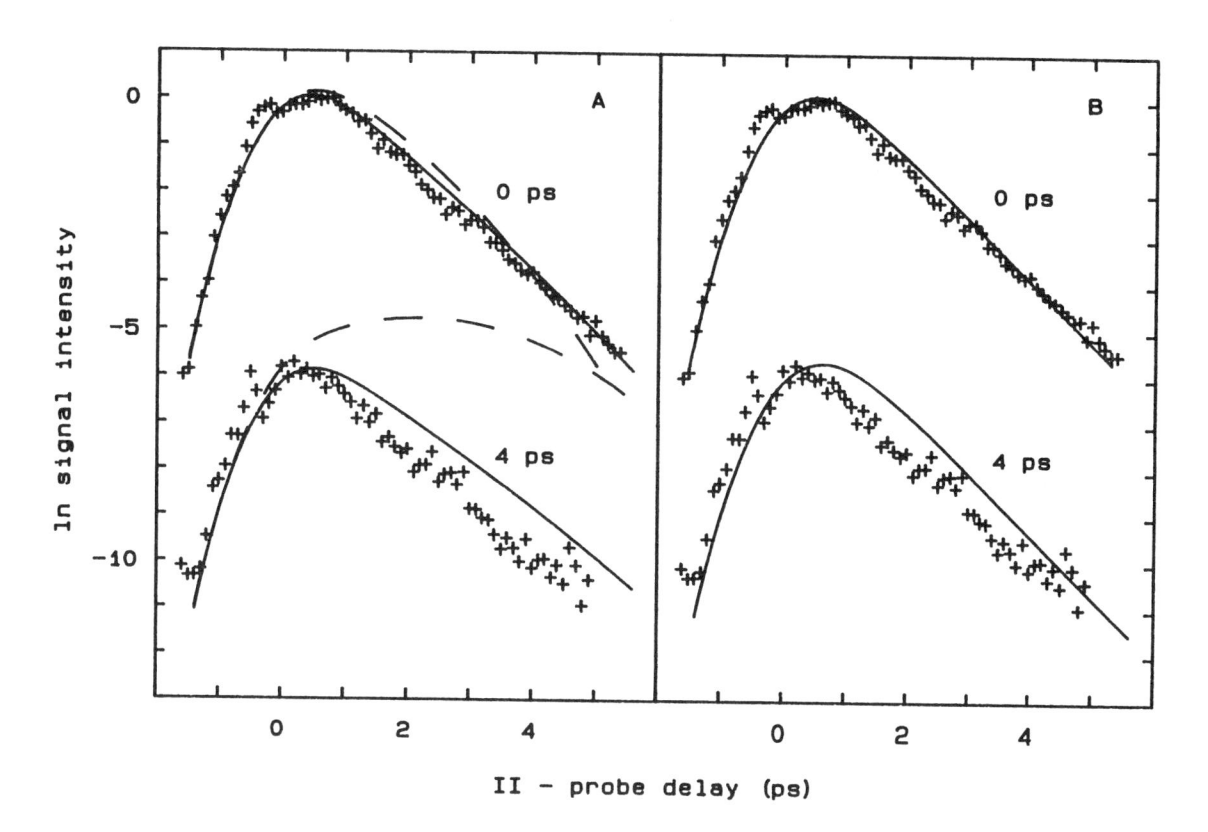

FIG. 5. Points: Natural log of the Raman echo intensity versus t_2 for $t_1 = 0$ and 4 ps. A) Comparison to a model including both homogeneous and inhomogeneous components. Dashed lines: Predicted decays for a vibrational line with a substantial inhomogeneous broadening [$T_2 = 3.4$ ps, $\sigma = 2.12$ cm^{-1} (5 cm^{-1} FWHM)] typical of theoretical predictions[13,14] and earlier TRCRS experiments.[9] Solid lines: Predicted decays for a vibrational line with 35% inhomogeneous broadening ($T_2 = 1.9$ ps, $\sigma = 1.3$ cm^{-1}) indicating the sensitivity of the technique. B) Comparison to a model of a single modulation process with a modulation time of $\tau = 0.5$ ps ($\sigma = 6.15$ cm^{-1}).

REFERENCES

1. D.W. Oxtoby, Adv. Chem. Phys. **40**, 1 (1979).
2. J. Schroeder, V.H. Schiemann, P.T. Sharko and J. Jonas, J. Chem. Phys. **66**, 3215 (1977).
3. J. Yarwood, R. Arndt and G. Doge, Chem. Phys. **25**, 387 (1977); G. Doge, A. Khuen and J. Yarwood, *ibid*. **42**, 331 (1979).
4. J. Yarwood, R. Ackroyd, K.E. Arnold, G. Doge and R. Arndt, Chem. Phys. Lett. **77**, 239 (1981).
5. K. Tanabe, Chem. Phys. **38**, 125 (1979).
6. K. Tanabe, Chem. Phys. Lett. **84**, 519 (1981).
7. A. Laubereau and W. Kaiser, Rev. Mod. Phys. **50**, 607 (1978).
8. S.M. George, H. Auweter and C.B. Harris, J. Chem. Phys. **73**, 5573 (1980).
9. S.M. George, A.L. Harris, M. Berg and C.B. Harris, J. Chem. Phys. **80**, 83 (1984).
10. G.M. Gale, P. Guyot-Sionnest and W.Q. Zheng, Opt. Commun. **58**, 395 (1986).
11. M. Fickenscher and A. Laubereau, J. Raman Spectrosc. **21**, 857 (1990).

12. W. Zinth, R. Leonhardt, W. Holzapfel and W. Kaiser, J. Quantum Electron. QE-24, 455 (1988).
13. K.S. Schweizer and D. Chandler, J. Chem. Phys. 76, 2296 (1982).
14. S.M. George and C.B. Harris, J. Chem. Phys. 77, 4781 (1982).
15. P.-O. Westlund and R.M. Lynden-Bell, Mol. Phys. 60, 1189 (1987); R.M. Lynden-Bell and P.-O. Westlund, *ibid.* 61, 1541 (1987).
16. W. Zinth, H.-J. Poland, A. Laubereau and W. Kaiser, Appl. Phys. B 26, 77 (1981).
17. R.F. Loring and S. Mukamel, J. Chem. Phys. 83, 2116 (1985).
18. S.R. Hartman, IEEE J. Quantum Electron. QE-4, 802 (1968); K.P. Leung, T.W. Mossberg and S.R. Hartman, Opt. Commun. 43, 145 (1982).
19. D. Vanden Bout, L.J. Muller and M. Berg, Phys. Rev. Lett., in press.
20. M. Muller, K. Wynne and J.D.W. van Voorst, Chem. Phys. 128, 549 (1988).
21. M. Levenson, *Introduction to Nonlinear Laser Spectroscopy*, Sec. 6.3 (Academic, NY, 1982).
22. C.P. Slichter, *Principles of Magnetic Resonance*, (Springer-Verlag, NY, 1990).
23. R.F. Loring and S. Mukamel, Chem. Phys. Lett. 114, 426 (1985).

Photophysics of rigidized 7-aminocoumarin laser dyes

M. R. V. Sahyun

3M Graphic Research Laboratory, St. Paul, MN 55144

D. K. Sharma

Canadian Centre for Picosecond Laser Flash Spectroscopy
Department of Chemistry, Concordia University
Montreal, Quebec, Canada H3G 1M8

ABSTRACT

An unusual slow rise of fluoresence intensity from Coumarins 102, 314 and 334, observed under conditions of high intensity laser excitation prompted further investigation of their photophysics by photoquenching, time-resolved transient absorption spectroscopy and time-resolved amplified stimulated emission. The slow rise was found to be the consequence of multi-photon excitation. Principal fate of coumarin molecules so excited was shown to be intersystem crossing to the non-emissive triplet manifold. Rate of intersystem crossing was estimated as ca. 10^{10} s^{-1}; energy matching between a higher excited singlet state, S_n, and a higher triplet, T_n, is implicated.

1. INTRODUCTION

Recently we observed a slow rise for the fluorescence of Coumarins 102, 314 and 334 in acetonitrile (AN), toluene, or methanol solution under conditions of high intensity laser excitation[1]. The experiments were carried out with 355 nm frequency-tripled Nd:YAG excitation, ca 0.7 - 2.7 mJ/pulse, 30 ps FWHM, focused to a 6 mm spot on the sample cuvette. A typical result obtained with a Hamamatsu streak camera is shown in Fig. 1. Curve fitting procedures allowed estimation of the rise time as ca. 100 ps; the radiative lifetime was similarly estimated as 3.5 ns, in agreement with the literature[2].

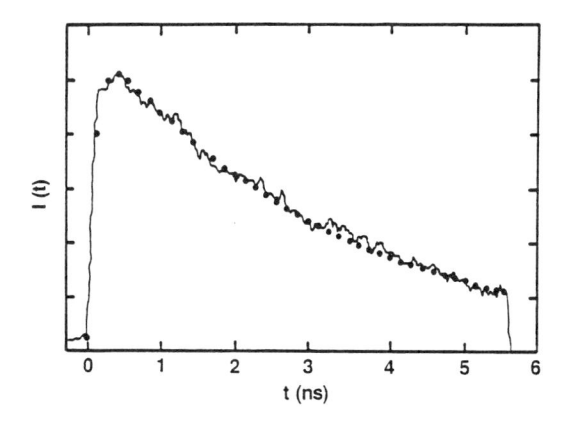

Fig. 1. Time resolved fluorescence of Coumarin 314 in AN.

Our results were at variance with other reports on time-resolved photoemission of these laser dyes, probed at lower excitation density[3,4]. All three of the dyes are characterized by incorporation of the amino function into a rigid julolidyl ring system to eliminate relaxation to a so-called TICT state[4,5]; multiple fluorescence was excluded, both in our work and in the low intensity studies[3,4]. We accordingly undertook the present work to identify the photophysical basis for our observations, and to infer limitations in the utility of these and structurally similar dyes for dye laser applications.

2. RESULTS AND DISCUSSION

2.1 Photoquenching

Photoquenching is the process by which emission efficiency of a fluorophore decreases with increasing energy of the exciting laser pulse, I_p, owing to increasing involvement of multi-photon processes[2]. Photoquenching experiments were carried out on the three coumarin laser dyes of interest. Raw fluorescence intensity data for Coumarin 314 in AN, again with 355 nm excitation, are shown in Fig. 2. A plot of reciprocal relative quantum efficiencies for fluorescence for the same system is shown in Fig. 3. Its linearity is diagnostic for photoquenching by two-photon excitation.

From the slope of the plot, a cross-section for photoexcitation of the coumarin in its lowest excited, emissive singlet state by absorption of a second photon to populate S_n, is estimated to be $(1.0 \pm 0.1) \times 10^{-17}$ cm^2. Similar values were observed for Coumarins 102 and 334. Our estimates differ significantly from literature values for the same dyes[2]. We infer that observation of photoquenching may depend significantly on experimental conditions.

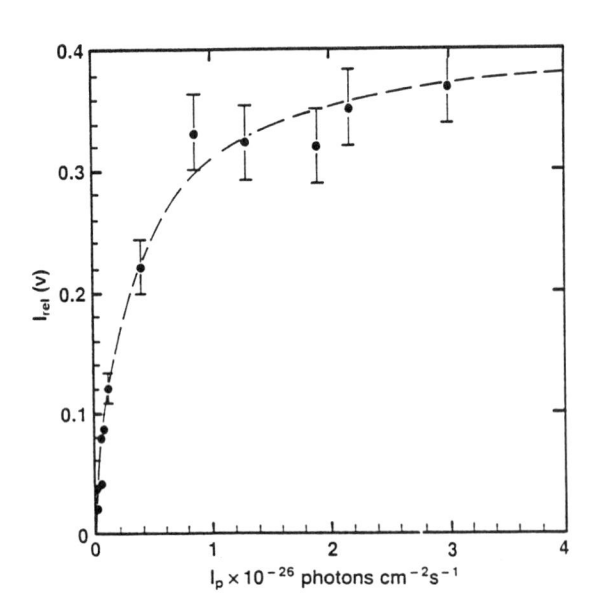

Fig. 2. Photoquenching of Coumarin 314 at 355 nm.

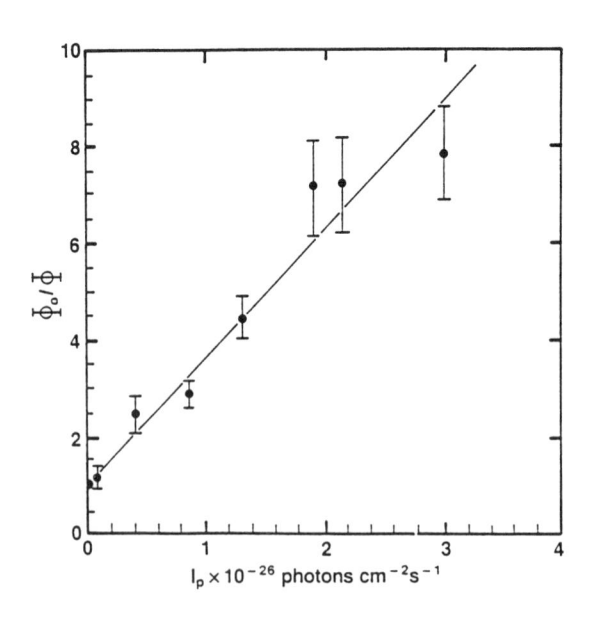

Fig. 3. Linearized representation of data of Fig. 2.

At the same time, we are led to identify the fluorescence rise time with the lifetime for internal conversion of S_n to emissive S_1. The high levels of photoquenching observed with the coumarins suggest, however, that this is not the principal fate of the S_n species. The usual model for photoquenching[2] is independent of the decay pathway for S_n, and the observation of photoquenching, in intself, provides no insight in this regard. We can infer that these julolidylcoumarins are efficient laser dyes only in applications involving relatively low energy pumping, i.e. $I_p \leq 10^{25}$ photons cm^{-2} s^{-1}.

2.2 Transient absorption spectroscopy

Transient absoprtion spectra were recorded as previously described[6] for Coumarins 314 and 334 in AN. Traces obtained 50 ps, 500 ps, 5 and 10 ns after 355 nm excitation (2.5 mJ pulse, 30 ps FWHM) are shown in Figs 4 and 5 for these two coumarins respectively. Three features are apparent in both sets of spectra. (1) Ground state absorption, centered at 427 and 452 nm for the two coumarins, respectively, is bleached. This is purely a transient effect, and no long term spectroscopically observable changes occur in the samples as a result of these experiments. (2) Amplified stimulated emission (ASE) is observed in the region of the usual lasing activity of these dyes (ca 500 nm); as expected, intensity of ASE decays at a rate comparable to the raditive lifetime of S_1, above, while ground state absorption is not significantly regenerated on the same time scale. (3) A broad, weak transient centered on 600 nm is observed for both coumarins. By comparison with independently generated absorption spectra of lowest triplet states of coumarins of this series[4], we assign the this absorption to triplet coumarin, which does not normally form under conditions of low intensity excitation.

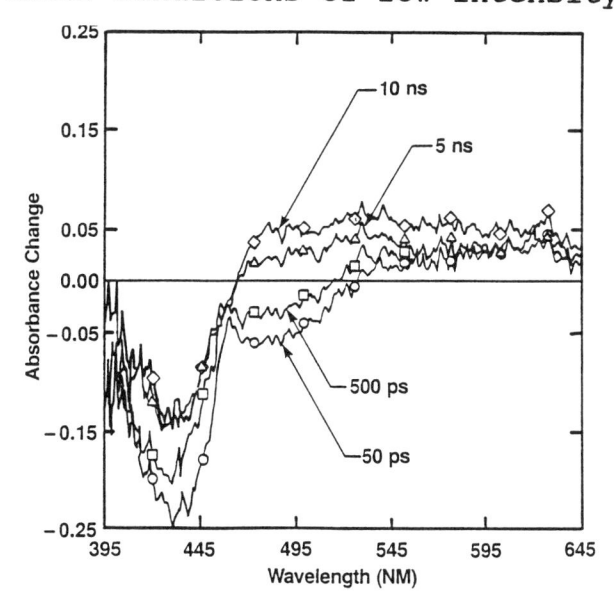

Fig. 4. Transient absorption spectra recorded 50 ps, 500 ps, 5 and 10 ns after 355 nm laser excitation of Coumarin 314 in AN.

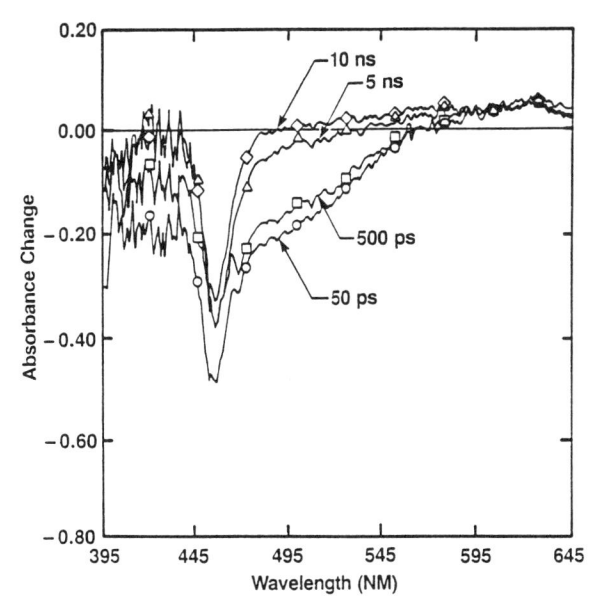

Fig. 5. As Fig. 4 for Coumarin 334.

3. CONCLUSIONS

The results obtained in the photoquenching, transient absorption, and time-resolved ASE studies of Coumarins 102, 314 and 334 can be accomodated by the scheme of Fig. 6. Two-photon excitation populates S_n which undergoes stepwise internal conversion, eventually to emissive S_1. One step, most likely S_2 to S_1, is rate determining, giving rise to the observed time constant for onset of fluorescence under our conditions. Lifetimes of ca 40 ps have been inferred by Speiser and Shakkour[2] for higher singlets of similar compounds in their photoquenching studies.

The intermediate state, e.g. S_2, is also, we propose, matched in energy with a higher triplet, T_m. Intersystem crossing is accordingly efficient and its rate actually limits S_2 lifetime, i.e. the observed rate of fluorescence rise corresponds to this intersystem crossing rate. A rate constant of 4×10^{10} s^{-1} has been reported for intersystem crossing of S_2 acenaphthylene[7].

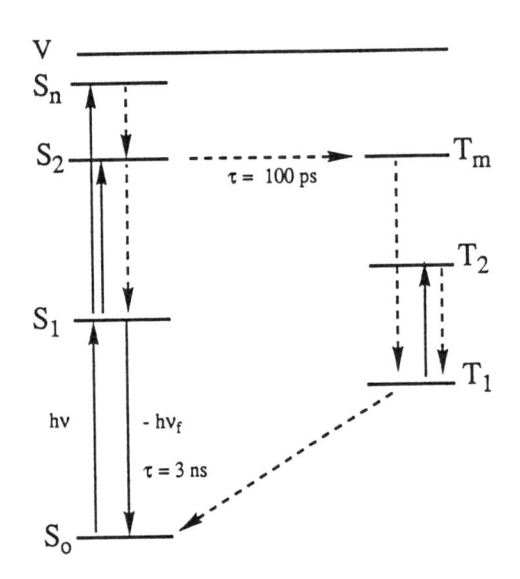

Fig. 6. Proposed energy level diagram for julolidylcoumarin laser dyes.

From photoquenching we can infer that these julolidylcoumarins are efficient laser dyes only in applications involving relatively low energy pumping, i.e. $I_p \leq 10^{25}$ photons cm^{-2} s^{-1}. The cross section estimated in our photoquenching studies is quite different from Speiser and Shakkour's value[2]; we would accordingly also caution other workers that photoquenching parameters may be very sensitive to experimental conditions. Finally the efficient, rapid population of the coumarin triplet manifold on two-photon excitation commends two-photon excitiation to photochemists intent on preparation and study of otherwise difficult to obtain triplet species.

4. REFERENCES

1. M. R. V. Sahyun and D. K. Sharma, "Photophysics of rigidized 7-aminocoumarin laser dyes: efficient intersystem crossing from higher excited state," Chem. Phys. Lett., in press.
2. S. Speiser and N. Shakkour, "Photoquenching parameters for commonly used laser dyes," Appl. Phys. B **38**, 191-197 (1985).
3. W. Jarzeba, M. Kahlow and P. F. Barbara, "Ultrafast fluorescence measurements in chemistry," J. Imaging Sci. **33**, 53-7 (1989).
4. R. W. Yip and Y. X. Wen, "Photophysics of 7-diethylamino-4-methylcoumarin: picosecond time-resolved absorption and amplified emission study," J. Photochem. Photobiol., A: Chem. **54**, 263-70 (1990).
5. G. Jones II, W. R. Jackson, C.-y. Choi, and W. R. Bergmark, "Solvent effects on emission yield and lifetime for coumarin laser dyes," J. Phys. Chem. **89**, 294-300 (1985) and references cited therein.
6. N. Serpone, D. K. Sharma, J. Moser, and M. Graetzel, "Reduction of acceptor relay species by conduction band electrons of colloidal TiO_2," Chem. Phys. Lett. **136**, 47-51 (1987); see also U. P. Wild, "Kurzzeit-photochemie," Chimia **27**, 421-7 (1973).
7. A. Samanta, "Direct evidence for intersystem crossing involving higher excited states of acenaphthylene," J. Amer. Chem. Soc. **113**, 7427-9 (1991).

OPTICAL METHODS FOR
TIME- AND STATE-RESOLVED CHEMISTRY

SESSION 5

Optical Techniques and Molecular Spectroscopy

Chair
Edward R. Grant
Purdue University

Population transfer by stimulated Raman scattering with delayed pulses: the concept, some problems and experimental demonstration

K. Bergmann, S. Schiemann, and A. Kuhn

Fachbereich Physik der Universität, Postfach 3049, W–6750 Kaiserslautern

ABSTRACT

We discuss recent progress in the attempt to gain complete control over the internal state population of atoms and molecules for collision dynamic studies and spectroscopy. We describe a method which relies on stimulated Raman scattering with pulses delayed in a counter–intuitive way. This method (STIRAP) allows highly efficient transfer of population between atomic or molecular levels. It allows, in particular selective population of highly excited vibrational levels.

1. INTRODUCTION

We discuss recent progress in the attempt to gain complete control over the internal state population for collision dynamics and spectroscopy.

Early experiments with state selection involving the lowest or first excited vibrational level in the electronic ground state of molecules employed infrared excitation, optical pumping depletion labelling or off–resonance stimulated Raman scattering. For a review see ref. (1).

Recently, the interest is shifting to experiments involving also highly vibrationally excited states. These states are thermally not populated, neither in molecular beams nor in cells and population transfer methods are required. Obviously, lasers need to be involved. Ideally, such a technique should allow efficient and selective transfer of population to a predetermined level. It should also be robust with respect to small variations of laser frequencies or intensities.

Successful techniques which are in routine use for studies in collision dynamics or spectroscopy of highly vibrationally excited states since many years are the so–called "Franck–Condon pumping" method FCP[1-6]), which relies on population transfer by spontaneous emission following laser excitation and "stimulated emission pumping" (SEP)[7-10,14].

The main advantage of the FCP method is its simplicity: only one laser is needed to excite molecules to a level of the electronically excited state. Vibrationally excited levels in the electronic ground state are populated by subsequent spontaneous emission. An obvious disadvantage is the lack of selectivity. Selectivity and efficiency are significantly enhanced when a second laser is used to stimulate a fraction of the population in the electronically excited level into a predetermined vibrational level of the electronic ground state, as realized in SEP.

The recently developed "stimulated Raman scattering involving adiabatic passage" technique (STIRAP[11-24]) which will be discussed in the remainder of this report holds great promise for a major step forward in terms of efficiency and selectivity of the population transfer process. In fact, as we shall see, it combines features of a stimulated Raman process with the concept of adiabatic passage, well known in nuclear magnetic resonance[25] but also applied in the infrared and visible part of the spectrum[26-29].

2. THE STIRAP CONCEPT

Consider an open three level system consisting of states 1, 2 and 3, coupled by two lasers, the pump laser connecting the levels 1 and 2 and the Stokes laser connecting the levels 2 and 3. The level system is considered "open" since the population in level 2 may decay by spontaneous emission to (many) levels in addition to the levels 1 and 3. Levels 1 and 3 may be vibrational levels of a molecule in the electronic ground state typically with the vibrational quantum numbers v = 0 and v >> 1, respectively, while level 2 is a level in an electronically excited state. Alternatively, all levels can be electronically excited states of an atom or a molecule, typically with level 3 and level 1 being metastable[15]. Consider furthermore particles in a molecular beam crossing the two laser beams at right angle. Alternatively, consider particles in a beam or a cell interacting with radiation from two pulsed lasers.

It is the goal to achieve complete population transfer from level 1 to level 3. Obviously strong coupling between these levels is required. This calls for the tuning of both laser frequencies to resonance with their respective transitions 1↔2 and 2↔3. We emphasize, however, that any transient population established in level 2 may decay to levels other than level 1 and 3.

Because strong radiative coupling, given by the Rabi frequency [30]

$$\Omega = (\vec{\mu}\vec{E})/\hbar \tag{1}$$

is required. The application of perturbation theory is not appropriate. Rather, the dressed state approach [31] is needed. Here μ is the transition dipol moment and E is the electric field of the laser radiation. Invoking the rotating wave approximation[30] which is well justified for the experimental conditions relevant here, and neglecting (at this stage of the discussion) collisional or radiative decay, the eigenfunctions of the Hamiltonian including the interaction with the radiation and written as linear combination of the bare system states $|1>$, $|2>$ and $|3>$ are [12,13]

$$|a^+> = \sin\theta \sin\Phi \,|1> - \sin\Phi \,|2> + \cos\theta \cos\Phi \,|3> \tag{2a}$$

$$|a^0> = \cos\theta \,|1> - \sin\theta \,|3> \tag{2b}$$

$$|a^-> = \sin\theta \sin\Phi \,|1> + \cos\Phi \,|2> + \cos\theta \sin\Phi \,|3> \tag{2c}$$

with

$$\cos\theta = \frac{\Omega_s}{\sqrt{\Omega_s^2 + \Omega_p^2}} \quad , \quad \sin\theta = \frac{\Omega_p}{\sqrt{\Omega_s^2 + \Omega_p^2}} \,. \tag{2d}$$

The indices s and p refer to the Stokes and pump laser, respectively. The angle Φ depends on the Rabi frequencies as well as on the detuning Δ of the lasers from their respective resonances[13]. For simplicity, we restrict most of the discussion to the case $\Delta = 0$, for which $\Phi = \pi/4$ is valid.

A more complete notation would also include the photon numbers of both radiation fields, which are reduced or increased by one when a stimulated absorption or emission process, respectively, occurs. Since we are interested only in the molecular part of the dressed states, we do not keep track of the photon numbers in the equations.

The dressed eigenstate $|a^0>$ is of particular interest since it contains no contribution of level 2, which may be subject to radiative decay. Unlike the states $|a^+>$ and $|a^->$, the state $|a^0>$ is a linear combination of the initial and final state alone. We may speculate at this stage that $|a^0>$ provides the route for efficient and selective transfer to level 3. Closer inspection of eq. (2b) reveals the recipe for achieving complete population transfer. Before the interaction with the laser fields, the particles are in state $|1\rangle$. This state emerges into state $|a^0\rangle$ in the region of strong radiative coupling if

$$|\langle 1|a^0\rangle|^2 \approx 1 \ , \tag{3a}$$

or $\cos^2\theta \approx 1$ at early times of the interaction. Efficient population transfer is achieved, if we also have

$$|\langle a^0|3\rangle|^2 \approx 1 \tag{3b}$$

or $\sin^2\theta \approx 1$, at late times of the interaction. From eq. (2d) we see that eqs. (3a) and (3b) are valid for

$$\Omega_s \gg \Omega_p \text{ at early times} \quad \text{and} \quad \Omega_p \ll \Omega_s \text{ at late times.} \tag{3c}$$

It is the surprising ("counter- intuitive") consequence of eq. (3c) that complete population transfer can in fact be achieved if the interaction begins with the Stokes laser and ends with the pump laser, provided there is a region of overlap between the two.

If the lasers are spatially or temporally shifted to assure this sequence of interaction (the "STIRAP-configuration"), the mixing angle θ, see eq. (2), is initially zero. During the interaction θ increases for smooth pulse shapes monotonically to $\pi/2$. Thus, the molecular part of the dressed state eigenfunction changes smoothly from $|1\rangle$ to $|3\rangle$. No transient population is established in state $|2\rangle$. Therefore, spontaneous emission is eliminated despite the fact that the lasers are tuned to resonance with their respective transitions.

For overlapping pulses and $\Omega_p = \Omega_s$ we find $\cos^2\theta = 0.5$ and $\sin^2\theta = 0.5$, meaning that only a fraction of the population of level 1 emerges into the "dark state" $|a^0\rangle$. In fact a reduction of spontaneous emission by about 25% has been observed in this configuration some time ago[32] and $|a^0\rangle$ has been called the "trapped state", because its population does not leak out of the open three-level system.

The first observation of highly efficient population transfer is reported in ref. (11). A more extensive set of data and a more complete discussion is given in ref. (13). Historically, Hioe, Oreg and Eberly[33] were the first to recognize, in the context of their discussion of coherent phenomena in N-level systems, that the counter-intuitive sequence of interaction may lead to complete population transfer. Hioe and Carroll elaborated on that further[34,35]. They concentrate in particular on analytically tractable solutions for specific pulse shapes $\Omega(t)=f(t)\Omega^0$.

3. CONDITIONS FOR ADIABATIC EVOLUTION DURING THE STIRAP PROCESS

We note that the eqs. (2) are eigenfunctions of the dressed state Hamiltonian at any given time. However, since the interaction with the pump laser is delayed with respect to the interaction with the Stokes laser, i.e.

$$\Omega_s(t) = f(t+\tau)\Omega_s^0 \quad \text{and} \quad \Omega_p(t) = f(t) \Omega_p^0 \ , \tag{4}$$

with $f(t) = 0$ for very early and late times, the ratio of Rabi-frequencies and thus the Hamiltonian is explicitely time dependent. The time dependent state vector $|\Psi\rangle$, describing the evolution of population in the various levels according to $|\langle i|\Psi\rangle|^2$ with $i = 1, 2, 3$, is written as a linear combination of the eigenfunctions of the dressed Hamiltonian, i.e.

$$|\Psi> = a^+|a^+> + a^0|a^0> + a^-|a^-> \qquad (5)$$

where both, the dressed eigenstates and the coefficients are in general time dependent. The time dependent Schrödinger equation is needed to determine of the evolution of $|\Psi>$. However, if the Hamiltonian changes sufficiently slow, the state vector $|\Psi>$ may follow adiabatically the evolution of the dressed eigenstates, eq. (2), in particular $|a^0>$, if the conditions given in eq. (3c) are satisfied.

Applying the criteria for adiabatic evolution given by Messiah[36] we find that

$$\Omega \Delta \tau \gg 1 \qquad (6a)$$

or alternatively

$$\Omega^2 \Delta \tau \gg \frac{1}{\Delta \tau} \, . \qquad (6b)$$

yields a sufficient criterion for complete population transfer for an interaction of duration $\Delta \tau$ with laser light showing negligibly small phase fluctuations during the interaction. The left hand side of eq. (6b) is proportional to the pulse energy. Here we have assumed that the maximum Rabi frequencies related to the interaction with the pump and Stokes laser are approximately equal. Equation 6 was derived[13] for the most critical moment during the interaction, namely in the middle of the overlap region, where the mixing angle θ changes most rapidly and where the non–adiabatic coupling to the radiatively decaying states $|a^+>$ and $|a^->$ is largest. For very early and very late times during the interaction, the splitting of the dressed eigenvalues, which are degenerate for $\Omega_p = \Omega_s = 0$ is small and adiabatic evolution is not guaranteed. Non–adiabatic evolution at very early and very late times has, however, no consequences, since – for optimum overlap – the mixing angle θ is zero or $\pi/2$, respectively. The time or the location in space beyond which eq. (6) is satisfied, may be called the "locking time" or "locking distance" in analogy to similar phenomena well known in atomic physics[37]. A more detailed discussion can be found in ref. 13.

We emphasize that eq. (6) is valid for optimum delay of the interaction of the particle with the pump and Stokes laser fields. If the delay is smaller than the optimum value, adiabatic evolution may still be assured. However, with decreasing delay an increasing population of the states $|a^+>$ and $|a^->$, which are subject to radiative decay, will occur because the mixing angle at early times is no longer zero. For zero delay, the mixing angle is 45 degrees. It is a consequence of this observation that – although details of the temporal evolution of the Rabi frequencies are not important – broad temporal or spatial wings underlying a narrower laser intensity profile may be detrimental to the success of the transfer process. Further details can be found in ref. 38 and will be published elsewhere.

Evaluation of the conditions for adiabatic evolution also reveals that it is not crucial to tune the pump and Stokes laser frequency to their respective one–photon resonance, $\Delta=0$. Adiabatic evolution and complete population transfer can still be achieved for $\Delta \gg \Delta \nu$, where $\Delta \nu$ is the natural linewidth, which is typically of the order of 10 MHz. With increasing Δ the required minimum peak Rabi frequency Ω^0, i.e. the required minimum laser power, increases. We have observed nearly complete population transfer for detunings as large as $\Delta = 1$ GHz. It is, however, crucial to maintain the two photon resonance

$$\hbar(\omega_s - \omega_p) = \Delta E_{31} \tag{7}$$

where ΔE_{31} is the energy gap between levels 1 and 3. A lower limit of the two–photon linewidth is given by the transit time broadening. Therefore, it is important to stabilize the frequency **difference** rather than the absolute frequency of the lasers. A successful relatively simple concept has been described in ref. 39. Quantitative evaluation of eq. (6) reveals furthermore that a sufficiently large Rabi frequency can be achieved with commercially available cw dye lasers only for molecules with an extraordinary large transition dipol moment, such as atomic systems, alkali molecules or some radicals. Many molecules of interest require the use of radiation in the UV or VUV region of the spectrum and pulsed lasers are needed. Unlike cw lasers, pulsed lasers may exhibit strong amplitude and phase fluctuations on the time scale of the interaction.

The effect of **amplitude fluctuations** on the transfer efficiency can be estimated based on the preceeding discussion. If eq. (3c) is valid, depite the fluctuations, and if eq. (6) is fulfilled, whith $\Delta\tau$ being the characteristic time scale of the fastest fluctuations, then the transfer efficiency will be unity.

A detailed analytical analysis of the consequences of **phase fluctuations**[18] showed that instead of eq. (6) we may write, subject to approximations,

$$\Omega^2 \Delta\tau \gg \frac{1}{\Delta\tau}(1 + N^2) \tag{8}$$

where N is the ratio of the actual bandwidth of the radiation, determined by the extend of the phase fluctuations, and the transform limited bandwidth related to the pulse width. Thus, the required pulse energy increases strongly with increasing bandwidth of the radiation. Obviously, nearly transform limited radiation is best suited for the implementation of the STIRAP process.

Extensive numerical studies[18] showed that the required minimum pulse energy is not solely given by the bandwidth of the radiation but also by the nature of the fluctuations. For a given bandwidth a higher pulse energy is required when the spectral distribution is more of lorentzian–shape as opposed to gaussian.

4. APPLICATION OF THE STIRAP CONCEPT TO MULTILEVEL SYSTEMS

The STIRAP concept can also be applied to polyatomic molecules with a high level density in the region of the intermediate and final level. It could be argued that the coherent nature of the process may lead to detrimental interference phenomena of two competing coupling channels if, for instance, the intermediate and/or final levels are split. It was shown analytically[17] that splitting of the intermediate level 2 has no detrimental effect at all. Selective population transfer to closely spaced levels 3 and 3' is achievable if the level splitting exceeds the two photon linewidth. If the splitting is smaller a transfer efficiency of unity is again possible if the sum of the population of levels 3 and 3' is considered.

The STIRAP concept also allows the transfer of population in an N–level system using N–1 lasers. It is shown in ref. (16) that appropriate timing of the (counter intuitive) interaction sequence of N–1 radiation fields with an N–level system, where N is an odd number, leads to highly efficient population transfer. However, some population is placed into levels 3, 5, 7 etc. while no transient population appears in levels 2, 4, 6 etc. Therefore the maximum transfer efficiency may be smaller than unity, depending on the radiative loss of population from the populated levels. This loss will be small for vibrational levels in the electronic ground state of a molecule coupled by infrared lasers. In fact, efficient population transfer between vibrational levels in polyatomic molecules using the STIRAP concept has been demonstrated experimentally by Reuss and coworkers[40,41].

Application of the STIRAP concept to N–level systems coupled by N–1 radiation fields with even N has been discussed in ref. (19). It is a qualitative difference of significant importance that, unlike for odd N, such level systems have no zero–eigenvalue dressed states throughout the interaction. For N = 4, efficient population transfer is achievable. The insensitivity to small variations of the laser intensities and frequencies is assured only if the laser frequencies are detuned from resonance with transitions to any intermediate level[19]. Of course, the (N–1)–photon resonance needs to be maintained. It was also found for N=4 that the time evolution of the intensity of the lasers coupling levels 2→3 and 3→4 or 1→2 and 2→3 may be identical. Thus, efficient transfer of population from level 1 to level 4 may be achievable with two lasers only instead of three.

5. ALIGNMENT AND ORIENTATION OF ATOMS OR MOLECULES USING THE STIRAP CONCEPT

We have explored the potential of the STIRAP concept to align and orient atoms or molecules[15,24]. Consider an open three level system with total angular momentum $j_1=0$, $j_2=1$ and $j_3=2$ such as realized for the two metastable levels of Ne[42] (levels 1 and 3) coupled via a short lived intermediate level 2. A convenient source of a beam of metastable Ne atoms is a cold cathode discharge[41]. For linearly polarized light and for $\hat{E}_s \parallel \hat{E}_p$, coupling is restricted to $m_1 = 0 \rightarrow m_2 = 0 \rightarrow m_3 = 0$. Thus, if the population in level 3 is depleted (by optical pumping for instance) prior to the application of the STIRAP process, level $m_3 = 0$ can be selectively populated. If, on the other hand, we have $\hat{E}_s \perp \hat{E}_p$ population transfer occurs to levels $m_3 = +1$ and $m_3 = -1$[44]. Quantitative agreement between experimental and theoretical results has been achieved[15]. We note that selective population (or selective depletion) of these latter levels (degenerate with $m_3 = 0$ and $m_3 = \pm 2$) by standard optical pumping techniques[1,45-47] is not possible.

Orientation of an atomic angular momentum i.e. selective population of a single m–level, has also been demonstrated[24] by removing the degeneracy of the m–levels in a magnetic field. Selective population depletion of single m–levels by standard optical pumping has been demonstrated before[47]. In these latter experiments, the difference of the transition frequencies between m–levels in level 3 and level 2 needs to exceed the saturation broadened linewidth. It is a particular attractive feature of the STIRAP concept that splitting of the m–levels on the final state needs only to be of the order of the two–photon linewidth. Therefore the required magnetic field may be one or even two orders of magnitude smaller than needed for the manipulation of the population in selected m–levels using optical pumping techniques. Therefore application of the STIRAP–concept to orient molecules appears also possible.

6. EFFECT OF RADIATIVE OR COLLISIONAL DAMPING AND FREQUENCY DETUNING ON THE EFFICIENCY OF THE STIRAP–PROCESS

Because of the coherent nature of the STIRAP–process damping induced by radiative or non–radiative processes may prohibit the achievement of a high transfer efficiency. We need to discuss such processes separately for level 2 and levels 1 and 3. At first glance we may conclude that damping of level 2, characterized by a rate constant γ_2, has little, if any, effect since the level carries no population during the transfer process. However, if radiative or collisional coupling to other levels or continua is strong, the preceeding discussion based on a three-level model breaks down. Detailed numerical studies[23] using the density matrix approach[48] revealed that the transfer efficiency is significantly reduced only if $\gamma_2/\Omega > 1$. Thus, the consequences of damping processes acting on level 2 can be overcome with sufficiently large Rabi frequencies. The transfer efficiency is, however, very sensitive to damping processes acting on the initial and final level. It is essential to avoid such processes, e.g. collisional quenching of these levels, during the interaction of the particle with the pump and Stokes laser fields.

The results of a detailed study of the sensitivity of the transfer efficiency on the detuning of the laser frequencies from the one-photon and the two-photon resonance are also given in ref. 23. In that work it is confirmed that the transfer efficiency is rather insensitive to detuning from the one-photon resonance, provided the Rabi frequency is sufficiently large, while any detuning of the frequencies from the two photon-resonance exceeding the two-photon linewidth is detrimental.

7. THE STIRAP PROCESS AND THE LANDAU-ZENER PICTURE

The transfer process by the STIRAP methods is analyzed from a different perspective in ref. 21. There, we start from two individual two-level systems consisting of levels 1 and 2 and levels 2' and 3. We introduce radiative coupling $\beta\Omega$ between levels 1 and 2' as well as levels 2 and 3 with $0 < \beta < 1$. For pulses delayed according to the STIRAP configuration and for small values of β the dressed states related to the two different two-level systems cross. Analytical treatment of these crossings in the Landau-Zener picture[29,49] yields a transfer efficiency in good agreement with numerical results. For $\beta \to 1$ a transfer efficiency of unity is recovered and the two coupled two-level systems reduce to the three-level system relevant here.

8. THE STIRAP PROCESS INDUCED BY FREQUENCY CHIRPING

Efficient population transfer in a two level system through adiabatic following induced by frequency chirping[25-30] is a well known phenomenon. We may ask the question whether or not the STIRAP-type interaction sequence, defined by eq. (3c), can also be induced by frequency chirping. It is shown in ref. 20 that a chirpe-induced STIRAP-process may indeed be successful in a ladder-type three-level system, such as one relevant for vibrational excitation by infrared lasers. The chirped frequency approach fails, however, for a lambda-type three-level system — typically needed for population transfer to highly excited vibrational levels — since the required interaction sequence cannot be imposed while maintaining the two-photon resonance throughout the process.

9. APPLICATION OF THE STIRAP PROCESS TO COLLISION EXPERIMENTS

The first application of the STIRAP process in a crossed beam reactive collision experiment is discussed in ref. 22. In that work, cw-lasers were used to study the effect of vibrational excitation on the chemiluminescent channel of the reaction of sodium dimers with chlorine atoms. This reaction yields Na atoms in electronically excited levels. A linear increase of the related rate constant of the order of 1% per vibrational quantum in the dimer molecule is found. A novel method for the stabilization of the frequency difference of the pump and Stokes laser was used[39]. It was shown that the condition imposed by eq. (7) can in fact be satisfied within $\Delta\nu \approx 1$ MHz for cw-lasers under realistic laboratory conditions for a period of 30 minutes or longer.

10. SUMMARY AND OUTLOOK

We have presented a status report on the investigation and experimental implementation of the recently developed STIRAP method. We have not commented on a series of publications by Band and Julienne where they discuss specific aspects of the population transfer by stimulated Raman scattering [50-53]. The inherent potential for further studies in collision dynamics and spectroscopy involving, in particular, highly vibrationally excited states should have become obvious. The current experimental effort in Kaiserslautern concentrates on the implementation of this technique with pulsed lasers. It is hoped that this effort will eventually allow successful completion of collision experiments under single collision conditions, preferentially involving crossed molecular beams. A better understanding of the collision dynamics of highly vibrationally excited molecules, the selective preparation of which is difficult with other techniques, should result.

Although an understanding of the underlying basic mechanism leading to the previously unexpected high transfer efficiency has been achieved by now, there are many important aspects which need further theoretical studies. It is, for instance, of interest to study the STIRAP process when the coupling (level 2) occurs via a continuum or a predissociating level. Under very high laser intensity, available from pico-second or femto-second lasers, the Rabi splitting may be of the order of the level separation of the bare particle. Under such conditions the current treatment is no longer valid. Since in most cases levels with angular momenta $j \neq 0$ are involved, a more rigorous treatment of the transfer process involving degenerate levels is also desirable. Such questions are currently persued, to the best of our knowledge, in particular by F.T. Hioe, J. Oreg and B.W. Shore and their coworkers.

The STIRAP concept may have an impact in other areas as well. When the transfer is induced by two counter-propagating lasers, the two-photon resonance condition, essential for the success of the transfer process, is satisfied only for particles with velocities in a well defined very narrow range. Furthermore, the momentum of both photons is transferred in the direction of propagation of the pump laser. Most importantly – unlike photon-atom interaction involving a two-level system – the momentum transfer is non-dissipative. In the STIRAP configuration it can be repeatitively imposed even in an open three level system. Zoller and coworkers showed [55] that this interesting feature, together with the coherent nature of the process, can be used to create, for instance, atomic beam splitters.

11. ACKNOWLEDGEMENT

We wish to acknowlege the significant contributions to this work made by the students, postdocs and visiting scientist, which have cooperated with us on these problems in the past, in particular G.W. Coulston, P. Dittmann, U. Gaubatz, G. Z. He, F.T. Hioe, E.Konz, M. Külz, I.C.M. Littler, J. Martin, J. Oreg, B. W. Shore, S. Rosenwaks, P.Rudecki, H.-G. Rubahn, and W. S. Warren. We also thank E. Arimondo, C. Cohen-Tannoudji, J. Reuss and P. Zoller for valuable comments on specific aspects. The interest of Y. Band and P. S. Julienne in these problems is appreciated as well.

One of us (K.B.) thanks C. Y. Ng for arranging a very fruitful meeting on "Time and State Resolved Chemistry".

This work has been supported by the Deutsche Forschungsgemeinschaft under SFB 91 "Energy Transfer in Atomic and Molecular Collisions".

12. REFERENCES

1. K. Bergmann in: "Atomic and Molecular Beam Methods", edited by G. Scoles (Oxford University Press, New York, 1988, chapter 12)

2. K. Bergmann, U. Hefter, and J. Witt, J. Chem. Phys. 72, 4777 (1980)

3. M. Fuchs and J. P. Toennies, J. Chem. Phys. 85, 7062 (1986)

4. H.-G. Rubahn and J. P. Toennies, J. Chem. Phys. 89, 287 (1988)

5. H.-G. Rubahn and K. Bergmann, Ann. Rev. Phys. Chem. 41, 735 (1990)

6. H.-M. Keller, M. Külz, R. Setzkorn, G. Z. He, and K. Bergmann, J. Chem. Phys. 96, 000 (1992)

7. C. H. Hamilton, J. L. Kinsey, and R. W. Field, Ann. Rev. Phys. Chem. 37, 493 (1986)

8. X. Yang and A. Wodtke, J. Chem. Phys. 92, 116 (1990)

9. X. Yang, E. H. Kim, and A. M. Wodtke, J. Chem. Phys. 93, 4483 (1990)

10. H. L. Dai in: "Advances in Multi-Photon Processes and Spectroscopy", Vol. 7, edited by S. H. Lin (World Scientific London, 1991)

11. U. Gaubatz, P. Rudecki, M. Becker, S. Schiemann, M. Külz, and K. Bergmann "Population Switching Between Vibrational Levels in Molecular Beams", Chem. Phys. Lett. 149, 463-8 (1988)

12. J.R. Kuklinski, U. Gaubatz, F.T. Hioe, and K. Bergmann, "Adiabatic Population Transfer in a Three Level System Driven by Delayed Laser Pulses", Phys. Rev. A 40, 6741 – 6744 (1989)

13. U. Gaubatz, P.Rudecki, S. Schiemann, and K. Bergmann,"Population Transfer Between Molecular Vibrational Levels by Stimulated Raman Scattering with Partially Overlapping Laser: A New Concept and Experimental Results", J. Chem. Phys. 92, 5363 – 5376 (1990)

14. G.Z. He, A. Kuhn, S. Schiemann, and K. Bergmann, "Population Transfer by Stimulated Raman Scattering with Delayed Pulses and by the SEP Method: A Comparative Study", J. Opt. Soc. Am. B7, 1960–1969 (1990)

15. H.-G. Rubahn, E. Konz, S. Schiemann, and K. Bergmann, "Alignment of Electronic Angular Momentum by Stimulated Raman Scattering with Delayed Pulses", Z. Phys. D – Atoms, Molecules and Clusters, 22, 401–406 (1991)

16. B.W. Shore, K. Bergmann, J. Oreg, and S. Rosenwaks, "Multilevel Adiabatic Population Transfer", Phys. Rev. A 44, 7442–7447 (1991)

17. G. Coulston and K. Bergmann, "Population Transfer by Stimulated Raman Scattering with Delayed Pulses: Analytical Results for Multilevel Systems", J. Chem. Phys. 96, 000 (1992)

18. A. Kuhn, S. Schiemann, G.Z. He, G. Coulston, W.S. Warren, and K. Bergmann "Population Transfer by Stimulated Raman Scattering with Delayed Pulses using Spectrally Broad Light", J. Chem. Phys. 96, 000 (1992)

19. J. Oreg, K. Bergmann, B.W. Shore, and S. Rosenwaks, "Population Transfer with Delayed Pulses in Four-State Systems", Phys. Rev. A 45, 000 (1992)

20. B.W. Shore, K. Bergmann, A. Kuhn, S. Schiemann, J. Oreg, and J.H. Eberly "Laser-induced Population Transfer in Multistate Systems: A comparative study, Phys. Rev. A 45, 000 (1992)

21. B.W. Shore, K. Bergmann, and J. Oreg, "Coherent Population Transfer: Stimulated Raman Adiabatic Passage and the Landau-Zener Picture", Z. Phys. D – Atoms, Molecules and Clusters, 22, 000 (1992)

22. P. Dittmann, J. Martin, G. Coulston, G.Z. He, and K. Bergmann, "Effect of Initial Vibrational Excitation on the Chemiluminescent Channels of the Reaction $Na_2(v")+ Cl \rightarrow NaCl + Na(3p)$", (in preparation)

23. G. Coulston, U. Gaubatz, A. Kuhn, S. Schiemann, and K. Bergmann "Population Transfer by Stimulated Raman Scattering with Delayed Pulses: Numerical Results" (in preparation)

24. H.-G. Rubahn, J. Martin, T. Kraft, H. Hotop and K. Bergmann, "Optical preparation of single m-levels by stimulated Raman scattering with delayed pulses", (in preparation)

25. A. Abragan "The Principles of Nuclear Magnetism" (Oxford University Press, London, 1970)

26. L. Allen and J. H. Eberly, "Optical Resonances and Two-Level Atoms" (Wiley, New York, 1974)

27. S. Avrillier, J. M. Raimond, Ch. J. Bordé, D. Bassi, and G. Scoles, Opt. Commun. 39, 311 (1981)

28. A. G. Adams, T. E. Gough, N. R. Isenor, and G. Scoles, Phys. Rev. A 32, 1451 (1985)

29. J. P. C. Kroon, H. A. J. Senhorst, H. C. W. Beijerinck, B. J. Verhaar, and N. F. Verster, Phys. Rev. A31, 3724 (1985)

30. B.W. Shore, "The Theory of Coherent Atomic Excitation" (Wiley, New York (1990)

31. C. Cohen-Tannoudji, and S. Reynaud, J. Phys. B 10, 345 and 365 (1977)

32. G. Alzetti, A. Gozzini, L. Moi, and G. Orriols, Nuova Cimento B 36, 5 (1976)

33. J. Oreg, F. T. Hioe, and J. H. Eberly, Phys. Rev. A 29, 690 (1984)

34. C. E. Carroll and F. T. Hioe, J. Phys. B: At. Mol. Opt. Phys. 22, 2633 (1989)

35. C. E. Carroll and F. T. Hioe, Phys. Rev. A 42, 1522 (1990)

36. A. Messiah "Quantum Mechanics" (North Holland, Amsterdam, 1962), Vol II, P. 744

37. I. V. Hertel, H. Schmidt, A. Bähring, and E. Meyer, Rep. Prog. Phys. 48, 375 (1985)

38. A. Kuhn, Diplom Thesis, University of Kaiserslautern (1990)

39. I. C. M. Littler, P. Jung, and K. Bergmann, Opt. Commun. 87, 61 (1992)

40. C. Liedenbaum, S. Stolte, and J. Reuss, Phys. Rep. 178, 1 (1989)

41. N. Dam, L. Oudejans, and J. Reuss, Chem. Phys. 140, 217 (1990)

42. K. Hart, M. Raab, and H. Hotop, Z. Phys. D − Atoms, Molecules and Clusters 7, 213 (1987)

43. H. Hotop, E. Kolb, J. Lorenzen, Electron spectrosc. relat. phenomena 23, 347 (1981)
    We thank H. Hotop for providing such a source for this experiment

44. S. Schiemann, Diplom Thesis, University of Kaiserslautern (1989)

45. W. Bussert, Z Phys. D − Atoms, Molecules and Clusters 1, 321 (1986)

46. U. Hefter, G. Ziegler, A. Mattheus, A. Fischer, and K. Bergmann, J. Chem. Phys. 84, 756 (1986)

47. C. Bender, W. Beyer, H. Haberland, D. Hausamann, and H. P. Ludescher, J. Phys. (Paris) Collq. 46, C1 75 (1985)

48. K. Blum "Density Matrix Theory and Application" (Plenum Press, New York, 1981)

49. L. D. Landau, Z. Phys. 2, 46 (1936)

50. Y. B. Band and P. S. Julienne, J. Chem. Phys. 94, 5291 (1991)

51. Y. B. Band, Phys. Rev. A (submitted)

52. Y. B. Band and P. S. Julienne, J. Chem. Phys. 95, 5681 (1991)

53. Y. B. Band and P. S. Julienne, J. Chem. Phys. 96, 3339 (1992)

54. A. Mattheus, A. Fischer, G. Ziegler, E. Gottwald, and K. Bergmann, Phys. Rev. Lett. 56, 712 (1986)

55. P. Marte, P. Zoller, and J. L. Hall, Phys. Rev. A 44, 4118 (1991)

Orientation and molecular orbital dependences in electronic relaxation collisions studied through van der Waals complexes

L. Lapierre, P. Y. Cheng, S. S. Ju, Y. M. Hahn

Department of Chemistry
University of Pennsylvania
Philadelphia, PA 19104-6323 USA

and

H. L. Dai

Department of Chemistry
University of Pennsylvania
Philadelphia, PA 19104-6323

and

Department of Chemistry
National Taiwan University
Taipei, Taiwan, ROC

ABSTRACT

The orientation dependence in the spin changing collisions $C_2H_2O_2(S_1)+Ar \rightarrow C_2H_2O_2(T_1)+Ar$ and $C_2H_2(S_1)+Ar \rightarrow C_2H_2(T)+Ar$ has been examined by time-resolved laser induced fluorescence studies of the intersystem crossing rates in the $C_2H_2O_2$-Ar and C_2H_2-Ar complexes with different isomeric structures. It was found that when Ar interacts primarily with the n(O) orbital the $S_1 \rightarrow T_1$ transition rate is about two orders of magnitude faster than that induced by Ar interacting primarily with the $\pi^*(CO)$ orbital. On the other hand, the studies in acetylene show that the Ar induced intersystem crossing rate is nearly identical for both the π and π^* orbitals.

I. INTRODUCTION

Since the 1960s, the combination of molecular beam and laser spectroscopic techniques has provided powerful means for examining molecular collision and reaction dynamics. Molecular beam methods have enabled the control of translational energy and temperature and ensured that collisions can be studied with single collision resolution. Lasers can be used to prepare collision partners or reactants in a well-defined quantum state distribution and to detect the products with quantum state resolution. Plenty of examples now exist, some of them reviewed in this proceeding, showing that intermolecular potentials and collision dynamics can be deduced with great success from this type of study[1]. However, in such studies, the control of the relative orientation between the collisional partners remains one of the most difficult problems in the experiment. As a result, in our understanding of bimolecular collisions, the anisotropicity is often neglected.

Very recently, several experimental methods have been developed to control, or at least provide, a specific orientation in a bimolecular encounter. For example, optical excitation with a polarized light source provides a well-defined distribution of the transition moment, and thus the molecular orientation, in space. Bimolecular reactions of reactants excited at different molecular orbitals have been studied this way[2]. An inhomogeneous magnetic field has been successfully used to align molecules with nonzero dipole moment[3].

Another way to set the orientation is to use the van der Waals complex formed between the collisional partners in a bimolecular encounter.

In a weakly bound molecular complex with a well-defined structure, the interaction between the constituent species is mainly localized near the minimum of the intermolecular potential surface. Dynamic processes occurring in the complex resemble the corresponding collisional processes occurring within a narrow range of impact parameter and orientation. The distribution of the initial collisional conditions is essentially determined by the intermolecular motion within the complex, with the narrowest distribution defined by the zero-point fluctuation. The relative collisional energy in the complex is very low. However, when excitation of van der Waals vibrational motions is possible, collisional energy close to room temperature conditions could be achieved. The recent studies of reactions in van der Waals (vdW) complexes have clearly demonstrated some of these advantages. Wittig and coworkers have shown that by photodissociating the H-X bond in an X-H...Y complex, the H + Y reaction can be studied at a predetermined scattering angle[4]. Furthermore, Breckenridge, Soep, and others have demonstrated the effect of molecular orbital alignment in reactions occurring in a complex[5].

One can easily imagine that if a complex formed by the two collisional partners has more than one isomeric structure, each providing a specific interaction orientation, the orientation dependence in the collisional process can be examined through the corresponding dynamical process in the complex. Furthermore, selective excitation of the van der Waals vibrational modes can be used to fine-tune the collisional energy and direction. Here we present the first examples of the orientation dependence in electronic spin changing and energy transfer collisions studied through the corresponding complexes.

II. $C_2H_2O_2(S_1)+Ar \rightarrow C_2H_2O_2(T_1)+Ar$

In the glyoxal molecule[6], the first excited electronic state in the singlet system is 21,953 cm^{-1} above the ground state zero-point level. 2,776 cm^{-1} below the S_1 origin is the first triplet state T_1. The T_1 state vibrational level density near the S_1 zero-point level is estimated to be approximately 0.1 per cm^{-1}. The S_1-T_1 coupling matrix elements have been measured by Lombardi and coworkers to be on the order of 10^{-2} cm^{-1}[7]. Thus, near the S_1 origin, occurrence of S_1-T_1 coupling relies on accidental degeneracy. However, it was found that in the presence of high pressure Ar gas [8,9] the initially excited S_1 state may cross into the T_1 state, as was indicated by the decreased S_1 fluorescence lifetime and increased phosphorescence yield with Ar pressure. Theoretically[10], it can be understood that collision with Ar would increase the glyoxal S_1-T_1 coupling matrix element through the External Heavy Atom Effect[11] as well as the linewidth of both the S_1 and T_1 levels through lifetime broadening. Consequently, efficient S_1 to T_1 transition occurs when the combined magnitude of the matrix element and the linewidths are comparable to the T_1 level spacings.

In the glyoxal·Ar complex, the energies of the glyoxal electronic states are lowered by the van der Waals interactions which are on the order of a few hundreds of cm^{-1}. The energy levels of the bare molecule and the complex are schematically shown in Fig. 1. The constant presence of Ar at the van der Waals radii distance from glyoxal would induce a larger S_1-T_1 matrix element, again because of the External Heavy Atom Effect. Most importantly, the three van der Waals vibrational modes with frequencies from 10 to 30 cm^{-1}[12] would dramatically increase the $\bar{T}_1$ level density near the $\bar{S}_1$ origin by many orders of magnitude. Thus, we have a situation in which a single, zero-order, optically active $\bar{S}_1$ rovibronic level couples to a dense manifold of optically dark $\bar{T}_1$ rovibronic levels. As a result of this coupling, the stationary complex molecular-eigenstate would be predominantly of the $\bar{T}_1$ nature but consist of a small fraction of the $\bar{S}_1$ character. A laser with a coherent linewidth wider than the $\bar{S}_1$ character distribution--indicated by the dashed line on the righthand side of the figure--among the molecular-eigenstates can prepare an $\bar{S}_1$ wavefunction through a

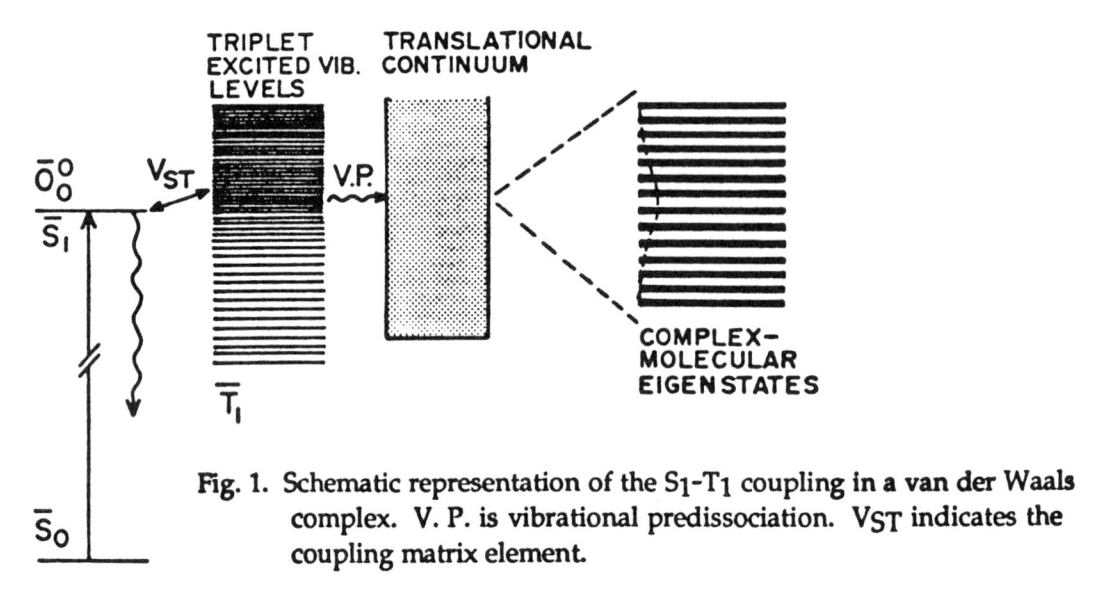

Fig. 1. Schematic representation of the S_1-T_1 coupling in a van der Waals complex. V. P. is vibrational predissociation. V_{ST} indicates the coupling matrix element.

coherent superposition of the molecular-eigenstates excited by the $\bar{S}_1 \leftarrow \bar{S}_0$ optical transition. Each of the zero-order complex $\bar{T}_1$ levels is broadened through coupling to the translational continuum associated with the molecular T_1 state, i.e., the $\bar{T}_1$ levels have enough energy to vibrationally predissociate.

It has been shown theoretically[13] that the fluorescence ($\bar{S}_1 \rightarrow \bar{S}_0$) decay following such an excitation will be biexponential:

$$I(t) = A_1 \exp(-k_1 t) + A_2 \exp(-k_2 t) \qquad (1).$$

Here, the ratio A_1/A_2 equals the number of molecular-eigenstates excited. The first decay is the dephasing of the coherently excited $\bar{S}_1$ wavefunction into an incoherent superposition of the molecular-eigenstates. Since the molecular-eigenstate is predominantly T_1, this decay is essentially the $S_1 \rightarrow T_1$ (intersystem crossing ISC) transition. The second exponential is the population decay of the molecular-eigenstates. In this case, k_2 is primarily from vibrational predissociation of the $\bar{T}_1$ levels. Thus, by observing k_1 in the fluorescence decay following the optical excitation by a short enough laser pulse, one can directly monitor the Ar induced glyoxal S_1 to T_1 transition rate in the complex. The complex molecular-eigenstates involved here each consists of glyoxal S_1 and T_1, intermolecular vibrational mode, as well as translational continuum characters. The complex intersystem crossing followed by vibrational predissociation thus mimics half of the collision induced $S_1 \rightarrow T_1$ transition process.

Three isomeric structures have been identified for the 1:1 glyoxal·Ar complex[12]. From rotational band contour analysis of the laser induced fluorescence spectra, the approximate structures of these isomers have been determined. In the $\tilde{X}$ state, the Ar atom is either on the C_2 axis 3.6 Å above the glyoxal plane (the top isomer), in the HCCO gulf of the glyoxal plane 4.3 Å away from the glyoxal center of mass (the side isomer), or in the HCO gulf of the glyoxal plane 4.8 Å away form the glyoxal center of mass (the front isomer). As the glyoxal molecule is excited to the S_1 state, the Ar atom moves away from oxygen by more than 1 Å in all three isomers. The three glyoxal(S_1)·Ar isomeric structures are shown in Fig. 2. In these different structures, Ar interacts with glyoxal(S_1) in three different directions. In this case, the three different directions are nearly orthogonal to each other. The effect of orientations in the collision Ar + $C_2H_2O_2(S_1) \rightarrow$ Ar + $C_2H_2O_2(T_1)$ can be explored over the entire molecular frame by examining the intersystem crossing rate in the three different glyoxal(S_1)·Ar isomers.

The van der Waals complexes were generated by expanding a 100 psi gas mixture that contained 99% He, 1% Ar, and 500 mTorr glyoxal through a 250 mm diameter pulsed nozzle (General Valve) into a vacuum chamber. A pulsed dye laser (Lambda-Physik FL2002) pumped by an Excimer laser (Lambda-Physik 201MSC) was used for the laser induced fluorescence experiment. The laser beam intercepted the molecular beam perpendicularly at 8 mm distance from the nozzle. Fluorescence was detected at a right angle to the laser and molecular beam axes by a photomultiplier tube with f1 imaging optics and masking. A shortpass filter with a cut-off at 470 nm was used for reducing the scattered light. Near the glyoxal monomer $\tilde{A}^1A_2 \leftarrow \tilde{X}^1A_1$, 0_0^0 band origin of 21,953 cm^{-1}, thirty-some spectral features that can be assigned to the glyoxal$_n$·Ar$_m$ complexes were observed[12]. Of these spectral features, about twenty are from the three 1:1 glyoxal·Ar isomers and their associated van der Waals vibrational level progressions in the $\tilde{A}$ state[12]. The deduction of the isomeric structure and van der Waals excitation is based on rotational bandshape, energy, and symmetry considerations. The dye laser was scanned and locked at the peak of each of these glyoxal·Ar features and the fluorescence decay trace was recorded by a transient digitizer/oscilloscope (LeCroy 9400).

The 125 MHz oscilloscope has 256 points on the x-axis. For slower decay traces a 10 nsec per point time resolution was used. The fluorescence traces were typically averaged over 10^4 laser pulses at about a 15 Hz repetition rate. For faster decay traces, interleaving of the digital scope was used to achieve a 0.8 nsec per point resolution. Since interleaving requires 100 times more pulses, only 200 averages were performed for the faster decay traces. A blank reference trace was recorded for each fluorescence trace by setting the laser about 1 cm^{-1} away from the spectral feature. Subtraction of the reference trace from the signal one removes background noise which arises from the scattered light and the monomer fluorescence. Each decay trace was then fitted to a biexponential decay according to Eqn. 1 along with a floating baseline. For traces in which the faster decay lifetime approaches the laser pulse length so that excitation and decay had to be deconvoluted, I(t) in the fitting was multiplied by a laser pulse shape function. The latter was determined from the scattered light pulse to be a Gaussian function with a FWHM of 8.8 nsec.

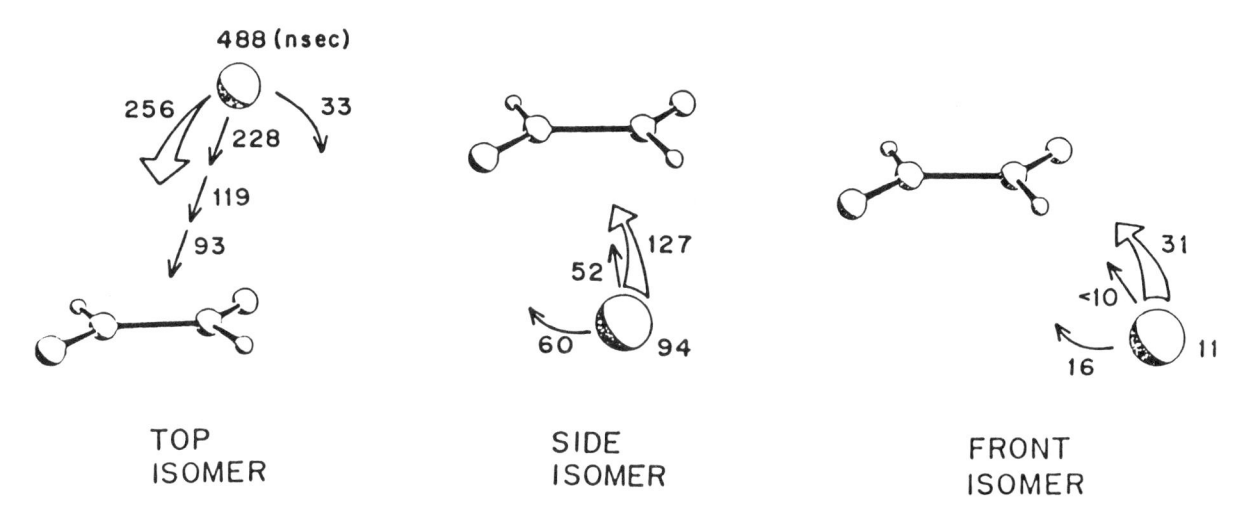

Fig. 2. The structures of the three glyoxal·Ar isomers with glyoxal excited in the S_1 state, and a graphic display of the intersystem crossing rates in the three glyoxal(S_1)·Ar isomers with various van der Waals mode excitations. Each arrow indicates one quantum of excitation in the pointed vibrational motion. The zero-point levels are represented by the Ar atoms themselves. The numbers are the associated lifetimes. The uncertainty from the nonlinear least-squares fit is smaller than one unit of the last significant digit.

All of the decay traces display distinct biexponential decay behavior, with the first decay varying dramatically with the isomeric structure as well as the van der Waals excitation. As examples, several decay traces were shown in Fig. 3. The assignment of the first exponential decay to intersystem crossing can be confirmed by comparing this decay rate in the glyoxal·Ar complex with those in the glyoxal·Ne and glyoxal·Xe complexes. From a straightforward Fermi Golden Rule argument[13], the intersystem crossing rate increases with the T_1 state level density and the square of the S_1-T_1 matrix element. Both quantities increase with the inert gas atom mass: the former because heavier reduced mass in the van der Waals vibrational modes decreases the vibrational frequencies and the latter because a heavier external atom induces a stronger spin-orbit coupling. Thus, one would expect a faster intersystem crossing rate in the direction of heavier inert gas atom mass. Indeed, similar fluorescence decay experiments on the glyoxal·Ne and glyoxal·Xe complexes have revealed such a trend. Similar to the Ar-complex situation, dozens of spectral features have been observed in the Ne- and Xe-complex fluorescence excitation spectra. They all display biexponential behaviour in the fluorescence decay. The first decay rate spans over a range of more than one order of magnitude. Since we have not determined the structures of these complexes, we cannot compare the decay rate associated with a specific structure or van der Waals excitation among different complexes. Thus, the decay lifetime distribution is compared here. The ranges of the first decay lifetime observed for the Ne and Xe complexes are 80-320 nsec and 10-100 nsec respectively, while the one for the Ar complexes is 10-500 nsec. The most probable lifetimes for the lifetime distributions of the three systems are 160, 110, and 50 nsec in the direction of Ne, Ar, and Xe.

The intersystem crossing rates of the three isomers with various degrees of van der Waals vibrational excitations are shown in Fig. 2. One notices immediately that the in-plane front isomer with Ar in the HCO gulf has a rate 44 times faster than that of the out-of-plane top isomer. The Ar atom in the zero-point level undergoes only zero-point motion near the equilibrium position. These dramatically different intersystem crossing rates associated with different structures thus unambiguously indicated a strong orientation dependence. This observation is further supported by the effect of van der Waals excitations on the intersystem crossing rate.

There are one stretching and two bending van der Waals vibrational modes in each complex. In the top isomer, excitation of the stretching mode increases the collisional energy between Ar and glyoxal and hence decreases the decay rate. Further excitation in this mode displays the anticipated results. Excitation of both bending vibrations also shortens the decay lifetime. In particular, as the Ar atom moves in the direction of the HCO gulf, the rate increases by 15 times! In the side isomer, excitation in the stretching and the in-plane bending modes increases the decay rate. However, the out-of-plane bending excitation that reduces the Ar-glyoxal in-plane interaction decreases the decay rate. A similar trend is also observed for the front isomer. Results associated with a few combination levels have also been obtained. The numbers observed for the combination levels are consistent with the observations on the fundamental and overtone levels.

Both the structure and the van der Waals excitation dependences in the decay rate suggest that the Ar interaction in the molecular plane is much more effective in inducing the glyoxal $S_1 \rightarrow T_1$ transition than the out-of-plane interaction. The orientation dependence can be related to the spatial distribution of the molecular orbitals in glyoxal.

Both the S_1 and T_1 states arise from the single electron $\pi^*(CO) \leftarrow n(O)$ excitation[14]. The triplet state has a totally symmetric spin wavefunction with S=1, while the singlet state has an S=0 asymmetric spin wavefunction. Classically, a transition from the singlet to the triplet state corresponds to a change from an originally 180° out-of-phase spinning of the two electrons (paired spin) to an in-phase spinning (parallel spin). This can be achieved by changing the spin of either the electron in the π^* orbital or the one in the n orbital. The CO π^* orbital sticks out perpendicularly to the molecular plane, while the oxygen nonbonding orbital lies in the molecular plane. The Ar atom in the planar isomers interacts primarily with the n(O) orbital, while in the top isomer Ar interacts with the π^* orbital. Thus, the difference between the in-plane and out-of-plane interactions is a result of the Ar interactions with the two different orbitals.

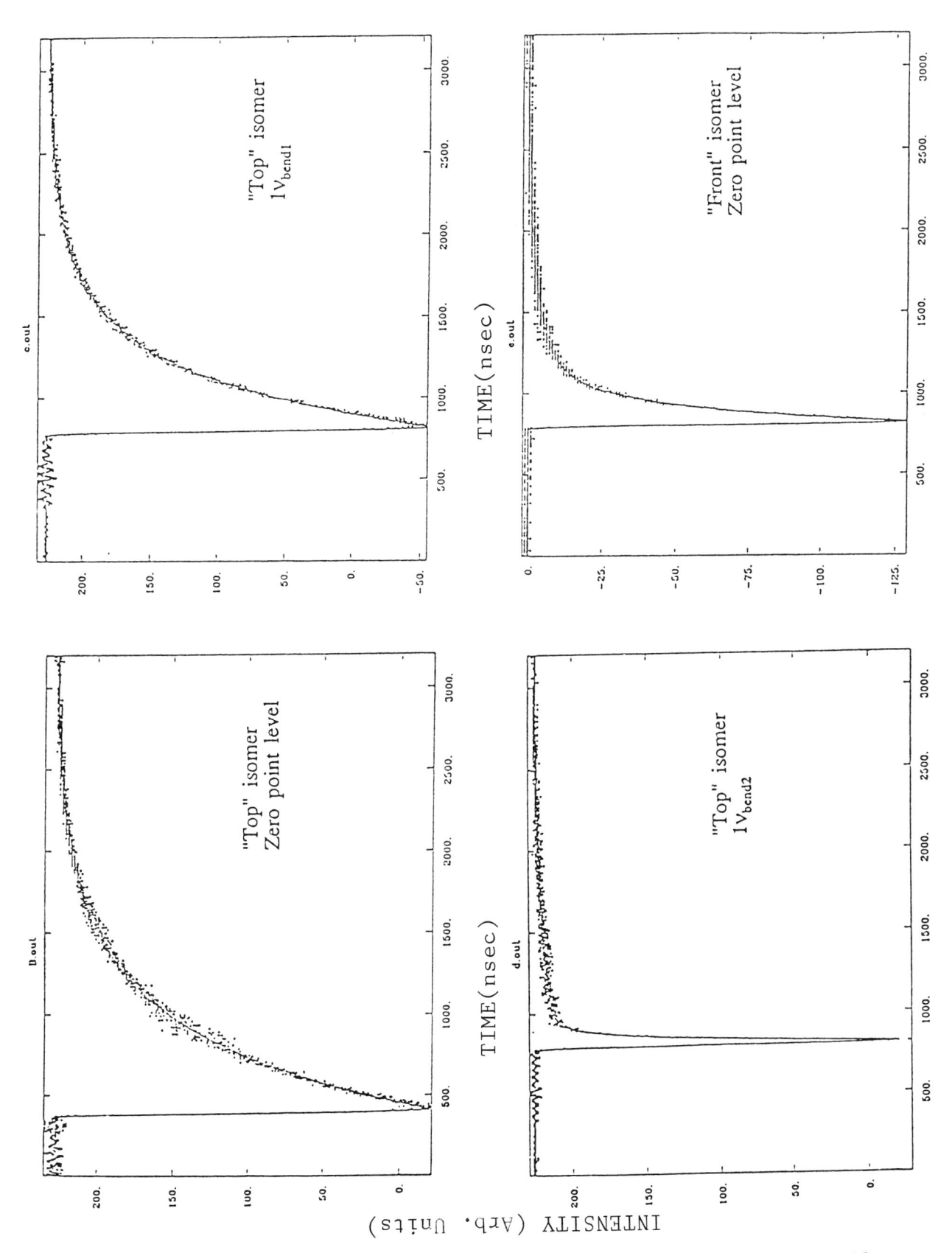

Fig. 3. Fluorescence decay traces of the glyoxal·Ar complexes in several isomeric structures with different van der Waals mode excitations.

A quantitative understanding of the results requires a sophisticated calculation on the spin-orbit coupling induced by an external heavy atom. However, we find that the spin-orbit coupling constant of the CO molecule is several times smaller than that of the oxygen atom[15]. The molecular orbitals may simply be reflecting the properties of their component atomic orbitals.

III. $C_2H_2(S_1)+Ar \rightarrow C_2H_2(T)+Ar$

The equilibrium structure of the ground state acetylene, $S_0(\tilde{X}^1\Sigma_g^+)$ is linear, whereas that of the first excited singlet state, $(\tilde{A}^1\Pi_u)$ is trans-bent planar [16]. The ground state of acetylene-Ar has been studied by molecular-beam electric resonance [17] and by FT-microwave spectroscopy [18]. Both studies indicated that the complex has a T-shaped, C_{2v} symmetry, structure with the Ar atom located about 4 Å away from the center of mass of C_2H_2.

Rotational contour analysis of the laser induced fluorescence features observed in association with the acetylene $S_1 \leftarrow S_0$ transitions revealed that there are two isomeric structures for acetylene$(S_1)\cdot$Ar: one with the argon atom positioned at a distance about 3.7 Å above the acetylene molecular plane near the C_2 symmetry axis (out-of-plane isomer), and the other with the argon atom located in the C_2H_2 molecular plane (in-plane isomer) with a distance between Ar and the CM of C_2H_2 about 3.8 Å. These two complex isomers exhibit very different rotational band structures due to their different transition types. The two structures are schematically shown in Figure 4. Fluorescence decay following the $S_1 \leftarrow S_0$, 3_0^n (n=1-4) band excitation of the

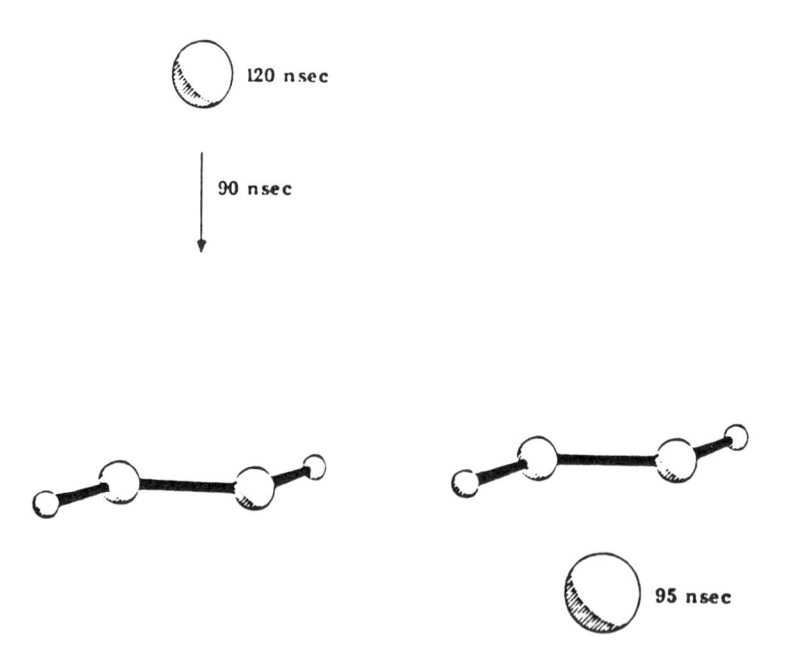

Fig. 4 Pictorial representations of the two $C_2H_2(S_1)\cdot$Ar isomers. The numbers in nsec are the observed fluorescence decay lifetimes of the corresponding isomers.

acetylene·Ar complexes indicates that rapid vibrational predissociation occurs for the vibrationally excited acetylene(S_1)·Ar. However, ISC dominates relaxation of the zero-point level of acetylene(S_1)·Ar. Through monitoring the fluorescence decay behavior of the two excited state C_2H_2·Ar isomers, electronic relaxation dynamics influenced by the Ar atom along two different directions have been characterized.

The acetylene-Ar complex was formed in a supersonic expansion of a 100 PSI gas mixture (C_2H_2 0.5%, Ar 20%, He 79.5%) through a pulsed valve (General Valve) with a 350 mm orifice. The output of an excimer pumped dye laser (Lambda-Physik 201MSC and FL2002) was frequency doubled through a BBO crystal to obtain the UV laser radiation in the 220-250 nm region. The laser beam crossed the supersonic jet 9 mm down stream from the nozzle. The fluorescence in the 280-700 nm spectral range was collected. For time-resolved measurements, the digital storage scope at 10 nsec/channel was used to digitize and average the fluorescence decay traces. The fwhm of the overall instrumental time-response function is ≤20 nsec. Typically, 1000-5000 laser shots were averaged. Background fluorescence was also recorded at laser wavelengths right next to each vdW complex peak and was removed from the major complex decay traces.

Without vibrational excitation in the C_2H_2 moiety, the $C_2H_2(S_1)$·Ar complexes decay exponentially with lifetimes of only about 100 nsec; much shorter than the C_2H_2 0^0 lifetimes observed under the same conditions. This lifetime shortening in the vibrationless level of acetylene should be due to enhancement of nonradiative decay processes upon complexation.

Strong singlet-triplet interaction in $C_2H_2(S_1)$ with coupling matrix elements on the order of 10^{-2} cm^{-1} have been observed for the v_3'=2-4 rovibronic levels by Zeeman quantum beat spectroscopy [20]. The v_3'=0-3 rotationless levels are also found to couple with neighboring triplet states by Zeeman anticrossing spectroscopy and a coupling strength of about 10^{-4} cm^{-1} was measured for the v_3'=0 level [21]. Ab initio calculations [22] have placed the lowest potential minima of the two lowest triplet states at energies between S_0 and S_1. They both are predicted to have two planar minima in the cis and trans-bent configuration lying at about 14,000 (cis T_1), 11,200 (trans T_1), 8,500 (trans T_2) and 5,500 (cis T_2) cm^{-1} below the trans-bent S_1 minimum. In addition, minima in the vinylidene configuration are also predicted for both T_1 and T_2 surfaces [23]. Electron energy loss spectra [24] have confirmed the existence of two triplet states in this region.

The intersystem crossing rate is determined by the density of triplet states and the S-T coupling matrix element which is induced by the spin-orbit coupling term in the molecular hamiltonian. The $C_2H_2(S_1)$ intersystem crossing rate could be enhanced in the complex for reasons similar to the glyoxal·Ar case: (1) The complexation of a small molecule such as acetylene with an argon atom should drastically increase the level density of the triplet manifolds that couple to the S_1 0^0 level. (2) The presence of the argon atom can also enhance the spin-orbit coupling as a result of the "external heavy atom effect." Based on this reasoning and the previous experimental and theoretical results, it is reasonable to attribute the observed complex lifetime shortening to the $C_2H_2(S_1)$ intersystem crossing enhanced by the presence of the Ar atom. Internal conversion (IC) of $C_2H_2(S_1)$·Ar to the highly excited vibrational levels of $C_2H_2(S_0)$·Ar is excluded here because the $S_1 \rightarrow S_0$ in the acetylene monomer is not a dominant channel [21] and the presence of the Ar atom is not expected to enhance the S_0-S_1 matrix element.

The acetylene·Ar fluorescence decays show only a single exponential, although bi-exponentials are expected, just as in the glyoxal·Ar complexes. This situation may arise if either the A_1/A_2 ratio is very large or if the second decay is very long or short in comparison with the first decay. Most likely the former condition is satisfied for the acetylene·Ar complexes. Since there are two triplet states below the S_1 state in acetylene with one of the two ~7,000 cm^{-1} lower in energy, the density of triplet levels coupling with the S_1 level could be much higher, inducing a much larger A_1/A_2 ratio.

The fluorescence lifetimes of the 0^0 levels of the two $C_2H_2(S_1)\cdot Ar$ isomers are not very different. This indicates that the ISC rates of acetylene induced by interacting with Ar along the two different directions are similar. The different interaction orientations can be related to the spatial distribution of the molecular orbitals in acetylene. The acetylene S_1 state is a $(\pi,\pi*)$ excitation with one electron in the out-of-plane π orbital and one in the $\pi*$ orbital in-plane orbital [25]. ISC of C_2H_2 from the singlet state with paired electron spins to the triplet state with parallel electronic spins can be regarded as simply flipping the spin of one of the electrons in either the π or the $\pi*$ orbital. According to this excited state orbital configuration, the argon atom in the out-of-plane isomer would interact mostly with the π orbital, whereas the one in the in-plane isomer interacts with the $\pi*$ orbital. The lack of a significant difference in the ISC rates suggests that Ar collision with either the π or the $\pi*$ orbital is similarly effective in enhancing the spin-orbit coupling and causing the electron spin to flip.

Brus [26] has reported fluorescence lifetimes of 80 nsec for vibrationally relaxed $C_2H_2(S_1)$ in both Ne and Ar matrices at 4 K. This value is very close to what we have measured in the acetylene complexes, even though only one argon atom is attached to the acetylene molecule. Jortner et al. have proposed that the heavy atom enhancement of ISC in the tetracene-Kr_n/Xe_n complexes essentially originates from the single pair interaction, i.e., from the interaction with the first rare gas atom [27]. The similarity of vibrationless $C_2H_2(S_1)$ fluorescence lifetimes observed in the argon matrix and in the single argon atom complex seems to support their conclusion and the assignment of the fluorescence decay to ISC.

IV. CONCLUDING REMARKS - SEARCHING FOR VAN DER WAALS COMPLEXES WITH ISOMERIC STRUCTURES

One of the key elements in the study of orientational dependence in collisional processes through van der Waals complexes is the existence of the isomeric structures. Determination of the complex structure through spectral analysis, particularly for the electronically excited states, is a tedious and nontrivial process. In order for this kind of study to be widely applicable to many molecular systems, it would be much more efficient to have some simple and physically intuitive way to guide the search for the isomers. One very promising approach is the pair-potential model calculation of complex structures.

In the pair-potential model, the potential energy of the complex is calculated as a sum of the individual pair interaction potential between the constituent atoms of the two molecules. The potential between a pair of atoms is usually predetermined experimentally or theoretically through quantum chemical calculations. The complex structures are the local minima on the calculated potential energy surface.

In the case of glyoxal·Ar, the Ar-H and Ar-CO interaction potentials have been determined through molecular beam scattering [28] and matrix isolation [29] techniques. The pair-potential calculation of the glyoxal·Ar potential surface indeed produced three local minima, shown in Fig. 5, that resemble the spectrally determined structures. The calculated structures are for the electronic ground state. In the S_1 state, the Ar position in the complex shifts about 1Å away from the oxygen atom[12]. However, when the pair-potential parameters are properly adjusted for the electronic excitation, such shifts in Ar positions can be reproduced[30].

The pair-potential parameters for acetylene(S_1)·Ar are borrowed from the ones theoretically calculated and experimentally tested for the ethene·Ar complex[31]. The trans-bent structure of acetylene in the S_1 state resembles the one of ethene minus two H atoms in the trans positions. The atomic orbital hybridization and bond order are also similar for the two molecules in two different electronic states. Thus, we simply took the ethene(S_0)·Ar pair-potential parameters less two H·Ar pairs and used them for acetylene(S_1)·Ar. This calculation generated two potential minima that resemble the structures of the two experimentally determined isomers, Fig. 6.

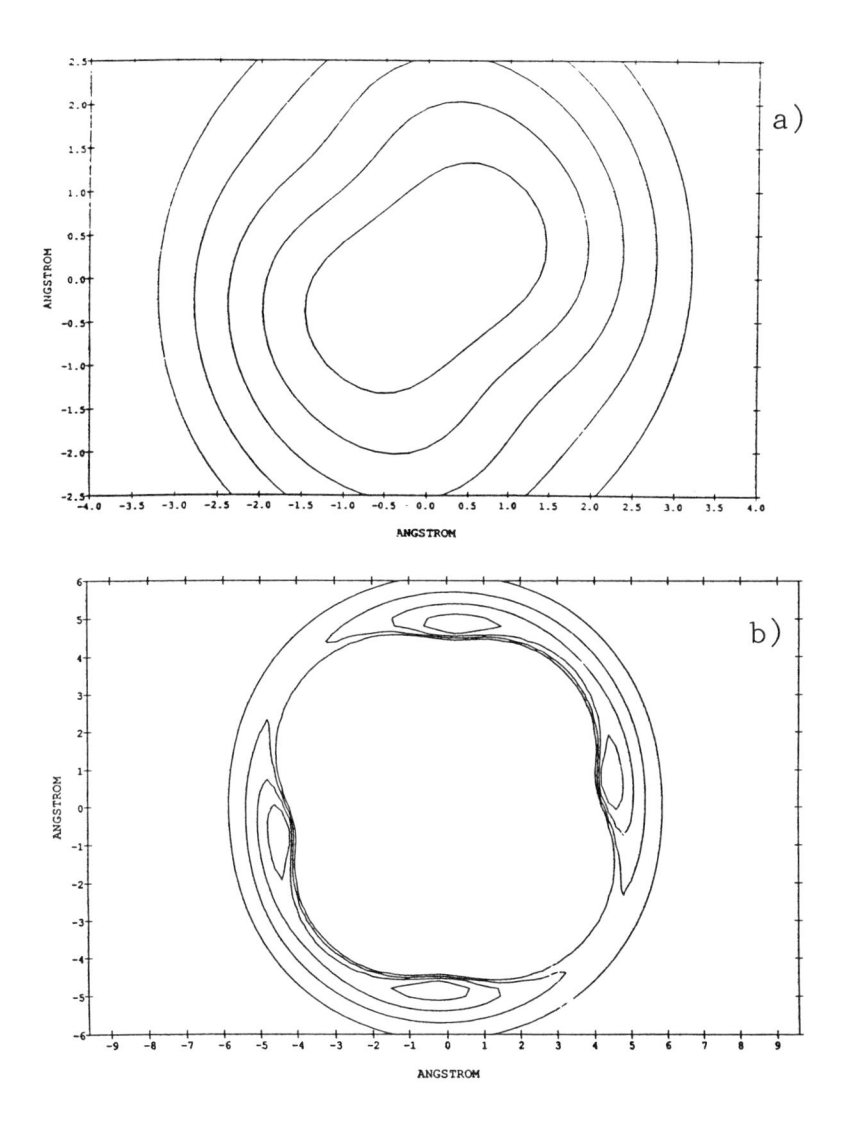

Fig. 5. Cross sections showing local minima on the potential energy surface of glyoxal·Ar from the pair-potential calculation. a) 3.6 Å above the glyoxal plane, the top isomer. b) The two in-plane front and side isomers.

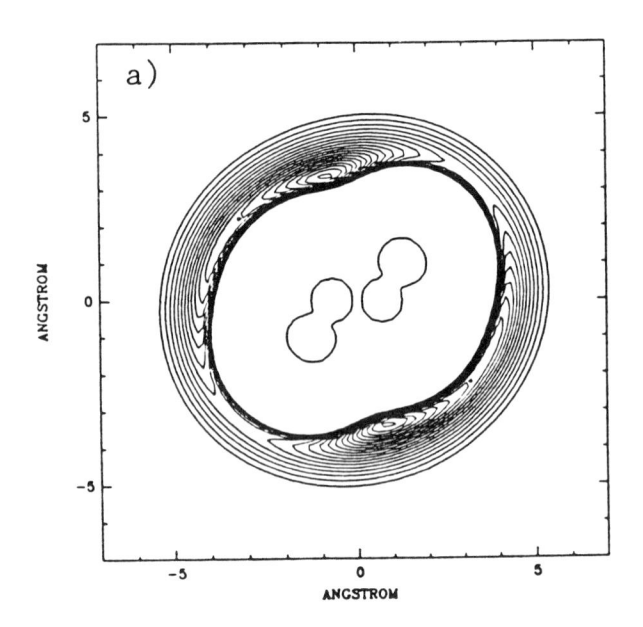

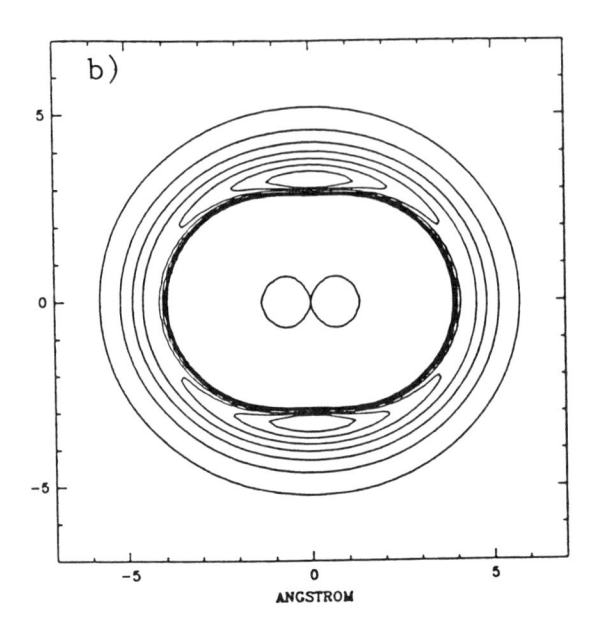

Fig. 6. Pair-potential calculation of acetylene(S_1)·Ar complex structures. a) The in-plane isomer. b) The out-of-plane isomer.

These two cases, along with others, have exemplified the transferability of pair-potential parameters from molecules to molecules and the usefulness of this simple model for structural calculation. It seems reasonable to speculate that a sufficient number of pair-potential parameters can be established through similar experimental and theoretical approaches on small molecular complexes. They can then be generally used for structural calculations of complexes and provide guidance for the search of complex isomeric structures.

ACKNOWLEDGEMENTS:

This work is supported by the National Science Foundation (CHE88-20177). LL thanks the Quebec Ministry of Education for a predoctoral fellowship. HLD acknowledges the receipt of an Alfred P. Sloan Fellowshp and a Camille and Henry Dreyfus Teacher-Scholar Award.

REFERENCES

1. R. D. Levine and R. B. Bernstein, "Molecular Reaction Dynamics and Chemical Reactivity" (New York: Oxford University Press, 1987).
2. See, for example,
    a) G. E. Busch and K. R. Wilson, *J. Chem. Phys.* **56**, 3626 (1972).
    b) M. A. O'Halloran, H. Joswig, and R. N. Zare, *J. Chem. Phys.* **87**, 303 (1987).
    c) R. L. Robinson, L. J. Kovalenko, C. J. Smith, and S. R. Leone, *J. Chem. Phys.* **92**, 5260 (1990).
3. S. R. Gandhi and R. B. Bernstein, *J. Chem. Phys.* **88**, 1472 (1988).
4. a) C. Wittig, S. Sharpe, and R. A. Beaudet, *Acc. Chem. Res.* **21**, 341 (1988).
    b) S. K. Shin, Y. Chen, S. Nickolaisen, S. W. Sharpe, R. A. Beaudet, and C. Wittig, *Adv. Photochem.* Vol. 16 (Wiley & Sons, New York, 1990) p. 249; and references therein.
5. a) W. H. Breckenridge, *Acc. Chem. Res.* **22**, 21 (1989).
    b) C. Jouvet, M. Boivineau, M. C. Duval, and B. Soep, *J. Phys. Chem.* **91**, 5416 (1987).
    c) W. H. Breckenridge, C. Jouvet, and B. Soep, *J. Chem. Phys.* **84**, 1443 (1986).

6. a) F. W. Birss, J. M. Brown, A. R. H. Cole, A. Lofthus, S. L. N. G. Krishnamachari, G. A. Osborne, J. Paldus, D. A. Ramsay, and I. Watmann, *Can. J. Phys,* **48**, 1230 (1970).
   b) L. H. Spangler, Y. Matsumoto, and D. W. Pratt, *J. Phys. Chem.,* **87**, 4781 (1983).
7. a) M. Lombardi, R. Jost, C. Michel, and A. Tramer, *Chem. Phys.* **46**, 273 (1980); ibid, **57**, 341 (1981).
   b) P. Dupre, R. Jost, and M. Lombardi, *Chem. Phys.* **91**, 355 (1984).
8. C. Michel, M. Lombardi, and R. Jost, *Chem. Phys.* **109**, 357 (1986).
9. a) L. G. Anderson, C. S. Parmenter, and H. M. Poland, *Chem. Phys.* **1**, 401 (1973).
   b) K. W. Holtzclaw, D. B. Moss, C. S. Parmenter, and G. W. Lodge, *J. Phys. Chem.* **87**, 4495 (1983).
   c) R. A. Beyer, and W. C. Lineberger, *J. Chem. Phys.* **62**, 4024 (1975).
10. K. F. Freed, *Chem. Phys. Lett.* **37**, 47 (1976); *J. Chem. Phys.* **64**, 1604 (1976).
11. a) J. B. Birks, "Photophysics of Aromatic Molecules" (Wiley-Interscience, 1970).
    b) S. P. McGlynn, T. Azumi, and M. Kinoshita, "Molecular Spectroscopy of the Triplet State" (Prentice-Hall, 1969).
12. L. Lapierre and H. L. Dai, *J. Chem. Phys.*, in press.
13. a) M. Bixon and J. Jortner, *J. Chem. Phys.* **48**, 715 (1968).
    b) J. Kommandeur, W. A. Majeuski, W. L. Meerts, and D. W. Pratt, *Ann. Rev. Phys. Chem.* **38**, 433 (1987).
    c) F. Lahmani, A. Tramer, and C. Tric, *J. Chem. Phys.* **60**, 4431 (1974).
14. D. C. Moule and A. D. Walsh, *Chem. Rev* **75**, 67 (1975).
15. H. Lefebrve-Brion and R. W. Field, "Perturbations in the Spectra of Diatomic Molecules" (Academic Press, New York, 1986), and references therein.
16. a) C. K. Ingold and G. W. King, *J. Chem. Soc.* 2725 (1953).
    b) K. K. Innes, *J. Chem. Phys.* **22**, 863 (1954).
17. R. L. Deleon and J. S. Muenter, *J. Chem. Phys.* **72**, 6020 (1980).
18. Y. Ohshima, M. Iida, and Y. Endo, *Chem. Phys. Lett.* **161**, 202 (1989).
19. S. S. Ju, P. Y. Cheng, M. Y. Hahn, and H. L. Dai, to be published.
20. N. Ochi and S. Tsuchiya, *Chem. Phys.* **152**, 319 (1991).
21. P. Dupre, R. Jost, M. Lombardi, P. G. Green, E. Abramson, and R. W. Field, *Chem. Phys.* **152**, 293 (1991).
22. a) H. Lischka and A. Karpfen, *Chem. Phys.* **102**, 77 (1986).
    b) M. Peric, R. J. Buenker, and S. D. Peyerimhoff, *Mol. Phys.* **53**, 1177 (1984).
    c) R. W. Wetmore and H. F. Schaefer III, *J. Chem. Phys.* **69**, 1648 (1978).
    d) J. S. Binkley, *J. Am. Chem. Soc.* **106**, 603 (1984).
23. M. P. Conrad and H. F. Schaefer III, *J. Am. Chem. Soc.* **100**, 7820 (1978).
24. D. G. Wilden, P. J. Hicks, and J. Comer, *J. Phys. B* **10**, L403 (1977).
25. R. Ditchfield, J. D. Bene, and J. A. Pople, *J. Am. Chem. Soc.* **94**, 4806 (1972).
26. L. E. Brus, *J. Mol. Spectrosc.* **75**, 245 (1979).
27. A. Amirav, U. Even, and J. Jortner, *Chem. Phys. Lett.* **67**, 9 (1979), and references therein.
28. J. P. Toennies, W. Welz, and G. Wolf, *J. Chem. Phys.* **71**, 614 (1979).
29. K. Mirsky, *Chem. Phys.* **46**, 445 (1980).
30. L. Lapierre, Ph.D. Thesis, University of Pennsylvania (1991).
31. A. C. Pect, D. C. Clary, and J. M. Hutson, *J. Chem. Soc. Faraday Trans. 2* **83**, 1719 (1987).

Advances in high repetition rate, ultra-short, gigawatt laser systems for time-resolved spectroscopy

Louis F. DiMauro

Chemistry Department, Brookhaven National Laboratory, Upton, NY 11973

1. INTRODUCTION

The evolution of lasers from their early conception to a general laboratory tool has been earmarked by periodic quantum jumps in the technology. Such advances include tunability, high peak power, and ultra-short temporal pulses. Since the mid-70's the advent of easily usuable high peak-power pump lasers has made the generation of intense tunable radiation from 0.2 to 1 μm routine practice in many laboratory settings. The general philosophy has been the use of powerful pump lasers to efficiently produce radiation at wavelengths appropriate for the task. The use of this intense tunable radiation has opened the opportunity for studies involving the use of nonlinear probes, viz. resonantly enhanced multiphoton ionization (REMPI), pump-probe techniques and the generation of plasma continuum. Even the ability to produce tunable VUV radiation below 0.2μm by third harmonic generation in various gases[1,2], although inefficient, becomes a viable spectroscopic technique with high peak power systems. Likewise, the use of intense pulses of radiation has led to the understanding of new fundamental phenomena[3,4] in the area of nonlinear optics. These are just a few examples where high peak power lasers have made an impact and it is beyond the scope of this review to cover all applications. However, the one thing that both the laser industry and the scientific community have been striving towards over the last decade is the development of a class of megawatt (MW) peak power lasers with kilohertz repetition rates. The ability to generate short pulses with low peak-power, 0.1-10 kilowatts (kW), at 50-100 megahertz (MHz) repetition rates has been available from mode-locked and colliding pulse (CPM) laser systems. Consequently, the availability of MHz source lasers has geared the approach towards the development of high peak power lasers as high repetition-rate pump amplifiers that would complement the existing technology. A schematic representation of a high-repetition rate, tunable, short pulse laser system is shown in Fig. 1. The low peak-power output of the mode-locked (ML) or cw-pumped CPM lasers are seeded into a tunable amplifier, i.e. dye or solid state. The tunable amplifier is in turn pumped by the output of a fixed-frequency amplifier pump operating in the kilohertz regime. This amplifier can be electrically and/or optically synchronized to the master oscillator, as indicated by the dashed line in Fig. 1. The objective of this article is to emphasize the current advances in the development of high-repetition rate amplifier pumps. Various approachs to this problem will be reviewed and contrasted along with our own developments here at Brookhaven National Laboratory (BNL). Although this review will highlight amplifier pump development, any recent data from achieved outputs via the tunable amplifier section will also be discussed albeit without detail. The first section will describe desirable parameters attributable to the pump amplifier while the rest of the article will deal with specific examples for various options. The pump amplifiers can be characterized into two distinct classes; those achieving operation in the hundred hertz regime and those performing at repetition rates ≥ 1 kHz.

In general, most experiments would benefit from an increase in repetition rate, assuming that the other characteristics of the laser are not dramatically compromised. Exceptions would of course be studies in which some dynamical decay or recovery time of the system under investigation is slow compared to the repetition rate of the laser. The most basic concept of signal averaging predicts that the signal-to-noise ratio (SNR) for an experiment is proportional to the square root of the number of counts. In principle, one could attain arbitrarily high precision by counting for long times but this is often not practical because of long term instabilities in the experimental system or finite sample quantities which introduce new sources of uncertainty to the signal. Consequently, high-repetiton rate laser systems are advantageous in their ability to accumulate a statistically meaningful signal in a relatively short time period. The systems discussed in this article typically have a 100 fold increase in duty cycle compared to standard 10 Hz laser systems. A more subtle but important improvement in the signal-to-noise ratio of an experiment can be realized with high repetition rate lasers by application of signal-enhancement techniques based on bandwidth reduction[5]. Phase-sensitive or lock-in amplifier detection is one such technique that although widely used in various spectroscopy applications has not found wide spread use with low repetition rate pulsed laser systems. Although the ability to detect and process a signal against a white noise environment is significantly enhanced by such techniques the low-duty cycle associated with low repetition rate lasers makes application essentially impossible (inefficient). Such is not the case for laser systems approaching the kilohertz regime where the higher duty factor makes application of signal-enhancement techniques an effective detector. Finally, the need for high repetiton rate systems can also be dictated by the specific details of an experiment. Thus, for certain experimental studies to achieve a minimal acceptable count rate there exists an important need for high peak power pulses for either driving a particular weak transition and/or efficiently up- or down-converting radiation in a spectral region of interest, but one finds that there is indeed a practical upper limit for usable peak power. That is, the signal does not continue to increase with increasing laser power. For instance, an experiment designed to perform transient Raman spectroscopy requires high peak power lasers to efficiently pump the inherently weak transition and overcome the low transient densities; however too much pump power leads to unwanted secondary processes, i.e. multiphoton ionization and dissociation, that tend to mask or diminish the desired signal. Thus, the ability to work at lower peak power dictates the need for higher repetition rates in order to successfully perform the experiment. The use of such laser systems also provides the ability to perform coincidence experiments, i.e. electron-ion, in molecular and atomic beams that were virtually impossible at lower repetiton rates but now come into the realm of possibilities.

One of our interests at BNL is the coupling of high-repetition rate laser systems to accelerator-based facilities, i.e. National Synchrotron Light Source (NSLS). The NSLS is a facility that provides tunable VUV radiation from 10-1000 eV with high average power and megahertz repetition rates, but the energy per pulse is extremely low, 10^5 photons/pulse. Consequently, to pursue synchrotron-laser pulse-probe studies it is necessary that the laser system perform with a repetition rate closely matched with that of the synchrotron, high duty cycle. One approach to this problem is to use low peak-power cw or mode-locked lasers. However, if the specifics of an experiment require high peak-power from the laser then currently kilohertz based systems become the only viable alternative. We are also exploring the construction of a UV-FEL amplifier facility operating in the wavelength range from visible to VUV with millijoule output and kilohertz repetition rates[6]. High-powered, high-repetition rate lasers will be an integral part of this proposed facility. The lasers will not only drive the photocathode RF gun for the LINAC but will also provide the megawatt seed pulses for injection in the FEL amplifier. A second interest in my group involves the use of neutral molecular beam detection techniques to study time-resolved vibrational relaxation and dissociation dynamics of medium sized molecules and clusters. Specifically, our beam machine will employ bolometric detection which directly detects the change in energy content of a neutral species. This technique has been used with great success in CW high resolution laser studies but has alluded application to time-resolved studies. The main reason for this shortcoming has been the low-duty cycle associated with conventional short pulse lasers. However, the high-repetition rate lasers discussed in this article makes application of this technique feasible and such a program in currently underway at BNL.

2. DESIRABLE AMPLIFIER PARAMETERS

The purpose of this section is to quantify the desirable output parameters of the "ideal" amplifier pump. Of course, this laser does not exist and the choices have to be based on the individual needs of the experiment. The *pulse repetition rate* is strongly coupled to the *energy per pulse* since the laser's average power is fixed. This is shown in Fig. 2 where the existing amplifier systems are clustered about the 1 and 10 Watt average power lines. The nanosecond green/uv amplifier pump sources are shown in Fig.2(a) while sub-nanosecond green sources are plotted in Fig.2(b). The available sub-nanosecond amplifiers are solely based on the second harmonic output of either Nd:YAG or Nd:YLF. The researcher must essentially choose between peak power or repetition rate in their application. However, it is our view that amplifiers operating in the 1-5 kHz regime represent the best compromise. Since this dynamic range offers peak power performance similar to commecical 10 Hz systems but with at least a 100-fold increase in repetition rate. The specifics of these sources are discussed in more detail in the next sections.

The temporal *pulse width* of the amplifier's output is crucial in determining the dye amplifier design. Since the gain lifetime of the dyes is of the order of a nanosecond[7], a clear distinction exists between the group of amplifiers plotted in Fig.2(a) and (b). The nanosecond amplifier pumps require a multipass cell arrangement[8], sometimes referred to as "butterfly" or "bow-tie", in order to maximize the use of the pump energy. However, the dye amplifier can be simplified to a single pass longitudinal pumping arrangement with sub-nanosecond amplifier pumps. Conversion efficiencies up to 15% can be realized for the shorter pump pulses as compared to 1% for nanosecond pump pulses. The amount of amplified spontaneous emission (ASE) present in the amplified tunable output is mainly determined by the amplifier pump duration. The use of sub-nanosecond pump pulses typically achieve 10 times less ASE than nanosecond pumps. However, if the tunable amplifier media is Ti:Sapphire with gain lifetimes of the order of a few microseconds then green nanosecond amplifier pump pulses are desirable in order to minimize the energy density. In particular, Q-switched CW pumped Nd:YLF amplifier pump pulses with 300 nsec width have been used in a Ti:Sapphire amplifier[9].

The *wavelength* of the amplifier pump should ideally coincide with the maximum in the absorption peak of the tunable media for good conversion efficiency. Likewise, the frequency or energy difference between the absorption and emission of the tunable amplifier should be minimized to reduce the heat load. Considering that the host of tunable short pulse generation techniques operate at wavelengths longer than 550nm, it is preferrable to have an amplifier pump operating in the visible region. Furthermore, when considering Ti:Sapphire as the tunable media, one finds that the narrow absorption band limits the pump wavelength exclusively to the green ($\lambda_{max} \sim 500$nm).

Figure 2 does show the energy per pulse for various sources without reflecting the *beam quality*. Good beam quality means that the energy spatial distribution is described by a simple gaussian function, TEM$_{00}$ mode, and with no beam pointing instabilities. In order to optimize the conversion efficiency between the amplifier pump and amplified tunable pulse, it is essential to achieve good spatial overlap of the two beam volumes (mode-matching). This is most easily accomplished with diffraction limited gaussian pump beams. Furthermore, good beam quality is necessary if standard up- and/or down-nonlinear conversion techniques are to be employed to extend the wavelength coverage to the UV or near IR. This is illustrated in Fig. 1 by the dashed box labelled mixer where the amplified tunable output is mixed with the fundamental or harmonic frequency of the amplifier pump. Here the regenerative amplifier alternatives standout because of their low timing jitter, appropriate wavelength, and excellent beam quality.

The *timing jitter* between the dye oscillator and the amplifier pump must be maintained within a fraction ($\leq 20\%$) of the dye's storage time. This requirement is easily achieved with amplifier schemes that are optically synchronized to the master oscillator, i.e. regenerative amplifiers. In the case of nanosecond amplifier pumps this criterion is somewhat relaxed but a typical jitter requirement should be ± 1 ns. The *energy stability* or shot-to-shot energy fluctuations presents an obvious choice; the more stable the better, since the final amplified tunable output (saturated) will be limited by the shot-to-shot noise of the amplifier pump. Finally, from a practical standpoint the *utility requirements* can also be a criterion for choice of amplifier pump, as well as the *cost of consumables* which can be substantial, viz. flash-lamps, rare gas consumption.

3. REGENERATIVE AMPLIFIER PUMP SOURCES

In recent years there has been work directed toward the development of a unique class of solid state optical amplifiers capable of producing high gain and repetition rates in the kilohertz regime. This class of lasers has become known as regenerative amplifiers. The pioneering work of Lowdermilk and Murray[10] laid the foundations for amplifying picosecond IR pulses to nearly a millijoule with a 10 Hz regenerative amplifier. In their scheme, a single pulse from a mode-locked Nd:YAG laser was seeded into a pulsed flashlamp pumped Nd:YAG regenerative amplifier, where it made several passes through the gain medium before being cavity dumped. The single pass gain within the amplifier cavity was kept low allowing the seed pulse to make ≈ 100 passes before reaching the saturation limit. It is the multipass operation of this system which makes it distinctive from conventional traveling wave amplifiers. In this section, we will examine two different pumping schemes (*cw or pulsed*), as well as different gain media. Since the emphasis of this article is on high repetition rate sources, we will not discuss low repetition-rate (≤ 100Hz) chain amplifiers which can pump short pulses with very high energy per pulse.

3.1. CW-pumped regenerative amplifiers

A new arrival in the multi-kilohertz amplifier pump arena is the cw-pumped solid-state regenerative amplifier[11] (RGA). The early work on cw RGA were limited in their repetition rate to 1 kHz and energies of 1 mJ. However, substantial work on improving the resonator design and electro-optics have resulted in multi-kilohertz performance with higher output energies[12]. Our work at BNL has produced a cw-pumped YLF RGA capable of producing 4 watts average power at repetition rates of 5 kHz. Our system, shown in Fig.3, utilizes a convex-concave stable resonator design which produces excellent beam quality (TEM$_{00}$) with 4 mJ pulse output in 50 psec at a 1 kHz repetition rate while minimizing the energy density on critical optical components. This allows us to produce 2.5 watts of 40 psec green light in order to pump our dye amplifier. Wang et al.[13] using an improved electro-optic driving scheme and proper acoustic damping in a BNL resonator have reported an output power of 6.6 watt with up to 10 kHz operation.

The RGA operation can be visualized as four simple steps, (1) trapping a single pulse from the mode-locked seed train in the amplifier resonator, (2) simultaneously Q-switching the amplifier resonator, (3) letting the trapped pulse build up gain by undergoing N round trips, where $N \geq 2$, and (4) at saturation, cavity dumping the amplified pulse. For our RGA operating in the low gain single pass limit the number of roundtrips is typically 75 to 100. This is shown in Fig. 4(a) where the single trapped pulse builds up gain with each successive roundtrip through the amplifier. At the Q-switched build-up time the cavity dumping step is applied extracting all the energy from the amplifier which results in a single pulse output as shown in Fig. 4(b). This large number of roundtrips in our resonator contributes to better stability, 1% shot-to-shot energy fluctuations, and excellent beam quality. The advantages of cw-pumped RGA include the absence of any pulsed power supplies which are inherently cumbersome, gaussian beam quality, large dynamic range of operation (especially true for YLF), multi-kilohertz performance, short pump pulses, low energy fluctuations and timing jitter (optically synchronized). Damage threshold becomes a limitation with any RGA resonator and extraction of average powers in excess of 10 watts seems unlikely.

A recent development at the University of Chicago[14] has resulted in the operation of a 100 kHz cw RGA using acousto-optic switching. This system yields a gaussian beam with $8\mu J$ of green energy at 100 kHz and 5% conversion efficiency upon pumping a dye amplifier. Both YAG and YLF systems are now commercially available with multi-kilohertz outputs. The use of the second harmonic of a cw RGA system for an amplifier pump at 1 kHz has been reported to give conversion efficiencies between 1-5% for 7 psec[15] and 150 fsec[16] tunable pulses. Using 527nm radiation from our BNL regenerative amplifier, we have achieved 15% conversion efficiencies for 250 fsec TEM_{00} pulses using a three-stage longitudinally pumped dye amplifier scheme[17]. Three dye cells with 1, 2, and 10 mm thickness produced gains of 200, 35, and 50, respectively. The pump geometry consisted of counter-propagating pump/seed beams to ensure good spatial overlap in the dye cells. The green 527 nm light was distributed among the dye cells using 14%, 26%, and 60% pump energies, respectively. Using 1 mJ green energy at 1 kHz repetition rate we have observed over 140 μJ output at 590 nm and 250 fsec with less than 1% ASE. We believe that a combination of *excellent mode quality* from our RGA and dye oscillator and *good spatial overlap* in our dye lasers results in our high conversion efficiencies, which are the highest reported to date for any kilohertz picosecond dye system.

3.2. Pulsed pumped regenerative amplifiers

The first solid state regenerative amplifier was a pulsed flash-lamp pumped Nd:YAG amplifier[10]. The principle of operation is similar to that described in the section on cw RGA but the details of operation are quite different. First, there is the need to have pulsed power supplies which become an obvious limit on repetition rate performance. Commercial YAG systems are available that will operate at discrete repetition rates up to 1 kHz but multi-kilohertz rates have not been achieved. A second difference is that they operate in the high gain single pass limit, thus the number of roundtrips in the amplifier is about 5. A consequence of this limited time in the resonator results in poorer mode quality (near-gaussian) and greater shot-to-shot energy fluctuations (7-10%). However, the advantages of pulsed RGA are still provided by optical coupling to the oscillator and short pump pulse generation. The pulsed RGA provide output energies similar to their cw counterpart since both resonators are damage threshold limited. However, the commercial systems offer the possibility of further amplification via single pass high gain amplifiers after the RGA. Of course, such a scheme can be applied to cw-based RGA. This scheme can produce 15 mJ/pulse of green amplifier pump at 540 Hz operation. The conversion efficiencies reported[18] for short pulse dye amplification are similar to those given by cw RGA's.

4. ALTERNATIVE PUMP SOURCES

There is a number of alternate pump laser sources available to the researchers with widely varying characteristics, as discussed above. The choice of amplifier system depends upon the desirable characteristics necessary for each individual experiment. For example, if high peak power is desirable than a compromise on high kilohertz performance must be made. However, it should be noted that certain systems are inherently more flexible than others. The following section will describe some different schemes for amplifying tunable light.

4.1. Excimers

This group of pulsed, high pressure gas lasers emits radiation in discrete lines in the uv region. Commercially available excimer lasers use electronically excited diatomic rare gas halide molecules (XeF*, XeCl*) as the lasing medium. Even though their power supplies are pulsed circuitry and they still use thyratrons for electronically switching high voltages to initiate a discharge, they have the advantage of well established commercial technology, high energy per pulse, and the flexibility to pump a wide range of dyes. The XeCl and XeF transitions at 308nm and 351nm, respectively, are the best suited excimers for pumping dye amplifiers. However, for amplifying short pulses two major difficulties exist. First, amplification of short red pulses ($\lambda \geq 550$nm) with UV pump pulses results in a large frequency mismatch and consequently present problems associated with a large heat load. Second, the long nanosecond pulses are temporally mismatched with short seed pulses ($\tau \leq 5$ps), even when accounting for the dye's lifetime. This will lead to low conversion efficiencies and high ASE. Lasers with 308nm output with about 50 mJ of energy per pulse at 500 Hz in 20 nsec are commercially available . The problem associated with the nonuniform pumping of dye media because of the rectangular spatial distribution of excimer beams can be overcome by transverse pumping of prism or Bethune[19] cells. Rolland and Corkum[20] have demonstrated an excimer pumped dye amplifier with an output of 550 uJ in 70 fsec pulses and a maximum repetition rate of 125 Hz. The conversion efficiency for their four stage amplifier is 1% and the amount of ASE was about 10%. This same scheme could easily be used with commercial 500 Hz excimer lasers but this appears to be the upper limit in high repetition rate performance.

4.2. "All-in-one" solid-state laser

The mode-locked, Q-switched, and cavity dumped lasers or "all in ones" as they are sometimes called span the moderately high repetition rate range with about 1 mJ of energy per pulse. These solid-state (YAG or YLF) lasers have all three mechanisms combined in one resonator. Briefly, these lasers are operated in a prelasing mode such that there is a low level of cw mode-locked operation between Q-switched bursts. Once mode-locked operation is achieved, the resonator is Q-switched and ample time is allowed for the photon density to reach a maximum, at which time cavity dumping is initiated. For pulsewidths in the range of 75 to 120 psec repetition rates of up to 1 kHz have been reported[21]. Extensive research has been conducted on both YAG[22] and YLF[23] lasers in order to achieve high repetition rate. Currently, 1 kHz seems to be the performance limit because there is about a 1 msec recovery time after each cavity dumping step for these resonators to return to their mode-locked steady-state. Advantages of "all in ones" include stand-alone resonator operation which can perform with any choice of oscillator/dye laser system, optical independence from any master oscillator which is a desirable feature in adjusting the arrival times of pump and seed pulses at the dye amplifier. Disadvantages include a limited repetition rate and critical alignment of several active optical components in the cavity which are necessary for achieving good mode and short pulse operation. Newell et al.[21] have reported 1 kHz amplification of 150-300 fsec pulses in the range of 560-680 nm with a maximum energy of 50 μJ. The conversion efficiency for their system was 7%.

4.3. Copper vapor lasers

Copper vapor lasers (CVL) are pulsed lasers with output distributed between two wavelengths, 511 and 578 nm. A major advantage of a copper vapor laser is its repetition rate. Presently these lasers are available with repetition rates ranging between 2 and 32 kHz. CVL is inherently pulsed because of the metastable lower laser levels and therefore the optimum repetition rate is dependent on the lifetime of these levels. Typical commercial copper vapor laser outputs are 6 mJ at an optimized rate of 6.5 kHz in pulses which are 20 to 40 nsec wide. High ASE problems associated in pumping with CVL can be minimized by utilizing the well known bow-tie arrangement[8] shown in Fig. 3(a). Pulse energies of over 1 uJ were obtained in 100 fsec pulses at 5 kHz which corresponds to a conversion efficiency less than 1%. Since the two emission lines are visible nanosecond pulses any sum or difference frequency mixing techniques is not possible with CVL.

5. CONCLUSION

The recent advances in research on high powered, high repetition-rate laser systems, as discussed in this article, provide the impetus for a new generation of laser systems that will evolve into a general laboratory tool in the 1990's. Likewise, the potential for novel applications driven by the advantages provided by higher repetition rates will lead to new innovations and discoveries in science and technology.

6. ACKNOWLEDGEMENTS

I would like to thank the members of my group Joseph Dolce, Muhammad Saeed, and Baorui Yang for their contributions to the developments in my laboratory. The research carried out at Brookhaven National Laboratory under Contract No. DE-AC02-76CH00016 with the U.S. Department of Energy and supported, in part, by its Division of Chemical Sciences, Office of Basic Energy Sciences.

References

[1]   R.H. Page, R.J. Larkin, A.H. Kung, Y.R. Shen, and Y.T. Lee, *Rev. Sci. Instrum.* **58**, 1616 (1987).

[2] C.T. Rettner, E.E. Marinero, R.N. Zare, and A.H. Kung, *J. Phys. Chem.* **88**, 4459 (1984).

[3] *"Short Wavelength Coherent Radiation: Generation and Applications"*, eds. D.T. Attwood and J. Bokor, A.I.P. Conference Proceedings No. 147, articles therein, (1986).

[4] *"Atomic and Molecular Processes with Short Intense Laser Pulses"*, ed. A. Bandrauk, Plenum Press, NY and London (1988).

[5] J.H. Moore, C.C. Davis, and M.A. Coplan, *"Building Scientific Apparatus: A Practical Guide to Design and Construction"* Addison-Wesley Pub. Co., London, Chapter 6.8(1983).

[6] I. Ben-Zvi, L.F. DiMauro, S. Krinsky, M.G. White, and L.H. Yu, *Nuclear Inst. and Methods*, in press (1990).

[7] K.H. Drexhage, in *"Dye Lasers"* ed. F.P. Schäfer, Chapt. 5, Springer-Verlag, New York (1990).

[8] W.H. Knox, M.C. Downer, R.L. Fork, and C.V. Shank, *Opt. Lett.* **9**, 554 (1984).

[9] G. Vaillancourt, T.B. Norris, J.S. Coe, P. Bado, and G. Mourou, *Opt. Lett.* **15**, 317 (1990).

[10] J. E. Murray and W. H. Lowdermilk, *J. Appl. Phys.* **51**, 3548 (1980).

[11] P. Bado, M. Bouvier, and J. Scott Coe, *Opt. Lett.* **12**, 319 (1987).

[12] M. Saeed, Dalwoo Kim, and L.F. DiMauro, *Appl. Opt.* **29**, 1752 (1990).

[13] X.D. Wang, P. Basseras, R.J. Miller, J. Sweetser, and I.A. Walmsley, *Opt. Lett.*, in press (1990).

[14] A.J. Ruggiero, N. F. Scherer, G. M. Mitchell, G. R. Fleming, and J. N. Hogan, *J. Opt. Soc. Am. B* **8**, 2061 (1991).

[15] T. Juhasz, J. Kuhl, and W.E. Bron, *Opt. Lett.* **13**, 577 (1988).

[16] I.N. Duling III, T. Norris, T. Sizer II, P. Bado, and G.A. Mourou, *J. Opt. Soc. Am. B* **2**, 616 (1985).

[17] J. Dolce, B. Yang, M. Saeed, G. Scoles, and L. F. DiMauro, to be published.

[18] A.J. Taylor, J.P. Roberts, T.R. Gosnell, and C.S. Lester, *Opt. Lett.* **14**, 444 (1989).

[19] D. S. Bethune, *Appl. Opt.* **20**, 1897 (1981).

[20] C. Rolland and P. B. Corkum, *Opt. Comm.* **59**, 64 (1986).

[21] V. J. Newell, F. W. Deeg, S. R. Greenfield, and M. D. Fayer, *J. Opt. Soc. Am. B* **6**, 257 (1989).

[22] I. N. Dulling and T. Sizer, *Proc. Opt. Soc. Amer.* annual meeting Seattle, WA, Oct.23, 1986 paper I - PD6

[23] L. Min, Q. Bao and R. Miller, *Opt. Comm.* **68**, 427 (1988).

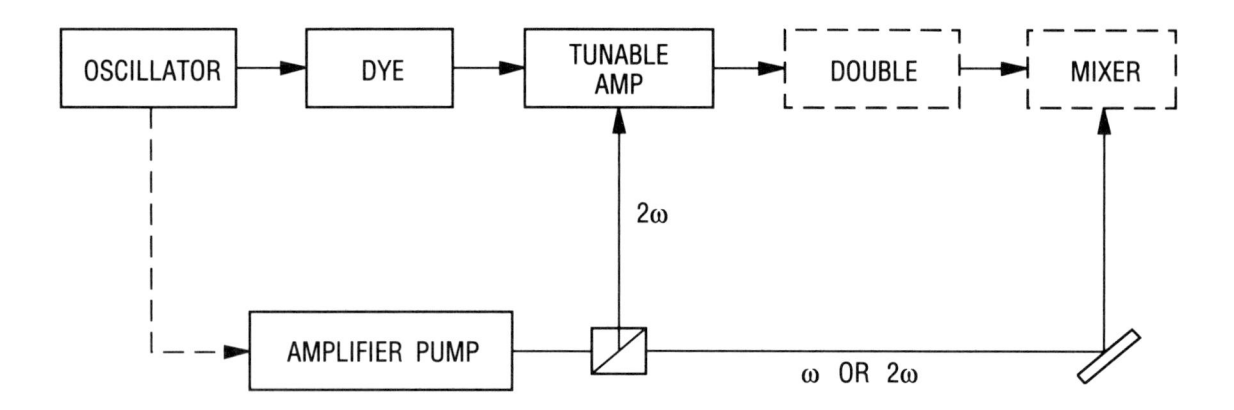

Fig. 1. Schematic representation of a high repetition-rate, tunable, short pulse laser system. The dashed line between the oscillator and amplifier pump exists for optically coupled systems, i.e. regenerative amplifiers. The dashed boxes indicated optional extensions of some of these systems.

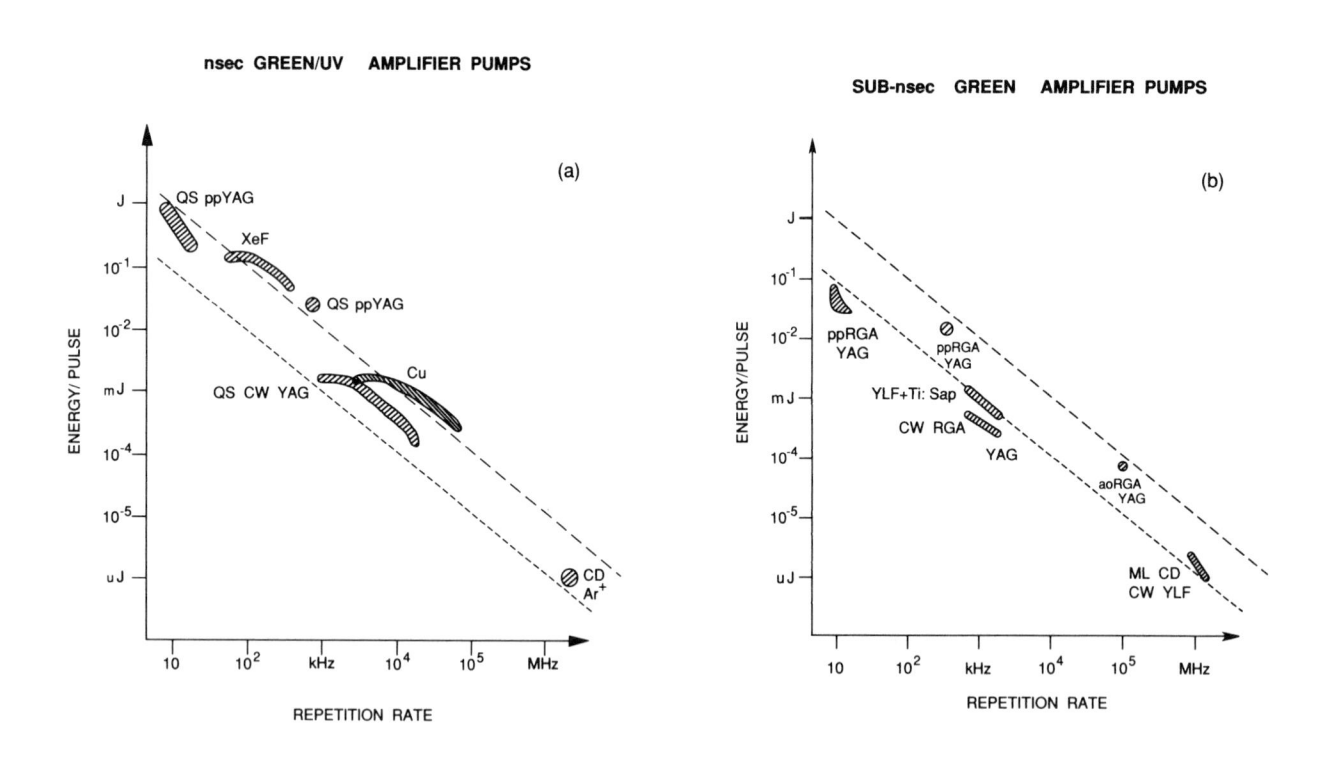

Fig. 2. Log-log plot of energy per pulse versus repetition rates for (a) nanosecond and (b) sub-nanosecond amplifier pumps. The small dashed line is for 1 watt average power and the large dashed line is for 10 watt. The abbreviations in the plot are; QS–Q-switched, pp–pulsed pumped, CD–cavity dumped, ao–acousto-optic, and RGA–regenerative amplifier.

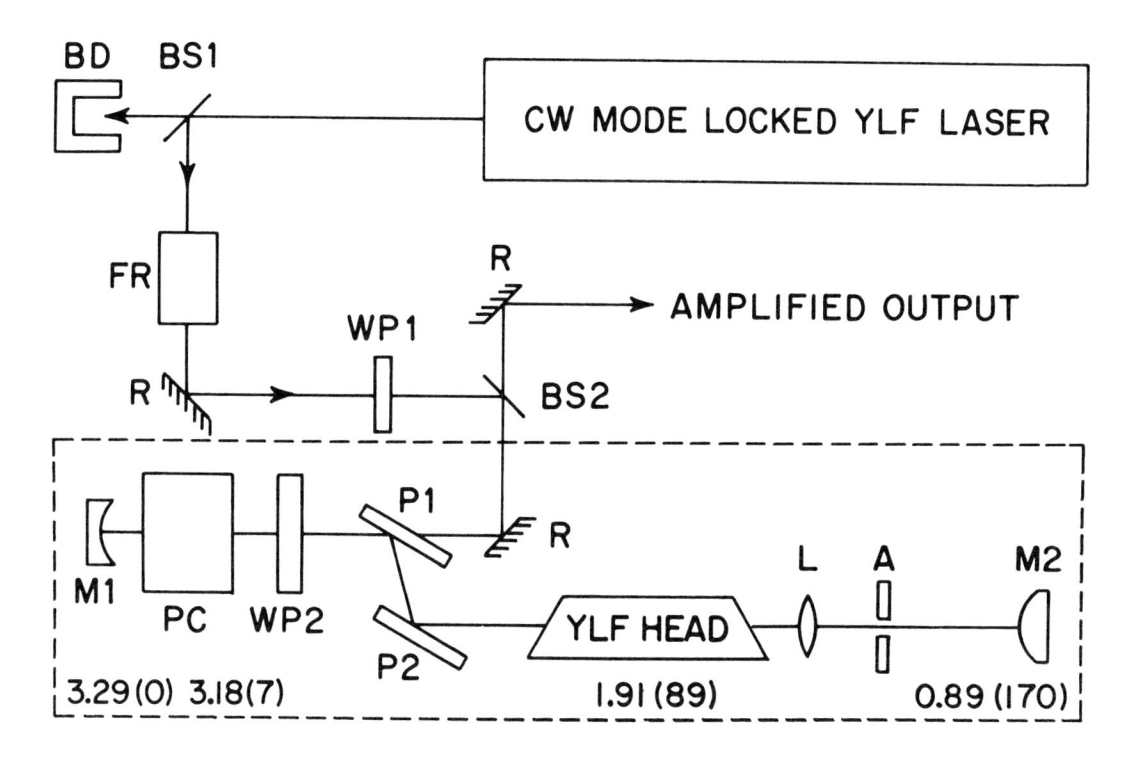

Fig. 3. A block diagram of the BNL regenerative amplifier system. M1 and M2 are concave and convex resonator mirrors, PC Pockels cell, L cylindrical lens, A mode selector, P1 and P2 Brewster angle polarizers, WP1 and WP2 half- and quarter-wave plates, BS1 and BS2 5% beam splitters. The e^{-2} diameters in mm are indicated at various positions (the positions in cm are in parenthesis) in the amplifier.

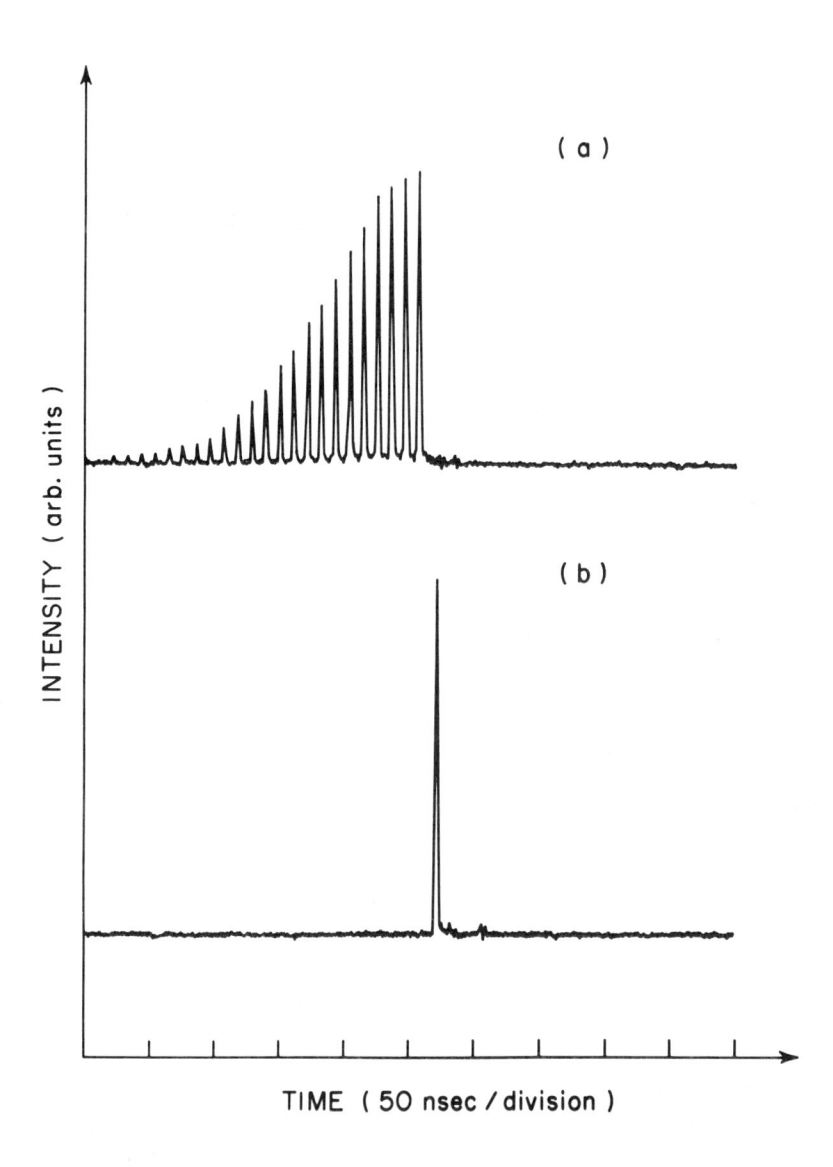

Fig. 4. Fast photodiode output during regenerative amplifier operation including cavity dumping. Trace *a* is output from a photodiode viewing 1% resonator leakage through mirror M1 and trace *b* is photodiode output of amplifier output. Note that at the time of cavity dumping all the energy is extracted from the cavity (trace *a*) and emerges as a single pulse (trace *b*).

Ultraviolet resonance Raman spectroscopy: studies of depolarization dispersion and strong vibronic coupling

Bruce S. Hudson

University of Oregon, Department of Chemistry
Eugene, Oregon 97403

ABSTRACT

The application of ultraviolet resonance Raman spectroscopy to the electronic excitations of the simple π-electron systems butadiene, cyclopentadiene, benzene, acetylene, dimethylacetylene (2-butyne) and diacetylene (1,3-butadiyne) is described. Raman scattering resonant with electronic excitations of these species provides new information that permits a check on the interpretation of the corresponding absorption transitions. The determination of the depolarization ratio for Raman scattering of totally symmetric modes and its variation with excitation wavelength is shown to be a useful way to demonstrate that an electronic band consists of two or more transitions with orthogonal polarization components. Raman spectra of acetylene, dimethylacetylene and benzene show evidence of strong vibronic coupling. A quantitative analysis has been developed for the benzene case where pseudo-Jahn-Teller distortion is observed. The general utility of resonance Raman spectroscopy using ultraviolet radiation as a tool in molecular spectroscopy is illustrated with these studies. Highly excited vibrational levels not seen by other methods are often observed with high intensity, overlapping electronic transitions can be detected and the nature of Franck-Condon displacements and vibronic coupling mechanisms can be determined.

1. INTRODUCTION

Resonance Raman scattering provides a mapping onto the ground electronic state of the dynamics of nuclear motion following electronic excitation. The performance of such experiments with excitation in the far ultraviolet region[1-5] permits resonance with higher excited electronic states of polyatomic molecules and with the excitations of small molecules formed from atoms with few electrons. The absorption spectra of polyatomic molecules at higher energy are usually difficult to understand due to their diffuse nature and overlapping valence and Rydberg excitations.[6] Electronic excitations of small molecules often result in very large geometry changes and thus electronic absorption spectra which, even if resolved, are often difficult to interpret. In some cases, such as that of ammonia,[7] water[8-10] and hydrogen sulfide,[11] excitation even to the lowest excited singlet state results in photodissociation and thus intrinsically broadened spectra. In all of these cases resonance Raman scattering can provide useful information as to the nature of the electronic excitation. The Franck-Condon active modes are directly reported in the resonance Raman spectrum reflecting the geometry change associated with excitation. The observation of fundamental or overtone transitions of non-totally symmetric modes permits the specification of the vibrational modes involved in vibronic coupling if that is important for the excitation mechanism. Because of the large geometry changes that are often associated with electronic excitation of small molecules, resonance Raman spectra obtained with ultraviolet excitation often have transitions that involve long progressions of the Franck-Condon active modes. While the intensity distribution associated with such progressions contains information about the excited electronic state potential energy surface, the frequencies of these overtones can be used to specify the ground state potential energy surface in regions far from the equilibrium position. This has been applied to the cases of ethylene and allene[12-17].

2. RAMAN INTENSITIES AND DEPOLARIZATION RATIOS

The Raman scattering process can be described within either time-dependent[18] or time-independent formalisms[19]. In the time-dependent formalism the interaction of a molecule with radiation creates a wavepacket describing the nuclear wavefunction on the upper state Born-Oppenheimer surface. This wavepacket is, in general, not an eigenstate of the upper state surface and thus propagates in time. This time-dependent wavepacket has an overlap with the various ground electronic state stationary wavefunctions which varies in time. The absorption spectrum as a function of frequency and the resonance Raman spectrum as a function of excitation frequency are given by Laplace transforms of these time-dependent overlap integrals with the initial wavepacket (for absorption) or that describing the final vibrational state (for a Raman transition). In the Condon (constant transition dipole moment) approximation the initial wavepacket is the starting ground state vibrational wavefunction. If the transition moment varies in space then this "transition moment surface" multiplies the the vibrational wavefunctions. If the harmonic approximation is added to the Condon approximation, then the wavepacket is a Gaussian function and remains so for all times with its mean position describing classical harmonic motion.

This picture of Raman scattering places emphasis on the region of the upper state potential energy surface that is directly above the peak of the initial ground state wavefunction (the "Franck-Condon region") and permits an expansion of the overlap integral in a time series and thus a "short time approximation" and higher time-series expansion terms. This treatment makes it clear that in certain limits (such as the "far-from-resonance" case and when the upper state is dissociative) the short time behavior dominates and Raman scattering is related to the slope of the upper state potential energy surface in the Franck-Condon region along the displacement directions that best describe the ground state modes of motion. In this same limit the various overtone transitions observed in the Raman spectrum represent increasingly longer time motion on the upper state potential surface.

For resonance Raman scattering the wavepacket description is perhaps best suited to the case of dissociative upper state potential surfaces. In that case the wavepacket moves smoothly away from the Franck-Condon region and never returns. This implies a broad diffuse absorption spectrum (a continuous distribution of eigenstates) and, in the same approximation, a resonance Raman spectrum that does not change in form as the excitation wavelength is changed within a given electronic transition. When an absorption spectrum exhibits structure or the resonance Raman spectrum changes in relative intensities as the excitation wavelength is changed, the short time limit terminology does not apply. In the wavepacket description, the time-dependent overlap integral has recurrances and the regions of the upper electronic state surface far from the Franck-Condon surface are being probed.

In the time-independent picture of resonance Raman scattering the stationary states of the upper excited electronic surface are used as a basis set with an oscillatory time dependence that is related to the degree to which they are off-resonance. If the upper state surface has a high density of states (or if the excitation is well off-resonance) then contributions from many intermediate states must be summed to describe the virtual state. The relative contribution of each stationary state is related to a product of vibrational overlap integrals. If the harmonic approximation is made for the upper electronic state then the major feature of the upper state oscillators that determines the overlap integrals is (for symmetric modes) the degree of displacement of these oscillators from their equilibrium position in the ground electronic state. The slope of the potential surface of the upper state at the

Franck-Condon region which dominates the short time behavior is proportional to the product of the displacement and the force constant. This means that a force constant change has only a second order effect if there is no change in the equilibrium position but a shift in the equilibrium without a change in the force constant has a first order effect. It is interesting to note that within the time independent picture two parameters (the shift and the force constant) are needed to specify the slope which is the only paramter apparing in the fiorst-order time dependent picture.

The inclusion of the non-Condon terms (the dependence of the transition moment between electronic states on nuclear position) is usually treated within the stationary state method by a Taylor series expansion leading to linear, quadratic, etc. corrections to the value of the transition moment at the equilibrium nuclear position. If this value is zero by symmetry, the transition is said to be "forbidden" and a non-zero linear or higher term is necessary in order for the transition to be observed. In this case the transition moment will depend only on excursions of the nuclei along nontotally symmetric coordinates. Symmetry considerations impose the restriction that in all electronic states the potential energy surface as a function of displacement along a nontotally symmetric coordinate must have only even terms in its Taylor series representation. This means that electronic excitation will, at most, change the second and higher order even derivatives of the potential surface with respect to these vibronically active coordinates; the derivative of the potential with respect to these displacements will remain zero and the *overall* symmetry of the potential surface will be retained. Similar considerations impose the constraint that the variation in the transition moment with respect to nonsymmetric displacements can have only odd or only even terms. The dominant nuclear coordinate dependent vibrational integrals will involve the first power of the displacement between vibrational wavefunctions leading to the restriction within the harmonic approximation (and ignoring very large changes in the harmonic force constant) of a change of one in vibrational quantum number. This means that resonance Raman spectra obtained under conditions of resonance with symmetry forbidden electronic transitions will have a characteristic enhancement of a two-quantum overtone transition of the mode or modes that are vibronically active. Behavior of this type has been observed for resonance Raman studies of benzene in the region of the symmetry forbidden transition to the excited state of B_{1u} symmetry near 200 nm.[20-22]

In the case of an electronic transition which is not forbidden by symmetry, the transition moment may still depend on displacements along either totally symmetry or nontotally symmetric coordinates. Symmetry again requires that the variation of the transition moment with respect to excursions of the nuclei along non-totally symetric paths away from the symmetry point must have only odd terms in its Taylor series representation. This is consistent with a nonvanishing value of the moment at the symmetry point only if the vibronically induced component is vectorially orthogonal to that present in the absence of the displacement. This vibronic activity within an allowed electronic transition gives rise to intensity in fundamental transitions of nontotally symmetric vibrations. One of the two electric dipole transitions contributing to the Raman process is allowed (involving only a vibrational overlap integral) while the other is induced by linear vibronic coupling (involving the first power of the displacement coordinate). The fact that these allowed and induced transition dipoles have orthogonal polarizations leads to the well-known result that nontotally symmetric Raman transitions are depolarized. Resonance with a weakly allowed electronic transition that has a significant amount of vibronically induced intensity results in enhancement of these fundamental transitions of the vibronically active modes. This has been the interpretation of the spectra observed for substituted benzenes with excitation in the ca. 200 nm region[23,24].

Within the time independent formalism, the intensity of a Raman transition from the ith vibrational level of the ground electronic state g to the jth vibrational level in the ground electronic state g is:

$$I_{gi,gj} \propto I_0 \nu^4 \Sigma_{\rho,\sigma} |(\alpha_{\rho\sigma})_{gi,gj}|^2 \qquad (1)$$

where $I_{gi,gj}$ is the intensity of the Raman scattered radiation for the above specific Raman transition. $(\alpha_{\rho\sigma})_{gi,gj}$ is the polarizability tensor element possessing polarization indices ρ and σ for the $|gi\rangle \to |gj\rangle$ vibronic transition. Within the Kramers-Heisenberg-Dirac formulation the polarizability tensor elements are:

$$(\alpha_{\rho\sigma})_{gi,gj} = \Sigma_{ev} \langle gi|\mu_\sigma|ev\rangle \langle ev|\mu_\rho|gj\rangle [E_{ev} - E_\ell + i\Gamma_{ev}]^{-1} + \text{anti-resonant term} \qquad (2)$$

μ_σ and μ_ρ are the electronic dipole operators of σ and ρ polarization, respectively; E_{ev} is the energy of the $|ev\rangle$ vibronic level; E_ℓ is the energy of the laser source, and Γ_{ev} is the linewidth of the $|ev\rangle$ vibronic state.

If the vibronic wavefunctions are described by Born-Oppenheimer wavefunctions, $\alpha_{\rho\sigma}$ becomes:

$$(\alpha_{\rho\sigma})_{gi,gj} = \Sigma_{ev} \langle i|(M_\sigma)_{ge}|v\rangle \langle v|(M_\rho)_{eg}|j\rangle [E_{ev} - E_\ell + i\Gamma_{ev}]^{-1} \qquad (3)$$

where $(M_\sigma)_{ge} = \langle g|\mu_\sigma|e\rangle = (M_\sigma(Q))_{ge}$ is the nuclear coordinate dependent electric dipole transition moment between the ground electronic state and excited electronic state e. The dependence of $(M_\sigma)_{ge}$ on the nuclear coordinates and be described by the Taylor series expansion in terms of displacements from the equilibrium position:

$$(M_\sigma)_{ge} = (M_\sigma^o)_{ge} + (Q-Q_o)(\partial(M_\sigma)_{ge}/\partial Q)_o = (M_\sigma^o)_{ge} + (Q-Q_o)(M_\sigma')_{ge} \qquad (4)$$

Substitution of this expression into the Kramers-Heisenberg-Dirac polarizability expression gives three terms designated the A, B, and C terms:

$$(\alpha_{\rho\sigma})^A_{gi,gj} = \Sigma_{ev} (M_\sigma^o)_{ge} (M_\rho^o)_{eg} \langle i|v\rangle \langle v|j\rangle [E_{ev} - E_\ell + i\Gamma_{ev}]^{-1} \qquad (5a)$$

$$(\alpha_{\rho\sigma})^B_{gi,gj} = \Sigma_{ev} (M_\sigma')_{ge} (M_\rho^o)_{eg} \langle i|Q|v\rangle \langle v|j\rangle [E_{ev} - E_\ell + i\Gamma_{ev}]^{-1}$$
$$+ \Sigma_{ev} (M_\sigma^o)_{ge} (M_\rho')_{eg} \langle i|v\rangle \langle v|Q|j\rangle [E_{ev} - E_\ell + i\Gamma_{ev}]^{-1} \qquad (5b)$$

$$(\alpha_{\rho\sigma})^C_{gi,gj} = \Sigma_{ev} (M_\sigma')_{ge} (M_\rho')_{eg} \langle i|Q|v\rangle \langle v|Q|j\rangle [E_{ev} - E_\ell + i\Gamma_{ev}]^{-1} \qquad (5c)$$

The superscript o indicates the value at the equilibrium position; the prime indicates derivative with respect to normal coordinate Q. By definition, M^o is non-zero for an allowed transition and zero for a forbidden transition. For an allowed electronic transition the terms involving M^o usually dominate the electric dipole transition moment and thus the absorption spectrum. This means that under conditions of resonance with an allowed transition the major term is the A-term. This gives rise to fundamental, overtone and combination intensity in those modes that are Franck-Condon active in the transition. A long series of overtone transitions can be seen in resonance Raman spectra for resonance with allowed transitions that are to upper states that have very different geometries than for the ground state. If a transition is allowed but has a relatively significant contribution from vibronic coupling then the B-term can be important. As noted above this gives rise to activity in the fundamental transitions of non-totally symmetric vibrations. (A special case of this which is particularly important is the Jahn-Teller effect in which the two vibronically coupled transitions are degenerate.) Clearly, for a symmetry forbidden electronic transition, the only non-zero term in the polarizability expression is the C term. Each of the matrix elements in this term are dominated by the contribution from those vibrational levels that differ from each other by one quantum. This gives rise to the dominance of two-quantum Raman vibrations under conditions of resonance with a forbidden transition. The observation of such transitions, which are otherwise very weak, can be used as an indicator of the presence of an excited electronic state that gives rise to a forbidden transition.

The depolarization ratio for a Raman transition for 90° scattering is defined as the ratio of the intensity scattered with polarization perpendicular to that of the excitation divided by the intensity with polarization parallel to the excitation polarization. The depolarization ratio is a sensitive measure of the extent to which the transition dipoles giving rise to Raman activity are colinear. We are interested here in a particular special case in which a totally symmetric vibration gains intensity due to resonance with one or more nearby transitions whose transition dipole orientations are prescribed by symmetry to lie along mutually orthogonal axes. For A-type Raman scattering for the case of two overlapping transitions with perpendicular transition dipoles ρ is given by

$$\rho = \{[\alpha_{xx}]^2 - \alpha_{xx}\alpha_{yy} + [\alpha_{yy}]^2\}/\{3[\alpha_{xx}]^2 + 2\alpha_{xx}\alpha_{yy} + 3[\alpha_{yy}]^2\} \qquad (6)$$

For the case to be considered below the two coordinate axes are parallel and perpendicular to the C_2 axis of the molecule. We will arbitrarily designate the y axis as that corresponding to the B_2 transition (in-plane, perpendicular to the C_2 axis) and the x-axis as that parallel to the C_2 axis corresponding to a transition to an A_1 excited state. The depolarization ratio ρ can be thought of as a function of the ratio of the two polarization components, α_{xx}/α_{yy}. If only the transition to the B_2 state were present then $\alpha_{xx} = 0$ and $\rho = 1/3$. A small contribution of an x-polarized transition to the scattering of a mode relative to a dominant y-polarized transition can have an easily measurable effect on ρ causing it to deviate appreciably from 1/3.

The limiting case of $\alpha_{xx}/\alpha_{yy} = 1$ corresponds to the situation which occurs for resonance with a doubly degenerate allowed electronic transition. In this case of two equal polarizability components the depolarization ratio is 1/8. When two electronic transitions overlap but are displaced in energy then the relative sign of the two contributing polarizabilities will change when the excitation energy exceeds a certain value. The ratio α_{xx}/α_{yy} then becomes negative and the depolarization rises above 1/3 reaching a limiting value of 3/4. Thus, for two energetically displaced electronic transitions with perpendicular polarizations one expects the value of ρ to be 1/3 at low excitation frequency (where only one state contributes), drop toward (or possibly to) 1/8, rise through 1/3 to a higher value (as high as 3/4) and then drop back to 1/3 when only the higher energy transition contributes.

3. BUTADIENE AND CYCLOPENTADIENE

Linear conjugated polyenes have an electronic structure which results in a low energy electronic excitation that has no simple description within the standard molecular orbital picture.[25,26] (In the following we restrict ourselves exclusively to singlet electronic states.) In the C_{2h} point group of the *all trans* linear polyene the ground electronic state has A_g symmetry and is designated the $1A_g$ state. The excitation corresponding to promotion of an electron from the highest occupied to the lowest unoccupied orbital (HOMO to LUMO) has B_u symmetry. This simple orbital picture is a good representation of the $1B_u$ state. The second state of A_g symmetry (the $2A_g$ state) has been shown to be lower in energy than the $1B_u$ excited state in polyenes with four or more conjugated double bonds. Because this excited electronic state is the lowest exitation within the singlet manifold, it plays a particularly important role in the photochemistry of these species. This $2A_g$ excited state is a mixture of singly and doubly excited configurations within the molecular orbital picture. The presence of such an excitation at low energies represents one of the major failures of the simplest implementation of molecular orbital theory. The energetic location of this electronic excited state in short polyenes is of considerable interest because these species should be amenable to reliable treatment by modern *ab initio* quantum chemical methods. Attempts to locate the $2A_g$ state in butadiene using a variety of methods have failed to provide any evidence for its presence at low energy. In the present section we describe how resonance Raman methods have been used to detect this excitation in both the unsubstituted butadiene itself (which is predominantly in the s-trans conformation) and in cyclopentadiene where the molecule is constrained to an s-cis conformation.

The basis for the method used in the study of butadiene has been outlined above. Under conditions of resonance with a (proposed) forbidden electronic transition, the C-term contribution to the resonance intensity will become activated. In order for the vibronic coupling matrix element, M', in the C term for the $1^1A_g \rightarrow 2^1A_g$ transition to be non-zero the normal coordinate Q must have b_u symmetry for this C_{2h} point group. Therefore, b_u symmetry vibrations will be enhanced in a $1^1A_g \rightarrow 2^1A_g$ resonance enhanced Raman transition. In the harmonic approximation the vibrational integrals given in the C term dictate the approximate selection rule $\Delta v = \pm 1$ for each vibrational integral. Since the Raman transition starts in the i = 0 vibrational level, the above selection rule implies j = 2. That is, the first overtone of b_u vibrations will be enhanced for resonance Raman scattering resonant with the 2^1A_g state. Also, since the vibrational wavefunctions in the above expressions are products of wavefunctions for normal modes active in the electronic transition, combination bands involving the overtone of the promoting mode plus other Franck-Condon modes will also be enhanced.

From one-photon fluorescence excitation studies of linear unsubstituted and alkyl substituted polyenes, it has been found that the lowest frequency in-plane b_u symmetry chain deformation mode is efficient at lowering the symmetry of these molecules, thereby promoting the $1^1A_g \rightarrow 2^1A_g$ transition in these systems. In butadiene, this mode is ν_{24} with a fundamental frequency of 300 cm^{-1}. The experiment aimed at detection of the $2A_g$ state is simply to monitor the intensity of the $2\nu_{24}$ transition at ca. 600 cm^{-1} as a function of excitation wavelength. The results of such a study are shown in figures 1 and 2. The observation of intensity in the expected two-quantum transition with excitation in the energy region slightly lower than that of the strongly allowed transition to the $1B_u$ state demonstrates that the $2A_g$ state is present in this energy region. Alternative explanations for this intensity in $2\nu_{24}$ have been eliminated by studies of deuterated butadiene which also confirm the assignment of the vibrational bands. The very strong transitions corresponding to totally symmetric vibrations obtain most of their intensity from the $1B_u$ state via an A-term mechanism. The special utility of this Raman methodology is that it provides symmetry-specific information for a forbidden electronic transition even though this transition is overlapped by a much stronger allowed transition.

In the case of the constrained s-cis diene cyclopentadiene the C_{2v} symmetry produces A_1 and B_2 excited electronic states that are the counterparts of the A_g and B_u states of the C_{2h} trans structure.[29] Because of the lower symmetry transitions are allowed to both of these excited states. The multiconfigurational character of the A_1 excited state results in a low intensity for this transition. The polarization of these transitions is parallel to the C_2 axis for the A_1 excited state and perpendicular to this axis in the molecular plane for the B_2 state. These two allowed

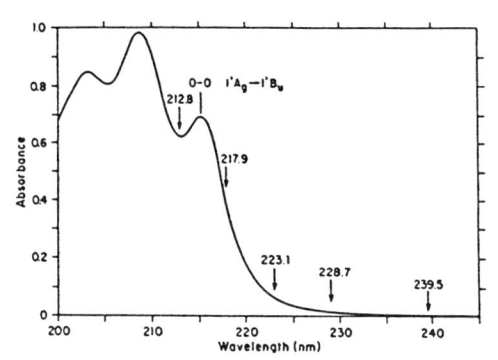

Figure 1A Butadiene absorption spectrum.

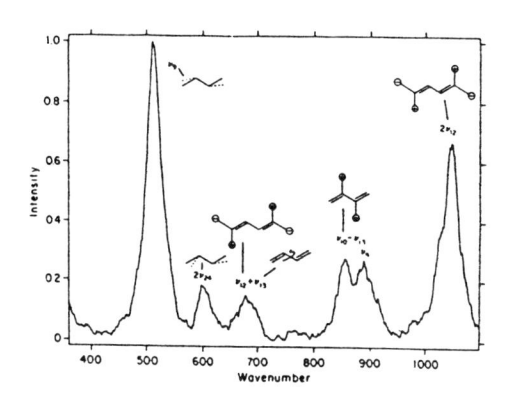

Figure 1B. Raman spectrum of butadiene with 217.9 nm excitation.

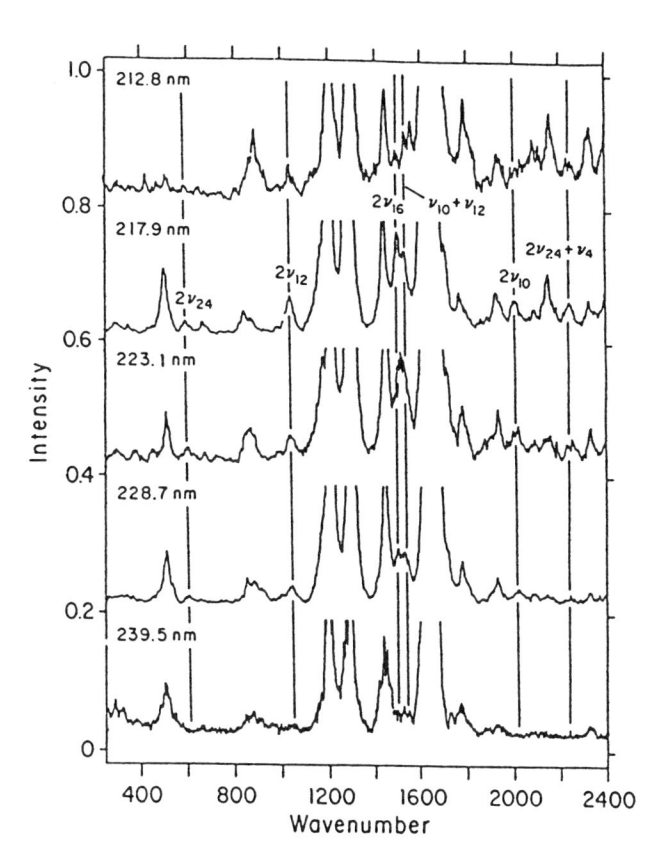

Figure 2. Raman spectra of butadiene at the indicated excitation wavelengths.

transitions are thus polarized in relatively perpendicular directions. These transition orientations are maintained during vibrational excursions along totally symmetric modes of nuclear motion. Because of this resonance with one or the other of these transitions results in enhancement of the α_{xx} or α_{yy} polarizability components via the A-term mechanism of resonance Raman scattering for totally symmetric modes.

The absorption spectrum of cyclopentadiene is shown in Figure 3. The features at shorter wavelength have been identified with a Rydberg transition. The rest of the absorption spectrum has been interpreted as being due to a single transition[30] or, alternatively, as due to two overlapping bands,[31] one relatively weak and structured (due to the transition to the A_1 state) and the other stronger, broad component (due to the transition to the B_2 state). Resonance Raman scattering depolarization ratio measurements provide a simple and unambiguous method to distinguish between these two explanations. The wavelengths at which resonance Raman spectra have been taken are indicated in Figure 3 as are the resulting depolarization ratios for three totally symmetric vibrations. The deviation of the values from 1/3 is conclusive proof that there are two contributing electronic transitions with perpendicular polarization components. We identify these as the A_1 and B_2 excited states. The variation of the depolarization ratio makes sense if we associate one transition with the low energy structured component and the other with the stronger broad component. The success of this simple method depends on the Franck-Condon activity of one or more vibrations in each of two excited states that give rise to perpendicularly polarized electronic excitations.

Figure 3. The absorption spectrum of cyclopentadiene. The arrows indicate wavelengths used for Raman excitation. Values of ρ for several bands are given at the top.[29]

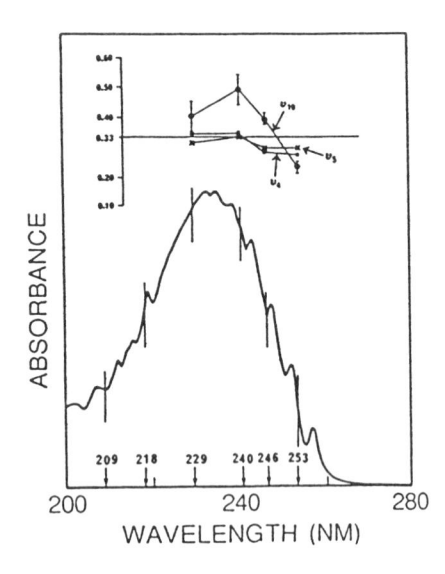

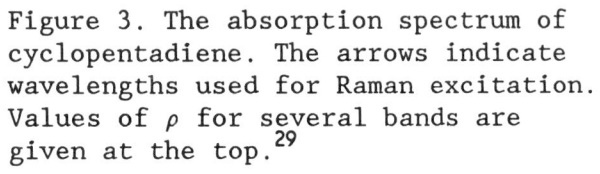

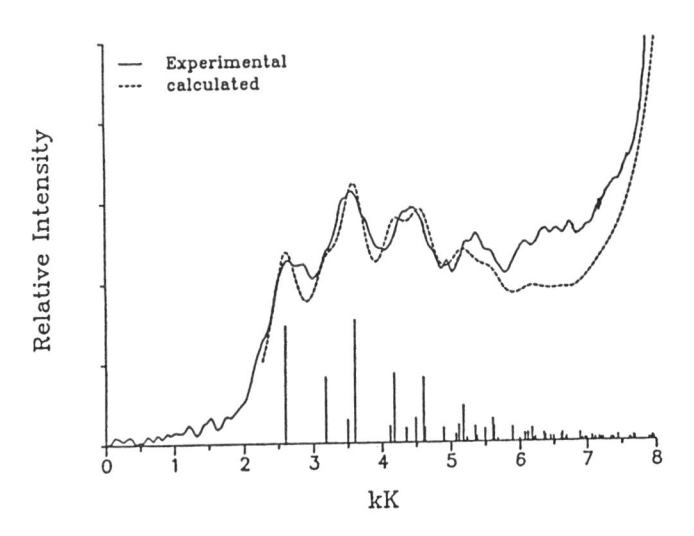

Figure 4. The absorption spectrum of benzene in the 210 to 180 nm region compared with the results of a vibronic calculation.[33]

4. THE B_{1u} EXCITED STATE OF BENZENE

The electronic spectroscopy of benzene has been the subject of extensive study for many years.[32] Excitations from the degenerate HOMO to degenerate LUMO levels results in states of B_{2u}, B_{1u} and E_{1u} symmetry in the 260 to 170 nm region. The spectral region near 200 nm includes the B_{1u} excitation which is relatively strong for a symmetry forbidden electronic transition due to strong coupling with the nearby allowed E_{1u} state. Previous resonance Raman studies of benzene in this spectral region[20-23] have been interpreted in terms of a first order linear perturbative treatment of the vibronic coupling of the B_{1u} state with the E_{1u} state due to the e_{2g} vibrations ν_8, ν_9 and ν_7 in decreasing order of importance. The experimental observation of greatest interest which is explained by this treatment is the strong enhancement of the binary overtones and combinations of these vibronically active modes. This simple treatment correctly predicts depolarization ratios and the effect of alkyl substitution on the intensity of fundamantal transitions of these modes[23,24]. Similar treatments[32] have been applied to the weakly structured absorption spectrum in this region (Figure 4). However, when these treatments are quantitatively compared it is found that the description that appears to be appropriate to the Raman scattering data does not adequately describe the absorption spectrum. In very recent work[33] we have used semiempirical quantum chemical methods and exactly solvable models for potential energy surfaces to develop a non-perturbative vibronic coupling model for the B_{1u} state. The result is very similar to that originally proposed[34] by Moffitt and Liehr and also to the similar situation deduced directly from *ab initio* calculations for the triplet manifold[35]. The absorption spectrum that results from this model is shown in figure 4 superimposed on the experimental results. The Raman spectra predicted on the basis of this model are shown in Figure 5. (The peaks in the experimental data for which no calculated bars are shown are associated with electronic states other than the B_{1u} state.) Excellent agreement is observed between theory and experimental results.

A simplified version of the potential energy surface for the lower branch of the B_{1u}/E_{1u} surface that is deduced on the basis of this treatment is shown in Figure 5. The most stable configuration is distorted with respect to the D_{6h} symmetry point along an anti-quinoidal displacement direction (two long bonds and four short bonds). The stabilization of these minima relative to the symmetry point is about 1000 cm^{-1}. The local maxima in the 3-fold symmetric surface correspond to quinoidal geometries. These are estimated to be about 300 cm^{-1} above the absolute minima. This small differential stabilization results from the (amplified) Jahn-Teller effect within the degenerate E_{1u} state. This effect does not seem to occur in the triplet manifold so in that case the potential has full cylindrical symmetry.[35]

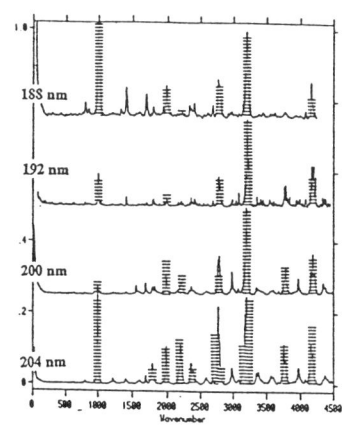

Figure 5. Resonance Raman spectra of benzene obtained at the indicated excitation wavelengths compared with the results of a full vibronic model.[33]

Figure 6. The potential energy surface of the B_{1u} excited state of benzene as a function of displacements along the two orthogonal ν_8 coordinates.[33]

5. ACETYLENE, DIMETHYLACETYLENE AND DIACETYLENE

The low lying $\tilde{A}$ and $\tilde{B}$ states of acetylene have Σ_u^- and Δ_u symmetry in the linear geometry. The electric dipole transition from the ground electronic state is forbidden in each case. It is well established from the detailed studies of Price and Walsh that the $\tilde{A}$ state is strongly bent[36,37]. The situation for the $\tilde{B}$ state is not so clear. Theoretical potential energy surfaces and transition dipole variations are shown in Figure 7. Note the Renner-Teller splittting of the degenerate $\tilde{B}$ state. This pattern of energy levels is similar to that for benzene but both of these states give rise to transitions that are forbidden at the equilibrium geometry. According to the calculations summarized in Figure 7 there is a strong variation of the transition dipole from its zero value as the molecule is deformed from linearity. This must be associated (in Herzberg-Teller terminology) with coupling to some higher lying excited states with dipole intensity. Such states have Π_u and Σ_u^+ symmetry in $D_{\infty h}$.

The resonance Raman spectrum of acetylene obtained with excitation at 204 nm is shown in Figure 8.[38] At this excitation significant contributions from both the $\tilde{A}$ and $\tilde{B}$ states are expected. The high activity for transitions involving even overtones of the gerade bending vibration is consistent with a *trans* bent geometry for an upper state with dipole intensity for the $\tilde{A}$ state. The activity in the symmetric C≡C stretch is also anticipated. The unexpected observation in this spectrum and others obtained at other wavelengths is the resonance enhanced activity seen for the fundamental transition of the bending vibration. As outlined above, fundamental transitions of non-totally symmetric vibrations are expected to obtain resonance enhancement via vibronic coupling within a resonant allowed electronic transition according to the B-term mechanism. Odd quantum transitions are not expected for non-totally symmetric modes under conditions of resonance with a forbidden transition. In the case of the

B_{1u} state of benzene, for example, very little activity was seen in the fundamental transitions of the vibronically active e_{2g} vibrations. The mechanism by which this odd-quantum vibration obtains intensity has recently been worked out[38]. As might be anticipated, a vibronic mechanism that involves both symmetry excited states with allowed transitions from the ground state is required and for one of these the coupling must be quadratic in the vibrational displacement.

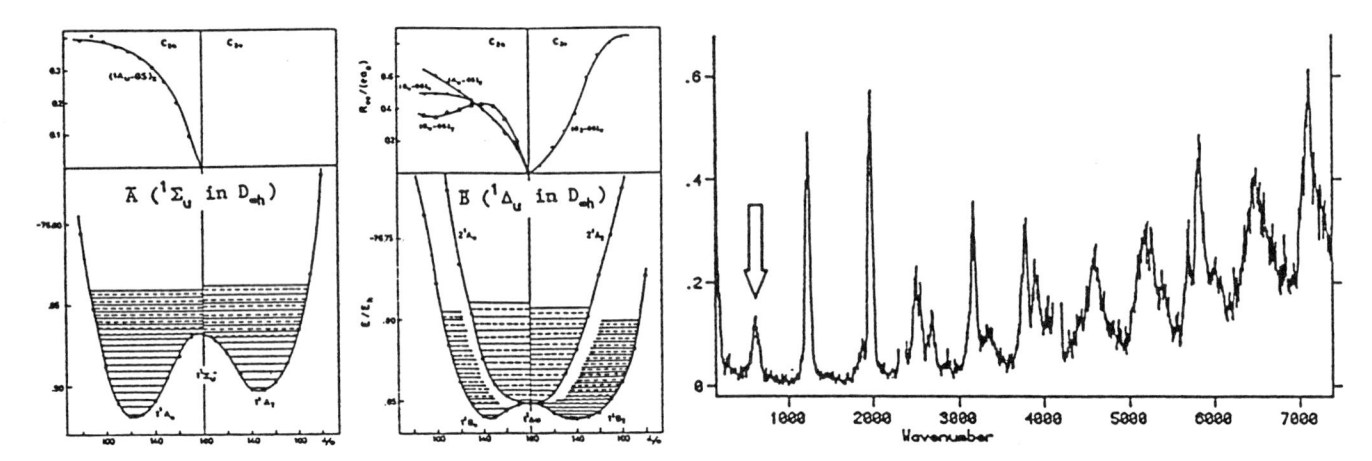

Figure 7. Potential energy surfaces and transition dipole moment variation for the A and B states of acetylene.[36]

Figure 8. The resonance Raman spectrum of acetylene obtained with excitation at 204 nm.

Similar studies of dimethylacetylene (2-butyne) and diacetylene (1,3-butadiyne = HC≡C-C≡CH) have recently been initiated. Space and the preliminary nature of these results does not permit a more detailed presentation but several interesting features have become clear. In the case of dimethylacetylene there is activity in the fundamental of the bending vibration as seen for acetylene but very little intensity for the even overtones. There is also interesting intensty in a band near 2950 cm^{-1} which may be the fundamental of a nonsymmetric doubly degenerate CH stretching vibration. The spectrum of diacetylene poses several unresolved problems[39-41]. The resonance Raman spectra obtained so far are explicable in terms of a slightly bent excited state with significantly different CC bond lengths than in the ground state. One feature of interest is the observation of bands that are also seen at certain excitation wavelengths for acetylene. These are probably due to the photogeneration of the HCC radical. This is one of several examples of the creation and detection of radical species under the nanosecond pulse conditions of these experiments.

6. ACKNOWLEDGEMENTS

The contributions of my associates, the coauthors of the publications listed in the references, is acknowledged. This work was supported by grants from the National Science Foundation (CHE-8816698) and the Petroleum Research Fund.

7. REFERENCES

1. B. Hudson, "Vacuum Ultraviolet Resonance Raman Spectroscopy", in *Recent Trends in Raman Spectroscopy*, S.S. Jha and S.B. Banerjee, editors (World Scientific Press, Singapore, 1989) pgs. 368-385.

2. B. Hudson and R. J. Sension, "Far Ultraviolet Resonance Raman Spectroscopy: Methodology and Applications" in *Vibrational Spectra and Structure*, Vol. 17A, H. D. Bist, J. R. Durig and J. F. Sullivan, eds. (Elsevier, Amsterdam, 1989) pgs 363-390.

3. B. Hudson, P. B. Kelly, L. D. Ziegler, R. A. Desiderio, D. P. Gerrity, W. Hess and R. Bates, "Far Ultraviolet Resonance Raman Studies of Electronic Excitations", in *Advances in Laser Spectroscopy*, Volume 3, edited by B. A. Garetz and J. R. Lombardi, (John Wiley & Sons, New York, 1986) pages 1-32.

4. B. Hudson, "Resonance Raman Spectroscopy in the Far-Ultraviolet Region", Spectroscopy **1**, 22-31 (1986).

5. B. Hudson, "Ultraviolet Resonance Raman Studies of Electronic Excitations", in *Time-Resolved Vibrational Spectroscopy*, edited by A. Laubereau and M. Stockburger (Springer-Verlag, Berlin, 1985) pages 170-174.

6. M. B. Robin *Higher Excited States of Polyatomic Molecules* (Academic Press, New York, 1974) volumes I, II and III.

7. L. D. Ziegler and B. Hudson, "Resonance Rovibronic Raman Scattering of Ammonia", J. Phys. Chem. **88**, 1110-1116 (1984); L. D. Ziegler, P. B. Kelly and B. Hudson, "Resonance Rovibrational Raman Scattering as a Probe of Unimolecular Subpicosecond Dynamics", J. Chem. Phys. **81**, 6399-6400 (1984).

8. R. J. Sension, R. J. Brudzynski and B. Hudson, "Resonance Raman Studies of the Low-Lying Dissociative Rydberg/Valence States of H_2O, D_2O and HDO", Phys. Rev. Lett. **61**, 694-697 (1988).

9. R. J. Sension, R. J. Brudzynski, B. Hudson, J. Zhang and D. G. Imre, "Resonance Emission Studies of the Photodissociating Water Molecule", Chem. Phys. **141**, 393-400 (1990).

10. V. Engel, V. Staemmler, R. L. Vander Wal, F. F. Crim, R. J. Sension, B. Hudson, P. Andresen, K. Weide and R. Schinke, "The Photodissociation of Water in the First Absorption Band: A Prototype Example for Dissociation on a Repulsive Potential Energy Surface" J. Phys. Chem., invited review article.

11. R. J. Brudzynski, R. J. Sension and B. Hudson, "Resonance Raman Study of the First Absorption Band of H_2S", Chem. Phys. Lett. **165**, 487-493 (1990).

12. L. D. Ziegler and B. Hudson, "Resonance Raman Scattering of Ethylene: Evidence for a Twisted Geometry in the V State", J. Chem. Phys. **79**, 1197-1202 (1983).

13. R. J. Sension, L. C. Mayne and B. Hudson, "Far Ultraviolet Resonance Raman Scattering: Highly Excited Torsional Vibrations of Ethylene", J. Am. Chem. Soc. **109**, 5036-5038 (1987).

14. R. J. Sension and B. Hudson, "Vacuum ultraviolet resonance Raman studies of the excited electronic states of ethylene," J. Chem. Phys. **90**, 1377-1389 (1989).

15. R. J. Brudzynski and B. Hudson, "Determination of the Torsional Potential of Allene from Highly Excited Torsional Vibrations Observed by Ultraviolet Resonance Raman Spectroscopy: The Torsional Barrier of Cumulenes", J. Amer. Chem. Soc. **112**, 4963-4965 (1990).

16. R. Wallace, "The Torsional Energy Levels of Ethylene: A Re-evaluation", Chem. Phys. Letters **159**, 35 (1989).

17. L. Giroux, M.H. Back and R.A. Back, "A comment on the rotational isomerization of ethylene", Chem. Phys. Letters **154**, 610 (1989).

18. A. C. Albrecht, "On the Theory of Raman Intensities", J. Chem. Phys. **34**, 1476 (1961).

19. S.-Y. Lee and E. J. Heller, "Time-dependent theory of Raman scattering", J. Chem. Phys. **71**, 4777 (1979); E. J. Heller, R. L. Sundberg and D. L. Tannor, "Simple aspects of Raman scattering", J. Phys. Chem. **86**, 1822 (1982); D. J. Tannor and E. J. Heller, "Polyatomic Raman scattering for general harmonic potentials", J. Chem. Phys. **77**, 202 (1982).

20. L. D. Ziegler and B. Hudson, "Resonance Raman Scattering of Benzene and Benzene-d6 with 212.8 nm Excitation", J. Chem.Phys. **74**, 982-992 (1981).

21. D. P. Gerrity, L. D. Ziegler, P. B. Kelly, R. A. Desiderio and B. Hudson, "Ultraviolet Resonance Raman Spectroscopy of Benzene Vapor with 220-184 nm Excitation", J. Chem. Phys. **83**, 3209-3213 (1985).

22. Roseanne J. Sension, Richard J. Brudzynski and Bruce S. Hudson, "Vacuum ultraviolet resonance Raman studies of the valence excited electronic states of Benzene and Benzene-d_6: The E_{1u} state and a putative A_{2u} state," J. Chem. Phys. **94**, 873-882 (1991).

23. L. D. Ziegler and B. Hudson, "Vibronic Coupling Activity in the Resonance Raman Spectra of Alkyl Benzenes", J. Chem. Phys. **79**, 1134-1137 (1983).

24. S. Li and B. Hudson, "Resonance Raman Studies of the ^1L_a State of 1,2,3-Trisubstituted Benzene Derivatives: Lack of an Induced Transition Moment", Chem. Phys. Lett. **148**, 581-585 (1988).

25. B. Hudson, B. E. Kohler and K. Schulten, "Linear Polyene Electronic Stucture and Potential Surfaces", Excited States, Vol. **6**, E. C. Lim, editor (New York: Academic Press, 1982), pgs 1-95.

26. B. Hudson and B. E. Kohler, "Electronic Structure and Spectra of Finite Linear Polyenes", Synthetic Metals **9**, 241-252 (1984).

27. R. R. Chadwick, D. P. Gerrity and B. Hudson, "Resonance Raman Spectroscopy of Butadiene: Demonstration of a 2^1A_g State Below the 1^1B_u State", Chem. Phys. Lett. **115**, 24-28 (1985).

28. Richard R. Chadwick, Gary D. Strahan, Marek Z. Zgierski and B. Hudson, "Resonance Raman scattering of butadiene: Vibronic activity of the b_u modes demonstrates the presence of a low lying A_g state", J. Chem. Phys. **95**, 7204-7211 (1991).

29. Quan-yuan Shang and Bruce S. Hudson, "Resonance Raman Depolarization Ratios for Cyclopentadiene Demonstrate the Presence of Two Overlapping Electronic Transitions with Perpendicular Polarizations in the Low Energy Absorption Band: The $1B_2$ and $2A_1$ States", Chem. Phys. Lett. **183**, 63-68 (1991).

30. A. Sabljic and R. McDiarmid, "Analysis of the absorption spectrum of the NV_1 transition of cyclopentadiene", J. Chem. Phys. **93**, 3850 (1990).

31. Reference 6, volume II, page 173.

32. L. D. Ziegler and B. Hudson, "The Vibronic Spectroscopy of Benzene: Old Problems and New Techniques", Excited States, Vol. **5**, E.C. Lim, editor (New York: Academic Press, 1982), pgs 41-139.

33. Roseanne J. Sension, Richard J. Brudzynski, Shijian Li, Bruce S. Hudson, Francesco Zerbetto and Marek Z. Zgierski, "Resonance Raman spectroscopy of the B_{1u} region of benzene: analysis in terms of pseudo-Jahn-Teller distortion", J. Chem. Phys., in press (15 February 1992 issue).

34. W. Moffitt and A.D. Liehr, "Configurational Instability of Degenerate Electronic States", Phys. Rev. **106**, 1195 (1957); A.D. Liehr, "Interaction of the Vibrational and Electronic Motions of Some Simple Conjugated Hydrocarbons", Z. Naturfors. **16a**, 641 (1961).

35. W. J. Buma, J. H. van der Waals and M. C. van Hemert, "Conformational Instability of the Lowest Triplet State of Benzene: The Result of ab Initio Calculations", J. Am. Chem. Soc. **111**, 86 (1989); "Conformational instability of the lowest triplet state of the benzene nucleus. I. The unsubstituted molecule", J. Chem. Phys. **93**, 3733 (1990).

36. M. Peric, S. D. Peyerimhoff and R. J. Buenker, "Theoretical study of the U. V. spectrum of acetylene III. *Ab initio* investigation of the valence-type singlet electronic states", Mol. Phys. **62**, 1339 (1987).

37. Reference 6, volume II, pages 106-108.

38. J. Berryhill, S. Pramanick, M. Zgierski and B. Hudson, "Resonance Raman spectroscopy of acetylene", in preparation.

39. Reference 6, volume II, page 330.

40. J. L. Hardwick and D. A. Ramsey, "The Near Ultraviolet Band System of Diacetylene", Chem. Phys. Lett. **48**, 399 (1977).

41. A. Karpfen and H. Lischka, "Ab initio calculations on the excited states of π-systems. II. Valence excitations in diacetylene", Chem. Phys. **102**, 91 (1986).

Vibrational spectroscopy and picosecond dynamics of gaseous trienes and tetraenes in S_1 and S_2 electronic states

Hrvoje Petek, Andrew J. Bell, Ronald L. Christensen,* and Keitaro Yoshihara

Institute for Molecular Science,
Myodaiji, Okazaki 444 Japan
and *Department of Chemistry
Bowdoin College
Brunswick, Maine 04011

ABSTRACT

Fluorescence excitation and emission spectra of the S_1 and S_2 states of model trienes and tetraenes are measured in free jet expansions. The barriers to *cis-trans* isomerization in the S_1 state are <200 cm^{-1} for trienes and ~2000 cm^{-1} for tetraenes. <250 fs nonradiative decay of the S_2 state of tetraenes is deduced from the observed Lorentzian linewidths.

1. INTRODUCTION

Nature has harnessed polyenes as transducers for converting light into chemical energy by exploiting their ability to undergo photochemical *cis-trans* isomerization.[1,2] Thus polyenes can be found to perform vital functions in processes such as vision and plant and bacterial photosynthesis. Furthermore, their unique electronic structure makes polyenes attractive for potential applications as organic conductors and nonlinear optical materials with large third-order susceptibilities.[3] There is a growing effort to exploit the unique electronic structure and reactivity of polyenes as chemical sensors, switches, logic devices, etc. In order to incorporate polyenes into useful functional devices it is imperative to understand such fundamental properties as the dependence of the electronic structure on the conjugation length and the effect of substitution on the structure and reactivity of the chromophore.[1] Most of the interesting photochemical transformations occur in the three low energy electronic states - the 1^1A_g ground state (S_1), 2^1A_g first excited state (S_1), and 1^1B_u second excited state (S_2). Since the $S_2 \leftarrow S_0$ transition is strongly allowed, the S_2 state is the chromophore in simple polyene systems. Because the $S_1 \leftarrow S_0$ ($2^1A_g \leftarrow 1^1A_g$) transition is forbidden by symmetry, it is usually produced by $S_2 \leftarrow S_0$ excitation followed by subpicosecond timescale $S_2 \rightarrow S_1$ internal conversion. Since both the S_1 and S_2 states play important roles in the biological activity of polyenes, we have made a systematic study of the electronic structure and dynamics of the S_1 and S_2 states of short linear polyenes (trienes[4] and tetraenes[5]) under molecular beam conditions and in solution phase.

2. EXPERIMENTAL

The experimental setup used to measure fluorescence excitation spectra under collision free conditions have been described.[5] The apparatus consists of; (i) a vacuum chamber together with a pulsed valve, for production of the molecular beam; (ii) an excimer pumped dye laser for excitation of the molecule under study; and (iii) emission collection optics and electronics.

All-trans-2,4,6,8-decatetraene (DT), *all-trans*-2,4,6,8-nonatetraene (NT), and octatriene were prepared respectively from the Wittig reaction between hexadienal (Aldrich) and crotyltriphenylphosphonium bromide, allyltriphenylphosphonium bromide, and ethyltriphenylphosphonium bromide (Fluka), as described previously.[6] The samples were purified by column chromatography followed by recrystallization in hexane for tetraenes. This procedure was sufficient to get >95% pure *all-trans*-

tetraenes. The analysis of octatriene by GC/MS and HPLC showed it to contain >90% *all-trans*-octatriene together with small amounts of *cis*-isomers. The hexatriene sample, consisting of a mixture of isomers, was obtained from Aldrich Chemical Company (cat. no.# H1,258-7) and used as obtained without separating the isomers. The samples were introduced into the vacuum chamber through the 500 μm orifice of the pulsed valve. Tetraene samples were heated to 40 - 70° C, and triene samples were used at room temperature. The samples were coexpanded with He carrier gas at a total pressure of 100 - 3000 Torr. Pressures of <300 Torr were used in the $S_2 \leftarrow S_0$ fluorescence excitation measurements because the spectra were sensitive to the presence of clusters at higher pressures.

The excitation source consisted of an excimer pumped dye laser system (Lambda-Physik 104/2003). The bandwidth of the laser was ~0.3 cm^{-1}, with a pulsewidth of ~15 ns. The experiment was performed with a laser repetition rate of 10 Hz and the dye laser wavelength was calibrated by measuring the optogalvanic spectrum of neon using a neon filled hollow cathode lamp. The laser beam intersected the molecular beam 15 mm from the nozzle orifice. At this distance the polyenes were found to be collision free, on the time scale of the experiment. The laser intensity was carefully adjusted to optimize the signal while avoiding saturation of the transitions. Fluorescence was collected by a single f/1 quartz lens and detected by a photomultiplier tube (Hamamatsu Photonics H3177). The signal was captured with a gated integrator (SRS 250) and then passed to a microcomputer along with a monitor of the dye laser intensity for normalization. For the measurement of lifetimes the fluorescence decay signal was digitized and averaged using a digital storage oscilloscope (Le Croy 7400A) and transfered to a microcomputer for analysis by fitting to a single exponential decay profile.

3. S_1 STATE SPECTRA OF TRIENES

The fluorescence excitation spectrum measured with a mixture of *cis*- and *trans*-hexatrienes is shown in Fig. 1. The positions and relative intensities of the peaks are given within the figure. Based on the assignment of previously measured Resonance Enhanced Multiphoton Ionization (REMPI) spectrum by Buma et al.,[7] the spectrum in Fig. 1 is assigned to the S_1 state of the *cis*-hexatriene. Even though the concentration of the *trans*-isomer is at least as great as the *cis*-isomer, its spectrum could not be observed. The origin region consists of two major peaks separated by ~5.7 cm^{-1}. This doublet shows further structure due to partially resolved rotational features. Changing the expansion conditions affects the rotational structure, however, the relative integrated intensities of the two bands are not changed. The rotational lineshapes of the origin bands are consistent with a predominantly parallel transition of *cis*-hexatriene. The $^1A_1 \leftarrow {}^1A_1$ transition is allowed for a molecule with a C_{2v} symmetry, however, a perpendicular band is expected. This spectrum appears to derive its oscillator strength by intensity borrowing from the $S_2 \leftarrow S_0$ transition. This is a common feature to all linear polyene spectra presented here: the symmetry forbidden $2^1A_g \leftarrow 1^1A_g$ transition gains intensity by vibronic coupling between 1^1B_u and 2^1A_g states. Although the presence of a *cis*-linkage or asymmetric substitution by a methyl group breaks the molecular symmetry, the dominant effect which determines the oscillator strength for the $S_1 \leftarrow S_0$ spectra of linear polyenes is vibronic coupling. For that reason g and u symmetry labels will be retained throughout this text, even though the labels may not apply strictly to individual molecules.

The ratio of the fluorescence excitation spectrum intensity relative to the REMPI spectrum drops rapidly between the peaks at 71.3 and 157.7 cm^{-1}. Since the hexatriene lifetimes are shorter than the experimental time resolution, these intensity ratios provide indirect evidence for a nonradiative decay channel with an activation energy between 71.3 and 157.7 cm^{-1}, as will be explained below. Due to a rapid decrease in the fluorescence quantum yields, the highest energy peaks which could be detected are at 247 - 263 cm^{-1} (not shown), however, the REMPI spectrum is seen to increase in intensity for >4000 cm^{-1} above the origin.

The fluorescence excitation spectrum of octatriene, and the fluorescence lifetimes measured at several peaks are shown in Fig. 2. Qualitatively, the fluorescence excitation spectrum is similar to a previously measured REMPI spectrum of a *cis*-octatriene isomer.[8] The smaller fluorescence yields of other isomers that are present in the sample, such as *all-trans*-octatriene which comprises ~90% of the sample, precludes their detection in our experiments. The origin, which could not be determined with certainty in the REMPI study is found to be at at 33648 cm^{-1}. As in *cis*-hexatriene, there are considerable differences between the fluorescence excitation and REMPI spectra at high excess energy, where the fluorescence excitation spectrum rapidly drops in intensity to below the detection limit, while the REMPI spectrum has further vibrational structure followed by a rising continuous absorption.[8] Since *cis*-octatriene isomers could not be isolated in either REMPI or the present study, the true structure of the octatriene responsible for the spectrum in Fig. 2 has not yet been determined.[4,8]

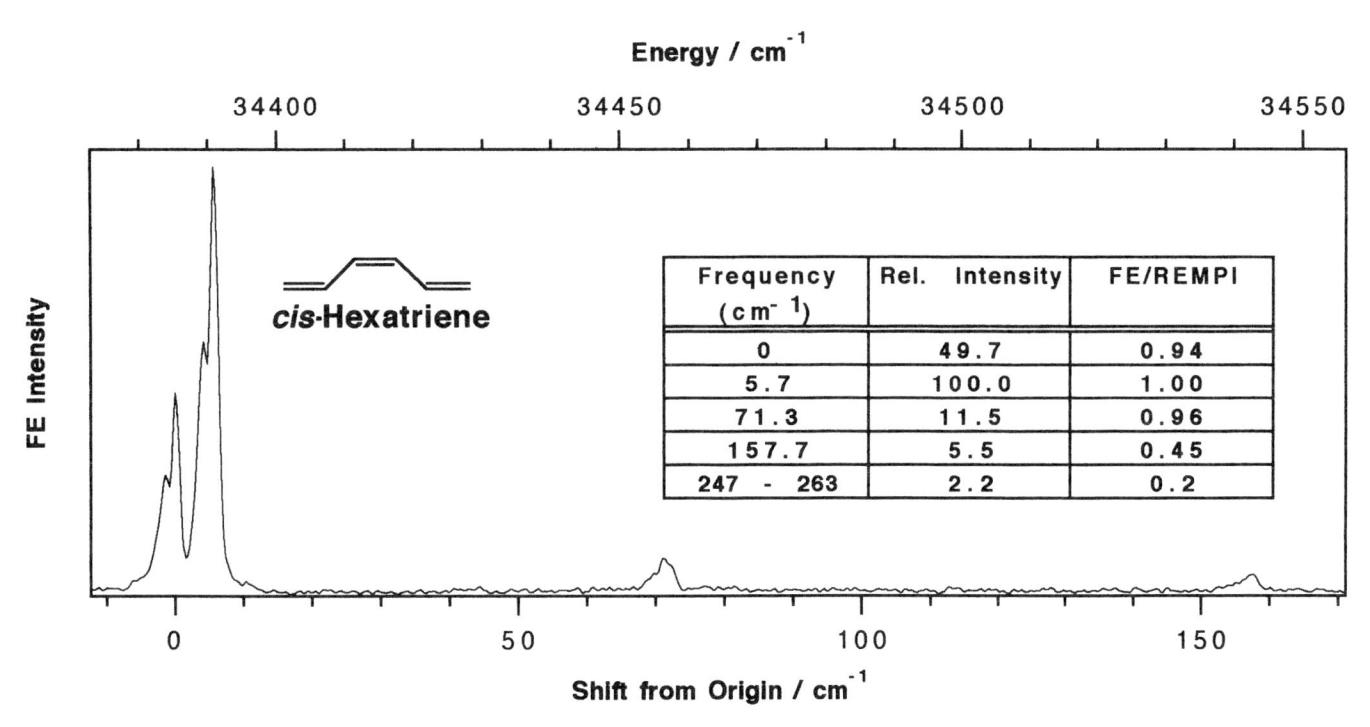

Frequency (cm^{-1})	Rel. Intensity	FE/REMPI
0	49.7	0.94
5.7	100.0	1.00
71.3	11.5	0.96
157.7	5.5	0.45
247 - 263	2.2	0.2

Figure 1. The fluorescence excitation spectrum of hexatriene. The abscissa shows both absolute energy and energy shift from the origin. The molecular beam was formed by coexpanding hexatriene with helium at a pressure of 575 Torr. The origin is at 34382 cm^{-1}. The table gives the ratios of integrated intensities for individual peaks measured by REMPI[7] and fluorescence excitation.

The octatriene fluorescence lifetimes measured for states with <200 cm^{-1} excess energy show a reproducible scatter about an average of 63 ns. This may be due the vibrational dependence of radiative and nonradiative decay rates, or possibly due to the presence of two or more octatriene isomers. The marked decrease in the fluorescence lifetimes shown in Fig. 2 coincides with the drop in fluorescence excitation spectral intensity relative to the REMPI spectrum.[8] This implies a sudden decrease in the fluorescence quantum yields as is also observed in hexatriene at low excess energies.

Even though the S_1 states of hexatriene and octatriene have similar electronic structures, their spectra are remarkably different. The hexatriene spectrum is simpler, but the splitting of most lines, including the origin, into two or more peaks, indicates that the S_1 state surface has some complex features. Buma *et al.* made a proposal supported by *ab initio* calculations that this splitting is due to out-of-plane distortion of the terminal hydrogens which gives rise to two distinct geometries in the S_1 state.[7] The presence of the methyl groups make the spectrum of octatriene significantly more complex than that of hexatriene. The phenomenally large number of lines observed in the first 250 cm^{-1} of the octatriene spectrum implies a large geometrical change upon $S_1 \leftarrow S_0$ excitation, most likely torsion of the methyl groups. In addition, torsion of the CC bonds and out-of-plane bending distortions of methyl groups, that are analogous to the hydrogen bending proposed for hexatriene, also may contribute.[4]

The 15 ns upper limit for the *cis*-hexatriene S_1 state lifetime is considerably shorter than the ~350 ns decays observed for isolated *all-trans*-decatetraene[5] and *all-trans*-nonatetraene (*vide infra*). This indicates that, even at the origin, the fluorescence quantum yield of hexatriene is significantly less than unity. Relative lifetimes as a function of vibronic energy can be deduced from the comparison between relative integrated peak intensities observed in fluorescence excitation and REMPI spectra tabulated in Fig. 1. The intensity of lines in the REMPI spectra mainly are dictated by the absorption of the species (assuming that the ionization rate is faster than the nonradiative decay rate and that the $S_1 \leftarrow S_0$ transition is not saturated), whereas the fluorescence excitation spectra are controlled by the product of the absorption cross section and fluorescence quantum yield.[4,7] The decrease of the intensity ratios from unity above 71.3 cm^{-1} starting with the peak at 157.7 cm^{-1}, implies that a nonradiative decay channel opens up in this energy interval. Thus in hexatriene there are at least two processes which lead to nonradiative decay: one appears to be energy independent, and the other has an activation energy of <157.7 cm^{-1}. Further evidence for a low barrier on the S_1 state surface may be the continuous absorption in the REMPI spectrum. This continuum may be due to a rapid increase in the density of vibrational states due to the anharmonicity of the S_1 state surface, or due to the coupling between the S_1 and another dark state.[4]

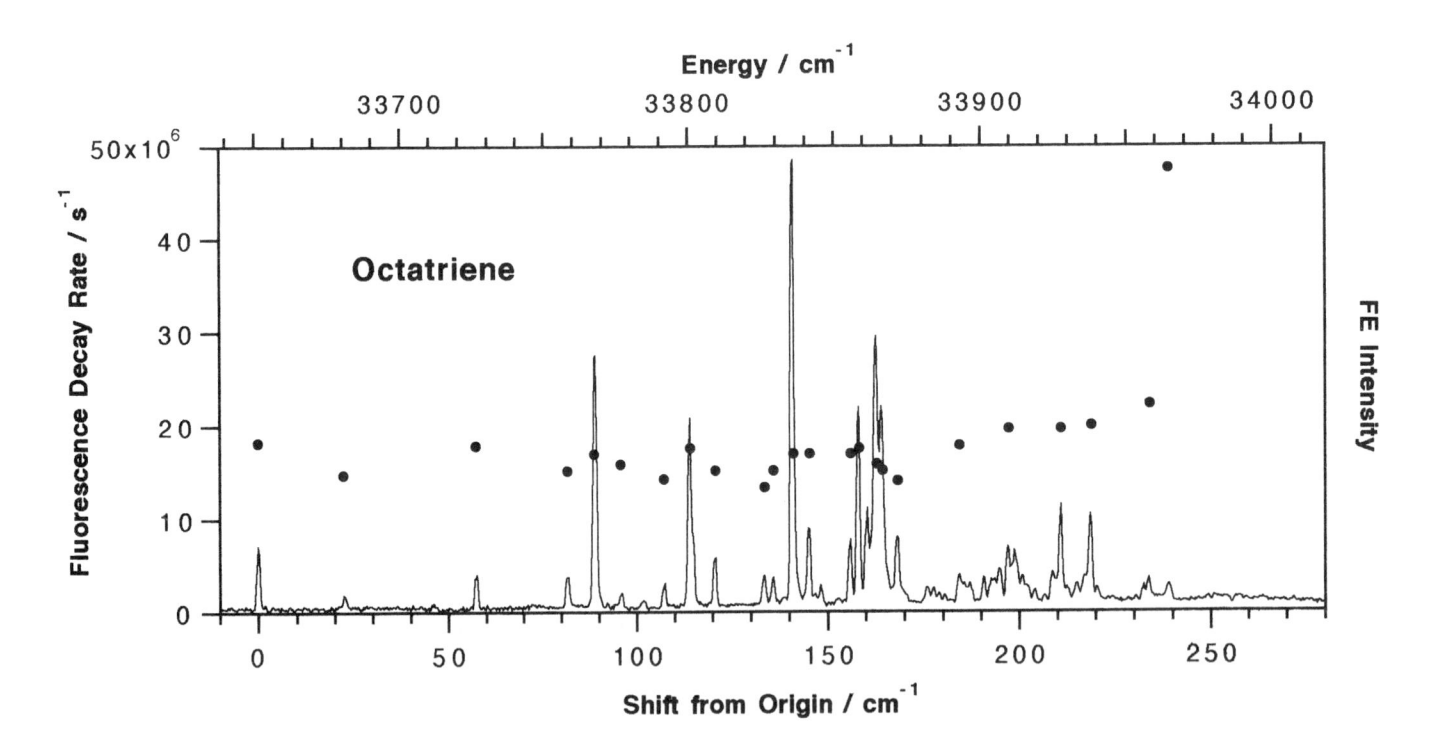

Figure 2. The fluorescence excitation spectrum of octatriene. The abscissa shows both absolute energy and energy shift from the origin. The ordinate shows the fluorescence decay rate (left) and fluorescence excitation intensity (right). Also shown are lifetimes for the major peaks. The error bars correspond to one standard deviation determined from the fit of the fluorescence decays to single exponentials. The origin is at 33648 cm^{-1}.

Although the octatriene lifetimes are significantly longer than those of hexatriene, their shortness (relative to the tetraenes)[6] and the variation in lifetimes in the <200 cm^{-1} energy region may be due to nonradiative decay processes that have a weak dependence on energy and quantum state. The decrease in octatriene lifetimes and the loss of the fluorescence excitation spectrum intensity above 200 cm^{-1} from the origin both are indicative of nonradiative decay by barrier crossing. The transition from discrete vibrational structure to the continuous absorption seen in the REMPI spectrum at higher energies probably implies an abrupt increase in the density of states above the barrier. As in hexatriene, there is evidence for two distinct nonradiative decay mechanisms.[4]

4. FLUORESCENCE EXCITATION SPECTRA OF TETRAENES

The fluorescence excitation spectra of DT and NT are measured from their $S_1 \leftarrow S_0$ origins at ~345 nm to 265 nm. The spectra of both molecules show the presence of two electronic states of greatly different character. The S_2 state with origins at ~287.47 nm (34786 cm^{-1}) for DT and 284.97 nm (35091 cm^{-1}) for NT, corresponds to the intense near uv-visible $\pi^* \leftarrow \pi$ transition seen in the solution phase absorption spectra of polyenes.[1] Previously, we have shown that the excitation of tetraenes to the S_2 state under collision free conditions results in emission from both S_1 and S_2 states.[5,6] Below the S_2 state, there is a considerably weaker $S_1 \leftarrow S_0$ spectrum, which can only be seen in emission from intermediate length polyenes in solution. Under supersonic expansion conditions the $S_1 \leftarrow S_0$ fluorescence excitation spectra of DT and NT are characterized by sharp vibrational structures, a complex vibrational developments, and maxima in intensity at the C = C stretching fundamentals, which are at ~1793 cm^{-1} above the origins. Above the C = C stretch the sharp structure gradually is replaced by a broad continuum.

4.1. $S_1 \leftarrow S_0$ Fluorescence Excitation Spectra

Decatetraene. The measured fluorescence excitation spectra of the $S_1 \leftarrow S_0$ transition and the fluorescence decay rates for DT are shown in Fig. 3a. The origin of the DT $S_1 \leftarrow S_0$ spectrum has been assigned to the intense triplet of peaks with the center of the lowest energy feature at 29022 cm^{-1}.[5] Since the $2^1A_g \leftarrow 1^1A_g$ transition is strictly forbidden by symmetry, this is probably a false origin, as will be discussed further. In Fig. 3 it can be seen that the low energy region is rather complicated due to excitation of many low frequency modes which involve skeletal bending vibrations and methyl torsions. Many low frequency modes derive intensity from Herzberg-Teller coupling between the 2^1A_g and 1^1B_u states, which makes the $2^1A_g \leftarrow 1^1A_g$ symmetry forbidden transition vibronically allowed.[1,5] There are at least two types of low frequency b_u symmetry modes, methyl torsion and in-plane bending, which break the molecular symmetry and make the spectrum weakly allowed. The characteristic signature of methyl torsion induced transitions is a triplet structure with a 3.4 cm^{-1} splitting (doublet in case of nonatetraene, which has only one methyl); for the transitions which are induced by b_u symmetry in-plane bending the signature is a singlet structure. High resolution fluorescence excitation spectra of DT and NT $S_1 \leftarrow S_0$ origin regions are shown in Fig. 4. The ~3.4 cm^{-1} splitting is due to the tunneling of two equivalent methyl groups through a low barrier on the S_1 state surface. Assuming that the surface for methyl torsion is given by a simple sinusoidal potential and that the two methyl groups are not coupled, a barrier of ~40 cm^{-1} reproduces the observed splitting and it also successfully predicts the line positions of a number of observed low frequency lines in both DT and NT, which are due to excitation of higher quanta of methyl torsion.[9]

At higher energies we can identify two prominent features, the C - C stretch at 1231 cm^{-1} and C = C stretch at 1793 cm^{-1}. The CC stretches have the largest Franck-Condon factors due to the well known bond order inversion between the S_1 and S_0 states, which is common to all linear polyenes.[1] The frequencies of CC stretches in the S_1 states of symmetric (or nearly symmetric) polyenes are unusually high. This is characteristic of the S_1 states of polyenes, and can be ascribed to strong vibronic coupling between the S_1 and S_0 states.[10]

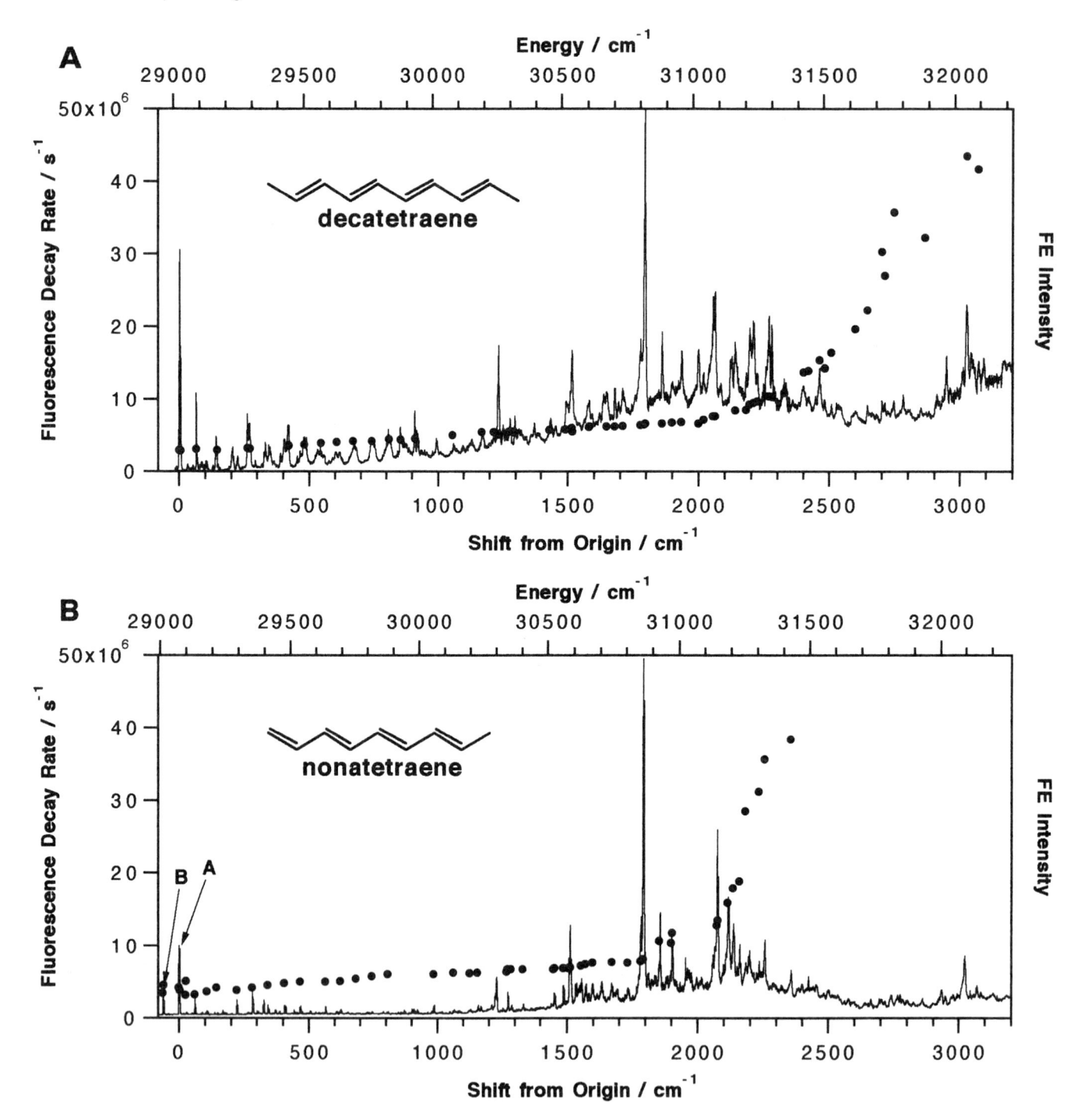

Figure 3. The $S_1 \leftarrow S_0$ fluorescence excitation spectra and single vibronic level fluorescence decay rates (•) of DT (a) and NT (b). The two origins seen in the NT spectrum are indicated by "A" and "B". The most intense feature in both spectra is assigned to the C = C stretch fundamental. The discontinuous change in the decay rates at ~2000 cm^{-1} excess energy is due to onset of nonradiative decay as described in the text.

The commonly observed increase in intensity in the $2^1A_g \leftarrow 1^1A_g$ absorption spectra of polyenes in solutions with increase in energy can be attributed to two causes: i) the vibronic coupling between the 2^1A_g and 1^1B_u states; and ii) the Franck-Condon factors for the C - C and C = C stretching vibrations.[1] The increase in vibronic coupling with excess energy due to a decrease in the energy gap between the 2^1A_g and 1^1B_u states is responsible for the modest increase in the absorption and emission rates for states with <2000 cm^{-1} excess energy seen in Fig 3. From the absorption spectra of octatetraene in solution[11] and emission spectra of DT and NT measured in this work (Fig. 5), we expect the maximum in the $S_1 \leftarrow S_0$ absorption spectra of tetraenes to be at a considerably higher energy than the intensity maximum observed in the DT and NT fluorescence excitation spectra. The maximum in emission intensity upon exciting the origin of DT and NT occurs for a combination band with two quanta of C = C stretch and one quantum of C - C stretch at ~4,500 cm^{-1} above the origin (Fig. 5). The Franck-Condon maximum is expected to be at an even higher energy in $S_1 \leftarrow S_0$ absorption due to the steeper slope of the S_1 state potential energy surface in the CC stretching coordinates as compared to the S_0 state. In fact, *ab initio* and semi-empirical calculations predict that the vertical excitation from the ground state geometry is at a higher energy for the 2^1A_g state than for the 1^1B_u state.[2] The discrepancy between the expected and observed maxima in the $S_1 \leftarrow S_0$ fluorescence excitation spectrum is due to a nonradiative decay process which has an activation energy of ~2000 cm^{-1}. We will discuss this further in the context of lifetime measurements.

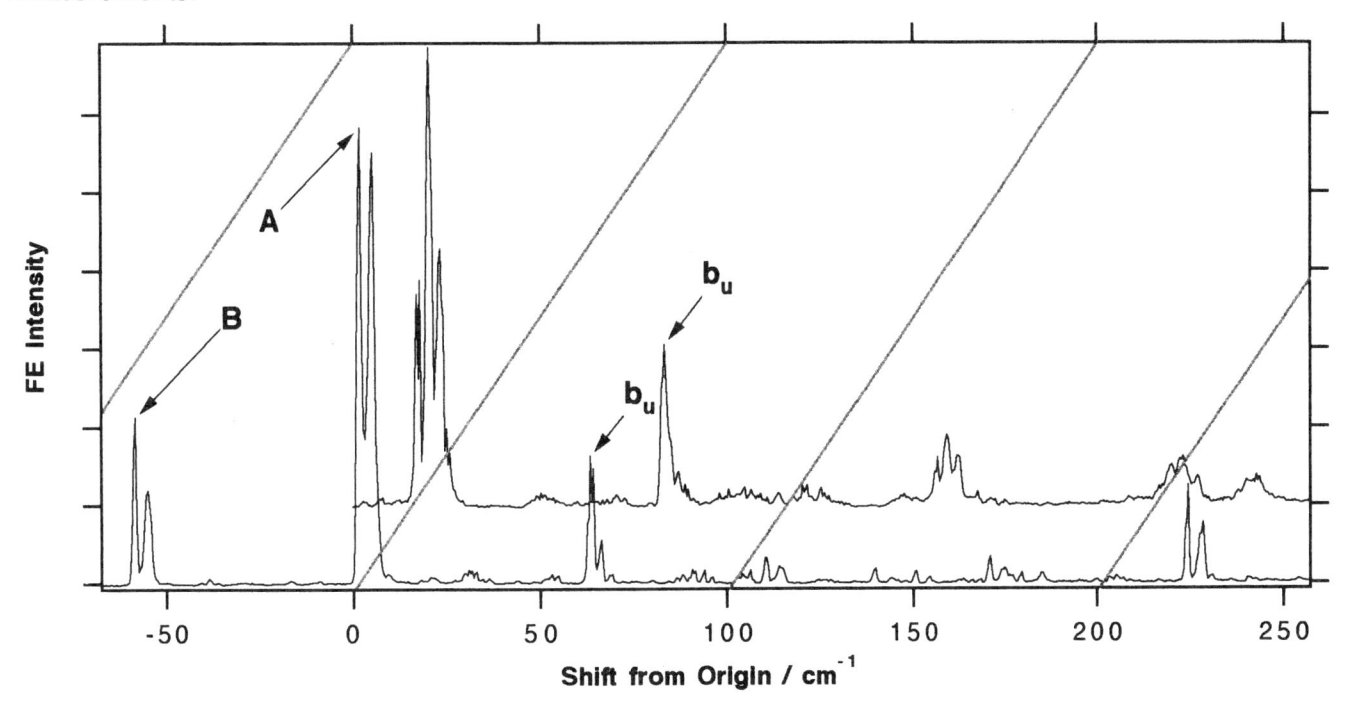

Figure 4. High resolution fluorescence excitation spectra of DT (top) and NT (bottom) $S_1 \leftarrow S_0$ origins. The two origins observed in NT are indicated by "A" and "B". Most of the lines are split into triplets (DT) and doublets (NT) with a 3.4 cm^{-1} splitting due to the tunneling of the methyl groups. However, the b_u symmetry modes appear as singlets, for example the in-plane bending fundamentals indicated bu "b_u".

Nonatetraene Although DT and NT have very similar structures, their $S_1 \leftarrow S_0$ spectra exhibit large differences. Again the spectrum (Fig. 3b) consists of many low frequency modes due to methyl torsion and in-plane bending. However, the spectrum is much sharper at higher energies, because the density of vibrational states grows with energy more slowly than for DT.

The origin of the NT spectrum cannot be trivially assigned to the lowest energy feature. In the origin region (Fig. 3b and 4) there are two pairs of lines which might be assigned to the origin. Both pairs are split by ~3.4 cm^{-1} due to the tunneling of a single methyl group. This splitting is an exquisitely sensitive probe of the electronic structure in the vicinity of the methyl group, so the fact that both pairs of lines show the same splitting, which is essentially the same as the splitting in DT, indicates that the methyl groups are experiencing very similar repulsive and attractive interactions.

The C = C stretching fundamental, which is expected to be the most intense peak by analogy with DT, is a useful guide for locating the origin. If the most intense line in the spectrum is assigned to the C = C stretching fundamental and the most intense line in the origin region at 29072 cm^{-1} (we will call it origin "A") is taken to be the origin, then the C = C stretching frequency is 1793 cm^{-1}, as in DT. However, origin A is not the lowest energy feature in the spectrum. The lowest energy feature at 29012 cm^{-1} (we will call it origin "B") has integrated intensity ~1/4 as large as the origin A independent of the sample preparation, the temperature of the sample reservoir in the pulsed nozzle (which is also the same temperature as the nozzle), and the He stagnation pressure. However, the only line in the C = C stretching region which can be associated with the origin B has <1/20 of the intensity of the C = C stretch assigned to the origin A.

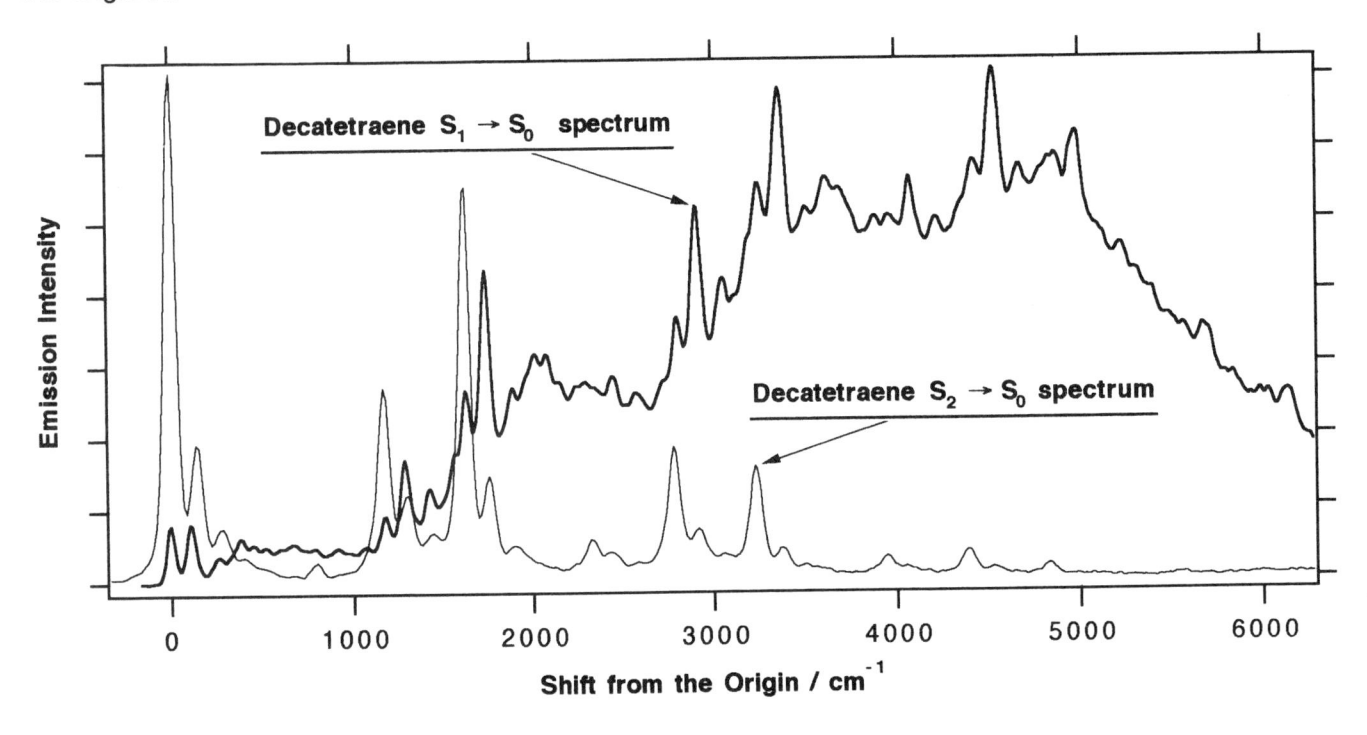

Figure 5. The emission spectra from the origins of S_1 (thicker line) and S_2 states. The very broad intensity distribution in the $S_1 \rightarrow S_0$ spectrum is due to the bond order reversal for CC stretching in the S_1 state. By contrast, the S_2 state has much smaller displacement in CC stretching with respect to the S_0 state.

Further information on the nature of origins A and B can be gained from the rotational bandshapes, emission lifetimes, and emission spectra. In the low excess energy region (<350 cm^{-1} above B) there appear to be two progressions with nearly identical frequency intervals but different lifetimes: the one associated with B has relatively longer lifetimes of 350 - 300 ns and narrower rotational bandshapes; the other, which is associated with A, has shorter lifetimes of 300 - 270 ns and broader rotational bandshapes. The difference in the lifetimes and the rotational profiles suggests that these two sets of lines belong to molecules which have different molecular structures at least in the S_1 state. However, the emission spectra from A and B show that the vibrational structure due to the C - C and C = C stretching vibrations in the S_0 state is identical to within ± 5 cm^{-1} up to 6000 cm^{-1} excess vibrational energy. If the molecular structure were different for the species that give rise to A and B in the S_0 state this would not be the case. For instance, a comprehensive study by Tasumi and co-workers of the S_0 state vibrational spectra of hexatriene isomers and rotamers shows that the C - C stretching frequency is very sensitive to the molecular structure: the C - C stretching for *cis*-hexatriene is 56 cm^{-1} higher than for *trans*-hexatriene.[12] These observation point to the possibility that there are two nearly isoenergetic conformations on the S_1 state surface of *all-trans*-nonatetraene, which can be accessed by optical excitation from the vibrationally and rotationally cold NT. Above 350 cm^{-1} excess energy in the S_1 state the lifetimes do not show quantum state dependence suggesting that at these energies the two conformations are no longer distinct. At this energy the emission spectra are nearly structureless due to intramolecular vibrational redistribution (IVR) which is apparently faster than the emission rate. The IVR results in a loss of identity between A and B species. There is *no* evidence in the measured fluorescence lifetimes that the B species disappears above 350 cm^{-1} excess energy due to opening up of a nonradiative decay channel.

4.2 S_1 state fluorescence decay measurements

The measured fluorescence decay rates as a function of excess energy for DT and NT are displayed in Fig. 3. For both DT and NT the fluorescence lifetimes are seen to decrease from ~350 ns at the origin by approximately 1/2 upon increasing the excess energy to 2000 cm^{-1}. The increase of the radiative decay rate as a function of S_2 - S_1 separation is commonly observed in the S_1 states of polyenes in solution and is attributed to the intensity borrowing between S_2 and S_1 states.[1] For NT there is a large quantum state dependence in the decay rates in the low energy region, as discussed above.

Above 2000 cm^{-1} excess energy there is a rapid increase in the decay rates in both DT and NT. Such a discontinuous change in the decay rate is evidence for an energy activated nonradiative decay channel such as isomerization. As discussed above, the absorption maximum for $S_1 \leftarrow S_0$ spectrum is implied to be >4500 cm^{-1} above the origin by the large Stokes shift in the emission. This lack of mirror symmetry between fluorescence excitation and emission spectra above the C = C stretch fundamental is further evidence for nonradiative decay above 2000 cm^{-1} excess energy. Although there have been no measurements on *trans-cis* isomerization of DT and NT under isolated conditions or in solution, Kohler and coworkers have extensively studied *cis-trans* isomerization of several octatetraene isomers in low temperature mixed crystals.[13] Since isomerization of octatetraene is well established, we tentatively assign this nonradiative decay process to *cis-trans* isomerization. Since the IVR rate is much faster than the fluorescence decay rate at 2000 cm^{-1} excess energy, it is a good assumption that the energy is statistically distributed in all of the available vibrational degrees of freedom, and that the barrier to isomerization is located where the decay rate begins to increase rapidly. The faster fluorescence decay rate and greater loss of intensity in NT as compared with DT at comparable energies above the barrier is due to the lower density of states in NT.

5. S_2 STATE SPECTRA OF TETRAENES

The fluorescence excitation spectra of the S_2 state of DT and NT are shown in Fig. 6. The spectra were found to be extremely sensitive to the excitation laser fluence and the expansion conditions. The laser was attenuated to the point where the fluorescence intensity had a linear dependence on the laser power. At low stagnation pressures (<200 Torr He) and low tetraene concentration, the emission decay was too fast to resolve with the instrumental time resolution of ~15 ns. At higher He stagnation pressures and tetraene concentrations there was an additional long lived decay component and several

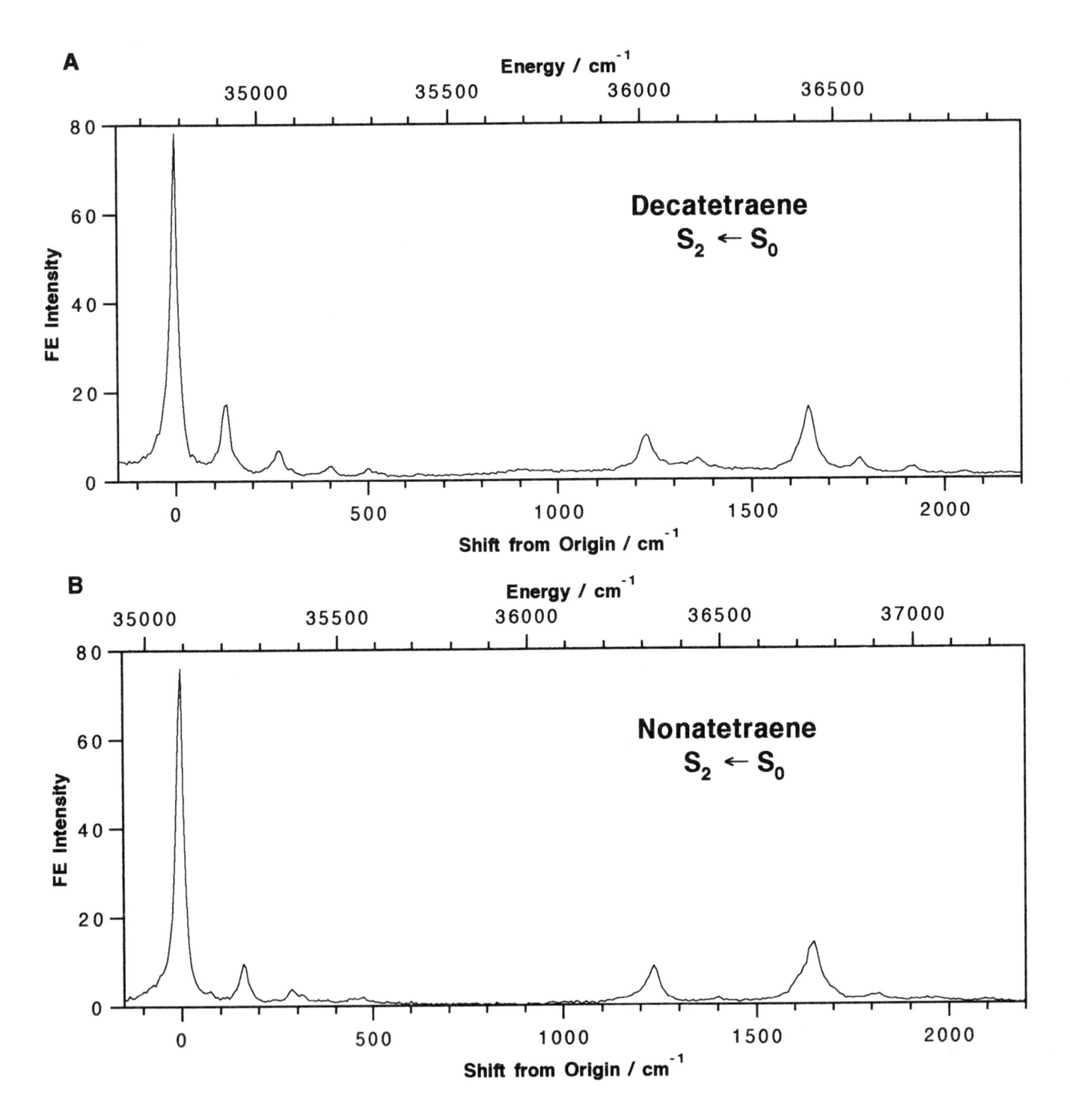

Figure 6. The $S_2 \leftarrow S_0$ fluorescence excitation spectra of DT (a) and NT (b). The origins are at 34786 and 35091 cm^{-1}, respectively. The $\geq$21 cm^{-1} Lorentzian linewidths are due to strong $S_2 - S_1$ coupling which induces internal conversion on <250 fs timescale.

broad and intense peaks could be observed below the origin. This long lived decay and the associated structure is attributed to excitation to He:tetraene clusters and tetraene dimers and higher oligomers. The tetraene clusters can be observed with very high sensitivity because the fluorescence quantum yield of tetraenes in the S_2 state or in the isoenergetic highly vibrationally levels of the S_1 state is very small for isolated molecules. Apparently the dissociation of tetraene clusters can compete with internal conversion to the S_0 state. Dissociation of a cluster leaves the electronically excited tetraene with much lower internal energy and therefore higher fluorescence quantum yield than a monomer, for which this energy dissipation pathway is not available. By this mechanism a small concentration of clusters makes a significant contribution to the fluorescence excitation spectra in the region extending to ~1000 cm^{-1} below the S_2 state origin.

The linewidths in the S_2 state were found to be much broader than can be attributed to rotational structure or to methyl torsion. The line profiles were fitted to Lorentzian lineshapes with full-widths at half-maximum of ≥21 cm^{-1} at the origin and >40 cm^{-1} at CC stretches. There was no additional structure that could be resolved with the 0.3 cm^{-1} laser resolution. Although Lorentzian lineshapes qualitatively described every peak, there were consistent deviations on most lines. These deviations could be of trivial nature such as the presence of a small amount of impurities such as *cis*-isomers tetraene complexes, and sequence bands from vibrationally excited molecules, which may be present due to the low He stagnation pressures; or more interestingly, the deviations may be caused by the $S_2 - S_1$ interaction. The true cause for these deviations could not be determined because the $S_2 \leftarrow S_0$ spectra of DT and NT are extremely sensitive to the experimental conditions. Due to these difficulties the reported linewidths represent an upper limit to the true Lorentzian linewidths of individual states. The observed linewidths imply that the S_2 state internal conversion to the S_1 state occurs on <250 fs timescale.[9]

The increase in linewidths with excess energy in the S_2 state implies an increase in nonradiative decay rates and a decrease in the emission quantum yields. The absorption spectrum of octatetraene has been observed in a molecular beam.[14] The comparison between the intensities of the CC stretches in the absorption spectrum of octatetraene and the fluorescence excitation spectra of DT and NT confirms that there is a gradual decrease in the fluorescence quantum yield with excess energy in the S_2 states of tetraenes.

6. EMISSION SPECTRA FROM THE S_1 AND S_2 STATES OF TETRAENES

The observed $S_2 \to S_0$ and $S_1 \to S_0$ emission spectra measured by exciting the respective origins of decatetraene are shown in Fig. 5. The very different nature of the $S_2 - S_0$ and $S_1 - S_0$ optical transitions is very clearly demonstrated in these spectra. The $S_2 \to S_0$ spectrum shows very simple structure which is dominated by CC stretches and weaker transitions due to the even quanta of in-plane bending. In the $S_1 \to S_0$ spectrum there are many more states with significant transition moments. The overall envelope and the large Stokes shift are due to progressions in CC stretches, which have the largest Franck-Condon factors as discussed above. However, the comparison with the $S_2 \to S_0$ spectrum shows that the most intense transitions are due to combinations bands of CC stretches and other low frequency modes. These low frequency modes can readily be assigned to the b_u symmetry in-plane bending modes which promote this symmetry forbidden transition.

7. CONCLUSIONS

Linear polyenes show several interesting trends in structure and reactivity as a function of conjugation length. Triene S_2 and S_1 states decay very rapidly by internal conversion to the S_0 state. Only the *cis*-isomers have observable fluorescence in the S_1 state, and even these decay by an activated nonradiative decay process with a <200 cm^{-1} barrier. The spectra of *cis*-isomers show that the structures in the S_1 state are nonplanar, and probably, that the methy torsional angle changes between S_1 and S_0 states. Tetraenes are significantly more stable with respect to internal conversion in both S_2 and S_1 states. The structures appear to be planar, but in nonatetraene there is evidence for existence for two inequivalent minima on the S_1 state surface, which arise from distortion in the in-plane bending coordinate. This may be a common feature in longer polyenes. Both tetraenes show evidence for an activated nonradiative channel with a ~2000 cm^{-1} barrier, which is tentatively assigned to *cis-trans* isomerization. The S_2 states of tetraenes are emissive and have well resolved structure with ≥ 21 cm^{-1} Lorentzian linewidths. These linewidths are independent of methyl substitution. The homogeneous broadening in the S_2 states of tetraenes is attributed to internal conversion to the S_1 state in <250 fs. The study of the structure and dynamics of short linear polyenes in free jet expansions provides important insights into the photochemistry of biologically and technologically important polyenes.

8. ACKNOWLEDGEMENTS

We are grateful to Royal Society/Japan Society for the Promotion of Science for support of AJB, the donors of the Petroleum Research Fund, administered by the American Chemical Society, and the DuPont Fund grant to Bowdoin College for partial support of research by RLC, B. Tounge for the preparation of octatriene, and I. Ohmine for invaluable discussions.

9. REFERENCES

1. B. S. Hudson, B. E. Kohler, K. Schulten, "Linear polyene electronic structures and potential Surfaces," in Excited States, ed. E. C. Lim, vol 6, pp 1-95, Academic Press, New York (1982).

2. G. Orlandi, F. Zerbetto, and M. Z. Zgierski, "Theoretical analysis of spectra of short polyenes," *Chem. Rev.* **91**, 867 (1991).

3. B. E. Kohler, "Electronic properties of linear polyenes," in Conjugated Polymers: The Novel Science and Technology of Conducting and Nonlinear Optically Active Materials, eds. J.L. Breads and R. Silby, pp 405- 434, Kluver Press, Dodrecht (1991).

4. H. Petek, A. J. Bell, R. L. Christensen, and K. Yoshihara, "Fluorescence excitation spectra of the S_1 states of isolated trienes," *J. Chem. Phys.* (in press).

5. H. Petek, A. J. Bell, K. Yoshihara, and R. L. Christensen, "Spectroscopic and dynamical studies of the S_1 and S_2 states of decatetraene in supersonic molecular beams," *J. Chem. Phys.* **95**, 4739 (1991).

6. W. G. Bouwman, A. C. Jones, D. Phillips, P. Thibodeau, C. Friel, and R. L. Christensen, "Fluorescence of gaseous tetraenes and pentaenes," *J. Phys. Chem.* **94**, 7429 (1990).

7. W. J. Buma, B. E. Kohler and K. Song, "Lowest energy excited singlet state of isolated *cis*-hexatriene," *J. Chem. Phys.*, **94**, 6367 (1991).

8. W. J. Buma, B. E. Kohler, K. Song, "Lowest energy excited singlet states of isomers of alkyl substituted hexatrienes," *J. Chem. Phys.* **94**, 4691 (1991).

9. H. Petek, A. J. Bell, R. L. Christensen, and K. Yoshihara, "Dynamical Study of Methyl Substituted Tetraenes in the S_1 and S_2 Electronically Excited States," (in preparation).

10. M. Aoyagi, I. Ohmine, and B. Kohler, "Frequency increase of the C = C a_g stretch mode of polyenes in the $2^1A_g^-$ state: *Ab initio* MCSCF study of butadiene, hexatriene, and octatetraene,"*J. Phys. Chem* **94**, 3922 (1990).

11. M. F. Granville, G. R. Holtom. B. E. Kohler, "High resolution one and two photon excitation spectra of *trans, trans*-1,3,5,7-octatetraene," *J. Chem. Phys.* **72**, 4671 (1980).

12. H. Yoshida, Y. Furukawa, and M. Tasumi, "Structural studies and vibrational analysis of stable and less stable conformers of 1,3,5-hexatriene based on *ab initio* calculations." *J. Mol. Struct.* **194**, 279 (1989).

13. B. E. Kohler, P. Mitra, and P. West, "Barriers in the excited 2^1A_g state of *cis, trans*-octatetraene: General features and the excited state potential," *J. Chem. Phys.* **85**, 4436 (1986).

14 D. G. Leopold, V. Vaida, and M. F. Granville, "Direct absorption spectroscopy of jet-cooled polyenes. I. The $^1B_u^+ \leftarrow {}^1A_g^-$ transition of *trans, trans*-octatetraene," *J. Chem. Phys.* **81**, 4210 (1984).

Coherent effects in laser chemistry

Moshe Shapiro

Chemical Physics Department
Weizmann Institute of Science
Rehovot, Israel 76100

and

Paul Brumer

Chemical Physics Theory Group
Department of Chemistry
University of Toronto
Toronto, Ontario M5S 1A1, Canada

ABSTRACT

We describe the current status of coherent radiative control, a quantum-interference based approach to controlling molecular processes by the use of coherent radiation. In addition to providing an overview of proposed laboratory scenarios, ongoing experimental studies and recent theoretical developments, we call attention to recent theoretical results on symmetry breaking in achiral systems.

1. INTRODUCTION

This paper is concerned with new phenomena which emerge when one manipulates the quantum phase of material systems. We discuss a theory, based upon quantum interference effects, which predicts virtually total control over branching ratios in isolated molecular processes.

Control over the yield of chemical reactions is the *raison d'etre* of practical chemistry and the ability to use lasers to achieve this goal is one primary thrust of modern Chemical Physics. The theory of coherent radiative control of chemical reactions, which we have developed over the last five years[1-16], affords a direct method for controlling reaction dynamics using coherence properties of weak lasers, with a large range of yield control expected in laboratory scenarios. In addition, the theory of coherent control provides deep insights into the essential features of reaction dynamics, and of quantum interference, which are necessary to achieve control over elementary chemical processes. Below we provide a schematic overview of coherent radiative control including an example which emphasizes the general principles (Section 2), a survey of several experimental scenarios which implement the principles of coherent control (Section 3), an interesting application to symmetry breaking in achiral systems (Section 4) and a summary of the current status of theory and experiment. A qualitative introduction to coherent control is provided in Ref. 9. An alternative approach, based upon the use of shaped light pulses has been advocated by Tannor and Rice[17] and Rabitz[18].

2. INTERFERENCE AND CONTROL: AN EXAMPLE

Consider a chemical reaction which, at total energy E produces a number of distinct products. The total Hamiltonian is denoted $H = H_q^0 + V_q$, where H_q^0 is the Hamiltonian of the separated products in the arrangement channel labeled by $q, (q = 1, 2, ...)$ and V_q is the interaction between products in arrangement q. We denote eigenvalues of H_0^q by $|E, n, q^0\rangle$, where n denotes all quantum numbers other than E; eigenfunctions of H, which correlate with $|E, n, q^0\rangle$ at large product separation, are labeled $|E, n, q^-\rangle$. By definition of the minus states, a state prepared experimentally as as superposition $|\Psi(t=0)\rangle = \sum_{n,q} c_{n,q} |E, n, q^-\rangle$ has probability $|c_{n,q}|^2$ of forming product in channel q, with quantum numbers n. As a consequence, control over the probability of forming a product in any asymptotic state is equal to the probability of initially forming the appropriate minus state which correlates with the desired product. The essence of control lies, therefore, in forming the desired linear combination at the time of preparation. The essential *modus operandi* of coherent radiative control is to utilize phase and intensity properties of laser excitation to alter the character of the prepared state so as to enhance production of the desired product.

Many of the proposed coherent control scenarios rely upon a simple way of achieving active control over the prepared state and product. Specifically, active control over products is achieved by driving an initially pure molecular state through two or more independent coherent optical excitation routes. The resultant product probability displays interference terms between these two routes, whose magnitude and sign depend upon laboratory parameters. As a consequence, product probabilities can be manipulated directly in the laboratory.

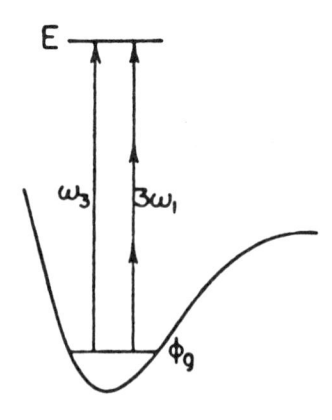

Figure 1: One-photon plus three-photon coherent control scenario.

The scheme outlined above has, as a well-known analogy, the interference between paths as a beam of either particles or of light passes through a double slit. In that instance source coherence leads to either constructive or destructive interference, manifest as patterns of enhanced or reduced probabilitites on an observation screen. In the case of coherent control the overall coherence of a pure state plus laser source allows for the constructive or destructive manipulation of probabilities in product channels.

We note, as an aside, that practitioners of "mode-selective chemistry" often advocate the use of lasers to excite specific bonds or modes in the reactant molecule in order to drive a reaction in a desired direction. The brief remarks above, elaborated upon elsewhere[10], make clear that the proper modes to excite in order to produce product in arrangement q are the eigenfunctions $|E, n, q^-\rangle$, the system's "natural modes", if you will. Thus the extent to which excitation of some zeroth order state $|\chi\rangle$ is successful in promoting reaction to the desired product q depends entirely on the extent to which $|\chi\rangle$ overlaps $|E, n, q^-\rangle$.

Consider, as an example of coherent control, a specific scenario for unimolecular photoexcitation[5,13] (Figure 1) where a system, initially in pure state ϕ_g (or $|E_i\rangle$ below), is excited to energy E, by simultaneous application of two CW frequencies ω_3 and $3\omega_1$ ($\omega_3 = 3\omega_1$), providing two independent optically driven routes from ϕ_g to $|E, n, q^-\rangle$.

Straightforward perturbation theory, valid for the weak fields under consideration, gives the probability $P(E, q; E_i)$ of forming product at energy E in arrangement q as:

$$P(E, q; E_i) = P_3(E, q; E_i) + P_{13}(E, q; E_i) + P_1(E, q; E_i) . \tag{1}$$

with terms defined as follows:

$$P_3(E, q; E_i) = (\frac{\pi}{\hbar})^2 \epsilon_3^2 \sum_n |\langle E, n, q^- | (\hat{\epsilon}_3 \cdot \underset{\sim}{\mu})_{e,g} | E_i \rangle|^2 . \tag{2}$$

Here $\underset{\sim}{\mu}$ is the electric dipole operator, and

$$(\hat{\epsilon}_3 \cdot \underset{\sim}{\mu})_{e,g} = \langle e | \hat{\epsilon}_3 \cdot \underset{\sim}{\mu} | g \rangle , \tag{3}$$

where $|g\rangle$ and $|e\rangle$ are the ground and excited electronic state wavefunctions, respectively. The second and third terms in Eq. (1) are

$$P_1(E, q; E_i) = (\frac{\pi}{\hbar})^2 \epsilon_1^6 \sum_n |\langle E, n, q^- | T | E_i \rangle|^2 , \tag{4}$$

with

$$T = (\hat{\epsilon}_1 \cdot \underset{\sim}{\mu})_{e,g}(E_i - H_g + 2\hbar\omega_1)^{-1}(\hat{\epsilon}_1 \cdot \underset{\sim}{\mu})_{g,e}(E_i - H_e + \hbar\omega_1)^{-1}(\hat{\epsilon}_1 \cdot \underset{\sim}{\mu})_{e,g} . \tag{5}$$

and

$$P_{13}(E, q; E_i) = -2(\frac{\pi}{\hbar})^2 \epsilon_3 \epsilon_1^3 \cos(\theta_3 - 3\theta_1 + \delta_{13}^{(q)})|F_{13}^{(q)}| \tag{6}$$

with the amplitude $|F_{13}^{(q)}|$ and phase $\delta_{13}^{(q)}$ defined by

$$|F_{13}^{(q)}|\exp(i\delta_{13}^{(q)}) = \sum_n \langle E_i | T | E, n, q^- \rangle \langle E, n, q^- | (\hat{\epsilon}_3 \cdot \underset{\sim}{\mu})_{e,g} | E_i \rangle . \tag{7}$$

The branching ratio $R_{qq'}$ for channels q and q', can then be written as

$$R_{qq'} = \frac{P(E,q;E_i)}{P(E,q';E_i)} = \frac{\epsilon_3^2 F_3^{(q)} - 2\epsilon_3\epsilon_1^3 \cos(\theta_3 - 3\theta_1 + \delta_{13}^{(q)})|F_{13}^{(q)}| + \epsilon_1^6 F_1^{(q)}}{\epsilon_3^2 F_3^{(q')} - 2\epsilon_3\epsilon_1^3 \cos(\theta_3 - 3\theta_1 + \delta_{13}^{(q')})|F_{13}^{(q')}| + \epsilon_1^6 F_1^{(q')}}, \quad (8)$$

where

$$\begin{aligned} F_3^{(q)} &= \left(\frac{\hbar}{\pi}\right)^2 \frac{P_3(E,q;E_i)}{\epsilon_3^2}, \\ F_1^{(q)} &= \left(\frac{\hbar}{\pi}\right)^2 \frac{P_1(E,q;E_i)}{\epsilon_1^6}, \end{aligned} \quad (9)$$

with $F_3^{(q')}$ and $F_1^{(q')}$ defined similarly.

The numerator and denominator of Eq. (8) each display what we regard as the canonical form for coherent control. That is, independent contributions from two routes, modulated by an interference term. Since the interference term is controllable through variation of laboratory parameters (here the relative intensity and relative phase of the two lasers), so too is the product ratio R.

This 3-photon + 1-photon scenario has now been experimentally implemented[19] in studies of HCl ionization through a resonant bound Rydberg state. Specifically, HCl is excited to a selected rotational state in the $^3\Sigma^-(\Omega^+)$ manifold using $\omega_1 = 336$ nm; ω_3 is obtained by third harmonic generation in a Krypton gas cell. The relative phase of the light fields was then varied by passing the beams through a second Krypton cell and varying the cell gas pressure. The population of the resultant Rydberg state was interrogated by ionizing to HCl$^+$ with an additional photon. This REMPI type experiment showed that the HCl$^+$ ion probability depended upon both the relative phase and intensity of the two exciting lasers, in accord with the theory described above. Note that in this case, control is achieved by simultaneous excitation to a nondegenerate bound state. As a consequence there is no sum over n in Eqs. (2), (4) and (7) and hence $|F_3^{(q)} F_1^{(q)}| = |F_{13}^{(q)}|^2$. Satisfaction of this Schwartz equality suffices to ensure that the probability of forming the HCl Rydberg state can be varied over the full range of zero to unity[1]. Specifically, setting $\theta_3 - 3\theta_1 = -\delta_{13}^{(q)}$ and $\epsilon_3^2/\epsilon_1^6 = F_1^{(q)}/F_3^{(q)}$ gives zero probability in channel q.

A similar phase control experiment has been performed on atoms[20], in the simultaneous 3-photon + 5-photon ionization of Hg. Although the effect of the relative laser intensity was not studied, the Hg$^+$ ionization probability was shown to be a function of relative phase of the two lasers. Since in this case excitation is nonresonant, this experiment shows the feasibility of control over continuum channels via interfering optical excitation routes.

These experiments clearly show that coherent control of simple molecular processes through quantum interference of multiple optical excitation routes is both feasible and experimentally observable. Further experimental studies designed to show control over processes with more than one product channel are in progress by a number of experimental groups. Further, our recent computational results[13] on photodissociation of IBr, to produce I + Br and I + Br* suggest that such experiments should show huge control over the product ratio. For example, 1-photon plus 3-photon photodissociation of the ground vibrational state in the ground electronic state of IBr showed variations in the Br*/Br ratio of 45% to 95% with variations of the relative laser intensity and phase. Similar results were obtained for higher J, including $J = 42$ where extensive M_J averaging was required.

3. COHERENT CONTROL SCENARIOS

The 3-photon plus 1-photon case is but one example of a scenario which embodies the essential principle of coherent control, i.e. that *coherently driving a pure state through multiple optical excitation routes to the same final state allows for the possibility of control*. Given this general principle, numerous scenarios may be proposed to obtain control in the laboratory. Such proposals must, however, properly account for a number of factors which reduce or eliminate control. Amongst these are the need to (a) adhere to selection rule requirements, (b) minimize extraneous and parasitic uncontrolled satellites, (c) insure properly treated laser spatial dependence and phase jitter and (d) allow maintainance of coherence over time scale associated with any true relaxation processes, e.g. collisions. Briefly:

(a) control requires that the interference term (e.g. $F_{13}^{(q)}$ above) arising between optical routes is nonzero. For example[21], consider the 1-photon + 3-photon case discussed above. Here excitation from, e.g. $J = 0$, where J is the rotational angular momentum, yields $J = 1$ for the 1-photon route and $J = 1$ and $J = 3$ for the three photon route. In this case selection rules are such that only final states with the same J, here $J = 1$, interfere;

(b) the $J = 3$ state created by three photon absorption in this scenario is an example of a "parasitic" uncontrolled state. Effective scenarios must insure that such contributions are small compared to the controlled component;

(c) since the ability to accurately manipulate the laser relative phase is crucial to coherent control, one must account for all laboratory features which affect the phase. For example, scenarios must be designed so as to eliminate effects due to the $\mathbf{k} \cdot \mathbf{R}$ spatial dependence of the laser phase. Not doing so would results in the dimunition of control resulting from the variation of this term over the interaction volume.

(d) proposed scenarios must maintain phase coherence in the face of possibly dephasing effects (e.g. collisions or partial laser coherence). Studies indicate[7,22] that control can survive moderate levels of such phase destructive processes.

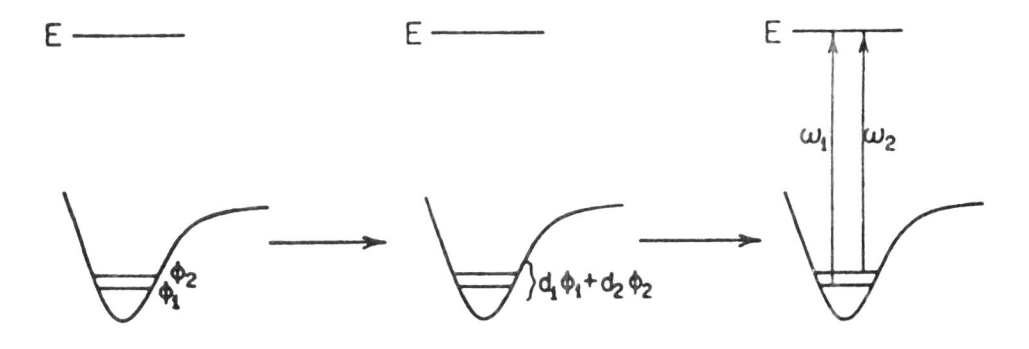

Figure 2: Basic control scenario.

We have, thus far, proposed a number of different scenarios for coherent control. Below we comment on some of them, with the intent of providing only the briefest of "roadmaps" for the proposed schemes. Detailed discussions are provided in the literature, along with extensive computations. The original coherent control scenario is shown in Figure 2 wherein a superposition of two bound states is subjected to two cw frequencies which lift the system to energy E. An analysis of this very basic scenario shows that the control parameters are the relative intensity and relative phase of the two indicated electromagnetic fields. This basic mechanism has been examined in the gas phase, both in the presence and in the absence of collisions and this scenario has been adapted to control currents in semiconductors. In all cases control was extensive.

Developments in pulsed laser technology may be used to good advantage in a straightforward modification of the above scenario. Specifically, the superposition state preparation and subsequent excitation with frequencies ω_1, ω_2 may be both carred out using pulsed lasers in accord with Figure 3:

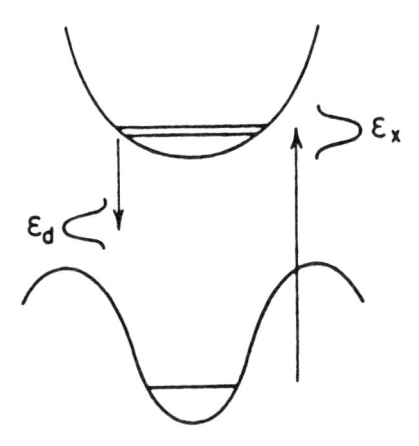

Figure 3: Pump-Pump scenario

In this instance the frequencies required to excite to energy E are contained within the second excitation pulse, which also creates product, and associated interference contributions over a wider energy range. Multiple excitation paths are contained within this overall pump-pump excitation scheme and serve to introduce the necessary interfering coherent paths. The straightforward analysis[8,11] of this scenario consists of a perturbation theory treatment of the molecule in the presence of two temporally separated, sequential pulses. The first pump step, in which the system is excited from ground state $|E_g\rangle$, yields a superposition state $\sum_k c_k |E_k\rangle$ with c_k proportional to $\langle E_k |\mu| E_g \rangle$; the subsequent pulse causes dissociation. Perturbation theory gives the probability, $P(E, m_j, q)$, required in the section below, of forming product in arrangement channel q at energy E with total fragment angular momentum projection m_j along a space fixed axis as,

$$P(E, m_j, q) = \sum_n{}' |B(E, n, q|t = \infty)|^2$$
$$= (2\pi/\hbar^2) \sum_n{}' | \sum_{k=1,2} c_k \langle E, n, q^- |\mu| E_k \rangle \epsilon_d(\omega_{EE_k})|^2 \qquad (10)$$

where $\omega_{EE_k} = (E - E_k)/\hbar$. Here $\epsilon_d(\omega_{EE_k})$ is the amplitude of the electric field of the second pulse at the frequency ω_{EE_k} and the prime denotes summation over all quantum numbers n (including scattering angles) other than m_j.

Expanding the square, for the case where c_k is nonzero for two states, and using a gaussian pulse shape gives:

$$P(E, m_j, q) = (2\pi/\hbar^2)[|c_1|^2 \mu_{1,1}^{(q)} \varepsilon_1^2 + |c_2|^2 \mu_{2,2}^{(q)} \varepsilon_2^2 + 2|c_1 c_2^* \varepsilon_1 \varepsilon_2 \mu_{1,2}^{(q)}| \cos(\omega_{2,1}(t_d - t_x) + \alpha_{1,2}^{(q)}(E) + \phi)] \quad (11)$$

where $(t_d - t_x)$ is the temporal delay between pulses, $\varepsilon_i = |\epsilon_d(\omega_{EE_i})|$, $\omega_{2,1} = (E_2 - E_1)/\hbar$ and the phases ϕ, $\alpha_{1,2}^{(q)}(E)$ are defined by

$$\langle E_1 |\mu| E_g \rangle \langle E_g |\mu| E_2 \rangle \equiv |\langle E_1 |\mu| E_g \rangle \langle E_g |\mu| E_2 \rangle| e^{i\phi}$$

$$\mu_{i,k}^{(q)}(E) \equiv |\mu_{i,k}^{(q)}(E)| e^{i\alpha_{i,k}^{(q)}(E)} = {\sum_n}' \langle E, n, q^- |\mu| E_i \rangle \langle E_k |\mu| E, n, q^- \rangle \quad (12)$$

Note the generic form of Eq. (11) in which there are two direct routes plus a term representing interference between them. Integrating over E to encompass the width of the second pulse, and forming the ratio $Y(m_j) = P(q = 1, m_j)/[P(q = 1, m_j) + P(q = 2, m_j)]$, gives the controllable ratio of products in fixed m_j states. This particular quantity is of interest in the case of chirality control discussed in Section 4. Of greater interest in general is the ratio of products in each arrangement channel, i.e.
$Y = P(q = 1)/[P(q = 1) + P(q = 2)]$ where $P(q) = \sum_{m_j} P(q, m_j)$.

These results show that in this case the laboratory control parameters which affect the relative yield are the time delay between the pulses and the detuning of the initial excitation pulse which, in turn, determines the relative population and phase of the intermediate superposition state. We have applied this scenario to both model DH_2^8 and to IBr photodissociation[1]. In both cases relatively small changes in the detuning or in the time delay, which is in the convenient picosecond domain, resulted in extremely large variations of the yield (e.g. variations in the Br yield in IBr dissociation case over a range of 4% to 96%).

We have also applied this pump-dump type scheme to control bimolecular reactions[14]. In this case the initial excitation is from the continuum ground electronic state to a bound excited electronic state, followed by a dump back down to the ground electronic state. Significant control results only in the energy domain which is essentially below reaction threshold. Above threshold the interference term does not effectively compete with the energetically allowed reaction.

It is important to note that the scenarios involving two bound levels described above are not "restricted-two-state problems". Rather, we recognize that two levels are sufficient to carry the necessary phase information to allow control over the relative yield. That is, lasers with narrow frequency profiles (i.e. long pulses in time) are sufficient. Indeed as the molecules increase in size, and the density of states increases, longer and longer pulses could be used, and two intermediate levels retained. There appears little advantage to the inclusion of more than two levels, a situation which would result if the pulses were considerably shorter in time.

4. SYMMETRY BREAKING AND CHIRALITY CONTROL

The results above make clear that two arrangement channels, whose photodissociation amplitudes differ, can be controlled through manipulation of external laser fields. A particularly interesting example results from consideration of the photodissociation of a molecule schematically written as :

$$(q = 2), B' + AB \stackrel{h\nu}{\leftarrow} BAB' \stackrel{h\nu}{\rightarrow} B + AB', (q = 1) \tag{13}$$

where q defines the product arrangement channel and BAB' denotes a molecule, or a molecular fragment, which is symmetric with respect to reflection σ across the plane which interchanges the enantiomers B and B' (e.g., Ref. 23). The existence of this plane implies that the point group of the reactant is at least C_s.

Consider then the molecule $B'AB$ subjected to the two pulse scenario sketched above. The probability of forming product in arrangement channel q with products in the m_j state is given by Eq. (11). Active control over the products $B + AB'$ vs. $B' + AB$, attained by manipulating the time delay or the pulse detuning will result as long as $P(q = 1, m_j)$ and $P(q = 2, m_j)$ have different functional dependences on laboratory parameters.

It is surprising that $P(q = 1, m_j)$ and $P(q = 2, m_j)$ differ in the $B'AB$ case[24]. To see this consider the behavior of c_k and $\mu_{i,k}^{(q)}$, for $|E_1\rangle$ and $|E_2\rangle$ of different symmetry with respect to the reflection σ. For simplicity we restrict attention to BAB' belonging to point group C_s, the smallest group possessing the required symmetry plane. Further, we focus upon transitions between electronic potential energy surfaces of similar species, e.g. A' to A' or A'' to A'' and assume the ground vibronic state to be of species A'. Similar arguments apply for larger groups containing σ, to ground vibronic states of odd parity, and to transition between electronic states of different species.

Excitation Coefficients c_k: Components of μ lying in the symmetry plane, denoted μ_s, transform as A' and are symmetric with respect to reflection whereas the component perpendicular to the symmetry plane, denoted μ_a, is antisymmetric (A''). Hence $\langle E_k|\mu|E_g\rangle = \langle E_k|\mu_a + \mu_s|E_g\rangle$, and hence c_k, is nonzero for transitions between vibronic states of the same symmetry due to the μ_s component, and nonzero for transitions between vibronic states of different symmetry due to the μ_a component. The latter is common in IR spectroscopy. In electronic spectroscopy such transitions result from the non Franck-Condon Herzberg-Teller intensity borrowing[25] mechanism. Thus, the excitation pulse can create an $|E_1\rangle, |E_2\rangle$ superposition consisting of two states of different reflection symmetry and hence a state which no longer displays the reflection symmetry of the Hamiltonian.

Cumulative matrix elements $\mu_{i,k}^{(q)}$: In contrast to the bound states, the continuum states of interest $|E, n, q^-\rangle$ are neither symmetric nor antisymmetric. Rather, $\sigma|E, n, q = 1^-\rangle = |E, n, q = 2^-\rangle$ and vice-versa. Such a choice is possible because of the exact degeneracies which exist in the continuum. To examine $\mu_{i,k}^{(q)}$ we introduce symmetric and antisymmetric continuum eigenfunctions of σ, $|\psi^s\rangle = (|E, n, q = 1^-\rangle + |E, n, q = 2^-\rangle)/2$ and $\psi^a = (|E, n, q = 1^-\rangle - |E, n, q = 2^-\rangle)/2$. Assuming $|E_1\rangle$ is symmetric and $|E_2\rangle$ antisymmetric, rewriting $|E, n, q^-\rangle$ as a linear combination of $|\psi^a\rangle$ and $|\psi^s\rangle$, and adopting the notation $A_{s2} = \langle \psi^s|\mu_a|E_2\rangle$, $S_{a1} = \langle \psi^a|\mu_s|E_1\rangle$, etc. we have, after elimination of null matrix elements,

$$\mu_{11}^{(q)} = {\sum}' \left[|S_{s1}|^2 + |A_{a1}|^2 \pm A_{a1}S_{s1}^* \pm A_{a1}^*S_{s1}\right]$$

$$\mu_{22}^{(q)} = {\sum}' \left[|A_{s2}|^2 + |S_{a2}|^2 \pm A_{s2}S_{a2}^* \pm A_{s2}^*S_{a2}\right]$$

$$\mu_{12}^{(q)} = \sum{}' [S_{s1}A_{s2}^* + A_{a1}S_{a2}^* \pm S_{s1}S_{a2}^* \pm A_{a1}A_{s2}^*] \qquad (14)$$

where the plus sign applies for $q = 1$ and the minus sign for $q = 2$. Equation (14) displays two noteworthy features:

(1) $\mu_{kk}^{(1)} \neq \mu_{kk}^{(2)}$. That is, the system displays *natural symmetry breaking* in photodissociation from state $|E_k\rangle$, with right and left handed product probabilities differing by $2\sum{}' \text{Re}(S_{s1}^* A_{a1})$ for excitation from $|E_1\rangle$ and $2\sum{}' \text{Re}(A_{s2}S_{a2}^*)$ for excitation from $|E_2\rangle$. Note that these symmetry breaking terms may be relatively small since they rely upon non Franck-Condon contributions. However:

(2) $\mu_{12}^{(1)} \neq \mu_{12}^{(2)}$. Thus laser controlled symmetry breaking, which depends upon $\mu_{12}^{(q)}$ in accordance with Eq. (11), is therefore possible, allowing enhancement of the enantiomer ratio for the m_j polarized product.

To demonstrate the extent of control we considered a model of the enantiomer selectivity, i.e. HOH photodissociation in three dimensions, where the two hydrogens are assumed distinguishable. The computation is in accord with state-of-the art formalism and computational machinery and details are provided elsewhere[24]. Typical results showed that it was possible to vary the product ratio, in this completely symmetric case, over a range of 61% to 39%. As expected the control disappears if the products are not m_j selected. Specifically, the natural symmetry breaking between right and left handed products seen with fixed m_j is lost upon m_j summation, both channels $q = 1$ and $q = 2$ having equal photodissociation probabilities. In addition, control over the enantiomer ratio is lost since the interference terms no longer distinguish the $q = 1$ and $q = 2$ channels.

5. CONCLUSION

We have provided a capsule summary of the current status of coherent control of chemical reactions and molecular processes. Three essential points should be emphasized:

a) theoretical studies show that control over molecular processes, via simple quantum interference, is possible;

b) computational results show that such control can be extensive, with achievable variations in yield close to the full possible range of 0% to 100%;

c) experimental studies have shown that the proposed coherent control scenarios are feasible in the laboratory.

6. ACKNOWLEDGMENTS

We acknowledge support from the U.S. Office of Naval Research under contract numbers N00014-87-J-1204 and N00014-90-J-1014. Work carried out in conjunction with R.J. Gordon was supported by NSF PHY-8908161.

7. REFERENCES

1. P. Brumer and M. Shapiro, "Control of Unimolecular Reactions Using Coherent Light", Chem. Phys. Lett. **126**:541 (1986).

2. P. Brumer and M. Shapiro, "Coherent Radiative Control of Unimolecular Reactions: Three Dimensional Results", Faraday Disc. Chem. Soc. **82**:177 (1987).

3. M. Shapiro and P. Brumer, "Laser Control of Product Quantum State Populations in Unimolecular Reactions", J. Chem. Phys. **84**:4103 (1986).

4. C. Asaro, P. Brumer and M. Shapiro, "Polarization Control of Branching Ratios in Photodissociation", Phys. Rev. Lett. **60**:1634 (1988).

5. M. Shapiro, J.W. Hepburn and P. Brumer, "Simplified Laser Control of Unimolecular Reactions: Simultaneous (ω_1, ω_3) Excitation", Chem. Phys. Lett. **149**:451 (1988).

6. G. Kurizki, M. Shapiro and P. Brumer, "Phase Coherent Control of Photocurrent Directionality in Semiconductors", Phys. Rev. **B39**:3435 (1989).

7. M. Shapiro and P. Brumer, "Laser Control of Unimolecular Decay Yields in the Presence of Collisions", J. Chem. Phys. **90**:6179 (1989).

8. T. Seideman, M. Shapiro and P. Brumer, "Coherent Radiative Control of Unimolecular Reactions: Selective Bond Breaking with Picosecond Pulses", J. Chem. Phys. **90**:7132 (1989).

9. For an introductory discussion see P. Brumer and M. Shapiro, "Coherence Chemistry: Controlling Chemical Reactions With Lasers", Accounts Chem. Res. **22**:407 (1989).

10. P. Brumer and M. Shapiro, "One Photon Mode Selective Control of Reactions by Rapid or Shaped Laser Pulses: An Emperor Without Clothes?", Chem. Phys. **139**:221 (1989).

11. I. Levy, M. Shapiro and P. Brumer, "Two-Pulse Coherent Control of Electronic States in the Photodissociation of IBr: Theory and Proposed Experiment", J. Chem. Phys. **93**:2493 (1990).

12. M. Shapiro and P. Brumer (to be published).

13. C.K. Chan, P. Brumer and M. Shapiro, "Coherent Radiative Control of IBr Photodissociation via Simultaneous (ω_1, ω_3) Excitation", J. Chem. Phys. **94**: 2688 (1991).

14. J. Krause, M. Shapiro and P. Brumer, "Coherent Control of Bimolecular Reactions", J. Chem. Phys. **92**:1126 (1990).

15. M. Shapiro and P. Brumer (to be published).

16. P. Brumer, X-P. Jiang and M. Shapiro (to be published).

17. See, e.g., S.H. Tersegni, P. Gaspard and S.A. Rice, " On Using Shaped Laser Pulses to Control the Selectivity of Product Formation in a Chemical Reaction", J. Chem. Phys. **93**:1670 (1990) and references therein.

18. See, e.g., S. Shi and H. Rabitz, "Optimal Control of Selective Vibrational Excitation of Harmonic Molecules: Analytic Solution and Restricted Forms for the Optimal Field", J. Chem. Phys. **92**:2927 (1990) and references therein.

19. S.M. Park, S-P. Lu and R.J. Gordon, "Coherent Laser Control of the Resonant Enhanced Multiphoton Ionization of HCl", J. Chem. Phys. **94**:8622 (1991)

20. C. Chen, Y-Y. Yin and D.S. Elliott, "Interference Betweeen Optical Transitions", Phys. Rev. Lett. **64**:507 (1990).

21. As another interesting case consider the use of two components of laser polarization to excite a bound state $|E_i\rangle$ to the continuum[4]. These two components will interfere constructively and destructively and allow control over product ratios in the differential cross section. However, the interference contribution to the *total* cross section vanishes since these two routes reach different values of M_J.

22. X-P. Jiang, P. Brumer, and M. Shapiro (to be published)

23. An example, although requiring a slight extension of the BAB' notation, is the Norrish type II reaction: $D(CH_2)_3CO(CH_2)_3D'$ dissociating to $DCHCH_2 + D'(CH_2)_3COCH_3$ and $D'CHCH_2 + D(CH_2)_3COCH_3$ where D and D' are enantiomers.

24. M. Shapiro and P. Brumer, "Controlled Photon Induced Symmetry Breaking: Chiral Molecular Products from Achiral Precursors", J. Chem. Phys. (in press)

25. J.M. Hollas, "High Resolution Spectroscopy", (Butterworths, London, 1982)

Vibrational Energy Transfer in the Chemical Energy Regime using Stimulated Emission Pumping

Xueming Yang[a], Eun H. Kim[b], Alec M. Wodtke[c], Department of Chemistry, University of California, Santa Barbara, CA 93106

a) Present address: Dept. of Chemistry, Princeton University, Princeton, NJ 08544
b) Present address: Dept. of Chemistry, University of California, Berkeley CA 94720
c) National Science Foundation Presidential Young Investigator

Abstract

The dependence of vibrational energy transfer on vibrational excitation has been studied using the stimulated emission pumping technique to efficiently prepare a large range of specific vibrational states of the Nitric Oxide molecule in its ground electronic state. Laser induced fluorescence was used to detect collisionally relaxed NO. The self-relaxation rate constants of NO(v>>1) were up to two-hundred times larger than that of NO(v=1). Multiquantum relaxation was found to be important at high energy and was quantified at 3.8 eV. Theoretical explanations of our experimental results were attempted and it is shown that at vibrational energy up to ≈ 3 eV the qualitative trends observed in these experiments such as: the mass effect and the linear dependence of the relaxation constant on v can be explained by Schwartz-Slawsky-Herzfeld theory. A simple explanation of the anomalously high NO self-relaxation rate is given. The large acceleration of the vibrational relaxation rate above 3.0 eV is coincident with the energetic onset of high energy $(NO)_2$ isomer-complexes. More theoretical and experimental work is needed to explain the quantitative aspects of these observations.

1) Introduction

Exothermic chemical reactions, ultraviolet photodissociation, and electrical discharges are three common examples of means by which highly vibrationally excited molecules can be produced. Such excited states may have unique and interesting chemical and physical properties. Slow, normally unimportant chemical reactions may become rapid when large amounts of vibrational energy are found in the reaction coordinate.[1,2] Infrared emission may be strongly affected due to the fact that vibrationally excited molecules may sample a different part of the dipole moment function.[3,4] Vibrational energy transfer may be much faster as has been observed experimentally for hydrogen halides,[5] or it may be much slower as has been predicted for highly vibrationally excited O_2.[6]

In order to understand such phenomena, one needs a means for state specifically producing and detecting highly vibrationally excited species. Overtone pumping[7] and stimulated emission pumping[8] are presently the two methods of choice. While overtone pumping works well for anharmonic hydride bonds, stimulated emission pumping offers, at least in principle, a more general route to the production of highly vibrationally excited molecules.

One particularly favorable application of stimulated emission pumping is to the preparation of highly vibrationally excited Nitric Oxide (NO).[9,10,11] In this example, the vibrational energy becomes essentially a "dial-in" experimental quantity for nearly every vibational state up to $v = 25$, which contains more than 4.6 eV of vibrational energy. The unique success of this example relies on using a suitable laser to access the B state of NO which has very large Franck-Condon factors to many vibrational states of the ground electronic state. Using this approach vibrational state specific energy transfer properties of NO from $v = 8$ to 25 have been investigated giving an uncommonly complete picture of the vibrational dependence of the energy transfer.[11,12] This study reveals that the vibrational energy transfer below 24,000 cm^{-1} is qualitatively different than that above this energy value. In the low energy regime it is found that the relaxation rate constant is linearly dependent on the vibrational quantum number, which is the first observation of the simple prediction of the SSH theory of vibrational energy transfer. In the high energy regime the dependence on the vibrational quantum number is much stronger than linear and $\Delta v = 2$ transitions are observed to play as important a role as $\Delta v = 1$ transitions. It is suggested that in the low energy regime SSH theory gives a qualitatively correct picture of the vibrational energy transfer if one adapts the theory to account for an increased repulsive interaction due to the attractive well of the NO dimer. In the high energy regime it is suggested that the interaction is more "chemical" in nature, leading to a stronger coupling of the two NO oscillators.

2) Experimental

A schematic diagram of the experimental set-up is shown in Figure 1. Highly vibrationally excited NO was prepared by stimulated emission pumping and laser induced fluorescence was used to probe the collisionally relaxed NO molecules as a function of time after the stimulated emission pumping event. A tunable ArF laser (Model EMG150TMSC) was used to pump the NO B(7)-X(0) vibronic band and a XeCl excimer pumped dye laser (with frequecy doubling where necessary) stimultaed emission in the B(7)-X(v") vibronic bands for a large range of v". A detailed description of this experimental arrangement can be found in our recent work[9,10,13] and references therein. Another XeCl excimer pumped dye laser was used to probe either the prepared vibrational state or the collisionally populated ones using laser induced fluorescence through either the A-X or B-X systems. By varying the delay between the DUMP and the PROBE, the pressure dependent vibrational state specific decay and/or appearance could be measured. A gated PMT (Thorn-EMI9816B/GB1001B) with a risetime of 0.5 μs was used to discriminate against fluorescence background coming from the SEP preparation. Great care was taken with the spatial overlap of the three laser beams to eliminate the influence of molecular diffusion.

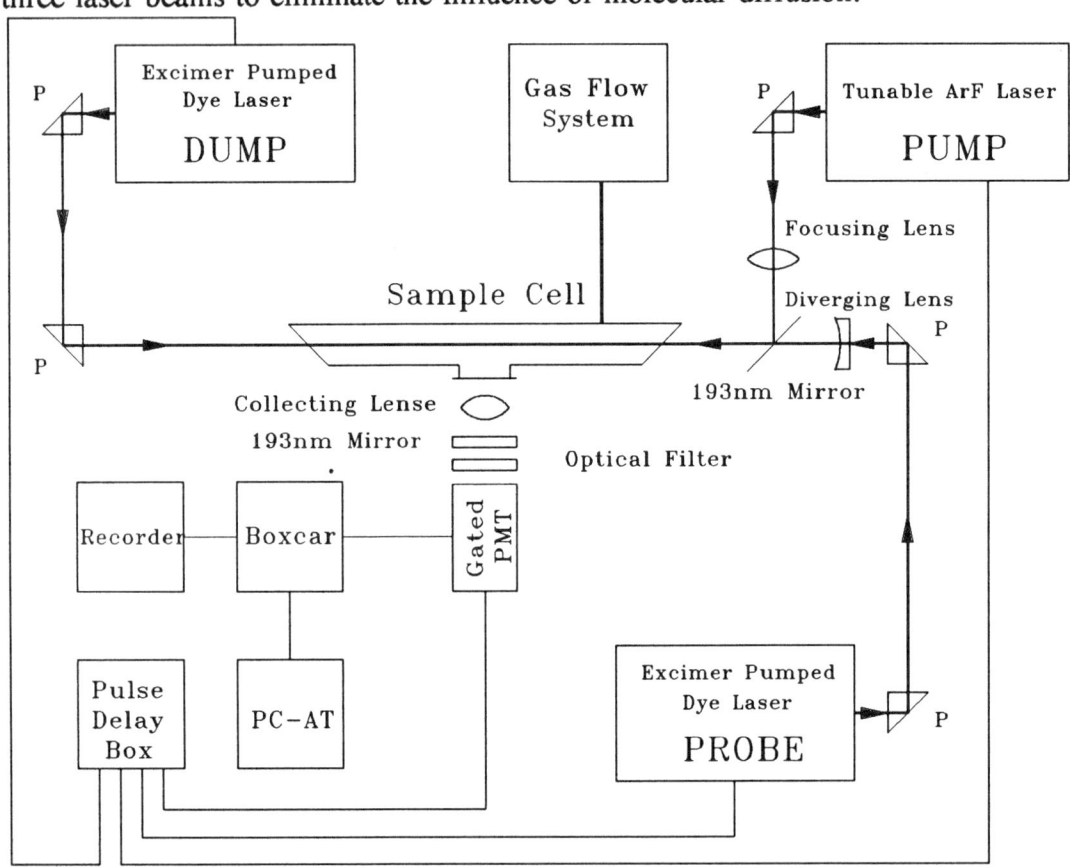

Figure 1 Experimental set-up for measuring vibrational relaxation rate constants.

3) Results

The described experimental apparatus allowed the determination of vibrational state specific total removal rate constants for vibrational levels up to v = 25. In addition, the determination of branching ratios between Δv = 1 and 2 relaxation was possible.

3.1) Vibrational Relaxation Rate Constants

Figure 2 shows the fully rotationally relaxed laser induced fluorescence spectrum of the (1,16) B-X band obtained by scanning the probe laser wavelength at a pressure of 0.4 torr, 8 μs after stimulated emission pumping. The observed signal is very strong. Under these conditions, where the PMT (Hamamatsu R212) was set at 800 Volts and a 200mV high 1 μs transient was observed on the oscilloscope using a 50 Ohm terminator and a ten times preamplifier. Figure 3 shows the decay curves for a typical experiment. Here, $^2\Pi_{3/2}$, v = 16, J = 25.5 has been prepared by stimulated emission pumping. The vibrational relaxation which proceeds after rotational equilibration is observed by monitoring the time dependence of a single rotational state of v = 16 by LIF. The solid line is the best exponential fit to the data. From such data collected at several pressures a Stern-Volmer plot can be constructed and a bimolecular relaxation rate constant can be extracted for the sum of all process that give rise to the decay of the prepared vibrational state. Several Stern-Volmer plots are shown in Figure 4. The slope of the curves is proportional to the rate constant.

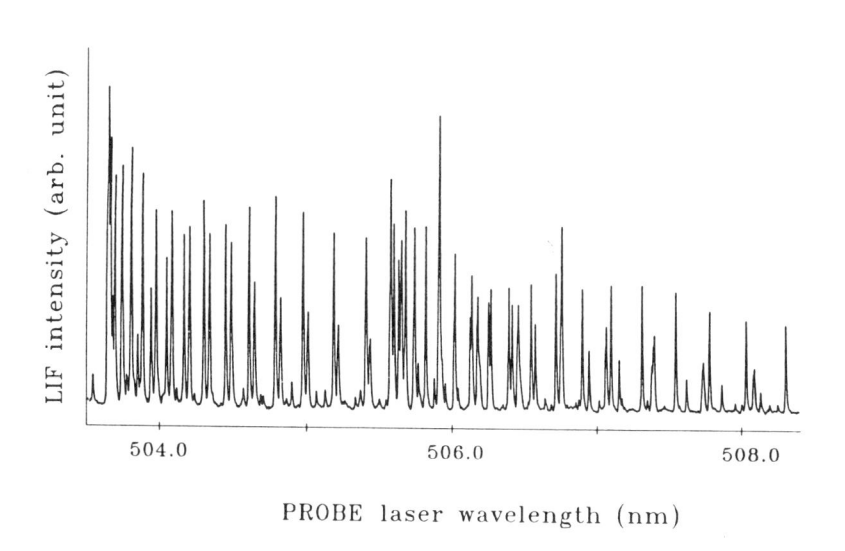

Figure 2 LIF spectrum of B(1)-X(16) band of $^{14}N^{16}O$ after SEP

The vibrational state dependent self-relaxation rate constants for both isotopes ($^{14}N^{16}O$ and $^{15}N^{18}O$) are tabulated in tables 1 and 2. Rotational and spin-orbit state-to-state energy transfer, which we have also measured and will be discussed in another paper, is typically fifty times larger than the largest vibrational decay rate constant we have observed. Consequently, the vibrational state specific rate constants reported here are an average deactivation rate constant that has contributions from all the populated rotational and spin-orbit states. The effect of radiative loss is negligible since calculations of the radiative lifetimes of high vibrational states are approximately milliseconds,[14] while our measured collisional relaxation lifetimes are always near ten μs.

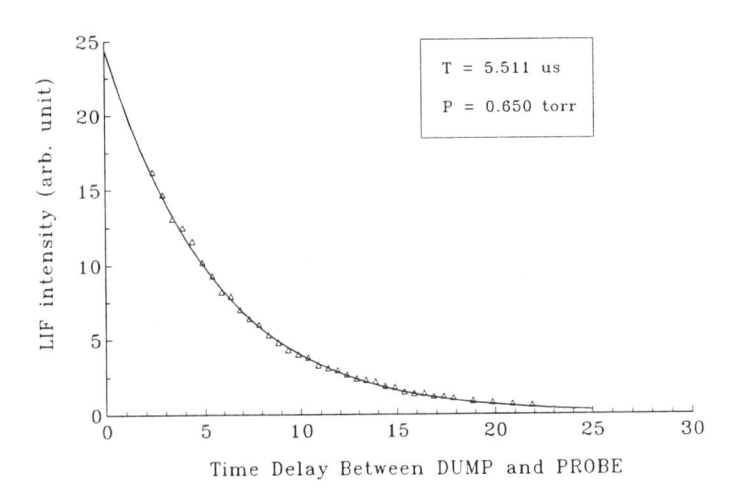

Figure 3 Exponential decay of NO(v=16)

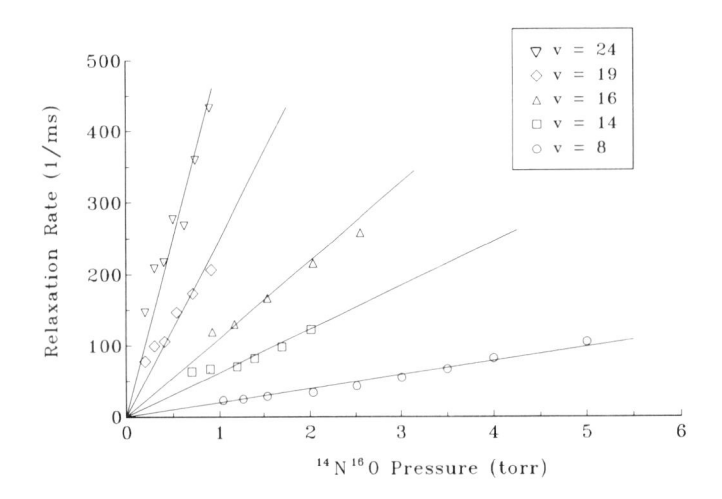

Figure 4 Pressure dependence of collisional lifetime

The dependence of the relaxation rate on vibrational excitation is very strong, about 200 times faster for the highest vibrational levels in comparison to v = 1. As pointed out before there appears to be a region at low vibrational energy where the relaxation rate constant increase linearly with vibrational quantum number. Our more complete set of data verifies this as is shown in Figure 5. This figure shows the measured rate constants divided by the vibrational quantum number as a function of vibrational quantum number. The region over which the graph is constant indicates the linear part of the vibrational quantum number dependence.

As will be discussed below we believe this linear region is dominated by $\Delta v = -1$ collisional propensity rules. Similarly we believe the region where the dependence is much stronger than linear has significant

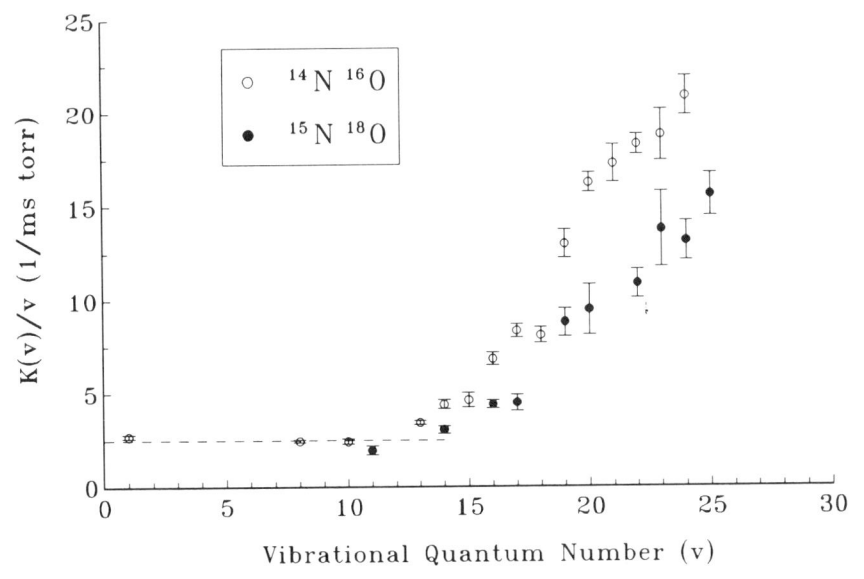

Figure 5 k_v/v as a function of vibrational excitation

contribution from multiquantum relaxation and the following section describes the strong experimental evidence for this conclusion. It is easy to notice that the vibrational relaxation of $^{15}N^{18}O$ is always slower than that of $^{14}N^{16}O$ for the same vibrational quantum level. This behavior is still observed even if the rate constants are plotted vs. vibrational energy instead of vibrational quantum number. This is the opposite behavior one would expect for an exponential energy gap law, since heavy NO has more closely spaced vibrational quantum levels. We believe this is the observation of the well known SSH mass effect between collision partners.

3.2) Multiquantum Relaxation

In order to determine the importance of multiquantum relaxation, it is necessary to measure the relative population of both the prepared state and collisionally populated states. The multiquantum relaxation is characterized by the following kinetic scheme

$$NO(v) + NO \xrightarrow{k_{v,v-1}} NO(v-1) + NO + \Delta E$$

$$\xrightarrow{k_{v,v-2}} NO(v-2) + NO + \Delta E$$

$$\xrightarrow{k_{v,v-n} \ (n>2)} NO(v-n) + NO + \Delta E$$

The kinetic theory of multiquantum relaxation has been described previously[15]. Assuming the simple kinetic scheme shown above but neglecting relaxation for n > 2, the kinetic equations can be integrated analytically to give the vibrational state specific population as a function of time. These are shown below.

$$N(v) = N_0(v) \exp(-k_v p t), \qquad (1)$$

$$N(v-1) = N_0(v) \; k_{v,v-1} / (k_v - k_{v-1}) \; [\exp(-k_{v-1} p t) - \exp(-k_v p t)] \qquad (2)$$

$$N(v-2) = N_0(v) [\alpha \exp(-k_{v-2} p t) - \beta \exp(-k_{v-1} p t) - \tau \exp(k_v p t)] \qquad (3)$$

where

$$\alpha = \frac{k_{v,v-1} \, k_{v-1,v-2}}{(k_v - k_{v-2})(k_{v-1} - k_{v-2})} + \frac{k_{v,v-2}}{(k_v - k_{v-2})}$$

$$\beta = \frac{k_{v,v-1} \, k_{v-1,v-2}}{(k_v - k_{v-2})(k_{v-1} - k_{v-2})}$$

$$\tau = \frac{k_{v,v-1} \, k_{v-1,v-2}}{(k_v - k_{v-2})(k_{v-1} - k_{v-2})} - \frac{k_{v,v-2}}{(k_v - k_{v-2})}$$

$N_0(v)$ is the initial population of state v

p is the pressure in the cell

t is time

k_v is the total removal rate constant for state v

$k_{v,v-n}$ is the state to state relaxation rate constant between states v and v-n

The state to state relaxation rate constants are found by integrating the population equations (1), (2), and (3) over all time as shown in equations (4) and (5).

$$\frac{k_{v,v-1}}{k_v} = \frac{A_{v-1} \, k_{v-1}}{A_v \, k_v} \qquad (4)$$

and

$$\frac{k_{v,v-2}}{k_v} = \frac{A_{v-2} \, k_{v-2} \, k_{v,v-1}}{A_v \, k_{v-1} \, k_v} - \frac{k_{v-1,v-2} \, k_{v,v-1}}{k_{v-1} \, k_v} \qquad (5)$$

in these equations, A_v is the integral over the population equation for state v.

Because a relative measurement of three different vibrational states is required there is ample room for systematic error in the measurement and we have been very careful to obtain the best data possible. Since optical filters were used in the experiment one must be very careful when different probing schemes are employed. Saturation of all optical transitions must also be ensured. Figure 6 is a typical measurement of the time evolutions and relative intensities of the three related states v = 19, 18, 17 made by initially preparing v = 19 of NO.

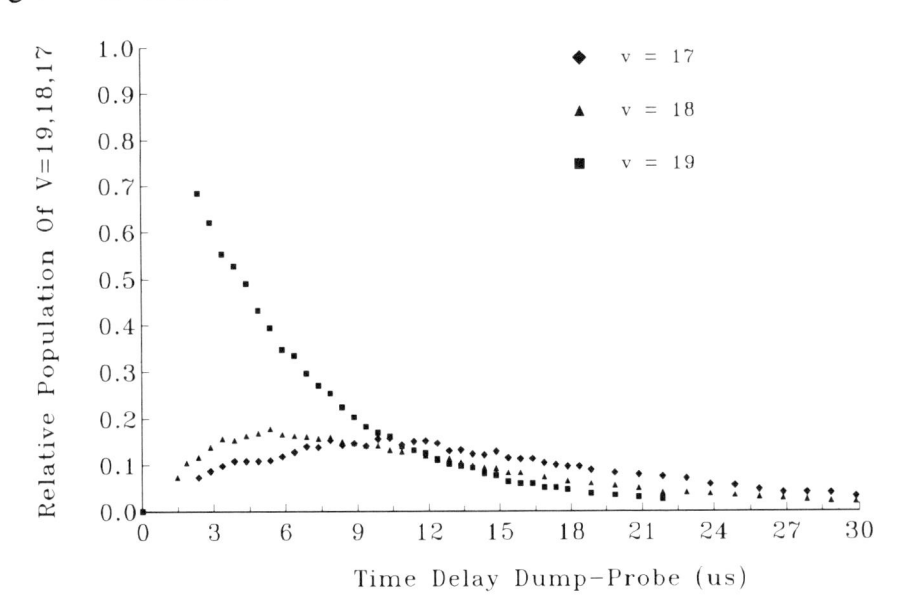

Figure 6 Multiquantum relaxation

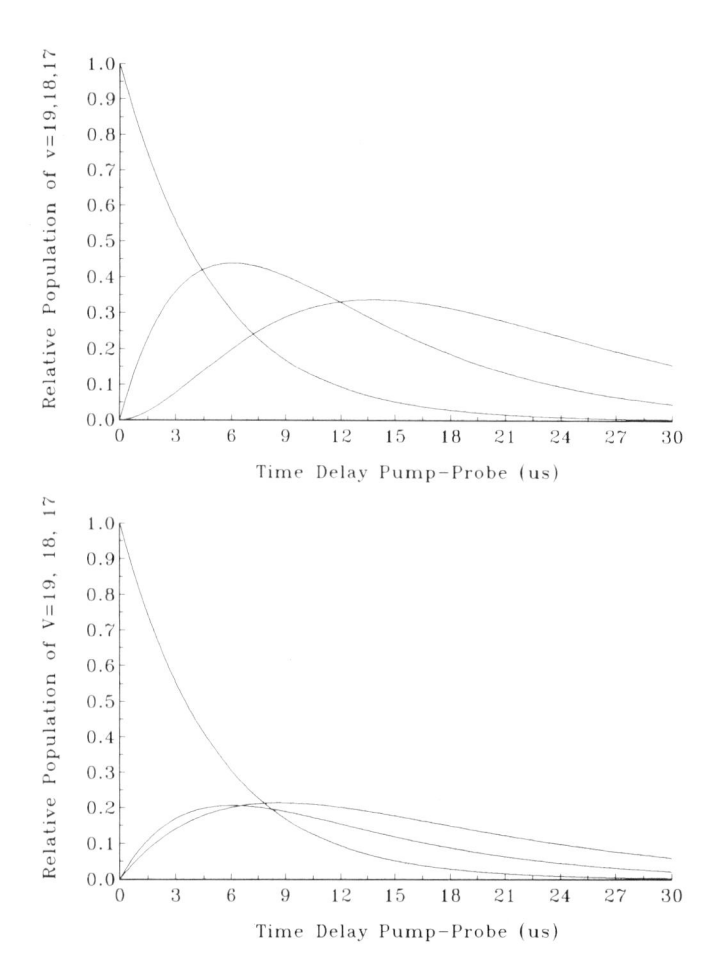

Figure 7 calculated time profiles: no multiquantum = top, multiquantum included = bottom

Figure 7 shows the calculated influence of multiquantum relaxation on the time profiles. Notice specifically the area under the time profile curves of v = 18 and 17. These are the curves which first appear then disappear. Figure 7(bottom), which includes multiquantum relaxation is clearly in much better agreement with the observed time profiles.

Analyzing our experimental data yields the first experimental evidence of multiquantum relaxation for highly vibrationally excited molecules.

$$k_{19,17} / k_{19} = 0.33 \pm 0.19$$
$$k_{19,18} / k_{19} = 0.47 \pm 0.12$$
$$k_{18,17} / k_{18} = 0.48 \pm 0.12$$

These rate constants can also be used to determine what fraction of the total vibrational deactivation process we have accounted for in this experiment.

$$k_{19,17} / k_{19} + k_{19,18} / k_{19} = 0.80 \pm 0.29$$

One can see that we have directly measured the large majority of the deactivation process. It is arguable that as much as 20% of the total deactivation remains unaccounted for. This could be made up by $\Delta v > 2$ collisions or chemical reaction to form $N_2O + O$. This result supports our earlier work which concluded that chemical reaction plays a minor role in the deactivation of vibrationally excited NO under our experimental conditions.

These results show that single quantum relaxation accounts for only about half of the total relaxation out of states $v = 19$ of NO, and two-quantum relaxation makes significant contribution (33%) to the vibrational relaxation at this vibrational energy (3.7 eV). There is a clear qualitative difference in the vibrational energy transfer of highly vibrationally excited NO and this observation has strong implications that the vibrational energy transfer mechanism may be significantly different for molecules that have vibrational energy in the "chemically interesting regime".

4) Discussion

In this work many features of nitric oxide vibrational energy transfer have been observed including: the vibrational dependence of the relaxation rate constants, multiquantum relaxation at high vibrational energy as well as the related mass and isotope effects. It is of obvious significance to determine to what extent vibrational energy transfer theories derived from experimental investigations at low vibrational energy can be used in the "chemical energy regime". One of the simplest quantum mechanical theories of vibrational energy transfer is that of Schwartz, Slawsky, Herzfeld (SSH).[16] Although designed for the general case this theory has had the most success reproducing vibrational energy transfer in diatomic molecules with atoms and other diatoms. The reader is referred to the literature for a detailed discussion of SSH theory. Only the assumptions and central ideas of the theory will be discussed here.

SSH theory assumes that the interaction between the two diatoms can be modelled as an exponential repulsive wall and the problem is reduced to a single dimension. The rationale for this is the recognition that vibrational energy transfer happens most efficiently at that point where the interaction potential is changing most rapidly with distance, i.e on the repulsive wall of the interaction potential. This is where the impulsive collision can most efficiently couple the high frequency harmonic oscillator of the diatomic bond to low frequency translational motion. Consequently, all details of the attractive part of the potential are assumed unimportant. The probability of vibrational energy transfer can then be calculated assuming harmonic oscilators for the internal degrees of freedom and perturbation theory like assumptions are used to calculate the transition probability.

One very important qualitative prediction of the SSH theory, that is in fact observed in this experiment is the linear dependence of V-T vibrational energy transfer on vibrational quantum number. This is a result of the first order of perturbation theory being the largest contribution to the relaxation probability.

4.1) Magnitude of NO's vibrational self relaxation rate

For many diatomic collision systems it has been found that the exponential fall-off constant for the repulsive interaction potential, α, can be determined from gas transport properties and the relaxation rates and temperature dependences can be determined using no adjustable parameters. But NO is a special case for which one cannot determine the NO/NO potential from its gas transport properties because of the unusually strong NO/NO interaction.[17] One of the most notorious problems encountered with SSH theory is its inability to explain why NO self vibrational relaxation is about 10^5 times faster than that of other comparable molecules such as CO in v = 1. This is not specifically a problem with the SSH theory since there is no clear explanation of the NO anomaly in the literature. Suggestions concerning "attractive forces" as well as electronic non-adiabatic vibrational energy transfer have been made.[18]

We have found that the SSH theory can be applied in a modified but logical fashion to at least partially explain this large discrepancy. Consider Figure 8 which compares a Lennard-Jones potential for CO/CO with a Lennard-Jones potential for NO/NO. The Lennard-Jones potential for NO-NO has been constructed using the formulae from Atkins[19] and the N-N bond length of 2.24 Å[20] and the dimer bond energy of 560 cm-1.[21] SSH theory operates from the principle that it is the repulsive part of the potential which induces vibrational relaxation. To find the appropriate value of the exponential fall off constant for NO/NO one must consider the part of the potential which induces vibrational energy transfer, the repulsive part at a total energy close to kT (see dashed line of Figure 8).

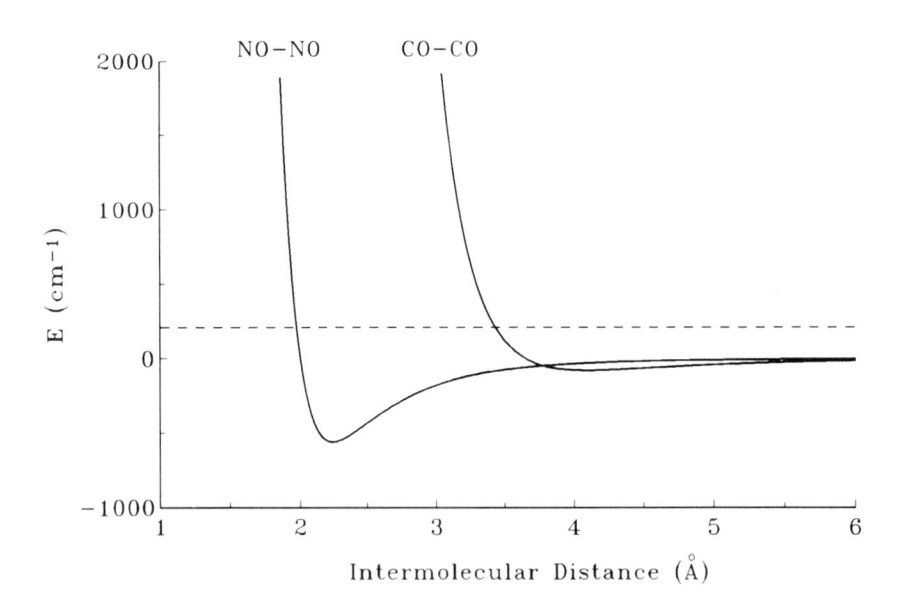

Figure 8 Potentials for NO-NO and CO-CO interaction. kT at 300K (dashed line)

Fitting the points near this position of the potential to an exponential one derives an an exponential fall off constant that is much larger than that of CO. The SSH theory is extremely sensitive to the steepness of the repulsive wall and using a steeper potential consistent with figure 8 gives rise to 25,000 fold increase in the relaxation rate of NO/NO relative to CO/CO. This simple calculation implies that the attractive forces are the most important factor in accelerating NO vibrational relaxation. However, the effect is subtle. Notice that because of the presence of the small (560 cm^{-1}) attractive well in the NO-dimer, the repulsive part of the potential is significantly steeper than that of the weakly interacting CO-CO case. The strong dependence of the vibrational relaxation on the steepness of the repulsive wall amplifies this small difference in the interaction potential. Similar arguments have been made for ion-neutral vibrational relaxation independently by Tanner and Maricq.[22] The ion-induced dipole interaction in these systems is directly analogous to the shallow well of $(NO)_2$. It could be that the true potential may be slightly different, thus making the agreement between experiment and calculation more quantitative. This analysis, however, awaits a more precise potential surface for the NO-dimer.

It appears that SSH theory can predict the correct order of magnitude of the $v = 1$ rate and it is consistent with the linear dependence on vibrational quantum number observed up to $v = 14$. Consequently, we conclude that SSH theory qualitatively explains the V-T contribution of the relaxation rate below $v = 14$.

4.4) Vibrational Dependence of Relaxation Rate Constants and Multiquantum Relaxation

SSH theory predicts very simple behavior for the vibrational quantum number dependence of the vibrational relaxation as can be seen from the form of the vibrational matrix elements in equations in references 16. $\Delta v = 1$ energy transfer is predicted to depend linearly on vibrational quantum number. If the collisional energy behaves as a perturbation to the system, $\Delta v = 1$ transitions are to be expected. Figure 5 shows clearly that there is a very large region, up to 24,000 cm-1 where a linear dependence on vibrational quantum number is observed. This observation is partially accidental because no measurement in the region where V-V energy transfer is important were made, $2 < v < 8$. This is a very dramatic success of the SSH theory and suggests two conclusions be drawn.

1) Up to 24,000 cm^{-1}, $\Delta \& v = 1$ energy transfer dominates the V-T energy transfer.
2) In this energy range SSH theory gives a qualitatively correct picture of the V-T component of the energy transfer.

Above 24,000 cm^{-1} the quantum number dependence is clearly non-linear. Such behavior has been observed when the perturbation approximation inherent to the SSH theory is invalid, for example in I_2 where the vibrational spacing is significantly less than kT.[23] However, we do not believe that this experiment is similar to the I_2 results. Even at these high energies the vibrational spacing is still much bigger than kT. For example the vibrational spacing in $^{14}N^{16}O$ between $v = 18$ and $v = 19$ is 1350 cm^{-1} and that between $v = 24$ and 25 is 1155 cm^{-1}. For the harmonic bending vibrations of CF_2, $\Delta v > 1$ transitions were observed above $v = 6$ for CF_2 colliding with SF_6.[24] However, the harmonic frequency in this case is also relatively low, 494 cm^{-1}.

At high vibrational energy, the electronic structure of NO is much different than that of NO at its equilibrium geometry. Ab initio calculation of the vibrational state specific dipole moments show this conclusively.[3] All diatomics that separate to neutral atoms must asymptotically have a zero dipole moment. This means that NO's dipole moment, which is essentially zero at its equilibrium geometry, rises slowly and turns over going back to zero at infinite separation. Although this interesting effect has not yet been observed experimentally, this turn-over apparently occurs near the outer classical turning point of v = 20.[3] Recent calculations of Gordon[25] also show the possibility of extremely unique interactions which are possible for high energy $(NO)_2$ molecules. Three $(NO)_2$ complexes are shown in Figure 9.

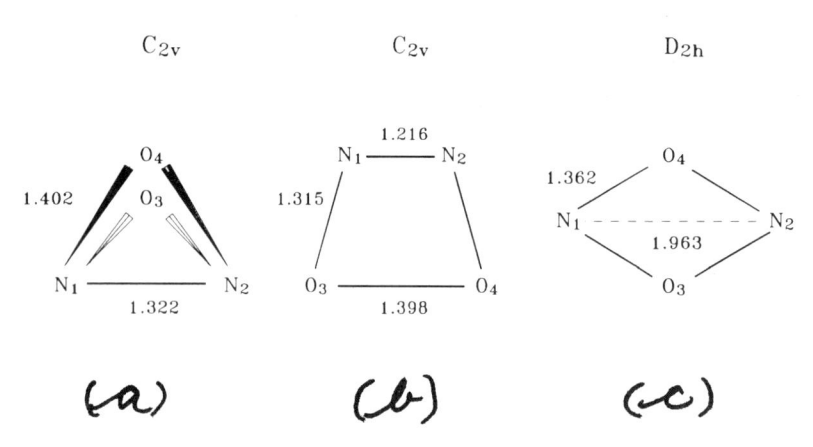

Figure 9 Calculated high energy structures of NO-NO (Ref. 24)

Figure 10 shows an apparent correlation between the calculated energies for complex (a) and (c) in fig. 9 and energies at which vibrational relaxation accelerates.

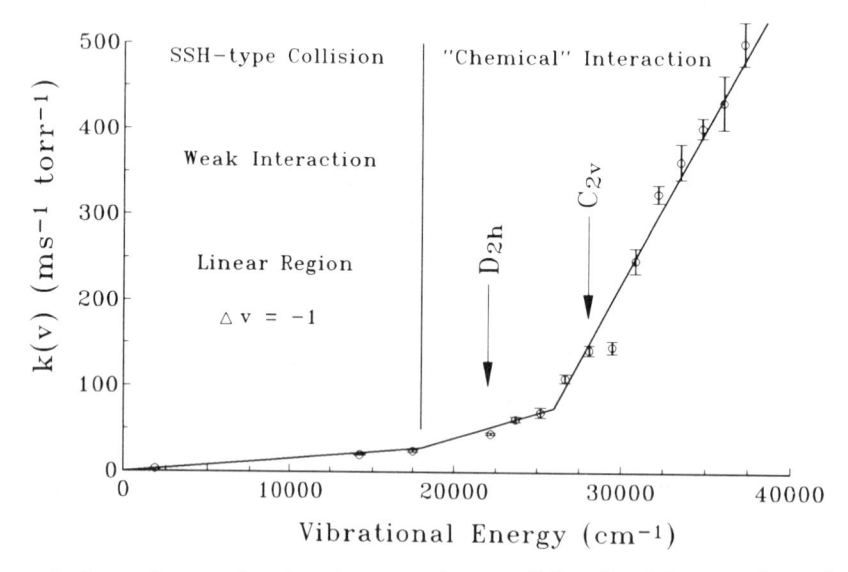

Figure 10 Correlation of energies for the complexes of fig. 9 with energies where relaxation accelerates

There is one other empirical piece of evidence which argues in this direction. The threshold for the bimolecular chemical reaction NO + NO → N_2O + O is very close to 24,000 cm^{-1}. This chemical reaction is clearly accessing a part of the $(NO)_2$ potential where the electronic structure is strongly dependent on nuclear displacement. Access to this part of the potential could, for example, enhance v-v energy transfer.

All of these ideas taken together summarize the thoughts behind the interpretation of the observations. That is, due to the high vibrational excitation, the electronic structure of the $(NO)_2$ transition state for energy transfer is severely altered, resembling transient "chemical" interactions. This leads to a much more efficient rate of vibrational energy transfer and more quanta being transferred per collision. More theoretical work is required to fully understand this problem.

5) Conclusion

Collisional vibrational energy transfer dynamics of very highly vibrationally excited NO was studied by a PUMP-DUMP-PROBE method. This method proved to be a very efficient way to study collisional dynamics of highly vibrationally excited molecules. Self-relaxation of NO was studied for vibrational energies as high as 4.6 eV. Vibrational self-relaxations for v = 8, 10, 11, 13, 14, 15, 16, 17, 18, 19, 20, 21, 22, 23, 24 of $^{14}N^{16}O$ and for v = 11, 14, 16, 17, 19, 20, 22, 23, 24, 25 of $^{15}N^{18}O$ were investigated. The main results are the following. a) NO self-relaxation rate constants for $^{14}N^{16}O$ and $^{15}N^{18}O$ for vibrational states up to v = 25 were measured very accurately. b) A simple explanation of the NO(v=1) relaxation anomaly was given, using a modified application of SSH theory. c) Up to v = 14, the V-T component of the self-relaxation rate constants are almost linearly dependent on the vibrational quantum number. This suggests that up to that vibrational energy, collisional relaxation occurs through one-quantum energy transfer. Multi-quantum relaxation plays a negligible role. d) Above this energy regime, state-to-state vibrational energy transfer rates of NO v= 19 to 18, 17 and v= 18 to 17 were measured. It was found that $\Delta v = 2$ relaxation is as fast as $\Delta v = 1$. This clearly shows that multiquantum relaxation plays a very important role in the self-relaxation process at high energy. The multiquantum relaxations at high energy is accompanied by a dramatic increase in the overall relaxation rates. e) The unusually fast increase of the vibrational relaxation rate at high vibrational energy may be due to the recently predicted high energy $(NO)_2$ isomers which are accessible in this energy regime.

Along with HF-HF, NO-NO appears to be one of the best characterized systems in terms of vibrational energy transfer. This provides a prototype non-hydride system for detailed theoretical studies of vibrational relaxation at vibrational energy up to 4.6 eV. More experimental work is also needed to fully understand vibrational energy transfer processes of this unique system.

Acknowledgements

This work was partially supported by Petroleum research Fund Grant No. 21762-G6, National Science

Foundation Presidential Young investigator award CHE-8957978 and National Science Foundation Atmospheric Chemistry
division ATM-8922214. In addition, a small grant from the University-wide Energy Research Group from the University of California is gratefully acknowledged. This work was also made possible by the Santa Barbara Laser pool under NSF grant number CHE-8411302. Special thanks go to Professor J. William Rich at the Ohio State University for his help with models of vibrational energy transfer as well as many useful discussions.

REFERENCES

1. A. Sinha, M. C. Hsiao, and F. F. Crim, J. Chem. Phys. 92(10), 6333 (1990)

2. K. Kleinermanns, J. Wolfrum, Appl. Phys. B34, 5(1984)

3. F. P. Billingsley II, J. Chem. Phys. 62, 864(1975); F. P. Billingsley II, J. Chem. Phys. 63, 2267 (1975)

4. C. Amiot, J. Mol. Spectrosc. 94, 150(1982)

5. L. S. Dzelzkalns and F. Kaufman, J. Chem. Phy. 79(8),3836 (1983); L. S. Dzelzkalns and F. Kaufman, J. Chem. Phys. 77(7), 3508 (1982)

6. T. G. Slanger, L. E. Jusinski, G. Black, G. E. Gadd, Science, 241, 945 (1988)

7. F. F. Crim, Ann. Rev. Phys. Chem. 35, 657(1984)

8. C. E. Hamilton, J. L. Kinsey, and Robert W. Field, Ann. Rev. Phys. Chem. 37, 493(1986)

9. X. Yang and A. M. Wodtke, J. Chem. Phys. 92(1), 116(1990)

10. X. Yang, D. McGuire, A. M. Wodtke, J. Mol. Spectrosc. (submitted)

11. X. Yang, E. H. Kim and A. M. Wodtke, J. Chem. Phys. 93, 4483 (1990)

12. X Yang, E. H. Kim, A. M. Wodtke, J. Chem. Phys. (accepted)

13. X. Yang, C. A. Rogaski and A. M. Wodtke, J. Opt. Soc. Am. B, 1835 (1990)

14. C. H. Kuo, C. G. Beggs, P. R. Kemper, M.T. Bowers, D. J. Leahy and R. N. Zare, Chem. Phys. Lett. 163, 291(1989); M. Henniger, S. Fenistein, M. Durup-Ferguson, E. E. Ferguson, R. Marx and G. Mauclaire, Chem. Phys. Lett. 439(1986); G. Chambaud and P. Rosmus, Chem. Phys. Lett. 165, 429(1990)

15. G. M. Jursich and F. F. Crim, J. Chem. Phys. 74, 4455 (1981)

16. R. N. Schwartz, Z. I. Slawsky, and K. F. Hertzfeld, J. Chem. Phys. 20, 1591 (1952); R. N. Schwartz and K. F. Hertzfeld, J. Chem. Phys. 22, 767(1954); F. I. Tanczos, J. Chem. Phys. 25, 439(1956)

17. J. O. Hirschfelder, C. F. Curtiss and R. B. Bird, Molecular theory of gases and liquids, John Wiley & Sons, Inc, 1964, p1111

18. E. G. Nikitin, Optics Spectroscopy 9, 8(1960)

19. P. W. Atkins, Physical Chemistry, Oxford University Press, 1978

20. C. M. Western, P. R. R. Langridge-Smith, B. J. Howard, and S. E. Novick, Mol. Phy. 44, 145(1981); S. G. Kukolich, J. Mol. Spectrosc. 98, 80(1983)

21. J. T. Yardley, Introduction to Molecular Energy Transfer, Academic Press, New York, 1980

22. J. J. Tanner and M. M. Mariq, submitted

23. J. I. Steinfeld, W. Klemperer, J. Chem. Phys. 42, 3475 (1965); R. B. Kerzel, J. I. Steinfeld, J. Chem. Phys. 53, 3293 (1970)

24. D. L. Akins, D. S. King, J. C. Stephenson, Chem. Phys. Lett. 65, 257 (1979)

25. M. Gordon, North Dakota State University (private communication)

Tables

Table 1 Collisional relaxation rate constants (in torr-1 ms-1) and cross sections (Å²) of $^{15}N^{18}O$

v	energy(cm-1)	rate constant(a)	cross section
11	18318	21.4(2.6)	0.093
14	22771	42.8(2.9)	0.186
16	25617	70.3(3.4)	0.306
17	27001	76.5(7.1)	0.332
19	29685	167.1(14.0)	0.726
20	30986	189.5(27.2)	0.824
22	33502	239.2(17.0)	1.039
23	34722	316.5(46.0)	1.375
24	35914	315.9(25.2)	1.373
25	37078	390.6(28.2)	1.698

(a) Numbers in the parentheses are 1σ errors.

Table 2 Collisional relaxation rate constants (in torr-1 ms-1) and cross section (Å²) of $^{14}N^{16}O$

v	energy(cm-1)	rate constants(a)	cross section
8	14223	19.7(0.5)	0.086
10	17499	24.3(1.3)	0.106
13	22203	44.6(1.4)	0.194
14	23714	61.6(5.8)	0.268
15	25196	69.6(5.8)	0.302
16	26650	109.4(5.3)	0.475
17	28074	142.1(5.9)	0.618
18	29469	146.0(7.5)	0.635
19	30833	247.4(14.5)	1.075
20	32167	324.9(10.2)	1.412
21	33470	362.4(20.7)	1.575
22	34740	402.1(11.9)	1.748
23	35979	432.2(31.4)	1.878
24	37183	501.0(25.2)	2.177

(a) Numbers in the parentheses are 1σ errors.

OPTICAL METHODS FOR
TIME- AND STATE-RESOLVED CHEMISTRY

Volume 1638

SESSION 6

State-Selected and State-to-State Reaction Dynamics

Chair
Cheuk-Yiu Ng
Ames Laboratory/USDOE and Iowa State University

Reactive scattering of "hot" H atoms with CO_2 and OCS

Harry E. Cartland

Photonics Research Center, U.S. Military Academy
West Point, NY 10996

Scott L. Nickolaisen and Curt Wittig

Chemistry Department, University of Southern California
Los Angeles, CA 90089

ABSTRACT

Time-resolved infrared diode laser spectroscopy has been used to state selectively monitor CO produced in the "hot" atom reaction of H with CO_2 and OCS. For center of mass (CM) kinetic energies of 2.4 and 1.4 eV, respectively, CO internal excitations are substantially colder than predicted by statistical theory, with energy preferentially channeled into CM translation. In the collision energy regime of these experiments, $\approx 10,000$ cm^{-1} above the barrier, statistical rate theory is not applicable even with an intermediate.

1. INTRODUCTION

Hydrogen atom reactions with CO_2 and OCS

$$H + CO_2 \rightarrow CO + OH \qquad \Delta H_{298} = 8960 \text{ cm}^{-1}$$

$$H + OCS \rightarrow CO + SH \qquad \Delta H_{298} = -3930 \text{ cm}^{-1}$$

are of interest because of their role in combustion and atmospheric chemistry. In addition, their relative simplicity makes them amenable to computational studies, allowing a rigorous comparison between experiment and current dynamical theories.

The two reactions present an interesting pair for comparison. Though the first is endothermic and the second exothermic, under the conditions employed here the barrier is crossed with $\approx 10,000$ cm^{-1} of excess energy in both cases. As we shall see, both reactions yield qualitatively similar non-statistical CO product state distributions.

1.1. $H + CO_2 \rightarrow CO + OH$

The $H + CO_2$ reaction has already received considerable attention, mostly focussing on laser-induced fluorescence (LIF) detection of the OH fragment. Quick and Tiee detected OH in up to v=2 at a CM collision energy of 2.6 eV.[1] Kleinermanns et al. measured OH(v=0) rotational distributions for CM energies of 1.9 and 2.6 eV, and found them to be colder than statistical.[2] Jacobs et al. determined a reaction cross section of 0.4±0.2 Å^2 at a collision energy of 1.9 eV.[3] Radhakrishnan et al.,[4] Rice et al.,[5] and Scherer et al.[6] have studied the reaction with complexed precursors, the last using a picosecond pump-probe technique to demonstrate the presence of an intermediate. Finally, Rice et al. have used VUV LIF to measure CO(v=0,1) rotational distributions under bulk and complexed conditions.[7] We consider the CO distributions of Rice et al., as well as the ab initio HOCO potential energy surface of Schatz et al.,[8] later in this paper.

1.2. H+OCS → CO+SH

The H+OCS reaction has also been studied previously, but not nearly to the same extent as H+CO$_2$. Tsunashina et al.[9] and Lee et al.[10] have measured the reaction rate constant, establishing an activation barrier of 1350 cm^{-1}. Using 193 nm photolysis of DBr as a D atom source, Hauesler et al. measured SD(v=0,1) rotational distributions under bulk and complexed conditions, and found them to be non-statistical;[11] predissociation prevents a similar analysis of the analogous H atom reaction. We return to the results of Hauesler et al. later in the discussion.

2. EXPERIMENTAL SECTION

The UV laser pump/IR diode laser probe technique has been described elsewhere in detail.[12] The apparatus used for the study of H+CO$_2$ was not the same as that used for H+OCS, but the differences were for the most part minor. We provide a general description of the technique, followed by a discussion of details specific to each experiment.

Translationally "hot" H atoms were generated in a 2 m pressure stabilized flow cell by a UV laser photolysis pulse. The pulse energy was measured during the course of the experiments to allow normalization of the measured nascent product distributions.

A collinear CW IR diode laser (Laser Analytics) then state selectively probed the CO(v,J) reaction product. The output of the IR diode was collimated and passed through a monochromator to select a single laser mode with a nominal linewidth of ≈ 0.0003 cm^{-1}. Because of anharmonicity, virtually any CO(v,J) level is accessible by temperature and current tuning of the diode output, though more than one diode is usually required to cover the wavelength range of interest. The distributions reported here are based on a sampling of even and odd J states in both the P and R branches.

Prior to entering the sample cell, $\approx 10\%$ of the IR beam was split off and passed through a reference flow cell. The reference cell and a Ge etalon permitted precise wavelength calibration. Passing a discharge through the cell generated a reference population in higher vibrational states (v>0), allowing excited product states to be probed in the sample cell.

Standard line locking techniques were used to prevent drift of the diode laser. The reference detector signal was input to a lock-in amplifier. By using phase sensitive detection, a small dither in the diode output produced an error signal which was fed back into the diode current controller to drive the laser output to line center.

The sample and reference beams passed through long wavepass filters ($\lambda > 2.5$ μm) before being focussed onto N$_2(\ell)$ cooled InSb detectors. The signal level was boosted by impedance matched preamplifiers and line drivers. The reference signal was input directly to the lock-in amplifier, while the sample signal underwent another stage of amplification and filtering prior to being input to a transient digitizer (LeCroy). The digitizer sampled the absorption transients at 10^8 samples s^{-1}, and hard wire averaged them in the CAMAC crate controller. The averaged data was stored, processed and displayed on a PC AT computer (IBM) using original software written in ASYST.

2.1. H+CO$_2$ → CO+OH

Excimer laser (Lambda Physik) photolysis of H$_2$S at 193 nm produced H atoms with a CM kinetic energy of 2.4 eV (19,200 cm^{-1}). Although 193 nm photolysis actually produces H atoms with a distribution of kinetic energies corresponding to varying internal excitation of the SH fragment, the strong dependence of reaction probability on CM collision energy means that 96% of the reactive H atoms have this energy.[12a]

For this experiment, the sample mixture was 20% H_2S in CO_2 at a total pressure of 200 mTorr. Typically 1500 transient signals were averaged at a repetition rate of 10 Hz. Detection system response time was 200 μs.

2.2. H+OCS → CO+OH

Hydrogen atoms were generated by 278 nm photolysis of HI. The 278 nm photons were produced by doubling the output of a Nd:YAG pumped dye laser (Quantel) using Rhodamine 590 dye. Pulse energies were typically 10-12 mJ. At this wavelength, 90% of the I atoms are produced in the $^2P_{1/2}$ ground state,[13] yielding a H atom CM energy of 1.4 eV (11,100 cm^{-1}).

Normally 250 transient signals were averaged at a repetition rate of 2 Hz. The sample mixture was 20% HI in OCS with both components requiring distillation before use. Sample pressure was typically 200 or 400 mTorr, but ranged from 50 to 1000 mTorr during pressure studies. Using a photodissociation process that produced CO, the detection system risetime was measured at $\approx$ 80 ns.

For time-resolved CO(v,J) lineshape measurements, a fraction of the IR beam was directed to a vacuum spaced confocal etalon (Laser Analytics). Instead of dithering the diode laser wavelength, the etalon scanner driver was modulated thereby dithering the fringe positions in time. The line locking technique described above was used to slave the laser output to an etalon fringe, which was subsequently stepped across the absorption feature of interest. Simultaneous acquisition of a room temperature CO reference lineshape served to calibrate the step size in terms of wavelength.

3. RESULTS

The raw data were converted to absorbance units by subtracting the AC coupled signal from the DC background, dividing the difference by the background, and taking the natural logarithm. Some typical v=0 signals from the H+OCS reaction are shown in Fig. 1. These transient signals clearly illustrate the reaction and CO relaxation dynamics. The two component rise for low J is evidence of a small nascent population, increased by rotational relaxation, and to a lesser extent vibrational relaxation, as equilibrium is reached. In contrast, high J levels show substantial initial populations which rapidly decay. Relaxation of a large nascent population to some non-zero value as thermal equilibrium is established is characteristic of intermediate J.

In order to interpret the data, we develop the following simple model where X designates either a O or S atom (X=O,S).

$$H^* + M \rightarrow H + M \quad (1)$$

$$H^* + OCX \rightarrow CO(v,J) + XH \quad (2)$$

$$H^* + OCX \rightarrow CO(v',J') + XH \quad (2')$$

$$CO(v,J) + M \rightarrow CO(v',J') + M \quad (3)$$

$$CO(v',J') + M \rightarrow CO(v,J) + M \quad (3')$$

Reaction 1 is collisional deactivation, but effectively includes all processes that remove translationally "hot" H atoms, such as diffusion out of the probed volume and various radical reactions. Reactions 2 and 2' represent reactive scattering into the monitored state CO(v,J), and all other states CO(v',J'). Modelling the CO manifold as a quasi two level system consisting of the probed state and a "bath" state, with which energy is exchanged via reactions 3 and 3', yields an analytic expression for the time dependent CO(v,J) population [CO(v,J,t)]. As the fits in Fig. 1 show, the transient absorption signals are well modelled by this simple scheme.

The derivation of the analytic expression for [CO(v,J,t)], in various forms, is discussed elsewhere in detail.[12] Here we consider only two limiting cases. First, in the long time limit we find

$$[CO(v,J,t)]_{t\to\infty} = [H_o^*] \frac{(k_2\chi_{OCX} + k_{2'}\chi_{OCX})}{(k_1 + k_2\chi_{OCX} + k_{2'}\chi_{OCX})} \frac{k_{3'}}{k_3 + k_{3'}}$$

where $[H_o^*]$ is the initial "hot" H atom number density and χ_{OCX} is the mole fraction of the OCX (X=O,S) collision partner. This expression has a straight forward interpretation: the concentration of CO(v,J) after attainment of equilibrium is a product of the initial "hot" H atom concentration, the branching ratio for reactive scattering, and the equilibrium fraction of molecules in that state. Of more interest, though, is the short time limit

$$[CO(v,J,t)]_{t\to 0} = k_2\chi_{OCX}[H_o^*][M]t$$

where [M] is the total sample number density. Initially, the CO(v,J) density is linear in time, with a proportionality constant dependent on the state specific rate constant k_2. Therefore, a plot of the initial slopes of the transient signals gives a relative distribution of state specific rates, which is equivalent to plotting the actual nascent CO(v,J) number densities.

Confidence in the distributions obtained in this manner requires that the model incorporate all essential features of the kinetics. Elsewhere, we explore the effect of processes specific to the individual systems.[12] In the H+OCS system for example, the unusually high relaxation rate of vibrationally excited CO in OCS ($P \approx 0.1$ for CO(v=4)) must be considered. Inclusion of additional processes in the model of course improves the fit. In no case will any of these secondary processes change the t→0 limit, but they may influence the time interval over which the linear approximation is valid. Though relaxation rates may in some cases be significant, even gas kinetic, we note that our absorption measurements are sensitive to net changes in population. Net population changes will be small initially, especially near the peak of the distribution, since the relaxation rates into and out of a given state will be comparable. Thus the distributions reported here may reflect some relaxation, but not enough to change any major conclusions.

The validity of the model is supported by the plot in Fig. 2 which is derived from H+OCS data. Multi-parameter iterative fitting can produce ambiguous results, but certain features of the transient signals, such as the time to peak amplitude, are not subject to misinterpretation. For CO(v=0,J=35), $k_{3'} \approx 0$ since this level falls on the tail of the v=0 rotational distribution. In this instance the model predicts an inverse pressure dependence for the time required to reach maximum amplitude, which the plot in Fig. 2 verifies. The time constants derived from iterative fitting also show the predicted pressure dependencies.[12b,c]

The time-resolved absorption technique used in our experiments is not as sensitive as MPI or LIF, but it does present one advantage. Distinction between gain and absorption clearly indicates the presence of population inversion. Thus we are assured, at least qualitatively, of the accuracy of our distributions. The possibility of gain raises an important point. Absorption techniques measure population differences. This means that the initial slope of the absorption signal is proportional to the difference of the upper and lower state specific rate constants. By probing successively higher vibrational levels until no further absorption is observed, the nascent rotational distribution of the highest populated vibrational level is directly determined. Working backwards, the difference measurements yield the distributions in lower vibrational levels. It is important to make these corrections, particularly for H+OCS where excitation of CO is observed up to v=4.

Using the initial slope method, CO rotational and vibrational distributions were determined for the title reactions. Distributions for $H+CO_2$ are shown in Fig. 3, while those for H+OCS are summarized in Table I.

The surprisal analysis procedure developed by Levine, Bernstein, and coworkers[14] was used to compare the observed distributions to statistical distributions based on random sampling of the available phase space. The prior distribution is given by

$$P^\circ(v,J) = (2J_{CO}+1) \sum_{v_{XH}=0}^{v^*} (1-f_V^{XH}-f_V^{CO}-f_R^{CO})^{\frac{3}{2}}$$

where the V and R subscripts refer to vibration and rotation, f is the fraction of available energy in the designated degree of freedom, and v^* is the highest energetically accessible vibrational level of the HX fragment. Here we assume continuous functions for translation and rotation, and count vibrations explicitly.[12b] The insets in Fig. 3 show rotational surprisal plots for CO(v=0,1) from H+CO$_2$. The solid lines overlying the rotational distributions represent back-calculated distributions derived from the surprisal analysis. The surprisal parameters listed in Table I summarize the H+OCS results.

Time-resolved sub-Doppler lineshape measurements on the CO(v=0) product from H+OCS indicate substantial partitioning of energy into CM translation. A total of six transitions from J=7 to J=30 were examined: R(7), P(10), R(16), R(19), R(22), and R(30). A 3D plot for R(19) is shown in Fig. 4. Cross sections at 100 ns intervals were fit to Gaussian lineshapes, from which E_{CM} was determined. Since the S/N ratio is initially poor due to the small amount of product present, a preliminary estimation of the nascent E_{CM} was made by fitting a plot of E_{CM} vs time to an exponential function, and extrapolating back to t=0. Such a plot for R(19) is shown in Fig. 4. In all cases the exponential fit appears to have underestimated the nascent CM kinetic energy. We take the average E_{CM} thus determined, 7180 cm^{-1}, as a lower limit for CO(v=0).

4. DISCUSSION

Because of the large difference in mass, it is unlikely that the H atom will simply abstract an O or S atom in passing. While the required slowing of the H atom implies the existence of an intermediate, it does not guarantee statistical distribution of available energy among product degrees of freedom. In the energy regime of these experiments, ≈10,000 cm^{-1} at the transition state, RRKM estimates of reaction rates are on the order of 10^{13} s^{-1}.[15] Since IVR may not be complete on this time scale, the applicability of statistical theory to these experiments is questionable.

4.1. H+CO$_2$ → CO+OH

With a total of 10,000 cm^{-1} available to the products, the average energy channeled into CO internal degrees of freedom, 1500 cm^{-1} (15%), is surprisingly small. The v=0 and 1 rotational distributions are colder than would be predicted statistically, with rotational surprisal parameters of Θ_R=7 and 9, respectively.

Likewise, our measured vibrational distribution is cold with a [v=1]/[v=0] ratio of 0.4. No signal was detected for 3←2 transitions. We conclude that little or no population of v>1 occurs since it is unlikely that the v=1 and 2 populations are equal.

Our results are consistent with those of Weston and coworkers,[16] but disagree with the study of Rice et al. who found a [v=1]/[v=0] ratio of 1.[7] Using VUV LIF, Rice et al. also reported an inverted v=1 rotational distribution peaked at J=23 which should have manifested itself as gain in our experiments: none was observed. While our observations do not quantitatively support those of Rice et al., we both conclude that CO internal excitation is less than statistical.

Previous studies of nascent OH distributions have shown that they too are colder than statistical.[2-4] Using 193 nm photolysis of HBr as the H atom source, an average OH(v=0) rotational energy of 1350 cm^{-1} was determined.[4] Under similar conditions, Chen et al. were able to measure a CM kinetic energy of ≈9800 cm^{-1} for OH(v=0, low J) using sub-Doppler techniques.[17] Although the average energy partitioned into CM translation will be less than 9800 cm^{-1}, all results indicate a propensity for CM translation at the expense of product internal excitation.

Some insight into the dynamics can be gained by examining the HOCO potential energy surface. H and O atom energy contours were generated from analytic functions fit to the surface of Schatz et al..[8] For CO_2 in its equilibrium geometry, a shallow H atom well exists along a C_{2v} approach to the C atom. As the CO_2 molecule is bent and stretched, prominent minima develop in both the cis and trans positions. The geometry shown in Fig. 5 produces the minimum H atom barrier, which is found along a cis approach of the H atom. However, those collisions leading to the distortion of the CO_2 equilibrium geometry shown in Fig. 5 will leave the H atom in the trans position. The resulting dynamical barrier is consistent with the low reaction probability of 0.024 for E_{CM}=2.4 eV,[16] and the strong dependence of reaction probability on collision energy.[12a,17] Similar contours, generated by fixing the H atom at the minima shown in Fig. 5 and moving the O atom, reveal steep gradients forcing the O atom away from the CO.

Evidence for the role of a HOCO intermediate can be found in the oriented precursor work of Zewail and coworkers.[6] They demonstrate that the appearance of OH following photoexcitation of CO_2-HI complexes is delayed by several picoseconds. Insofar as these experiments can be interpreted in terms of H+CO_2 collisions, they demonstrate the importance of a HOCO intermediate. But in this case the effect of the nearby halogen atom on reaction mechanism remains to be established.

4.2. H+OCS → CO+SH

Although the energy in the transition state, 9700 cm^{-1}, is comparable to that for H+CO_2, the total amount available to the products, 15,000 cm^{-1}, is much greater due to a substantial reaction exothermicity of 11.3 kcal mol^{-1}. It is not surprising, then, to find significantly more internal excitation of the CO product, 1920 cm^{-1} (13%), with measurable population in levels up to v=4. Comparison with statistical theory, however, again indicates a strong bias against channeling of energy into CO internal excitation. The distribution of population over vibrational levels is somewhat cold (Boltzmann temperature of 3890 K) with a vibrational surprisal parameter Θ_V=3.6. The rotational surprisal parameters Θ_R offer a striking comparison, increasing monotonically from 12.9 for v=0 to >50 for v=4 (Boltzmann temperatures from 1400 K down to 175 K). The strong dependence of rotational temperature on vibrational quantum number suggests that CO internal excitation is conserved, that is vibrational excitation occurs at the expense of rotational excitation.

In contrast to OH, predissociation precludes the use of LIF for monitoring the SH fragment. However, the technique has been successfully applied to the SD(v=0,1) product following 193 nm photolysis of DBr.[11] Though the results are not directly comparable to the present studies due to the large difference in available energy, 23,000 cm^{-1} vs 15,000 cm^{-1}, the SD rotational distributions were cold and non-statistical in nature.

Our CO(v=0) transient lineshape measurements for six rotational levels between J=7 and 30 yield an average nascent CM translational energy of 7180 cm^{-1}. This value is an upper limit in that S/N limitations prevented similar measurements over the entire ensemble of CO states, i.e. v>0. Nevertheless, it again underscores preferential partitioning of available energy into CM translation at high collision energies.

Conservation of energy leaves a minimum of 5,900 cm^{-1} for SH internal excitation. To the extent that the D+OCS experiments mentioned earlier are relevant, we can expect that most of this energy will appear as product vibration. Intuitively, this is not surprising since the HS bond undergoes a large change in amplitude during the reactive scattering process.

For H+OCS, there is no potential energy surface to aid in the elucidation of reaction mechanism. Fortunately, the results of our experiments provide a number of clues. We begin by considering CO vibrational excitation. Mapping of OCS ν_1 (C-O stretch) harmonic oscillator wave functions onto those of the free CO molecule produces a vibrational distribution of only 1020 K. Hence a simple Franck-Condon analysis based upon a sudden crossing from the OCS to the CO surface will not suffice to explain the observed 3890 K CO vibrational distribution. Energy must be coupled into the CO stretch of the HSCO intermediate during the reaction.

Assuming a short-lived HSCO intermediate (incomplete IVR), two limiting approach geometries present themselves, side-on and end-on. For side-on approach of the H atom, little excitation of the CO stretch would be expected since the force of the collision is directed along an axis normal to the C-O bond. In addition, large impact parameters expected with side-on collisions would introduce substantial orbital angular momentum, which should be conserved as product rotation. In contrast for collinear approach, compression of the S-C and C-O bonds would serve to slow the H atom, ultimately resulting in some CO vibrational excitation. The small impact parameter inherent in the collinear geometry also predisposes the products to little rotational excitation. We conclude that our experiments suggest a collinear geometry in the reactive process.

We note in closing the work of Flynn and coworkers who reached similar conclusions in their inelastic H+OCS scattering studies.[18] They found a negative correlation between vibrational and rotational excitation. Those collisions which produced excitation of the OCS ν_1 mode caused statistically little rotational excitation.

5. CONCLUSION

We have studied reactive scattering of H atoms with CO_2 and OCS at $\approx 10,000$ cm^{-1} above the barrier through state resolved monitoring of the CO product. The CO rotational and vibrational distributions obtained under arrested relaxation conditions are significantly colder than those predicted by statistical theory. Our results, coupled with earlier studies, indicate incomplete
IVR in the HOCO and HSCO intermediates, with preferential partitioning of available energy into CM frame translational degrees of freedom.

6. ACKNOWLEDGEMENT

H.E.C. wishes to thank the Flynn Group for their assistance. In particular, helpful discussion of the data with George Flynn, suggestions concerning experimental technique from Tom Kreutz, and the use of John Herschberger's line search program proved invaluable.

This work was supported by the Army Research Office and the Office of Naval Research.

7. REFERENCES

1. C.R. Quick and J.J. Tiee, Chem. Phys. Lett. 100, 223 (9183).

2. K. Kleinermanns, E. Linnebach, and J. Wolfrum, J. Phys. Chem. 89, 2525 (1985).

3. A. Jacobs, M. Wahl, R. Weller, and J. Wolfrum, Chem. Phys. Lett. 158, 161 (1989).

4. G. Radhakrishnan, S. Buelow, and C. Wittig, J. Chem. Phys. 84, 727 (1986).

5. J. Rice, G. Hoffman, and C. Wittig, J. Chem. Phys. 88, 2841 (1988).

6. N.F. Scherer, L.R. Khundkar, R.B. Bernstein, and A.H. Zewail, J. Chem. Phys. 92, 5239 (1990).

7. (a) J.K. Rice, Y.C. Chung, and A.P. Baronavski, Chem. Phys. Lett. 167, 151 (1990); (b) J.K. Rice and A.P. Baronavski, J. Chem. Phys. 94, 1006 (1991).

8. G.C. Schatz, M.S. Fitzcharles, and L.B. Harding, Faraday Disc. Chem. Soc. 84, 359 (1987).

9. S. Tsunashina, T. Yokota, I. Safarik, H.E. Gunning, and O.P. Strausz, J. Phys. Chem. 79, 775 (1975).

10. J.H. Lee, L.J. Stief, and R.B. Timmons, J. Chem. Phys. 67, 1705 (1977).

11. D. Hauesler, J. Rice, and C. Wittig, J. Chem. Phys. 91, 5413 (1987).

12. (a) S.L. Nickolaisen, H.E. Cartland, and C. Wittig, J. Chem. Phys., submitted for publication; (b) S.L. Nickolaisen, PhD dissertation, University of Southern California (1991); (c) S.L. Nickolaisen and H.E. Cartland, J. Chem. Phys., in preparation.

13. H. Okabe, Photochemistry of Small Molecules (John Wiley & Sons, New York, 1978) p. 165.

14. R.D. Levine and R.B. Bernstein, Acc. Chem. Res. 7, 393 (1974).

15. G. Hoffman, D. Oh, Y. Chen, and C. Wittig, Israel J. Chem. (1989).

16. R.E. Weston, private communication.

17. Y. Chen, G. Hoffman, D. Oh, and C. Wittig, Chem. Phys. Lett. 159, 426 (1989).

18. L. Zhu, J.F. Hershberger, and G.W. Flynn, J. Chem. Phys. 92, 1687 (1990).

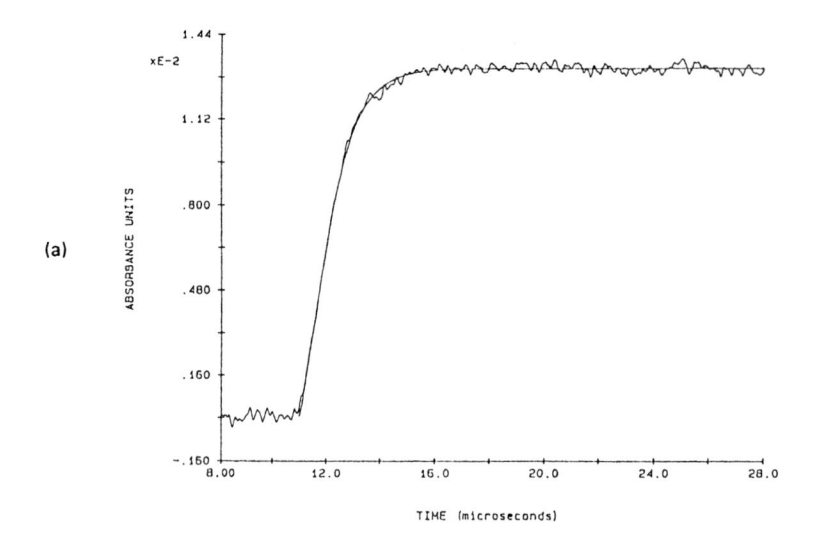

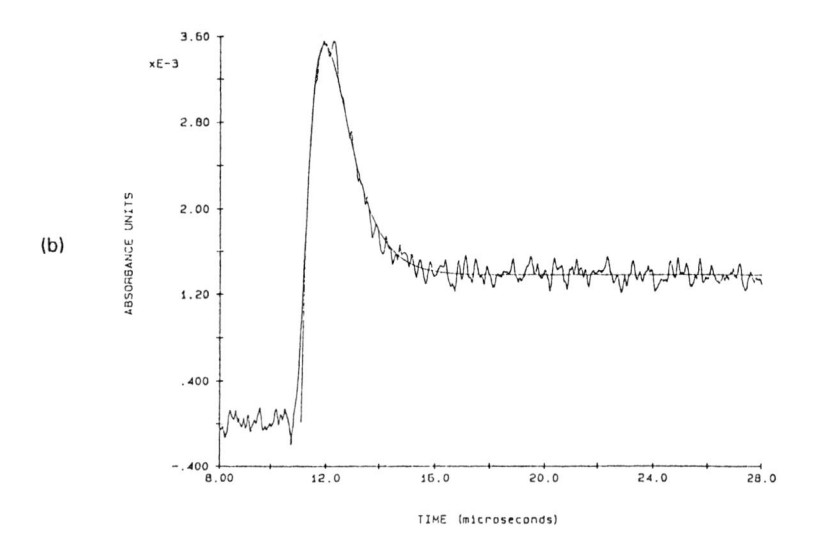

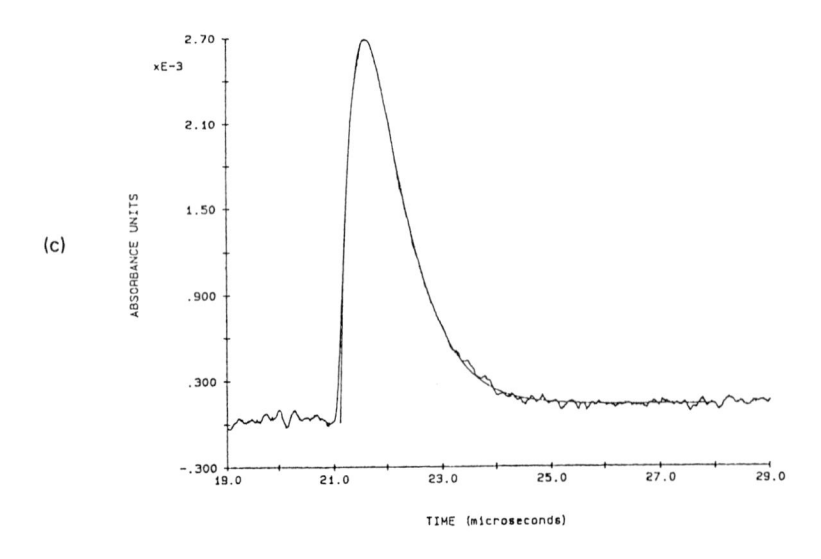

Figure 1. CO(v = 0) transient absorption signals from H + OCS→CO + SH; (a) P(10), (b) R(20), (c) P(35).

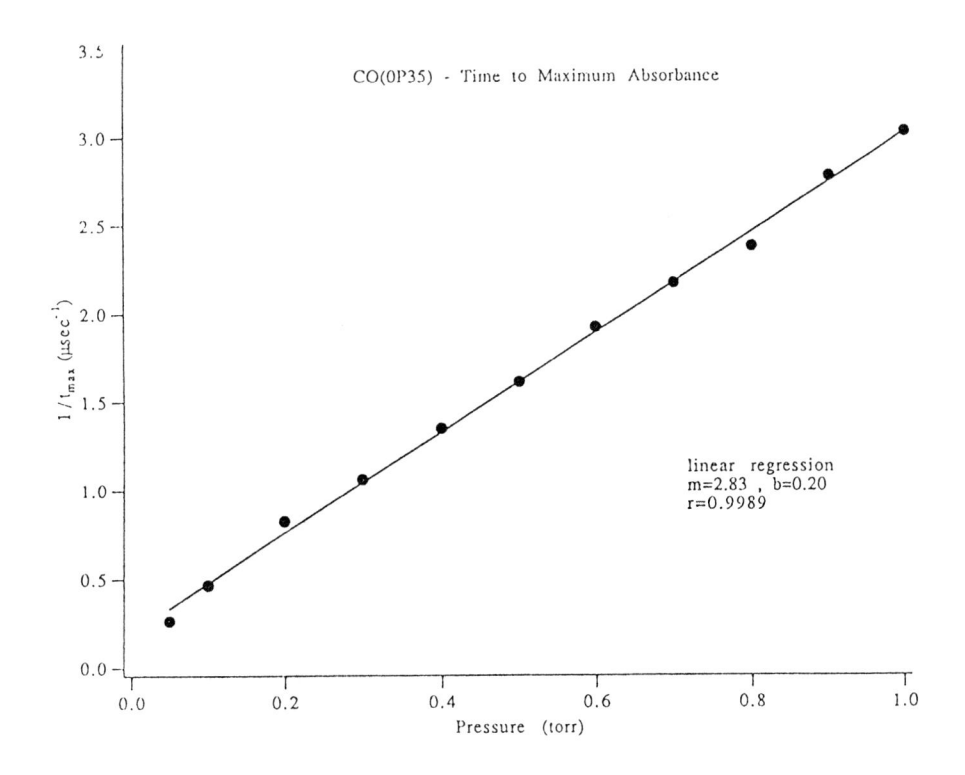

Figure 2. The time to maximum signal amplitude for CO(v = 0, J = 35) from H + OCS → CO + SH shows the pressure dependence predicted by the model.

Table I. Nascent CO product rotational temperatures T_R and surprisal parameters Θ_R from H + OCS → CO + SH. The vibrational temperature $T_V = 3890 \pm 830$ K and surprisal parameter $\Theta_V = 3.6 \pm 1.1$.

CO vibrational quantum number	Boltzmann temperature T_R (K)	Surprisal parameter Θ_R
0	1400 ± 140	12.9 ± 1.4
1	780 ± 45	21.3 ± 1.3
2	590 ± 100	24.0 ± 3.8
3	405 ± 20	28.6 ± 1.5
4	175 ± 20	51.9 ± 5.6

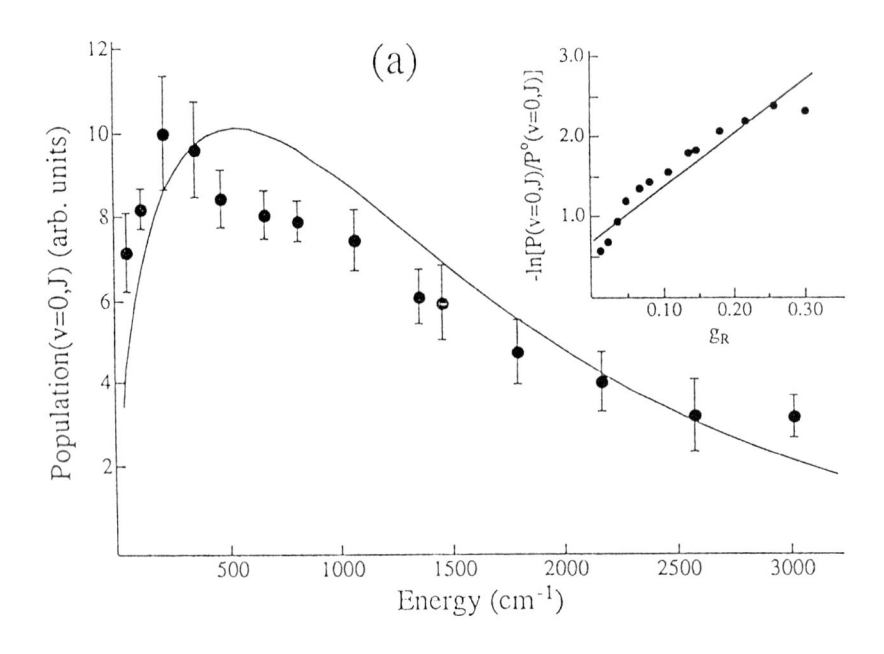

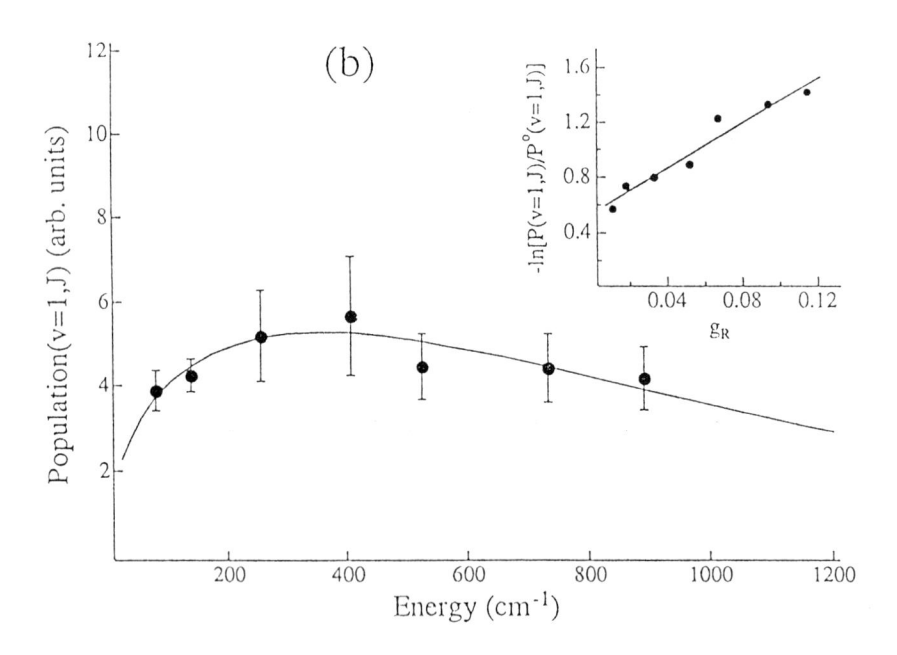

Figure 3. CO rotational distributions from H + CO_2 + CO + OH; (a) v = 0, (b) v = 1. Solid lines are from surprisal analyses (insets); Θ_R = 7 and 9 for v = 0 and 1, respectively.

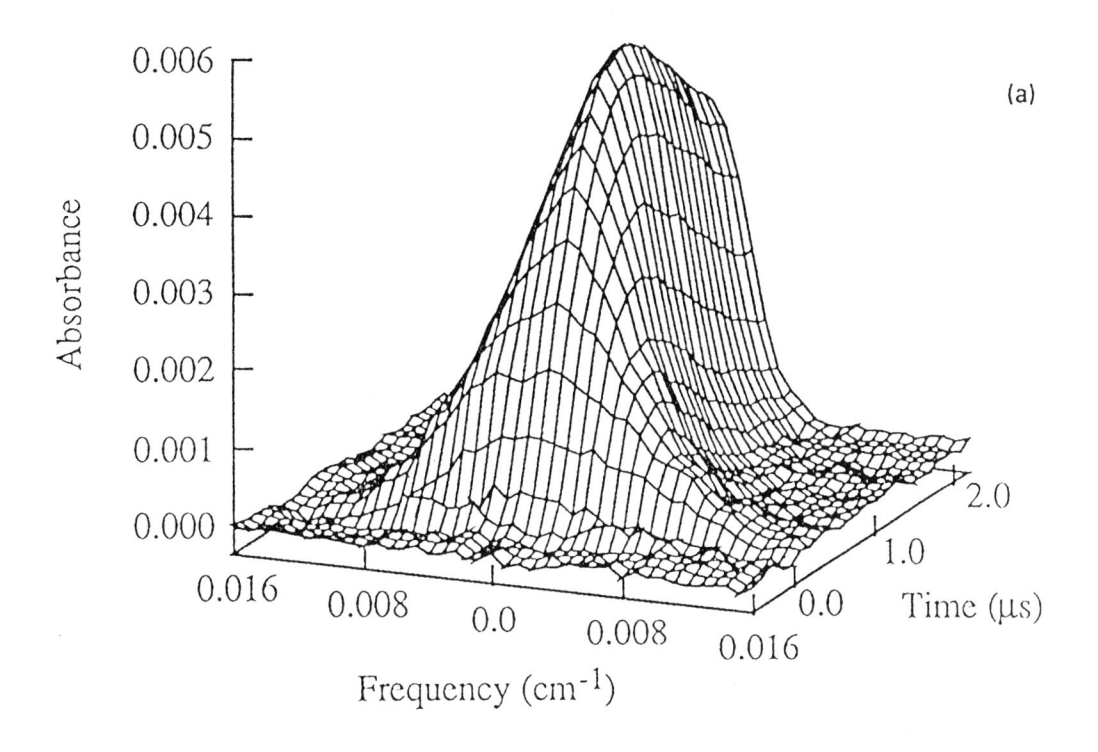

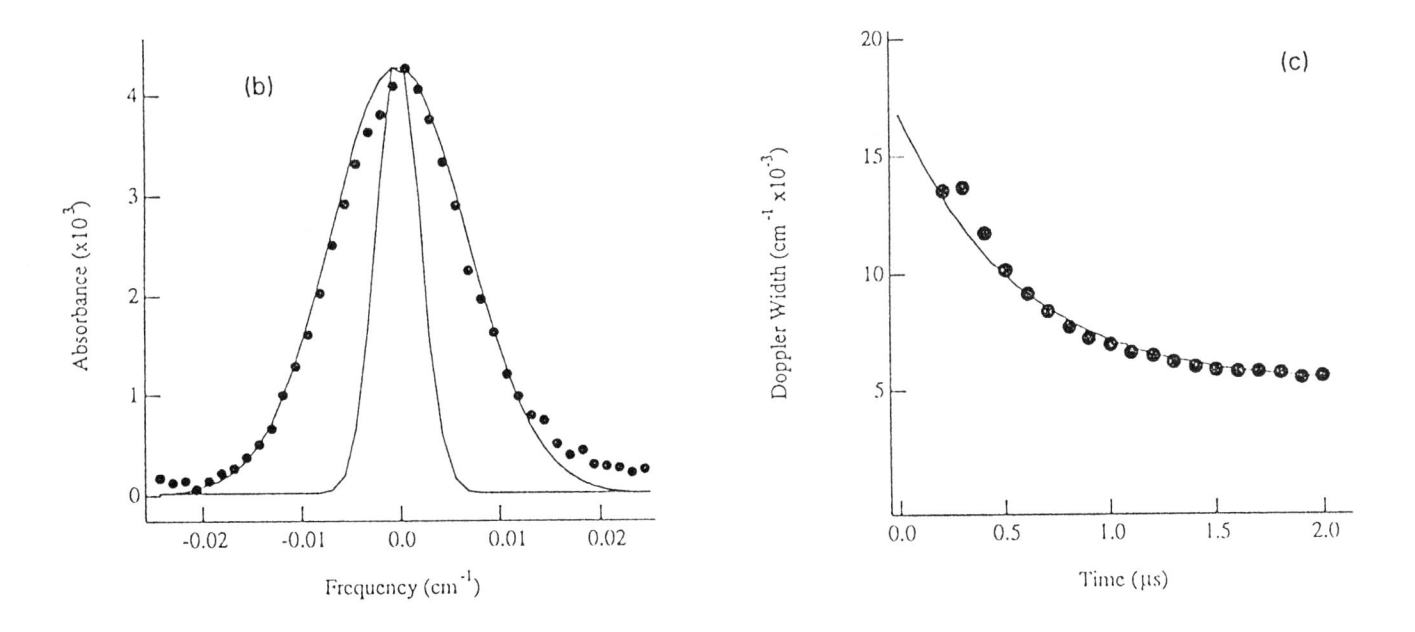

Figure 4. Time-resolved sub-Doppler spectroscopy of CO(v=0, R19) from H+OCS→CO+SH; (a) absorbance as a function of frequency and time, (b) lineshape, Gaussian fit, and 298 K reference lineshape at 500 ns following the photolysis pulse, (c) Doppler width versus time extrapolates to 0.0168 cm^{-1} at t=0.

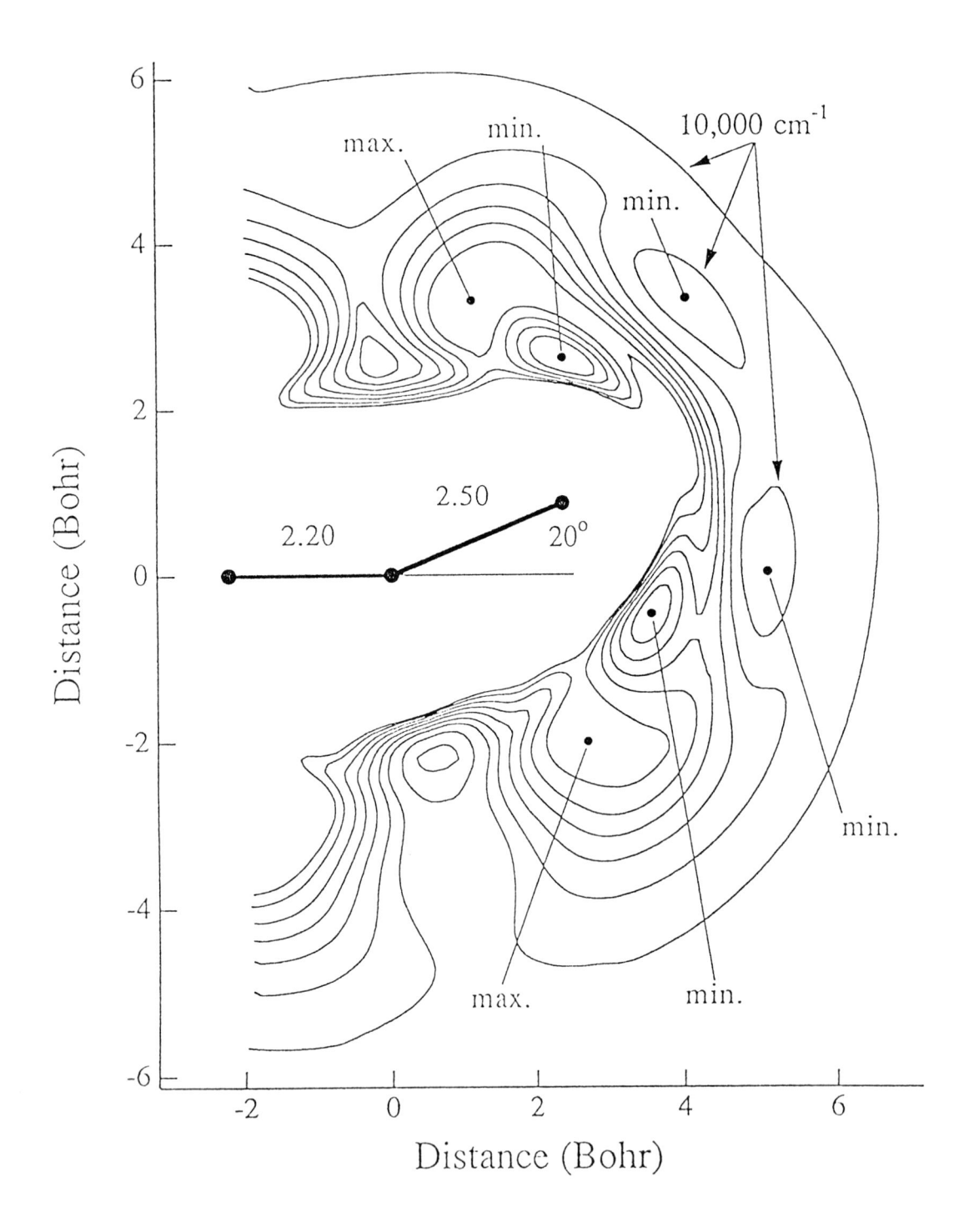

Figure 5. H atom energy contours for H+CO$_2$ at the cis saddle point geometry. Contours are separated by 2500 cm^{-1}; the highest is at 25,000 cm^{-1}.

Chemiluminescent reactions of laser-generated metal atoms with oxidants

Kuo-mei Chen

Department of Chemistry, National Sun Yat-sen University
Kaohsiung, Taiwan, ROC

ABSTRACT

Crossed beam chemical reactions of excimer laser-generated metal atoms (Al, Ba, La, Ce, Gd and Dy) with oxidants are studied under the condition of high collisional energy. Electronically excited metal monoxides are positively identified by their chemiluminescent spectra. From time- and spectrally-resolved emission studies, the dominant role of high speed, ground state metal atoms in these chemiluminescent reactions is revealed. State-resolved excitation functions at high collisional energy are obtained for Ba + O_2 reaction. To the exclusion of Al + O_2 reaction, all the systems display anisotropic angular distributions of chemiluminescence. A Newton diagram analysis of the observed chemiluminescent spatial patterns infers direct reactions with backward scatterings for these metal atom + O_2 reactions at high collisional energy.

1. INTRODUCTION

Chemiluminescent reactions of metal atoms with oxidants have received much attention in the past two decades.[1-4] In particular, the dependence of reaction cross sections upon translational energy of reactants[5-21] and the angular distribution of products[22-27] have provided indispensable evidences to elucidate reaction dynamics of these important elementary reactions. Combining laser vaporization and pulsed nozzle techniques,[28-30] Costes and co-workers[31,32] have studied reactive collisions of carbon and aluminum cooled beams with a variety of oxidants. Wicke and co-workers[33,34] have exploited laser vaporization techniques in their studies on chemiluminescent reactions of Pb and Zn with nitrous oxide. Utilizing similar techniques, Levy[35] has investigated the Mn + N_2O chemiluminescent reaction by time-resolved emission spectrometry.

Chemical reactions of metal atoms with high collisional energy have technological applications in the fields of high-temperature combustion and chemical laser. In this paper, we report the first investigation on crossed beam reactions of excimer laser-generated metal atoms, including Al, Ba, La, Ce, Gd and Dy under the condition of high collisional energy, with a variety of oxidants. In section 2, laser-generated pulsed metal beam techniques, experimental procedures in acquiring time- and spectrally-resolved emission waveforms and the photographic recording of chemiluminescences are described. Experimental results and their implications to reaction dynamics are presented in section 3.

2. EXPERIMENTAL

To perform crossed beam chemical reactions of laser-generated metal atoms with oxidants, a differentially pumped, stainless steel vacuum system which composed of a source and a main chamber, was built. A schematic diagram of the experimental apparatus is depicted in Fig. 1. Inside the source chamber, a pulsed beam value (NRC, BV-100V) was mounted on motor-driven positioning devices such that it was aligned with respect to a skimmer. Without gas loads, the source chamber which was pumped by a turbo molecular pump maintained a 1×10^{-4} mbar pressure, while the main chamber could be evacuated to a 6×10^{-6} mbar pressure by an oil diffusion pump. To improve pumping speed, a liquid nitrogen cold baffle was installed inside the main chamber.

Metal targets were mounted on a platform which was driven to a new position by a stepping motor after each laser shot to avoid deep crater formation. To deplete oxide films, laser ablation techniques were employed in situ to clean the metal surface. By inspecting the abalated grooves on the target surfaces at the end of each experiment, it was evident that oxide formation from reaction with background gases was negligible.

The experimental details for laser-generated pulsed metal beams have been published elsewhere.[36] In essence, photon pulses ($\approx$ 0.15 J/pulse, 15 ns pulse duration) at 248 nm from a KrF excimer laser (Lambda Physik, EMG 53 MSC) were focused onto the metal target along the vertical axis to generate fast atomic species. Pulsed oxidant beams (400 μs pulse duration, 5 bar stagnation pressure), including O_2, N_2O and CO_2, crossed the metal beam perpendicularly at 3.3 cm above the target. The divergent angles of the metal beam were quite large ($\approx$ 45°). On the other hand, pulsed oxidant beam with divergent angles less than 5° was collimated by the skimmer. The delay time of the laser-generated metal beam with respect to the pulsed oxidant beam was searched until chemiluminescent intensities were maximized. A delay generator (LeCroy, 2323A) was utilized to trigger the excimer laser and the beam valve driver properly.

Crossed beam chemiluminescences were optically dispersed by a monochromator (Pacific, MP-1018B) and detected by a thermoelectrically cooled photomultiplier tube (Hamamatsu, R943-02 and C2761). To register chemiluminescent waveforms, a modular, two-channel transient digitizer (LeCroy, 6102 amplifier, 8828c transient recorder and 8104 memory) with a 200 MHz sampling rate capability was employed. The transient digitizer was triggered by the current pulse from a fast phototube (Hamamatsu, R1193U-02) which detected a small portion of the excimer laser pulse by utilizing a beam splitter. Both the laser pulse and the chemiluminescent waveforms were digitized and displayed by the transient digitizer. The reference point of the time frame was set at the raising edge of the laser pulse. Energy fluctuations of excimer laser pulses during the experiment were less than ±2%. When the beam valve was fired, the pressure of the main chamber raised to a level around 5×10^{-5} mbar. Thus, crossed beam reactions could be executed in a single-collision environment and the observed chemiluminescences were from nascent products.

Velocity distributions of pulsed oxidant beams were measured by time-of-flight (TOF) mass spectrometric techniqus.[37] To register the velocity distributions of excited and ground state metal atoms, the time- and spatially-resolved emission and LIF techniques[36] were employed. A schematic diagram of the experimental apparatus is depicted in Fig. 2. Viewing through the slit, arrival times of laser-generated atomic species were registered. From TOF data, one can calculate the velocity and kinetic energy distributions. The convolution formula and the numerical algorithms have been reported[37] and will not be reproduced here. A pulsed tunable dye laser (Lambda Physik LPX 200 and LPD 3000) with coumarin 153 dyes has been utilized in the measurement of velocity distributions of ground state Ba atoms.

To register the spatial distribution patterns of chemiluminescences, photographic recording was chosen in this preliminary study. The shutter of the camera was kept open during the 10 to 20 single-shot experiments in order to reach an acceptable exposure.

The purities of the reagents are: Al (Koch, 99.99%), Ba (Merck, 99.9%), La, Ce, Gd, Dy (Koch, 99.9%) and O_2 N_2O, CO_2 (Matheson, 99.98%).

3. RESULTS AND DISCUSSION

Chemiluminescent spectra of excimer laser-generated metal atoms, including Al, Ba, La, Ce, Gd and Dy, with O_2, N_2O and CO_2 pulsed oxidant beams were obtained according to procedures described in section 2. The spectrally-resolved chemiluminescent waveforms at a chosen wavelength were averaged by the transient digitizer after 10 single-shot experiments. Representative spectra of these chemiluminescent reactions are shown in Figs. 3-8. Chemiluminescent spectra of rare earths Ce, Gd and Dy with oxidants are reported for the first time. The general features of these spectra clearly display vibrational structures. As a matter of fact, transitions can be assigned to emissions from electronically excited metal monoxides according to the term formula of diatomic molecules and their spectroscopic constants.[38] Spectral assignments are given in Table 1 for a representative reaction Ba + O_2. An $A^1\Sigma^+ \rightarrow X^1\Sigma^+$ transition of BaO* is assigned to chemiluminescences in the 400-800 nm spectral region and the tentative vibrational quantum numbers of the A and X states are given for prominent spectral features. It should be reminded that only chemiluminescence along the oxidant beam was effectively imagined onto the slit of a monochromator. It turns out that emitting species with long lifetimes which imply weak emission intensity, can escape the observation zone and defy spectroscopic detection. On the other hand, those long-lifetime emitting species can be detected photographically by adopting a longer exposure time. The BaO ($A'^1\Pi$) products were identified accordingly (see later discussion).

In the present investigation, endothermic reactions $Al(^2P) + O_2(X^3\Sigma_g^-) \rightarrow AlO^* + O(^3P)$ and $Ba(^1S) + O_2(X^3\Sigma_g^-) \rightarrow BaO^* + O(^3P)$ are worthy of a special attention. Energy deficiencies for these two reactions are compensated by either electronic excitations of atomic species or the relative kinetic energy of reactants. Time-resolved chemiluminescence studies are suitable to tackle this problem. Combining information on time- and spectrally-resolved chemiluminescences, time- and spatially-resolved emissions of excited atoms as well as LIF probe on ground state atoms, state-specific excitation function $\sigma(E_T)$, which represents a final-state-resolved, chemiluminescence cross section as a function of relative kinetic energy E_T, can be estimated.

It is well-known that the relative cross section σ is proportional to[15,19,35]

$$\sigma(\lambda, E_T) \propto \frac{I_{CL}(\lambda, t)}{N_{OX} N_M(t) \langle v \rangle}, \tag{1}$$

where $\sigma(\lambda, E_T)$ denotes a final-state-resolved excitation function at emission wavelength λ, $I_{CL}(\lambda, t)$ represents the time-resolved chemiluminescence intensities at wavelength λ, N_{OX} is the number density of the oxidant beam which can be assumed to be a constant for its long pulse duration in the present experiment, $N_M(t)$ is the number density of the excited state atoms or ground state atoms as a function of arrival times and $\langle v \rangle$ is the relative velocity. After calculating σ from eq. (1), σ in the temporal frame t can be transformed into $\sigma(\lambda, E_T)$ as a function of $E_T = \frac{1}{2}\mu \langle v \rangle^2$, where μ is a reduced mass.

In Fig. 9, time-resolved experimental results on Ba + O_2 → BaO* + O reaction are presented. Panel (a) is the TOF data of Rydberg state Ba atoms by a time-, spectrally- and spatially-resolved emission study at 553.5 nm (Ba ^1P → ^1S transition). Panel (b) is the TOF data of ground state Ba atoms by LIF probing techniques at the 553.5 nm resonance line. Panels (c) and (d) are time- and spectrally-resolved BaO* chemiluminescences at 470 and 583 nm, respectively. Judging from the raising edges of these four waveforms, it is quite safe to assume that ultra-high speed ground state and Rydberg state Ba atoms (arrival times <2 μs) can only lead to ionization and dissociation channels such that no chemiluminescence is observed. In the time interval where chemiluminescence intensity raises rapidly, the branching ratio of Rydberg state atoms should still favor the formation of ions and fragments since they have 2-3 eV more energy than ground state atoms with the same speed. Thus, the dominant role of high speed, ground state Ba atoms in Ba + O_2 reaction by laser-generated metal beam techniques can be established. Invoking eq. (1), final-state-resolved excitation functions for Ba(^1S) + O_2($X^3\Sigma_g^-$) → BaO*(A) + O(^3P) reaction at 470 and 583 nm are shown in Fig. 10. State-resolved excitation functions for other metal atom oxidation reactions at high collisional energy can be found be employing similar experimental techniques.

The spatial distribution patterns of chemiluminescences from the metal atom + O_2 reactions have been recorded photographically. Typical results are shown in figs. 11-14. Bluish AlO* chemiluminescence along the O_2 beam is shown vividly in Fig. 11. For collimated Al beam, AlO* chemiluminescence (see Fig. 12) exhibits a more symmetric angular distribution. To the contrary, both BaO* and LaO* display anisotropic angular distribution (see Figs. 13 and 14, respectively).

To analyze the anisotropic angular distribution and the spatial extension of chemiluminescence outside the crossing region, a schematic diagram of the chemiluminescence pattern of Ba + O$_2$ reaction is depicted in Fig. 15 with a dimension scale. It is important to recognize that lifetimes of emitting species, speed of reaction products and spatial extension of chemiluminescence all correlate with each other. Judging from the length of spatial extension of chemiluminescence in Fig. 15, the emitting species which escape outside the crossing region in Ba + O$_2$ reaction should have lifetime on the order of a few μs and speed in the laboratory frame around 2000 m/s. Since the BaO A'$^1\Pi$ state has a 10 μs lifetime,[39] this state can be identified as the desired reaction product.

Assuming that the speed of BaO A'$^1\Pi$ products is around 1500 m/s in the laboratory frame, a Newton diagram for Ba + O$_2$ reaction can be constructed (see Fig. 16). From the Newton diagram and the anisotropic angular distribution, a direct reaction with backward scattering for Ba + O$_2$ reaction can be identified. In principle, Ba + O$_2$ reaction at high collisional energy can lead to the following reaction channels:

$$\begin{aligned} \text{Ba}(^1S) + O_2 &\longrightarrow \text{BaO}(X^1\Sigma^+) + O(^3P) &\text{(2-a)}\\ &\ \text{BaO}(A^1\Sigma) + O(^3P) &\text{(2-b)}\\ &\ \text{BaO}(A'^1\Pi) + O(^3P). &\text{(2-c)} \end{aligned}$$

The ground state product can be probed by LIF techniques and is under investigation in author's laboratory. The non-adiabatic transition at high collisional energy which leads to the formation of A$^1\Sigma$ and A'$^1\Pi$ state products via a direct reaction mechanism should be noted.

4. CONCLUSION

Studies on crossed beam chemical reactions of laser-generated metal atoms with oxidants probe into many unexplored fields of reaction dynamics, especially in the high collisional energy region. The laser vaporization techniques can be applied in generating high speed neutral and ionized species from refractory materials rountinely. Advantages of time-, spectrally- and spatially-resolved experimental techniques have been demonstrated in this work.

Specifically, state-resolved excitation functions at high collisional energy are found for prototypical chemiluminescent reactions. In addition, chemiluminescent reactions of metal atoms with O_2 at high collisional energy can be classified to be a direct mode.

5. ACKNOWLEDGMENT

This research was supported by the National Science Council of the Republic of China.

6. REFERENCES

1. Ch. Ottinger and R.N. Zare, "Crossed Beam Chemiluminescence," *Chem. Phys. Letters*, vol. 5(4), pp. 243-248, April 1970.
2. C.D. Jonah, R.N. Zare and Ch. Ottinger, "Crossed-beam Chemiluminescence Studies of Some Group IIa Metal Oxides," *J. Chem. Phys.*, Vol. 56(1), pp. 263-274, January 1972.
3. M. Menzinger, "Electronic Chemiluminescence," *Adv. Chem. Phys.*, vol. 42, pp. 1-62, 1980.
4. A. Fontijn ed., *Gas-phase Chemiluminescence and Chemi-ionization*, North-Holland, Amsterdam, 1985.
5. R. Dirscherl and K.W. Michel, "Dependence of Reaction Cross Sections of Eu + O_2 on Collisional Energy," *Chem. Phys. Letters*, vol. 43(3), pp. 547-551, November 1976.

6. C.B. Cosmovici, E. D'anna, A. D'innocenzo, G. Leggieri, A. Perrone and R. Dirscherl, "The Reaction Yb + $O_2 \rightarrow$ YbO + O in a Crossed Molecular Beam Experiment," *Chem. Phys. Letters*, vol. 47(2), pp. 241-244, April 1977.

7. D.M. Manos and J.M. Parson, "Chemiluminescent Reactions of Group IIIb Atoms with O_2: Spectral Simulations and Extended Energy Dependence," *J. Chem. Phys.*, vol. 69(1), pp. 231-236, July 1978.

8. A. Siegel and A. Schultz, "Laser-induced Fluorescence Study of Reactions of Ba with HCl and HBr at Different Collision Energies," *J. Chem. Phys.*, vol. 72(11), pp. 6227-6236, June 1980.

9. A. Gupta, D.S. Perry and R.N. Zare, "Effect of Reagent Translation on the Dynamics of the Exothermic Reaction Ba + HF," *J. Chem. Phys.*, vol. 72(11), pp. 6237-6249, June 1980.

10. T. Munakata, Y. Matsumi and T. Kasuya, "Collision Energy Dependence of Vibrational/Rotational Distribution of BaBr Produced in the Crossed Beam Reaction Ba + CH_3Br," *J. Chem. Phys.*, vol. 79(4), pp. 1698-1707, August 1983.

11. H.-J. Meyer, U. Ross and Th. Schulze, "Observation of Activation Energies in the Chemiluminescent Reactions of Sr with Cl_2 and Br_2 at Collision Energies below 4 eV," *J. Chem. Phys.*, vol. 82(6), pp. 2644-2649, March 1985.

12. H. Jalink, F. Harren, D. van den Ende and S. Stolte, "The Influence of the Internal State and Translational Energy of the Molecular Reactant upon the Chemiluminescent Reaction Ba + $N_2O \rightarrow$ BaO* + N_2," *Chem. Phys.*, vol. 108, pp. 391-402, 1986.

13. H. Jalink, D.H. Parker, K.H. Meiwes-Broer and S. Stolte, "Translational Energy Dependence of the Steric Effect: Oriented N_2O + Ba $\rightarrow$ BaO* + N_2 Reaction," *J. Phys. Chem.*, vol. 90, pp. 552-554, 1986.

14. C. Alcaraz, P. de Pujo, J. Cuvellier and J.M. Mestdagh, "Collision Energy Dependence of the Chemiluminescent Reaction: Ba + $N_2O \rightarrow$ BaO + N_2," *J. Chem. Phys.*, vol. 89(4), pp. 1945-1949, August 1988.

15. J. Cuvellier, P. de Pujo, J.M. Mestdagh, P. Meynadier, J.P. Visticot, J. Berlande and A. Binet, "Chemiluminescent Reactions of Electronically Excited Alkaline-earth Atoms. I. Energy Dependence in Ba(6s6p $^1P_1$) + $CO_2 \rightarrow$ BaO (A, A') + CO," *J. Chem. Phys.*, vol. 90(12), pp. 7050-7054, June 1989.

16. J.M. Mestdagh, P. Meynadier, P. de Pujo, O. Sublemontier, J.P. Visticot, C. Alcaraz, J. Berlande and J. Cuvellier, "Energy Dependence of the Chemiluminescent Ba(6s5d $^1D_2$) + $O_2 \rightarrow$ BaO* + O Reaction," *Chem. Phys. Letters*, vol.

164(1), pp. 5-11, December 1989.

17. J.M. Parson and C.C. Fang, "Effects of Initial Electronic and Translational Energy on Chemiluminescent Reactions of Cu with F_2," *J. Chem. Phys.*, vol. 92(8), pp. 4823-4832, April 1990.

18. B.S. Cheong, J.H. Wang, C.C. Fang and J.M. Parson, "Product Vibrational Distributions and Collision Energy Dependence of Chemiluminescent Reactions of Group IVA Elements with O_2, N_2O and NO_2," *J. Chem. Phys.*, vol. 92(8), pp. 4839-4848, April 1990.

19. E. Verdasco and A. González Ureña, "Reaction Cross Section by Time Profile Measurements under Crossed-beam Conditions: The Ca(^3P, ^1D) + $N_2O \to$ CaO* + N_2 Reaction," *J. Chem. Phys.*, vol. 93(1), pp. 428-433, July 1990.

20. Q. Xu, R.S. Mackay, F.J. Aoiz, M.A. Quesada, P.J. Grunberg and R.B. Bernstein, "Measurement of the Translational Energy Dependence of the Cross Section for the Reaction of Sr + $CH_3I \to$ SrI + CH_3 from 0.1-1.0 eV," *Chem. Phys. Letters*, vol. 176(6), pp. 499-503, February 1991.

21. J.P. Visticot, C. Alcaraz, J. Berlande, J. Cuvellier, T. Gustavsson, J.M. Mestdagh, P. Meynadier, P. de Pujo and O. Sublemontier, "Chemiluminescent Reactions of Electronically Excited Alkaline Earth Atoms. II. Energy Dependence in Ba* + $O_2 \to$ BaO* + O," *J. Chem. Phys.*, vol. 94(7), pp. 4913-4920, April 1991.

22. P.J. Dagdigian and R.N. Zare, "Primitive Angular Distribution Studies of Internal States in Crossed-beam Reactions using Laser Fluorescence Detection," *J. Chem. Phys.*, vol. 61(6), pp. 2464-2465, September 1974.

23. G.P. Smith and R.N. Zare, "Angular Distribution of Product Internal States using Laser Fluorescence Detection: The Ba + KCl Reaction," *J. Chem. Phys.*, vol. 64(6), pp. 2632-2640, March 1976.

24. C.A. Mims and J.H. Brophy, "Angular Distributions of Chemiluminescence from Ba + Cl_2*," *J. Chem. Phys.*, vol. 66(3), pp. 1378-1379, February 1977.

25. T.P. Parr, A. Freedman, R. Behrens, Jr. and R.R. Herm, "Crossed Molecular Beams Kinetics: BaO Recoil Velocity Spectra from Ba + N_2O," *J. Chem. Phys.*, vol. 67(5), pp. 2181-2190, September 1977.

26. A. Siegel and A. Schultz, "Direct Observation of the Angular Distribution of the Chemiluminescence from Ba + $N_2O \to$ BaO + N_2," *Chem. Phys.*, vol. 28, pp. 265-271, 1978.

27. R. Kiefer, A. Siegel and A. Schultz, "Two Chemiluminescence Studies of the Reaction Ba + $CCl_4 \to BaCl_2 + CCl_2(A)$," *Chem. Phys. Letters*, vol. 59(2), pp. 298-

302, November 1978.

28. D.E. Powers, S.G. Hansen, M.E. Geusic, A.C. Pulu, J.B. Hopkins, T.G. Dietz, M.A. Duncan, P.R.R. Langridge-Smith and R.E. Smalley, "Supersonic Metal Cluster Beams: Laser Photoionization Studies of Cu_2," *J. Phys. Chem.*, vol. 86, pp. 2556-2560, 1982.

29. J.B. Hopkins, P.R.R. Langridge-Smith, M.D. Morse and R.E. Smalley, "Supersonic Metal Cluster Beams of Refractory Metals: Spectral Investigations of Ultracold Mo_2," *J. Chem. Phys.*, vol. 78(4), pp. 1627-1637, February 1983.

30. D.E. Powers, S.G. Hansen, M.E. Geusic, D.L. Michalopoulos and R.E. Smalley, "Supersonic Copper Clusters," *J. Chem. Phys.*, vol. 78(6), pp. 2866-2881, March 1983.

31. G. Dorthe, M. Costes, C. Naulin and J. Joussot-Dubien, C. Vaucamps and G. Nouchi, "Reactive Scattering using Pulsed Crossed Supersonic Molecular Beams. Example of the C + NO → CN + O and C + N_2O → CN + NO reactions," *J. Chem. Phys.*, vol. 83(6), pp. 3171-3172, September 1985.

32. M. Costes, C. Naulin, G. Dorthe, C. Vaucamps and G. Nouchi, "Dynamics of the Reactions of Aluminium Atoms Studied with Pulsed Crossed Supersonic Molecular Beams," *Faraday Discuss. Chem. Soc.*, vol. 84, pp. 75-87, 1987.

33. B.G. Wicke, S.P. Tang and J.F. Friichtenicht, "Velocity Dependence of the Pb + N_2O → PbO (B, v=0) + N_2 Chemiluminescent Reaction," *Chem. Phys. Letters*, vol. 53(2), pp. 304-308, January 1978.

34. B.G. Wicke, "Dynamics of the Chemiluminescent Oxidation of Zinc Atoms by Nitrous Oxide," *J. Chem. Phys.*, vol. 78(10), pp. 6036-6044, May 1983.

35. M.R. Levy, "Chemiluminescence and Energy Transfer in Mn + N_2O Collisions at High Translational Energy," *J. Phys. Chem.*, vol. 93, pp. 5195-5203, 1989.

36. K. Chen, C. Kuo, C. Lin, T. Tseng, C. Pan, M. Lin, G. Wang and C. Pei, "Characteristics of Excimer Laser-generated Pulsed Na Beams," *Chem. Phys. Letters*, in print.

37. K. Chen, T. Chein and C. Pei, "Convolution of Velocity Distribution in a Pulsed Molecular Beam," *J. Chin. Chem. Soc.*, vol. 36(2), pp. 131-134, 1989.

38. K.P. Huber and G. Herzberg, *Molecular Spectra and Molecular Structure. IV. Constants of Diatomic Molecules*, Van Nostrand Reinhold, New York, 1979.

39. J.G. Pruett and R.N. Zare, "Lifetime-separated Spectroscopy: Observation and Rotational Analysis of the BaO $A'^1\Pi$ state," *J. Chem. Phys.*, vol. 62(6), pp. 2050-2059, March 1975.

Table 1. Vibrational assignments of the $A^1\Sigma^+ - X^1\Sigma^+$ transition of BaO*

wavelength (nm)	(v'', v')	wavelength (nm)	(v'', v')
394	(20, 1)	584	(9, 6)
411	(19, 2)	587	(2, 1)
416	(17, 1)	596	(11, 8)
424	(16, 1)	600	(19, 14)
431	(15, 1)	605	(5, 4)
441	(14, 1)	609	(17, 13)
448	(16, 3)	632	(13, 11)
453	(14, 2)	636	(6, 6)
459	(12, 1)	651	(12, 11)
466	(17, 5)	669	(7, 8)
472	(12, 2)	691	(10, 11)
483	(11, 2)	727	(2, 6)
497	(7, 0)	739	(16, 17)
512	(17, 8)	751	(5, 9)
524	(16, 8)	761	(2, 7)
532	(7, 2)	767	(7, 11)
537	(8, 3)	780	(4, 9)
549	(3, 0)	800	(2, 8)
522	(18, 11)	812	(12, 16)
559	(12, 7)	818	(4, 10)
580	(12, 8)	824	(0, 7)

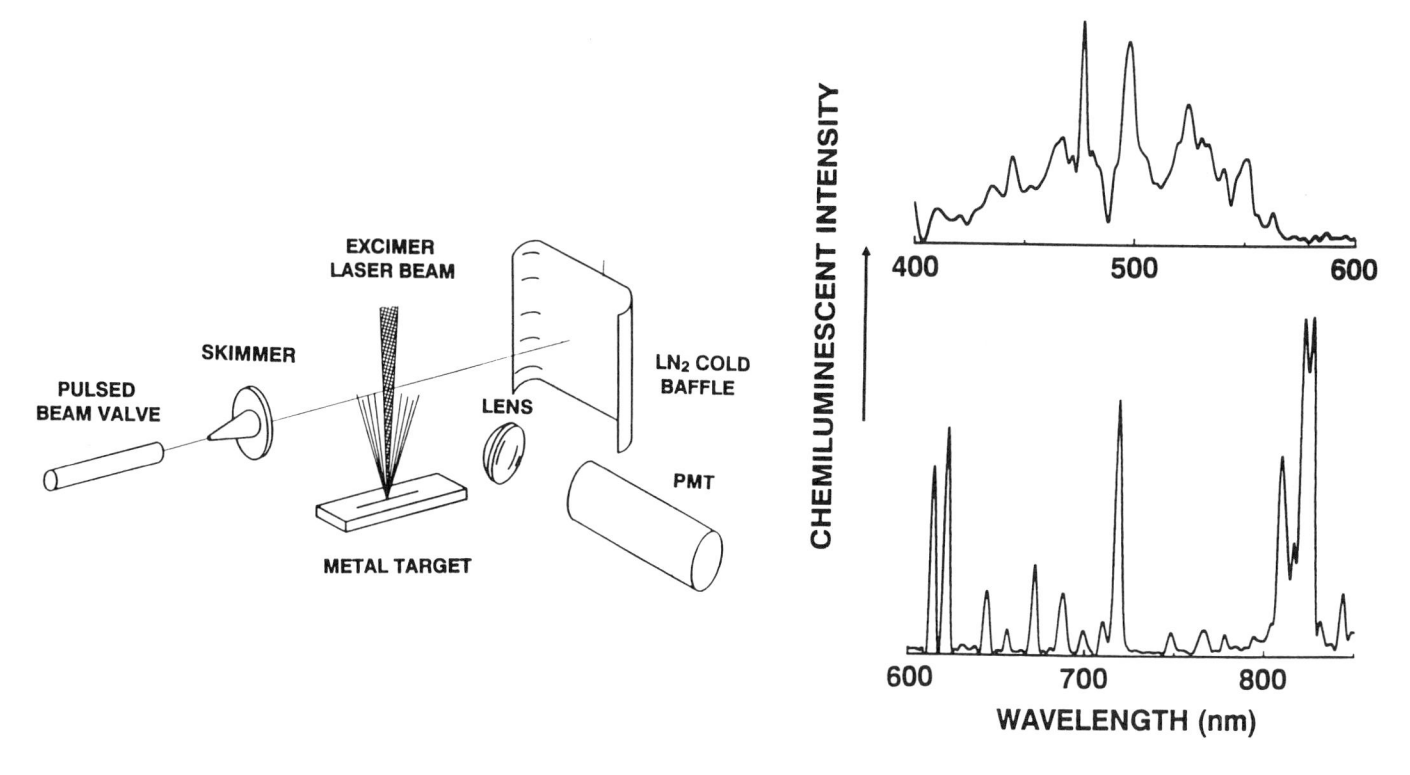

Fig. 1. A shematic diagram of the chemiluminescence experimental apparatus.

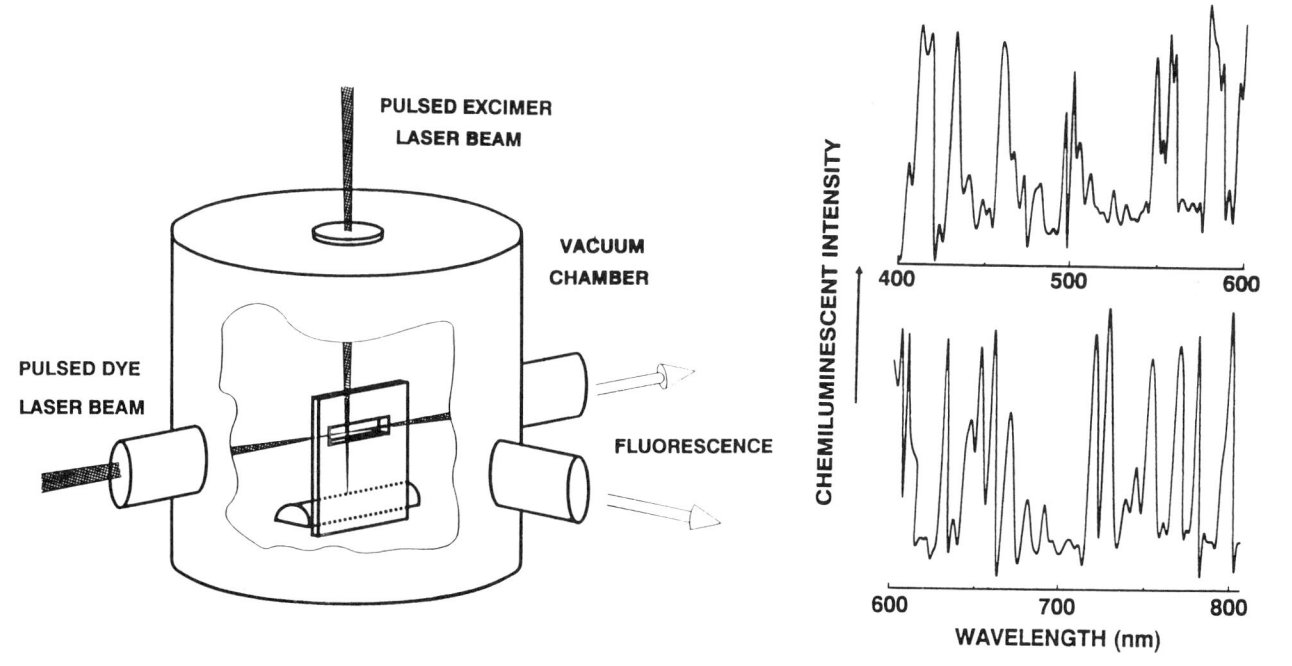

Fig. 2. A schematic diagram of the TOF experiment utilizing spatially-resolved emission and LIF techniques.

Fig. 3. The chemiluminescent spectrum of Al + O_2 reaction.

Fig. 4. The chemiluminescent spectrum of Ba + O_2 reaction.

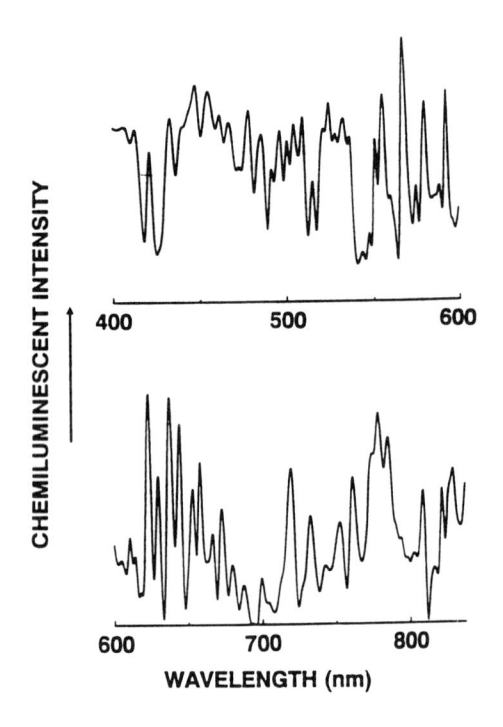

Fig. 5. The chemiluminescent spectra of La + O_2 reaction.

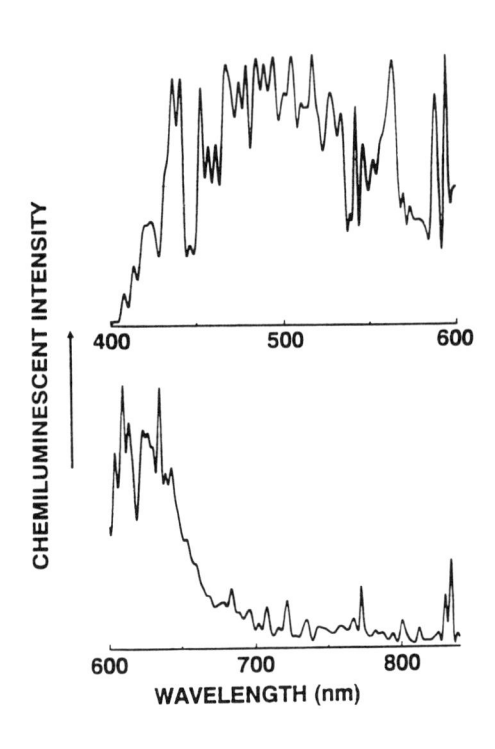

Fig. 7. The chemiluminescent spectrum of Gd + CO_2 reaction.

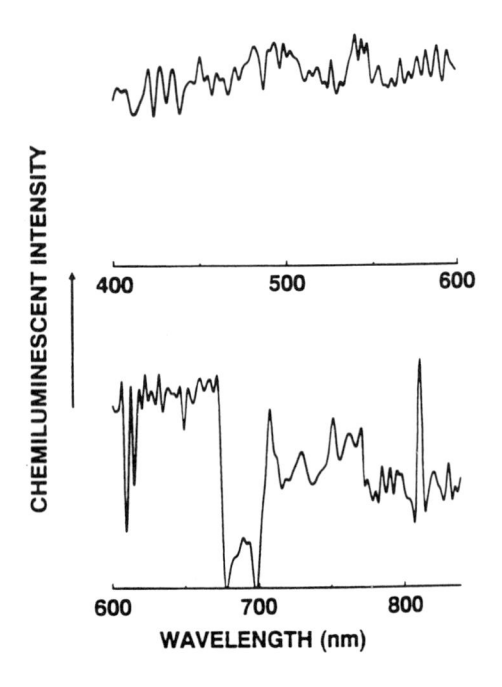

Fig. 6. The chemiluminescent spectrum of Ce + CO_2 reaction.

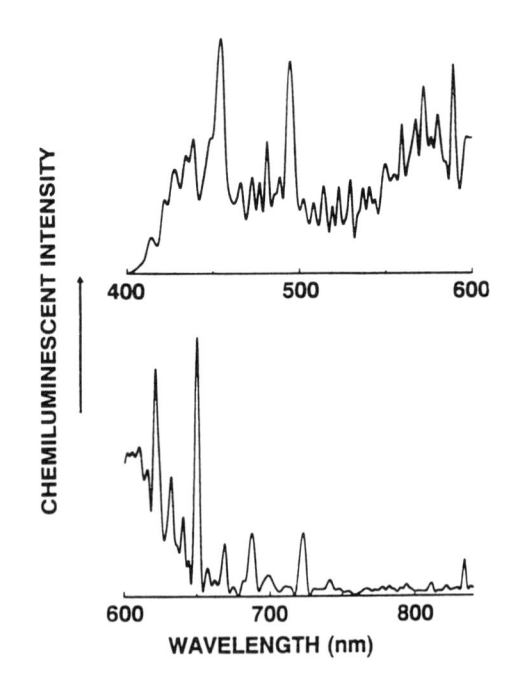

Fig. 8. The chemiluminescent spectrum of Dy + N_2O reaction.

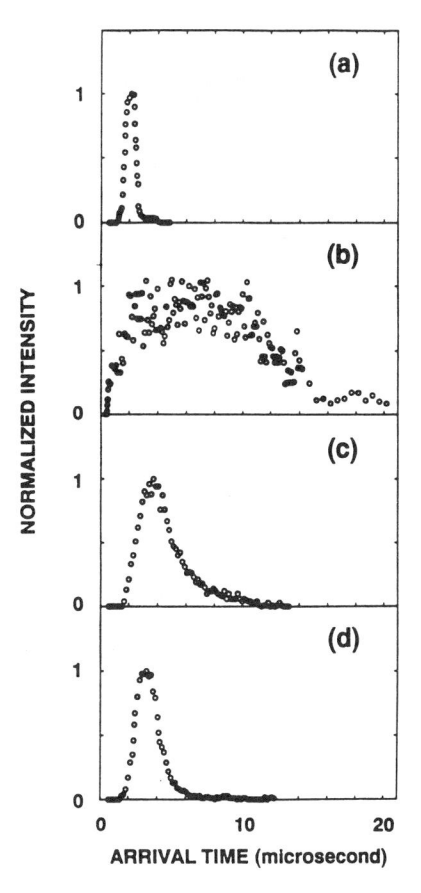

Fig. 9. Time-resolved experimental results of Ba + O_2 reaction. (a). The emission profile of Rydberg state Ba atoms at 553.5 nm. The viewing slit is 3.3 cm above the target. (b). TOF of ground state Ba atoms probed by LIF techniques at 553.5 nm. Absolute LIF signals are two orders of magnitude stronger than emission intensities in (a). (c). BaO* emission profile at 470 nm. (d). BaO* emission profile at 583 nm.

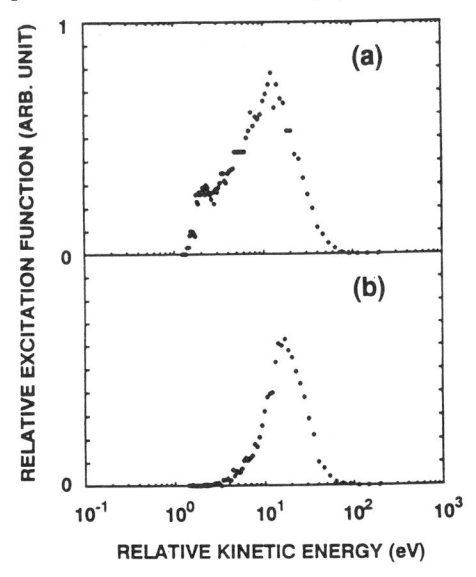

Fig. 10. State-resolved excitation functions of Ba + O_2 reaction at 470 nm (panel (a)) and 583 nm (panel (b)).

Fig. 11. AlO* chemiluminescence from uncollimated Al + O₂ reaction.

Fig. 13. BaO* chemiluminescence from collimated Ba + O₂ reaction.

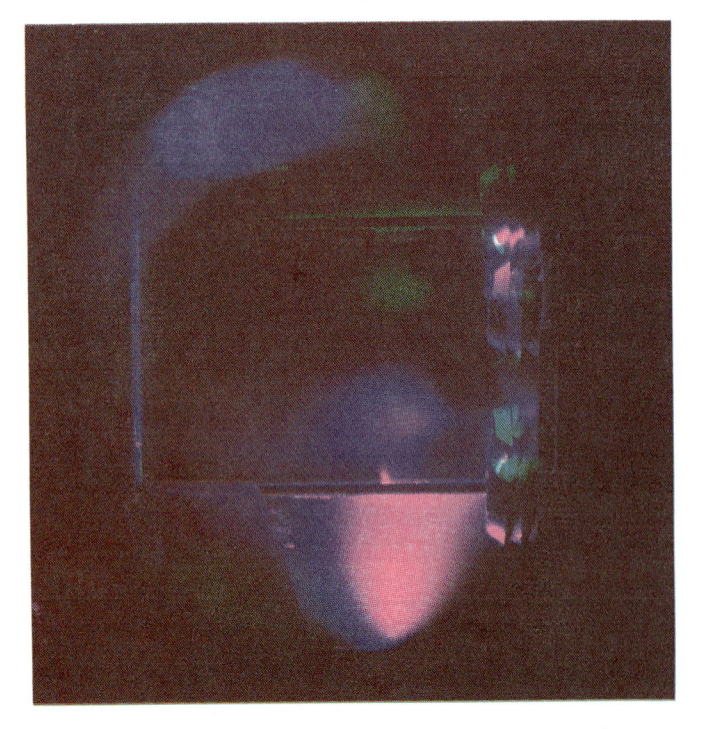

Fig. 12. AlO* chemiluminescence from collimated Al + O₂ reaction.

Fig. 14. LaO* chemiluminescence from collimated La + O₂ reaction.

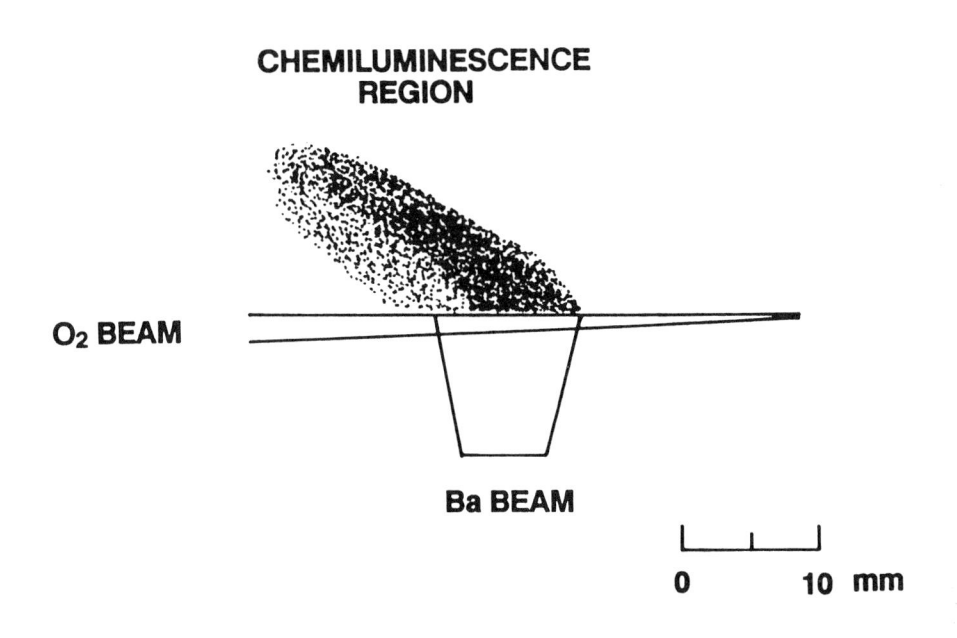

Fig. 15. A schematic diagram of the chemiluminescence spatial pattern of Ba + O$_2$ reaction.

SPECIES	SPEED IN LABORATORY FRAME (m/s)
Ba	12,000
O$_2$	700
BaO	1,500
O	35,000

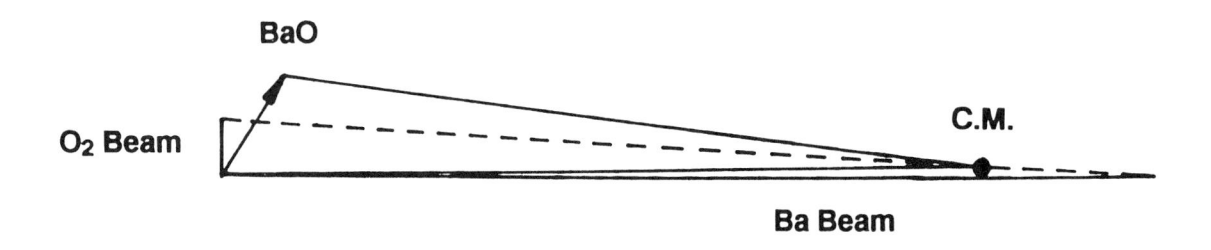

Fig. 16. A Newton diagram for Ba + O$_2$ reaction.

State-to-state Collision Dynamics of Molecular Free Radicals

R. Glen Macdonald and Kopin Liu
Argonne National Laboratory
Chemistry Division
Argonne, Illinois 60439

ABSTRACT

State-to-state collision dynamics of molecular radicals were investigated by the laser-induced fluorescence technique in a pulsed, crossed-beam apparatus. Dramatically different product state distributions were observed for two prototypical radicals, NCO($\widetilde{X}^2\Pi$) and CH($X^2\Pi$). Based on a quantum scattering formalism and general considerations of the potential energy surfaces these observations were interpreted as generic features for the inelastic scattering of $^2\Pi$ radicals. The differences observed for NCO and CH are the results of well-known Hund's coupling classification of linear molecules.

1. INTRODUCTION

The vast majority of chemical reactions have at least one of the reactants and/or products as an open-shell atom or radical. Radical reactivity and energy transfer properties also play a central role in the chemistry of combustion and the atmosphere as well as astrophysics. Despite the importance of radical chemistry to these processes, the experimental and theoretical progress in a detail understanding of the collision dynamics of molecular radicals has been slow. The experimental challenge arises from the transient nature of free radicals, while the theoretical challenge arises because several potential energy surfaces (PES) are involved in describing their collision dynamics.

Radicals are characterized by the existence of unpaired electrons. The unpaired electron endows radicals with fine-structure energy levels induced by the additional couplings due to the unquenched electron orbital and/or spin angular momenta. The spectroscopic consequences of these additional electronic degrees of freedom has long been well-understood. Hund's coupling schemes provide a powerful means of classifying the different spectroscopic features of different radicals. During the past few years, we have been engaged in the systematic study of the collision dynamics of radicals. From these studies, some dynamical consequences of the various Hund's coupling schemes have just begun to emerge. In this paper we present a brief account of these investigations, focussing on one particular feature of the collision dynamics of radicals: How do the product state distributions differ for a Hund's case (a) vs a case (b) Π radical for rotationally inelastic processes ? and Why ?

2. EXPERIMENTAL

The experiments were conducted in a pulsed, crossed-beam apparatus described previously.[1] A cross sectional view of the apparatus is shown in Fig. 1. The NCO radical beam was generated by photolyzing mixtures of C_2N_2(~4%)/O_2(~10%) /N_2(~15%)/H_2(~70%) with ArF laser light at 193 nm directly in front of a pulsed molecular beam nozzle. The photolytically generated CN radical reacted almost completely with O_2 to produce NCO. The NCO was subsequently cooled both vibronically and rotationally to a temperature of about 2 K in the strong supersonic expansion. After colliding with He the product state distribution of NCO($\widetilde{X}^2\Pi_{3/2,1/2},00^10$) was interrogated by the saturated laser-induced fluorescence (LIF) technique using the NCO $(00^00)\widetilde{A}^2\Sigma \leftrightarrow (00^10)\widetilde{X}^2\Pi$ transition near 440 nm.[2]

The CH($X^2\Pi$) radical beam was produced by 193 nm multiphoton dissociation of mixtures of CH_3I(~0.5%)/Xe(~3%) and H_2(~94%) directly in front of the pulsed molecular beam nozzle. After photolysis and the supersonic expansion approximately 93% of the CH radicals were in the lowest N=1 rotational level. The collisionally induced product state distribution was again interrogated by the saturated LIF technique using the CH $A^2\Delta \leftrightarrow ^2\Pi$ transition near 430 nm.[3]

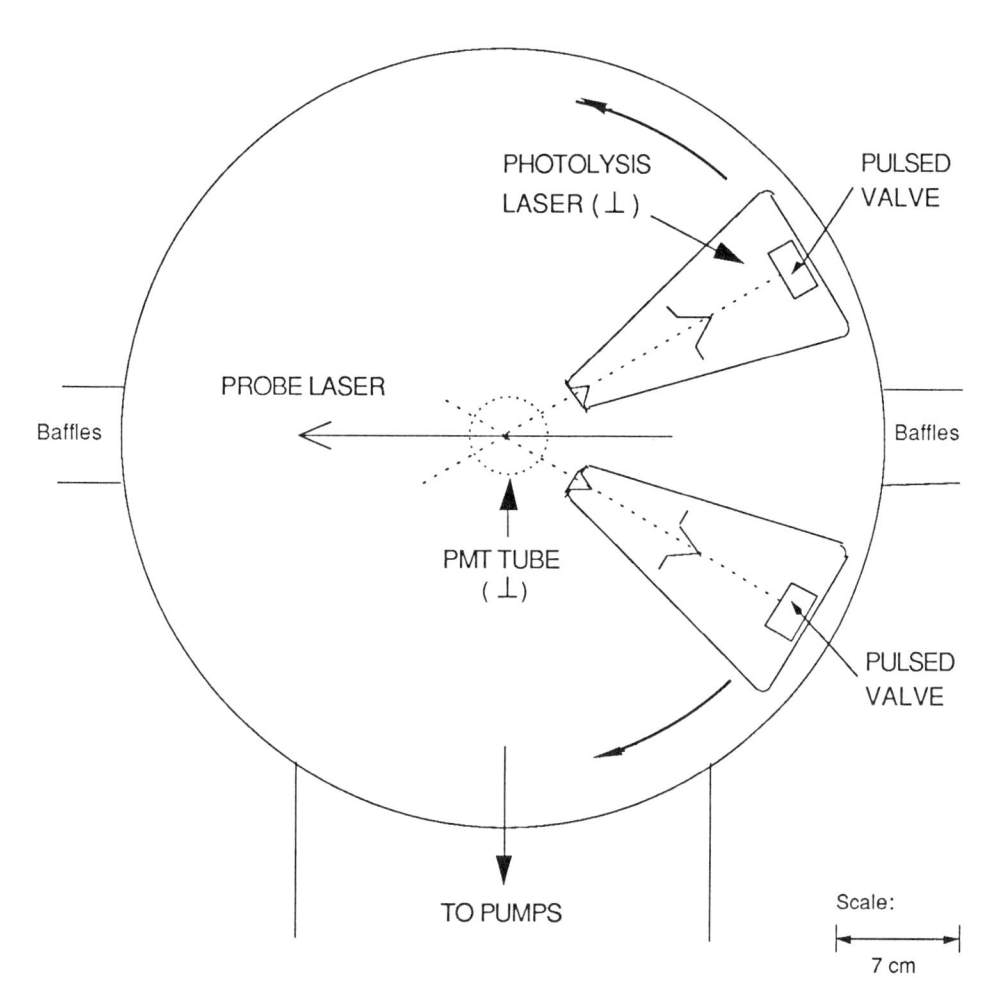

Figure 1. A cross sectional view of the pulsed-molecular beam apparatus. The photomultiplier tube (PMT), recording the LIF, fits in a hollow cylinder in the rotating lid and remains fixed as the beam source rotates. Each pulsed molecular beam source is attached to rotatable lids which cap a cylinder 180 cm long and 60 cm in diameter.

As indicated in Fig. 1, one unique feature of this crossed-beam apparatus is that the state specific detector remains fixed while the two beam sources can be independently rotated; thus, without any change in the expansion conditions, the relative collision energy can be varied merely by changing the intersection angle between the two beams.

3. RESULTS

Figures 2 and 3 show portions of the NCO LIF spectrum for spin-orbit conserving and spin-orbit changing collisions, respectively, at an initial collision energy (E_o) of 3.74 kcal mol^{-1}.[4] The experiments were conducted by scanning the wavelength of the dye laser while the He beam was alternating on and off. Only the difference spectra, indicating the change in NCO state population induced by collisions with He, are shown. The negative signals for $J \leq 4.5$ in Fig. 2 result from attenuation of the primary NCO beam. The spectra correspond to the inelastic scattering process,

$$\text{NCO}(\widetilde{X}^2\Pi_{3/2}, 00^10, J \leq 4.5) + \text{He} \rightarrow \text{NCO}(\widetilde{X}^2\Pi_i, 00^10, J, e/f) + \text{He} \qquad (1)$$

where i=3/2 represents spin-orbit conserving transitions, and i=1/2 corresponds to spin-orbit changing transitions. For spin-orbit conserving transitions, Fig. 2, the distribution shows a monotonic decline with increasing final J, while for spin-orbit changing collisions, Fig. 3, the distribution is bell-shaped, centered around $J \sim 20.5$. Thus, dramatically different product state distributions for these two types of collision processes are anticipated just from a casual inspection of the LIF spectra.

$NCO(^2\Pi_{3/2}) + He \rightarrow NCO(^2\Pi_{3/2}, J, e/f) + He$

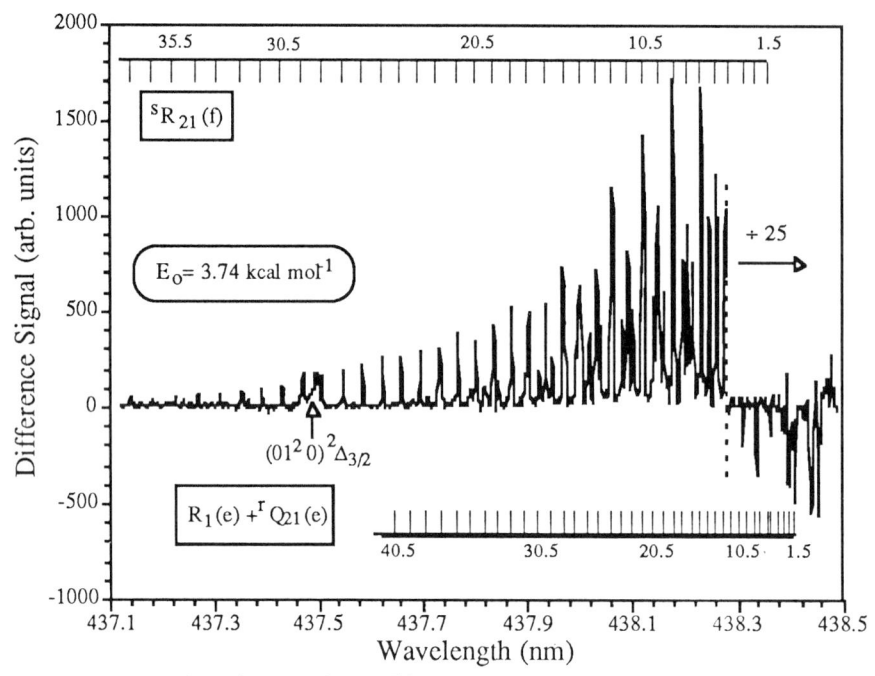

Figure 2. A difference spectrum of the $^SR_{21}$ and $R_1 + {}^rQ_{21}$ branches of NCO, corresponding to spin-orbit conserving collisions at an initial translational energy of 3.74 kcal mol^{-1}. The monotonically declining rotational state distribution is clearly evident. The negative going portion of the spectrum to the right of the dotted line results from attenuation of the lowest J states of NCO.

$NCO(^2\Pi_{3/2}) + He \rightarrow NCO(^2\Pi_{1/2}, J, e) + He$

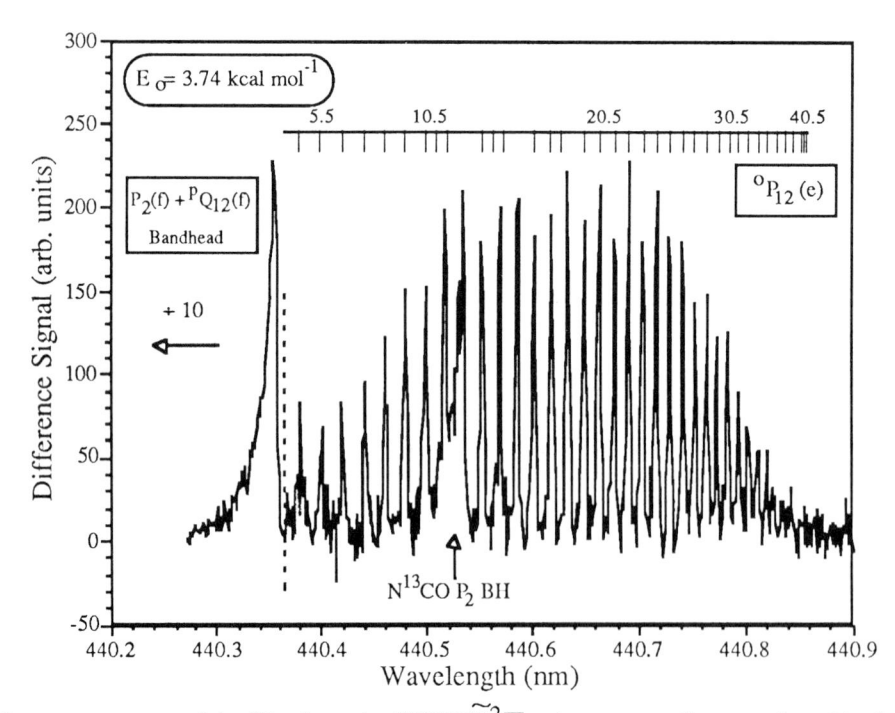

Figure 3. A difference spectrum of the $^OP_{12}$ branch of $NCO(\widetilde{X}^2\Pi_{1/2})$ corresponding to spin-orbit changing collisions at a translational energy of 3.74 kcal mol^{-1}. A bell-shaped rotational state distribution is clearly seen.

This anticipation is quantitatively borne out after the complete data analysis. The final rotational state distributions are shown in Fig. 4 for spin-orbit conserving transitions, by the circles, and for spin-orbit changing transitions, by the squares. All the data shown in Fig. 4 have been normalized to one another. It can readily be estimated that the ratio of total cross section for spin-orbit conserving to spin-orbit changing collisions is about 2.2 at $E_O = 3.74$ kcal mol^{-1}. Yet the most remarkable result is that the two processes yield such dramatically different rotational state distributions. In particular, the low J's population for spin-orbit changing collisions are greatly suppressed. It should be mentioned that the monotonic decline in final rotational population within the $^2\Pi_{3/2}$ manifold is typical for a rotational energy transfer process; however, the bell-shaped distribution in the $^2\Pi_{1/2}$ manifold is quite abnormal.

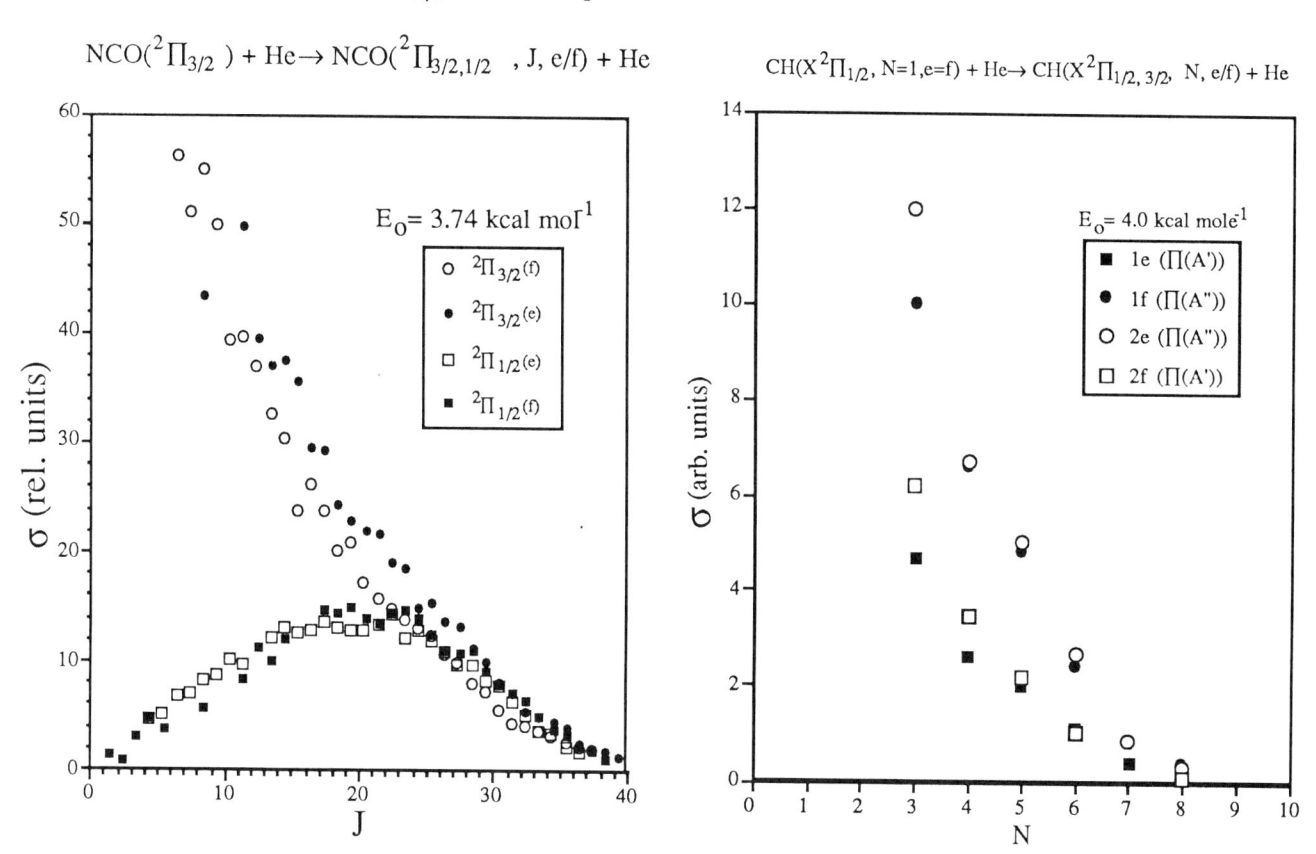

Figure 4. Final rotational state population of NCO($\widetilde{X}^2\Pi_{3/2,1/2}$, 00^{1}0) from process (1) at a collision energy of 3.74 kcal mol^{-1}.

Figure 5. Final rotational state population of CH(X$^2\Pi$) from process (2) at a collision energy of 4.0 kcal mol^{-1}.

Figure 5 shows the results for a similar inelastic scattering process for CH(X$^2\Pi$) + He.[1]

$$CH(X^2\Pi_{1/2}, N=1, e=f) + He \rightarrow CH(X^2\Pi_{1/2,3/2}, N, e/f) + He \qquad (2)$$

As can be seen the product state distribution for a given fine-structure manifold always displays a monotonic decline with increasing final N in this case. However, there appears to be two groups of distributions depending on the nature of the fine-structure states. A comparison of Fig. 4 and 5 clearly shows the sharp contrast in the rotational and fine-structure state distributions for CH + He compared to NCO + He.

4. DISCUSSION

To understand the origin of the dramatically different behavior of the product state distributions between process (1) and (2) displayed in Fig. 4 and 5, it should be first realized that the NCO($\widetilde{X}^2\Pi$) and CH(X$^2\Pi$) radicals are two prototypical examples of different Hund's cases. The NCO radical has a large spin-orbit constant (A = -95 cm^{-1}) and a small rotational constant (B = 0.39 cm^{-1}), i.e. Y = A/B = -244; thus, it is an excellent example of an Hund's case (a) radical. On the other hand, the CH radical exhibits nearly perfect Hund's case (b) character (small Y) with A = 28 cm^{-1} and B = 14 cm^{-1}, i.e. Y = 2. Physically, these two Hund's cases reflect the competition between the coupling of the electron spin angular momentum with the internuclear axis and that with the nuclear rotational angular momentum.

A typical $^2\Pi$ radical has an electronic configuration with an unpaired electron occupying a p orbital. Depending on the orientation of this unpaired p electron with respect to the collision plane, its interactions with a closed shell (^1S) atom are characterized by two non-degenerate PESs of A' (in the plane) or A" (perpendicular to the plane) in C_s symmetry. These two PESs can be regarded as a pair of Renner-Teller surface because in the collinear approach of the atom with the Π radical the A' and A" PESs are exactly degenerate. M. Alexander has formulated a rigorous quantum scattering theory for inelastic collisions between a Π radical and a ^1S atom.[5] In this diabatic representation, the collision occurs on neither $V_{A'}$ or $V_{A''}$ PESs, rather the average potential $1/2(V_{A'} + V_{A''})$ and the difference potential $1/2(V_{A''} - V_{A'})$ are used to describe the collision dynamics. From this formal quantum analysis, a number of general propensity rules have been derived and discussed[5,6] previously and will not be repeated here. Rather, we will outline the underlying physics and applications to the present results.

As usual, the intermolecular interaction is assumed to be purely electrostatic so that a collision can not directly interact with the electron spin of the unpaired electron. A collision can either cause rotational excitation within a given spin-orbit manifold (This interaction is governed by the average potential.) or change both nuclear rotation and the electronic orbital angular momentum motions causing a transition between spin-orbit manifolds and rotational excitation (This interaction is governed by the difference potential.). As shown in Fig. 6, these two different types of collisions represent distinct paths for a Hund's case (a) radical because of the dominance of the spin-orbit interaction. An entirely different situation arises for a Hund's case (b) radical. The strong spin-rotation and weak spin-orbit interaction allows for the decoupling of electron spin from the bond axis. As a result, both the average and difference potentials can contribute to spin-orbit conserving and changing collisions, as depicted in Fig. 6. Furthermore, the existence of two different scattering pathways results in interference of these different scattering amplitudes. One manifestation of this interference is unequal population in the product Λ-doublet states even though initially they were equally populated. For a Hund's case (b) radical Λ-doublet states have well-defined electron orbital alignment with respect to the plane of nuclear rotation; thus, unequal population in Λ-doublet states can be regarded as a preference for orbital alignment from the collision.

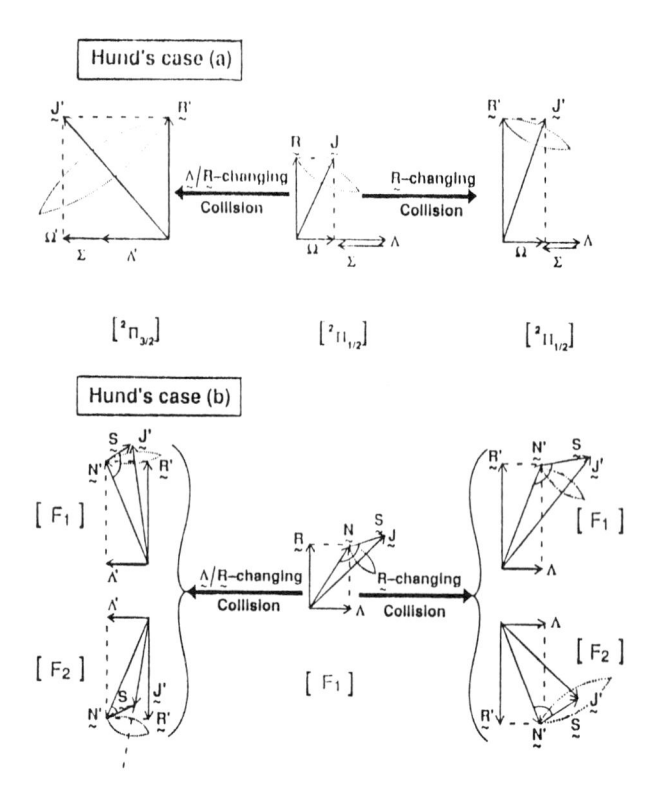

Figure 6. A pictorial representation of angular momenta coupling used to qualitatively visualize the inelastic scattering of a $^2\Pi$ radical and ^1S atom.

In general, the interaction of a molecule with a closed-shell atom is expected to be predominantly repulsive with a weak van der Waal's attraction. The presence of unpaired electrons in radicals will result in more anisotropic interactions. However, for the interaction of a $^2\Pi$ radical and a ^1S atom neither of the adiabatic $V_{A'}$ or $V_{A''}$ PESs are expected to be dramatically different from that expected for closed shell systems. In particular, the average potential is not likely to be very different from that for closed shell systems. For a Hund's case (a) radical such as NCO, as already mentioned, the rotational state distribution for spin-orbit conserving collisions is governed by this average potential and therefore should be similar to those for closed-shell systems, i.e. decline monotonically with increasing ΔJ, as observed in Fig. 4. On the other hand, because $V_{A'}$ and $V_{A''}$ are a pair of mainly repulsive Renner-Teller surfaces, a weak, short-ranged but very anisotropic potential is expected to be the general characteristic of the difference potential. This leads to the reason for the bell-shaped rotational state distribution observed for process (1) as illustrated in Fig. 4. Because the difference potential is short-ranged, only small impact parameter collisions can induce a spin-orbit change. However, large impact parameter collisions are largely responsible for small ΔJ transitions, the very type of collision which experiences only a weak difference potential. Hence the probability of inducing a change in spin-orbit manifold in these types of collisions is small; as a result, small ΔJ transitions are suppressed. Qualitatively, this rationalizes the "missing" population in the low J states for spin-orbit changing, $^2\Pi_{3/2} \rightarrow \,^2\Pi_{1/2}$, collisions as seen in Fig. 4.

As mentioned earlier, the simple correspondence for spin-orbit conserving collisions with the average potential and spin-orbit changing collisions with the difference potential is lost for collisions involving a Hund's case (b) radical, for example, process (2). For this case, because the CH($X^2\Pi$) radical has an electronic configuration of $^1\pi$, the repulsive interaction will be larger on the $V_{A'}$ surface, where the unpaired p orbital electron lies in the triatomic plane, due to an increase in electron repulsion energy. When the singly occupied p orbital is orientated such that it is perpendicular to the triatomic plane, i.e. the $V_{A''}$ PES, the electron repulsion energy is reduced compared to $V_{A'}$. This results in the difference potential, $1/2(V_{A''} - V_{A'})$ being negative. In the repulsive interaction region, where rotational transitions are largely induced, the average potential is always positive. In has been shown[6] that this difference in sign between the average and difference potentials leads to constructive interference for Λ-doublet states F_{2e} and F_{1f} labeled by ($\Pi(A'')$ symmetry), in which the unpaired electron is orientated perpendicular to the plane of rotation, and destructive interference for the other two Λ-doublet states F_{2e} and F_{1e} labeled by ($\Pi(A')$ symmetry) in which the unpaired electron lies in the plane of rotation. Accordingly, the preferential production of Λ-doublet states with symmetry $\Pi(A'')$ is expected for equal population in the initial Λ-doublet states. This is exactly what was observed experimentally as shown in Fig. 5.

Because the simple interpretations presented above are based on a quantum scattering formalism and only general considerations of PESs the observed different collisional behavior should be generic features for the inelastic scattering of Hund's case (a) and case (b) radicals with symmetric collision partners. In this regards, similar distributions though not as dramatic as depicted in Fig. 4 have previously been reported for NO($X^2\Pi$, Y= 73) with noble gas atoms.[7] Furthermore, for an intermediate case radical with a $^3\pi$ electronic configuration such as OH($X^2\Pi$, Y = -7.50), an opposite Λ-doublet preference to that observed for the CH radical can be predicted from the arguments presented above.[6] Indeed, this is exactly what has been observed experimentally for OH + H_2,[8] CO and N_2.[9]

5. SUMMARY and OUTLOOK

In summary, completely different types of collisional behavior have been observed in the inelastic scattering of $^2\Pi$ radicals. We interpreted these observations based on the well known Hund's coupling classification. Extensions of these basic concepts to even more complicated collisional processes such as rovibronic energy transfer and chemical reactions are currently underway. It is hoped that through these systematic investigations a simple classification and better understanding of radical reactivity can eventually be achieved, just as in the case of molecular spectroscopy.

6. ACKNOWLEDGEMENTS

This research was supported by the U.S. Department of Energy, Office of Basic Energy Sciences, Division of Chemical Sciences, under Contract No. W-31-109-ENG-38. Many helpful discussions with A. Wagner are also acknowledged

7.REFERENCES

1. R.G. Macdonald and K. Liu, "State-to-state integral cross sections for the inelastic scattering of CH($X^2\Pi$) + He: Rotational rainbow and orbital alignment," *J. Chem. Phys.* vol. 91, pp. 821-838, July 1989.

2. P.S.H. Boman, J.M. Brown, A. Carrington, I. Koop, and D.A. Ramsay, "A re-investigation of the $A^2\Sigma^+$ - $X^2\Pi_i$ band system of NCO," *Proc. R. Soc. London, Ser A.*, vol. 343, pp. 17- 44, April 1975.

3. I. Botterud, A. Lofthus and L. Veseth, "Term values and molecular parameters for CH and CH^+," *Physica Scrita*, vol. 8, pp. 218-224, 1973.

4. R.G. Macdonald and K. Liu, "Inelastic scattering of $NCO(X^2\Pi)$ + He: Prototypical rotational state distributions for Hund's case (a) radicals?" *J. Phys. Chem.* vol. 95, pp. 9630-9633, Dec. 1991.

5(a). M.H. Alexander,. " Rotationally inelastic collisions between a diatomic molecule in a $^2\Pi$ electronic state and a structureless target," *J. Chem. Phys.* vol. 76, pp. 5974-5988, June 1982; 5(b) M.H. Alexander, "Quantum treatment of rotationally inelastic collisions involving molecules in Π electronic states: New derivation of the coupling potential," *Chem. Phys.* vol.92, pp. 337-344, 1985.

6 P. J. Dagdigian, M.H. Alexander, and K. Liu, "The inelastic scattering of $^2\Pi$ [case (b)] molecules and an understanding of the differing Λ doublet propensities for molecules of π^1 vs π^3 orbital occupancy," *J. Chem. Phys.* vol. 91, pp. 839-848, July 1989.

7. H. Joswig, P. Andresen, and R. Schinke, "Electronic fine structure transitions and rotational excitation in NO rare gas collisions," *J. Chem. Phys.* vol. 85, pp. 1904-1914, August 1986.

8. P. Andresen, N. Aristov, V. Beushausen, D. Hausler, and H.W. Lulf, "Λ-doublet substate specific investigation of rotational and fine structure transitions in collisions of OH with H_2 and D_2," *J. Chem. Phys.* vol. 95, pp. 5763-5774, October 1991.

9. D.M. Sonnenfroh, R.G. Macdonald and K.Liu, "A crossed-beam study of the state-resolved integral cross sections for the inelastic scattering of $OH(X^2\Pi)$ with CO and N_2," *J. Chem. Phys.* vol. 94, pp. 6508-6518, May 1991.

State-selected ion-neutral reactive scattering

James E. Pollard

The Aerospace Corporation, P.O. Box 92957, M5/754
Los Angeles, California 90009

ABSTRACT

Molecular beam reactive scattering experiments are performed with H2+ prepared in selected vibrational state distributions via resonantly-enhanced multiphoton ionization. The technique allows a systematic exploration of the effects of reactant energy on product angular and energy distributions. Cross sections for the exoergic reaction H2+ + H2 → H3+ + H, differential in scattering angle and recoil energy, are measured at c.m. collision energies of 1-5 eV. The endoergic reaction He + H2+ → HeH+ + H is investigated at c.m. collision energies of 0.3-1.9 eV.

1. INTRODUCTION

During the last three decades molecular beam scattering has been one of the most valuable experimental methods for learning about bimolecular reactions. Previous investigations of ion-neutral reactive scattering have measured integral and differential cross sections as functions of collision energy. Reactant state-selection for integral cross section measurements has been achieved using photoionization techniques, but differential cross section studies have relied primarily on electron-bombardment ion sources that give a broad distribution of reactant internal energies and yield less definitive tests of theory.

We have developed a scattering technique[1-4] that uses resonantly-enhanced multiphoton ionization (REMPI) to prepare H2+ reactants in selected vibrational state distributions. The goal is to observe how reactant translational and vibrational energy influence the product recoil energy and angular distributions. For the exoergic proton-transfer reaction,

$$H_2^+ + H_2 \rightarrow H_3^+ + H, \quad \Delta E = -1.72 \text{ eV}, \qquad (1)$$

we have measured the differential cross section vs. scattering angle θ and recoil energy E_r at c.m. collision energies E_c = 1.5, 2.3, 3.5, and 5.3 eV. For the endoergic proton-transfer reaction,

$$He + H_2^+ \rightarrow HeH^+ + H, \quad \Delta E = +0.81 \text{ eV}, \qquad (2)$$

we have measured the differential cross section vs. recoil energy at c.m. scattering angles of θ = 0° and 180° for collision energies near threshold (E_c = 0.3-1.9 eV). The results provide insight on reaction mechanisms and product energy disposal, and will allow quantitative comparisons with theoretical predictions for these three-electron ion-neutral reactions.

2. EXPERIMENTAL METHODS

The apparatus shown in Fig.1 produces state-selected H2+ by 3+1 REMPI within a collimated H2 molecular beam. A 10-Hz Nd:YAG-pumped frequency-doubled dye laser excites the three-photon Q(1) resonances,

$$C^1\Pi_u(v'=0-2, J'=1) \leftarrow X^1\Sigma_g^+(v''=0, J''=1), \qquad (3)$$

in the wavelength range 288-303 nm and a fourth photon of the same color serves for ionization. A typical REMPI spectrum is shown in Fig.2. The ion vibrational and rotational state distributions are known from previously measured photoelectron spectra. Although this REMPI scheme has relatively high ionization efficiency, the state-selectivity is sufficient only for scattering experiments with the first three vibrational states of H2+.

Photoions are impulsively accelerated by a grid assembly that rotates about an axis perpendicular to the neutral beam and coaxial with the laser beam. One grid is held at ground potential, while a positive voltage pulse is applied to the other grid. Precise control of the H2+ velocity (and hence, the collision energy) is achieved by varying the amplitude in the range 100-600 V. The pulse is approximately triangular with a FWHM of 10 ns, and the leading edge occurs within 5 ns after the photoions are formed. A short pulse-duration is needed to ensure that very few collisions occur during the time that the electric field is applied.

Reactant ion velocity distributions are determined from the drift times measured with a microchannel plate detector (MCP) located above the interaction region. Collision energy resolution is controlled by the spread in H2+ velocities due to space-charge repulsion within the the ionization volume. The typical resolution of $\Delta E_c/E_c$ = 15%-18% corresponds to about 20,000 reactant ions per pulse. Laser power density at the ionization region is adjusted with a 500-mm focal-length lens for a compromise between energy resolution and product ion signal.

Product ions are selected by a quadrupole (QP) mass filter and counted with a conversion-dynode electron multiplier. The QP is positioned 60° from the neutral beam axis, in the vertical plane defined by the intersecting reactant ion and neutral beams. After leaving the interaction region, product ions drift through a field-free region, and are accelerated on passing through two grids at the QP entrance. The rods and housing are floated at −20 V relative to ground, which achieves a compromise between transmission and velocity resolution.

Angular distributions of product ions are measured by rotating the initial reactant ion velocity vector with the variable-angle impulsive electric field. Because the ionic and neutral reactants are stationary relative to each other prior to the acceleration pulse, the collision energy remains constant during the angular scan, even though the reactant ion velocity in the laboratory frame changes with grid angle. A complication of this approach is that the number of reactions per laser shot varies due to the changing length of the collision region. In addition one must account for the changing attenuations of the neutral beam and of the product ions as they pass through the grids at varying angles. Correction functions for all these effects have been measured and applied to the data. For experiments on H2+ + H2 the rotating grid assembly is driven by a stepping motor through a set of 20-26 programmed angles, pausing at each angle to record a product velocity spectrum for 1000 laser shots. A complete data set at a given reactant energy is typically the sum of 20 such angular scans. The maximum H3+ count rate is about 2 Hz at the peak of the angular distribution.

Due to the small cross section for He + H2+ (0.1-0.5 Å^2) an extended scattering path is required for the signal to be detectable. This is achieved by rotating the reactant ion velocity vector to within 2° of the neutral beam axis, as indicated in Fig.3. Collisions take place over the 80-mm distance between where the ions are formed and where they emerge from the neutral beam. The mass spectrometer is positioned to view the interaction region along the direction of the H2+ laboratory velocity. As shown by the Newton diagram in Fig.4, the viewing axis is nearly collinear with the reactant relative velocity axis, which allows the differential cross section at c.m. angles $\theta = 0°$ and 180° to be measured from product velocity spectra at this one laboratory angle. We designate $\theta = 0°$ as the c.m. velocity vector of the incident He. Thus the "fast" HeH+ in the laboratory frame corresponds to $\theta = 180°$, and the "slow" HeH+ corresponds to $\theta = 0°$. Velocity resolution is sufficient to separate the fast and slow components of the HeH+ signal but does not allow resolution of product vibrational states. The HeH+ count rate is rather low (0.1-0.5 Hz), but the ratio of signal to time-correlated background is nevertheless greater than 100:1.

3. RESULTS

Product flux measured by a detector subtending a constant solid angle in the laboratory frame is I_{lab}; flux measured by a detector of constant solid angle in the c.m. frame is I_{cm}. Transformation between these quantities and the Cartesian flux distribution I_{car} is accomplished using

$$I_{\text{lab}}(v,\Theta)/v^2 = I_{\text{car}}(u,\theta) = I_{\text{cm}}(u,\theta)/u^2. \qquad (4)$$

Solid-angle differential cross sections with respect to c.m. velocity u, c.m. angle θ, and product c.m. recoil energy E_r are related to the Cartesian flux distribution as follows:

$$d^2\sigma/d\omega du \propto I_{\text{cm}}(u,\theta) = I_{\text{car}}(u,\theta)\, u^2, \qquad (5)$$

$$d^2\sigma/d\omega dE_r = d^2\sigma/d\omega du\, |du/dE_r| \propto I_{\text{car}}(u,\theta)\, u, \qquad (6)$$

$$d\sigma/d\omega = \int_0^\infty (d^2\sigma/d\omega du)\, du \propto \int_0^\infty I_{\text{car}}(u,\theta)\, u^2\, du, \qquad (7)$$

$$d\sigma/dE_r = \int_0^\pi (d^2\sigma/d\theta dE_r)\, d\theta \propto u \int_0^\pi I_{\text{car}}(u,\theta)\, \sin\theta\, d\theta. \qquad (8)$$

3.1 $H_2^+ + H_2 \rightarrow H_3^+ + H$

The Cartesian flux distribution and differential cross sections for a typical data set are shown in Figs.5-8. The incident H2+ velocity vector defines the c.m. scattering angle $\theta = 0°$. Product ions appear in forward and backward scattered peaks whose positions and intensities vary with the choice of collision energy E_c and REMPI transition v'. Counting statistics in the forward hemisphere are generally 3-5 times better than for a peak of comparable intensity in the backward hemisphere. Coverage of product velocity space is sufficiently complete to allow the differential cross sections to be put on an absolute scale by normalizing to published integral cross sections. The H2+ + H2 reaction is one of the few for which the translational and vibrational dependence is known accurately enough to allow this procedure.

Forward/backward symmetry is seen in some of the data sets and is attributed to competing direct mechanisms rather than to long-lived complex formation, based on previous experiments with isotopic variants of H2+ + H2. We assume that H3+ is formed by a direct mechanism in which bond rupture can occur at either reactant if there is rapid equilibration of charge states. The nominal designation AT (atom transfer) refers to a collision in which two of the three nuclei in the product ion come from the reactant traveling toward $\theta = 0°$; the nominal designation PT (proton transfer) applies when two of the three nuclei come from the reactant traveling toward $\theta = 180°$. Assuming a direct mechanism, "stripping" encounters at large impact parameter favor forward scattered product from AT ($\theta \leq 90°$) and backward scattered product from PT ($\theta \geq 90°$).

Angular distributions (Fig.7) show that backward scattering (PT) falls off faster than forward scattering (AT) as E_c increases. This is replicated by trajectory calculations on H2+ + H2, but disagrees with what is seen in integral cross sections measured by isotopic substitution. Perhaps the competition between H3+ formation and collision-induced dissociation (CID) is affected enough by isotopic substitution to change the qualitative direction of this trend. Backward scattering (PT) in H2+ + H2 exhibits more vibrational enhancement than forward scattering (AT) at a given E_c, which agrees with both the trajectory calculations and the isotopic cross sections. Further efforts in both theory and experiment will probably be needed to produce a fully consistent picture.

A major trend in the energy distributions (Fig.8) is that the average recoil energy $<E_r>$ increases in proportion to E_c and is more-or-less independent of reactant vibration. The reaction is translationally endoergic despite the overall exoergicity. The fraction of available energy appearing as product recoil ranges from 31% to 39% for v'=0, from 28% to 38% for v'=1, and from 26% to 38% for v'=2. For the data sets recorded at reactant energies above the CID threshold (2.65 eV), a significant portion of the product ions appears at c.m. velocities corresponding to H3+ internal energies greater than the dissociation limit. Metastable ions must survive for at least 20 μs to be detected in this experiment. Theory shows that H3+ states at high total angular momentum can be classically bound with an internal energy up to 1.5 eV above the dissociation limit.

3.2 He + H_2^+ → HeH$^+$ + H

Cartesian flux distributions measured for He + H2+ are given in Fig.9, showing the two peaks corresponding to HeH+ scattering at $\theta = 0°$ and 180°. Fitting the data using a convolution model to account for instrumental averaging yields the differential cross section vs. recoil energy shown in

Fig.10. For convenience, negative and positive values of E_r are used to label the components at $\theta = 0°$ and $180°$ (all recoil energies are actually positive). An increase in reactant vibrational energy enhances the total cross section of this endoergic reaction and promotes scattering of HeH+ at $\theta = 0°$ (the direction of the incident He) through reactive encounters at large impact parameter. Vibrationally cold reactants require collisions at smaller impact parameters to reach the HeH+ product channel, which gives a more symmetric forward-backward distribution.

The dependence of the average recoil energy $\langle E_r \rangle$ on reactant translational and vibrational energy is shown in Fig.11, along with a dashed line defining spectator-stripping kinematics ($E_r/E_c = 2/5$). The measured recoil energies for $\theta = 0°$ are generally 0.1-0.2 eV lower than the dashed line, meaning that less of the reactant translational energy appears as product recoil than predicted by this simple model. The difference arises because energy is needed to overcome the 0.81 eV endoergicity and because the product can be vibrationally and rotationally excited. For $v'=2$ at $E_c = 1.5$ and 1.8 eV the measured $\langle E_r \rangle$ at $\theta = 0°$ closely matches the stripping value. Recoil energy is generally 10%-30% lower at $\theta = 180°$ than at $\theta = 0°$, showing that reactions at small impact parameter yield HeH+ with more internal excitation, due to the greater efficiency of momentum transfer.

We have also examined the energy dependence of the differential cross section at a resolution of $\Delta E_c = 0.15$ eV, but have found no definitive evidence for the narrow scattering resonances predicted by theory. A modified version of this experiment could achieve a resolution of $\Delta E_c = 0.05$ eV in a search for resonance phenomena in He + H2+.

4. ACKNOWLEDGEMENTS

This work was supported by the Aerospace Sponsored Research program and included contributions by L.K. Johnson, J.A. Syage, D.A. Lichtin, and R.B. Cohen.

5. REFERENCES

The following articles give a full description of the experiment along with citations of work by other investigators:

1. J.E. Pollard, D.A. Lichtin, and R.B. Cohen, "Differential cross sections for state-selected reactions in the H2+ + H2 system," Chem. Phys. Lett. **152**, 171-176 (1988).
2. J.E. Pollard, J.A. Syage, L.K. Johnson, and R.B. Cohen, "Collinear reactive scattering of state-selected H2+ + He → HeH+ + H: A method for probing dynamical resonances," J. Chem. Phys. **94**, 8615-8617 (1991).
3. J.E. Pollard, L.K. Johnson, D.A. Lichtin, and R.B. Cohen, "State-selected reactive scattering. I. H2+ + H2 → H3+ + H," J. Chem. Phys. **95**, 4877-4893 (1991).
4. J.E. Pollard, L.K. Johnson, and R.B. Cohen, "State-selected reactive scattering. II. He + H2+ → HeH+ + H," J. Chem. Phys. **95**, 4894-4904 (1991).

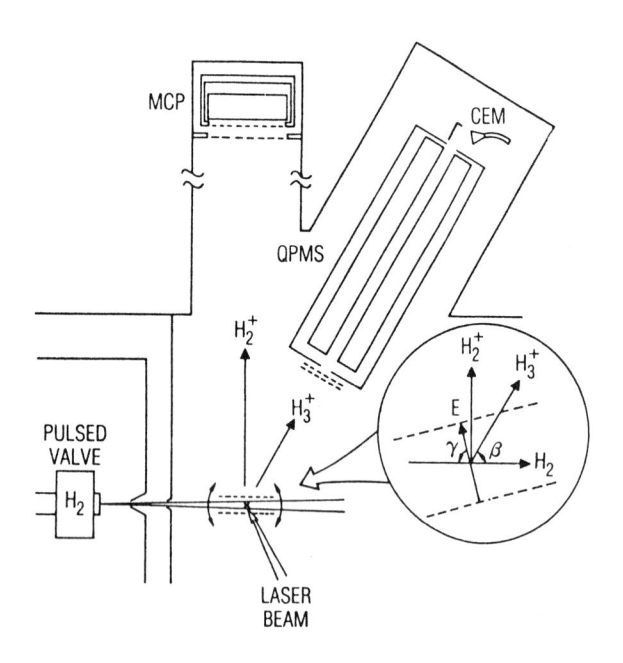

Fig.1. View of the apparatus showing the laser beam and molecular beam crossing between the rotatable grids. Reactant ions are detected by the MCP, and product ions are detected by the QPMS and CEM. Arrow E shows the direction of the impulsive electric field.

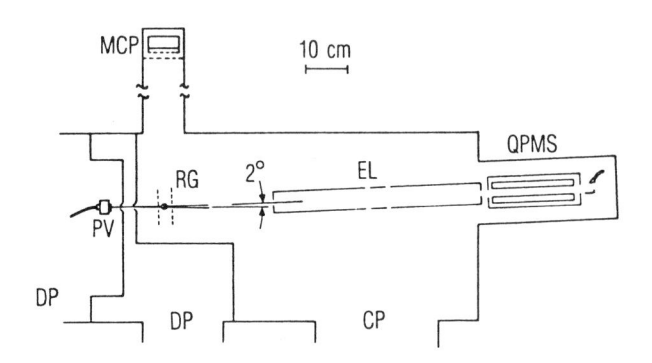

Fig.3. View of the apparatus configured for studies of He + H2+. Reactant ions are accelerated impulsively and collide with He in the beam downstream of the ionization point. Product ions are collected by the electrostatic lens and QPMS.

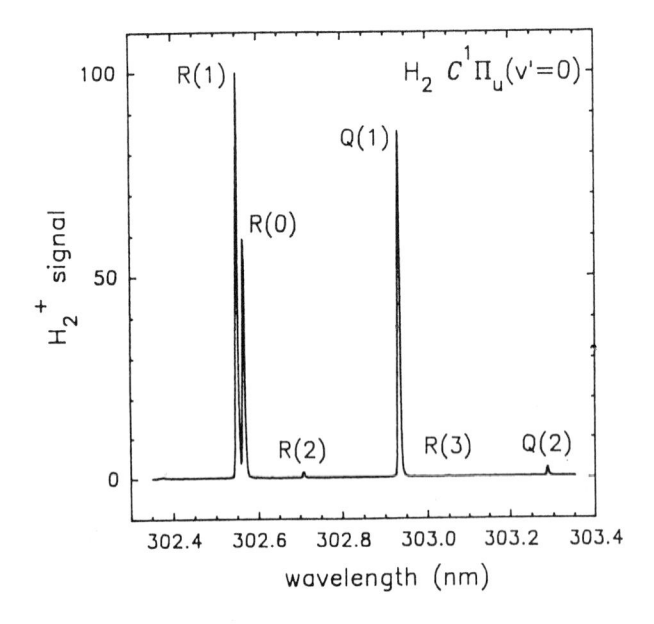

Fig.2. REMPI spectrum of H2 in the molecular beam. Reactive scattering is performed using the Q(1) transition. Rotational cooling is indicated by the weakness of the Q(2), R(2), and R(3) transitions.

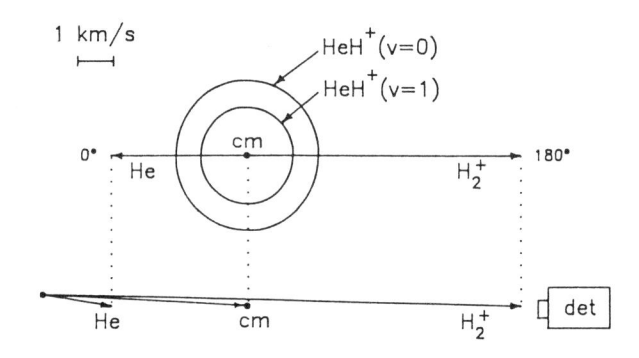

Fig.4. Newton diagram for He + H2+. The detector views the product flux scattered at c.m. angles of 0° and 180°. Circles show the maximum c.m. velocities for HeH+(v=0,1).

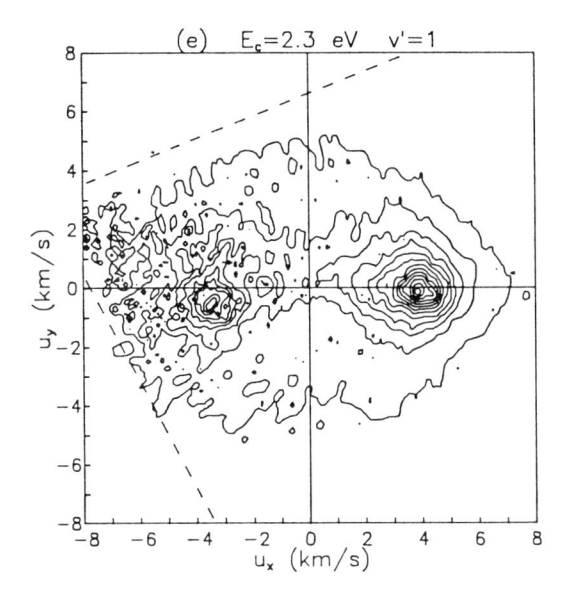

Fig.5. Cartesian flux vs. product c.m. velocity for H2+ + H2 at 2.3 eV collision energy with the REMPI transition v'=2. Incident H2+ velocity is at $\theta=0°$ (+x axis). Contours are evenly spaced in arbitrary flux units. Dashed lines indicate the angular range of the apparatus.

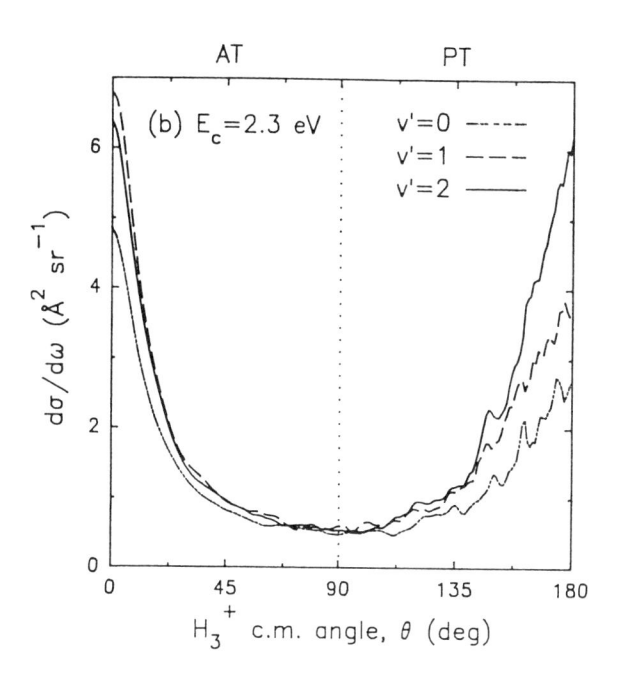

Fig.7. Differential cross section vs. c.m. scattering angle for H2+ + H2 at 2.3 eV collision energy with the REMPI transitions v'=0,1,2.

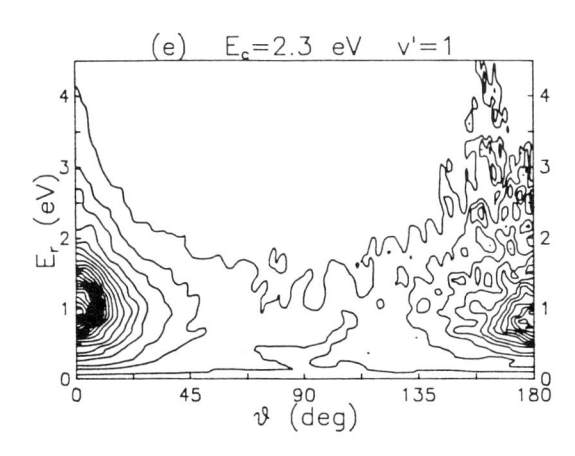

Fig.6. Differential cross section vs. c.m. scattering angle and recoil energy derived from Fig. 6. Incident H2+ velocity is at $\theta=0°$. Contour interval is 0.25 Å2 sr-1 eV-1.

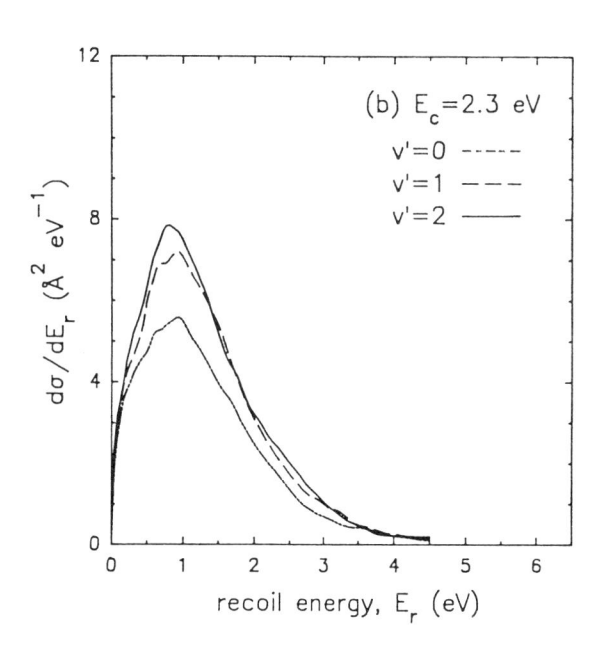

Fig.8. Differential cross section vs. product recoil energy for H2+ + H2 at 2.3 eV collision energy with the REMPI transitions v'=0,1,2.

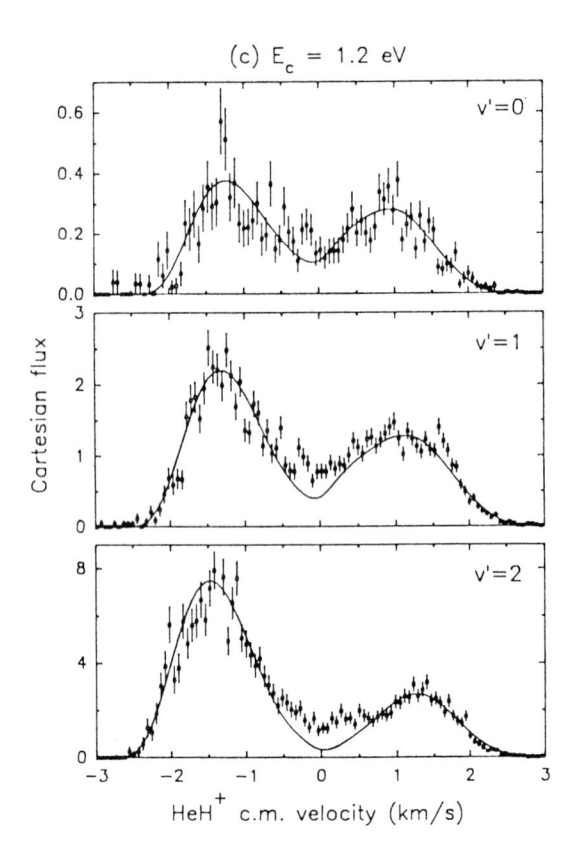

Fig.9. Cartesian flux vs. product c.m. velocity for He + H2+ at 1.2 eV collision energy with the REMPI transitions v'=0,1,2. Incident He velocity is at $\theta=180°$ (-x axis). Flux is in arbitrary units which indicate the observed relative intensities.

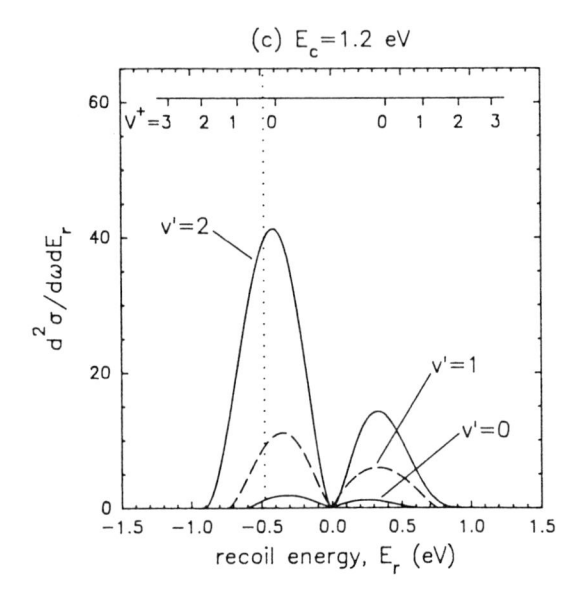

Fig.10. Differential cross section vs. product recoil energy for He + H2+ at 1.2 eV collision energy with the REMPI transitions v'=0,1,2. Incident He is at $\theta=180°$ (-x axis). Negative and positive recoil energies are used as a plotting convenience. Cross section is in arbitrary units.

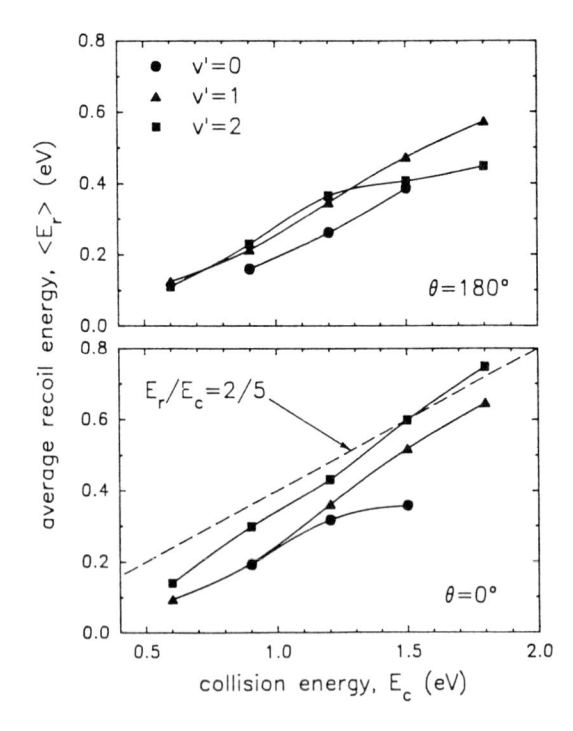

Fig.11. Average product recoil energy vs. collision energy at scattering angles $\theta=180°$ and $0°$. Dashed line defines spectator stripping kinematics.

Investigation of the roles of vibrational excitation and collision energy in the ion-molecule reaction
$NH_3^+(v_2) + ND_3$

Lynmarie A. Posey

Department of Chemistry, Vanderbilt University
Nashville, Tennessee 37235

Robert D. Guettler and Richard N. Zare

Department of Chemistry, Stanford University
Stanford, California 94305

ABSTRACT

The influence of vibrational excitation and collision energy on the reaction $NH_3^+(v_2) + ND_3$ has been investigated using a quadrupole-octopole-quadrupole mass spectrometer. The NH_3^+ reagent ions are prepared state-selectively with 0–7 quanta in the v_2 umbrella bending mode by (2 + 1) resonance enhanced multiphoton ionization. The mass-filtered reagent ion beam interacts with a thermal distribution of neutral ND_3 molecules at controlled center-of-mass collision energies (0.5-10.0 eV) within the octopole ion guide, enabling product ions to be collected independent of scattering dynamics. The reaction of NH_3^+ with ND_3 has three major product channels: (1) deuterium abstraction, (2) charge transfer, and (3) proton transfer. The product branching ratios and relative cross sections for each of these channels exhibit strong dependences on ion vibrational excitation and collision energy. Briefly, both deuterium abstraction and charge transfer are enhanced by vibrational excitation, whereas proton transfer is suppressed. As the collision energy is increased, the branching fraction for charge transfer increases sharply while proton transfer decreases. The branching ratio for deuterium abstraction does not exhibit a significant dependence on collision energy. The influence of ion vibrational excitation is discussed in terms of its relationship to the reaction coordinates for the three product channels. The behavior of this reaction points to a short-lived collision complex in which vibration and translation play inequivalent roles.

1. INTRODUCTION

Chemists have long been intrigued by the concept of manipulating the outcome of chemical reactions through judicious control of starting conditions. To this end, considerable effort has been devoted to understanding how initial conditions, such as reagent collision energy and internal excitation (electronic, vibrational, rotational), influence the outcome of chemical processes. From these state-selected studies as well as state-to-state measurements in which product translational and/or internal energies are probed, it is possible to infer what has happened during the course of reaction. These approaches, which have been widely applied to neutral reaction systems, are no less suitable for ion-molecule reactions. In fact, despite the strong potential interaction between ions and dipoles or induced dipoles and the accurate prediction of thermal collision rates by theories which treat the ion as a point charge (Langevin, ADO, AADO),[1] ion-molecule reactions can exhibit internal state-dependent behavior.

The reaction of the ammoniumyl ion, NH_3^+, with neutral ammonia, NH_3, was one of the first ion-molecule reactions for which the influence of reagent internal energy was studied.[2] When either the ionic or neutral reagent is labeled, three major product channels for this reaction are revealed: H/D atom abstraction, charge/electron transfer, and proton/deuteron transfer. In this work the system $NH_3^+(v_2) + ND_3$ was studied as a function of collision energy and vibrational excitation in the ν_2 umbrella bending mode; the three major product channels are given below:

$$NH_3^+(\nu_2) + ND_3 \rightarrow NH_3D^+ + ND_2 \quad \text{(deuterium abstraction)} \quad (1a)$$
$$\rightarrow ND_3^+ + NH_3 \quad \text{(charge/electron transfer)} \quad (1b)$$
$$\rightarrow ND_3H^+ + ND_2 \quad \text{(proton transfer)} \quad (1c)$$

Heavy particle transfer, which can proceed via either neutral or charged particle transfer, is exothermic by 0.92 ± 0.18 eV[3-10] and occurs at nearly the thermal collision rate limit predicted by AADO theory[5,11] ($k = 2.3 \pm 0.2 \times 10^{-9}$ cm^{3}s^{-1}molecule^{-1}).[5,6,12] The rate for the thermoneutral charge transfer process is inhibited at thermal collision energies by a barrier on the ground electronic surface that arises from the degeneracy of the system ($k \leq 4 \times 10^{-11}$ cm^{3}s^{-1}molecule^{-1}).[13] Photoionization measurements[7,14] and theoretical work[15-17] indicate that the entrance channel ion-neutral complex, $NH_3^+ \cdot NH_3$, is 1.11 ± 0.32 eV more stable than the separated reagents. The exit channel complex, $NH_4^+ \cdot NH_2$, is bound by 1.02 ± 0.15 eV with respect to the products according to theoretical calculations.[15-20] Ab initio calculations[16,17] predict that the barrier to forward reaction is as much as 0.99 eV below the entrance channel asymptote, suggesting that during reaction little time is spent sampling the potential energy surface in the region of the local minima corresponding to the entrance and exit channel complexes.

The $NH_3^+ + NH_3$ ion-molecule reaction has received considerable attention mainly as a result of a seminal state-selected study by Chupka and Russell[2] who used 1-photon VUV photoionization to produce vibrationally excited NH_3^+. While this technique did not prepare ions in a single vibrational level, it was possible to probe the effect of increased excitation in ν_2 on the reaction. Chupka and Russell discovered that the relative cross section for formation of NH_4^+ decreased by a factor of two in going from $NH_3^+(\nu_2 = 0)$ to $NH_3^+(\nu_2 = 0-10)$ with the distribution determined by Franck-Condon overlap between the neutral and ionic ground states. The following discussion of previous work on the ammoniumyl/ammonia system is not meant to be inclusive but to highlight work significant to the current study; a more extensive review is provided in a paper which is in preparation.[21] In subsequent state-selected work using photoion photoelectron coincidence techniques, Baer and Murray[22] found that the proton transfer channel decreased with increasing vibrational energy at 0.1–1.0 eV collision energy and that charge transfer was enhanced by increasing vibrational excitation at 1–100 eV collision energy. More recent state-selected studies of this system have employed deuterium labeling of either the ionic or neutral reagent to distinguish between H atom and proton transfer. These studies have included work by Conaway, Ebata, and Zare[9] who used (2 + 1) resonance enhanced multiphoton ionization (REMPI) to state-selectively prepare $NH_3^+(\nu_2)$ for reaction in a tandem quadrupole mass spectrometer with a static collision cell. Unfortunately, with this instrument geometry some product channels were collected more efficiently than others making it impossible to determine product branching ratios. These workers did find that atom abstraction was enhanced by as much as a factor of 6 in going from $\nu_2 = 0$ to 10, while proton/deuteron transfer decreased by a factor of 2. Charge transfer showed enhancement by a factor of 2 over the same range of vibrational excitation. Tomoda et al.[10] saw similar trends in a threshold electron secondary ion coincidence study of $NH_3^+(\nu_2) + ND_3$ and $ND_3^+(\nu_2) + NH_3$.

Below we report branching ratios and relative cross sections for the ion-molecule reaction $NH_3^+(\nu_2) + ND_3$ as a function of vibrational excitation ($\nu_2 = 0-7$) and collision energy (0.5–10.0 eV center-of-mass). These experiments were carried out using the quadrupole-octopole-quadrupole mass spectrometer described in the next section. This instrument takes advantage of radiofrequency (RF) guided ion beam techniques developed by Teloy and Gerlich,[23] which find widespread application in ion-molecule reaction studies.[24-29] Addition of an octopole ion guide to the collision cell region of the tandem quadrupole mass spectrometer used by Conaway et al.[9] removed the discrimination in product ion detection that hampered their work.

2. EXPERIMENTAL

Measurement of product branching ratios and relative cross sections for the $NH_3^+ + ND_3$ reaction was performed using the instrument shown in Fig. 1. The experiment can be divided into three phases: (1) preparation of a state-selected, mass-selected reactant ion beam, (2) reaction of the ions with a thermal distribution of neutral target gas, and (3) collection and mass-selective detection of product ions. The state-selected ion beam is produced by (2 + 1) REMPI of a skimmed and collimated molecular beam of 10% NH_3 seeded in He. Ionization through the $\tilde{B}$ or $\tilde{C}'$ Rydberg states of NH_3 allows the reactant ion beam to be prepared with 74–100% state selectivity.[30] After interaction of the molecular beam with the frequency-doubled output of a Nd:YAG-pumped tunable dye laser (10 Hz), the ions produced are swept onto the ion beam axis, which is oriented at 90° with respect to the molecular beam, by a static turning quadrupole.[31] The first quadrupole mass filter removes any fragment ions produced as byproducts of the ionization process before the ion beam is introduced to the RF octopole ion guide, which passes through the collision cell region of the instrument. The octopole ion guide was added to this instrument to remove the discrimination in product detection inherent in a static collision cell configuration. Although our collision cell geometry precludes direct measurement of the neutral target gas pressure, we have determined that reactions are carried out under single collision conditions by monitoring formation of the secondary product ion ND_4^+ produced by reaction of the charge transfer product ND_3^+ with the neutral reactant. The pressure of the neutral target gas is kept sufficiently low to ensure that no more than 2–3% of the product ions result from this process. Thin einzel lens assemblies at the entrance and exit of the octopole ion guide carry the ions across the interfaces between the octopole and quadrupoles. The second quadrupole mass filter is used to distinguish the unreacted ion beam from products and to detect product ions by mass. Typically, 50-150 reactant ions reach the detector per laser shot permitting analog detection, while less than 1 product ion per laser shot is collected necessitating ion counting.

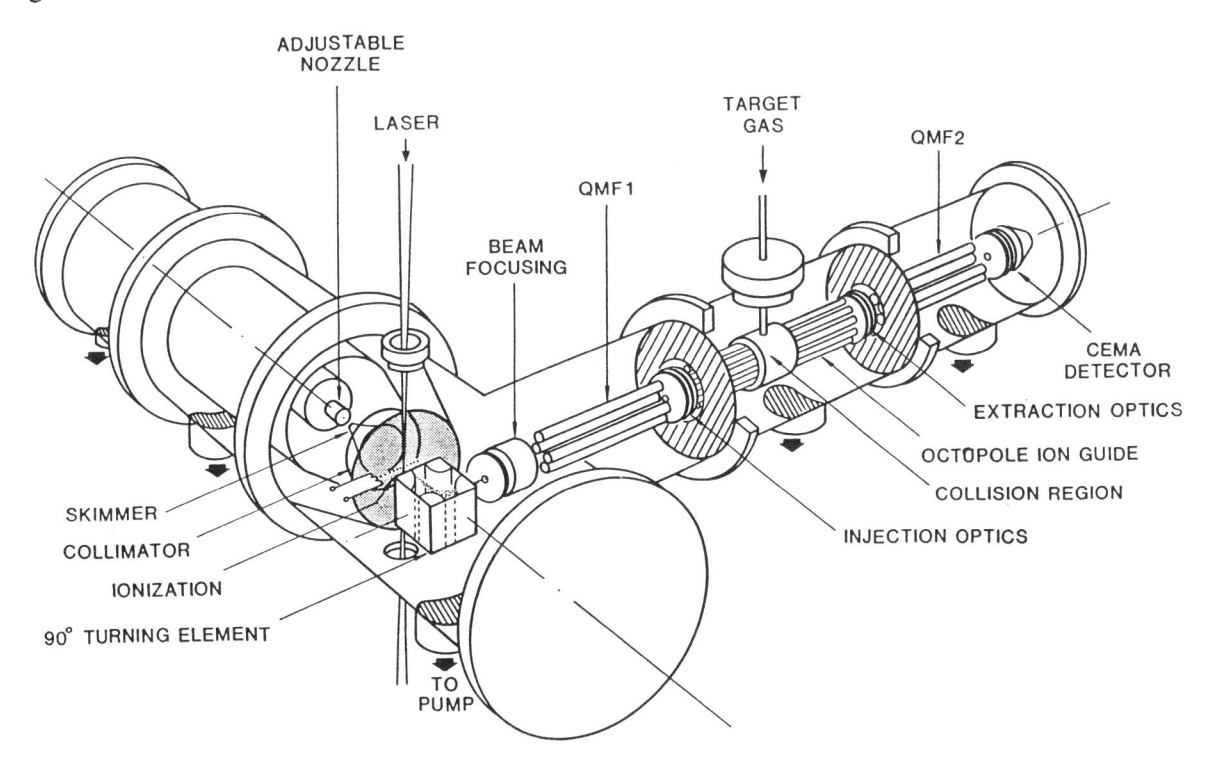

Figure 1. Quadrupole-octopole-quadrupole mass spectrometer used in ion-molecule reaction studies with state-selective preparation of the ionic reagents by resonance enhanced multiphoton ionization (QMF = quadrupole mass filter, MCP = microchannel plate).

3. RESULTS

We have investigated the influence of vibrational excitation and collision energy on relative reaction cross sections and branching ratios for $NH_3^+(v_2 = 0-7) + ND_3$. The effects of vibrational excitation on the relative cross sections for the three product channels, deuterium abstraction, charge transfer, and proton transfer, are shown in Fig. 2 for collision energies of 1, 5, and 10 eV. The effects of vibrational excitation are clearly most pronounced at low collision energy. At 1 eV deuterium abstraction is enhanced by a factor of 2.5 as vibrational excitation increases from $v_2 = 0$ to $v_2 = 7$ ($\Delta E_{int} = 0.85$ eV). Charge transfer shows some enhancement, although it is not as pronounced. At low collision energy the proton transfer channel decreases by a factor of 2. The effect of vibrational excitation on the relative cross sections is washed out at higher collision energy.

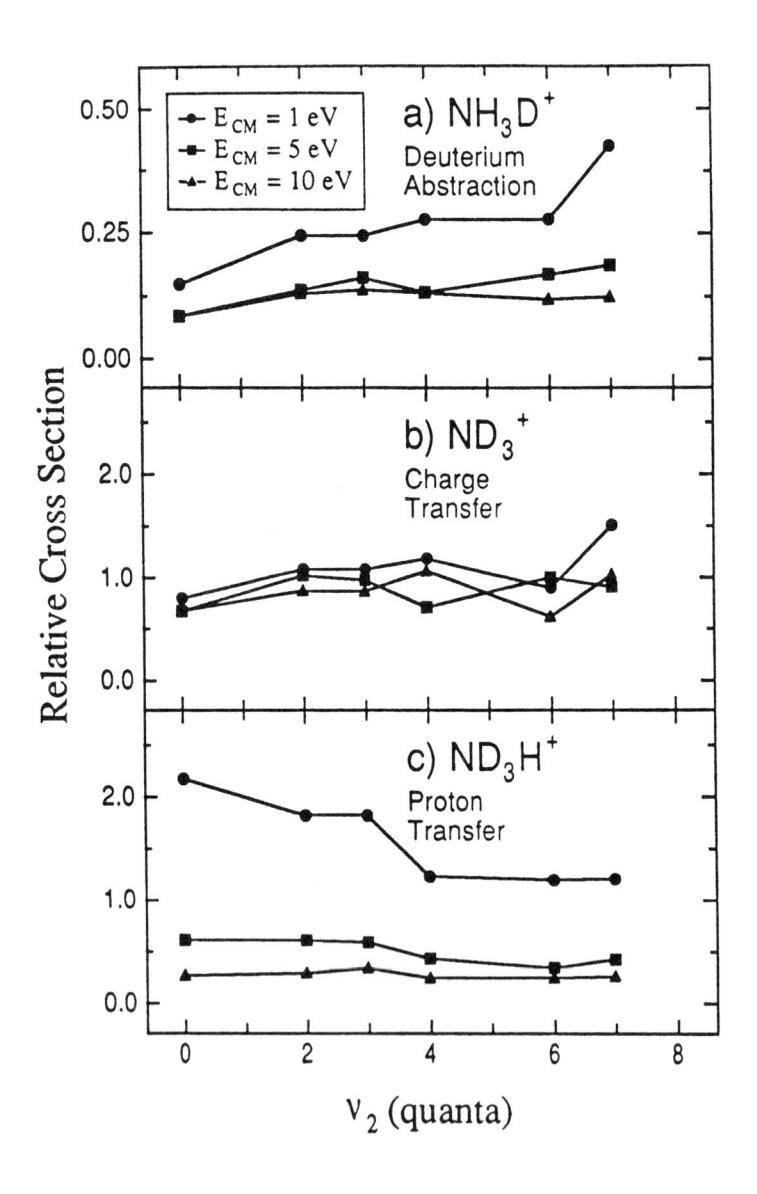

Figure 2. Relative cross sections plotted as a function of excitation in the v_2 umbrella bending mode of NH_3^+.

We installed the octopole ion guide in our mass spectrometer to remove discrimination in product ion detection arising from differences in the dynamics for formation of each product channel. Removal of product channel-dependent detection allows direct comparison of the influence of vibrational excitation and collision energy on the three major product channels through branching ratios (Fig. 3). Deuterium abstraction is always the smallest channel; the branching to this channel shows a slight increase with vibrational excitation. The other heavy particle transfer channel, proton transfer to form ND_3H^+, decreases with increased excitation in the umbrella mode of NH_3^+ as well as increased collision energy. Interestingly, under the conditions studied the branching ratios for proton transfer and deuterium abstraction are never equal as expected if rapid electron transfer between the ion and neutral during the interaction removes any distinction between the two reagents. The branching fractions for these two heavy particle transfer channels only approach each other at high collision energy (10 eV). Charge transfer becomes a dominant channel at high vibrational excitation and collision energy.

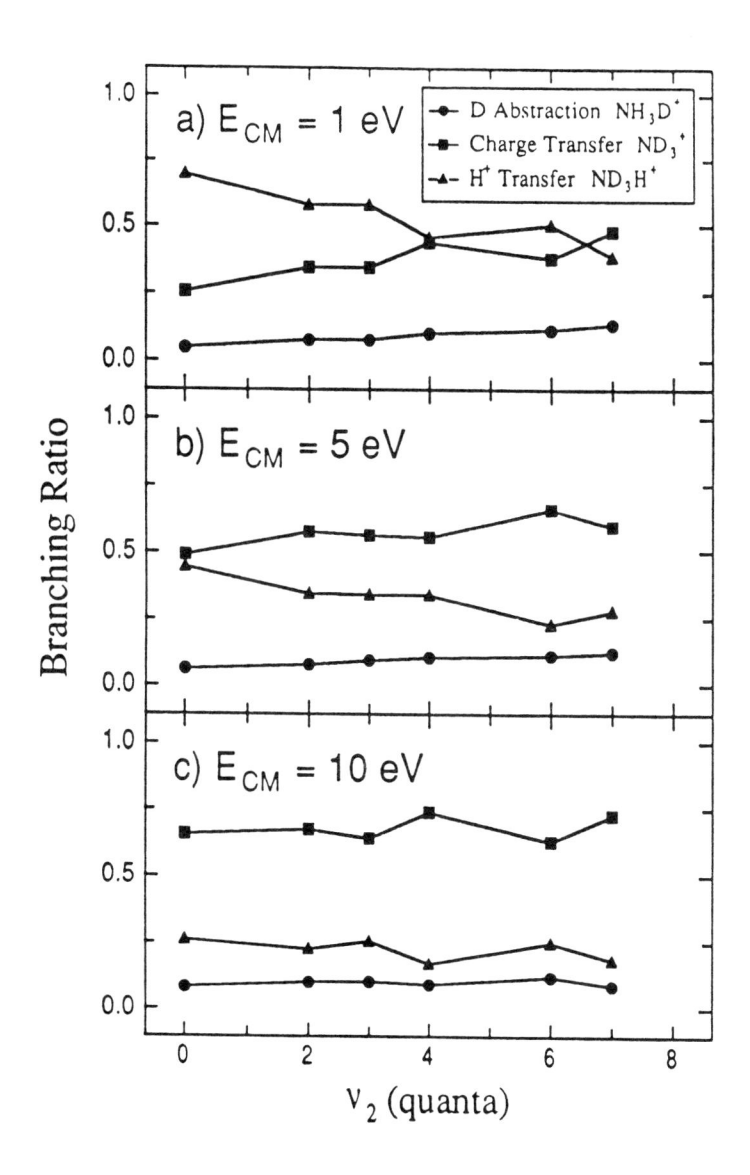

Figure 3. Product branching ratios plotted as a function of excitation in the ν_2 umbrella bending mode of NH_3^+.

Figure 4 shows the effects of collision energy and vibrational excitation on the total reaction cross section summed over the three major product channels. The total cross section depends most strongly on collision energy. It decreases with increasing collision energy in much the same way as the collision cross section is predicted to decrease by the Langevin model and its variants (ADO, AADO). These models treat the ion as a point charge and hence, assume that any internal excitation of the ion has no effect on the collision cross section. We observe that the propensity for reaction is independent of internal excitation, but the internal excitation controls the branching between product channels.

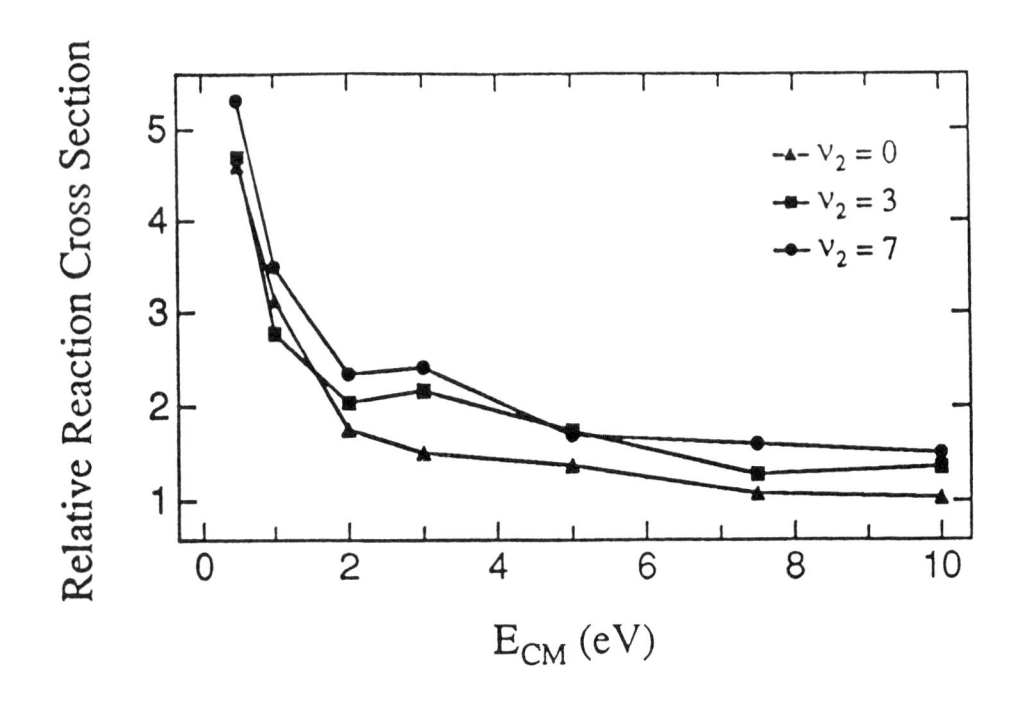

Figure 4. Total reaction cross sections at $v_2 = 0, 3, 7$ plotted as a function of the center-of-mass collision energy, E_{CM}.

4. DISCUSSION

In developing models to explain the observed dependence of the reaction of $NH_3^+(v_2)$ with ND_3, we return to earlier explanations for the role of vibrational excitation in this reaction. Chupka and Russell[2] offered a statistical argument to rationalize the decrease in NH_4^+ product formation in terms of competition with charge transfer. Chesnavich and Bowers[32] tested this hypothesis by performing statistical phase space calculations and found poor agreement with the experimental data. Chesnavich and Bowers proposed that proton transfer was inhibited by increased vibrational excitation in the ionic reagent because the motion induced was orthogonal to the reaction coordinate. Statistical analysis of relative cross sections for formation of NH_4^+ as a function of ion internal excitation at thermal collision energies by van Pijkeren et al.[8] indicated that fewer degrees of freedom than expected participated in redistribution of energy within the reaction complex. Franklin and Haney[3] found that the translational energy of the NH_4^+ product in the $NH_3^+ + NH_3$ reaction was independent of internal excitation in the reactant ion. All of this information is consistent with a short-lived, loosely coupled reaction complex, which is not surprising because it appears that the barrier for this exothermic reaction is significantly below the entrance channel asymptote. Additional data on the arrival time distributions for product ions at the detector, which will be reported elsewhere,[21] are also consistent with a picture of incomplete momentum transfer during a short-lived interaction.

The model developed to explain the behavior of the $NH_3^+ + ND_3$ system is illustrated in Fig. 5. We will first consider the heavy particle transfer channels, proton transfer and deuterium abstraction. Formation of ND_3H^+ requires that a proton be transferred from the vibrationally excited ion to the nitrogen center of the neutral ND_3. Figure 5b shows a likely approach geometry for this process with the dipole of the neutral directed toward the ion and one of the hydrogen atoms from NH_3^+ directed toward the ND_3 nitrogen. There is general agreement amongst experimentalists[9,10] and theoreticians[33] that proton transfer is a direct process. An explanation for the inhibition of proton transfer by excitation in the v_2 vibrational is that the motion induced in the ion is orthogonal to the reaction coordinate.

(a) Deuterium Abstraction

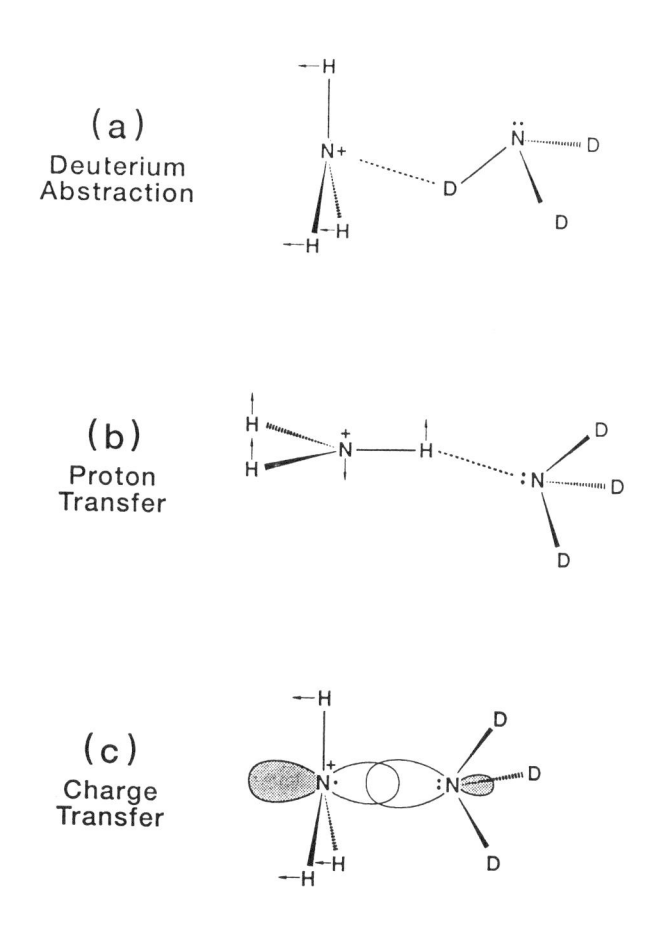

(b) Proton Transfer

(c) Charge Transfer

Figure 5. Schematic illustration of the model used to explain the influence of vibrational excitation and collision energy on the outcome of the reaction of $NH_3^+(v_2) + ND_3$.

There is less agreement on how to view the process with deuterium abstraction as the net result. Formation of NH_3D^+ has been described as either a direct one-step deuterium atom transfer[9] or a two-step process initiated by charge transfer and followed by deuteron transfer.[10,33] Recent molecular orbital calculations (HF/4-31G)[33] have found a minimum path on the potential energy surface from $NH_3^+ + NH_3$ reactants to the "atom abstraction" product that corresponds to stepwise charge transfer followed by proton transfer. Tomoda et al.[10] have also invoked a similar two-step process to explain their experimental data. The enhancement of atom abstraction is attributed to enhancement of the charge transfer process with increased vibrational excitation, although Baer and Murray[22] have noted that the cross section for formation of the charge transfer product does not increase as much as expected from Franck-Condon factors. A possible argument against this model is the observation that the cross sections for the two heavy particle transfer channels are not equivalent. The current data do not allow us to distinguish between the one- and two-step models; however, we favor the one-step model illustrated in Fig. 5a. To form the tetrahedral NH_3D^+ deuterium abstraction product, the planar NH_3^+ must distort to accommodate a fourth atom at the central nitrogen. The motion introduced by umbrella mode excitation distorts the NH_3^+ ion to accommodate the fourth atom. In other words, vibrational excitation in the v_2 mode corresponds to motion along the reaction coordinate for this process. The deuterium abstraction channel accounts for less than 20% of the products. The unfavorable approach geometry, which has the dipole directed away from the ion, may account for the low efficiency of the process. An important test of the notion that reagent motion directed along a reaction coordinate can promote formation of a specific product in ion-molecule reactions is currently underway; we are investigating the effect of symmetric stretch excitation on the proton transfer channel in the NH_3^+ + ND_3 reaction.[34]

The enhancement of charge transfer branching ratios by vibrational excitation and collision energy is particularly noticeable because this process, which is almost two orders of magnitude slower at thermal collision energies, dominates at high vibrational excitation and/or collision energy. At low collision energy, a barrier to electron transfer on the ground electronic surface resulting from exchange interaction impedes charge transfer. Collision energy and vibrational energy can supply enough energy to the system to surmount the barrier. In this study the vibrational frequency mismatch between the deuterium-labeled and unlabeled reagents eliminates resonant charge transfer. A possible approach geometry for this process is illustrated in Fig. 5c. The neutral dipole is directed toward the ion's nitrogen center, and the filled and half-filled nonbonding orbitals on the reagents overlap. The dominance of charge transfer at higher collision energies and vibrational excitation may simply reflect shortening of the interaction time between the reagents.

5. CONCLUSIONS

We have investigated the influence of vibrational excitation and collision energy on the reaction of NH_3^+, prepared state-selectively by REMPI, with ND_3 using a recently constructed quadrupole-octopole-quadrupole mass spectrometer. Measurement of the relative cross sections and branching ratios for the three major product channels, deuterium abstraction, charge transfer and proton transfer, have revealed that these two types of energy play inequivalent roles in the apparently short-lived reaction complex. Translational energy seems to control the total reaction cross section while vibrational excitation appears to play an important role in determining the branching between product channels. A simple model which views vibrational excitation in terms of its relationship to the reaction coordinates for this ion-molecule reaction system is consistent with the data presented.

6. ACKNOWLEDGMENTS

Support for this work was provided by the Air Force Office of Scientific Research under grant AFOSR-89-0264. LAP gratefully acknowledges the National Science Foundation for support provided by an NSF Postdoctoral Research Fellowship in Chemistry (CHE-8907493). We thank Beat A. Keller and Nicholas J. Kirchner for their roles in development of our quadrupole-octopole-quadrupole apparatus.

7. REFERENCES

1. T. Su and M. T. Bowers, "Classical Ion-Molecule Collision Theory," *Gas Phase Ion Chemistry*, M. T. Bowers, ed., vol. 1, pp. 83-118, Academic Press, New York, 1979.
2. W. A. Chupka and M. E. Russell, "Ion-Molecule Reactions of NH_3^+ by Photoionization," *J. Chem. Phys.*, vol. 48, pp. 1527-1533, 1968.
3. J. L. Franklin and M. A. Haney, "Translational Energies of Products of Exothermic Ion-Molecule Reactions," *J. Phys. Chem.*, vol. 73, pp. 2857-2863, 1969.

4. W. T. Huntress, Jr. and R. F. Pinizzotto, Jr., "Product Distributions and Rate Constants for Ion-Molecule Reactions in Water, Hydrogen Sulfide, Ammonia, and Methane," *J. Chem. Phys.*, vol. 59, pp. 4742-4756, 1973.

5. R. S. Hemsworth, J. D. Payzant, H. I. Schiff and D. K. Bohme, "Rate Constants at 297°K for Proton Transfer Reactions with NH_3. Comparisons with Classical Theories and Exothermicity," *Chem. Phys. Lett.*, vol. 26, pp. 417-421, 1974.

6. W. Lindinger, D. L. Albritton, F. C. Fehsenfeld, A. L. Schmeltekopf and E. E. Ferguson, "Flow-Drift Tube Measurements of Kinetic Energy Dependences of Some Exothermic Proton Transfer Rate Constants," *J. Chem. Phys.*, vol. 62, pp. 3549-3553, 1975.

7. S. T. Ceyer, P. W. Tiedemann, B. H. Mahan and Y. T. Lee, "Energetics of Gas Phase Proton Solvation by NH_3," *J. Chem. Phys.*, vol. 70, pp. 14-17, 1979.

8. D. van Pijkeren, J. van Eck and A. Niehaus, "Internal-Energy-Selected Measurements on the Relative Cross Section of the Ion-Molecule Reaction $NH_3^+(NH_3,NH_2)NH_4^+$," *Chem. Phys.*, vol. 95, pp. 449-457, 1985.

9. W. E. Conaway, T. Ebata and R. N. Zare, "Vibrationally State-Selected Reactions of Ammonia Ions. III. $NH_3^+(v) + ND_3$ and $ND_3^+(v) + NH_3$," *J. Chem. Phys.*, vol. 87, pp. 3453-3460, 1987.

10. S. Tomoda, S. Suzuki and I. Koyano, "State-Selected Ion-Molecule Reactions by a Coincidence Technique. XV. Hydrogen Atom Abstraction as an Electron Jump Followed by Proton Transfer in the $ND_3^+(v) + NH_3$ and $NH_3^+(v) + ND_3$ Reactions," *J. Chem. Phys.*, vol. 89, pp. 7268-7276, 1988.

11. T. Su, E. C. F. Su and M. T. Bowers, "Ion-Polar Molecule Collisions. Conservation of Angular Momentum in the Average Dipole Orientation Theory. The AADO Theory," *J. Chem. Phys.*, vol. 69, pp. 2243-2250, 1978.

12. N. G. Adams, D. Smith and J. F. Paulson, "An Experimental Survey of the Reactions of NH_n^+ Ions ($n = 0-4$) with Several Diatomic and Polyatomic Molecules at 300 K," *J. Chem. Phys.*, vol. 72, pp. 288-297, 1980.

13. W. T. Huntress, Jr., M. M. Mosesman and D. D. Elleman, "Relative Rates and Their Dependence on Kinetic Energy for Ion-Molecule Reactions in Ammonia," *J. Chem. Phys.*, vol. 54, pp. 843-849, 1971.

14. W. Kamke, R. Herrmann, Z. Wang and I. V. Hertel, "On the Photoionization and Fragmentation of Ammonia Clusters Using TPEPICO," *Z. Phys. D*, vol. 10, pp. 491-497, 1988.

15. J. C. Greer, R. Ahlrichs and I. V. Hertel, "Proton Transfer in Ammonia Cluster Cations: Molecular Dynamics in a Self Consistent Field," *Z. Phys. D*, vol. 18, pp. 413-426, 1991.

16. W. J. Bouma and L. Radom, "Hydrazinium Radical Cation ($NH_3NH_3^{+\bullet}$) and Dication ($NH_3NH_3^{2+}$): Prototypes for the Ionized Forms of Medium-Ring Bicyclic Compounds," *J. Am. Chem. Soc.*, vol. 107, pp. 345-348, 1985.

17. P. M. W. Gill and L. Radom, "Structures and Stabilities of Singly Charged Three-Electron Hemibonded Systems and Their Hydrogen-Bonded Isomers," *J. Am. Chem. Soc.*, vol. 110, pp. 4931-4941, 1988.

18. S. Tomoda, "Ionization of the Ammonia Dimer: Proton Transfer in the Ionic State," *Chem. Phys.*, vol. 110, pp. 431-445, 1986.

19. S. Tomoda, "Proton Transfer in the Ionic State of Simple Hydrogen-Bonded Dimers $(H_2O)_2^+$, $(NH_3)_2^+$, and $(HF)_2^+$: An Elementary Process of Protonation," *Faraday Discuss. Chem. Soc.*, vol. 85, pp. 53-63, 1988.

20. S. Tomoda, "Proton Transfer in Ionic States of Hydrogen-Bonded Dimers. A Photoelectron Spectroscopic Approach," *Vacuum Ultraviolet Photoionization and Photodissociation of Molecules*, C. Ng, ed., World Scientific, River Edge, NJ, 1991.

21. L. A. Posey, R. D. Guettler, N. J. Kirchner, B. A. Keller and R. N. Zare, "Influence of Vibrational Excitation and Collision Energy on the Ion-Molecule Reaction $NH_3^+(v_2) + ND_3$," in preparation.

22. T. Baer and P. T. Murray, "Cross Sections for Symmetric Charge Transfer and Proton Transfer Reactions of Internally Energy Selected $NH_3^+(v)$," *J. Chem. Phys.*, vol. 75, pp. 4477-4484, 1981.

23. E. Teloy and D. Gerlich, "Integral Cross Sections for Ion-Molecule Reactions. I. The Guided Beam Technique," *Chem. Phys.*, vol. 4, pp. 417-427, 1974.

24. S. L. Anderson, F. A. Houle, D. Gerlich and Y. T. Lee, "The Effects of Vibration and Translational Energy on the Reaction Dynamics of the $H_2^+ + H_2$ System," *J. Chem. Phys.*, vol. 75, pp. 2153-2162, 1981.

25. K. M. Ervin and P. B. Armentrout, "Translational Energy Dependence of $Ar^+ + XY \rightarrow ArX^+ + Y$ (XY = H_2, D_2, HD) from Thermal to 30 eV c.m.," *J. Chem. Phys.*, vol. 83, pp. 166-189, 1985.

26. J. D. Shao and C. Y. Ng, "A State-Selected Study of the Reaction $H_2^+(v_0') + H_2(v_0''=0) \rightarrow H_3^+ + H$," *Chem. Phys. Lett.*, vol. 118, pp. 481-485, 1985.

27. D. Gerlich, "Low Energy Ion Reactions Measured with Guided Beams," *Electronic and Atomic Collisions*, D. C. Lorents, W. E. Meyerhof and J. R. Peterson, eds., pp. 541-553, Elsevier, Amsterdam, 1986.

28. D. Gerlich, R. Disch and S. Scherbarth, "$C^+ + H_2(j) \rightarrow CH^+ + H$: The Effect of Reagent Rotation on the Integral Cross Section in the Threshold Region," *J. Chem. Phys.*, vol. 87, pp. 350-359, 1987.

29. D. Gerlich, "Inhomogeneous RF Fields: A Versatile Tool for the Study of Processes with Slow Ions," *Adv. in Chem. Phys., State-Selected and State-to-State Ion-Molecule Reaction Dynamics: Experiment*, C. Y. Ng and M. Baer, eds., vol. 82, part 1, Wiley, New York, in press.

30. W. E. Conaway, R. J. S. Morrison and R. N. Zare, "Vibrational State Selection of Ammonia Ions Using Resonant 2 + 1 Multiphoton Ionization," *Chem. Phys. Lett.*, vol. 113, pp. 429-434, 1985.

31. H. D. Zeman, "Deflection of an Ion Beam in the Two-Dimensional Electrostatic Quadrupole Field," *Rev. Sci. Instrum.*, vol. 48, pp. 1079-1085, 1977.

32. W. J. Chesnavich and M. T. Bowers, "Statistical Phase Space Theory of Polyatomic Systems: Application to the Reactions $NH_3^+ + NH_3 \rightarrow NH_4^+ + NH_2$ and $NH_3^+ + H_2O \rightarrow NH_4^+ + OH$," *Chem. Phys. Lett.*, vol. 52, pp. 179-183, 1977.

33. A. Tachibana, S. Kawauchi, Y. Kurosaki, N. Yoshida, T. Ogihara and T. Yamabe, "Role of Charge Transfer for the Vibrational-Mode-Specific Chemical Reaction of $NH_3^+(v)$ and NH_3," *J. Phys. Chem.*, vol. 95, pp. 9647-9653, 1991.

34. R. D. Guettler and R. N. Zare, unpublished results.

State-to-state studies of intramolecular dynamics

Thomas R. Rizzo, Xin Luo, and Rebecca D. F. Settle

Department of Chemistry, University of Rochester
Rochester, New York 14627

ABSTRACT

Infrared-optical double resonance incorporating vibrational overtone excitation of a light atom stretch mode prepares reactant molecules in single quasibound rovibrational states at energies above their decomposition threshold on the ground potential surface. When combined with laser induced fluorescence detection of the dissociation fragments, this approach permits fully quantum state resolved studies of unimolecular dissociation reactions. This report describes the application of this technique to study the unimolecular dissociation of hydrogen peroxide from a variety of vibrational and rotational states. We emphasize here the spectroscopy associated with the reactant excitation process and demonstrate how highly resolved excitation spectra provide detailed information about the dynamics of the ensuing dissociation. We then describe a new spectroscopic technique for detecting vibrational overtone excitation in low pressure environments based upon selective CO_2 laser infrared multiphoton dissociation of highly vibrationally excited molecules. We show preliminary results applying this approach to CH_3OH.

1. INTRODUCTION

The development of quantitative models to predict the decomposition dynamics of an isolated molecule on a single potential energy surface has challenged experimentalists and theoreticians for over 70 years.[1,2] Statistical theories of unimolecular reactions such as Rice-Ramsperger-Kassel-Marcus (RRKM) Theory,[1,2] Phase Space Theory (PST),[3] and the Statistical Adiabatic Channel Model (SACM),[4] have enjoyed considerable success in predicting both dissociation rates and product energy partitioning by making the assumption that internal energy is rapidly (and statistically) redistributed among all the modes of the reactant molecule on a timescale that is short relative to chemical bond breaking. Statistical theories further assume that all product channels that conserve total energy energy and angular momentum are equally probable, although in some cases additional constraints are introduced.[4]

Experimental studies that prepare reactant molecules in a wide distribution of initial quantum states do not provide a stringent test of statistical theories insofar as agreement with experiment can arise in part from averaging over the initial reactant state distributions. Advances in laser and molecular beam techniques over the last several years have permitted increasingly detailed studies of unimolecular reactions to the point that true state-to-state studies are now feasible, and this has brought the assumptions of statistical models under closer scrutiny. The most difficult part of achieving full state-to-state specificity in unimolecular reaction studies is in preparing reactant molecules with well defined energy and angular momentum (*i.e.*, in single rovibrational states).[5] Optical excitation techniques typically project the room temperature distribution of thermally populated reactant states to higher energy, causing reaction to occur with a wide distribution of total energy and angular momentum. Two methods have been used to overcome this problem. In one approach, a supersonic expansion cools reactant molecules prior to excitation, narrowing the initial distribution of reactant states. This method has been used in conjunction with electronic excitation/internal conversion[6, 7] and direct overtone excitation[8, 9] to prepare reactant molecules in well defined initial states for unimolecular reaction studies. Another approach to selective reactant preparation uses double resonance excitation from a room temperature sample.[10, 11] By requiring reactant molecules to satisfy two resonance requirements, spectral congestion can in some cases be completely eliminated, allowing the preparation of reactant molecules in single rovibrational states. In general, the double resonance approach has the advantage of being able to access a wide variety of reactant states, whereas supersonic molecular beam techniques are limited, by the low temperatures in the expansion and the rotational selection rules, to low J, K states of the reactant.

We have recently implemented an infrared-optical double resonance technique in which infrared excitation of an light atom stretch fundamental is combined with overtone excitation of the same vibrational mode to selectively prepare highly excited reactant molecules.[11-13] When combined with LIF detection of the nascent dissociation fragments, this approach permits fully state-resolved studies of unimolecular dissociation reactions. Over the last few years, we have used this technique to investigate the unimolecular dissociation of HOOH,[11-13] and NH$_2$OH,[14]. Information about the dissociation dynamics comes from highly resolved vibrational overtone excitation spectra of the reactant and from the product quantum state distributions. In this report, we summarize the reactant excitation spectra of HOOH and explore the rotational and vibrational state dependence of the dissociation dynamics revealed by such spectra. These results have important implications for defining the limit of statistical theories and provides information on how these theories should treat rotational degrees of freedom. We then describe a new experimental technique that permits the measurement of highly resolved vibrational overtone spectra of more strongly bound molecules. This new approach is based upon selective CO$_2$ laser infrared multiphoton dissociation (IRMPD) of vibrationally excited molecules as a means to detect vibrational overtone excitation in low pressure environments. We describe initial results on CH$_3$OH to demonstrate the feasibility of this approach.

2. INFRARED-OPTICAL DOUBLE RESONANCE

Several recent publications describe our infrared-optical double resonance scheme, and we refer the reader to these sources for the experimental details.[11-13] Here we briefly describe the general approach and summarize its various implementations. Figure 1 shows three different configurations of the infrared-optical double resonance excitation scheme that allow access to a range of reactant states differing both in energy and in vibrational character.

An infrared pulse from a Nd:YAG pumped optical parametric oscillator excites HOOH molecules from the ground vibrational state to either v=1 of an OH stretch (schemes (a) and (b)) or a combination level containing one quantum of OH stretch and one quantum of lower frequency vibration (scheme (c)). In excitation scheme (a), a pulse from a Nd:YAG pumped dye laser further excites the parent molecules from $v_{OH}=1$ to a level above the unimolecular dissociation threshold *via* vibrational overtone excitation of the same local OH stretch mode. In the case shown in Fig. 1, this scheme prepares highly excited reactant molecules in single rotational states of the $6v_{OH}$ level. Once above the dissociation threshold, the energized reactants decompose to form two OH radicals which are detected by laser induced fluorescence LIF in the A-X band. In scheme (b), subsequent to excitation of one local OH stretch oscillator to the v=1 level, the pulsed dye laser excites a $\Delta v=4$ vibrational overtone transition of the *other* local OH stretch. This scheme produces energized reactant molecules in the (4,1) local-local combination level (designated $4v_{OH}+v_{OH'}$) which is 88 cm^{-1} below the unimolecular dissociation threshold. Addition of rotational energy in access of this deficit prepares reactant molecules in single rotational levels extremely close to threshold. In the scheme labeled (c), the infrared pulse prepares reactant molecules in a combination level containing one quantum of OH stretch and one quantum in a low frequency vibrational mode, designated $v_{OH}+v_x$. The pulsed dye laser then excites a $\Delta v_{OH}=4$ transition to access the local-normal combination level designated $5v_{OH}+v_x$. We have used this approach to prepare reactant molecules in single rotational states of the $5v_{OH}+v_{OO}$ and $5v_{OH}+v_{OOH}$ combination levels. Using the $6v_{OH}$, $4v_{OH}+v_{OH'}$, $5v_{OH}+v_{OO}$ and $5v_{OH}+v_{OOH}$ levels for reactant state preparation allows us to vary the energy from just above the dissociation threshold to ~2100 cm^{-1} in excess of threshold. Moreover, these different excitation schemes vary the vibrational character of the reactant states.

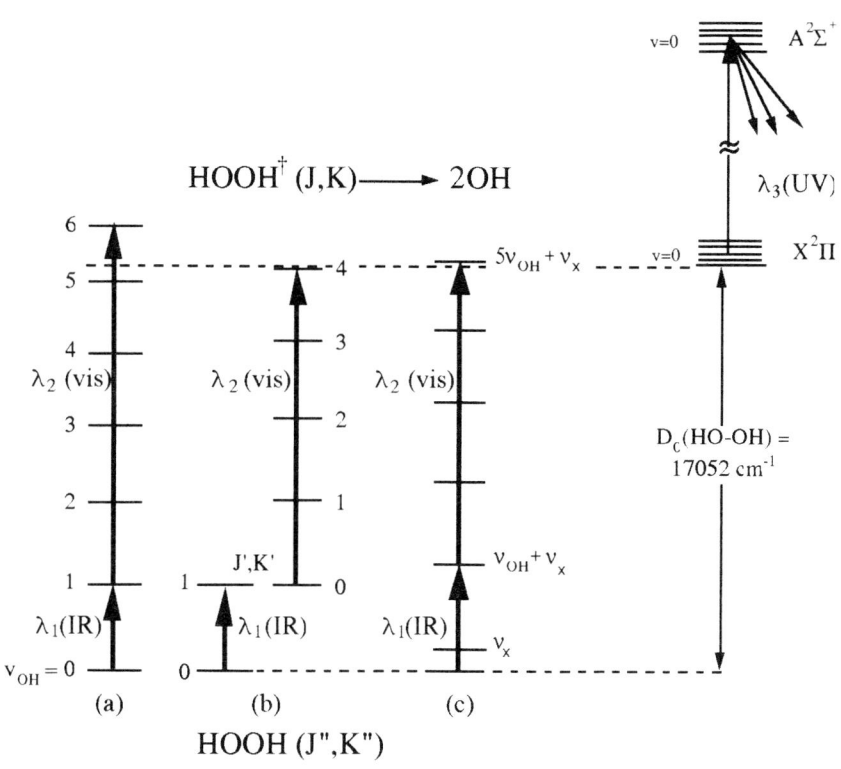

Figure 1. Energy level diagram for infrared-optical double resonance excitation of HOOH molecules to single rovibrational states above the O-O dissociation threshold. The excitation schemes labeled (a), (b) and (c) show different implementations of the double resonance approach which allow us to vary the vibrational character of the reactant states.

We distinguish the rotational quantum numbers J and K of the HOOH parent in the ground, intermediate and terminal state of the double resonance scheme by using double primes, single primes, and no primes respectively. The use of J and K implies that HOOH is a symmetric top. With an asymmetry parameter $\kappa=-0.992$, HOOH is very nearly a prolate symmetric top, and we observe no asymmetry splitting with the limited (0.11 cm^{-1}) resolution of our overtone excitation laser.

3. ROTATIONALLY RESOLVED VIBRATIONAL OVERTONE SPECTRA

3.1 Survey spectra

Figure 2 shows a series of rotationally resolved vibrational overtone excitation spectra in the (4,1) local-

local combination band of HOOH.[13] Each spectrum is obtained by fixing the infrared frequency on a particular transition in the 1←0 OH stretch fundamental and scanning the overtone excitation laser with the LIF probe laser fixed to monitor the $Q_1(1)$ transition of the OH fragments. Because detection overtone absorption is achieve by monitoring the products, such spectra represent the absorption of those molecules which ultimately dissociate and produce a fragment in the probed level.

Figure 2. Rotationally resolved vibrational overtone excitation spectra in the (4,1) local-local combination band of HOOH.

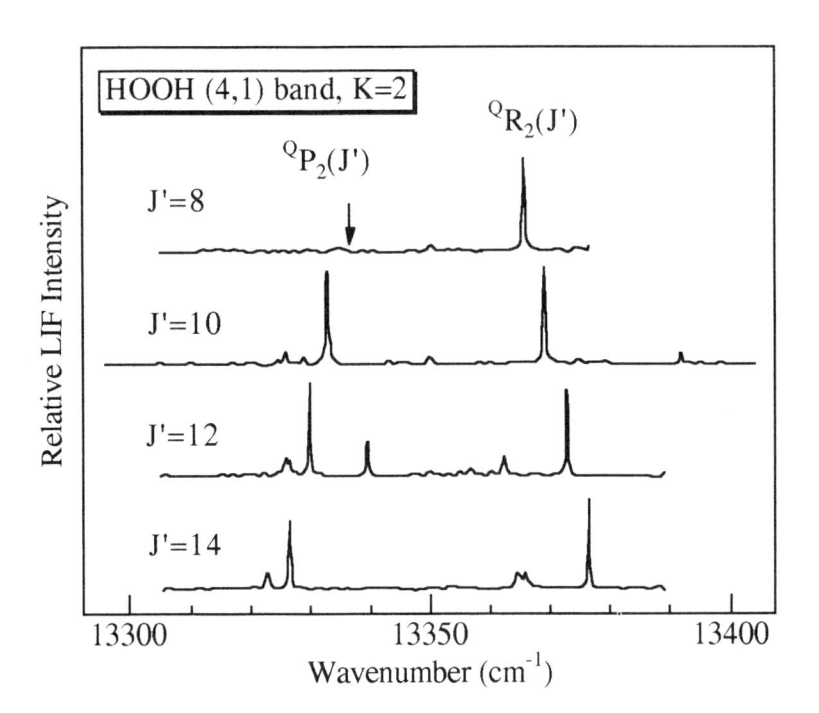

Each spectrum is labelled by J', the total rotational quantum number in the intermediate state. As is indicated by their labels, the major features in these spectra correspond to P and R branch transitions of a parallel band of a prolate symmetric top. Because it is a parallel band (i.e., ΔK=0) the K quantum number of the intermediate and terminal level of the double resonance scheme are the same, and we denote only the K quantum number of the dissociating molecule. The smaller features in the spectrum come from transitions with other J and K which overlap with our target transition in the infrared. Although we don't have perfect state selection in the first step, it is sufficient to generate fully resolved transitions in the second step.

Note that the P branch is missing from the J'=8 spectrum of Fig. 2. Because the (4,1) combination level is 88 cm^{-1} below the dissociation threshold, only states that provide at least this much energy will lead to dissociation and hence be detected as a feature in the vibrational overtone excitation spectra. The missing feature indicates that the J=7, K=2 level is below the unimolecular dissociation threshold. By observing a number of missing transitions, we determine the dissociation threshold to be 17,051.83 ± 3.4 cm^{-1}.[13]

The spectra of Fig. 2 provide important information about the dynamics of the dissociation process. Although these transitions terminate on states above the dissociation threshold, they are completely assignable to zeroth-order J and K quantum numbers. The quantum number J must be conserved in the absence of external fields; however K also appears to be a good quantum number insofar as the positions of each feature can be predicted from knowing J and K. Recall that K is the projection of the total angular momentum J on the symmetric top axis. The motion associated with K corresponds to the rotational motion about the top axis, whereas the motion associated with J corresponds to the precession of the top axis about the J vector. For spectra such as those of Fig. 2 to be assignable to K as observed, the molecule must be maintaining a fixed projection for at least several rotational periods. From a chemist's point of view, the important question is: how does the timescale for conservation of K compare to the lifetime of the molecule? That is, does the molecule maintain a fixed projection of J throughout the course of its dissociation, or does it undergo a more statistical type of rotational motion in which it samples different projections? Physically, this would correspond to energy being transferred between vibrations and K rotation. If a reactant molecule does maintain a fixed projection as it dissociates, then models of the dissociation dynamics such as RRKM theory or phase space theory must incorporate conservation of K rotation, which makes a substantial difference in their predictions near threshold. In their usual applications, these theories do not conserve K but treat it statistically on the timescale of dissociation. To address this question further, we now show single rovibrational transitions of the reactant molecule under higher resolution and examine their dependence upon the rotational and vibrational quantum numbers.

3.2 High resolution spectra - clump structure and rotational state dependence

Figure 3 shows a series of transitions to individual zeroth-order rotational states of HOOH in the (4,1) local-local combination level under higher resolution. Figure 3(a) shows how these transition lineshapes change with increasing K quantum number, while 3(b) shows their J dependence at fixed K.

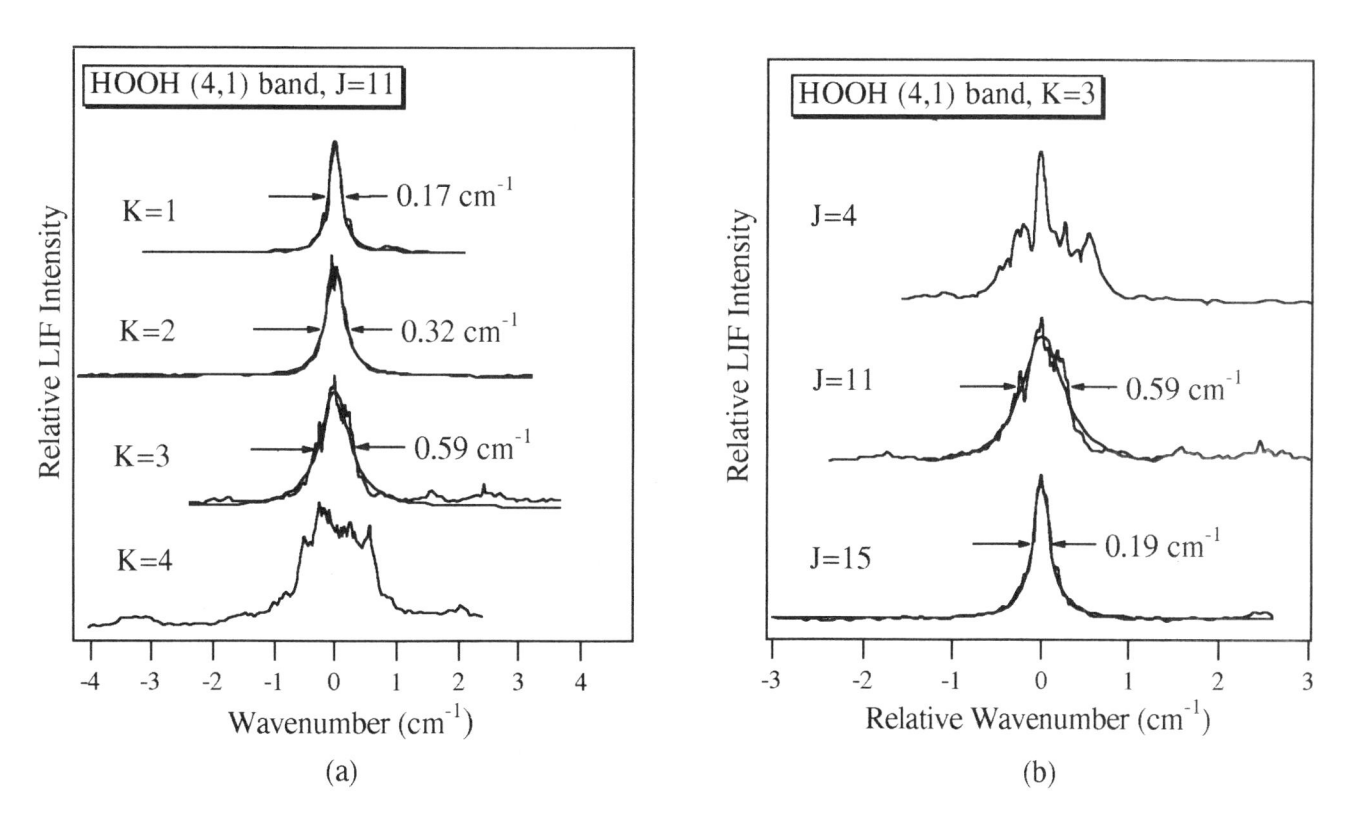

Figure 3. Rotational state dependence of the single rovibrational transitions to states above the unimolecular dissociation threshold. (a) K dependence with J fixed at 11; (b) J dependence with K fixed at 3. Where included, the cited linewidths come from Lorentzian fits to the data.

Before drawing any conclusions from spectral lineshapes, one ensure that the spectra are not contaminated by residual spectral overlap. To determine that the structure we observe in an individual transition represents splitting of the excited state level, we access every level by at least two different routes (*i.e.*, different combinations of P and R branch transitions). Since the spectral overlap will differ for different routes, the observation of common structure in spectra generated by different paths to the same state verifies that it results from excited state splittings. We have verified that each spectrum in Fig. 3 represents a transition which derives oscillator strength from a state with a single value of J and K (*i.e.*, a single zeroth-order J, K level).

Note that some of the transitions in Fig. 3 appear as clumps of lines rather than a single Lorentzian feature. This clump structure is completely reproducible among transitions that terminate on the same excited state level. Consider the origin of this spectral structure, shown schematically in Fig. 4.

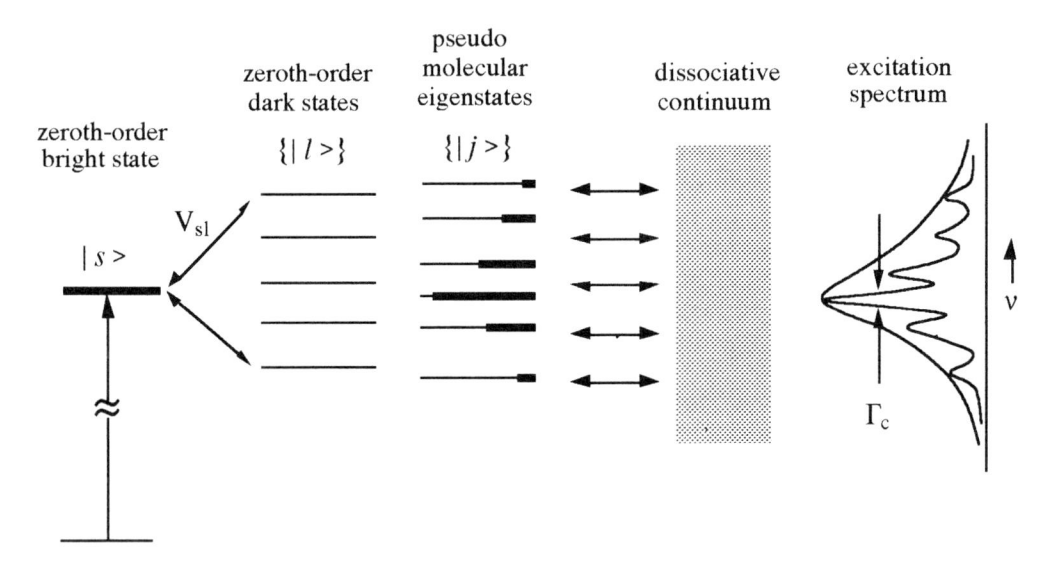

Figure 4. Schematic of coupling model which accounts for the spectral structure of single zeroth-order rovibrational transitions to quasibound states of the HOOH reactant molecules.

Rotationally resolved vibrational overtone spectra of molecules excited above their dissociation threshold potentially contain information about both IVR and the unimolecular decomposition rate. The vibrational overtone level is the zeroth order optically active state, $|s>$ (*i.e.*, the state that carries oscillator strength from the ground state). For a molecule the size of HOOH, there are other nearly isoenergetic states, $\{|l>\}$, that do not carry oscillator strength from the ground state, referred to as zeroth-order "dark" states. We refer to these states as *zeroth-order* states inasmuch as that they are not eigenstates of the full molecular Hamiltonian, but rather of a Hamiltonian that neglects the coupling between vibrational modes, indicated in Fig. 4 by the matrix elements V_{sl}. From a perturbation theory point of view, including coupling terms in the Hamiltonian causes the molecular eigenstates $\{|j>\}$ to be mixtures of the zeroth-order bright state, $|s>$, and the dark states, $\{|l>\}$. Each eigenstate in a frequency range determined by the matrix elements coupling $|s>$ and $\{|l>\}$ will contain some component of $|s>$ and hence contribute to the absorption spectrum. This will give rise to a "clump" of lines in absorption, the overall width of which reflects the coupling matrix elements V_{sl}. We must also consider, however, that our spectra involve transitions to states that are above the dissociation threshold. Each of these states $\{|j>\}$ are in turn coupled to the dissociative continuum which imposes a finite width, Γ_c, upon them determined by the dissociation lifetime. If the dissociation were fast enough, the clump might appear as a single broadened line. Depending upon the timescale for reaction, the overall linewidth might represent vibrational state mixing or the unimolecular decomposition rate, or a convolution of the two.

The observation of transitions to single zeroth-order states that break up into discrete components (see Fig. 3) indicates that the coupling to the continuum is much weaker than the coupling among zeroth-order vibrational states (*i.e.*, $V_{sl} \gg \Gamma_c$). In this case, the overall clump width results from coupling of the bright state to the dark states and hence defines a timescale for intramolecular vibrational energy redistribution (IVR) subsequent to coherent excitation. The width of a single component results from coupling to the dissociative continuum and represents the dissociation rate. Although we cannot completely resolve the individual components of a clump, we can conclude that IVR occurs faster than unimolecular dissociation. However, the timescales do not appear to be completely separable as statistical theories assume. Since the K quantum number is well defined for the overall clump and is preserved for a time which is not insignificant with respect to the dissociation time, these results suggest that statistical models of the dissociation process may have to incorporate at least partial conservation of K to quantitatively predict the results of state-resolved experiments.

Given that the overall clump widths are determined predominantly by vibrational state mixing (IVR), the rotational state dependence to these clump widths can elucidate the nature of the coupling between the zeroth-order bright state and the zeroth-order dark states. The spectra of Fig. 3(a) illustrate that the overall clump width of an individual J, K transition, and hence the IVR rate, increases sharply with increasing K. This K dependence suggests that the coupling between the bright state and zeroth-order dark states results predominantly from Coriolis interactions induced by a-axis rotation.[15, 16] The matrix elements for a-axis Coriolis coupling are linearly proportional to the quantum number K, while those for b- and c-axis Coriolis coupling decrease with increasing K.[15, 16] In a case where the density of coupled states is large, one expects the overall clump width to be proportional to the square of the coupling matrix elements; however, in a small molecule like HOOH where the states are relatively sparse, the dependence can be somewhere between linear and quadratic.[17] Moreover, lumpiness in the state density will cause fluctuations about a smooth trend. Given these considerations, the observed increase in clump widths with increasing K is clearly consistent with a mechanism that involves a-axis Coriolis coupling.

The J dependence of the overall clump widths is somewhat more intriguing. What appear to be clumps of individual features at low J coalesce into a smooth Lorentzian profile with increasing J. We attribute this unusual behavior to J dependent interactions among zeroth-order dark states that breaks down the goodness of the K quantum number and increases the effective density of states. We have previously shown, using a simple model which assumes constant average coupling matrix elements between the bright and dark states, that the overall width of a clump of transitions will initially narrow with increasing J before reaching an asymptotic J independent value. Indeed our measured spectra show no significant change after ~J=15. For the effective density of states to increase with J, b- or c-axis Coriolis coupling between bath states must be invoked insofar as these mechanisms destroy the goodness of the K quantum number. As K breaks down, the effective density of states can increase by as much as a factor of 2J+1. There have been many cases in which rotational involvement in IVR has been invoked to explain an effective density of states that appears higher than the calculated vibrational state density.[18, 19] These spectra seem to provide an example in which these J dependent interactions *among bath states* have direct spectral consequences.

3.3 High resolution spectra - vibrational state dependence

The ability to prepare excited HOOH molecules in different vibrational states above the unimolecular dissociation threshold allows us to examine the vibrational state dependence of the coupling between the bright state and zeroth-order dark states. Figure 5 shows transitions to single rotational levels in three different vibrational bands: $6\nu_{OH}$, $5\nu_{OH} + \nu_{OO}$, and $4\nu_{OH} + \nu_{OH'}$. Each of these transitions arises from a single zeroth-order bright state with well defined J and K.

Figure 5. Comparison of single rotational transitions to the $6\nu_{OH}$, $5\nu_{OH} + \nu_{OO}$, and $4\nu_{OH} + \nu_{OH'}$ vibrational levels.

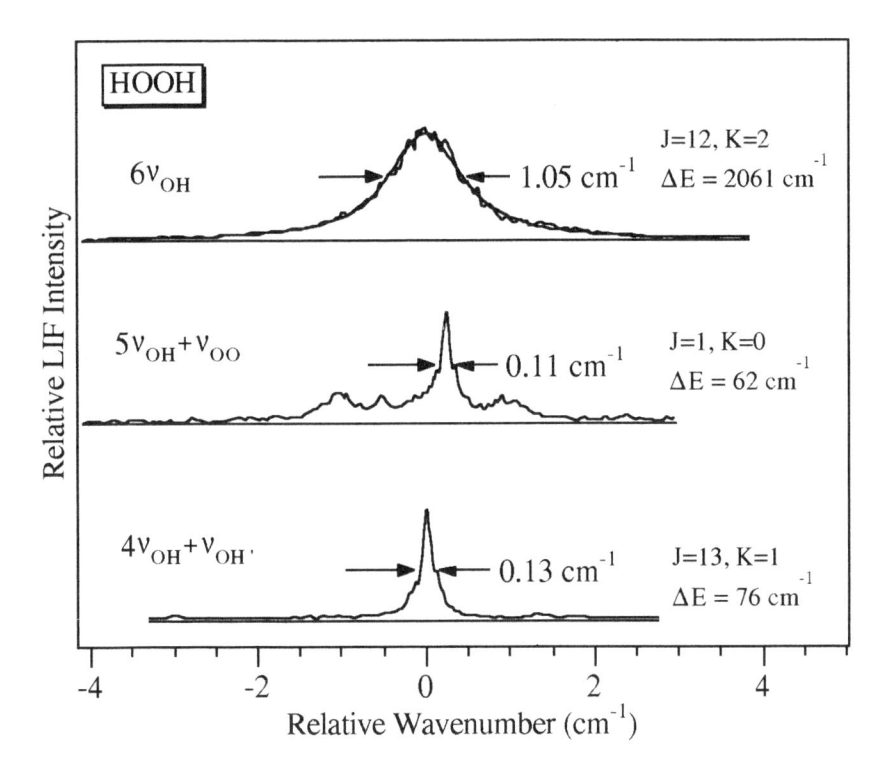

Although the $5\nu_{OH} + \nu_{OO}$ level is ~2000 cm^{-1} *below* the $6\nu_{OH}$ level, the overall clump widths of transitions to the former is on the same order as the $6\nu_{OH}$ transitions. If we assume that the lifetime broadening component at $6\nu_{OH}$ is at least as much as at the $5\nu_{OH} + \nu_{OO}$ level (and it is almost certainly greater), then the component of the $6\nu_{OH}$ transitions resulting from vibrational state mixing is clearly smaller than that at the combination level, even though the energy is 2000 cm^{-1} higher!

The comparison between transitions to the $5\nu_{OH} + \nu_{OO}$ level and those to $4\nu_{OH} + \nu_{OH'}$ are particularly enlightening in that the excess energies are very nearly the same. The $4\nu_{OH} + \nu_{OH'}$ transitions show what appear to be single lines that have widths comparable to the single components of the $5\nu_{OH} + \nu_{OO}$ transitions. The widths of the $4\nu_{OH} + \nu_{OH'}$ transitions are at least a factor of 5 smaller than the overall clump width of the $5\nu_{OH} + \nu_{OO}$ transitions at equivalent excess energy above the dissociation threshold.

The spectra of Fig. 5 suggest that the vibrational coupling of the $5\nu_{OH} + \nu_{OO}$ level to the zeroth-order dark states is significantly stronger than that of the $6\nu_{OH}$ and $4\nu_{OH} + \nu_{OH'}$ levels. The $5\nu_{OH} + \nu_{OO}$ level contains one quantum in the low frequency O-O stretch mode whereas the others have all the energy in high frequency OH stretch modes. These results seem to confirm the general expectation that high frequency light atom stretch vibrations are more poorly coupled to the rest of the molecule than modes with lower frequency vibrational components. In this case the difference may be as much as a factor of 10. In cases in which the

IVR and unimolecular dissociation times are comparable, this result indicates that the assumption of statistical theories that only the total energy and angular momentum determines the dissociation dynamics may not be valid.

4. CO_2 LASER ASSISTED VIBRATIONAL OVERTONE SPECTROSCOPY

Although the infrared-optical double resonance technique described in the preceding sections are providing new and detailed views of the intramolecular dynamics of vibrationally excited molecules, this approach is limited to a relatively small number of molecules. The major limitation to its general applicability is the small oscillator strengths for vibrational overtone transitions. These transitions gain oscillator strength because of the slight breakdown of the harmonic oscillator selection rules caused by anharmonicity in the vibrational potential and non-linearity in the dipole moment function. A general rule of thumb is that vibrational overtone transitions become an order of magnitude weaker for each additional change in vibrational quantum number greater than 1. As one goes to higher vibrational overtone levels, experiments become progressively more difficult, and because we must break chemical bonds for spectroscopic detection, this limits us to studying molecules that have dissociation energies less than about 60 kcal/mole, which corresponds to 7 quanta in a typical OH stretch. Unfortunately most molecules have significantly higher bond dissociation energies than this. For example, most C-O and C-C bond energies are ≥ 80 kcal/mol,[20] and this is simply too high to reach directly from the ground state or in double resonance using vibrational overtone excitation.

One way to overcome the limitations of the previously described spectroscopic techniques is to relax the requirement that transitions terminate upon a state above the dissociation threshold and and use an alternative means to supply additional energy for dissociation. In our infrared-optical double resonance scheme, vibrational overtone excitation could still be detected by monitoring dissociation products if the dissociation were assisted by an additional photoexcitation process. We have implemented a new detection scheme for weak vibrational overtone transitions that should be applicable to a wide variety of molecules in both static gas cells (using infrared-optical double resonance) and molecular beams (using direct overtone excitation). After excitation of a light atom stretch overtone transition, we dissociate vibrationally excited molecules *via* CO_2 laser multiphoton excitation. Detection of the overtone excitation is ultimately achieved by LIF detection of the dissociation fragments.

For this approach to be successful, we must selectively dissociate vibrationally excited molecules without dissociating the large excess of ground state molecules. Two characteristics of CO_2 laser infrared multiphoton dissociation (IRMPD) lend themselves to achieving such selectivity. IRMPD of ground state molecules exhibits a threshold fluence[21] which has been associated with pumping through the vibrational quasicontinuum.[22] Vibrational overtone excitation of a light atom stretch puts molecules half way to typical bond dissociation energies where they should require a significantly lower fluence to dissociate. Adjusting the CO_2 laser fluence just below the threshold for dissociating ground state molecules permits selective dissociation of the molecules which have undergone a vibrational overtone transition. A second factor which enhances this selectivity is the red-shift of the CO_2 laser absorption with increasing vibrational energy.[21,23] Evidence for such a red-shift comes from comparing photofragment yield spectra with linear absorption spectra at CO_2 laser wavelengths[21] and from two-color CO_2 laser multiphoton absorption studies.[23] Adjusting the CO_2 laser frequency to the red of the ground state absorption discriminates against IRMPD of ground state molecules while still efficiently dissociating overtone excited molecules.

We have demonstrated the feasibility of this new experimental scheme using it to detect the 5←0 OH stretch transition of CH_3OH in a low pressure gas cell. Figure 6(a) shows a schematic energy level diagram for this test experiment. A 40 mJ pulse from a Nd:YAG pumped dye laser first excites the 5←0 OH stretch overtone transition of methanol molecules in a 50 mTorr static gas sample. After a 50 ns delay, a pulse from a CO_2 TEA laser promotes molecules initially prepared in $v_{OH}=5$ to an energy above the 92.6 kcal/mol dissociation threshold,[20] producing CH_3 and OH. The CO_2 laser operates on the P(20) line of the 00^01-10^00 transition at 944 cm^{-1}, which is outside the rotational band contour of the v_8 (C-O stretch) transition of the ground state molecule.[24] We focus the CO_2 beam into the cell with a 50 cm ZnSe lens to give an average laser fluence of 1-5 J cm^{-2} in the center of the cell, which is less than the 10 J cm^{-2} reported for methanol.[21] Approximately 320 nanoseconds later, an ultraviolet pulse from a frequency doubled dye laser electronically excites the OH dissociation products *via* the A-X band, and a photomultiplier tube collects the resulting fluorescence from a 5 cm region in the center of the glass fluorescence cell. Scanning the vibrational overtone excitation laser while collecting the total OH fluorescence produces the 5←0 OH stretch overtone excitation spectrum of CH_3OH shown in Fig. 6(b). Figure 6(c) shows a photoacoustic spectrum of the same overtone transition taken at a pressure of ~40 Torr.

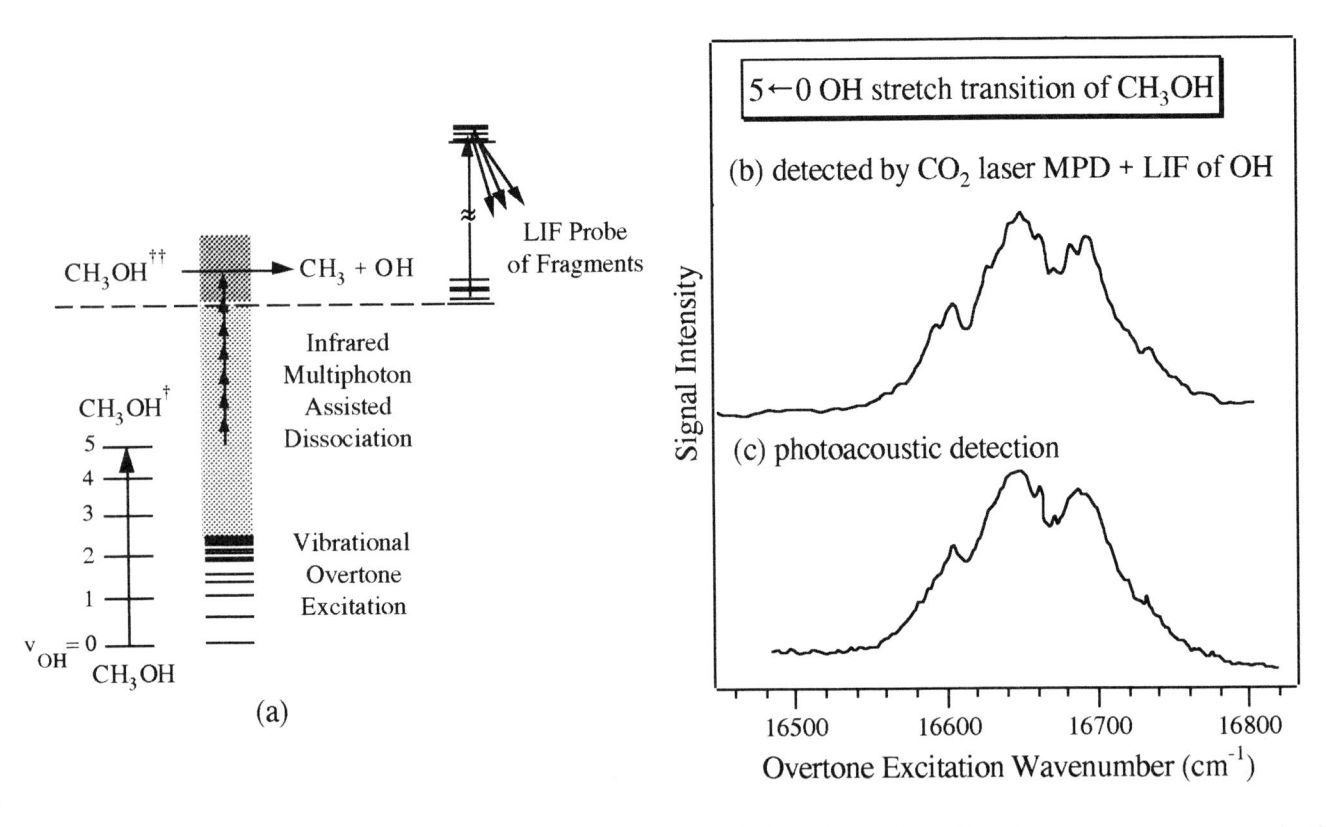

Figure 6. (a) Energy level scheme for CO_2 laser assisted detection of 0←5 OH stretch overtone excitation in methanol; (b) vibrational overtone excitation spectrum; (c) photoacoustic spectrum.

This preliminary experiment establishes several important points about the use of CO_2 laser IRMPD for detection of weak vibrational overtone transitions. At the laser frequency and fluence used here, *no* OH LIF signal was observed when the overtone laser was blocked, although tuning the CO_2 laser to the peak of the ground state methanol absorption and increasing the fluence produced copious amounts of signal. This demonstrates that we can discriminate against dissociating ground state molecules by at least a factor of 10^5 by using the CO_2 laser fluence and frequency. The laser fluence required for dissociation of the vibrationally excited molecule was 2-10 times smaller than that of the ground state molecules, verifying our assumption of a lower threshold fluence for the vibrationally excited molecules. Moreover, the spectrum of Fig. 7(b) is virtually identical to the photoacoustic spectrum of Fig. 7(c) indicating that the CO_2 laser photodissociation step does not distort the vibrational overtone (absorption) spectrum. There is no reason to believe the same will not be true for the overtone spectrum generated by infrared-optical double resonance. We are currently applying this IRMPD/LIF detection scheme in conjunction with infrared optical double resonance overtone excitation and direct overtone excitation in a supersonic molecular beam.

5. ACKNOWLEDGEMENTS

We gratefully acknowledge the support of this work by the Office of Basic Energy Sciences of the Department of Energy and the donors of the Petroleum Research Fund, administered by the ACS.

6. REFERENCES

1. P. J. Robinson and K. A. Holbrook, Unimolecular Reactions, p. (Wiley, London, 1972).
2. R. G. Gilbert and S. C. Smith, Theory of Unimolecular and Recombination Reactions, p. (Blackwell, 1990).
3. P. Pechukas, J. C. Light, and C. Rankin, J. Chem. Phys. **44**, 794 (1966).
4. M. Quack and J. Troe, "Specific rate constants of unimolecular processes II. Adiabatic channel model", Ber. Bunsenges. Phys. Chem. **78**, 240 (1974).
5. F. F. Crim, "Selective excitation studies of unimolecular reaction dynamics", Ann. Rev. Phys. Chem. **34**, 657 (1984).
6. W. F. Polik, D. R. Guyer, W. H. Miller, and C. B. Moore, "Eigenstate-resolved unimolecular reaction dynamics: Ergodic character of S_0 formaldehyde at the dissociation threshold.", J. Chem. Phys. **92**, 3471 (1990).
7. C. X. W. Qian, M. Noble, I. Nadler, H. Reisler, and C. Wittig, "NCNO → CN + NO: Complete NO (E,V,R) and CN(V,R) nascent population distributions from well-characterized monoenergetic unimolecular reactions.", J. Chem. Phys. **83**, 5573 (1985).
8. L. J. Butler, T. M. Ticich, M. D. Likar, and F. F. Crim, "Vibrational overtone spectroscopy of bound and predissoicative states of hydrogen peroxide cooled in a supersonic expansion.", J. Chem. Phys. **85**, 2331 (1986).
9. B. R. Foy, M. P. Casassa, J. C. Stephenson, and D. S. King, "Overtone-excited HN_3 (X^1A'): Anharmonic resonance, homogeneous linewidths, and dissociation rates", J. Chem. Phys. **92**, 2782 (1990).
10. E. Abramson, R. W. Field, D. Imre, K. K. Innes, and J. L. Kinsey, "Stimulated emission pumping of acetylene: Evidence for quantum chaotic behavior near 27,900 cm^{-1}.", J. Chem. Phys. **80**, 2298 (1984).
11. X. Luo and T. R. Rizzo, "Rotationally resolved vibrational overtone spectroscopy of hydrogen peroxide at chemically significant energies", J. Chem. Phys. **93**, 8620 (1990).
12. X. Luo and T. R. Rizzo, "Unimolecular dissociation of hydrogen peroxide from single rovibrational states near threshold", J. Chem. Phys. **94**, 889 (1991).
13. X. Luo and T. R. Rizzo, "Vibrational overtone spectroscopy of the $4\nu_{OH}+\nu_{OH'}$ combination band of HOOH via sequential local mode-local mode excitation", J. Chem. Phys. (in press). (1991).

14. X. Luo, P. R. Fleming, T. A. Seckel, and T. R. Rizzo, "Broad vibrational overtone linewidths in the $7\nu_{OH}$ band of rotationally selected NH_2OH.", J. Chem. Phys. **93**, 9194 (1990).
15. A. E. W. Knight, "Rotational Involvement in Intramolecular Vibrational Redistribution", in <u>Excited States</u>, edited by E. C. Lim and K. K. Innes; p. 1; (Academic Press, San Diego, 1988).
16. H. L. Dai, C. L. Korpa, J. L. Kinsey, and R. W. Field, "Rotation-induced vibrational mixing in X^1A_1 formaldehyde: Non-negligible dynamical consequences of rotation", J. Chem. Phys. **82**, 1688 (1985).
17. M. Bixon and J. Jortner, "Intramolecuar Radiationless Transitions", J. Chem. Phys. **48**, 715 (1968).
18. C. S. Parmenter, "Vibrational redistribution within excited electronic states of polyatomic molecules", Faraday Discuss. Chem. Soc. **75**, 7 (1983).
19. K. W. Holtzclaw and C. S. Parmenter, "Chemical Timing 3. The picosecond dynamics of intramolecuar vibrational redistribution from 11 levels in S_1 p-difluorobenzene vapor.", J. Chem. Phys. **84**, 1099 (1986).
20. D. F. McMillen and D. M. Golden, "Hydrocarbon bond dissociation energies", Ann. Rev. Phys. Chem. **33**, 493 (1982).
21. V. N. Bagratashvili, V. S. Letokhov, A. A. Makarov, and E. A. Ryabov, <u>Multiple Photon Infrared Laser Photophysics and Photochemistry</u>, p. (Harwood, Amsterdam, 1985).
22. M. C. Gower and T. K. Gustafson, Sov. Phys. Optics Commun. **23**, 69 (1977).
23. P. Mukherjee and H. S. Kwok, "Dynamical temporal evolution of molecular IR absorption spectra observed with picosecond CO_2 laser pulses", J. Chem. Phys. **87**, 128 (1987).
24. D. S. Bomse, S. Dougal, and R. L. Woodin, "Multiphoton Ionization Studies of IR Multiphoton Dissociation: Direct C-H Bond Cleavage in Methanol", J. Phys. Chem. **90**, 2640 (1986).

Photoelectron spectroscopy with cm^{-1} resolution: VO$^+$, benzyl$^+$, toluene$^+$

James C. Weisshaar

Department of Chemistry, University of Wisconsin-Madison
Madison, Wisconsin 53706

ABSTRACT

Detection of nominal "zero kinetic energy electrons" (ZEKEs) in resonant two-photon ionization with ω_1 fixed and ω_2 scanned permits photoelectron spectroscopy with 1-5 cm^{-1} resolution. We report vibrational bands from 0-650 cm^{-1} for benzyl$^+$-h$_7$, benzyl$^+$-αd$_2$, and benzyl$^+$-d$_7$. The exocyclic C-C bond of benzyl cation has substantial double bond character. Band assignments from *ab initio* frequencies illuminate the mechanism of vibronic coupling in the 1^2A_2-2^2B_2 system of neutral benzyl. In toluene$^+$, we observe internal rotor frequencies that fix the V$_6$ barrier height of the CH$_3$ rotation at 20 ± 5 cm^{-1}, which is remarkably low. In VO → VO$^+$ + e$^-$, we measure state-to-state relative photoionization cross sections from the intermediate spin-rotational state N'J' to different cation spin-rotation states N$^+$J$^+$. Remarkably strong lines corresponding to large changes in angular momentum (ΔN = -6 to +7 and ΔJ = -6.5 to +6.5) are observed. Accurate values of the adiabatic ionization potentials of each molecule are reported.

1. INTRODUCTION

Recent advances in threshold photoelectron spectroscopy permit measurement of cation vibrational spectra with ~1 cm^{-1} resolution. The breakthrough experiments are due to Müller-Dethlefs, Schlag, and co-workers,[1] who developed "zero kinetic energy" photoelectron spectroscopy (ZEKE-PES). The technique promises to provide a much more comprehensive *overview* of the vibrational structure of polyatomic cations than either traditional photoelectron spectroscopy[2] or high resolution infrared absorption.[3] Recent variations on this theme include pulsed-field ionization (PFI)[4] and mass-analyzed threshold ionization (MATI),[5] which detect only the electrons or cations from field ionization.

We have extended ZEKE-PES to the study of the vibrational spectroscopy of cations derived from *unstable* neutrals. We previously reported the adiabatic ionization potential of Al$_2$ and the vibrational spectrum of the X$^2\Sigma_g^+$ state of Al$_2^+$.[6] Here we report vibrational spectra of the X^{1}A$_1$ electronic state of the polyatomic, closed-shell benzyl cation (C$_6$H$_5$-CH$_2^+$); spectra of the low frequency, nearly free CH$_3$ rotation in toluene cation (C$_6$H$_5$-CH$_3^+$); and rotationally resolved spectra of vanadium monoxide cation (VO$^+$). The ZEKE-PES technique improves the resolution of early[7] and more recent[8] photoelectron spectra of organic radicals by ~200 and ~20, respectively.

2. EXPERIMENTAL SECTION

Pulsed beams of the species of interest are skimmed and intersected by two tunable dye lasers in a high vacuum chamber. Toluene is seeded at 0.1-1% in 1-5 atm Ar for the toluene$^+$ studies. To form benzyl radical, we intersect the toluene expansion with a pulsed excimer laser 2 mm downstream; hot photolysis products, including benzyl radical, cool internally during the subsequent expansion, as described by Heaven and Miller.[11] VO is formed by laser ablation of a vanadium target rod in the high density portion of an expansion of Ar with 5% O$_2$. The dye lasers are 5 ns fwhm pulses in the visible or near-UV derived from Nd:YAG pumped dye lasers.

The ZEKE-PES device is modeled after that of Schlag and co-workers.[1] The molecular beam and the two dye lasers overlap in space and time between a pair of magnetically shielded stainless steel electron extraction plates spaced by 1.7 cm. The plates are held at ground potential during the laser pulses. At a time delay of 2.3 μs following the laser pulses, we apply a fast voltage step to the lower plate (risetime 50 ns; typical electric field strength 0.2-1.5 V/cm). Extracted electrons travel in the 1.5 V/cm field until they reach a mesh in the upper plate, where they accelerate to 6.5 eV kinetic energy and are subsequently detected by a microchannel plate chevron 50 cm from the interaction region. Gated integration of the resulting pulse of electrons (60 ns gate width) vs ω_2 at fixed ω_1 produces a nominal ZEKE spectrum.

The delayed extraction and time-gated detection strongly discriminate against electrons with > 1 cm^{-1} kinetic energy. It is a subtle question whether our nominal ZEKE experiment detects *any* genuine threshold electrons, since stray electric fields on the order of 1 mV/cm would remove such electrons during the 2.3 μs "field-free" delay. We definitely detect electrons due to laser excitation of high Rydberg states which live at least 2.3 μs and are field ionized by the extraction pulse. The signature of pulsed field ionization is an asymmetric broadening of the observed bands towards lower energy as the strength of the extraction field increases. In benzyl$^+$, the bands broaden from 4 cm^{-1} FWHM at 0.6 V/cm to 7 cm^{-1} FWHM at 3.4 V/cm, in rough accord with the classical field ionization equation $\Delta E(\text{cm}^{-1}) = 6[F(\text{V/cm})]^{1/2}$, where F is electric field strength. In VO$^+$, individual rotational lines broaden much more slowly than predicted by the classical equation.

3. BENZYL CATION: VIBRATIONAL FREQUENCIES AND NEUTRAL VIBRONIC MIXING MECHANISM

Internally cold benzyl radical is probed by two-color, resonant two-photon ionization (R2PI) through vibronically mixed 1^2A_2-2^2B_2 excited states near 450 nm.[9,10,11] For each isotopomer, benzyl$^+$-h$_7$, benzyl$^+$-αd$_2$, and benzyl$^+$-d$_7$, we report vibrational frequencies in the range 0-650 cm^{-1}. Measurement of vibrational spectra to 2500 cm^{-1} is in progress; we have already observed some 40 bands below 2000 cm^{-1} in benzyl$^+$-h$_7$.

ZEKE spectra of the three isotopomers of benzyl$^+$ by R2PI through specific intermediate states are displayed in Fig. 1. The origin bands provide the following adiabatic ionization potentials: 58465 ± 5 cm^{-1} = 7.2487 ± 0.0006 eV for benzyl-h$_7$; 58410 ± 5 cm^{-1} for benzyl-αd$_2$; and 58382 ± 5 cm^{-1} for benzyl-d$_7$. The benzyl-h$_7$ value refines our previous measurement[12] of 58456 ± 14 cm^{-1} from extrapolation of cation yield curves to zero field as well as the earlier value[7a] of 58100 ± 160 cm^{-1}. The vibronic mixing in the intermediate states[13,14] allows us to observe benzyl$^+$ vibrational states of both a$_1$ and b$_1$ symmetry (C$_{2v}$ point group) in the ZEKE spectrum. In addition to the origin, we observe several low-frequency states for each isotopomer. Table I collects the measured frequencies for all three isotopomers in the range 0-650 cm^{-1}. The relative intensities of these bands depend sensitively on the neutral intermediate state, as shown in Fig. 2 for benzyl$^+$-h$_7$.

The bands in Table I are assigned as either fundamentals or combinations by comparison with harmonic frequencies from *ab initio* calculations. Vibronic coupling of the two low-lying, nearly degenerate excited states (1^2A_2 and 2^2B_2) qualitatively explains the irregular absorption spectrum. Assuming C$_{2v}$ symmetry in the neutral benzyl ground and excited states, two types of mixed state can carry oscillator strength from the vibrationless ground state, which has $^eB_2 \times {}^va_1 = {}^{ev}B_2$ vibronic symmetry. These are:

$$|{}^eA_2 \times {}^vb_1\rangle \text{ mixed with } |{}^eB_2 \times {}^va_1\rangle \qquad \text{(type-A bands)} \qquad (1a)$$

and $\quad |{}^eA_2 \times {}^va_1\rangle \text{ mixed with } |{}^eB_2 \times {}^vb_1\rangle \qquad \text{(type-B bands).} \qquad (1b)$

In the simplest approximation, both the 1^2A_2 and 2^2B_2 electronic states can photoionize to X^1A_1 benzyl$^+$ in a dipole-allowed, one-electron transition. All observed bands are accommodated by the calculated a_1 and b_1 vibrational states, including low frequency combinations. This provides indirect evidence that the intermediate state geometry is planar. Agreement within several percent between experimental vibrational intervals and harmonic frequencies is typical for MP2/6-31G* theory.[15]

Our data presently allow several conclusions about the vibronic coupling mechanism. First, observation of both $\nu_{13}(a_1)$ and $\nu_{28}(b_1)$ in the same ZEKE spectrum (Figs. 1 and 2) is direct evidence of vibronic mixing in the A^1, A^2, and A^3 states. According to the mixed-state description of Eq. 1a, observation of both a_1 and b_1 modes indicates that both the $^2A_2$ and $^2B_2$ electronic components have comparable photoionization cross sections, i.e., the ZEKE-PES experiment detects *both* types of components in each vibronically mixed state. Observation of ν_{28} (b_1) confirms its importance in the vibronic mixing mechanism. Observation of the combination bands $\nu_{36} + \nu_{17}$ and $\nu_{36} + \nu_{35}$ strongly suggests their inclusion in future models of the vibronic coupling in benzyl. The *absence* of ν_{29} from ZEKE spectra through the A^1, A^2, and A^3 states indicates that ν_{29} is *ineffective* at mixing $^2A_2$ with $^2B_2$. While we might worry that the Rydberg states converging to ν_{29} in the cation have unusually short vibrational autoionization lifetimes resulting in anomalously weak pulsed-field-ionization signals for ν_{29}, the dispersed fluorescence measurements of Fukushima and Obi[10] confirm the minor contribution of ν_{29} to the mixed states. The activity in ν_{13} (a_1) indicates substantial geometry change from intermediate to cation along this normal coordinate.

Finally, we briefly compare corresponding vibrational frequencies in X^1A_1 benzyl$^+$ and X^2B_2 neutral benzyl. The most useful comparisons at present use *ab initio* harmonic frequency sets for both cation and neutral.[17] Our experimental data for benzyl$^+$ and the existing data for the neutral ground state give us confidence in the essential accuracy of the calculated values. We compare harmonic frequencies for benzyl$^+$ from MP2/6-31G* calculations with those for neutral benzyl from UHF/3-21G calculations. We correct the latter by the usual factor of 0.89 to improve agreement with experiment. For five experimentally known benzyl frequencies below 650 cm^{-1},[16] the corrected UHF/3-21G frequencies agree within 2% for four modes and within 9% for the fifth.

The lowest-frequency motions, which involve deformation of both the ring and the -CH$_2$ group, have quite similar frequencies in the cation and neutral: 157 cm^{-1} (cation) vs 188 cm^{-1} (neutral) for m_{36} (b_2, out-of-plane deformation); 323 cm^{-1} vs 359 cm^{-1} for ν_{17} (a_2, torsion); and 358 cm^{-1} vs 353 cm^{-1} for ν_{29} (b_1, in-plane rock). More telling are the modes that nearly isolate motions of the exocyclic -CH$_2$ group, ν_{16} (a_2, -CH$_2$ torsion) and ν_{30} (b_2, quasi-umbrella). These have much higher frequencies in the cation than in the neutral: 627 cm^{-1} vs 481 cm^{-1} for ν_{16} and 1103 cm^{-1} vs 676 cm^{-1} for ν_{30}. In particular, the benzyl cation torsional frequency of 627 cm^{-1} is similar to those of alkenes (575 cm^{-1} for propene and 770 cm^{-1} for 1,3-butadiene);[17] the torsion of the radical is some 30% lower. The comparison clearly indicates substantially greater double-bond character in the exocyclic C-C bond of benzyl$^+$ than in that of the neutral free radical.

4. TOLUENE CATION: NEARLY FREE INTERNAL ROTATION OF CH$_3$ GROUP

We have used the resolution of the ZEKE-PES technique to study the internal rotational motion of the CH$_3$ group in toluene cation. Previous work[18] has shown that neutral toluene has very small barriers to internal rotation in its S_0 and S_1 electronic states. Fits of observed band positions to a model V_6 potential yield estimates of the barrier height of 10 cm^{-1} for S_0 and 25 cm^{-1} for S_1. Figure 3 shows a ZEKE-PES spectrum of toluene$^+$ taken with ω_1 tuned to the S_1-S_0 origin. The beam contains both the $0a_1'$ ground vibronic state and the $1e''$ first excited rotor state. In Fig. 3, ω_1 excites both the

0-0 and 1-1 bands of the neutral. Assignments of cation rotor states are as shown. Additional ZEKE spectra obtained by tuning ω_1 to other S_1-S_0 bands confirms our assignments.

The adiabatic ionization potential of toluene is $71,199 \pm 8$ cm^{-1} = 8.8276 ± 0.0010 eV. The cation rotor states occur at the following energies: $0a_1'$, 0 cm^{-1}; $1e''$, 5 cm^{-1}; $2e'$, 20 cm^{-1}; $3a_2''$, 45 cm^{-1}; $3a_1''$, 54 cm^{-1}; $4e'$, 80 cm^{-1}. State energies are deemed accurate to ± 2 cm^{-1}.

We have not yet fit these energies to the states of a V_6 potential, but the entire set of frequencies is nearly identical to those within S_1. The V_6 model barrier height for the cation is thus 20 ± 5 cm^{-1}. The CH$_3$ group rotates nearly freely in the cation as it does in S_1 and S_0. This is surprising in view of the natural bond orbital analysis of rotational barriers due to Weinhold and co-workers. In neutral toluene S_0, they find the small barrier to be due to a *cancellation* of relatively strong σ and π donor-acceptor interactions between the ring and the CH bonds and anti-bonds. These vary in out-of-phase fashion as the CH$_3$ group rotates. We expected removal of a π electron to drastically alter the π component of the ring-CH$_3$ interaction. The results of *ab initio* calculations on the cation internal rotation potential will thus be interesting.

In addition, we have obtained preliminary low frequency spectra of the cations derived from the van der Waals species toluene·Ar and toluene·Ar$_2$, which also appear in our beam at high Ar stagnation pressures.

5. VANADIUM OXIDE CATION: PROPENSITY FOR LARGE CHANGES IN ROTATIONAL STATE ON IONIZATION

We ionize VO through the (3-0) band of the $C^4\Sigma^- \leftarrow X^4\Sigma^-$ transition (ω_1 = 19935-19960 cm^{-1}). The state-specific resonant two-photon ionization process of interest is:

$$\text{VO } [X^4\Sigma^-, v''=0, N''J''] \rightarrow \text{VO}^* [C^4\Sigma^-, v'=3, N'J'] \rightarrow \text{VO}^+ [X^3\Sigma^-, v^+=0, N^+J^+] + e^-. \quad (2)$$

Here double primes label the neutral ground state, primes label the resonant excited state, and '+' labels the cation state. The angular momentum quantum numbers are $\mathbf{N} = \mathbf{L} + \mathbf{R}$ (sum of orbital angular momentum and diatomic rotation) and $\mathbf{J} = \mathbf{N} + \mathbf{S}$ (where $\mathbf{S}$ is electron spin) from Hund's case b. Merer and co-workers[19] have carried out beautiful high resolution studies of the C ← X transition in VO. By tuning ω_1 judiciously within the 3-0 band, we can selectively excite a single spin-rotation level (N' S') of the C state. However, the 0.1 cm^{-1} linewidth of ω_1 excites many hyperfine levels ($\mathbf{F} = \mathbf{I} + \mathbf{J}$) due to the ^{51}V nuclear spin of 7/2.

Figure 4 shows a ZEKE spectrum obtained through the intermediate state N' =7, J' =8.5 at ω_1 = 19937.0 cm^{-1}. The lines are ~1.5 cm^{-1} fwhm. For sufficiently large values of N^+J^+, this resolves all cation quantum numbers except hyperfine. The key to obtaining narrow lines is to minimize the charge density produced in the ionization region and to lower the electron extraction field to 0.2 V/cm. The positions of all of the strong lines can be fit to within 0.1 cm^{-1} using the $^3\Sigma^-$ model Hamiltonian due to Schlapp.[20] From the fit to the line positions, we obtain the spectroscopic constants summarized in Table II. It is perhaps unprecedented to obtain spin-rotation and spin-spin coupling constants from a photoelectron technique.

The adiabatic ionization potential of 7.2386 ± 0.0004 eV is in reasonable agreement with the previous experimental result of 7.25 ± 0.01 eV from Dyke and co-workers.[21] The average bond length $r_0 = 1.561 \pm 0.003$ Å disagrees with Dyke's estimate of 1.54 ± 0.01 Å from a Franck-Condon analysis but agrees well with the value 1.56 Å from an *ab initio* calculation due to Carter and Goddard.[22]

The most surprising results are the relative intensities of the different N^+J^+ cation states. Figure 5 is a histogram of cation state intensities for ionization from $N' = 7$, $J' = 8.5$; the corresponding ZEKE spectrum is shown in Fig. 4. This one example shows $\Delta N = N^+ - N'$ ranging from -5 to +7, and $\Delta J = J^+ - J'$ ranging from -5.5 to +4.5. The intensity vs N^+ remains rather constant to $N^+ = 14$ and then falls off sharply. The distribution of intensity among the three spin states F_1, F_2, and F_3 associated with each N^+ varies substantially with N^+. Distributions vs N^+ (summed over the three J^+ values that contribute to each N^+) and distributions vs J^+ (summed over the three N^+ values contributing to each J^+) decay less abruptly at the high end. We have obtained similar cation state distributions for five other intermediate states $N'J'$; the results are qualitatively similar. The total range of ΔN and ΔJ observed is -6 to +7 and -6.5 to +6.5, respectively.

The mechanism behind the large positive ΔN and ΔJ is not clear. We very likely detect absorption to high-n Rydberg states converging to each cation level (Fig. 6) and subsequent field ionization by the delayed extraction pulse. It is worth attempting to fit our spectra by calculating transition strengths from the Hund's case b intermediate state to a set of Rydberg states in Hund's case d', for which the good quantum numbers are $N^+S^+J^+;\ell sj;J$. Here ℓ, s, and j are the orbital, spin, and total angular momenta of the Rydberg electron and **J** = **J**$^+$ + **j** is the total angular momentum neglecting nuclear spin. Müller-Dethlefs has calculated such intensities in a recent paper.[23]

In a case d' ← case b transition, the selection rules are $\Delta N = \ell + 1, ..., -(\ell + 1)$ and $\Delta J = \ell + 1.5, ..., -(\ell + 1.5)$. Within that model, observation of $\Delta N = -6$ to $+7$ implies Rydberg electron orbital angular momentum components as large as 6 ("i-waves"). A crucial set of input data is the decomposition of the initial orbital (nominal $3d\sigma$ orbital on V) into its ℓ'-components about the center of mass of the diatomic. Presumably the same distribution of ℓ' should fit all ZEKE spectra obtained with the C state as intermediate. However, the fact that the distribution of intensity among the F_1, F_2, and F_3 states associated with each N^+ varies substantially with N^+ indicates that the simple case d' ← case b model is inadequate. We can still hope that Hund's case d can describe the intensities into each N^+, summed over the contributing J^+. Another interesting possibility is that the Born-Oppenheimer approximation fails badly for the high-n Rydberg states of interest, leading to strong electron-rotational coupling. There is a wealth of information in the ZEKE spectra that we have just begun to analyze.

The complexity of the problem is underscored by the following preliminary but reproducible result. We have excited two different subsets of hyperfine levels associated with $N' = 7$, $J' = 8.5$. The one-photon energies differ by about 0.2 cm^{-1}. The two resulting N^+J^+ distributions are very different from each other. This strongly suggests that hyperfine interactions, which degrade N' and J' as quantum numbers in the intermediate state, play an important role in the state-to-state photoionization cross sections.

6. ACKNOWLEDGMENTS

I thank the National Science Foundation and the donors of the Petroleum Research Foundation for generous support of this research. My skillful graduate students Dr. Joel Harrington, Dr. Gregory Eiden, and Ms. Kueih-Tzu Lu carried out all of the experimental work described here and provided valuable conceptual insights as well. I also thank Professors Phil Certain, Frank Weinhold, and Claude Woods for stimulating and useful discussions.

7. REFERENCES

[1] a) L.A. Chewter, M. Sander, K. Muller-Dethlefs, and E. W. Schlag, J. Chem. Phys. **86**, 4737 (1987). b) K. Muller-Dethlefs, M. Sander, and E. W. Schlag, Chem. Phys. Lett. **112**, 291 (1984).

[2] a) K. Kimura, *Handbook of HeI Photoelectron Spectra of Fundamental Organic Molecules: Ionization energies, ab initio assignments, and valence electronic structure for 200 molecules* (Halsted, New York, 1981). b) J. Berkowitz, *Photoabsorption, Photoionization and Photoelectron Spectroscopy* (Academic Press, New York, 1979).

[3] a) M. W. Crofton, M. F. Jagod, B. D. Rehfuss, W. A. Kreiner, and T. Oka, J. Chem. Phys. **88**, 666 (1988). b) M. W. Crofton, M. F. Jagod, B. D. Rehfuss, and T. Oka, J. Chem. PHys. **91**, 5139 (1989). c) C. S. Gudeman, M. H. Begemann, J. Pfaff, and R. J. Saykally, Phys. Rev. Lett. **50**, 727, (1983).

[4] a) M. Braunstein, V. McKoy, S. N. Dixit, R. G. Tonkyn, and M. G. White, J. Chem. Phys. **93**, 5345 (1990). b) R. G. Tonkyn, J. W. Winniczek, and M. G. White, Chem. Phys. Lett. **164**, 137 (1989).

[5] L. Zhu and P. M. Johnson, J. Chem. Phys. **94**, 5769 (1991).

[6] J. E. Harrington and J. C. Weisshaar, J. Chem. Phys. **93**, 854 (1990).

[7] a) F. A. Houle and J. L. Beauchamp, J. Am. Chem. Soc. **100**, 3290 (1978). b) J. M. Dyke, N. Jonathan, A. E. Lewis, J. D. Mills, and A. Morris, Mol. Phys. **50**, 77 (1983).

[8] a) X. Song, E. Sekreta, and J.P. Reilly, J. Chem. Phys. **91**, 6062 (1989) and references 1-19 therein. b) D.W. Minsek and P. Chen, J. Phys. Chem. **94**, 8399 (1990).

[9] a) G. Porter and B. Ward, J. Chim. Phys. **61**, 1517 (1964).
b) C. Cossart-Magos and S. Leach, J. Chem. Phys. **54**, 1534 (1972). c) C. Cossart-Magos and W. Goetz, J. Mol. Spectrosc. **115**, 366 (1986). d) J. H. Miller and L. Andrews, J. Mol. Spectrosc. **90**, 80 (1981).

[10] a) M. Fukushima and K. Obi, J. Chem. Phys. **93**, 8488 (1990). b) J. I. Selco and P. G. Carrick, J. Mol. Spectrosc. **137**, 13 (1989).

[11] M. Heaven, L. Dimauro, and T. A. Miller, Chem. Phys. Lett. **95**, 347 (1983).

[12] G. C. Eiden and J. C. Weisshaar, J. Phys. Chem., in press.

[13] C. Cossart-Magos and S. Leach, J. Chem. Phys. **64**, 4006 (1976).

[14] F. Negri, G. Orlandi, F. Zerbetto, and M. Z. Zgierski, J. Chem. Phys. **93**, 600 (1990).

[15] R. D. Amos, J. F. Gaw, N. C. Handy, and S. Carter; J. Chem. Soc., Faraday Trans. 2, **84**, 1247 (1988).

[16] M.E. Jacox, J. Phys. Chem. Ref. Data **13**, 945 (1984); ibid. **17**, 269 (1988).

[17] T. Shimanouchi, H. Matsuura, Y. Ogawa, and I. Harada, J. Phys. Chem. Ref. Data **9**, 1149 (1980).

[18] P.J. Breen, J.A. Warren, E.R. Bernstein, and J.I. Seeman, J. Chem. Phys. **87**, 1917 (1987).

[19] A.S-C. Cheung, R.C. Hansen, and A.J. Merer, J. Mol. Spec. **91**, 165 (1982).

[20] G. Herzberg, *Spectra of Diatomic Molecules* (Van Nostrand Reinhold, New York, 1950), p. 223.

[21] J.M. Dyke B.W.J. Gravenor, M.P. Hastings, and A. Morris, J. Phys. Chem. **89**, 4613 (1985).

[22] E.A. Carter and W.A. Goddard III, J. Phys. Chem. **92**, 2109 (1988).

[23] K. Muller-Dethlefs, J. Chem. Phys. **95**, 4821 (1991).

Table I. Experimental and Calculated Vibrational Levels of X^1A_1 State of Benzyl Cation.[a]

Level[b]	symm	old label[c]	benzyl+-h7 expt	calc	Δ	benzyl+-αd2 expt	calc	Δ	benzyl+-d7 expt	calc	Δ
$\nu_{36}+\nu_{17}$	b_1	--	487	480	-1.4	456	452	-0.9	423	418	-1.2
$\nu_{36}+\nu_{35}$	a_1	--	--	552	--	--	540	--	488	494	+1.2
ν_{13}	a_1	ν_{6a}	526	537	+2.1	504	524	+4.0	500	511	+2.2
ν_{28}	b_1	ν_{6b}	598	613	+2.5	596	612	+2.7	575	588	+2.3

[a] Calculated harmonic frequencies using Gaussian-90, MP2/6-31G*. Δ is the percentage error, (calc - expt)/expt.

[b] Assignments with new mode labels. Approximate mode descriptions (op = out-of-plane, ip = in-plane): ν_{36} = ring and -CH$_2$ op wag; ν_{35} = op CH ring wag; ν_{28} = ip ring deform; ν_{17} = op ring + -CH$_2$ torsion; ν_{13} = ip CCC bend; ν_{12} = ring CC str + C-CH$_2$ str.

[c] Benzene mode labels of Wilson.

Table II. Spectroscopic constants[a] of VO^+ $X^3\Sigma^-$, $v^+ = 0$.

IP (cm^{-1})	58,383 ± 5
B_0 (cm^{-1})	0.568 ± 0.002
λ (cm^{-1})	6.9 ± 0.2
γ (cm^{-1})	0.06 ± 0.05
r_0 (Å)	1.561 ± 0.003

[a] IP is adiabatic ionization potential; B_0 is $v^+=0$ rotational constant; λ is spin-rotation coupling constant; γ is spin-spin coupling constant; and r_0 is the average bond length in $v^+ = 0$. See Ref. 20.

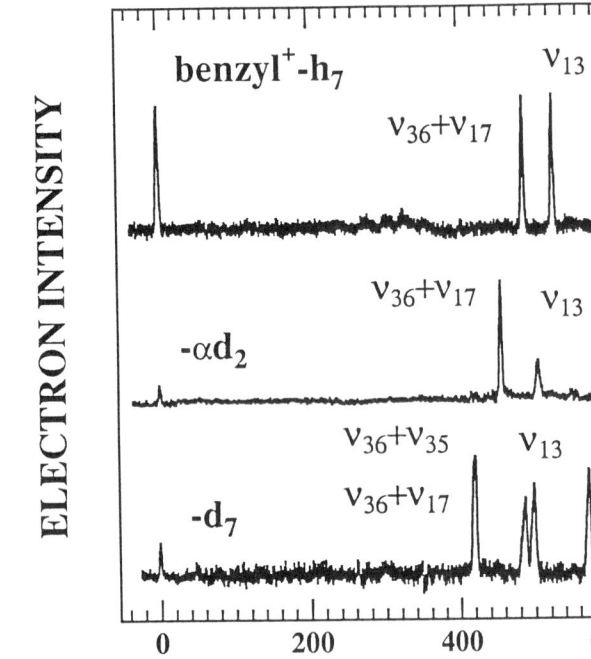

Figure 1. ZEKE spectra of three benzyl[+] isotopomers. Excitation through A[1] band.

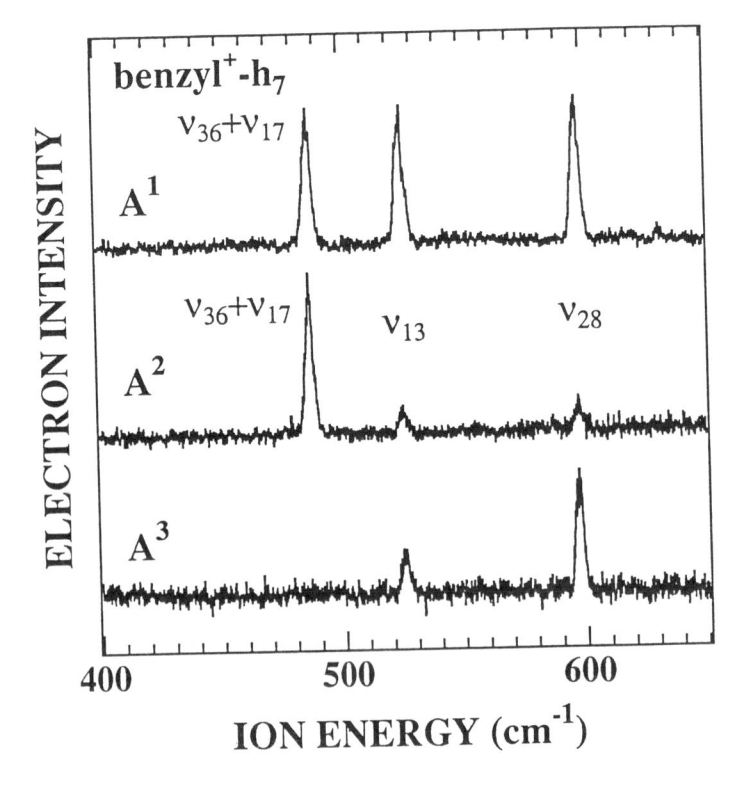

Figure 2. ZEKE spectra of benzyl[+]-h7 through three vibronically mixed intermediate states as shown.

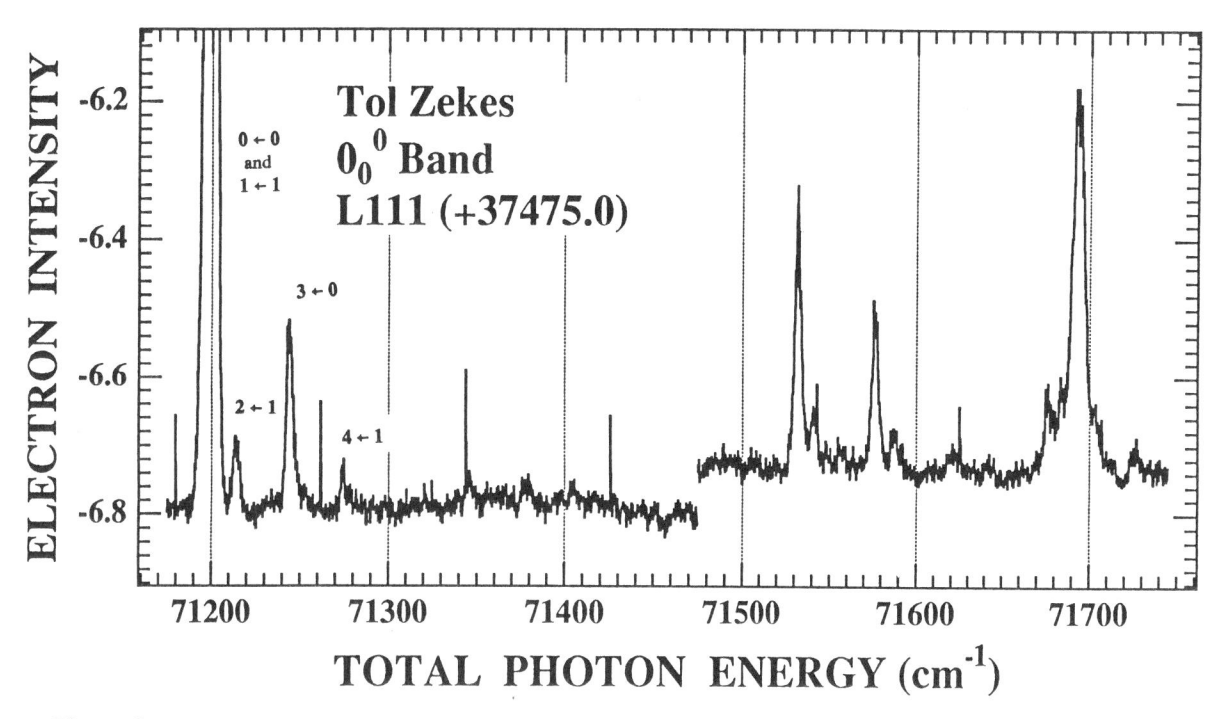

Figure 3. ZEKE spectrum of toluene cation excited through the S_1-S_0 origin.

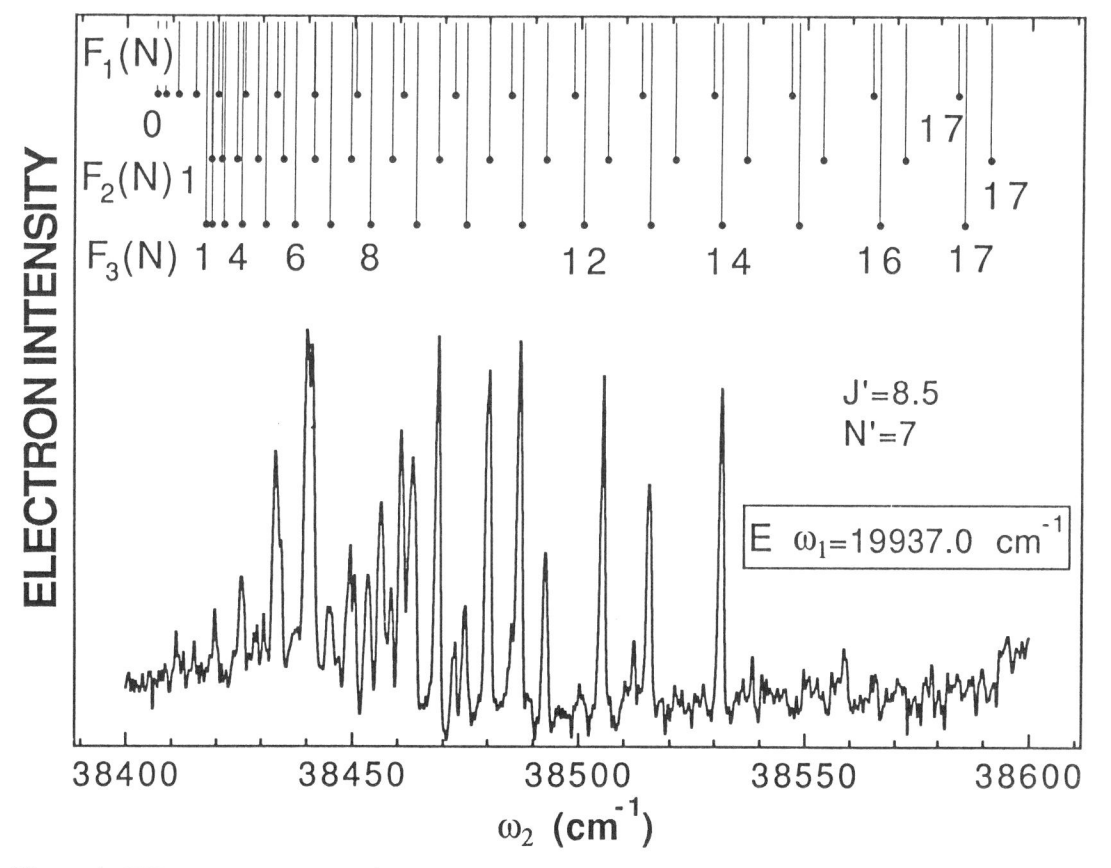

Figure 4. ZEKE spectrum of VO^+ through the $J' = 8.5$, $N' = 7$ intermediate; assignments as shown.

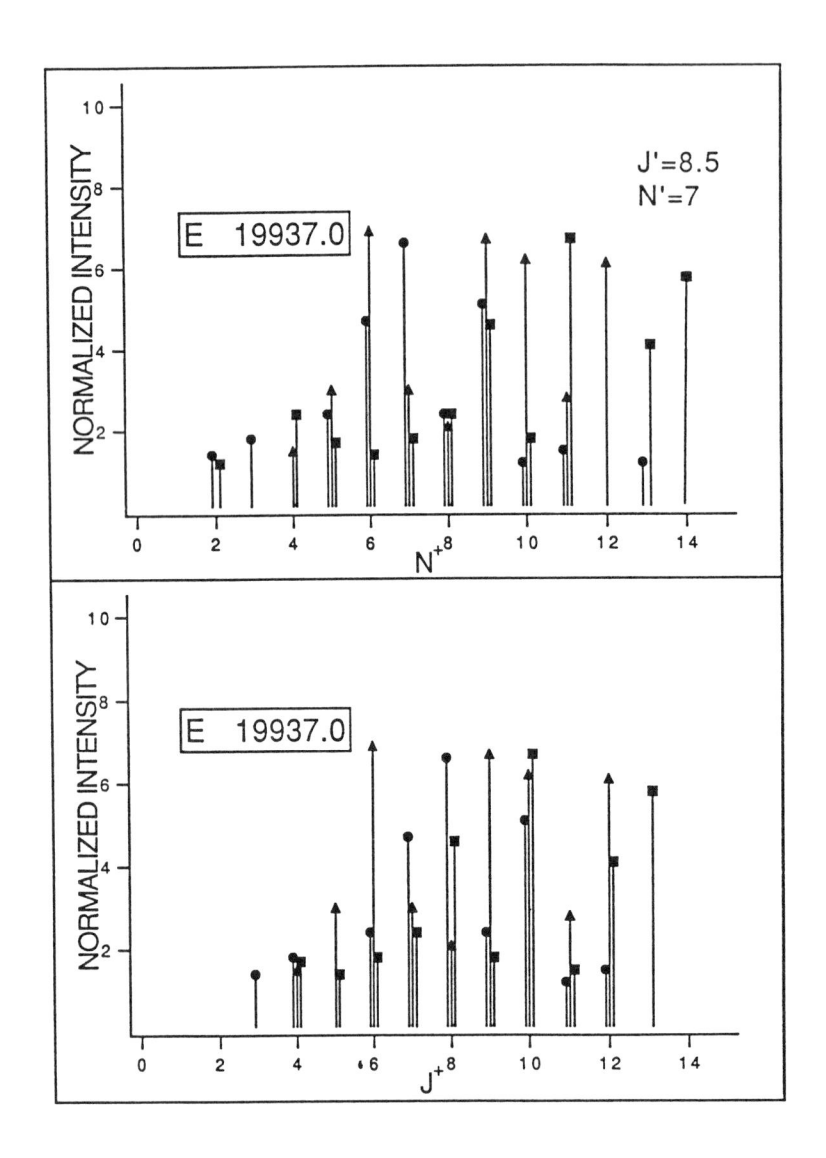

Figure 5. Distribution of intensities vs N^+ and J^+ in ZEKE spectrum of Fig. 4. Circles, triangles, and squares are F_1, F_2, and F_3 spin components, respectively.

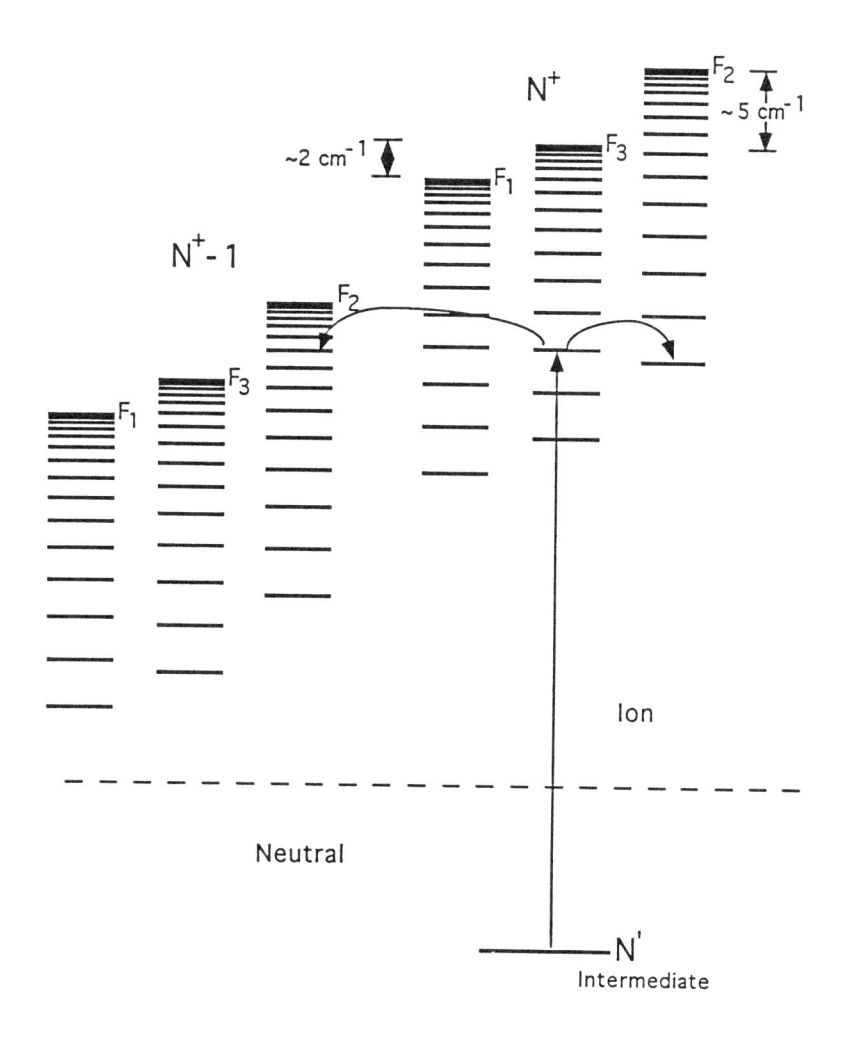

Figure 6. Schematic of Hund's case d′ Rydberg series converging to each spin-rotation level of VO$^+$ for two different values of N$^+$.

Addendum

The following papers were presented at this conference, but the manuscripts supporting the oral presentations are not available.

[1638-03] **High-resolution spectroscopy of high-pressure molecular solids by time-resolved photon echo spectroscopy**
E. L. Chronister, R. A. Crowell, Univ. of California/Riverside.

[1638-04] **Statistical mechanical studies of electronic line broadening in liquids**
R. F. Loring, N. E. Shemetulskis, A. M. Walsh, Cornell Univ.

[1638-50] **Vacuum UV photodissociation dynamics of the chlorofluorocarbons**
M. Yen, P. Johnson, M. G. White, Brookhaven National Lab.

Author Index

Achiba, Yohji, 44
Åkesson, Eva, 21
Apkarian, V. A., 2
Ashfold, Michael N., 144
Ashjian, P., 2
Baltzer, P., 127
Barbara, Paul F., 21
Bell, Andrew J., 345
Bengali, Ashfaq A., 106
Berg, Mark, 12, 282
Bergmann, Klaas, 298
Bieske, E. J., 254
Bishea, Gregory A., 74
Blake, N., 2
Blush, Joel A., 118
Broström, Lars, 228
Brum, Jeffrey L., 156
Brumer, Paul, 357
Burns, W. A., 170
Carlsson-Göthe, M., 127
Cartland, Harry E., 132, 386
Casey, Sean M., 106, 234
Chen, Kuo-mei, 399
Chen, Peter, 118, 210
Cheng, Chun-Lin, 106
Cheng, P. Y., 310
Christensen, Ronald L., 345
Chronister, Eric L., Addendum
Clauberg, Horst, 210
Crowell, Robert A., Addendum
Cyr, Donna M., 74
Dai, Hai-Lung, 310
Danylichev, A., 2
Deshmukh, Subhash, 156
DiMauro, Louis F., 322
Dick, Jonathan P., 106
Donnelly, Stephen G., 56
Dunn, Robert C., 29
Dvorak, M. A., 170
Earvolino, Patrick, 12
Farrar, James M., 56
Felker, Peter M., 63
Fenn, P. T., 106
Flanders, Bret N., 29
Ford, R. S., 170
Friedmann, A., 254
Guettler, Robert D., 431
Hahn, Y. M., 310
Han, Chau-Chung, 74
Hartland, G. V., 63
Henson, B. F., 63
Hill, M., 2
Hoffman, G. J., 2
Hsu, C.-W., 245

Hudson, Bruce S., 333
Hunziker, Heinrich E., 102
Imre, D., 2
Izawa, Yasukazu, 164
Johnson, Alan E., 21
Johnson, Mark A., 74
Johnson, P., Addendum
Ju, Shan S., 310
Karlsson, L., 127
Kim, Eun H., 368
Kimura, Katsumi, 216
Kondow, Tamotsu, 85
Koplitz, Brent D., 156
Kuhn, A., 298
Lambert, Ian R., 144
Lapierre, L., 310
Larsson, Mats, 127, 228
Lawrence, William G., 2
Lee, Yuan T., 179, 197
Leopold, Doreen G., 106, 234
Leopold, Kenneth R., 170
Levinger, Nancy E., 21
Liao, Chung-Lin, 245
Lineberger, W. C., 264
Lisy, James M., 92
Liu, Kopin, 416
Loring, Roger F., Addendum
Lovas, F. J., 170
Luo, Xin, 441
Macdonald, R. G., 416
Maier, John P., 254
Mannervik, Sven, 228
Maxton, P. M., 63
Meijer, Gerard, 102
Minsek, David W., 118, 210
Mordaunt, David H., 144
Moriwaki, Taro, 44
Morley, Gregory P., 144
Muller, Laura J., 282
Mullin, Amy S., 264
Nagata, Takashi, 85
Nakashima, Nobuaki, 164
Ng, Cheuk-Yiu, 245
Nickolaisen, Scott L., 132, 386
Nonose, Shinji, 85
Ojima, Yuichi, 164
Petek, Hrvoje, 345
Pollard, James E., 423
Posey, Lynmarie A., 74, 431
Prather, Kimberly A., 179
Reeve, S. W., 170
Rizzo, Thomas R., 441
Sahyun, Melville R., 291
Schiemann, S., 298

(continued)

Sekreta, E., 2
Settle, Rebecca D., 441
Shapiro, Moshe, 357
Sharma, D. K., 291
Shemetulskis, Norah E., Addendum
Shimada, Tetsuya, 164
Shiromaru, Haruo, 44
Shobatake, Kosuke, 185
Simon, John D., 29
Soliva, A. M., 254
Steadman, Jhobe, 36
Stolow, Albert, 197
Suenram, R. D., 170
Sundström, Goran, 228
Syage, Jack A., 36
Szaflarski, Diane M., 264
Tabayashi, Kiyohiko, 185
Takahashi, Masahiko, 216
Vanden Bout, David, 282
Veney, David, 132
Venturo, V. A., 63
Villalta, Peter W., 106
Walker, Gilbert C., 21
Walsh, Alan M., Addendum
Wang, Zhongrui, 156
Wannberg, B., 127
Weisshaar, James C., 453
Wendt, H. R., 102
Western, Colin M., 144
White, Michael G., Addendum, 273
Wiedmann, Ralph T., 273
Wittig, Curt, 132, 386
Wodtke, Alec M., 368
Xu, Xiaodong, 156
Yamanaka, Chiyoe, 164
Yang, Xueming, 368
Yen, Mei-Wen, Addendum
Yen, Yu-Fong, 156
Yokoyama, Kazushige, 264
Yoshihara, Keitaro, 345
Yu, Jongwan, 12
Zadoyan, Ruben, 2
Zare, Richard N., 431
Zoval, J., 2
de Vries, Mattanjah S., 102